Lecture Notes in Computer Science 4004

Commenced Publication in 1973
Founding and Former Series Editors:
Gerhard Goos, Juris Hartmanis, and Jan van Leeuwen

Editorial Board

Serge Vaudenay (Ed.)

Advances in Cryptology – EUROCRYPT 2006

24th Annual International Conference on the Theory
and Applications of Cryptographic Techniques
St. Petersburg, Russia, May 28 – June 1, 2006
Proceedings

 Springer

Volume Editor

Serge Vaudenay
EPFL, I&C, LASEC, Station 14
INF Building, 1015 Lausanne, Switzerland
E-mail: serge.vaudenay@epfl.ch

Library of Congress Control Number: 2006925895

CR Subject Classification (1998): E.3, F.2.1-2, G.2.1, D.4.6, K.6.5, C.2, J.1

LNCS Sublibrary: SL 4 – Security and Cryptology

ISSN   0302-9743
ISBN-10   3-540-34546-9 Springer Berlin Heidelberg New York
ISBN-13   978-3-540-34546-6 Springer Berlin Heidelberg New York

Springer is a part of Springer Science+Business Media

springer.com

© International Association for Cryptologic Research 2006
Printed in Germany

Typesetting: Camera-ready by author, data conversion by Scientific Publishing Services, Chennai, India
Printed on acid-free paper   SPIN: 11761679   06/3142   5 4 3 2 1 0

Preface

The 2006 edition of the Eurocrypt conference was held in St. Petersburg, Russia from May 28 to June 1, 2006. It was the 25th Eurocrypt conference. Eurocrypt is sponsored by the International Association for Cryptologic Research (IACR). Eurocrypt 2006 was chaired by Anatoly Lebedev, and I had the privilege to chair the Program Committee.

Eurocrypt collected 198 submissions on November 21, 2005. The Program Committee carried out a thorough review process. In total, 863 review reports were written by renowned experts, Program Committee members as well as external referees. Online discussions led to 1,114 additional discussion messages and about 1,000 emails. The review process was run using e-mail and the iChair software by Thomas Baignères and Matthieu Finiasz. Every submitted paper received at least three review reports. The Program Committee had a meeting in Lausanne on February 4, 2006. We selected 33 papers, notified acceptance or rejection to the authors, and had a cheese fondue. Authors were then invited to revise their submission. The present proceedings include all the revised papers. Due to time constraints the revised versions could not be reviewed again.

We delivered a "Eurocrypt Best Paper Award." The purpose of the award is to formally acknowledge authors of outstanding papers and to recognize excellence in the cryptographic research fields. Committee members were invited to nominate papers for this award. A poll then yielded a clear majority. This year, we were pleased to deliver the Eurocrypt Best Paper Award to Phong Q. Nguyen and Oded Regev for their brilliant paper "Learning a Parallelepiped: Cryptanalysis of GGH and NTRU Signatures."

The Program Committee invited two speakers: David Naccache and Kevin McCurley. The current proceedings include papers about their presentation.

I would like to thank Anatoly Lebedev for organizing the conference. I would like to thank the IACR Board for honoring me by asking me to chair the Program Committee. The Program Committee and external reviewers worked extremely hard. I deeply thank them for this volunteer work. Acknowledgments also go to the authors of submitted papers, the speakers, and the invited speakers. I am grateful to Thomas Baignères and Matthieu Finiasz for their hard work developing the iChair software and constantly adding features. I also thank Shai Halevi and Amr Youssef, who participated in the software testing. Finally, I heartily thank Christine and Emilien, my family, for letting me spend some time on Eurocrypt.

This year, we celebrated the 30th anniversary of the publication of the Diffie-Hellman seminal paper "New Directions in Cryptography." As cryptography was becoming a new academic research area, this pioneer paper invented public-key cryptography. My wish is that research in cryptography will lead us to 30 more years of fun.

March 2006

Serge Vaudenay
Lausanne

Eurocrypt 06

May 28–June 1, 2006, Saint Petersburg, Russia

General Chair

Anatoly Lebedev, LAN Crypto
Moscow, Russia

Program Chair

Serge Vaudenay, EPFL
Lausanne, Switzerland

Program Committee

Feng Bao .... Institute for Infocomm Research
Eli Biham .... Technion
Alex Biryukov .... Katholieke Universiteit Leuven
Alexandra Boldyreva .... Georgia Institute of Technology
Colin Boyd .... Queensland University of Technology
Jean-Sébastien Coron .... University of Luxembourg
Yevgeniy Dodis .... New York University
Matt Franklin .... University of California Davis
Eiichiro Fujisaki .... NTT Laboratories
Juan Garay .... Bell Labs — Lucent Technologies
Martin Hirt .... ETH Zurich
Tetsu Iwata .... Ibaraki University
Pil Joong Lee .... Pohang University of Science and Technology
Antoine Joux .... DGA and University of Versailles
Jonathan Katz .... University of Maryland
Arjen Lenstra .... Bell Labs – Lucent Technologies
and Technische Universiteit Eindhoven
Helger Lipmaa .... Cybernetica AS and University of Tartu
Javier Lopez .... University of Malaga
Stefan Lucks .... University of Mannheim
Philip MacKenzie .... DoCoMo USA Labs
Mitsuru Matsui .... Mitsubishi Electric
Alexander May .... University of Paderborn
Willi Meier .... FH Aargau
Atsuko Miyaji .... JAIST
Kaisa Nyberg .... Helsinki University of Technology and Nokia
Kenny Paterson .... Royal Holloway University of London
Greg Rose .... Qualcomm
Berry Schoenmakers .... Technische Universiteit Eindhoven
Serge Vaudenay (Chair) .... EPFL
Michael Wiener .... Cryptographic Clarity
Robert Zuccherato .... Entrust, Inc.

External Reviewers

Michel Abdalla
Masayuki Abe
Carlisle Adams
Luis von Ahn
Koichiro Akiyama
Elena Andreeva
Kazumaro Aoki
Seigo Arita
Frederik Armknecht
Tomoyuki Asano
Gildas Avoine
Thomas Baignères
Elad Barkan
Don Beaver
Zuzana Beerliová
Mihir Bellare
Vicente Benjumea
Dan Bernstein
John Black
Daniel Bleichenbacher
Johannes Blömer
Jean Christian Boileau
Xavier Boyen
Harry Buhrman
Jan Camenisch
Ran Canetti
Juyoung Cha
Liqun Chen
Rafi Chen
Kookrae Cho
Sherman Chow
Carlos Cid
Scott Contini
Yang Cui
Reza Curtmola
Ivan Damgård
Vanesa Daza
Alex Dent
Claus Diem
Yan Zong Ding
Martin Döring
Orr Dunkelman
Stefan Dziembowski
Daniela Engelbert

Nelly Fazio
Serge Fehr
Matthieu Finiasz
Marc Fischlin
Matthias Fitzi
Pierre-Alain Fouque
Felix Freiling
Jun Furukawa
Soichi Furuya
Martin Gagne
Steven Galbraith
David Galindo
Ran Gelles
Mark Gondree
Daniel Gottesman
Louis Goubin
Ignacio Gracia
Safuat Hamdy
Goichiro Hanaoka
Phil Hawkes
Ryotaro Hayashi
Javier Herranz
Florian Hess
Shoichi Hirose
Dennis Hofheinz
Thomas Holenstein
Nick Howgrave-Graham
Yong Ho Hwang
Yuval Ishai
Stanislaw Jarecki
Jorge Jiménez
Ellen Jochemsz
Pascal Junod
Senny Kamara
Akinori Kawachi
John Kelsey
Aggelos Kiayias
Joe Killian
Mehmet Kiraz
Kazukuni Kobara
Vladimir Kolesnikov
Chiu-Yuen Koo
Matthias Krause
Volker Krummel

Caroline Kudla
Ulrich Kühn
Simon Künzli
Kaoru Kurosawa
Tanja Lange
Joseph Lano
Peeter Laud
Sven Laur
Jung Wook Lee
Reynald Lercier
Christina Lindenberg
Moses Liskov
Yi Lu
Christoph Ludwig
Anna Lysyanskaya
Greg Maitland
John Malone-Lee
Keith Martin
Sebastiá Martín
Natsume Matsuzaki
Lorenz Minder
Serge Mister
Payman Mohassel
Jean Monnerat
Paz Morillo
Tim Moses
Siguna Mueller
Frederic Muller
Sean Murphy
Toru Nakanishi
Deholo Nali
Anderson Nascimento
Gregory Neven
Phong Nguyen
Antonio Nicolosi
Jesper Nielsen
Wakaha Ogata
Kazuo Ohta
Koji Okada
Takeshi Okamoto
Tatsuaki Okamoto
Rafail Ostrovsky
Raphael Overbeck
Michael Paddon

Table of Contents

Hash Functions

Oblivious Transfer

Numbers and Lattices

Foundations

Invited Talk II

Block Ciphers

Cryptography Without Random Oracles

Multiparty Computation

Cryptography for Groups

Security Analysis of the Strong Diffie-Hellman Problem

Jung Hee Cheon

ISaC and Dept. of Mathematics, Seoul National University, Republic of Korea
jhcheon@snu.ac.kr
http://www.math.snu.ac.kr/~jhcheon

Abstract. Let g be an element of prime order p in an abelian group and $\alpha \in \mathbb{Z}_p$. We show that if g, g^α, and g^{α^d} are given for a positive divisor d of $p - 1$, we can compute the secret α in $O(\log p \cdot (\sqrt{p/d} + \sqrt{d}))$ group operations using $O(\max\{\sqrt{p/d}, \sqrt{d}\})$ memory. If g^{α^i} $(i = 0, 1, 2, \ldots, d)$ are provided for a positive divisor d of $p + 1$, α can be computed in $O(\log p \cdot (\sqrt{p/d} + d))$ group operations using $O(\max\{\sqrt{p/d}, \sqrt{d}\})$ memory. This implies that the strong Diffie-Hellman problem and its related problems have computational complexity reduced by $O(\sqrt{d})$ from that of the discrete logarithm problem for such primes.

Further we apply this algorithm to the schemes based on the Diffie-Hellman problem on an abelian group of prime order p. As a result, we reduce the complexity of recovering the secret key from $O(\sqrt{p})$ to $O(\sqrt{p/d})$ for Boldyreva's blind signature and the original ElGamal scheme when $p - 1$ (resp. $p + 1$) has a divisor $d \leq p^{1/2}$ (resp. $d \leq p^{1/3}$) and d signature or decryption queries are allowed.

Keywords: Discrete logarithm, Diffie-Hellman, strong Diffie-Hellman, ElGamal encryption, blind signature.

1 Introduction

Let g be an element of prime order p in an abelian group and $\alpha \in \mathbb{Z}_p$. The ℓ-Strong Diffie-Hellman (ℓ-SDH) problem asks to find $g^{\alpha^{\ell+1}}$ given $g, g^\alpha, \ldots, g^{\alpha^\ell}$. Recently, many cryptographic schemes including encryption, signature, and key management schemes are proposed on the basis of the Strong Diffie-Hellman (SDH) problem [MSK02, BB04e, BB04s], or its variants such as the Bilinear Diffie-Hellman problem [BBS04, DY05] and the Bilinear Diffie-Hellman Exponent (BDHE) problem [BBG05, BGW05]. A lower bound on the computational complexity of the SDH problem or its variants for generic groups are known in the sense of Shoup [Sho97], but it does not guarantee the security for specific parameters.

In this paper, we analyze the security of the SDH problem. More precisely, we show that if g, g^α and g^{α^d} are given for a positive divisor d of $p - 1$, the secret $\alpha \in \mathbb{Z}_p$ can be computed in $O(\log p \cdot (\sqrt{p/d} + \sqrt{d}))$ group operations using $O(\max\{\sqrt{p/d}, \sqrt{d}\})$ memory. If g^{α^i} $(i = 0, 1, 2, \ldots, d)$ are provided for a positive

S. Vaudenay (Ed.): EUROCRYPT 2006, LNCS 4004, pp. 1–11, 2006.

divisor d of $p+1$, it can be computed in $O(\log p \cdot (\sqrt{p/d} + d))$ group operations using the same size of memory. This implies that the strong Diffie-Hellman problem and its related problems have computational complexity reduced by $O(\sqrt{d})$ from that of the discrete logarithm problem for such primes. Hence it is necessary to increase by the size of d the key size of the cryptographic schemes based on the ℓ-SDH problem or its variants if the base group has such a prime as its order.

We investigate some known elliptic curve parameters and find that either $p-1$ or $p+1$ has many small divisors for the largest prime divisor p of its order for each elliptic curve in [NIST, BLS01, KM05, MIRACL]. For example, if we use the curve E^+ over $GF(3^{155})$ [BLS01] for the broadcast encryption [BGW05], the secret key can be computed in $O(2^{59})$ exponentiations (resp. $O(2^{42})$ exponentiations) when the number of users is 2^{32} (resp. 2^{64}), rather than $O(2^{76})$ group operations.

Moreover, we apply this algorithm to the schemes based on the Diffie-Hellman problem on an abelian group of prime order p. As a result, we show the complexity of recovering the secret key is reduced from $O(\sqrt{p})$ to $O(\sqrt{p/d})$ for Boldyreva's blind signatures [Bol03] when d signature or decryption queries are allowed and $p-1$ has a divisor $d \leq p^{1/2}$ or $p+1$ has a divisor $d \leq p^{1/3}$. Similar results hold for the original ElGamal scheme [ElG85] with decryption oracles and the conference keying protocol by Burmester-Desmedt [BD94] with key issuing oracles.

The rest of the paper is organized as follows: In Section 2, we introduce the SDH related problems and some schemes based on them. In Section 3, we present our algorithms. In Section 4, we exploit our algorithms to attack several protocols based on the Diffie-Hellman problem. In Section 5, we investigate some known elliptic curve parameters in order to check if our algorithms are applicable for these parameters. We conclude in Section 6.

2   Strong Diffie-Hellman Problems and Their Variants

Let G be an abelian group of prime order p and g a generator of G. The **Discrete Logarithm (DL) Problem** in G asks to find $a \in \mathbb{Z}_p$ given g and g^a in G. Many cryptosystems are designed on the basis of the DL problem, but most of them have the security equivalent to a weaker variant of the DL problem rather than the DL problem itself. Two most important weaker variants are as follows:

> **The Computation Diffie-Hellman (CDH) Problem.** Given (g, g^a, g^b), compute g^{ab}.
> **The Decisional Diffie-Hellman (DDH) Problem.** Given (g, g^a, g^b, g^c), decide whether $c = ab$ in $\mathbb{Z}_p$.

Recently, some weakened variants of the CDH problem are introduced and being used to construct cryptosystems for various functionalities or security without random oracles. One characteristic of these problems is to disclose $g, g^\alpha, \ldots, g^{\alpha^\ell}$ for the secret α and some integer ℓ.

The ℓ-weak Diffie-Hellman (ℓ-wDH) Problem. Given g and g^{α^i} in G for $i = 1, 2, \ldots, \ell$, compute $g^{1/\alpha}$. This problem was introduced by Mitsunari, Sakai, and Kasahara for a traitor tracing scheme [MSK02].

The ℓ-Strong Diffie-Hellman (ℓ-SDH) Problem. Given g and g^{α^i} in G for $i = 1, 2, \ldots, \ell$, compute $g^{\alpha^{\ell+1}}$. This problem is considered as a weaker version of ℓ-wDH problem. It was first introduced by Boneh and Boyen to construct a short signature scheme, that is provably secure in the standard model (without random oracles) [BB04s], and later a short group signature scheme [BBS04].

The SDH problem is generalized into a group with bilinear maps. We further assume that $e : G \times G \rightarrow G'$ is an admissible bilinear map between two abelian groups G and G' with prime order p.

The ℓ-Bilinear Diffie-Hellman Inversion (ℓ-BDHI) Problem. Given g and g^{α^i} in G for $i = 1, 2, \ldots, \ell$, compute $e(g, g)^{1/\alpha} \in G'$. This problem was introduced by Boneh and Boyen to construct an identity-based encryption that is secure in the standard model [BB04e]. It is also used to construct verifiable random functions [DY05].

The ℓ-Bilinear Diffie-Hellman Exponent (ℓ-BDHE) Problem. Given g, h, and g^{α^i} ($i = 1, 2, \ldots, \ell-1, \ell+1, \ldots, 2\ell$) in G, compute $e(g, h)^{\alpha^\ell} \in G'$. This problem was introduced by Boneh, Boyen, and Goh [BBG05] to construct a hierarchical identity-based encryption scheme with constant size ciphertext, and later used for a public key broadcast encryption scheme with constant size transmission overhead [BGW05].

Given two problem instances A and B, we denote by $A \geq B$ if the problem B can be solved in polynomial time with polynomially many queries to the oracle to solve the problem A. Then we can easily deduce the following relations among the DL related problems [BBG05]:

$$DL \geq CDH \geq DDH \geq \ell\text{-wDH} \geq \ell\text{-SDH} \geq \ell\text{-BDHI}, (\ell+1)\text{-BDHE}.$$

3 Main Results

Theorem 1. *Let g be an element of prime order p in an abelian group. Suppose that d is a positive divisor of $p - 1$. If g, $g_1 := g^\alpha$ and $g_d := g^{\alpha^d}$ are given, α can be computed in $O(\log p \cdot (\sqrt{(p-1)/d} + \sqrt{d}))$ group operations using $O(\max\{\sqrt{(p-1)/d}, \sqrt{d}\})$ memory.*

Proof. Note that $\mathbb{Z}_p^*$ is a cyclic group with $\phi(p-1)$ generators, where $\phi(\cdot)$ is the Euler totient function. Since a random element in $\mathbb{Z}_p^*$ is a generator with probability

$$\frac{\phi(p-1)}{(p-1)} > \frac{1}{6 \log\log(p-1)},$$

which is large enough [MOV, p.162], we can easily take a generator of $\mathbb{Z}_p^*$. Let ζ_0 be a generator of $\mathbb{Z}_p^*$. Then we compute $\zeta = \zeta_0^d$ that is an element of order $(p-1)/d$ in $\mathbb{Z}_p^*$.

Since $(\alpha^d)^{(p-1)/d} = 1$ and ζ generates all $(p-1)/d$-th roots of unity in $\mathbb{Z}_p^*$, there exists a non-negative integer i less than $(p-1)/d$ such that $\alpha^d = \zeta^i$. If we take $d_1 = \lceil \sqrt{(p-1)/d} \rceil$, we must have

$$(\alpha^d)\zeta^{-u} = \zeta^{d_1 v}$$

for some $0 \le u, v < d_1$. It is equivalent to

$$g_d^{\zeta^{-u}} = g^{\zeta^{d_1 v}}. \tag{1}$$

We compute and store the left-hand side terms and compare them with each of right-hand side terms in Baby-Step Giant-Step style. Note that each of terms in both sides can be computed by repeated exponentiations by either ζ^{-1} or ζ^{d_1}. Thus we can find all non-negative integers u and v less than d_1 satisfying (1) in $O(d_1 \cdot \log p)$ group operations using $O(d_1)$ memory. For u and v which satisfies (1) and $u + d_1 v$ is smallest, we put $k_0 = u + d_1 v$. Then k_0 is a non-negative integer less than $(p-1)/d$.

Let $\alpha = \zeta_0^k$ for $0 \le k < p-1$. Then we have $dk \equiv dk_0 \mod (p-1)$ and so $k \equiv k_0 \mod (p-1)/d$. There exists a non-negative integer j less than d such that $k = k_0 + j(p-1)/d$. If we take $d_2 = \lceil \sqrt{d} \rceil$, we must have

$$\alpha \zeta_0^{-u'(p-1)/d} = \zeta_0^{k_0 + d_2 v'(p-1)/d}$$

for some $0 \le u', v' < d_2$. It is equivalent to

$$g_1^{\zeta_0^{-u'(p-1)/d}} = g^{\zeta_0^{k_0 + d_2 v'(p-1)/d}}. \tag{2}$$

By the same method as above, we can find non-negative integers u' and v' less than d_2 satisfying (2) in $O(d_2 \cdot \log p)$ group operations and $O(d_2)$ memory. This completes the proof. $\square$

We remark that the memory requirement of the above algorithm can be reduced by using Pollard's lambda techniques [Pol78]. We use the notation of Theorem 1 to sketch the idea: First we consider a function $F : \mathbb{Z}_p \to \mathbb{Z}_p$ with $F(x) = x\zeta^{f(g^x)}$ for a pseudo-random function $f : \langle g \rangle \to \mathbb{Z}_{(p-1)/d}$. For $\beta \in \mathbb{Z}_p$ and $t \ge 1$, $g^{F^t(\beta)}$ can be computed from g and g^β in $O(t \log p)$ group operations by using

$$g^{F(\beta)} = \left(g^\beta\right)^{\zeta^{f(g^\beta)}} \quad \text{and} \quad g^{F^i(\beta)} = \left(g^{F^{i-1}(\beta)}\right)^{\zeta^{f(g^{F^{i-1}(\beta)})}} \quad \text{if } i \ge 2.$$

If we find u, v such that $g^{F^u(\alpha^d)} = g^{F^v(1)}$, we have $F^u(\alpha^d) = F^v(1)$ in $\mathbb{Z}_p$ and so

$$\alpha^d \zeta^{\sum_{i=1}^{u} f(g^{F^{i-1}(\beta)})} = \zeta^{\sum_{j=1}^{v} f(g^{F^{i-1}(1)})}.$$

Hence if we store only distinguished points [Tes98], α^d can be computed in $O(\sqrt{(p-1)/d})$ exponentiations using small memory with some probability. The second part to compute α from g^α and α^d can be done using similar technique.

If we know $g^{\alpha^{(p-1)/d}}$ for many small d, we can do even better:

Corollary 1. *Let g be an element of prime order p in an abelian group. Suppose that $p-1 = d_1 d_2 \cdots d_t$ for pairwise prime d_i's. If g and $g_{(p-1)/d_i} := g^{\alpha^{(p-1)/d_i}}$ for $1 \le i \le t$ are given, α can be computed in $O(\log p \cdot \sum_{i=1}^{t} \sqrt{d_i})$ group operations using $O(\max_{1 \le i \le t} \sqrt{d_i})$ memory.*

Proof. Let ζ be a generator of $\mathbb{Z}_p^*$ and $\alpha = \zeta^k$. Since $(\alpha^{(p-1)/d_i})^{d_i} = 1$, there must be a non-negative integer k_i less than d_i satisfying $\alpha^{(p-1)/d_i} = (\zeta^{(p-1)/d_i})^{k_i}$. Hence by checking

$$g_{(p-1)/d_i} = g^{(\zeta^{(p-1)/d_i})^{k_i}} \quad \text{for } 0 \le k_i < d_i$$

or

$$\left(g_{(p-1)/d_i}\right)^{(\zeta^{(p-1)/d_i})^{-u_i}} = g^{(\zeta^{(p-1)/d_i})^{\lceil \sqrt{d_i} \rceil v_i}} \quad \text{for } 0 \le u_i, v_i < \lceil \sqrt{d_i} \rceil.$$

we can compute k_i in $O(\log p \cdot \sqrt{d_i})$ group operations using $O(\sqrt{d_i})$ memory. Since k satisfies $k \equiv k_i \mod d_i$, we can compute k by performing the above step for $1 \le i \le t$ and using Chinese Remainder Theorem. The total complexity is $O(\log p \cdot \sum_{i=1}^{t} \sqrt{d_i})$ using $O(\max_{1 \le i \le t} \sqrt{d_i})$ memory. $\square$

Next, we use an imbedding of $\mathbb{Z}_p$ into $\mathbb{F}_{p^2}$ to generalize Theorem 1.

Theorem 2. *Let g be an element of prime order p in an abelian group. Suppose that d is a positive divisor of $p+1$ and $g_i := g^{\alpha^i}$ for $i = 1, 2, \ldots, 2d$ are given. Then α can be computed in $O(\log p \cdot (\sqrt{(p+1)/d} + d))$ group operations using $O(\max\{\sqrt{(p+1)/d}, \sqrt{d}\})$ memory.*

Proof. Let a be a quadratic non-residue in $\mathbb{Z}_p$ and θ be a root of $x^2 = a$ in an algebraically closed field of $\mathbb{Z}_p$. Then $\mathbb{Z}_p[\theta] \cong \mathbb{F}_{p^2}$. Let H be a subgroup of order $p+1$ of $\mathbb{F}_{p^2}$. Since $\beta \in H$ is equivalent to $\beta^{p+1} = 1$, we see that $\beta_0 + \beta_1 \theta$ is an element of H for $\beta_0 = (1 + a\alpha^2)/(1 - a\alpha^2)$ and $\beta_1 = 2\alpha/(1 - a\alpha^2)$ from $\theta^p = -\theta$ and

$$\beta^{p+1} = (\beta_0 + \beta_1 \theta)(\beta_0 + \beta_1 \theta^p) = \beta_0^2 - a\beta_1^2. \tag{3}$$

Let ζ_0 be a generator of H (for example, the $(p+1)$-th power of a generator of $\mathbb{F}_{p^2}^*$). Then $\zeta := \zeta_0^d$ generates all the $(p+1)/d$-th roots of unity and so there must be some $k \in \mathbb{Z}$ such that $\beta^d = \zeta^k$ and $0 \le k < (p+1)/d$. For convenience, we denote $\zeta^i = s_i + t_i \theta$ for some $s_i, t_i \in \mathbb{Z}_p$ where the index i is defined modulo $(p+1)/d$. Also we denote

$$\beta^d = (\beta_0 + \beta_1 \theta)^d = \frac{1}{(1 - a\alpha^2)^d}(f_0(\alpha) + f_1(\alpha)\theta),$$

where f_i's are polynomials of degree $2d$. Then we must have

$$\beta^d \zeta^{-u} = \zeta^{d_1 v} \tag{4}$$

for some $0 \le u, v < d_1 := \lceil \sqrt{(p+1)/d} \rceil$. It is equivalent to

$$(f_0(\alpha)s_{-u} + af_1(\alpha)t_{-u}) + (f_0(\alpha)t_{-u} + f_1(\alpha)s_{-u})\theta = (1 - a\alpha^2)^d (s_{d_1 v} + t_{d_1 v}\theta). \tag{5}$$

Hence we compute $(g^{f_0(\alpha)s_{-u} + af_1(\alpha)t_{-u}}, g^{f_0(\alpha)t_{-u} + f_1(\alpha)s_{-u}})$ for all $0 \le u < d_1$ and store them. By comparing them with $(g^{(1-a\alpha^2)^d s_{d_1 v}}, g^{(1-a\alpha^2)^d t_{d_1 v}})$ for each $0 \le v < d_1$, we can find the (unique) non-negative integers u and v less than d_1 satisfying (4) and $u + d_1 v < (p+1)/d$. We put $k_0 = u + d_1 v$. Note that $g^{f_0(\alpha)}, g^{f_1(\alpha)}$ and $g^{(1-a\alpha^2)^d}$ can be computed from $g, g_1, \ldots, g_{2d}$ in $6d$ exponentiations. Hence k_0 can be found in $O(\log p \cdot (6d + \sqrt{(p+1)/d}))$ group operations with $O(\sqrt{(p+1)/d})$ memory.

Let $\beta = \zeta_0^k$ for $0 \le k < p+1$. Then we have $k \equiv k_0 \mod (p+1)/d$. There exists a non-negative integer j less than d such that $k = k_0 + j(p+1)/d$. If we take $d_2 = \lceil \sqrt{d} \rceil$, there must exist non-negative integers u', v' less than d_2 such that

$$\beta \zeta_0^{-u'(p+1)/d} = \zeta_0^{k_0 + d_2 v'(p+1)/d}. \tag{6}$$

We denote $\zeta_0^{-i(p+1)/d} = s_i' + t_i'\theta$ and $\zeta_0^{k_0 + d_2 i(p+1)/d} = s_i'' + t_i''\theta$ for some $s_i', t_i', s_i'', t_i'' \in \mathbb{Z}_p$ where the index i is defined modulo $(p+1)$. Then (6) is equivalent to

$$((1 + a\alpha^2)s_{u'} + 2a\alpha t_{u'}) + ((1 + a\alpha^2)t_{u'} + 2\alpha s_{u'})\theta = (1 - a\alpha^2)(s_{v'} + t_{v'}\theta). \tag{7}$$

Hence we compute $(g^{(1+a\alpha^2)s_{u'} + 2a\alpha t_{u'}}, g^{(1+a\alpha^2)t_{u'} + 2\alpha s_{u'}})$ for all $0 \le u' < d_2$ and store them. By comparing them with $(g^{(1-a\alpha^2)s_{v'}}, g^{(1-a\alpha^2)t_{v'}})$ for each $0 \le v' < d_2$, we can find non-negative integers u' and v' satisfying (6). That is, $\beta = \zeta_0^{k_0 + (u' + d_2 v')(p+1)/d}$ can be found in $O(\log p \cdot \sqrt{d})$ group operations and $O(\sqrt{d})$ memory. This completes the proof. $\square$

We remark that if $d \le p^{1/3}$, then Theorem 2 says that the secret can be computed in $O(\log p \cdot \sqrt{p/d})$ group operations using $O(\sqrt{p/d})$ memory.

Remark 1. We may consider that our proof utilizes Diffie-Hellman oracles in a very restricted way [Boe88, MW99]. That is, in our situations we can use the Diffie-Hellman oracle $DH(g^x, g^y) = g^{xy}$ only when x is fixed and $y = x^\ell$ for some small ℓ. This restriction is an obstacle when we try to generalize the proposed algorithm into other extension fields of $\mathbb{F}_p$ or elliptic or hyperelliptic curves over $\mathbb{F}_p$.

4   Analysis of Cryptographic Schemes Based on the Diffie-Hellman Problem

4.1   Blind Signature Based on the GDH Assumption

The Gap-Diffie-Hellman (GDH) group is an abelian group on which there is an polynomial time algorithm to solve the decisional Diffie-Hellman problem and there is no polynomial time algorithm to solve the computation Diffie-Hellman problem.

Boldyreva proposed a blind signature scheme on a Gap-Diffie-Hellman group [Bol03]. The scheme is as follows: Let G be a GDH group of prime order p and g a generator of G. Let $H : \{0,1\}^* \to G$ be a full domain hash function [BLS01]. A signer has a private key $x \in \mathbb{Z}_p$ and the corresponding public key $y = g^x$. In order to blindly sign a message $m \in \{0,1\}^*$, a user picks a random $k \in \mathbb{Z}_p^*$, computes $M' = H(m)g^k$, and sends it to the signer. The signer computes $\sigma' = (M')^x$ and sends it back to the user. Then the user computes the signature $\sigma = \sigma'/y^k (= H(m)^x)$ of the message m.

This scheme is shown to be secure against one-more forgery under chosen message attacks in the random oracle model [Bol03], that is the standard security notion for blind signature schemes. However, since the signer does not have any information on the message to be signed, we may use this blind signing phase as a Diffie-Hellman oracle and so reduce the security of this scheme under chosen message attacks: A chosen-message attacker $\mathcal{A}$ takes a random $\gamma_1 \in \mathbb{Z}_p$ and requests a signature on the message $y \cdot g^{\gamma_1}$. From the signature $\sigma_1 = (y \cdot g^{\gamma_1})^x$, $\mathcal{A}$ obtains $g_2 := g^{x^2} = \sigma_1/y^{\gamma_1}$. Second, $\mathcal{A}$ takes another random $\gamma_2 \in \mathbb{Z}_p$ and requests a signature on the message $g_2 \cdot g^{\gamma_1}$. From the signature $\sigma_2 = (g_2 \cdot g^{\gamma_2})^x$, $\mathcal{A}$ obtains $g_3 := g^{x^3} = \sigma_2/y^{\gamma_2}$. If ℓ signature queries are allowed, $\mathcal{A}$ repeats this procedure ℓ times to obtain $g_1, g_2, \ldots, g_{\ell+1}$ ($g_i := g^{x^i}$). By Theorem 1 and 2, if $p-1$ has a divisor $d \leq \min\{\ell+1, p^{1/2}\}$ or $p+1$ has a divisor $d \leq \min\{(\ell+1)/2, p^{1/3}\}$, the secret key x can be computed in $O(\sqrt{p/d})$. That is, the security of the scheme is reduced by $O(\sqrt{d})$ from that of the GDH assumption.

We note that the attack does not imply that the security proof of the scheme is wrong, but that more quantitative analysis on security reduction is required. In fact, the security proof of BLS signatures on which the Boldyreva's blind signature scheme is based shows that the advantage of an adversary can be increased by q_S when q_S signature queries are allowed [BLS01].

This method can be applied similarly to schemes which respond by its secret key power for an unknown message. For example, the conference keying protocol by Burmester-Desmedt has this property [BD94]. Thus, in this case, we need to take the order carefully or raise the security parameter.

4.2 Original ElGamal Encryption Scheme

We briefly introduce the original ElGamal encryption scheme in a generalized form: Let G be an abelian group of prime order p and g a generator of G. Suppose the secret key and the public key of the recipient is $x \in \mathbb{Z}_p$ and g^x, respectively. To encrypt a message $m \in G$, a sender takes a random $k \in \mathbb{Z}_p$ and sends a ciphertext $(c_1, c_2) := (g^k, mg^x)$ to the recipient. The recipient recovers the message m by computing c_2/c_1^x.

The ElGamal encryption is known not to satisfy non-malleability under chosen ciphertext attacks (Refer to the appendix in [ABR98]). That is, given a decryption oracle any target ciphertext can be decrypted without feeding itself to the decryption oracle. Here we show that the decryption oracle enables not only a decryption of any target ciphertext without the secret key, but also a reduction of the complexity to compute the secret key in some cases.

As in the previous subsection, first a chosen ciphertext attacker $\mathcal{A}$ takes random numbers $k_1, k_2 \in \mathbb{Z}_p$, requests a decryption of the ciphertext $(c_1, c_2) := (y^k, y^{k'})$ to the decryption oracle, and obtains $c_2/c_1^x = g^{xk'} \cdot g^{x^2 k}$. Since he knows k, k' and g^x, $\mathcal{A}$ can compute $g_2 := g^{x^2}$. By taking different random pairs (k, k') and replacing y by g_2, $\mathcal{A}$ can obtain $g_3 := g^{x^3}$ similarly. By repeating this procedure ℓ times, $\mathcal{A}$ can obtain $g_1, g_2, \ldots, g_\ell$ ($g_i := g^{x^i}$) when ℓ decryption queries are allowed. By Theorem 1 and 2, if $p - 1$ has a divisor $d \leq \min\{\ell, p^{1/2}\}$ or $p + 1$ has a divisor $d \leq \min\{\ell/2, p^{1/3}\}$, the secret key x can be computed in $O(\sqrt{p/d})$.

We might imagine a situation that this attack is harmful: One uses the original ElGamal encryption scheme, to encrypt not so important messages, with another cryptosystem having the same secret key. Then the secret key may be revealed from the original ElGamal encryption scheme and so the other system can be insecure. This shows that the original ElGamal scheme must not share the same secret key with another system.

5   Practicality of the Proposed Algorithm

In this section, we discuss the potential of the proposed algorithms. The algorithm in Theorem 1 has complexity $O(\log p \cdot (\sqrt{(p-1)/d} + \sqrt{d}))$ for a divisor d of $p - 1$. The complexity achieves the minimum value $O(\log p \cdot p^{1/4})$ when $d = O(p^{1/2})$. The algorithm in Theorem 2 has complexity $O(\log p \cdot (\sqrt{(p-1)/d} + d))$ for a divisor d of $p+1$. The complexity achieves the minimum value $O(\log p \cdot p^{1/3})$ when $d = O(p^{1/3})$. Hence the security of the ℓ-SDH problem on an abelian group of order p can be reduced up to $O(\log p \cdot p^{1/4})$ (resp. $O(\log p \cdot p^{1/3})$) for large ℓ if $p - 1$ (resp. $p + 1$) has a divisor $d = O(p^{1/2})$ (resp. $d = O(p^{1/3})$).

Now we give an example in which security reduction due to our algorithm yields a serious security problem.

Example 1. We consider the situation that $E^+(\mathbb{F}_{3^{97}})$ [BLS01] is used for the broadcast encryption scheme [BGW05]. $E^+(\mathbb{F}_{3^{97}})$ has a subgroup G of 151 bit prime order p. Let g be a generator of G and $\alpha \in \mathbb{Z}_p$ be the system secret. The scheme assuming n users publishes g and $g_i := g^{\alpha^i}$ for $0 \leq i \leq 2n$, $i \neq n$. Using a non-degenerate bilinear map e on G, we can compute $e(g, g)^{\alpha^i}$ for all non-negative integers $i \leq 4n$. Using Pollard ρ method [Pol78], the secret key can be found in $O(2^{76})$ group operations. But if we apply the proposed algorithm, it is reduced to about $O(2^{59})$ exponentiations or $O(2^{67})$ group operations for $n = 2^{32}$. Furthermore, if we use $n = 2^{64}$ as in the file sharing application [BGW05], the complexity is reduced to $O(2^{42})$ exponentiations or $O(2^{50})$ group operations.

We remark that in order to give 2^{80} security for the system with 2^{64} users, it is recommended to take the group of about 220 bit prime order unless p is of a special form.

Most cryptosystems based on SDH-related problems make use of bilinear maps. For practice, we investigate some known elliptic curve parameters and show that either $p - 1$ or $p + 1$ has many small divisors for the largest prime divisor p of the order for each elliptic curve in [NIST, BLS01, KM05, MIRACL].

NIST curves. NIST suggested several elliptic curves for federal government use [NIST]. They consist of three categories: Pseudo-random curves over a prime field, a pseudo-random curve over a binary field, and a Koblitz curve over a binary field. For most of them, the largest prime divisor p has the property that either $p - 1$ or $p + 1$ has enough small divisors. We present some of them:

- B-163: $p - 1 = 2 \cdot 53 \cdot 383 \cdot 21179 \cdot$ (a 132 bit prime), which is a 163 bit integer.
- K-163: $p - 1 = 2^4 \cdot 43 \cdot 73 \cdot$ (a 16 bit prime) $\cdot$ (an 18 bit prime) $\cdot$ (a 112 bit prime), which is a 163 bit integer.
- P-192: $p - 1 = 2^4 \cdot 5 \cdot 2389 \cdot$ (an 83 bit prime) $\cdot$ (a 92 bit prime), which is a 192 bit integer.

We note that P-192 gives the smallest security loss, that is 8 bits, if the parameter ℓ in the SDH problem is less than 83 bits. Otherwise, however, the security loss for P-192 can be more than 40 bits.

Elliptic curves with embedding degree 6. Boneh, Lynn and Shacham suggested two families of elliptic curves with embedding degree 6 for short signatures [BLS01]: $E^+ : y^2 = x^3 + 2x + 1$ and $E^- : y^2 = x^3 + 2x - 1$ over $\mathbb{F}_3$. We consider E^+ or E^- over $\mathbb{F}_{3^\lambda}$. We denote by p the largest prime factor of $E^\pm(\mathbb{F}_{3^\lambda})$.

- $E^+(\mathbb{F}_{3^{97}})$: $p - 1 = 2 \cdot 3^{49} \cdot 24127552321 \cdot 21523361 \cdot 76801$, which is a 151 bit integer.
- $E^+(\mathbb{F}_{3^{121}})$: $p - 1 = 2 \cdot 3 \cdot 11^2 \cdot 683 \cdot 6029 \cdot$ (a 123 bit prime), which is a 155 bit integer.

Koblitz-Menezes curves. Koblitz and Menezes [KM05] suggested seven supersingular elliptic curve parameters for pairing based cryptography. If we denote by p the order of the group to be used in cryptosystems, either $p + 1$ or $p - 1$ has divisor 2^i for $i \geq 60$ in all cases except one. The exceptional case is $p = 2^{160} + 2^3 - 1$. In this case, however, $p - 1 = 2 \cdot 29 \cdot 227 \cdot 27059 \cdot$ (a 37 bit prime) $\cdot$ (a 94 bit prime).

Elliptic curves in MIRACL library. MIRACL library [MIRACL] provides a sample parameter for pairing-friendly elliptic curves. The order of the group is $p = 2^{159} + 2^{17} + 1$. Then $p - 1$ has the following prime factorization: $p - 1 = 2^{17} \cdot 5 \cdot 569 \cdot$ (a 27 bit prime) $\cdot$ (a 32 bit prime) $\cdot$ (a 32 bit prime) $\cdot$ (a 39 bit prime).

We can see that our algorithm can be applied for all the examples above. We note that our algorithm is more plausible for pairing-friendly curves including Koblitz-Menezes curves and MIRACL library curves because a curve with an order of small Hamming weights in signed binary form admits efficient implementation of Weil or Tate pairing. In most cases, however, it is necessary and seems hard to find a prime p such that both of $p - 1$ and $p + 1$ have no small divisor greater than $(\log p)^2$. We may consider Gordon's algorithm [Gor84] to generate strong primes which resist against the proposed algorithms. Basically, the algorithm is to find a prime of the form $p = 2(p_1^{p_2-2} \bmod p_2)p_1 - 1 + p_1 p_2 k$ where p_1 and p_2 are primes of equal size and k is an integer. Then we have $p_1|p+1$ and $p_2|p-1$. But this algorithm usually yields a prime much larger than p_1 and p_2. It would be an interesting problem to find elliptic curve parameters for which the security loss of the SDH is minimized.

6   Conclusion and Further Studies

In this paper, we proposed a novel algorithm to solve the SDH-related problems. More precisely, given an element g of prime order p in an abelian group and a secret $\alpha \in \mathbb{Z}_p$, if g^{α^i} $(0 \le i \le \ell)$ are published for the secret α, the complexity to recover α can be reduced by a factor of $\sqrt{d}$ from that of the DLP, where d is the maximum of the largest divisor of $p - 1$ not exceeding $\min\{\ell, p^{1/2}\}$ and the largest divisor of $p + 1$ not exceeding $\min\{\ell/2, p^{1/3}\}$. This algorithm can be used to attack cryptographic schemes that admit an oracle to return its secret key power upon an arbitrary input.

Hence, if a cryptographic scheme or protocol is based on a variant of ℓ-SDH problems or allows such an oracle by ℓ times, it is recommended to increase the key size or use a prime p such that both of $p + 1$ and $p - 1$ have no small divisor greater than $(\log p)^2$. However, we have no idea about the distribution of such primes.

We may try to generalize the proposed algorithms as in [MW99]. One problem is to find an embedding of $\mathbb{F}_p$ to some other groups including extension fields of $\mathbb{F}_p$ and elliptic or hyperelliptic curves over $\mathbb{F}_p$.

Acknowledgement. I am grateful to Dong Hoon Lee and Taekyoung Kwon for helpful discussions and JaeHong Seo for his implementation. I would also like to thank the anonymous reviewers for their valuable suggestions.

References

[ABR98]   M. Abdalla, M. Bellare, and P. Rogaway, "DHAES: An encryption scheme based on Diffie-Hellman problem," IEEE P1363a Submission, 1998, Available at http://grouper.ieee.org/groups/1363/addendum.html.

[BB04e]   D. Boneh and X. Boyen, "Efficient Selective-ID Secure Identity-Based Encryption Without Random Oracles," Eurocrypt 2004, LNCS 3027, Springer-Verlag, pp. 223-238, 2004.

[BB04s]   D. Boneh and X. Boyen, "Short Signatures Without Random Oracles," Eurocrypt 2004, LNCS 3027, Springer-Verlag, pp. 56-73, 2004.

[BBG05]   D. Boneh, X. Boyen, and E. Goh, "Hierarchical Identity Based Encryption with Constant Size Ciphertext," Eurocrypt 2005, LNCS 3494, Springer-Verlag, pp. 440-456, 2005.

[BBS04]   D. Boneh, X. Boyen, and H. Shacham, "Short Group Signatures," Crypto 2004, LNCS 3152, Springer-Verlag, pp. 41-55, 2004.

[BD94]   M. Burmester and Y. Desmedt, "A Secure and Efficient Conference Key Distribution System (Extended Abstract)," Eurocrypt 1994, LNCS 950, Springer-Verlag, pp. 275-286, 1994.

[BGW05]   D. Boneh, C. Gentry, and B. Waters. "Collution Resistant Broadcast Encryption with Short Ciphertexts and Private Keys," Crypto 2005, LNCS 3621, Springer-Verlag, pp. 258-275, 2005.

[BLS01]   D. Boneh, B. Lynn, and H. Shacham, "Short Signatures from the Weil Pairing," J. of Cryptology, Vol. 17, No. 4, pp. 297-319, 2004. Extended abstract in proceedings of Asiacrypt '01, LNCS 2248, Springer-Verlag, pp. 514-532, 2001.

[Boe88]   B. den Boer, "Diffie-Hellman is as Strong as Discrete Log for Certain Primes," Crypto '88, LNCS 403, Springer-Verlag, pp. 530-539, 1989.

[Bol03]   A. Boldyreva, "Threshold Signatures, Multisignatures and Blind Signatures Based on the Gap-Diffie-Hellman-Group Signature Scheme," Public Key Cryptography 2003, LNCS 2567, pp. 31-46, 2003.

[DY05]   Y. Dodis and A. Yampolskiy, "A Verifiable Random Function with Short Proofs and Keys," Public Key Cryptography 2005, LNCS 3386, pp. 416-431, 2005.

[ElG85]   T. Elgamal, "A Public Key Cryptosystem and a Signature Scheme based on Discrete Logarithms," IEEE Transactions on Information Theory, Vol. 31, no 4, pp. 469-472, 1985.

[Gor84]   J. Gordon, "Strong Primes are Easy to Find," Eurocrypt '84, LNCS 209, Springer-Verlag, pp. 216-223, 1984.

[KM05]   N. Koblitz and A. Menezes, "Pairing-based Cryptography at High Security Levels," IMA Conference of Cryptography and Coding 2005, pp. 13-36, 2005.

[MIRACL]   M. Scott, *Multiprecision Integer and Rational Arithmetic C/C++ Library*, Available at http://indigo.ie/~mscott/.

[MOV]   A. Menezes, P. van Oorschot, and S. Vanstone, *Handbook of Applied Cryptography*, CRC Press, 1996.

[MSK02]   S. Mitsunari, R. Sakai, and M. Kasahara, "A New Traitor Tracing," IEICE Trans. Fundamentals, Vol. E85-A, no. 2, pp. 481-484, 2002.

[MW99]   U. Maurer and S. Wolf, "The Relationship Between Breaking the Diffie-Hellman Protocol and Computing Discrete Logarithms," SIAM J. Comput., Vol. 28, no. 5, pp. 1689-1721, 1999.

[NIST]   *Recommended Elliptic Curves for Federal Government Use*, Available at http://csrc.nist.gov/CryptoToolkit/dss/ecdsa/NISTReCur.pdf, 1999.

[Pol78]   J. Pollard, "Monte Carlo Methods for Index Computation (mod p)," Mathematics of Computation, Vol. 32, pp. 918-924, 1978.

[Sho97]   V. Shoup, "Lower bounds for Discrete Logarithms and Related Problems," Eurocrypt '97, LNCS 1233, Springer-Verlag, pp. 256-66, 1997.

[Tes98]   E. Teske, "Speeding up Pollard's Rho Method for Computing Discrete Logarithms," Algorithmic Number Theory Symposium III, LNCS 1423, pp.541-554, 1998.

Cryptography in Theory and Practice: The Case of Encryption in IPsec*

Kenneth G. Paterson and Arnold K.L. Yau**

Information Security Group, Royal Holloway, University of London,
Egham, Surrey, TW20 0EX, United Kingdom
{kenny.paterson, a.yau}@rhul.ac.uk

Abstract. Despite well-known results in theoretical cryptography highlighting the vulnerabilities of unauthenticated encryption, the IPsec standards mandate its support. We present evidence that such "encryption-only" configurations are in fact still often selected by users of IPsec in practice, even with strong warnings advising against this in the IPsec standards. We then describe a variety of attacks against such configurations and report on their successful implementation in the case of the Linux kernel implementation of IPsec. Our attacks are realistic in their requirements, highly efficient, and recover the complete contents of IPsec-protected datagrams. Our attacks still apply when integrity protection is provided by a higher layer protocol, and in some cases even when it is supplied by IPsec itself.

Keywords: IPsec, integrity, encryption, ESP.

1 Introduction

The need for authenticated encryption is well understood in the cryptographic research community – see for example [4, 5, 14]. High-profile examples where the lack of strong integrity checks is known to lead to attacks or where inappropriate use of integrity mechanisms still leaves systems vulnerable are plentiful [3, 6, 7, 8, 28, 30]. However the process of adopting authenticated encryption in fielded systems is slower. Naturally, it takes time to translate theory into standards, standards into products and finally, for users to take up the latest versions of products. There is also resistance to change without clear and easily-absorbed evidence that such change is imperative. Attacks in the cryptographic literature can be rather technical and difficult for non-experts to understand. In some cases, it may also be that the attacks are not perceived by users as having a high impact. Theoreticians are rightly concerned about attacks on indistinguishability of ciphertexts, but users are perhaps less so. Attacks requiring huge numbers of

---

* The work described in this paper was partly supported by the European Commission under contract IST-2002-507932 (ECRYPT). An extended version is available [25].
** This author supported by EPSRC and Hewlett-Packard Laboratories Bristol through CASE award 01301027.

S. Vaudenay (Ed.): EUROCRYPT 2006, LNCS 4004, pp. 12–29, 2006.

chosen plaintexts are interesting to theoreticians, but may not unduly concern practitioners. Attacks on paper are easier to dismiss than fully demonstrated attacks that work in practice against deployed systems.

In this paper, our focus is on the use of integrity protection and encryption in IPsec, a widely-used suite of protocols providing security for IP. We provide a short introduction to IPsec in Section 2. Bellovin [6] was the first to point out that the lack of integrity protection in the first version of IPsec's encryption protocol ESP (Encapsulating Security Payload) [1] leads to security weaknesses. However, the attacks in [6] are actually quite limited in their practical impact. A close examination of [6] shows that the attacks presented in [6, Sections 3.1 and 3.2] only work in the rather unrealistic scenario where the attacker has access to accounts on the two network hosts performing the IPsec processing. The other concrete attack in [6] is contained in Section 3.8 and is attributed to Wagner. It recovers just a single byte of plaintext, from datagrams having special formats, and then only if 2^{24} ciphertexts matching chosen plaintexts are available to the attacker. Moreover, the attacks in [6] (and the related paper [22]) are really only sketches of what might be possible rather than fully implemented, working attacks: they are examples of "attacks in theory". Nevertheless, Bellovin's attacks are well-known in the cryptographic and IPsec standards communities, and are cited in subsequent versions of the ESP standards [16, 18]. The version in current use, [16], refers to [6] when warning of the dangers of using encryption without additional integrity protection, and requires support for integrity protection. However it also mandates that any implementation of ESP *must* include support for encryption-only processing. This surely illustrates the chasm that exists between the theory and practice of cryptography. Note that the developers of [16] did have good practical reasons (backward compatibility and performance) for mandating support for an encryption-only mode.

It is our belief that the availability of the encryption-only option in IPsec has led users into actually using it, in spite of Bellovin's work. After all, users do not typically read RFCs or research papers, and an inexperienced network administrator might reasonably believe that it is sufficient to use an encryption algorithm on its own to provide confidentiality for data, especially when selecting from amongst the myriad of IPsec options. (This point is also made in [10].) We have found several on-line tutorials showing how to configure IPsec VPNs using ESP for encryption with no additional integrity protection.[1] After the release of the vulnerability announcement [24] describing our attacks, we became aware that some vendors were aware of Bellovin's work and had taken steps to prevent the selection of encryption-only configurations, but others were much less well-informed, or less concerned.

1.1 Our Contribution

We present new attacks against the encryption-only configuration of IPsec that are as realistic and devastating as possible, with the aim of finally convincing

---

[1] See for example: `http://www.netbsd.org/Documentation/network/ipsec` and `http://lartc.org/howto/lartc.ipsec.tunnel.html`

users not to select it. In this respect, our attacks have several attractive features. Firstly, they are ciphertext-only attacks. Thus they do not require any special operating conditions under which, for example, the ciphertexts matching chosen plaintexts are generated. Nor do they require large amounts of ciphertext to be successful: the attacks can be mounted given only a single encrypted datagram. Secondly, the attacks merely require the attacker to be able to inject IP datagrams into the network and intercept certain responses. Some variants of our attacks even enable these responses to be sent directly to the attacker's machine. Thirdly, the attacks are very efficient. For example, one variant that we have implemented requires the injection of only a handful of datagrams to recover the complete contents of a datagram encrypted using AES. Fourthly, the attacks are flexible, with a range of variants being applicable in different circumstances. And finally, we have written an attack client which shows that the attacks work in practice against the native implementation of IPsec in Linux. For example, our client effectively allows a real-time cryptanalysis of encryption-only IPsec when AES is used as the encryption algorithm. In all these senses, our attacks improve on the pioneering work of Bellovin [6].

Our work also has consequences for the newly published version of ESP [18]. This RFC no longer requires mandatory support for encryption-only, and repeats the advice of [16] concerning the need for integrity protection, but then goes on to say: *"ESP allows encryption-only [...] because this may offer considerably better performance and still provide adequate security, e.g., when higher layer authentication/integrity protection is offered independently."* It is already known in theory that applying authentication followed by encryption to build an authenticated encryption scheme does not result in a generically secure construction [19]. We demonstrate that relying on higher layers for the provision of integrity in IPsec is inherently insecure in practice as well. Some of our attacks even apply to configurations using the IPsec protocol AH (Authentication Header) for integrity protection.

More generally, our attacks provide a stark illustration, should one still be required, of the general need to make appropriate use of authenticated encryption in fielded systems. We hope that this paper will also be of use to theoreticians in the field of authenticated encryption searching for convincing real-world examples to motivate their work.

A further theme of this paper is to illustrate the gaps that exist between cryptography as studied in theory, as defined in standards, as implemented by software engineers, and as actually consumed by users. For example, we have already commented on the differences in viewpoints of theoreticians and users, and how this can lead to the use of encryption-only ESP in practice. As another example, our attacks should in fact be prevented by any RFC-compliant implementation of IPsec, because of some seemingly innocuous post-processing checks specified in the architectural standard for IPsec [15]. Yet the native Linux version of IPsec fails to implement these checks. Drawing on our experiences with IPsec, we make some recommendations which we hope will help to bridge these gaps.

2   Background

2.1   IPsec

IPsec, as defined in RFCs 2401–2412, provides security at the IP layer. The interested reader is invited to consult [9,12] for accessible introductions to IPsec. Implementations of IPsec exist in Microsoft Windows XP, in the Linux kernel from release 2.6 onwards.[2] Various other open source projects are also developing IPsec implementations and IPsec is widely supported in commercial networking hardware. The IPsec protocols provide data confidentiality, integrity protection, data origin authentication and anti-replay services as well as supporting automated key management.

The IPsec protocols can be deployed in two basic modes: transport and tunnel. In tunnel mode, on which we focus here, cryptographic protection is provided for entire IP datagrams. In essence, a whole datagram plus security fields is treated as the new payload of an outer IP datagram, with its own header, called the outer header. The original, or inner, IP datagram is said to be *encapsulated* within the outer IP datagram. In tunnel mode, IPsec processing is typically performed at security gateways on behalf of endpoint hosts. The gateways could be perimeter firewalls or routers.

IPsec provides authentication and integrity protection and/or confidentiality services through the AH and ESP protocols. Our focus here is on the ESP protocol, as defined in [16,18]. ESP is normally invoked to provide confidentiality, and usually makes use of a block cipher algorithm operating in CBC mode. In tunnel mode, the entire inner IP datagram is encrypted and forms part of the payload of the outer IP datagram. The use in ESP of a variety of block ciphers has been specified, including DES [21], triple-DES [26] and AES [11]. ESP in tunnel mode inserts security information in the form of a header between the outer IP header and the encrypted version of the inner datagram. This ESP header indicates which algorithms and keys were used to protect the payload in a 32-bit field called the Security Parameters Index (SPI). The ESP header also contains a 32-bit sequence number to prevent packet replays; when ESP is used with encryption-only, this sequence number is simply ignored by IPsec implementations (as it is not protected in any way). ESP in tunnel mode may also append an authentication field after the encrypted portion. This contains a MAC value if ESP's optional integrity protection features are in use.

Further discussion of IPsec configuration and the combined usage of AH and ESP in tunnel and transport modes is beyond the scope of this paper. IPsec provides an automated key management service through the Internet Key Exchange (IKE) [13]. We will simply assume that key establishment for ESP has taken place, either manually or using IKE.

---

[2] All further references to Linux in this paper refer to official release 2.6.8.1 of the Linux kernel from `http://kernel.org`.

2.2   CBC Mode Encryption in ESP

We outline how CBC mode is used by ESP in tunnel mode. For more details, see [16, 21, 11, 26]. First of all, the original (inner) datagram that is to be protected is treated as a sequence of bytes. This sequence is padded and then a single Next Header byte is appended. It is permissible for the padding to be of variable length and to extend over multiple blocks. We assume throughout that the minimum amount of padding is used, though our attacks are easily modified to handle variable length padding. Let us assume that the byte sequence after padding consists of q blocks, each of n bits. We denote these blocks by $P_1, P_2, \ldots, P_q$. We use K to denote the key used for the block cipher algorithm and $e_K(\cdot)$ ($d_K(\cdot)$) to denote encryption (decryption) of blocks using key K. An n-bit initialization vector, denoted IV, is selected at random. Then ciphertext blocks are generated according to the equations:

$$C_0 = IV, \quad C_i = e_K(C_{i-1} \oplus P_i), \quad (1 \le i \le q).$$

The encrypted portion of the outer datagram is then defined to be the sequence of $q + 1$ blocks $C_0, C_1, \ldots, C_q$.

At the receiving security gateway, the payload of the outer datagram can be recovered using the equations: $P_i = C_{i-1} \oplus d_K(C_i), 1 \le i \le q$. Any padding and the Next Header byte can then be stripped off. At this point, Section 5.2 of the IPsec architectural RFC [15] mandates that implementations should check that the cryptographic processing performed to recover the inner datagram does in fact match that specified in local IPsec policies. Presumably, if the check fails, the datagram should be dropped, though this is not made explicit in [15].[3] In the Linux kernel implementation of IPsec, the inner datagram is passed directly to the IP software on the receiving gateway, without any policy checks being performed. This IP software usually just routes the inner datagram to the intended destination specified in the destination address of the inner datagram.

2.3   Bit Flipping Attacks

CBC mode has a well-known weakness, commonly known as the bit flipping vulnerability. Suppose an attacker captures a CBC mode ciphertext $C_0, C_1, \ldots, C_q$, then flips (inverts) a specific bit j in C_{i-1} and injects the modified ciphertext into the network. Upon receipt and decryption, this bit flip is transformed into a bit flip in position j in the plaintext block P_i. This can be seen by examining the decryption equation $P_i = C_{i-1} \oplus d_K(C_i)$. Thus an attacker can introduce controlled changes into the value of block P_i seen by the decrypting party, simply by flipping bits in C_{i-1} and injecting modified ciphertexts.

Of course, a problem for the attacker is that any modification to C_{i-1} typically results in a value of P_{i-1} that is effectively randomized. On the other hand, if the modification is made in C_0 (equal to IV), then no damage to plaintext blocks will result.

---

[3] Note that these checks are not specified in the ESP RFCs [16, 18]. The requirement to drop datagrams has now been made explicit in [17].

2.4 IP Datagram Headers

The execution of our attacks on ESP in tunnel mode depends in a detailed way on the structure of the headers of IP datagrams and on the order in which the fields of these headers are processed. We focus here only on IPv4 headers, as specified in detail in [20], and on describing those fields that are key to our attacks. The lay-out of the IP header is shown schematically in Figure 1.

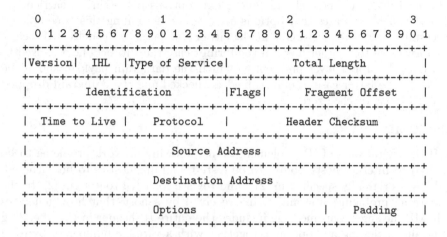

Fig. 1. Structure of IP header according to RFC 791, [20]

The IHL (Internet Header Length) field is 4 bits long and has a value between 5 and 15. This field indicates the length of the header in 32-bit words. The typical value is 5; larger values indicate that options bytes are present after the main header, in the Options field. This field can be up to ten 32-bit words (40 bytes) in length. It has a strict format; if the format is not followed, then IP implementations typically generate an ICMP (Internet Control Message Protocol) "parameter problem" message which is routed to the host indicated in the Source Address field. Experiments confirm that, upon receipt of a datagram with random bytes in the Options field, the implementation of IP in Linux generates an ICMP message with probability roughly 98.5%. We discuss ICMP in more detail below.

The Protocol field is 8 bits (1 byte) long and indicates which upper layer protocol is carried in the IP datagram payload. A minimal set of supported protocols include ICMP, TCP and UDP. When an IP datagram reaches its intended destination (as specified in the 32-bit Destination Address field), the protocol field is inspected. This value determines to which upper layer protocol the payload is passed. If the field contains a value corresponding to a protocol that is not supported at that host, then the local IP implementation should generate an ICMP "protocol unreachable" message.

The Header Checksum field is a 16-bit (2-byte) value that is formed by interpreting the header (including the Options field if present) as a sequence of

16-bit words, summing them using 1's complement arithmetic, and then taking the 1's complement of the result. If the Header Checksum fails, the datagram is discarded silently.

In Linux, the sequence of steps taken by IP when processing a datagram is as follows. First of all, basic checks are performed on the Version field and IHL field. The next action is to check the Header Checksum field. After this, a datagram length check is carried out using the Total Length field. The datagram is dropped if any of these checks fails. Next, options processing is carried out if the IHL field indicates that options are present. Assuming this is completed successfully, a routing decision is made: either the datagram is delivered locally or is forwarded to another host. In the former case the Protocol field is used to determine the upper layer protocol to which the datagram payload should be passed. In the latter case, the TTL field is checked and the datagram dropped if the TTL has reached zero.

2.5   ICMP

ICMP is a vital part of IP implementations, allowing network problems to be reported to Internet hosts, routes to be tested, and diagnostics to be gathered. ICMP was originally specified in [27], and revised for IPv4 routers in [2]. In the event of a "problem datagram" being received by a host, that host generates an ICMP message. This message includes the entire IP header of the offending datagram (including any options), together with a variable number of bytes of the datagram's payload. According to [27], 8 bytes of payload should be included. On the other hand, according to [2], the ICMP datagram should contain as much of the original datagram as possible without the length of the ICMP datagram exceeding 576 bytes. This is intended to aid fault diagnosis, and is how ICMP is implemented in the Linux kernel.

3   Attacks Based on Destination Address Rewriting

We are now ready to discuss our first group of attacks on encryption-only ESP in tunnel mode. We focus on the case where the block cipher used by ESP has 64-bit blocks. The two-phase attack we describe here serves as an introduction to the more sophisticated attacks to follow. We describe the attack in the context of a pair of security gateways communicating using encryption-only ESP in tunnel mode to protect the traffic between them. The attack also works in more general applications of this configuration of ESP.

We need to make one major assumption for the attack to work: that the attacker, controlling the host located at IP address AttAddr, knows the destination IP address DestAddr of the target inner datagrams. This assumption will be relaxed shortly.

3.1   The First Phase

Recall that the Destination Address field lies in the fifth 32-bit word of the IP header, and therefore forms the first 32 bits of plaintext block P_3 in the sequence

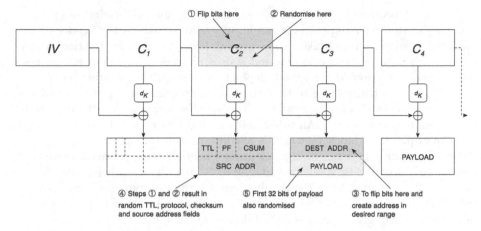

Fig. 2. Modifications to inner header fields in destination address rewriting attack, 64-bit case

of blocks to be encrypted in CBC mode by ESP. The second 32 bits of this block is the first 32 bits of the payload of the inner datagram. This phase proceeds as follows, with the attacker at `AttAddr` listening for IP datagrams during the attack (see also Figure 2):

1. Capture a target ESP-protected outer datagram from the network. Let $C_0, C_1, \ldots, C_q$ denote the encrypted portion of this datagram's payload.
2. Modify block C_2 in the first 32 bits by XORing it with the 32-bit mask $M = \mathtt{DestAddr} \oplus \mathtt{AttAddr}$ to obtain a block C_2'.
3. **Repeat:**
   – a. Modify block C_2', now in the last 32 bits, by setting these bits to a random 32-bit value R. Let C_2'' denote the modified block.
   – b. Prepare a modified datagram that is identical to the one captured in step 1, except that block C_2 of the encrypted portion is replaced with C_2''. Inject this modified datagram into the network.
   Until a datagram is received by the attacker at `AttAddr`.

To see why this phase might work, notice that each injected datagram now has `AttAddr` as the destination address of the inner datagram. So when the security gateway receives the modified outer datagram and decrypts the encrypted portion, it recovers an inner datagram that will then be routed directly to the attacker's machine (we are assuming here that datagrams are not checked after IPsec processing to see if the correct IPsec policies were applied; this is the case in the Linux kernel implementation, in contradiction to [15]). The inner datagram is in unencrypted form, and its payload will be identical to that of the original inner datagram except possibly in the first 32 bits (corresponding to the randomization of the second half of C_2). These payload bits can be recovered easily using the relation $P_3 = P_3' \oplus (M\|R)$ where P_3' is the third block in the received datagram, M is the address mask used in step 2 and R the random bits introduced in step 3.

Of course, because of the modifications made to block C_2 during the attack, block P_2 of the inner datagram is essentially randomized, so the header of the modified inner datagram is likely to be invalid. Block P_2 contains the time to live (TTL), protocol, header checksum and source address fields. Thus the success rate of each iteration of the attack depends on the combined probability that the TTL is sufficiently large so that the inner datagram reaches the attacker's machine, that the checksum is valid for the new header, and that the new inner source address is routable. All other fields in the header will be correct, since they lie in plaintext block P_1 which is not modified in the attack.

Based on our experience in implementing our other attacks, we estimate that this success probability should be roughly 2^{-17} per iteration, with the largest factor of 2^{-16} coming from the requirement for the random checksum to be a valid one. From this, it can be calculated that 2^{17} iterations of steps 3a and 3b of the attack will give a success probability of about 60%.

3.2   The Second Phase – Recovering Further Plaintext

An attacker who has conducted the first phase against an encrypted inner datagram of the form $C_0, C_1, \ldots, C_q$ does not need to repeat it in order to obtain decrypted versions of further inner datagrams. Instead, the contents of new datagrams can be recovered much more efficiently, as follows.

The attacker reuses the payload portion C_0, C_1, C_2'', C_3 of the outer datagram that was successful in the first phase, splicing onto it any $q - 6$ consecutive ciphertext blocks from the encrypted payload of the new target datagram, and finishing with the last three blocks C_{q-2}, C_{q-1}, C_q of the original target.[4] Dummy blocks can be used if necessary to ensure that a total of q blocks are present.

The attacker then uses this modified byte sequence as the encrypted payload of an outer datagram. This construction ensures that, upon decryption by the security gateway, the payload is correctly padded and is interpreted as an inner datagram with a valid header and a destination address equal to AttAddr. This datagram will be routed to the attacker's machine (for the same reasons that the successful datagram from the main attack was). From this datagram, a total of $64(q - 6)$ bits of plaintext from the new target datagram can be recovered (the first 64 bits are obtained using a similar to trick to that used to recover P_3 in the main attack; the remaining bits appear in clear in blocks 5 up to $q - 3$ of the datagram payload).

3.3   Relaxing the Address Assumption

Our main assumption that the attacker know the complete destination IP address of the inner datagram can be relaxed. It is enough that the attacker knows a significant portion of this IP address. The main idea is as follows. Instead of using a mask equal to DestAddr $\oplus$ AttAddr in step 2 of the attack, the attacker

---

[4] In fact, often only the last two blocks need to be preserved because the padding rarely extends over more than one block. Variable length padding of up to 255 bytes is allowed in [16]; our attacks are easily modified to handle this.

instead uses a mask which modifies that portion of the destination address known to the attacker so that it equals the corresponding portion of the address of his target machine. He then modifies the remaining bits of the destination address using a counter, and repeats the main attack for each counter value. One counter value will produce a destination address exactly matching that of the attacker; for this counter value, the attacker has the same probability as before (roughly 2^{-17}) of receiving a datagram from the gateway. After this effort, a more efficient second phase can again be used. Other variants are also possible [25].

3.4 Attack Implementation

As a proof of concept and as a precursor to our main attacks, we implemented a 128-bit version of the first phase of this attack against IP and IPsec as implemented in the Linux kernel. We found that roughly 2^{15} iterations were sufficient to produce the desired plaintext-bearing datagram, in line with a theoretical analysis of our 128-bit attack than can be found in [25]. This experiment confirmed the fact that the Linux implementation of IPsec does *not* carry out the policy checks described in Section 2.2 (otherwise the modified inner datagrams would be dropped because they would fail to match the IPsec policies used in their recovery).

4 Attacks Based on IP Options Processing

Our next set of attacks exploits the way in which IP implementations generate ICMP messages when processing incorrectly formatted options fields in IP headers. We focus on the case where the block cipher used by ESP has 64-bit blocks. We again describe the attack in the context of a pair of security gateways communicating using encryption-only ESP in tunnel mode.

We need to make some assumptions for the attack to work. As usual, we assume that the attacker is able to intercept ESP-protected datagrams and to inject modified datagrams into the network. We additionally assume that the attacker is able to monitor one of the gateways for ICMP messages not sent through the IPsec tunnel. A third-party network service provider is in a perfect position to mount this attack, for example. This would also be easily achievable if the IPsec traffic was being broadcast on a wireless network in which WEP (or an equivalent) was not in use. We will see later how this requirement can be relaxed in the 128-bit case, provided the attacker has (partial) information about inner source addresses.

4.1 The First Phase

As before, the attacker has captured an outer datagram and wishes to recover the plaintext version of the encrypted portion of its payload. Recall that the IHL field is located in the first byte of the IP header, and therefore lies in plaintext block P_1 in the sequence of blocks to be encrypted in CBC mode by ESP. The attacker modifies the contents of the IHL field of the inner datagram by flipping appropriate bits in IV, making the IHL equal a value greater than 5. When the inner datagram is subsequently processed by the IP software on the security gateway, the first word(s) of the payload (forming the contents of the second half of P_3 onwards) will be interpreted as options bytes. We randomize the values of these bytes (as seen by the security gateway) by placing a random value in the last 32 bits of C_2. Then with high probability, these bytes will be incorrectly formatted, resulting in the generation of an ICMP "parameter problem" message. The payload of this ICMP message will contain the header and a segment of the payload of the inner datagram. Thus, if it can be captured by the attacker, he can learn plaintext information from the inner datagram. However, randomizing bytes in C_2 has the additional effect of randomizing the contents of P_2 after decryption by the security gateway. So the inner datagram is likely to be dropped silently by the security gateway before any IP options processing takes place, because of an incorrect checksum value. Thus, in fact, the ICMP message will not often be generated. Moreover, the ICMP message, if generated, will be sent to the random source address now specified in P_2. This helps to ensure that the ICMP message is not sent through the IPsec tunnel between the security gateways, thus making it visible to the attacker, but also means that this address may not be routable. These problems can be overcome by iterating the attack sufficiently often and using new random bytes on each iteration. We will quantify the success rate for the Linux implementation of IP in Section 4.4 below.

This attack is illustrated in Figure 3 and formalized below.

1. Capture a target ESP-protected outer datagram from the network. Let $C_0, C_1, \ldots, C_q$ denote the encrypted portion of this datagram's payload.
2. Modify block $C_0 = IV$ in the first byte, XORing it with a mask which increases the IHL to a value greater than 5, obtaining a block C_0'.
3. **Repeat**:
   – a. Modify block C_2 in the last 32 bits, by setting these bits to a random 32-bit value R. Let C_2' denote the modified block.
   – b. Prepare a modified datagram that is identical to the one captured in step 1, except that blocks C_0 and C_2 of the encrypted portion are replaced with C_0' and C_2'. Inject this modified datagram into the network.
   Until an ICMP message is intercepted.

4.2   The Second Phase

Tricks similar to those introduced in Section 3.2 can be used in a second phase to speed up the recovery of all payload bytes from the remainder of the initial target datagram and further target datagrams. Once again, a successful header can be re-used and is guaranteed to always generate an ICMP message. The speed of recovery of plaintext in this second phase is limited only by the rate at which the security gateway is permitted to generate ICMP messages and by the number of payload bytes returned by ICMP.

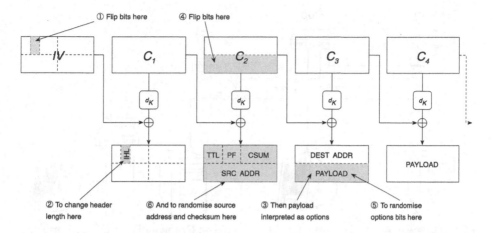

Fig. 3. Modifications to inner header fields in options processing attack, 64-bit case

4.3 The 128-Bit Case

A similar attack is possible when the block cipher used by ESP has 128-bit blocks. Now, however, the IHL field, Header Checksum field and Source Address field can all be manipulated by bit flipping in $C_0 = IV$. This allows the possible checksums to be tested systematically, which improves the success probability. The payload bytes which get interpreted as options bytes by the security gateway can be randomized by selecting a random value for C_2. Again, further plaintext can be recovered faster in a second phase which re-uses the successful header from the first phase. Moreover, if the attacker has some (or full) knowledge of the source address of the inner datagrams, then he can use similar ideas to those explored in Section 3.3 to direct the ICMP response to his own machine, this time by changing the source address in the inner header by manipulating the IV. This is an important variant, since it removes the most stringent requirement for our attack, namely that the attacker be able to monitor the security gateway for ICMP messages.

4.4 Attack Implementation

We have successfully carried out the two phases of our attack against IP and IPsec as implemented in the Linux kernel. We describe the main features and results of this implementation here.

Figure 4 shows the experimental set-up, with two Linux machines acting as security gateways for an ESP tunnel using either DES or AES as the encryption algorithm (the end host shown in this figure is not active during this attack). These machines are connected to a hub, as is the attack platform – this is simply to ease packet sniffing in the network. Also connected to this hub is a router, configured to act as the default router for the security gateways, thus ensuring that any ICMP messages can take at least a first hop towards their destinations.

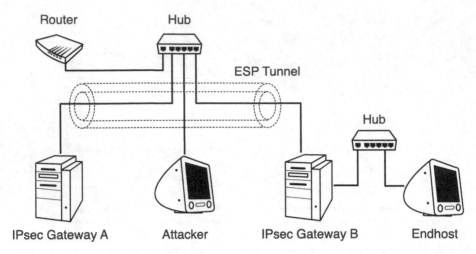

Fig. 4. Experimental set-up for attacks based on options processing and protocol field manipulation

We used a value of 6 for the modified IHL field, so as to maximise the number of plaintext bytes returned for each injected datagram in the second phase. We observed experimentally that presenting a datagram with a random source address and random options bytes to the IP implementation in Linux results in an ICMP "parameter problem" message with probability about 0.85. Moreover, the probability that a random 16-bit value represents the correct header checksum for the modified inner datagram is roughly 2^{-16}. Thus the expected success probability of the first phase of the attack in the 64-bit case is roughly 0.85×2^{-16} per iteration. For example, then, 2^{16} iterations should give a success rate of 57%.

We performed 100 runs of the first phase of the attack. An average of 77600 iterations (taking on average 2.64 minutes with our attack client) were needed to successfully generate an ICMP message. Linux is generous in providing 524 bytes of inner datagram payload in ICMP messages. As a consequence, the first phase and each injected datagram in the second phase yields 512 bytes of plaintext data (provided the encrypted payload in the target selected for the first phase is longer than 568 bytes, including the IV and encrypted inner header). Thus the second phase can rapidly recover the complete contents of inner datagrams. Our attack client, written in C, captures multiple ESP-protected datagrams, selects the one of optimum length for the first phase, conducts the first phase, and then runs the second, faster phase on remaining datagrams. Our attack client is also written to carry out the 128-bit variant of this attack.

5   Attacks Based on Protocol Field Manipulation

Our third class of attacks exploits the way in which IP implementations generate ICMP messages when faced with unsupported upper layer protocols. We focus on the case where the block cipher used by ESP has 128-bit blocks, as this is the more efficient case. We need to make the same assumptions as in Section 4 for the attack to work.

5.1   The First Phase

Recall that the protocol field is located in the second byte of the third 32-bit word of the IP header, and therefore lies in plaintext block P_1 in the sequence of blocks to be encrypted in CBC mode by ESP. The attacker modifies the contents of the protocol field of the inner datagram by flipping appropriate bits in IV, making the field equal a value corresponding to an upper layer protocol that is not supported by the end host receiving the inner datagram. Now, when the inner datagram arrives at the end host that is its final destination, an ICMP "protocol unreachable" message will be generated. The payload of this ICMP message will contain the header and a segment of the payload of the inner datagram. Thus, if it can be captured by the attacker, then he can learn plaintext information from the inner datagram. Note that, in contrast to the attack based on options processing, the end host, not the security gateway, generates the ICMP message.

An attacker must solve two problems here. Firstly, the attacker must alter the source address of the inner datagram, so that the ICMP response will not be routed through the IPsec tunnel and so that the attacker can intercept it. Secondly, the attacker must fix the header checksum so that it contains the correct value for the modified inner header. Fortunately, in the 128-bit case, both of these requirements can be met by further manipulating only IV, and in a systematic way that leads to a very efficient attack.

Consider an attacker who modifies the protocol field by forcing a flip in bit i of the field (where $0 \leq i < 8$) and who alters the inner source address by forcing a flip in bit j of the address (where $0 \leq j < 32$). These bit flips can both be induced by manipulating IV. To correct the inner header checksum, the attacker XORs it with two masks in sequence (one mask for each bit flip), again by flipping bits in IV. A detailed analysis of the checksum algorithm (see [25]) shows that one of only 17 possible masks will correct each bit flip. The attacker tries these pairs of masks in decreasing order of probability. A maximum of $17^2 = 289$ iterations will be needed, with an expected number much smaller than this because of the way mask probabilities are distributed. In fact, a simple analysis shows that when $i + 8 \neq j \bmod 16$, the expected number of iterations is slightly less than 7, and smaller still when $i + 8 = j \bmod 16$. This attack can be formalized just as with the earlier attacks.

In an important variant of this attack, now requiring on average 2^{15} iterations, the attacker can additionally exploit knowledge of the inner source address to rewrite this address, thus ensuring that any ICMP response is directed to a host he controls. This removes the requirement that the attacker be able to monitor the security gateway for ICMP messages.

5.2   The Second Phase

Just as with the attack in Section 4, once the first phase is complete, a second phase which recovers the complete contents of the remainder of the initial target datagram and further target datagrams can be invoked.

5.3   The 64-Bit Case

A similar, but less efficient, attack is possible when the block cipher used by ESP has 64-bit blocks, but now the protocol field is manipulated by randomizing the last 32 bits of block C_2. The success probability is now limited by the need for a random checksum to have the correct value, and for a random protocol field to represent an unsupported protocol. In practice, it is close to 2^{-16}, because, typically, only a handful of protocols are supported. Again, further plaintext can be recovered faster in a second phase which re-uses the successful header from the first phase.

5.4   Attack Implementation

We have successfully implemented the two phases of the 128-bit attack against the Linux kernel implementation of IP and IPsec in our attack client. The experimental set-up is shown in Figure 4. In our attack, we used values $i = 0$ and $j = 6$ (many other pairs worked equally well).

According to the probability analysis sketched in Section 5.1, the expected number of iterations of the first phase with these parameters is slightly less than 7. We performed 1000 runs of the first phase of the attack. An average of 6.53 iterations (taking 1.34 seconds with our attack client) was needed to successfully generate an ICMP "protocol unreachable" message containing plaintext information. Because of the way in which Linux implements ICMP, the first phase and each injected datagram in the second phase yields about 500 bytes of plaintext data. This means that our attack client is able to recover large amounts of plaintext easily in the second phase of the attack. Overall, because of the small number of trials needed, the attack effectively takes place in real time.

6   Impact

We have presented a number of attacks and variants on encryption-only ESP in tunnel mode, as implemented in the Linux kernel. The attacks are efficient and have been demonstrated to work under realistic network conditions. Perhaps surprisingly, ESP using a 128-bit block cipher such as AES may be more vulnerable to our attacks than one using a 64-bit block cipher. The underlying reason for this is that in the 128-bit case, more fields of the inner header can be manipulated by modifying IV, without any impact on the contents of plaintext blocks. A related point is that the complexity of the attacks does not depend on the key size of the block cipher employed by ESP: triple-DES is just as vulnerable as DES.

We note that, as with [23], our work demonstrates that the open source approach does not necessarily result in secure software: an encryption-only configuration was all too easy to select, the IPsec implementation did not carry out the post-processing checks mandated in the RFCs, and we found other flaws in the implementation, particularly in the handling of padding (c.f. [29]).

Concerning the real-world impact of our attacks, we have presented evidence in the introduction that encryption-only IPsec may still be in common use. But we have performed only limited experiments against other IP/IPsec implementations. We do know that several vendors attempt to disable encryption-only. However, disabling encryption-only configurations is not enough to prevent our attacks, as they still apply to some configurations where integrity-protection is supplied by IPsec itself. As just one instance, the attacks in Sections 3 and 4 still work if AH is applied in transport mode end-to-end and is tunnelled inside ESP from gateway-to-gateway. This is because the redirection or ICMP generation take place at the gateway, before any integrity checking occurs. We note too that our attacks are not prevented if integrity protection is offered independently of IPsec by a higher-layer protocol. This contradicts the statement made in [18] that we quoted in Section 1.

7 Conclusions

We believe that the dangers of encryption-only ESP that we have highlighted here, coupled with the difficulty of ensuring that security-unaware users pick strong configurations from amongst the myriad possibilities, means that a conservative approach is called for in the IPsec standards themselves. Unfortunately, ESPv3 [18] still permits the use of encryption-only ESP, though it is no longer mandatory to support it.

The complexity of the IPsec standards has been commented on before [10]. It certainly does not help an implementation team if processing checks important to the security of one module (ESP) are contained in another document altogether (RFC 2401, [15]). It is worrying that the security of the encryption-only mode depends completely on these checks being carried out: the security dangles from a very thin thread indeed, as our attacks on the native Linux implementation make clear. It would help, then, if the reasons why those checks need to be performed were spelled out in the standard: this would give an implementor a stronger motivation for getting things right.

We hope that this work will help in persuading users to migrate away from encryption-only IPsec configurations. We also hope that it serves as an instructive example to the theoretical community of the gaps that exist between theory and practice in cryptography, and that it helps to bridge these gaps.

Acknowledgements

We would like to thank Steve Kent and David Wagner for providing important information and context. We would also like to thank the members of the

NISCC Vulnerability Team for their assistance in evaluating the impact of our attacks and for helping us in working with the IPsec vendor and user communities ahead of their vulnerability advisory [24] concerning this work. Nessim Kisserli's assistance with lab-space and hardware issues was also invaluable.

References

1. R. Atkinson, "IP Encapsulating Security Payload (ESP)", RFC 1827, August 1995.
2. F. Baker, "Requirements for IPv4 Routers", RFC 1812, June 1995.
3. M. Bellare, T. Kohno and C. Namprempre, "Breaking and provably repairing the SSH authenticated encryption scheme: A case study of the Encode-then-Encrypt-and-MAC paradigm." *ACM TISSEC*, Vol. 7, No. 2, May 2004, pp. 206–241.
4. M. Bellare and C. Namprempre, "Authenticated Encryption: Relations among notions and analysis of the generic composition paradigm." In *T. Okamoto (ed.), Advances in Cryptology – ASIACRYPT 2000*, LNCS Vol. 1976, Springer-Verlag, 2000, pp. 531–545.
5. M. Bellare and P. Rogaway, "Encode-then-encipher encryption: How to exploit nonces or redundancy in plaintexts for efficient cryptography." In *T. Okamoto (ed.), Advances in Cryptology – ASIACRYPT 2000*, LNCS Vol. 1976, Springer-Verlag, 2000, pp.317–330.
6. S. Bellovin, "Problem Areas for the IP Security Protocols", in *Proceedings of the Sixth Usenix Unix Security Symposium*, pp. 1–16, San Jose, CA, July 1996.
7. N. Borisov, I. Goldberg and D. Wagner, "Intercepting Mobile Communications: The Insecurity of 802.11", in *Proc. MOBICOM 2001*, ACM Press, 2001, pp. 180–189.
8. B.Canvel, A.P. Hiltgen, S. Vaudenay and M. Vuagnoux, "Password Interception in a SSL/TLS Channel," in *D. Boneh (ed.), Advances in Cryptology – CRYPTO 2003*, LNCS Vol. 2729, Springer-Verlag, 2003, pp. 583–599
9. N. Doraswamy and D. Harkins. *IPsec: the new security standard for the Internet, Intranets and Virtual Private Networks (second edition)*, Prentice Hall PTR, 2003.
10. N. Ferguson and B. Schneier, "A cryptographic evaluation of IPsec." Unpublished manuscrip available from `http://www.schneier.com/paper-ipsec.html`.
11. S. Frankel, R. Glenn and S. Kelly, "The AES-CBC Cipher Algorithm and Its Use with IPsec", RFC 3602, Sept. 2003.
12. S. Frankel, K. Kent, R. Lewkowski, A.D. Orebaugh, R.W. Ritchey and S.R. Sharma, "Guide to IPsec VPNs", NIST Special Publication 800-77 (Draft), January 2005.
13. D. Harkins and D. Carrel, "The Internet Key Exchange (IKE)", RFC 2409, Nov. 1998.
14. J. Katz and M. Yung, "Unforgeable encryption and chosen ciphertext secure modes of operation." In *B. Schneier (ed.), FSE 2000*, LNCS Vol. 1978, Springer-Verlag 2001, pp. 284–299.
15. S. Kent and R. Atkinson, "Security Architecture for the Internet Protocol", RFC 2401, Nov. 1998.
16. S. Kent and R. Atkinson, "IP Encapsulating Security Payload (ESP)", RFC 2406, Nov. 1998.
17. S. Kent and K. Seo, "Security Architecture for the Internet Protocol", RFC 4301 (obsoletes RFC 2401), Dec. 2005.
18. S. Kent, "IP Encapsulating Security Payload (ESP)", RFC 4303 (obsoletes RFC 2406), Dec. 2005.

19. H. Krawczyk, "The Order of Encryption and Authentication for Protecting Communications (Or: How Secure Is SSL?)", in *J. Kilian (ed.), Advances in Cryptology – CRYPTO 2001*, LNCS Vol. 2139, Springer-Verlag 2001, pp. 310–331.
20. Internet Protocol, RFC 791, Sept. 1981.
21. C. Madson and N. Doraswamy, "The ESP DES-CBC Cipher Algorithm With Explicit IV", RFC 2405, Nov. 1998.
22. C.B. McCubbin, A.A. Selcuk and D. Sidhu, "Initialization vector attacks on the IPsec protocol suite." In *WETICE 2000*, IEEE Computer Society, pp. 171–175.
23. P.Q. Nguyen, "Can we trust cryptographic software? Cryptographic flaws in GNU Privacy Guard v1.2.3", in *C. Cachin (ed.), Advances in Cryptology – EUROCRYPT 2004*, LNCS Vol. 3027, Springer-Verlag 2004, pp. 555–570.
24. NISCC Vulnerability Advisory IPSEC - 004033, 9th May 2005. Available from `http://www.niscc.gov.uk/niscc/docs/al-20050509-00386.html?lang=en`.
25. K.G. Paterson and A.K.L. Yau, "Cryptography in Theory and Practice: The Case of Encryption in IPsec." Extended version of this paper available from `http://eprint.iacr.org/2005/416`.
26. R. Pereira and R. Adams, "The ESP CBC-Mode Cipher Algorithms", RFC 2451, Nov. 1998.
27. J. Postel, "Internet Control Message Protocol", RFC 792, Sept. 1981.
28. S. Stubblebine and V. Gligor, "On Message Integrity in Cryptographic Protocols", in *IEEE Security and Privacy*, May 1992, pp. 85–104.
29. S. Vaudenay, "Security flaws induced by CBC padding – applications to SSL, IPSEC, WTLS...", in *L.R. Knudsen (ed.), Advances in Cryptology – EUROCRYPT 2002*, LNCS Vol. 2332, Springer-Verlag 2002, pp. 534–545.
30. T. Yu, S. Hartman and K. Raeburn, "The perils of unauthenticated encryption: Kerberos version 4", in *Proc. NDSS 2004*, The Internet Society, 2004.

Polynomial Equivalence Problems: Algorithmic and Theoretical Aspects

Jean-Charles Faugère[1] and Ludovic Perret[2]

[1] LIP6, 8 rue du Capitaine Scott, F-75015, France
Jean-Charles.Faugere@lip6.fr
[2] UCL, Crypto Group, Microelectronic Laboratory, Place du Levant,
3 Louvain-la-Neuve, B 1348, Belgium
ludovic.perret@uclouvain.be

Abstract. The Isomorphism of Polynomials (IP) [28], which is the main concern of this paper, originally corresponds to the problem of recovering the secret key of a C* scheme [26]. Besides, the security of various other schemes (signature, authentication [28], traitor tracing [5], ...) also depends on the practical hardness of IP. Due to its numerous applications, the Isomorphism of Polynomials is thus one of the most fundamental problems in multivariate cryptography. In this paper, we address two complementary aspects of IP, namely its theoretical and practical difficulty. We present an upper bound on the theoretical complexity of "IP-like" problems, i.e. a problem consisting in recovering a particular transformation between two sets of multivariate polynomials. We prove that these problems are not NP-Hard (provided that the polynomial hierarchy does not collapse). Concerning the practical aspect, we present a new algorithm for solving IP. In a nutshell, the idea is to generate a suitable algebraic system of equations whose zeroes correspond to a solution of IP. From a practical point of view, we employed a fast Gröbner basis algorithm, namely F_5 [17], for solving this system. This approach is efficient in practice and obliges to modify the current security criteria for IP. We have indeed broken several challenges proposed in literature [28, 29, 5]. For instance, we solved a challenge proposed by O. Billet and H. Gilbert at Asiacrypt'03 [5] in less than one second.

Keywords: Public-Key Cryptography, Cryptanalysis, Isomorphism of Polynomials (IP), Gröbner bases, F_5 algorithm.

1 Introduction

Multivariate cryptography – which can be roughly defined as the cryptography using polynomials in several variables – offers a relatively wide spectrum of problems that can be used in public-key cryptography. The Isomorphism of Polynomials (IP) lies in this family [28]. Briefly, this problem consists in recovering a particular transformation between two sets of multivariate polynomials permitting to obtain one set from the other. It originally corresponds to the problem of recovering the secret key of a C* scheme [26]. Besides, the security of

S. Vaudenay (Ed.): EUROCRYPT 2006, LNCS 4004, pp. 30–47, 2006.

several other schemes is directly based on the practical difficulty of IP, namely the authentication/signature schemes proposed by J. Patarin at Eurocrypt'96 [28], and the traitor tracing scheme described by O. Billet and H. Gilbert at Asiacrypt'03 [5]. We also mention that IP is in a certain manner related to the security of Sflash [13] – the signature scheme recommended by the European consortium Nessie for low-cost smart cards [27] – and can be alternatively viewed as the problem of detecting affine equivalence between S-Boxes [6]. All in all, one can consider the hardness of IP as one of the major issues in multivariate cryptography. The goal of this paper is to provide new insights on the theoretical and practical complexity of IP and some of its relevant variants.

1.1   Previous Work

To the best of our knowledge, the most significant results concerning IP are presented in [11], where an upper bound on the theoretical complexity of IP is given. Nevertheless, we point out that the proof provided is actually not complete. Anyway, the upper bound presented in that paper is original and general. It is indeed based on a group theoretic approach of IP and actually dedicated to "IP-like" problems. A new algorithm for solving IP, called "To and Fro", is also described in [11]. This algorithm is however devoted to special instances of IP, namely the ones corresponding to a public key of C^* [26]. Thus, it can not be used for solving generic instances of IP. This is not the case for the algorithm presented here. Besides, we present in Section 4 experimental results demonstrating that our algorithm outperforms the "To and Fro" method. Finally, we would like to mention a result due to W. Geiselmann, R. Steinwandt, and T. Beth [23]. In the context of C^*, they showed how to easily recover the affine parts of a solution of IP. A similar property also holds in the context of HFE [20]. Such a kind of result does not exist for generic instances of IP. Nevertheless, it means that in the cryptographic context we can focus our attention on the linear variant of IP, called 2PLE here.

1.2   Organization of the Paper and Main Results

The paper is organized as follows. We begin in Section 2 by introducing our notation and defining essential tools of our algorithm, namely varieties and Gröbner bases. A recent algorithm (i.e. F_5 [17]) for computing these bases is also succinctly described. Finally, we define more formally the Isomorphism of Polynomials (IP) and two of its variants, namely the Isomorphism of Polynomials with one Secret (IP1S) [28], and the linear variant of IP that we name 2PLE. In Section 3, we show that these problems are actually particular instances of a more general problem that we call Polynomial Equivalence (PE). This problem provides a formal definition of an "IP-like" problem. Using classical results of group theory, we conclude this section by providing an upper bound on the theoretical hardness of PE. A new algorithm for solving 2PLE is presented in Section 4. The idea is to generate a suitable polynomial system of equations whose zeroes correspond to a solution of IP. In order to construct this system, we also provide some specific properties of 2PLE. From a practical point of view, we used the most recent (and efficient) Gröbner basis algorithm, namely F_5 [17], for solving this system. It is difficult to obtain a complexity bound really reflecting the practical behavior of the F_5 algorithm. We therefore carried out experimental results illustrating the practical efficiency of our approach. We have indeed broken several challenges proposed in literature [28, 29, 5]. For instance, we solved a challenge proposed by O. Billet and H. Gilbert at Asiacrypt'03 [5] in less than one second.

2   Preliminaries

The notation used throughout this paper is the following. We denote by $\mathbb{F}_q$ the finite field with $q = p^r$ elements (p a prime, and $r \geq 1$), and by $\mathcal{M}_{n,u}(\mathbb{F}_q)$ the set of $n \times u$ matrices whose components are in $\mathbb{F}_q$. As usual, $GL_n(\mathbb{F}_q)$ represents the set of invertible matrices of $\mathcal{M}_{n,n}(\mathbb{F}_q)$, and $AGL_n(\mathbb{F}_q)$ denotes the cartesian product $GL_n(\mathbb{F}_q) \times \mathbb{F}_q^n$. Finally, let $\mathbf{x} = (x_1, \ldots, x_n)$, and $\mathbb{F}_q[\mathbf{x}] = \mathbb{F}_q[x_1, \ldots, x_n]$, be the polynomial ring in the n indeterminates $x_1, \ldots, x_n$ over $\mathbb{F}_q$. By convention, a boldfaced letter will always refer to a row vector.

2.1   Gröbner Bases

We define now two essential notions of this paper, namely varieties and Gröbner bases. For a more thorough introduction to these tools, we refer to [1, 15].

Let $\mathbf{p} = (p_1, \ldots, p_s)$ be polynomials in $\mathbb{F}_q[\mathbf{x}]$. We shall call $\mathcal{I} = \langle p_1, \ldots, p_s \rangle = \left\{ \sum_{k=1}^{s} p_k u_k, u_1, \ldots, u_k \in \mathbb{F}_q[\mathbf{x}] \right\} \subset \mathbb{F}_q[\mathbf{x}]$ the *ideal generated by* $p_1, \ldots, p_s$, and denote by $V(\mathcal{I}) = \{ \mathbf{z} \in \mathbb{F}_q^n : p_i(\mathbf{z}) = 0, \forall i, 1 \leq i \leq s \}$ the *variety associated* to $\mathcal{I}$. Gröbner bases provide a method for computing this variety. Informally, a Gröbner basis of an ideal $\mathcal{I}$ is a computable generator set of $\mathcal{I}$ with "good" algorithmic properties. These bases are defined with respect to *monomial orders*. Here, we will use the lexicographical (LEX) and degree reverse lexicographical (DRL) orders, which are defined as follows:

Definition 1. *Let* $\alpha = (\alpha_1, \ldots, \alpha_n)$ *and* $\beta = (\beta_1, \ldots, \beta_n) \in \mathbb{N}^n$. *Then:*
$- x_1^{\alpha_1} \cdots x_n^{\alpha_n} \prec_{LEX} x_1^{\beta_1} \cdots x_n^{\beta_n}$, *if the left-most nonzero entry of* $\alpha - \beta$ *is positive.*
$- x_1^{\alpha_1} \cdots x_n^{\alpha_n} \prec_{DRL} x_1^{\beta_1} \cdots x_n^{\beta_n}$, *if* $\sum_{i=1}^{n} \alpha_i > \sum_{i=1}^{n} \beta_i$, *or* $\sum_{i=1}^{n} \alpha_i = \sum_{i=1}^{n} \beta_i$
and the right-most nonzero entry of $\alpha - \beta$ *is negative.*

To define Gröbner bases, we need to introduce the following definitions.

Definition 2. *For any* n-*tuple* $\alpha = (\alpha_1, \ldots, \alpha_n) \in \mathbb{N}^n$, *we denote by* $\mathbf{x}^\alpha$ *the* **monomial** $x_1^{\alpha_1} \cdots x_n^{\alpha_n}$. *We shall define the* **total degree** *of this monomial by the sum* $\sum_{i=1}^{n} \alpha_i$. *The* **leading monomial** *of a polynomial* $p \in \mathbb{F}_q[\mathbf{x}]$ *is the largest monomial (w.r.t some monomial ordering* $\prec$*) among the monomials of* p. *This leading monomial will be denoted by* $\mathrm{LM}(p, \prec)$. *The* **degree** *of* p, *denoted* $\deg(p)$, *is the total degree of* $\mathrm{LM}(p, \prec)$. *Finally, the* **maximal total degree** *of* p *is the maximal total degree of the monomials occurring in* p.

We are now in a position to define one of the main objects of this paper.

Definition 3. *A set of polynomials G is a* **Gröbner basis** *– w.r.t. a monomial ordering $\prec$ – of an ideal $\mathcal{I}$ in $\mathbb{F}_q[\mathbf{x}]$ if, for all $p \in \mathcal{I}$, there exists $g \in G$ such that* $\mathrm{LM}(g, \prec)$ *divides* $\mathrm{LM}(p, \prec)$.

Gröbner bases are a fundamental tool to study algebraic systems in theory and practice. They provide an algorithmic solution to several problems related to polynomial systems (see [1] for instance). We pay here particular attention to Gröbner bases computed for a lexicographical ordering. It offers a way of simplifying an algebraic system by giving an equivalent system with a structured shape. A lexicographical Gröbner basis of a zero-dimensional system (i.e. with a finite number of zeroes over the algebraic closure) is indeed always as follows:

$$\{f_1(x_1) = 0, f_2(x_1, x_2) = 0, \ldots, f_{k_2}(x_1, x_2) = 0, f_{k_2+1}(x_1, x_2, x_3) = 0, \ldots, \ldots\}$$

To compute the variety, we simply have to successively eliminate variables by computing zeroes of univariate polynomials and back-substituting results. However, computing a Gröbner basis w.r.t. a lexicographical order is in practice much slower than computing a Gröbner basis w.r.t. another monomial ordering. It is usually for a DRL order that the computation of Gröbner bases is the fastest in practice. Algorithms changing the monomial ordering of a Gröbner basis permit to handle efficiently this problem. The FLGM algorithm [19] allows to transform a Gröbner basis w.r.t. some monomial ordering into a lexicographical Gröbner basis in the zero-dimensional case and is polynomial-time.

The historical method for computing Gröbner bases is Buchberger's algorithm [9, 8]. Recently, more efficient algorithms have been proposed. The F_4 algorithm [16] is based on the intensive use of linear algebra methods. In short, the arbitrary choices – which limit the practical efficiency of Buchberger's algorithm – are replaced by computational strategies related to classical linear algebra problems (mainly the computation of a row echelon form).

In [17], a new criterion (the F_5 criterion) for detecting useless computations has been proposed. We mention that Buchberger's algorithm spends 90% of its time to perform these useless computations. Under some regularity conditions, it has been proved that all useless computations can be avoided. A new algorithm, called F_5, has then been built using this criterion and linear algebra methods. Briefly, it constructs incrementally the following matrices in degree d:

$$A_d = \begin{array}{c} \\ t_1 f_1 \\ t_2 f_2 \\ t_3 f_3 \\ \ldots \end{array} \overset{\displaystyle m_1 \succ m_2 \succ m_3 \ldots}{\begin{bmatrix} \ldots & \ldots & \ldots & \ldots \\ \ldots & \ldots & \ldots & \ldots \\ \ldots & \ldots & \ldots & \ldots \\ \ldots & \ldots & \ldots & \ldots \end{bmatrix}}$$

where the indices of the columns are monomials sorted for the admissible ordering $\prec$ and the rows are product of some polynomials f_i by some monomials t_j such that $\deg(t_j f_i) \leq d$. For a *regular system* ([17]) the matrices A_d are of full rank. In a second step, row echelon forms of theses matrices are computed, i.e.

$$
\begin{matrix}
& & m_1\ m_2\ m_3\ \ldots \\
A'_d = & \begin{matrix} t_1 f_1 \\ t_2 f_2 \\ t_3 f_3 \\ \ldots \end{matrix} & \begin{bmatrix} 1 & 0 & 0 & \ldots \\ 0 & 1 & 0 & \ldots \\ 0 & 0 & 1 & \ldots \\ 0 & 0 & 0 & \ldots \end{bmatrix}
\end{matrix}
$$

For d sufficiently large, A'_d contains a Gröbner basis of the ideal considered. Important parameters to evaluate the complexity of F_5 is the maximal degree d occurring in the computation and the size of the matrix A_d. The overall cost is thus dominated by $(\#A_d)^3$. Very roughly, $(\#A_d)$ can be approximated by $O(n^d)$. A more precise complexity analysis can be found in [3, 4].

From a practical point of view, the gap with other algorithms computing Gröbner basis is consequent. To date, F_5 is the most efficient method for computing Gröbner bases, and hence zero-dimensional varieties. In particular, it has been proved [2] – from both a theoretical and practical point of view – that XL [14] is less efficient than F_5. Due to the range of examples that become computable with F_5, Gröbner basis can be considered as a reasonable computable object in real scale applications. For systems arising in cryptography, F_5 has for instance given impressing results on HFE [18].

2.2   Isomorphism of Polynomials and Related Problems

Before defining formally IP, we briefly come back here to the origin of this problem. To do so, we describe the encryption scheme called C^* [26]. The public key of this system is a set of multivariate quadratic polynomials $\mathbf{b} = (b_1(\mathbf{x}), \ldots, b_n(\mathbf{x})) \in \mathbb{F}_q[\mathbf{x}]^n$. These polynomials are obtained by applying two bijective affine transformations $(S, \mathbf{V})$ and $(U, \mathbf{V})$ of $AGL_n(\mathbb{F}_q)$ to a particular set of polynomials $\mathbf{a} = (a_1(\mathbf{x}), \ldots, a_n(\mathbf{x})) \in \mathbb{F}_q[\mathbf{x}]^n$. That is:

$$
(b_1(\mathbf{x}), \ldots, b_n(\mathbf{x})) = (a_1(\mathbf{x}S + \mathbf{T}), \ldots, a_n(\mathbf{x}S + \mathbf{T}))U + \mathbf{V},
$$

denoted $\mathbf{b}(\mathbf{x}) = \mathbf{a}(\mathbf{x}S + \mathbf{T})U + \mathbf{V}$ in the sequel.

To encrypt, we simply evaluate a message $\mathbf{m} \in \mathbb{F}_q^n$ on $\mathbf{b}$, i.e. $(b_1(\mathbf{m}), \ldots, b_n(\mathbf{m}))$. To recover the correct plaintext, the legitimate recipient uses the bijectivity of the affine transformations combined with the particular structure of the polynomials of $\mathbf{a}$. How these polynomials are constructed is not relevant here. But, due to particular constraints, the polynomials of $\mathbf{a}$ are always considered as a public data. The secret key of C^* is constituted of $(S, \mathbf{T}), (U, \mathbf{V}) \in AGL_n(\mathbb{F}_q)$.

The first approach for attacking this scheme consists in trying to retrieve the message corresponding to a ciphertext $\mathbf{c} \in \mathbb{F}_q^n$, i.e. finding a zero of $\mathbf{b}(\mathbf{x}) = \mathbf{c}$. This corresponds to solving a particular instance of the so-called MQ problem, which is NP-Hard in general [10, 22]. We emphasize that such a kind of result uniquely guarantees the worst-case hardness and does not provide any information concerning the average-case difficulty. For instance, J.-C. Faugère and A. Joux proposed a polynomial-time algorithm for solving instances of MQ corresponding to the public key of HFE [18], which is an extension of C^*.

Another approach for breaking C^* consists in attempting to recover the affine transformations hiding the structure of $\underline{a}$. That is, extracting the secret key from the public key. This problem, introduced by J. Patarin at Eurocrypt'96 [28], is defined as follows:

Isomorphism of Polynomials (IP)
Input: $\mathbf{a} = (a_1, \ldots, a_u)$, and $\mathbf{b} = (b_1, \ldots, b_u)$ in $\mathbb{F}_q[\mathbf{x}]^u$.
Question: Find – if any – $(S, \mathbf{V}) \in AGL_n(\mathbb{F}_q)$ and $(U, \mathbf{V}) \in AGL_u(\mathbb{F}_q)$, s. t.:

$$\mathbf{b}(\mathbf{x}) = \mathbf{a}(\mathbf{x}S + \mathbf{V})U + \mathbf{V}.$$

More precisely, it is usually the linear variant of IP which is considered in practice [28, 5]. That is, when the vectors $\mathbf{T}$ and $\mathbf{V}$ are both equal to the null vector. This problem, that we call 2PLE is the following:
Input: $\mathbf{a} = (a_1, \ldots, a_u)$, and $\mathbf{b} = (b_1, \ldots, b_u)$ in $\mathbb{F}_q[\mathbf{x}]^u$.
Question: Find – if any – $(S, U) \in GL_n(\mathbb{F}_q) \times GL_u(\mathbb{F}_q)$, such that:

$$\mathbf{b}(\mathbf{x}) = \mathbf{a}(\mathbf{x}S)U.$$

However, it is without solving any of the two problems mentioned above that J. Patarin proposed a full cryptanalysis of C* [30]. This attack uses the very particular structure of the polynomials of $\underline{a}$. This result thus does not then affect at all the practical hardness of IP. The security estimate provided for this problem [29] is based on the complexity of the "To and Fro" (TF) algorithm [11, 12], which is $q^{n/2}$ for quadratic polynomials, and q^n otherwise.

In the rest of this paper, $\big(\mathbf{a} = (a_1, \ldots, a_u), \mathbf{b} = (b_1, \ldots, b_u)\big)$ will always denote an element of $\mathbb{F}_q[\mathbf{x}]^u \times \mathbb{F}_q[\mathbf{x}]^u$. We will always suppose that all the polynomials of $\mathbf{a}$ have the same maximal total degree noted D (in the practical applications, we have $2 \leq D \leq 4$). Note that, if $\mathbf{b}(\mathbf{x}) = \mathbf{a}(\mathbf{x}S)U$, for some $(S, U) \in GL_n(\mathbb{F}_q) \times GL_u(\mathbb{F}_q)$, then the polynomials of $\mathbf{b}$ must have the same maximal total degree than the ones of $\mathbf{a}$, i.e. D.

3   A Unified Point of View

The Isomorphism of Polynomials and 2PLE problems have actually a very similar formulation. An input of these problems is formed of two systems of multivariate polynomials and the question consists in recovering a particular transformation permitting to express one system in function of the other. All transformations have the same characteristic: inducing a group action on $\mathbb{F}_q[\mathbf{x}]^u$. Recall that a group $(G, \cdot)$, with identity element e, *acts* on $\mathbb{F}_q[\mathbf{x}]^u$ if there exists a map $\phi : G \times \mathbb{F}_q[\mathbf{x}]^u \to \mathbb{F}_q[\mathbf{x}]^u$ such that $\phi(e, \mathbf{p}) = \mathbf{p}$, for all $\mathbf{p} \in \mathbb{F}_q[\mathbf{x}]^u$, and:

$$\phi\big(g, \phi(g', \mathbf{p})\big) = \phi(g \cdot g', \mathbf{p}), \text{ for all } g, g' \in G, \text{ and for all } \mathbf{p} \in \mathbb{F}_q[\mathbf{x}]^u.$$

Remark 1. In order to simplify the notations, we will write G instead of $(G, \cdot)$.

For 2PLE, one can then easily check that $GL_n(\mathbb{F}_q) \times GL_u(\mathbb{F}_q)$ acts on $\mathbb{F}_q[\mathbf{x}]^u$ through:

$$\phi_{2PLE} : GL_n(\mathbb{F}_q) \times GL_u(\mathbb{F}_q) \times \mathbb{F}_q[\mathbf{x}]^u \to \mathbb{F}_q[\mathbf{x}]^u$$
$$((S, U), \mathbf{a}) \mapsto \mathbf{a}(\mathbf{x}S)U$$

Similarly for IP, $AGL_n(\mathbb{F}_q) \times AGL_u(\mathbb{F}_q)$ acts on $\mathbb{F}_q[\mathbf{x}]^u$ through:

$$\phi_{\text{IP}} : AGL_n(\mathbb{F}_q) \times AGL_u(\mathbb{F}_q) \times \mathbb{F}_q[\mathbf{x}]^u \to \mathbb{F}_q[\mathbf{x}]^u$$
$$((S, \mathbf{T}), (U, \mathbf{V}), \mathbf{a}) \mapsto \mathbf{a}(\mathbf{x}S + \mathbf{T})U + \mathbf{V}$$

This observation naturally leads to the introduction of the following problem. Let $(G, \cdot)$ be a group, and $\phi : G \times \mathbb{F}_q[\mathbf{x}]^u \to \mathbb{F}_q[\mathbf{x}]^u$ be an action of G on $\mathbb{F}_q[\mathbf{x}]^u$. Given $(\mathbf{a}, \mathbf{b}) \in \mathbb{F}_q[\mathbf{x}]^u \times \mathbb{F}_q[\mathbf{x}]^u$, the problem we call *Polynomial Equivalence*, with respect to $(G, \cdot)$ and ϕ – and denoted by $\text{PE}(G, \phi)$ – is the one of finding (if any) $g \in G$, verifying:

$$\mathbf{b} = \phi(g, \mathbf{a}),$$

denoted $\mathbf{a} \equiv_{(G,\phi)} \mathbf{b}$ in the sequel. This formulation is very convenient since it procures a unified description of IP and 2PLE. Indeed, $\text{PE}(GL_n(\mathbb{F}_q) \times GL_u(\mathbb{F}_q), \phi_{\text{2PLE}}) = 2\text{PLE}$, and $\text{PE}(AGL_n(\mathbb{F}_q) \times AGL_u(\mathbb{F}_q), \phi_{\text{IP}}) = \text{IP}$. More generally, PE provides a unified description of "IP-like" problems. In our mind, such a kind of problems consists in recovering a particular transformation between two sets of multivariate polynomials. For instance, the Isomorphism of Polynomials with one Secret (IP1S) – introduced at Eurocrypt'96 by J. Patarin [28] – falls into this new formalism. This problem, which can be used to design an authentication (resp. signature) scheme [28], is as follows. Given $(\mathbf{a}, \mathbf{b}) \in \mathbb{F}_q[\mathbf{x}]^u \times \mathbb{F}_q[\mathbf{x}]^u$, find – if any – $(S, \mathbf{T}) \in AGL_n(\mathbb{F}_q)$, such that $\mathbf{b}(\mathbf{x}) = \mathbf{a}(\mathbf{x}S + \mathbf{T})$. Using our formalism, we immediately obtain that $\text{PE}(AGL_n(\mathbb{F}_q), \phi_{\text{IP1S}}) = \text{IP1S}$, with $\phi_{\text{IP1S}} : AGL_n(\mathbb{F}_q) \times \mathbb{F}_q[\mathbf{x}]^u \to \mathbb{F}_q[\mathbf{x}]^u, ((S, \mathbf{T}), \mathbf{a}(\mathbf{x})) \mapsto \mathbf{a}(\mathbf{x}S + \mathbf{T})$. Finally, the following lemma justifies the use of the word equivalence in PE.

Lemma 1. *Let $(G, \cdot)$ be a group, and $\phi : G \times \mathbb{F}_q[\mathbf{x}]^u \to \mathbb{F}_q[\mathbf{x}]^u$ be an action of G on $\mathbb{F}_q[\mathbf{x}]^u$. Then, $\equiv_{(G,\phi)}$ is an equivalence relation on $\mathbb{F}_q[\mathbf{x}]^u$.*

3.1   Polynomial Equivalence Problems and Group theory

In the Graph Isomorphism context, the introduction of group theory concepts permitted to achieve significant advances from both a theoretical and algorithmic point of view [24, 21]. The formalism previously given permits to naturally extend these results to Polynomial Equivalence problems.

Definition 4. *Let $(G, \cdot)$ be a group. We shall call $\text{Aut}_{(G,\phi)}(\mathbf{a}) = \{g \in G : \phi(g, \mathbf{a}) = \mathbf{a}\}$, $\text{Aut}_{(G,\phi)}(\mathbf{b}) = \{g \in G : \phi(g, \mathbf{b}) = \mathbf{b}\}$, the automorphism groups of $\mathbf{a}$ and $\mathbf{b}$ w.r.t. (G, ϕ). We shall also set $S_{(G,\phi)}(\mathbf{a}, \mathbf{b}) = \{g \in G : \mathbf{b} = \phi(g, \mathbf{a})\}$.*

$\text{Aut}_{(G,\phi)}(\mathbf{a})$ and $.\text{Aut}_{(G,\phi)}(\mathbf{b})$ are also known as stabilizer of $\mathbf{a}$ (resp. $\mathbf{b}$) w.r.t. (G, ϕ). However, we will rather call these sets automorphism groups. This designation being indeed more usually used in the Graph Isomorphism context [24]. Anyway, the results that we are going to expose are classical results of group theory concerning the stabilizers and orbits, and then given without proofs.

Proposition 1. *Let $(G, \cdot)$ be a group, and $\phi : G \times \mathbb{F}_q[\mathbf{x}]^u \to \mathbb{F}_q[\mathbf{x}]^u$ be an action of G on $\mathbb{F}_q[\mathbf{x}]^u$. If there exists $g \in G$, such that $\mathbf{b} = \phi(g, \mathbf{a})$, then $S_{(G,\phi)}(\mathbf{a}, \mathbf{b})$ is a left (resp. right) coset – in G – of the automorphism group $Aut_{(G,\phi)}(\mathbf{a})$ $\bigl(resp.$ $Aut_{(G,\phi)}(\mathbf{b})\bigr)$. That is:*

$$\begin{cases} S_{(G,\phi)}(\mathbf{a}, \mathbf{b}) = \{g \cdot h : h \in Aut_{(G,\phi)}(\mathbf{a})\} = g \cdot Aut_{(G,\phi)}(\mathbf{a}), \\ S_{(G,\phi)}(\mathbf{a}, \mathbf{b}) = \{h \cdot g : h \in Aut_{(G,\phi)}(\mathbf{b})\} = Aut_{(G,\phi)}(\mathbf{b}) \cdot g. \end{cases}$$

Moreover, the automorphism groups $Aut_{(G,\phi)}(\mathbf{a})$ and $Aut_{(G,\phi)}(\mathbf{b})$ are conjugate, i.e. $Aut_{(G,\phi)}(\mathbf{b}) = g \cdot Aut_{(G,\phi)}(\mathbf{a}) \cdot g^{-1}$, and we have:

$$|S_{(G,\phi)}(\mathbf{a}, \mathbf{b})| = |Aut_{(G,\phi)}(\mathbf{b})| = |Aut_{(G,\phi)}(\mathbf{a})|.$$

3.2 A Generic Upper Bound on the Complexity of "IP-Like" Problems

Using the Polynomial Equivalence problem previously defined, we give in this part a general upper bound on the theoretical complexity of "IP-like" problems. To do so, Let us fix a group $(G, \cdot)$ acting on $\mathbb{F}_q[\mathbf{x}]^u$ through a map noted ϕ.

For simplicity, we suppose here that G is included in a finite set $\mathcal{E}$. We also suppose that the uniform distribution of the elements of $\mathcal{E}$ can be simulated in polynomial-time. These assumptions allows to facilitate the proofs, and are additionally well adapted to "IP-like" problems. Indeed, $AGL_n(\mathbb{F}_q) \subset \mathcal{M}_{n,n}(\mathbb{F}_q) \times \mathbb{F}_q^n$, $GL_n(\mathbb{F}_q) \times GL_u(\mathbb{F}_q) \subset \mathcal{M}_{n,n}(\mathbb{F}_q) \times \mathcal{M}_{u,u}(\mathbb{F}_q)$, $AGL_n(\mathbb{F}_q) \times AGL_u(\mathbb{F}_q) \subset \mathcal{M}_{n,n}(\mathbb{F}_q) \times \mathbb{F}_q^n \times \mathcal{M}_{u,u}(\mathbb{F}_q) \times \mathbb{F}_q^u$. To obtain our upper bound, we introduce:

Definition 5. *An **interactive proof** for a language L $\bigl(i.e.$ a subset of $\{0,1\}^*\bigr)$ is a two party protocol between a verifier $\mathcal{V}$ and a prover $\mathcal{P}$. At the end of the protocol, the verifier has to accept or reject a given input such that the following conditions hold:*

Efficiency. *The verifier strategy is a probabilistic polynomial time procedure.*

Completeness. *For all $x \in L$, $Pr[(\mathcal{V}, \mathcal{P})(x) \text{ accepts}] = 1$.*

Soundness. *For all $x \notin L$, and for any prover $\mathcal{P}^*$, $Pr[(\mathcal{V}, \mathcal{P}^*)(x) \text{ accepts}] \leq \frac{1}{2}$. The probabilities are taken over the random choices of the verifier.*

Let us analyse the following two party protocol:

Input: $(\mathbf{a_0}, \mathbf{a_1}) \in \mathbb{F}_q[\mathbf{x}]^u \times \mathbb{F}_q[\mathbf{x}]^u$
Protocol: PI(G, ϕ)
The verifier chooses uniformly at random $i \in \{0, 1\}$.
He also chooses uniformly at random $g \in \mathcal{E}$ and checks if $g \in G$. If after C trials the verifier does not obtain an element $g \in G$, he accepts directly.
Otherwise, he sends $\mathbf{a'} = \phi(g, \mathbf{a_i})$ to the prover.
The prover replies by sending $j \in \{0, 1\}$ to the verifier.
The verifier accepts if $i = j$ and rejects otherwise.

Efficiency. The efficiency of this protocol depends on the cost of computing $\phi(g, \mathbf{a_i})$, for all $g \in G$, and of the number of trials C.

Completeness. If $\mathbf{a_0} \not\equiv_{(G,\phi)} \mathbf{a_1}$, then a prover can always check if $\mathbf{a'} \equiv_{(G,\phi)} \mathbf{a_0}$ or $\mathbf{a'} \equiv_{(G,\phi)} \mathbf{a_1}$. In this situation, the verifier accepts with probability one.

Soundness. If $\mathbf{a_0} \equiv_{(G,\phi)} \mathbf{a_1}$, then by transitivity $\mathbf{a'} \equiv_{(G,\phi)} \underline{a_1}$ and $\mathbf{a'} \equiv_{(G,\phi)} \mathbf{a_0}$. In such a case, we will show that $\mathbf{a'} = \phi(g, \mathbf{a_i})$ yields no information about the bit i chosen by the prover. Let then ψ be a random variable uniformly distributed over $\{0,1\}$, and Σ be a random variable uniformly distributed over G.

Lemma 2. Let $\mathbf{a_0}, \mathbf{a_1}, \mathbf{a'} \in \mathbb{F}_q[\mathbf{x}]^u$. If $\mathbf{a_0} \equiv_{(G,\phi)} \mathbf{a_1}$ and $\mathbf{a'} \equiv_{(G,\phi)} \mathbf{a_0}$, then:

$$Pr[\psi = 0 \mid \mathbf{a}_\psi(\mathbf{x}\Sigma) = \mathbf{a'}] = Pr[\psi = 1 \mid \mathbf{a}_\psi(\mathbf{x}\Sigma) = \mathbf{a'}] = \frac{1}{2}.$$

Proof. We have $\Pr[\phi(\Sigma, \mathbf{a}_\psi) = \mathbf{a'} \mid \psi = 0] = \Pr[\phi(\Sigma, \mathbf{a_0}) = \mathbf{a'}] = \Pr[\Sigma \in S_{(G,\phi)}(\mathbf{a_0}, \mathbf{a'})]$. Moreover, according to Proposition 1:

$$|S_{(G,\phi)}(\mathbf{a_0}, \mathbf{a'})| = |Aut_{(G,\phi)}(\mathbf{a'})| = |S_{(G,\phi)}(\mathbf{a_1}, \mathbf{a'})|.$$

Therefore, $\Pr[\phi(\Sigma, \mathbf{a_0}) = \mathbf{a'}] = \Pr[\mathbf{a_1}(\mathbf{x}\Sigma) = \mathbf{a'}]$, and thus:

$$\Pr[\phi(\Sigma, \mathbf{a}_\psi) = \mathbf{a'} \mid \psi = 0] = \Pr[\phi(\Sigma, \mathbf{a}_\psi) = \mathbf{a'} \mid \psi = 1].$$

According to the Bayes formula:

$$\begin{aligned}
\Pr\psi = 0 \mid \phi(\Sigma, \mathbf{a}_\psi) = \mathbf{a'}] &= \frac{\Pr[\psi=0] \Pr[\phi(\Sigma,\mathbf{a}_\psi)=\mathbf{a'} \mid \psi=0]}{\Pr[\phi(\Sigma,\mathbf{a}_\psi)=\mathbf{a'}]} \\
&= \frac{\Pr[\psi=1] \Pr[\phi(\Sigma,\mathbf{a}_\psi)=\mathbf{a'} \mid \psi=1]}{\Pr[\phi(\Sigma,\mathbf{a}_\psi)=\mathbf{a'}]} \\
&= \Pr[\psi = 1 \mid \phi(\Sigma, \mathbf{a}_\psi) = \mathbf{a'}].
\end{aligned}$$

Finally:

$$\begin{aligned}
\Pr[\psi = 0 \mid \phi(\Sigma, \mathbf{a}_\psi) = \mathbf{a'}] &= \frac{\Pr[\psi=0] \Pr[\phi(\Sigma,\mathbf{a}_\psi)=\mathbf{a'} \mid \psi=0]}{\Pr[\phi(\Sigma,\mathbf{a}_\psi)=\mathbf{a'}]} \\
&= \frac{\Pr[\psi=1] \Pr[\phi(\Sigma,\mathbf{a_0})=\mathbf{a'}]}{\Pr[\phi(\Sigma,\mathbf{a}_\psi)=\mathbf{a'}]} \\
&= \frac{\Pr[\psi=1] \Pr[\Sigma \in S_{(G,\psi)}(\mathbf{a'},\mathbf{a_0})]}{\Pr[\Sigma \in S_{(G,\phi)}(\mathbf{a}_\psi,\mathbf{a'})]} = \frac{1}{2}.
\end{aligned}$$

$\square$

It follows that no prover – no matter what its strategy is – can guess i with probability greater than $\frac{1}{2}$. Finally, using a classical result of R. B. Boppana, J. Hastad, and S. Zachos [7], we get that:

Corollary 1. *If the polynomial hierarchy does not collapse then* IP, 2PLE, *and* IP1S *are not* NP-Hard.

Proof. We sketch the proof for IP1S. Note that for all $g \in AGL_n(\mathbb{F}_q)$, one can compute $\phi_{\text{IP1S}}(g, \mathbf{a'})$ in polynomial-time. Let L_{IP} be the language associated to IP1S (i.e. the set of instances of IP admitting a solution). We study now the number of trials in $\text{PI}(AGL_n(\mathbb{F}_q), \phi_{\text{IP1S}})$. Recall that more than $1/4$ of the matrices of $\mathcal{M}_{n,n}(\mathbb{F}_q)$ are invertible. Therefore for IP1S, we have $G = AGL_n(\mathbb{F}_q)$,

$\mathcal{E} = \mathcal{M}_{n,n}(\mathbb{F}_q) \times \mathbb{F}_q^n$, and $\Pr[g \in G \,|\, g \in \mathcal{E}] \geq \frac{1}{4}$. By setting $C = 10$, we get that no prover can guess i with probability greater than

$$\frac{1}{2} + \left(\frac{3}{4}\right)^{10} < \frac{1}{2} + \frac{1}{16} = \frac{9}{16},$$

where $\left(\frac{3}{4}\right)^{10} < \frac{1}{16}$ is the probability of not obtaining an element of $AGL_n(\mathbb{F}_q)$ after ten trials. By repeating the protocol two times, we obtain that no prover can fool the verifier into accepting $\mathbf{a_0} \not\equiv_{(AGL_n(\mathbb{F}_q), \phi_{\text{IP1S}})} \mathbf{a_1}$ with a probability greater than $\left(\frac{9}{16}\right)^2 < \frac{1}{2}$. The protocol $\text{PI}\big(AGL_n(\mathbb{F}_q), \phi_{\text{IP1S}}\big)$ is then an interactive proof for the complementary language of L_{IP1S} (i.e. $\{0,1\}^* \backslash \text{L}_{\text{IP1S}}$), where at most 4 messages are exchanged between the verifier and the prover. We do not detail the proof, but one can easily check that the same result holds for $\text{PI}\big(AGL_n(\mathbb{F}_q) \times AGL_u(\mathbb{F}_q), \phi_{\text{IP}}\big)$ and $\text{PI}\big(GL_n(\mathbb{F}_q) \times GL_u(\mathbb{F}_q), \phi_{\text{2PLE}}\big)$.

The corollary then follows from a result of [7], stating that if the complementary of a language admits a constant round interactive protocol, then this language can not be NP-Complete, unless the polynomial hierarchy collapses.   □

The new formalism introduced in this part allows to upper bound the theoretical hardness of IP, 2PLE, and IP1S. More generally, it provides a new insight on the complexity of "IP-like" problems. The previous corollary can be indeed easily adapted to any instance of the Polynomial Equivalence problem. An "IP-like" problem is then intrinsically not NP-Hard. Furthermore, we believe that our formalism is of independent interest. It indeed procures a general framework for studying "IP-like" problems. However, this is out of the scope of this paper. We investigate now another aspect of these problems.

4   An Algorithm for Solving 2PLE

We study here the practical hardness of a particular Polynomial Equivalence problem, namely 2PLE. Precisely, we present a new algorithm for solving this problem. We emphasize that – as explained in 1.1 – it is usually sufficient to consider this problem rather than its affine variant IP. Besides, any algorithm solving 2PLE can be transformed into an algorithm solving IP [11, 12].

4.1   A First Attempt: Evaluation and Linearization

Instead of directly describing the details of our method, we present the different steps that yielded to this algorithm. Anyway, most of the intermediate results that we are going to present will be used in our final algorithm, but differently. Our earliest idea for solving 2PLE was based on the following remark. If $\mathbf{b}(\mathbf{x}) = \mathbf{a}(\mathbf{x}S)U$, for $(S, U) \in GL_n(\mathbb{F}_q) \times GL_u(\mathbb{F}_q)$, then:

$$\mathbf{b}(\mathbf{p})U^{-1} = \mathbf{a}(\mathbf{p}S), \quad \text{for all } \mathbf{p} \in \mathbb{F}_q^n. \tag{1}$$

We hence obtain, for each $\mathbf{p} \in \mathbb{F}_q^n$, u non-linear equations in the $n^2 + u^2$ components of the matrices S and U^{-1}. We point out that the coefficients of U^{-1} only appear linearly in these equations. This is the advantage of considering the inverse of U rather than simply U in (1). The number of equations obtained is then significantly bigger than the number of unknowns. In this situation, one can simply use a linearization method (i.e. associating a new variable to each monomial) for solving the algebraic system. Unfortunately, our experiments rapidly revealed that the equations generated in this way are not all linearly independent. Besides, it also appeared that the number of unknowns is significantly bigger than the number of linearly independent equations. The use of a linearization method is then clearly no longer relevant. Let us explain this phenomenon.

Lemma 3. *Let* $\mathbf{y} = (y_{1,1}, \ldots, y_{1,n}, \ldots, y_{n,1}, \ldots, y_{n,n})$, *and* $\mathbf{z} = (z_{1,1}, \ldots, z_{1,u}, \ldots$ $, z_{u,1}, \ldots, z_{u,u})$. *For each* $i, 1 \leq i \leq u$, *there exists a subset* $S_i \subseteq \mathbb{F}_q^n$ *and polynomials* $p_{\alpha,i} \in \mathbb{F}_q[\mathbf{y}, \mathbf{z}]$, *such that the following equality holds:*

$$\left(\mathbf{b}(\mathbf{x})U^{-1} - \mathbf{a}(\mathbf{x}S)\right)_i = \sum_{\alpha \in S_i} p_{\alpha,i}(S, U^{-1})\mathbf{x}^\alpha, \tag{2}$$

$p_{\alpha,i}(S, U^{-1})$ *being the evaluation of* $p_{\alpha,i}$ *on* $S = \{s_{i,j}\}_{1 \leq i,j \leq n}$, $U^{-1} = \{u'_{i,j}\}_{1 \leq i,j \leq u}$.

Proof. The polynomial $\left(\mathbf{b}(\mathbf{x})U^{-1} - \mathbf{a}(\mathbf{x}S)\right)_i$ can be regarded as an element of:

$$\mathbb{F}_q[s_{1,1}, \ldots, s_{1,n}, \ldots, s_{n,1}, \ldots, s_{n,n}, u'_{1,1}, \ldots, u'_{1,u}, \ldots, u'_{u,u}, \ldots, u'_{u,u}][x_1, \ldots, x_n], \tag{3}$$

i.e. a polynomial with unknowns $x_1, \ldots, x_n$ and whose coefficients are polynomials in the components of S and U^{-1}. In this setting, the polynomials $p_{\alpha,i}$ exactly correspond to the coefficients of the monomials (in $x_1, \ldots, x_n$) occurring in $\left(\mathbf{b}(\mathbf{x})U^{-1} - \mathbf{a}(\mathbf{x}S)\right)_i$. Lastly $S_i = \{\alpha \in \mathbb{F}_q^n : p_{\alpha,i} \neq 0\}$. □

The cost of generating the polynomials $p_{\alpha,i}$ is proportional to the number of monomials occurring in $\left(\mathbf{b}(\mathbf{x})U^{-1} - \mathbf{a}(\mathbf{x}S)\right)_i$ viewed as a polynomial of (3), i.e. $O(n^{2D})$. Note also that each $p_{\alpha,i}$ is by construction the sum of a polynomial in $\mathbf{y}$, plus a linear polynomial in $\mathbf{z}$. Furthermore, the maximal total degree reached by a monomial in the variables $\mathbf{y}$ is equal to D.
From (2), we obtain that for all $i, 1 \leq i \leq u$:

$$\left(\mathbf{b}(\mathbf{p})U^{-1} - \mathbf{a}(\mathbf{p}S)\right)_i = \sum_{\alpha \in S_i} p_{\alpha,i}(S, U^{-1})p_1^{\alpha_1} \cdots p_n^{\alpha_n}, \text{for all } \mathbf{p} = (p_1, \ldots, p_n) \in \mathbb{F}_q^n.$$

It follows that, for all $\mathbf{p} \in \mathbb{F}_q^n$, the equations procured by (1) are linear combinations of the $p_{\alpha,i}(S, U^{-1})$. The number of polynomials $p_{\alpha,i}$ is limited by the number of monomials occuring in $\left(\mathbf{b}(\mathbf{p})U^{-1} - \mathbf{a}(\mathbf{p}S)\right)_i$. Thus, $u \cdot C_{n+D}^D$ bounds from above the number of linearly independent equations provided by linearizing (1). On the other hand, the number of unknowns in the linearized system is equal to the number of monomials in the variables $\mathbf{y}$ of degree smaller than D, plus the u^2 variables corresponding to $\mathbf{z}$. Using a rough bound, the linearization method yields a linear system of at most $O(u \cdot n^D)$ linearly independent equations with $O(u \cdot n^{2D})$ unknowns.

4.2   The 2PLE Algorithm

The linearization can thus not be employed for solving efficiently 2PLE. However, Gröbner basis procures another method for solving the algebraic system given by (1). From a practical point of view, this approach is quite promising. Indeed, the system obtained by evaluating $\mathbf{b}(\mathbf{x})U^{-1} = \mathbf{a}(\mathbf{x}S)$ on several vectors is overdetermined. Nevertheless, all the equations derived from $\mathbf{b}(\mathbf{p})U^{-1} = \mathbf{a}(\mathbf{p}S)$ are according to (2) linear combinations the polynomials $p_{\alpha,i}$. It is hence sufficient to only consider the system formed by these equations. Formally:

Proposition 2. *Let $\mathcal{I} = \langle p_{\alpha,i} : \text{for all } i, 1 \leq i \leq u, \text{ and for all } \alpha \in S_i \rangle \subset \mathbb{F}_q[\mathbf{y}, \mathbf{z}]$ be the ideal generated by the polynomials $p_{\alpha,i}$ defined as in Lemma 3, and $V(\mathcal{I})$ be the following variety:*

$$V(\mathcal{I}) = \left\{ \mathbf{s} \in \mathbb{F}_q^{n^2+u^2} : p_{\alpha,i}(\mathbf{s}) = 0, \text{for all } i, 1 \leq i \leq u, \text{ and for all } \alpha \in S_i \right\}.$$

If $\mathbf{b}(\mathbf{x}) = \mathbf{a}(\mathbf{x}S)U$, for some $(S, U) \in GL_n(\mathbb{F}_q) \times GL_u(\mathbb{F}_q)$, then:

$$\left(\phi_1(S), \phi_2(U^{-1}) \right) \in V(\mathcal{I}),$$

with:

$\phi_1 : \mathcal{M}_{n,n}(\mathbb{F}_q) \rightarrow \mathbb{F}_q^{n^2}, S = \{s_{i,j}\}_{1 \leq i,j \leq n} \mapsto (s_{1,1}, \dots, s_{1,n}, \dots, s_{n,1}, \dots, s_{n,n})$, and
$\phi_2 : \mathcal{M}_{u,u}(\mathbb{F}_q) \rightarrow \mathbb{F}_q^{u^2}, U^{-1} = \{u'_{i,j}\}_{1 \leq i,j \leq u} \mapsto (u'_{1,1}, \dots, u'_{1,u}, \dots, u'_{u,1}, \dots, u'_{u,u})$.

Proof. For all, $i, 1 \leq i \leq u$:

$$\left(\mathbf{b}(\mathbf{x})U^{-1} - \mathbf{a}(\mathbf{x}S) \right)_i = \sum_{\alpha \in S_i} p_{\alpha,i}(S, U^{-1})\mathbf{x}^\alpha = 0.$$

Thus, $p_{\alpha,i}(S, U^{-1}) = 0, \forall i, 1 \leq i \leq u$, and $\forall \alpha \in S_i$, i.e. $\left(\phi_1(S), \phi_2(U^{-1}) \right) \in V(\mathcal{I})$. □

In other words, if $\mathbf{b} = \mathbf{a}(\mathbf{x}S)U$, for some $(S, U) \in GL_n(\mathbb{F}_q) \times GL_u(\mathbb{F}_q)$, then the variety $V(\mathcal{I})$ contains the components of the matrices S and U^{-1}. The system associated to $\mathcal{I}$ has $n^2 + u^2$ variables and is of degree D. Once again, we recall that the variables of $\mathbf{z}$ only appear linearly in this system. The number of equations of the system is equal to the number of monomials occurring in the polynomials of $\mathbf{a}$, i.e. $O\left(u \cdot C_{n+D}^D \right)$. The system is then overdetermined.

Remark 2. In order to guarantee that $V(\mathcal{I}) \subseteq \mathbb{F}_q^{2n}$, we must generally join the field equations to the initial system. The fields considered in our case can be relatively large, leading then to a significant increase of the system's degree. This can artificially render impracticable the computation of a Gröbner basis. Fortunately, our systems are overdetermined and it is not necessary in practice to include the field equations. In our experiments the elements of $V(\mathcal{I})$ were indeed – without including these equations – all the times in $\mathbb{F}_q^{2n}$. It implies in particular that the hardness of 2PLE is not related to the size of the field. This is an important remark since the current security bound for 2PLE depends on this size.

The next proposition is fundamental to understand the practical behaviour of our approach. This result permits furthermore to improve the efficiency of our method.

Proposition 3. *Let d be a positive integer, and $\mathcal{I}_d \subset \mathbb{F}_q[\mathbf{y}, \mathbf{z}]$ be the ideal generated by the polynomials $p_{\alpha,i}$ of maximal total degree smaller than d. Let also $V(\mathcal{I}_d)$ be the variety associated to $\mathcal{I}_d$. If $\mathbf{b}(\mathbf{x}) = \mathbf{a}(\mathbf{x}S)U$, for some $(S, U) \in GL_n(\mathbb{F}_q) \times GL_u(\mathbb{F}_q)$, then:*

$$\left(\phi_1(S), \phi_2(U^{-1})\right) \in V(\mathcal{I}_d), \text{ for all } d, 0 \le d \le D,$$

ϕ_1 and ϕ_2 being defined as in proposition 2.

The proof is obviously deduced from the following result:

Lemma 4. *Let $(S, U) \in GL_n(\mathbb{F}_q) \times GL_u(\mathbb{F}_q)$. We have:*

$$\mathbf{b}(\mathbf{x}) = \mathbf{a}(\mathbf{x}S)U \iff \mathbf{b}^{(d)}(\mathbf{x}) = \mathbf{a}^{(d)}(\mathbf{x}S)U, \text{ for all } d, 0 \le d \le D,$$

$\mathbf{b}^{(d)}$ (resp. $\mathbf{a}^{(d)}$) being the homogeneous components of degree d (i.e. the sum of the terms of total degree d) of the polynomials of $\mathbf{b}$ (resp. $\mathbf{a}$).

The systems associated to $\mathcal{I}_1$ and $\mathcal{I}_0$ only contain linear equations in the components of S and U^{-1}. Indeed, let $\mathbf{0_n}$ be the null vector of $\mathbb{F}_q^n$, and $A \in \mathcal{M}_{n,u}(\mathbb{F}_q)$ (resp. $B \in \mathcal{M}_{n,u}(\mathbb{F}_q)$) be the matrix representation of $\mathbf{a}^{(1)}$ (resp. $\mathbf{b}^{(1)}$), i.e. $\mathbf{x}A = \mathbf{a}^{(1)}(\mathbf{x})$ (resp. $\mathbf{x}B = \mathbf{b}^{(1)}(\mathbf{x})$). According to Lemma 4:

$$\mathbf{b} = \mathbf{a}(\mathbf{x}S)U, \text{ for } (S, U) \in GL_n(\mathbb{F}_q) \times GL_u(\mathbb{F}_q) \implies \begin{cases} \mathbf{b}^{(0)}(\mathbf{0_n})U^{-1} = \mathbf{a}^{(0)}(\mathbf{0_n}), \\ BU^{-1} = SA. \end{cases}$$

That is, we get linear dependencies between the components S and U^{-1}. More precisely, we obtain $u(n+1)$ linear equations in the $n^2 + u^2$ components of the matrices solution. Anyway, we can not solve 2PLE just by using these equations. On the other hand, it is not necessary to consider the system formed by all the polynomials $p_{\alpha,i}$. According to Proposition 3, we can actually restrict our attention on $\mathcal{I}_{d_0}$, with d_0 being the smaller integer rendering the system overdetermined. This d_0 can be defined in function of $\mathbf{a}$. Indeed, $d_0 \approx \min\{d > 1 : \mathbf{a}^{(d)} \ne \mathbf{0}_u\}$. In practice, it is usually sufficient to take $d_0 = 2$. The hardness of an instance of 2PLE is then related to d_0 rather than to the maximal total degree D of this instance. It is also an important remark since the maximal degree of an instance is taken into account in the security estimate of 2PLE given by J. Patarin [28, 29]. Our algorithm for solving this problem is as follows:

Input: $(\mathbf{a}, \mathbf{b}) \in \mathbb{F}_q[\mathbf{x}]^u \times \mathbb{F}_q[\mathbf{x}]^u$
Let $d_0 = \min\{d > 1 : \mathbf{a}^{(d)} \ne \mathbf{0_u}\}$
Construct the polynomials $p_{\alpha,i}$ of max. total deg. smaller than d_0
Compute $V(\mathcal{I}_{d_0})$ using the F_5 algorithm
Find an element of $V(\mathcal{I}_{d_0})$ corresponding to a solution of 2PLE
Return this solution

The system associated to $\mathcal{I}_{d_0}$ is overdetermined by its very construction ($u^2 + n^2$ unknowns, and $O\left(u \cdot C_{n+d_0}^{d_0}\right)$ equations). The variety $V(\mathcal{I}_{d_0})$ is then very likely reduced to a solution of 2PLE (this has been indeed verified in our experiments). The complexity of this algorithm is (theoretically) dominated by the Gröbner basis computation. It is difficult to obtain a complexity bound really reflecting the practical behavior of the F_5 algorithm. We therefore carry out now experimental results illustrating the practical efficiency of our approach.

4.3 Experimental Results

We present in this part experimental results obtained with our algorithm. Before that, we provide the conditions of our experiments

Generation of the instances
We have only considered instances $(\mathbf{a}, \mathbf{b})$ of 2PLE admitting a solution. We constructed the instances in the following way:

(1) Choose the polynomials of $\mathbf{a}$
(2) Randomly choose $(S, U) \in GL_n(\mathbb{F}_q) \times GL_u(\mathbb{F}_q)$
(3) Return $\left(\mathbf{a}(\mathbf{x}), \mathbf{b}(\mathbf{x}) = \mathbf{a}(\mathbf{x}S)U\right)$

Precisely, we constructed the polynomials of $\mathbf{a}$ in two different ways. The first one simply consists in randomly choosing – w.r.t. a given maximal total degree D – the polynomials of $\mathbf{a}$. Precisely, each polynomial is a random linear combination of all the monomials of total degree smaller (or equal) to D. Note that we obtain in this way dense polynomials. We shall call *random instance*, an instance of 2PLE generated in this manner. In the second method, $\mathbf{a}$ corresponds to the public key of a C^* scheme [26]. An instance of 2PLE generated in this way will be named C^* *instance*.

Programming Language – Workstation
The experimental results have been obtained with an Opteron bi-processors 2.4 Ghz, with 8 Gb of Ram. The systems associated to an instance of 2PLE have been generated using the Magma software[25]. We used our own implementation (in language C) of F_5 for computing the Gröbner bases. However, for the sake of comparison, we sometimes used the last version of Magma (i.e. 2.12) for obtaining these bases. This version includes an implementation of the F_4 algorithm.

Table Notations
The following notations are used in the tables below:
– n, the number of variables,
– q, the size of the field,
– deg, the maximal total degree of the considered instance,
– T_{Gen}, the time needed to construct the system,
– T_{F_5}, the time of our algorithm for finding a solution of 2PLE (using the F_5 algorithm for computing the Gröbner bases,
– T, the total time of our algorithm, i.e. $T = T_{F_5} + T_{Gen}$,

– $T_{F_4/Mag}$, the time of our algorithm for recovering a solution of 2PLE, using Magma v. 2.12 for computing Gröbner bases,
– $q^{n/2}$ (resp. q^n), the security bound given in [11, 12] for instances of $deg = 2$ (resp. $deg > 2$).

Practical Results – Random Instances
We present here the results obtained on random instances of 2PLE. We emphasize that this family of instances is the one employed in the authentication and signature schemes based on 2PLE proposed by J. Patarin at Eurocrypt'96 [28, 29]. He suggested to use $u = n$ in practice. Since our main motivation is to study the security of these schemes, we can restrict our attention on the case $u = n$.

n	q	deg	T_{Gen}	T_{F_5}	$T_{F_4/Mag}/T_{F_5}$	T	$q^{n/2}$
8	2^{16}	2	0.35 s.	0.14 s.	6	0.49 s.	2^{64}
10	2^{16}	2	1.66 s.	0.63 s.	10	2.29 s.	2^{80}
12	2^{16}	2	7.33 s.	2.16 s.	16	9.49 s.	2^{96}
15	2^{16}	2	48.01 s.	10.9 s.	23	58.91 s.	2^{120}
17	2^{16}	2	137.21 s.	27.95 s.	31	195.16 s.	2^{136}
20	2^{16}	2	569.14 s.	91.54 s.	41	660.68 s.	2^{160}
10	65521	2	1.21 s.	0.44 s.	10	1.65 s.	$\approx 2^{80}$
15	65521	2	35.58 s.	8.08 s.	23	43.66 s.	$\approx 2^{120}$
20	65521	2	434.96 s.	69.96 s.	41	504.92 s.	$\approx 2^{160}$
23	65521	2	1578.6 s.	235.92 s.		1814 s.	$\approx 2^{184}$

Remark 3. Our implementation of F_5 is faster than the Gröbner basis algorithm available in Magma 2.12. For $n = 20$, F_5 is for instance 41 times faster than Magma. To fix ideas, $u = n = 8$, and $u = n = 16$ were two challenges proposed at Eurocrypt'96 [29]. We obtained exactly the same results as the ones quoted in the previous table for random instances of $deg > 2$. On the other hand, the security estimate for these instances is at least equal to $2^{128} (n = 8)$. The maximal total degree of the systems is indeed the same as for instances of $deg = 2$, i.e. d_0 is equal to 2 independently of D. In other words, increasing the maximal total degree of a random instance will not change its practical hardness. We observe the same behavior for the size of the field, that is increasing q does not really change the hardness of a random instance. This will indeed modify only the cost of the arithmetic operations in the different steps our algorithm.

Interpretation of the Results
In all these experiments, the varieties computed were reduced to one element, i.e. the components of the matrices solution of 2PLE. Furthermore, we observe in practice that the complexity of our algorithm is dominated by the time required to construct the system, and not by the Gröbner basis computation. This is surprising, but it clearly highlights that the systems considered here can be easily solved in practice. The generation of the systems being polynomial, we then conclude experimentally that our algorithm solves random instances of

2PLE in polynomial-time. This conclusion is supported by the fact that in all these experiments, the matrices generated by F_5 (see the Appendix) were of size at most equal to n^3. Experimentally, we deduce a complexity of $(n^3)^3 = n^9$ for our algorithm on random instances of 2PLE.

Practical Results – C* Instances

We now present the results obtained on C* instances $(\mathbf{a}, \mathbf{b})$ of degree D. We highlight that these instances are used in the traitor tracing scheme described in [5]. In this context, we also have $u = n$. The polynomials of $\mathbf{a}$ correspond to the public-key of a C* scheme [26]. Precisely, these polynomials are the "multivariate representation" of a univariate monomial (see [5] for details concerning the generation of this multivariate representation). The univariate monomial has the following shape: $m^{1+q^{\theta_1}+q^{\theta_2}+\cdots+q^{\theta_{D-1}}}$, with $\theta_1, \theta_2, \cdots, \theta_{D-1} \in \mathbb{N}^*$.

n	q	deg	T_{Gen}	T_{F_5}	$T_{F_4/Mag}/T_{F_5}$	T	q^n
5	2^{16}	4	0.2 s.	0.13 s.	45	0.33 s.	2^{80}
6	2^{16}	4	0.7 s.	1.03 s.	64	1.73 s.	2^{96}
7	2^{16}	4	1.5 s.	6.15 s.	90	7.65 s.	2^{112}
8	2^{16}	4	3.88 s.	54.34 s.	112	58.22 s.	2^{128}
9	2^{16}	4	5.43 s.	79.85 s.	145	85.28 s.	2^{144}
10	2^{16}	4	12.9 s.	532.33 s.	170	545.23 s.	2^{160}

Remark 4. $n = 5$, and $deg = 4$ is the first challenge proposed at Asiacrypt'03 [5]. Similarly to random instances, we observed that the size of the field does not really change the practical hardness of the C* instances. We can conclude that it is a general behaviour of 2PLE instances.

Interpretation of the Results and Future Work

Our algorithm is no longer polynomial for C* instances. The systems obtained for these instances are indeed harder to solve than the random ones. We believe that it is due to the fact that the systems are here sparser. The equality $\mathbf{b}(\mathbf{0_n}) = \mathbf{a}(\mathbf{0_n})U$ does not provide any information $(\mathbf{b}(\mathbf{0_n}) = \mathbf{a}(\mathbf{0_n}) = \mathbf{0_n}$ in the C* case). It is not clear yet but it seems that C* instances with $n = 19$ (the second challenge proposed in [5]), can not be solved with our approach.

More generally, we think that $d_0 = \min\{d \geq 0 : \mathbf{a}^{(d)} \neq \mathbf{0_u}\}$ provides a relevant measure of the practical hardness of 2PLE instances. It seems actually that this practical difficulty increases in function of d_0. Indeed, for random instances of 2PLE, $d_0 = 0$ and our algorithm solves 2PLE efficiently. For C* instances, $d_{min} = 1$ and there is a change of complexity class. We also checked that the practical complexity increases for homogeneous instances of degree 2, i.e. $d_0 = 2$. To summarize, for $d_0 = 0$ it is relatively clear that our algorithm solves 2PLE efficiently (likely in polynomial-time). For $d_0 \geq 1$, we conjecture that our algorithm is subexponential in n, and will depend on d_0. This anyway needs further investigations. It is an open problem to precisely determine, as a function of d_0, the asymptotic complexity of our algorithm. It could be possible that techniques presented in [3, 4] provide an answer.

Acknowledgements

We thank Françoise Levy-dit-Vehel and anonymous referees for numerous comments which improved the presentation of the results.

References

1. W.W. Adams and P. Loustaunau. *An Introduction to Gröbner Bases.* Graduate Studies in Mathematics, Vol. 3, AMS, 1994.
2. G. Ars, J.-C. Faugère, H. Imai, M. Kawazoe, and M. Sugita. *Comparison Between XL and Gröbner Basis Algorithms.* Advances in Cryptology – ASIACRYPT 2004, Lecture Notes in Computer Science, vol. 3329, pp. 338-353, 2004.
3. M. Bardet, J-C. Faugère, B. Salvy and B-Y. Yang. *Asymptotic Behaviour of the Degree of Regularity of Semi-Regular Polynomial Systems.* In MEGA 2005, Eighth International Symposium on Effective Methods in Algebraic Geometry, 15 pages, 2005.
4. M. Bardet, J-C. Faugère, and B. Salvy. *On the Complexity of Gröbner Basis Computation of Semi-Regular Overdetermined Algebraic Equations.* In Proc. of International Conference on Polynomial System Solving (ICPSS), pp. 71–75, 2004.
5. O. Billet, and H. Gilbert. *A Traceable Block Cipher.* Advances in Cryptology – ASIACRYPT 2003, Lecture Notes in Computer Science, vol. 2894, Springer–Verlag, pp. 331-346, 2003.
6. A. Biryukov, C. De Cannière, A. Braeken, and B. Preneel. *A Toolbox for Cryptanalysis: Linear and Affine Equivalence Algorithms.* Advances in Cryptology – EUROCRYPT 2003, Lecture Notes in Computer Science, vol. 2656, Springer–Verlag, pp. 33-50, 2003.
7. R. B. Boppana, J. Hastad, and S. Zachos. *Does co–NP Have Short Interactive Proofs?* Information Processing Letters, 25(2), pp. 127–132, 1987.
8. B. Buchberger. *Gröbner Bases : an Algorithmic Method in Polynomial Ideal Theory.* Recent trends in multidimensional systems theory. Reider ed. Bose, 1985.
9. B. Buchberger, G.-E. Collins, and R. Loos. *Computer Algebra Symbolic and Algebraic Computation.* Springer-Verlag, second edition, 1982.
10. N. Courtois. *La sécurité des primitives cryptographiques basées sur des problèmes algébriques multivariables: MQ, IP, MinRank, HFE.* Ph.D. Thesis, Paris, 2001.
11. N. Courtois, L. Goubin, and J. Patarin. *Improved Algorithms for Isomorphism of Polynomials.* Advances in Cryptology - EUROCRYPT 1998, Lecture Notes in Computer Science, vol. 1403, Springer-Verlag, pp. 84–200, 1998.
12. N. Courtois, L. Goubin, and J. Patarin. *Improved Algorithms for Isomorphism of Polynomials - Extended Version.* Available from http://www.minrank.org.
13. N. Courtois, L. Goubin, and J. Patarin. *SFLASH, a Fast Asymmetric Signature Scheme for low-cost Smartcards – Primitive Specification and Supporting Documentation.* Available at http://www.minrank.org/sflash-b-v2.pdf.
14. N. Courtois, A. Klimov, J. Patarin, and A. Shamir. *Efficient Algorithms for Solving Overdefined Systems of Multivariate Polynomial Equations.* Advances in Cryptology – EUROCRYPT 2000, Lecture Notes in Computer Science, vol. 1807, Springer–Verlag, pp. 392-407, 2000.
15. D. A. Cox, J.B. Little and, D. O'Shea. *Ideals, Varieties, and Algorithms: an Introduction to Computational Algebraic Geometry and Commutative Algebra.* Undergraduate Texts in Mathematics. Springer-Verlag. New York, 1992.

16. J.-C. Faugère. *A New Efficient Algorithm for Computing Gröbner Bases (F_4)*. Journal of Pure and Applied Algebra, 139(1-3), pp. 61–88, June 1999.
17. J.-C. Faugère. *A New Efficient Algorithm for Computing Gröbner Basis without Reduction to Zero:* F_5. Proceedings of ISSAC, pp. 75–83. ACM press, July 2002.
18. J.-C. Faugère, and A. Joux. *Algebraic Cryptanalysis of Hidden Field Equation (HFE) Cryptosystems using Gröbner bases*. Advances in Cryptology - CRYPTO 2003, Lecture Notes in Computer Science, vol. 2729, Springer-Verlag, pp. 44–60, 2003.
19. J. C. Faugère, P. Gianni, D. Lazard, and T. Mora. *Efficient Computation of Zero-Dimensional Gröbner Bases by Change of Ordering*. Journal of Symbolic Computation, 16(4), pp. 329–344, 1993.
20. P. Felke *On certain Families of HFE-type Cryptosystems*. Proceedings of WCC'05, International Workshop on Coding and Cryptography, March 2005.
21. S. Fortin. *The Graph Isomorphism problem*. Technical Report 96-20, University of Alberta, 1996.
22. M. R. Garey, and D. B. Johnson. *Computers and Intractability. A Guide to the Theory of NP-Completeness*. W. H. Freeman, 1979.
23. W. Geiselmann, R. Steinwandt, and T. Beth. *Attacking the Affine Parts of SFLASH*. Cryptography and Coding, 8th IMA International Conference, vol. 2260, Springer–Verlag, pp. 355-359, 2001.
24. M. Hoffman. *Group-theoretic algorithms and Graph Isomorphism*. Lecture Notes in Computer Science, vol. 136, Springer–Verlag, 1982.
25. http://magma.maths.usyd.edu.au/magma/
26. T. Matsumoto, and H. Imai. *Public Quadratic Polynomial-tuples for efficient signature-verification and message-encryption*. Advances in Cryptology – EUROCRYPT 1988, Lecture Notes in Computer Science, vol. 330, Springer–Verlag, pp. 419–453, 1988.
27. https://www.cosic.esat.kuleuven.be/nessie/deliverables/decision-final.pdf.
28. J. Patarin. *Hidden Fields Equations (HFE) and Isomorphisms of Polynomials (IP): two new families of Asymmetric Algorithms*. Advances in Cryptology – EUROCRYPT 1996, Lecture Notes in Computer Science, vol. 1070, Springer-Verlag, pp. 33–48, 1996.
29. J. Patarin. *Hidden Fields Equations (HFE) and Isomorphisms of Polynomials (IP): two new families of Asymmetric Algorithms – Extended Version*. Available from http://www.minrank.org/hfe/.
30. J. Patarin. *Cryptanalysis of the Matsumoto and Imai Public Key Scheme of Eurocrypt'88*. Advances in Cryptology – CRYPTO 1995, Lecture Notes in Computer Science, Springer-Verlag, vol. 963, pp. 248-261, 1995.

Alien *vs.* Quine, the Vanishing Circuit and Other Tales from the Industry's Crypt

Vanessa Gratzer[1] and David Naccache[1,2]

[1] Université Paris II Panthéon-Assas, Hall Goullencourt, casier 55,
12 place du Panthéon, F-75231, Paris, CEDEX 05, France
vanessa@gratzer.fr
[2] École Normale Supérieure, Équipe de Cryptographie,
45 rue d'Ulm, F-75230, Paris, CEDEX 05, France
david.naccache@ens.fr

Abstract. This talk illustrates the everyday challenges met by embedded security practitioners by five real examples. All the examples were actually encountered while designing, developing or evaluating commercial products.

This note, which is not a refereed research paper, presents the details of one of these five examples. It is intended to help the audience follow that part of our presentation.

1 Foreword

When I was asked to give this talk, I was delighted, but a bit concerned.

What in my brief decade in the card industry would be of interest to a group of practitioners far more experienced in security than myself?

What will my story be?

As I started to question ex-colleagues, competitors and suppliers, I quickly realized that the problem would be in deciding what to leave out rather than what to include. I was finally able to narrow my list to five examples.

The first ones will deal with an electronic circuit that mysteriously vanished into thin air, DES and RSA key-management in early-generation cards, a cryptographic watchdog chasing own tail and the story of the industry's first on-board sensors.

This note, which is not a refereed paper, presents the details of the fifth example – coauthored with one of my students. It is intended to help the audience follow that part of the talk – a talk that I dedicate to the memory of our friends and colleagues Prof. Dr. Thomas Beth (1949–2005) and Prof. Dr. Hans Dobbertin, (1952–2006).

David Naccache

2 Introduction

Aliens are a fictional bloodthirsty species from deep space that reproduce as parasites. Aliens lay eggs that release araneomorph creatures (facehuggers) when

S. Vaudenay (Ed.): EUROCRYPT 2006, LNCS 4004, pp. 48–58, 2006.

a potential host comes near. The facehugger slides a tubular organ down the victim's throat, implanting a larva in the victim's stomach.

Within a matter of hours the larva evolves into a chestburster and emerges, violently killing the host; chestbursters develop quickly and the cycle restarts.

Just as Aliens, rootkits, worms, trojans and viruses penetrate healthy systems and, once in, alter the host's phenotype or destroy its contents. Put differently, malware covertly inhabits *seemingly normal* systems until something triggers their awakening.

As illustrated recently [4], detecting new malware species may be a nontrivial task. In theory, the easiest way to exterminate malware is a disk reformat followed by an OS reinstallation from a trusted distribution CD. This relies on the assumption that computers can be *forced* to boot from trusted media.

However, most modern PCs have a flash BIOS. This means that the code-component in charge of booting has been recorded on a rewritable memory chip that can be updated by specific programs called *flashers* or, sometimes, by malware such as the CIH (Tchernobyl) virus.

Hence, a natural question arises:

> *How can we ascertain that malware did not re-flash the* BIOS *to derail disk reformatting attempts and simulate their successful completion?*

Flash smart cards[1] are equally problematic. Consider a SIM-card produced by Alice and sold empty to Bob. Bob keys the card. Alice reveals an OS code but flashes a malware simulating the legitimate OS. When some trigger-event occurs[2] the malware responds (to Alice) by revealing Bob's keys.

This note describes methods allowing Bob to check that SIMs bought from Alice contain no malware. Bob's only assumption is that his knowledge of the device's *hardware* specifications is correct.

In biology, the term *Alien* refers to organisms introduced into a foreign locale. Alien species usually wreak havoc on their new ecosystems – where they have no natural predators. In many cases, humans deliberately introduce matching predators to eradicate the alien species. This is the approach taken here.

Related topic. What we try to achieve differs fundamentally from program competitions for the control of a virtual computer, such as Core War. Here the verifier *cannot see* what happens inside a device and seeks to infer the machine's state given its behavior.

3   The Arena

We tested the approach on Motorola's 68HC05, a very common eight-bit micro-controller (more than five billion units sold). The chip's specifications were very slightly modified to better reflect the behavior of a miniature PC.

---

[1] *e.g.* SST Emosyn, Atmel AT90SC3232, Infineon SLE88CFX4000P, Electronic Marin's EMTCG, *etc.*

[2] *e.g.* a specific 128-bit challenge value sent during the GSM authentication protocol.

The 68HC05 has an accumulator A, an index register X, a program counter PC (pointing to the memory instruction being executed), a carry flag C and a zero flag Z indicating if the last operation resulted in a zero or not. We denote by $\zeta(x)$ a function returning one if $x = 0$ and zero otherwise (e.g. $\zeta(x) = \lfloor 2^{-x} \rfloor$).

The platform has $\ell \leq 2^{16} = 65536$ memory bytes denoted $M[0], \ldots, M[\ell - 1]$. Any address $a \geq \ell$ is interpreted as $a \bmod \ell$. We model the memory as a state machine insensitive to power-off. This means that upon shut-down, execution halts and the machine's RAM is backed-up in non-volatile memory. Reboot restores RAM, resets A, X, C and Z and launches execution at address 0x0002 (which *alias* is start).

The very first RAM state (digital *genotype*) is recorded by the manufacturer in the non-volatile memory. Then the device starts evolving and modifies its code and data as it interacts with the external world.

The machine has two I/O ports (bytes) denoted In and Out. Reading In allows a program to receive data from outside while assigning a value to Out displays this value outside the machine. In and Out are located at memory cells $M[0]$ and $M[1]$ respectively. Out's value is restored upon reboot (In isn't). If the device attempts to write into In, execute In or execute Out, execution halts.

The (potentially infested) system pretends to implement an OS function named Install(p). When given a string p, Install(p) installs p at start. We do not exclude the possibility that Install might be modified, mimicked or spied by malware. Given that the next reboot will grant p complete control over the chip, Install would typically require some cryptographic proof before installing p.

We reproduce here some of the 68HC05's instructions (for the entire set see [3]). β denotes the function allowing to encode short-range jumps [3].

EFFECT	lda i	sta i	bne k	bra k
new A ←	$M[i \bmod \ell]$			
new X ←				
new Z ←	$\zeta(\text{new A})$	$\zeta(\text{A})$		
EFFECT ON M		$M[i \bmod \ell] \leftarrow$ A		
new PC ←	$PC + 2 \bmod \ell$	$PC + 2 \bmod \ell$	$\beta(PC, Z, k, \ell)$	$\beta(PC, 0, k, \ell)$
OPCODE	0xB6	0xB7	0x26	0x20
CYCLES	3	4	3	3

---

[3] The seventh bit of k indicates if $k \bmod 128$ should be regarded as positive or negative, *i.e.*

$$\beta(PC, z, k, \ell) = \left(PC + 2 + (1 - z) \times \left(k - 256 \times \left\lfloor \frac{k}{128} \right\rfloor \right) \right) \bmod \ell.$$

EFFECT	inca	incx	lda ,X	ldx ,X
new A ←	A + 1 mod 256		M[X]	
new X ←		X + 1 mod 256		M[X]
new Z ←	ζ(new A)	ζ(new X)	ζ(new A)	ζ(new X)
EFFECT ON M				
new PC ←	PC + 1 mod ℓ	PC + 1 mod ℓ	PC + 1 mod ℓ	PC + 1 mod ℓ
OPCODE	0x4C	0x5C	0xF6	0xFE
CYCLES	3	3	3	3

EFFECT	ldx i	sta i,X	lda i,X	tst i
new A ←			M[i + X mod ℓ]	
new X ←	M[i mod ℓ]			
new Z ←	ζ(new X)	ζ(A)	ζ(new A)	ζ(M[i mod ℓ])
EFFECT ON M		M[i + X mod ℓ] ← A		
new PC ←	PC + 2 mod ℓ	PC + 2 mod ℓ	PC + 2 mod ℓ	PC + 2 mod ℓ
OPCODE	0xBE	0xE7	0xE6	0x3D
CYCLES	3	5	4	4

EFFECT	ora i	inc i	stx i
new A ←	A ∨ M[i mod ℓ]		
new X ←			ζ(X)
new Z ←	ζ(new A)	ζ(new M[i mod ℓ])	
EFFECT ON M		M[i mod ℓ] ← M[i mod ℓ] + 1 mod 256	M[i mod ℓ] ← X
new PC ←	PC + 2 mod ℓ	PC + 2 mod ℓ	PC + 2 mod ℓ
OPCODE	0xBA	0x3C	0xBF
CYCLES	3	5	4

4 Quines as Malware Predators

A Quine (named after the logician Willard van Orman Quine) is a program that prints a copy of its own code [1, 2]. Writing Quines is a tricky programming exercise yielding Lisp, C or natural language examples such as:

```
((lambda (x) (list x (list (quote quote) x)))
 (quote (lambda (x) (list x (list (quote quote) x)))))

char *f="char*f=%c%s%c;main(){printf(f,34,f,34,10);}%c";
main() {printf(f,34,f,34,10);}
```

Copy the next sentence twice. Copy the next sentence twice.

We start by loading a Quine into the tested computer. The device might be under the malware's total spell. The malware might hence neutralize the Quine or even analyze it and mutate (adapt its own code in an attempt to fool the verifier). As download ends, we start a protocol, called *phenotyping*, with whatever survived inside the platform.

Phenotyping will allow us to *prove* (Section 5) or assess the conjecture (Section 4) that the Quine survived and is now in full control of the platform. If the Quine survived we use it to reinstall the OS and eliminate itself; otherwise we *know* that the platform is infected. As we make no assumptions on the malware's malefic abilities, there exist extreme situations where decontamination *by software* is impossible. A trivial case is a malware controlling the I/O port and not letting anything new in. Under such extreme circumstances the algorithms presented in this note will only *detect* the malware but will be of no avail to eliminate it.

The underlying idea is that, upon activation, the Quine will (allegedly!) start dumping-out its own code plus whatever else found on board. We then *prove* or conjecture that the *unique* program capable of such a behavior, under specific complexity constraints, is *only* the Quine itself.

In several aspects, the setting is analogous to the scenario of *Alien vs. Predator*, where a group of humans (OS and legitimate applications) finds itself in the middle of a brutal war between two alien species (malware, Quine) in a confined environment (68HC05).

5   Space-Constrained Quines

We start by analyzing the simple Quine given below (`Quine1.asm`). This 19-byte program inspects $\ell = 256$ bytes platforms. `Quine1` is divided into three functional blocks separated by artificial horizontal lines. First, a primitive command dispatcher reads a byte from `In` and determines if the verifier wants to read the device's contents ($\texttt{In} = 0$) or write a byte into the RAM ($\texttt{In} \neq 0$).

As the program enters `print` the index register is null. `print` is a simple loop causing 256 bytes to be sent out of the device. As the loop ends, the device re-jumps to `start` to interpret a new command.

The `store` block queries a byte from the verifier, stores it in M[X] and re-jumps to `start`.

```
start:  ldx    In      ; X←In                    0xBE 0x00
        bne    store   ; if X≠0 goto store       0x26 0x09
-----------------------------------------------------------
print:  lda    M,X     ; A←M[X]                   0xE6 0x00
        sta    Out     ; Out←A                    0xB7 0x01
        incx           ; X++                      0x5C
        bne    print   ; if X≠0 goto print       0x26 0xF9
        bra    start   ; if X=0 goto start        0x20 0xF3
-----------------------------------------------------------
store:  lda    In      ; A←In                     0xB6 0x00
        sta    M,X     ; M[X]←A                   0xE7 0x00
        bra    start   ; goto start               0x20 0xED
```

The associated phenotyping ϕ_1 is the following:

1. Install(`Quine1.asm`) and reboot.
2. Feed `Quine1` with 235 random bytes to be `store`d at $M[21], \ldots, M[255]$.

3. Activate `print` (command zero) and compare the observed output to:

$$s_1 = \text{0x00 0x00 0xBE 0x00 0x26 0x09 0xE6 0x00 0xB7 0x01}$$
$$\text{0x5C 0x26 0xF9 0x20 0xF3 0xB6 0x00 0xE7 0x00 0x20}$$
$$\text{0xED M}[21], \ldots, \text{M}[255]$$

Is `Quine1.asm` the only nineteen-byte program capable of always printing s_1 when subject to ϕ_1?

We conjecture so although (unlike the variant presented in the next section) we are unable to provide a formal proof. To illustrate the difficulty, consider a *slight* variant:

```
start: ldx   In      ; X←In                    0xBE 0x00
       bne   store   ; if X≠0 goto store        0x26 0x0B
label: tst   label   ;                          0x3D 0x06
print: lda   M,X     ; A←M[X]                   0xE6 0x00
       ⋮     ⋮       ; same code as in Quine1
```

For all practical purposes, this modification (`Quine2.asm`)[4] has nearly no effect on the program's behavior: instead of printing s_1, this code will print:

$$s_2 = \text{0x00 0x00 0xBE 0x00 0x26 0x0B 0x3D 0x06 0xE6 0x00}$$
$$\text{0xB7 0x01 0x5C 0x26 0xF9 0x20 0xF1 0xB6 0x00 0xE7}$$
$$\text{0x00 0x20 0xEB M}[23], \ldots, \text{M}[255]$$

Let `Quine3` be `Quine2` where `tst` is replaced by `inc`.

When executed, `inc` will increment the memory cell at address `label` which is *precisely* `inc`'s own opcode. But since `inc`'s opcode is 0x3C, execution will transform 0x3C into 0x3D which is... the opcode of `tst`.

All in all, ϕ_2 does not allow to distinguish a `tst` from an `inc` present at `label`, as both `Quine2` be `Quine3` will output s_2.

The subtlety of this example shows that a microprocessor-Quine-phenotyping triple $\{\mu, Q, \phi\}$ rigorously defines a problem:

> Given a state machine μ find a state M (malware) that simulates the behavior of a state Q (legitimate OS) when μ is subject to stimulus ϕ (phenotyping).

Security practitioners can proceed by analogy to the assessment of cryptosystems which specifications are published and submitted to public scrutiny. If an M simulating Q with respect to ϕ is found, a fix can either replace Q or ϕ or both. Note the analogy: *Given a stream-cipher μ and a key Q (defining an observed cipher-steam ϕ), prove that the key Q has no equivalent-keys M.*

An alternative solution, described in the next section, consists in *proving* the Quine's behavior under the assumption that the verifier is allowed to count clock cycles (state transitions if μ is a Turing Machine).

---

[4] ϕ_1 should be slightly twitched as well (233 random values to write).

6   Time-Constrained Quines

Consider the following program loaded at address start:

```
start: ldx In    ; 3 cycles ; X←In              (instruction I₁)
       stx Out   ; 4 cycles ; Out←X             (instruction I₂)
       ⋮  ⋮     ;            ; other instructions
```

Latch a first value v_1 at In and reboot, as seven cycles elapse v_1 pops-up at Out. If we power-off the device before the eighth cycle and reboot, v_1 reappears on Out[5] immediately. Repeating the process with values v_2 and v_3, we witness two seven-cycle transitions $v_1 \rightsquigarrow v_2$ and $v_2 \rightsquigarrow v_3$.

It is impossible to modify two memory cells in seven cycles as all instructions capable of modifying a memory cell require at least four cycles. Hence we are assured that between successive reboots, the *only* memory changes are in Out. This means that no matter what the examined code is, this code has no time to mutate in seven cycles and necessarily remains invariant between reboots.

The instructions other than sta and stx capable of modifying directly Out are: ror, rol, neg, lsr, lsl, asl, asr, bset, bclr, clr, com, dec and inc. Hence, it suffices to select $v_2 \neq \text{dir}(v_1)$ and $v_3 \neq \text{dir}(v_2)$, where dir stands for any of the previous instructions[6], to ascertain that Out is being modified by an sta or an stx (we also need $v_1 \neq v_2 \neq v_3$ to actually *see* the transition).

$v_1 = \text{0x04}$, $v_2 = \text{0x07}$, $v_3 = \text{0x10}$ satisfy these constraints.

As reading or computing with a memory cell takes at least three cycles there are only four cycles left to alter the contents of Out; consequently, the only sta and stx instructions capable of causing the transitions fast enough are:

$$I_2 \in \boxed{\text{sta Out} \mid \text{stx Out} \mid \text{sta ,X} \mid \text{stx, X}}$$

To aim at Out (which address is 0x0001), sta ,X and stx ,X would require an X=0x01 but this is impossible (if the code takes the time to assign a value to X it wouldn't be able to compute the transition's value by time). Hence, we infer that the code's structure is:

```
start: ??? ???     ; 3 cycles ; an instruction causing • ←In
       st• Out     ; 4 cycles ; an instruction causing Out← •
       ⋮  ⋮       ;            ; other instructions
```

where • stands for register A or register X. The only possible code fragments capable of doing so are:

	adc In	adc ,X	add In	add ,X	eor In	eor ,X
I_1	sta Out	sta Out	sta Out	sta Out	sta Out	sta Out
$I_2 \in$	lda In	lda ,X	ora In	ora ,X	ldx In	ldx ,X
	sta Out	sta Out	sta Out	sta Out	stx Out	stx Out

---

[5] Out being a memory cell, its value is backed-up upon power-off.
[6] for ror and rol, consider the two sub-cases C = 0 and C = 1.

There is no way to further refine the analysis without more experiments, but one can already guarantee that as the execution of any of these fragments ends, the machine's state is either $S_A = \{A = v_3, X = 0x00\}$ or $S_X = \{A = 0x00, X = v_3\}$. Now assume that $Out = v_3 = 0x10$. Consider the code:

```
start: ldx In      ; 3 cycles ; X←In
       stx Out     ; 4 cycles ; Out←X
       lda ,X      ; 3 cycles ; A←M[X]    (instruction I₃)
       sta Out     ; 4 cycles ; Out←A     (instruction I₄)

       ⋮   ⋮       ;          ; other instructions
```

- Latch $In \leftarrow v_4 = 0x02$, reboot, wait fourteen cycles; witness the transition[7] $0x10 \rightsquigarrow 0x02 \rightsquigarrow 0xBE$; power-off before the fifteenth cycle completes.
- Latch $In \leftarrow v_6 = 0x04$, reboot, wait fourteen cycles; witness the transition[8] $0xBE \rightsquigarrow 0x06 \rightsquigarrow 0xF6$; power-off before the fifteenth cycle completes.

As $v_5 \neq \mathtt{dir}(v_4)$ and $v_7 \neq \mathtt{dir}(v_6)$ the second transition is, again, necessarily caused by some member of the **sta** or **stx** families and, more specifically[9] one of the following:

$$I_4 \in \boxed{\ \mathtt{sta\ Out}\ \mid\ \mathtt{stx\ Out}\ \mid\ \mathtt{sta\ ,X}\ }$$

I_3 cannot be an instruction that has no effect on X and A as this will either inhibit a transition or cause a transition to zero (remember: immediately before the execution of I_3 the machine's state is either S_A or S_X). This rules-out eighteen jump instructions as well as all **cmp**, **bit**, **cpx**, **tsta** and **tstx** variants. **lda** i and **ldx** i are impossible as both would have forced 0x02 and 0x04 to transit to the *same* constant value.

In addition, $v_5 \neq \mathtt{dir}(v_4)$ implies that I_3 cannot be a **dir**-variant operating on A or X, which rules-out **negx**, **nega**, **comx**, **coma**, **rorx**, **rora**, **rolx**, **rola**, **decx**, **deca**, **dec**, **incx**, **inca**, **clrx**, **clra**, **lsrx**, **lsra**, **lslx**, **lsla**, **aslx**, **asla**, **asrx** and **asra** altogether.

As no carry was set, we sieve-out **sbc** and **adc** whose effects will be strictly identical to **sub** i and **add** i (dealt with below).

add i, **sub** i, **eor** i, **and** i and **ora** i are impossible as the system

$$\begin{cases} 0x02 \star x = 0xBE \\ 0x06 \star x = 0xF6 \end{cases}$$

has no solutions when operator $\star$ is substituted by $+, -, \oplus, \wedge$ or $\vee$.
The only possible I_3 candidates at this point are:

$$I_3 \in \boxed{\ \mathtt{sub\ ,X}\ \mid\ \mathtt{and\ ,X}\ \mid\ \mathtt{eor\ ,X}\ \mid\ \mathtt{ora\ ,X}\ \mid\ \mathtt{add\ ,X}\ \mid\ \mathtt{lda,\ X}\ \mid\ \mathtt{ldx\ ,X}\ }$$

---

[7] $v_5 = 0xBE$ is the opcode of **ldx**, read from address 0x02.

[8] $v_7 = 0xF6$ is the opcode of **lda** ,X, read from address 0x06.

[9] taking timing constraints into account and ruling-out **stx** ,X who can only cause an $Out = 0x01$, a value never witnessed.

But before the execution of I_3 the machine's state is:

$$S_A = \{A = 0x06, X = 0x00\} \quad \text{or} \quad S_X = \{A = 0x00, X = 0x06\}$$

The ",X" versions of sub, and, eor, ora and add are impossible because:

– if the device is in state S_A we note that

$$0x06 \star 0x06 \neq 0xF6 \quad \text{for} \quad \star \in \{-, \vee, \oplus, \wedge +\}$$

– and if the device is in state S_X we note that

$$A - \text{opcode}(\text{sub}, X) = 0x00 - 0xF0 = 0x10 \neq 0xF6$$
$$A \wedge \text{opcode}(\text{and}, X) = 0x00 \wedge 0xF4 = 0x00 \neq 0xF6$$
$$A \oplus \text{opcode}(\text{eor}, X) = 0x00 \oplus 0xF8 = 0xF8 \neq 0xF6$$
$$A \vee \text{opcode}(\text{ora}, X) = 0x00 \vee 0xFA = 0xFA \neq 0xF6$$
$$A + \text{opcode}(\text{add}, X) = 0x00 + 0xFB = 0xFB \neq 0xF6$$

ldx ,X is impossible as it would have caused a transition to opcode(ldx, X) = 0xFE ≠ 0xF6 (if S_X) or to 0x06 (if S_A).

I_3 is hence identified as being necessarily lda ,X.

It follows immediately that $I_4 = $ sta Out and that the ten register-A-type candidates for $\{I_1, I_2\}$ are inconsistent.

The phenotyped code is thus one of the following two:

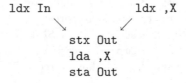

```
ldx In              ldx ,X
            ↘     ↙
          stx Out
          lda ,X
          sta Out
```

Only the leftmost is capable of causing the observed transition 0x02 ⤳ 0xBE.

All in all, we have built a *proof* that the device actually executed the fragment presented at the beginning of this section.

Extending the code further ahead to:

```
start: ldx  In      ; X←In            0xBE 0x00
       stx  Out     ; Out←X           0xBF 0x01
print: lda  ,X      ; A←M[X]          0xF6
       sta  Out     ; Out←A           0xB7 0x01
       incx         ; X ← X+1         0x5C
       bne  print   ; if X≠0 goto print  0x26 0xFA
```

and subjecting the chip to three additional experiments, we observe:

$$\text{In} \leftarrow 0x09 \quad \Rightarrow \quad 0xF6 \rightsquigarrow 0x09 \rightsquigarrow 0x5C$$

$$\text{In} \leftarrow 0x0A \quad \Rightarrow \quad 0x5C \rightsquigarrow 0x0A \rightsquigarrow 0x26$$

$$\text{In} \leftarrow 0x0B \quad \Rightarrow \quad 0x26 \rightsquigarrow 0x0B \rightsquigarrow 0xFA$$

Note that the identified code "happens to" allow the verifier to inspect *with absolute certainty* the platform's first 256 bytes. The rest is clear. The verifier does a last time measurement, allowing the Quine to print the device's first 256 bytes (power-off as soon as the last **bne** iteration completes, to avoid falling into the jaws of Aliens hiding beyond address 0x000B).

It remains to check the Quine's payload (code between 0x000C and 0x00FF) and unleash the Quine's execution beyond address 0x000B. Quine won the game.

7   Questions

This work raises a number of intriguing questions: Is it possible to prove security using only space constraints? In the negative, can we modify the assembly language to allow such proofs[10]? Can space-constrained Quines solve space-complete problems to flood memory instead of receiving random data?

Another interesting challenge consists in developing a time-constrained Quine whose proof does not require rebooting but the observation of one long succession of transitions. We conjecture that such programs exist. A possible starting point might be a code (not necessarily located at **start**) similar to:

```
loop: sta  Out
      lda  In
      sta  Out
      ldx  In
      stx  Out
      lda  ,X
      sta  Out
      bne  loop
```

Here the idea is that the verifier will feed the Quine with values chosen randomly in a specific set (to rule-out **dir**-variants) to repeatedly explore the code's immediate environment until some degree of certainty is acquired[11].

If possible, this would have the advantage of making the Quine a *function* automatically insertable into any application whose code needs to be authenticated. Moreover, if we manage to constrain the capabilities of such a Quine, *e.g.* not allow it read data beyond a given offset[12], we could offer the *selective* ability to audit critical program parts while preserving the privacy of others. For instance, the code of an accounting program could be audited while secret signature keys would provably remain out of the Quine's reach.

Finally, as time-constrained phenotyping is extremely quick (a few clock cycles), preserves nearly all the platform's data and requires only table lookups and comparisons, we currently try to extend the approach to more complex microprocessors and implement it between chips in motherboards.

---

[10] The approach would analogous to Java bytecode which is *purposely* shaped to fit type-inference.

[11] To exit the **bne** loop the verifier will purposely read a zero somewhere.

[12] *e.g.* the example above cannot read data beyond address 255.

References

1. J. Burger, D. Brill and F. Machi, *Self-reproducing programs*, Byte, volume 5, August 1980, pp. 74–75.
2. D. Hofstadter, *Godel, Escher, and Bach: An eternal golden braid*, Basic Books, Inc. New York, pp. 498–504.
3. Motorola Inc., 68HC(7)05H12 *General release specifications*, HC05H12GRS/D Rev. 1.0, November 1998.
4. T. Zeller, *The ghost in the* CD; *Sony* BMG *stirs a debate over software used to guard content*, The New York Times, C1, November 14, 2005.

Hiding Secret Points Amidst Chaff

Ee-Chien Chang and Qiming Li*

Department of Computer Science,
National University of Singapore
changec@comp.nus.edu.sg, qiming.li@ieee.org

Abstract. Motivated by the representation of biometric and multimedia objects, we consider the problem of hiding noisy point-sets using a secure sketch. A point-set X consists of s points from a d-dimensional discrete domain $[0, N-1]^d$. Under permissible noises, for every point $\langle x_1, .., x_d \rangle \in X$, each x_i may be perturbed by a value of at most δ. In addition, at most t points in X may be replaced by other points in $[0, N-1]^d$. Given an original X, we want to compute a secure sketch P. A known method constructs the sketch by adding a set of random points R, and the description of $(X \cup R)$ serves as part of the sketch. However, the dependencies among the random points are difficult to analyze, and there is no known non-trivial bound on the entropy loss. In this paper, we first give a general method to generate R and show that the entropy loss of $(X \cup R)$ is at most $s(d \log \Delta + d + 0.443)$, where $\Delta = 2\delta + 1$. We next give improved schemes for $d = 1$, and special cases for $d = 2$. Such improvements are achieved by pre-rounding, and careful partition of the domains into cells. It is possible to make our sketch short, and avoid using randomness during construction. We also give a method in $d = 1$ to demonstrate that, using the size of R as the security measure would be misleading.

1 Introduction

Many biometric data are noisy in the sense that small noises are introduced during acquisition and processing. Hence, two biometric samples that are different but close to each other, are considered to belong to the same identity. This poses technical challenges in applying classical cryptographic operations on them. Recently, new generic techniques such as fuzzy commitment [10], helper data [15] and secure sketch [7] are introduced to handle noisy data. These techniques attempt to remove the noise with the aid of some additional public data P. Here we follow Dodis et al. [7] and call such P a *sketch*. During registration, given original data X, a sketch P is constructed and made public. During reconstruction, given some other data Y and the sketch P, the original X can be reconstructed if Y is *close*[1] to X. In other words, the sketch aids in removing noise from noisy

---

* The author is currently with Department of Computer and Information Science, Polytechnic University.

[1] The formal definition of "closeness" will be given in Section 3.

S. Vaudenay (Ed.): EUROCRYPT 2006, LNCS 4004, pp. 59–72, 2006.

data Y. It is important that such sketch P should be *secure* in the sense that it reveals only limited information about the original X, so that the privacy of the original data can be sufficiently maintained. In other words, it is desirable to bound the *entropy loss* of X given P (Section 3 gives the definitions).

Not surprisingly, the design of a secure sketch is very much dependent on the definition of "closeness". Secure sketch for the following two main types of data have been proposed: (1) The data are from a vector space, and two sequences are close to each other if their distance (e.g., Hamming distance) is less than a threshold. (2) The data X and Y are subsets of a universe $\mathcal{U}$, where $|X| = |Y| = s$, and they are close with respect to a threshold t, if the set difference $s - |X \cap Y| \leq t$.

We observe that in many applications, a combination of the above is required. For example, a fingerprint template is typically represented as a set of minutiae points in a discrete 2-dimensional space, or even 3-dimensional if the less reliable orientation attribute is included [6]. Under noise, each points may be slightly perturbed, and a small number of points may be replaced.

We study secure sketch schemes for such *point-sets*. A point-set X is a set of s points from a discrete d-dimensional domain $[0, N-1]^d$. Under permissible *white noise*, for every point $\langle x_1, .., x_d \rangle \in X$, each x_i, $1 \leq i \leq d$, may be perturbed by at most δ. In addition, under *replacement noise*, at most t points in X may be replaced by randomly selected points. Hence, two point-sets X and Y are close to each other if we can find a subset $X' \subset X$, $|X'| \geq s - t$, such that for each $x \in X'$, there is a unique $y \in Y$ that satisfies $\|x - y\|_\infty \leq \delta$, where $\| \cdot \|_\infty$ is the infinity norm. We assume that a point-set X is always *well-separated*, that is, for any $x, x' \in X$, the distant $\|x - x'\|_\infty \geq 3\delta$. This assumption is reasonable in practice. For example, in a fingerprint template, two minutiae points cannot be too close to each other, otherwise they will be considered as false minutiae and should be corrected [11].

Clancy et al. [5] give the following construction of a two-part sketch for a point-set. The first part of the sketch is a *codebook* $\mathcal{C}$, which is a collection of points that are well-separated. We call each point in $\mathcal{C}$ a *codeword*, and we assume that all codewords are properly indexed in a pre-defined manner. The codebook $\mathcal{C}$ is the union of the original data X and a set of random *chaff* points R, i.e., $\mathcal{C} = (X \cup R)$. Consider another point-set Y that is a version of X corrupted only by white noise. For each point $y \in Y$, the codeword in $\mathcal{C}$ that is closest to y must be the corresponding $x \in X$. Thus, with $\mathcal{C}$, the white noise can be corrected. Hence we call $\mathcal{C}$ the *white noise sketch*. The second part of the sketch is constructed from the indices of the points in X, where the index of a point $x \in X$ is its location in the codebook $\mathcal{C} = (X \cup R)$. By using existing schemes for set difference, replacement of at most t points can be corrected. Hence we call it the *replacement sketch*. In this paper, we will focus on the construction of the white noise sketch. That is, we study how to hide the original points X amidst some chaff points R.

Clancy et al. propose the following method to generate R: The points in R are iteratively selected. During each iteration, a chaff point is chosen uniformly at random. If it is too close to any previously selected points or a point in X, then it is discarded. Otherwise it is selected. The iteration is repeated until sufficient points are selected or it is impossible to add more points. The above process of selecting a set of random points is essentially the online parking process which has intrinsic statistical properties [14, 13, 8].

Due to the dependencies among the selected points, the analysis of online parking process is difficult. This is especially so in higher dimensions. Many fundamental questions remain open, for example, the Palasti's Conjecture [13]. In our context of secure sketch, there is no known non-trivial bound of the entropy loss by revealing $(X \cup R)$. Furthermore, although the points generated seem to be "random", due to the dependencies, the original X may be statistically distinguishable from R. Indeed, an empirical study suggests a method to find X among $(X \cup R)$ [4].

Therefore, we propose another method of generating the points. First, many points are generated independently. Next, some points are removed so that among the remaining points, no two points are near to each other. In this way, we can eliminate the dependencies among the chaff points and give an upper bound $\mathcal{L}_H$ on the information revealed (i.e., the entropy loss) by the codebook $\mathcal{C} = (X \cup R)$. There are many ways to generate the points independently. The challenging issue now is to find a method whereby the randomness invested during generation is not much less than the number of bits required to represent the codebook.

For the second part of the sketch that corrects the replacement noise, we employ known techniques for set difference. Let $\mathcal{L}_{SD}(s, t, n)$ be the entropy loss of the sketch for set difference, where $n = |\mathcal{C}|$ is the size of codebook. There are sketch schemes such that $\mathcal{L}_{SD}(s, t, n)$ is in $O(t \log n)$ (e.g., those proposed by Juels and Wattenberg [9], Dodis et al. [7], and Chang et al. [3]).

In this paper, we propose a generic method to generate the white noise sketch and show that the upper bound of the entropy loss $\mathcal{L}_H < s(d \log \Delta + d + \log(e/2))$, where $\Delta = 2\delta + 1$, e is the base of natural logarithm and $\log(e/2) \approx 0.443$. The overall entropy loss is at most $\mathcal{L}_H + \mathcal{L}_{SD}(s, t, N^d/(4\delta + 1)^d)$. The bound is quite tight in the sense that there is a distribution of X such that the entropy loss of $\mathcal{C}$ is at least $\mathcal{L}_H - \epsilon$ where ϵ is a positive constant that is at most 3. When $t = 0$ (i.e., no replacement noise), a lower bound of the entropy loss is $sd \log \Delta$. Hence, the gap between our construction and the optimal is at most $s(d + \log(e/2))$. By pre-rounding and carefully partitioning the domain $[0, N - 1]$ into cells, we can improve the entropy loss in $d = 1$ to at most $s(1 + \log(\Delta - 1)) + \mathcal{L}_{SD}(s, t, N/(3\delta))$. We further apply the technique of partitioning to some special cases in two dimensions ($d = 2$) and obtain some improvements. Such technique probably can be extended to $d = 2$ in general, and to higher dimensions. In addition, we give two methods to reduce the size of the sketch. In one of them, we can avoid using randomness during sketch construction, thus some limited form of reusability can be achieved [2]. We also give another method in one dimension to demonstrate that, using the size of R as the security measure would be misleading.

2   Related Works

Recently, a few new cryptographic primitives for noisy data are proposed. Fuzzy commitment scheme [10] is one of the earliest formal approaches to error tolerance. The fuzzy commitment scheme uses an error correcting code to handle Hamming distance. The notions of *secure sketch* and *fuzzy extractor* are introduced by Dodis et al. [7], which gives constructions for Hamming distance, set difference, and edit distance. Under their framework, a reliable key is extracted from noisy data by reconstructing the original data with a given sketch, and then applying a normal extractor (such as pair-wise independent hash functions) on the data.

An important requirement of a secure sketch scheme is that the amount of information about X revealed by publishing the sketch P should be limited. Dodis et al. [7] propose a notion of entropy loss to measure the security of the sketch. They also provide a convenient way to bound the entropy loss for **any** distribution of X. Such worst case analysis is important in practice because typically, the actual distribution of the biometric data is not known.

The issue of *reusability* of sketches is addressed by Boyen [2]. It is shown that a sketch scheme that is provably secure may be insecure when multiple sketches of the same biometric data are obtained. It is also shown by Boyen that a sketch that can be constructed deterministically can achieve some limited form of reusability [2].

The set difference metric was first considered by Juels and Wattenberg [9], who gave a *fuzzy vault* scheme. Later, Dodis et al. [7] proposed three constructions. The entropy loss by all these schemes are roughly the same. They differ in the sizes of the sketches, decoding efficiency and also the degree of ease in practical implementation. The BCH-based scheme [7] has small sketches and achieves "sublinear" (with respect to the size of the universe) decoding by careful reworking of the standard BCH decoding algorithm. Chang et al. [3] gave a scheme for multi-sets, using the idea in set reconciliation [12].

A *fuzzy fingerprint vault* scheme is proposed by Clancy et al. [5], which is to be used in secure fingerprint verification using a smart card. The security of the scheme is analyzed by considering force attackers. Yang and Verbauwhede [16] employed similar approaches with different fingerprint representation.

3   Preliminaries

Entropy and entropy loss. We follow the definitions of entropy by Dodis et al. [7]. They propose to examine the *average min-entropy* of X given P, which gives the minimum length of an almost uniform secret key that can be extracted even if the sketch P is made public.

Let $\mathbf{H}_\infty(A)$ be the min-entropy of the random variable A, i.e., $\mathbf{H}_\infty(A) = -\log(\max_a \Pr[A = a])$. For two random variables A and B, the average min-entropy of A given B is defined as $\widetilde{\mathbf{H}}_\infty(A \mid B) = -\log(\mathbb{E}_{b \leftarrow B}[2^{-\mathbf{H}_\infty(A|B=b)}])$.

The entropy loss of X given sketch P is defined as $\mathcal{L} = \mathbf{H}_\infty(X) - \widetilde{\mathbf{H}}_\infty(X|P)$. When it is clear in the context, we simply call $\mathcal{L}$ the entropy loss of sketch P.

This definition is useful in the analysis of entropy loss, since for any ℓ-bit string B, we have $\widetilde{\mathbf{H}}_\infty(A \mid B) \geq \mathbf{H}_\infty(A) - \ell$. For any secure sketch scheme, let R be the randomness invested in constructing the sketch, it can be shown that when R can be recovered from X and P, then

$$\mathcal{L} = \mathbf{H}_\infty(X) - \widetilde{\mathbf{H}}_\infty(X \mid P) \leq |P| - \mathbf{H}_\infty(R). \tag{1}$$

Inequality (1) implies that the entropy loss can be bounded from above by the difference between the size of the sketch and the randomness we invested during construction. This gives a general method to find an upper bound of $\mathcal{L}$ that is independent of X, and hence it applies to any distribution of X. Therefore, $\mathcal{L}$ is an upper bound of entropy loss in the "worst-case".

Secure sketch. Let $\mathcal{M}$ be a set with a *closeness* relation $\mathsf{C} \subseteq \mathcal{M} \times \mathcal{M}$. When $(X,Y) \in \mathsf{C}$, we say the Y is close to X, or (X,Y) is a close pair. Similar to Dodis et al. [7], define

Definition 1. *A sketch scheme is a tuple* $(\mathcal{M}, \mathsf{C}, \mathsf{Enc}, \mathsf{Dec})$, *where* $\mathsf{Enc} : \mathcal{M} \to \{0,1\}^*$ *is an encoder and* $\mathsf{Dec} : \mathcal{M} \times \{0,1\}^* \to \mathcal{M}$ *is a decoder such that for all* $X, Y \in \mathcal{M}$, $\mathsf{Dec}(Y, \mathsf{Enc}(X)) = X$ *if* $(X,Y) \in \mathsf{C}$. *The string* $P = \mathsf{Enc}(X)$ *is to be made public and we call it the sketch. We say that the sketch scheme is $\mathcal{L}$-secure if for all random variable X over $\mathcal{M}$, the entropy loss of P is at most $\mathcal{L}$. That is,* $\mathbf{H}_\infty(X) - \widetilde{\mathbf{H}}_\infty(X \mid \mathsf{Enc}(X)) \leq \mathcal{L}$.

Closeness relations. For any two points x and y from the d-dimensional space $[0, N-1]^d$, we define the closeness C_δ, where $(x,y) \in \mathsf{C}_\delta$ if $\|x-y\|_\infty \leq \delta$. We further define the closeness $\mathsf{PS}_{\delta,s,t}$ for two point-sets.

Definition 2. *For any two sets of s points* $X = \{x_1, \ldots, x_s\}$ *and* $Y = \{y_1, \ldots, y_s\}$, *we say that* $(X,Y) \in \mathsf{PS}_{\delta,s,t}$ *if there exists a 1-1 correspondence f on $\{1, \ldots, s\}$ such that* $|\{i \mid (x_{f(i)}, y_i) \in \mathsf{C}_\delta\}| \geq s - t$.

A lower bound of the entropy loss. Here we give a lower bound $\mathcal{L}_0$ of the entropy loss. We say that $\mathcal{L}_0$ is a lower bound if, for any sketch scheme $(\mathcal{P}([0, N-1]^d), \mathsf{PS}_{\delta,s,t}, \mathsf{Enc}, \mathsf{Dec})$, there exists a distribution of X such that the entropy loss of $P = \mathsf{Enc}(X)$ is at least $\mathcal{L}_0$.

For any distribution of X, let $\mathcal{X}_b$ to be the set of all possible original point-sets given sketch $P = b$. We observe that

$$\max_a \Pr[X = a \mid P = b] \geq \frac{1}{|\mathcal{X}_b|}.$$

Substitute it into the definition, we have

$$\widetilde{\mathbf{H}}_\infty(X|P) \leq \max_{b, \Pr[P=b] \neq 0} \log |\mathcal{X}_b|. \tag{2}$$

Now, by considering X that is uniformly distributed over all well-separated sets of size s in $[0, N-1]^d$, using (2), we can show that (details omitted) when $s < (\frac{N}{2\Delta})^d$ and $t < (\frac{N}{2\Delta})^{\frac{d}{2}}$, $\mathcal{L}_0$ is in

$$sd \log \Delta + \Omega(td \log \frac{N}{2\Delta}). \tag{3}$$

Recall that $\Delta = 2\delta + 1$. An intuitive interpretation of the bound is that, it is the minimum number of bits needed to describe the noise. The first term in (3) is for the white noise, and the second term is for the replacement noise. When $t = 0$ (i.e., there is no replacement noise), the bound becomes $sd \log \Delta$.

4   The Basic Construction

Recall that our sketch consists of two parts $P_H P_S$, where P_H is the white noise sketch that removes the white noise. During encoding, a large number of points R is generated to form the codebook $\mathcal{C} = (X \cup R)$, and P_H is its description. During decoding, the points in Y are matched with the nearest codewords in $\mathcal{C}$, so that white noise can be removed. The sketch P_S for set difference is constructed using known schemes on $\mathcal{C}$ to correct the replacement noise. We also assume that X is well-separated.

Here we focus on the construction of P_H. We will first give our basic construction in one dimension ($d = 1$), and then show that it can be extended to higher dimensions.

The main idea of our construction is to first independently generate many points, but avoiding regions near the original X. We can also view the generation of these points as a two dimensional Poisson process. Next, remove some points so that among the remaining points, no two points are near to each other. The retained points form the codebook $\mathcal{C}$. Since the points are generated independently, it is easier to bound the entropy loss. To minimize the entropy loss, we need to find a way so that the size of the sketch is not much larger than the randomness we invested during the construction.

4.1   Construction of P_H in One Dimension ($d = 1$)

For any point $x \in [0, N - 1]$, call the set $S_1(x) = \{x + 1, x + 2, \ldots, x + 2\delta\}$ the *half-sphere* of x.

Given $X = \{x_1, \ldots, x_s\}$, the white noise P_H is constructed as below. We first construct a sequence $\langle h_0, h_1, \ldots, h_{N-1} \rangle$, where each $h_i \in [0, p_1 - 1]$, and p_1 is a parameter that is chosen to be $p_1 = |S_1(x)| + 1 = 2\delta + 1$ for optimal performance.

1. For each $x \in X$, set $h_x = 0$, and for each $a \in S_1(x)$, h_a is uniformly chosen at random from $\{1, \ldots, p_1 - 1\}$.
2. For each h_i that has not been set in step 1, uniformly choose its value from $\{0, \ldots, p_1 - 1\}$.

For each $w \in [0, N - 1]$, we select it to be in the codebook if and only if $h_w = 0$ and $h_a \neq 0$ for all $a \in S_1(w)$. Hence, if w is a codeword, there would be no other codeword in the half-sphere $S_1(w)$. The sequence $\langle h_0, \ldots, h_{N-1} \rangle$ is published as the white noise sketch P_H. Note that in practice, we can simply publish a description of the codebook $\mathcal{C}$ as the sketch. However, we choose to publish the entire sequence $\langle h_0, \ldots, h_{N-1} \rangle$ for the ease of analysis.

From the codebook $\mathcal{C}$, we can construct P_S, the second part of the sketch, using known schemes for set difference.

During decoding, given Y, each point $y \in Y$ is matched with its nearest codeword in $\mathcal{C}$. Suppose y is a noisy version of an $x \in X$, i.e. $|y - x| \leq \delta$, it is easy to verify that x is its closest point in $\mathcal{C}$. Hence, P_H can correct the white noise. Lemma 3 gives the entropy loss, and Lemma 4 shows that the bound is quite tight. Note that Lemma 3 and 4 still hold if we choose to publish a shorter description of the codebook instead of the entire sequence. In other words, publishing the entire sequence might seem to reveal more information about X, the "worst-case" entropy loss would not be much different.

Lemma 3. *The entropy loss of X given P_H is at most*

$$s \left(\log \Delta + (\Delta - 1) \log(1 + \frac{1}{\Delta - 1}) \right)$$

which is less than $s \left(\log \Delta + \log e \right)$, *where e is the base of natural logarithm.*

Proof. Since the randomness invested in constructing P_H can be recovered from X and P_H, we can apply (1) in Section 3. In particular, we look at the difference between the size of the sketch P_H, which is $N \log p_1$, and the randomness invested in constructing P_H. For any h_i in P_H, if it is not set in Step 1 of the above construction, then $|h_i| = \log p_1$, which equals to the invested randomness, and hence it does not contribute to the difference. For each h_x such that $x \in X$, it is set to 0, which contributes $\log p_1$ to the difference. For each h_a such that $a \in S_1(x)$ for some $x \in X$, we use $\log(p_1 - 1)$ bits of randomness, hence the difference introduced is $\log \frac{p_1}{p_1 - 1}$.

Therefore, the total difference (hence the entropy loss) is no greater than

$$s \left(\log p_1 + 2\delta \log \frac{p_1}{p_1 - 1} \right).$$

When $p_1 = 2\delta + 1$, and substituting $\Delta = 2\delta + 1$, we have

$$\mathcal{L}_H \leq s \left(\log \Delta + (\Delta - 1) \log(1 + \frac{1}{\Delta - 1}) \right).$$

Since $(1 + \frac{1}{\Delta - 1})^{\Delta - 1}$ approaches e from below when Δ approaches infinity, we have the above claimed bound. □

Lemma 4. *There exists a distribution of X, where the entropy loss of X given P_H is at least $s(\log \Delta + (\Delta - 1) \log(1 + \frac{1}{\Delta - 1})) - \epsilon$ for some positive constant ϵ.*

Proof. Consider the distribution $X = \{x_1, x_1 + 2\Delta, \cdots, x_1 + 2(s - 1)\Delta\}$, where x_1 is uniformly chosen from a set $A = \{a_1, \cdots, a_\lambda\}$ of λ points. Hence, $\mathbf{H}_\infty(X) = \log \lambda$. Recall that, given P_H, a point w is a codeword if and only if $h_w = 0$ and $h_b \neq 0$ for all $b \in S_1(w)$. Certainly, each point x_i in X itself *must* be a codeword. Hence, each point $a_i \in A$ is a possible candidate of the original point x_1 if and only if all the points in $\{a_i, a_i + 2\Delta, \ldots, a_i + 2(s - 1)\Delta\}$ are codewords in $\mathcal{C}$.

For any $a_i \neq x_1$, the probability that a_i is a possible candidate of x_1 is at most $\frac{1}{\Delta^s}(1 - \frac{1}{\Delta})^{(\Delta-1)s}$. Let C be the number of candidates of x_1 for a given P_H, then we have

$$\mathbb{E}[C] \leq 1 + \frac{\lambda - 1}{\Delta^s}(1 - \frac{1}{\Delta})^{(\Delta-1)s} \leq 1 + \frac{\lambda}{\Delta^s}(1 - \frac{1}{\Delta})^{(\Delta-1)s}.$$

Now by choosing

$$\lambda = 2^{s(\log \Delta + (\Delta-1)\log(1 + \frac{1}{\Delta-1}))}$$

we have $\mathbb{E}[C] \leq 2$. By Markov's Inequality, we have

$$\Pr[C \leq 4] \geq 1 - \mathbb{E}[C]/4 \geq 1/2.$$

We note that

$$\mathbb{E}_{b \leftarrow P_H}\left[2^{-\mathbf{H}_\infty(X|P_H=b)}\right]$$

$$=\mathbb{E}_{b \leftarrow P_H}\left[\max_a \Pr[X = a|P_H = b]\right]$$

$$\geq \frac{1}{4}\Pr[C \leq 4] \geq \frac{1}{8}.$$

Therefore, the left-over entropy $\widetilde{\mathbf{H}}_\infty(X|P) \leq -\log\frac{1}{8} = 3$. Considering that $\mathbf{H}_\infty(X) = \log\lambda = s\left(\log\Delta + (\Delta-1)\log(1 + \frac{1}{\Delta-1})\right)$, and let $\epsilon = 3$, we have the claimed bound. $\square$

4.2   Extension to Higher Dimensions

The construction in one dimension can be easily extended to higher dimensions by giving an appropriate notion of half-sphere. Let us first define a total order for the points in $[0, N-1]^d$. Define $\langle x_1, x_2, \ldots, x_d \rangle \succ \langle x'_1, x'_2, \ldots, x'_d \rangle$ if and only if there exists an i such that $x_i > x'_i$ and $x_j = x'_j$ for all $1 \leq j < i$. We define the half-sphere of x in d-dimensions $S_d(x) = \{y \mid 0 < \|y - x\|_\infty \leq 2\delta \text{ and } y \succ x\}$.

The sketch P_H is a set of N^d symbols. For each $h_y \in P_H$, we have $y \in [0, N-1]^d$ and $h_y \in \{0, \ldots, p_d - 1\}$ for some parameter p_d that is to be chosen later. We construct P_H as below.

1. For each $x \in X$, set $h_x = 0$. For every $a \in S_d(x)$, uniformly choose h_a at random from $\{1, \ldots, p_d - 1\}$.
2. For each h_y that is not set in step 1, choose its value uniformly at random from $\{0, \ldots, p_d - 1\}$.

From P_H we can determine the codebook $\mathcal{C}$ as follows. A point $x \in [0, N-1]^d$ is in $\mathcal{C}$ if and only if $h_x = 0$ and for every $a \in S_d(x)$, we have $h_a \neq 0$. We can then construct the second part P_S of the sketch for set difference. Suppose y is a noisy version of an $x \in X$, that is, $\|y - x\|_\infty \leq \delta$, it is not difficult to verify that its closest point in $\mathcal{C}$ is x.

In fact, this construction is essentially the same as the construction for $d = 1$, except that $S_d(x)$ is larger when $d > 1$. By simple counting we have

$$|S_d(x)| = \frac{(4\delta + 1)^d - 1}{2}.$$

Similar to the one-dimensional case, we choose $p_d = |S_d(x)| + 1$. By substituting $\Delta = 2\delta + 1$, we have

Theorem 5. *The entropy loss of X given sketch P_H is at most*

$$s\left(\log p_d + (p_d - 1)\log(1 + \frac{1}{p_d - 1})\right) \le s\left(d\log\Delta + d + \log\frac{e}{2}\right)$$

in d-dimensions, where $p_d = \frac{(4\delta+1)^d+1}{2}$, and e is the base of natural logarithm.

Similarly to the one-dimensional case, the above bound is tight. That is, there is a distribution of X such that the entropy loss is at least

$$s\left(\log p_d + (p_d - 1)\log(1 + \frac{1}{p_d - 1})\right) - \epsilon$$

for some positive constant ϵ. Taking into consideration the entropy loss of sketch for set difference, we have

Corollary 6. *In d-dimensions, the entropy loss of X given sketch $P_H P_S$ is at most $s\left(d\log\Delta + d + \log\frac{e}{2}\right) + \mathcal{L}_{SD}\left(s, t, \frac{N^d}{(2\delta+1)^d}\right)$.*

5   Improved Schemes

The generic construction in Section 4.2 can indeed be further improved in terms of entropy loss. We employ two techniques. The first is *pre-rounding*. That is, each point in X and Y is rounded prior to both encoding and decoding. We observe that, the effect of the white noise is reduced on the rounded points. The second technique is *partitioning*, where we carefully partition the domain into cells. Instead of selecting points independently from the space, in the improved scheme, at most one point is selected in each cell. Both techniques are useful in reducing the randomness required in constructing P_H.

5.1   Improvement in One Dimension ($d = 1$)

First, we give an improvement for $\delta = 1$ using partitioning, and we observe that this scheme can be extended to any $\delta > 1$ by pre-rounding.

We partition the domain $[0, N - 1]$ into cells of size 3, such that the i-th cell contains the 3 consecutive points $\{3i, 3i + 1, 3i + 2\}$. There are $n' = \lceil N/3 \rceil$ cells in total. We want to assign one bit h_i to the i-th cell for all $0 \le i \le n' - 1$, and construct P_H as the binary sequence $\langle h_0, h_1, \ldots h_{n'-1}\rangle$.

Our main idea is to use this binary sequence to describe the codewords in the cells. At the first glance, it seems impossible since each cell would have three different possible codewords, which cannot be described by one bit. However, since two codewords cannot be too close to each other, we can eliminate certain cases by considering each two consecutive cells together. In this way, we can use only two bits to describe the codewords in two consecutive cells.

Here is how the values in the binary sequence are determined: For each $x \in X$, it is in the $i = \lfloor x/3 \rfloor$-th cell, and $r = x \mod 3$ indicates the location of x in the i-th cell. We set two values h_i and h_{i+1} in P_H according to Table 1(a). Since there are s points in X, the above process sets the values for $2s$ bits in $\langle h_1, h_2, \dots h_{n'-1} \rangle$. For each h_i that is not set, we randomly assign a value from $\{0, 1\}$ to it.

Now, from $\langle h_0, h_1, \dots h_{n'-1} \rangle$, we determine a set of "potential codewords". For each i-th cell, the potential codeword in the cell is determined by h_i and h_{i+1} using Table 1(b). Next, for a potential codewords x, if there is another potential codeword x' such that $x' \in S_1(x)$, then x is removed. The retained points form the codebook $\mathcal{C}$. By the design of Table 1(a) & (b), each $x \in X$ will be a codeword.

Table 1. Improved Scheme for $d = 1$

	h_i	h_{i+1}
$r = 0$	0	0
$r = 1$	0	1
$r = 2$	1	1

(a)

	$h_{i+1} = 0$	$h_{i+1} = 1$
$h_i = 0$	$3i$	$3i + 1$
$h_i = 1$	$3i + 2$	$3i + 2$

(b)

Similar to the basic construction, in practice, we can publish a description of $\mathcal{C}$ as the sketch. However, for the ease of analysis, we choose to publish $\langle h_0, h_1, \dots, h_{n'-1} \rangle$. During decoding, each y is simply matched to the nearest codeword in $\mathcal{C}$.

Since we invested $n' - 2s$ bits of randomness, and the size of sketch is $2s$, the entropy loss is at most $2s$.

Extension to any δ. To extend this scheme to any δ, we employ rounding. The rounding is essentially a many-to-one mapping. For each point $w \in [0, N - 1]$, we map it to $\hat{w} = \lfloor w/\delta \rfloor$. Note that under white noise, the perturbed point w' can only be mapped to $\hat{w} - 1, \hat{w}$ or $\hat{w} + 1$. In other words, under the mapping, the white noise (that appears to be on $\hat{w}$) is reduced to -1, 0, or $+1$, which corresponds to white noise with unit strength. Since the mapping is many-to-one, for each $x \in X$, we keep the rounding error $x - \delta(\lfloor x/\delta \rfloor)$ and publish it as part of the sketch. Hence, the additional entropy loss due to the rounding is at most $\log \delta$ for each $x \in X$. In total, we have

Theorem 7. *The entropy loss for the above scheme is at most* $(2 + \log \delta)s + \mathcal{L}_{SD}(s, t, N/(3\delta))$.

5.2 Improvement for $d = 2$ and $\delta = 1$

For $\delta = 1$ in two dimensions, with a parameter $\alpha \in [0, 4]$, we partition the space such that every 5 points of the form $\{(w, 5k + \alpha), (w, 5k + \alpha + 1), (w, 5k + \alpha + 2),$

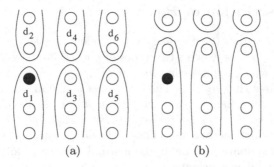

Fig. 1. Cells of size 5. For each scenario, the black point is a data point, the white points cannot be in the codebook.

$(w, 5k+\alpha+3), (w, 5k+\alpha+4)\}$ for some non-negative integer k, are grouped into a cell (Fig. 1). Each cell will be assigned a number $q \in [0, p_2 - 1]$ where p_2 is a constant to be decided later. If the assigned value q is less than or equal to 4, then we select the point $(w, 5k+q)$ to be a codeword in the cell, otherwise no codeword is selected in this cell.

There are five possible scenarios for a point $x \in X$, corresponding to the five different possible locations it occupies in a cell. Two of the five possible scenarios are illustrated in Fig. 1. Now we count the entropy loss for the scenario in Fig. 1(a). Same as in the basic construction, for any $x \in X$, all the points in the half-sphere $S_2(x)$ cannot be codewords. Therefore, all the white points in the figure cannot be codewords. Hence, for cell labeled d_1, there is only 1 choice for the value of the corresponding q, for d_3 and d_5, there are $p_2 - 3$ choices, and for d_2, d_4, and d_6, there are $p_2 - 2$ choices. Hence the entropy loss for this point is $\log p_2 + 2\log(p_2/(p_2 - 3)) + 3\log(p_2/(p_2 - 2))$.

Now we choose $p_2 = 14$, and the entropy loss for all five scenarios are as shown in Table 2.

Table 2. Entropy loss of the five scenarios

(a)	$\log p_2 + 2\log(p_2/(p_2 - 3)) + 3\log(p_2/(p_2 - 2)) < 5.1704$
(b)	$\log p_2 + 2\log(p_2/(p_2 - 4)) + 3\log(p_2/(p_2 - 1)) < 5.0990$
(c)	$\log p_2 + 2\log(p_2/(p_2 - 5))$ < 5.0823
(d)	$\log p_2 + 2\log(p_2/(p_2 - 4)) + 2\log(p_2/(p_2 - 1)) < 4.9921$
(e)	$\log p_2 + 2\log(p_2/(p_2 - 3)) + 2\log(p_2/(p_2 - 2)) < 4.9480$

Next, we choose a value for α, such that scenario (e) happens most often. By this choice of α, we can show that $\mathcal{L}_H \leq 5.0750s$, whereas in the basic construction in Section 4.2, the bound is at least $5.0861s$ for $\delta = 1$.

Although the improvement is small, this construction suggests that the basic construction can be further improved by partitioning. There are many ways to partition the 2-d domain, and it is interesting to find the optimal partition in terms of entropy loss.

6 Short Description of P_H

In the basic constructions (Section 4.2), we can view the sketch P_H as a random sequence of length $N^d \log p_d$ with two types of constraints: Type 0 constraint is of the form $(k, 0)$, which requires that $h_k = 0$, and type 1 constraint is of the form $(k, 1)$ which requires that $h_k \neq 0$. The main idea is as follows: *Find the seed of some pseudo-random generator, such that the generated sequence satisfies all the type 0 and 1 constraints, and use the seed as the sketch.* In this section, we give two methods. The first method has efficient decoding and encoding algorithms, but still requires randomness. The second method eliminates all randomness but there is no known efficient encoder.

Using a high degree polynomial. Let $n = N^d$, and assign each $x \in [0, N-1]^d$ a unique index $\mathtt{ind}(x)$ in $[0, n-1]$. Given a constraint set $S = \{(k_1, r_1), \ldots, (k_m, r_m)\}$, we construct a polynomial $f(x)$ of degree at most $m-1$ in $\mathbb{Z}_n$ as the following.

1. Uniformly choose $d_1, \ldots, d_m \in \mathbb{Z}_n$ at random such that for $1 \leq i \leq m$, if $r_i = 0$, then $d_i \equiv 0 \mod p_d$, otherwise $d_i \not\equiv 0 \mod p_d$.
2. Find the polynomial f of degree at most $m-1$ such that $f(\mathtt{ind}(k_i)) \equiv d_i \mod n$ for $1 \leq i \leq m$.

The m coefficients of f is published as the sketch. During decoding, each h_k in P_H can be recovered by computing $h_k = (f(\mathtt{ind}(k)) \mod n) \mod p_d$. Since for each point x we can have at most $|S_d(x)| + 1$ constraints, The polynomial f can be represented using $\frac{ds((4\delta+1)^d+1)}{2} \log N$ bits.

When p_d divides n, the entropy loss of this sketch is the same as the basic construction.

Using almost k-wise independence [1]. A sample space of n bits is almost k-wise independent if the probability distribution, induced on every k bit locations in a randomly chosen string from the sample space, is statistically close to uniform. The number of bits required to describe one sample is $(2+o(1))(\log \log n + 3k/2 + \log k)$. The sample space is pre-computed and made public.

We observe that this construction can be employed to make the sketch shorter. For instance, for $d = 1$ and $\delta = 1$ in our basic construction, we can construct such a sample space with $k = 3s$ and $n = N$. Given an original X, which in turn gives a set of constraints, we find the first sample that satisfies the constraints. The description of the sample is the sketch, whose size is in $o(s + \log \log N)$, which is also an upper bound for the entropy loss. In general, the size of the sketch would be in $o\left(s\Delta^d + \log \log(N^d)\right)$ in d-dimensional space. However, we are not aware of a better bound on the entropy loss other than the size of the sketch.

7 Entropy Loss of a Random Placement Method

Intuitively, it seems that it is better to have the codebook $\mathcal{C} = (X \cup R)$ as large as possible, since then a brute-force attacker will need to try more guesses to get X. In this section we give a seemingly natural random placement method to construct P_H with a large R in one dimension, and we show that the entropy loss is high for certain distributions of X.

The secure sketch P_H is a description of the sequence $\langle r_0, r_1, \ldots, r_{\lceil N/\Delta \rceil} \rangle$. Each r_i describes the gap between two consecutive codewords in $\mathcal{C}$ (except for r_0, which can be considered as the description of an "imaginary" starting codeword). Hence, instead of generating the codewords directly, we randomly choose the gaps between the codewords.

The sequence P_H is generated incrementally, starting from r_1. Most of the times the value of each gap can be chosen from Δ different values, but when a codeword w is close to a point $x \in X$, then the gap between w and the next codeword will be selected from a smaller interval (Steps 2 and 3).

1. Let $r_0 = -\delta$, $i = 1$.
2. If there is an $x \in X$ s.t. $x - r_{i-1} \in [2\delta, 4\delta]$ then let $r_i = x - r_{i-1}$.
3. If there is an $x \in X$ s.t. $x - r_{i-1} \in [4\delta + 1, 6\delta]$, uniformly choose r_i from $[\Delta - 1, x - r_{i-1} - \Delta]$. Otherwise, uniformly choose r_i from $[\Delta - 1, 2\Delta - 2]$.
4. Increase i by 1, and repeat from Step 2 until $i = \lceil N/\Delta \rceil + 1$.
5. Output $P_H = \langle r_1, \ldots, r_{\lceil N/\Delta \rceil} \rangle$.

The codewords can be recovered from P_H. In particular, the k-th codeword is $\sum_{i=0}^{k} r_i$, for $1 \leq k \leq \lceil N/\Delta \rceil$. If a codeword recovered in this process is greater than $N - 1$, it is removed. It is not necessary for P_H to have exactly $\lceil N/\Delta \rceil$ elements, and the extra padding is only for the ease of analysis.

Consider $X = \{x_1, x_1 + 2\Delta, \ldots, x_1 + 2(s-1)\Delta\}$, where x_1 is uniformly distributed. It can be shown that the entropy loss of X given P_H is at least $2s \log \Delta - \epsilon$ for some small positive constant ϵ. Comparing with other constructions in this paper, this method reveals the most information, even though it produces the largest number of codewords.

8   Conclusions and Discussions

In this paper, we investigate the technique of hiding a set of secret points by adding chaff points. Instead of considering brute force attackers as in known previous works, we give rigorous treatment under the secure sketch framework. We propose a construction of secure sketch for such point-sets, which can be extended to any dimension, and also some improvements for certain specific parameters. We give tight bounds of the entropy loss of our schemes.

Although we used infinity norm as the measure of closeness between any pair of points in the space, it is not difficult to extend our basic construction to any other closeness relations (e.g., using ℓ_2 norm). It seems that this is always possible as long as a total order can be defined on the points, so that the half-sphere of any given point is uniquely defined and is bounded.

On the other hand, the improvements in Section 5 are "ad-hoc" in the sense that they are specially designed for particular values of δ and d. We can also obtain improved schemes for another case where the white noise either leaves a coordinate unchanged or increased by one (we call this the 0-1 noise). An interesting question now is whether there is a generic method to find the "optimal" way of partitioning the space.

The proposed sketches are not suitable for large universe size N^d. The methods in Section 6 can reduce the sketch size, but the encoding and decoding algorithms can still be inefficient for large universe.

References

1. Noga Alon, Oded Goldreich, Johan Håstad, and René Peralta. Simple constructions of almost k-wise independent random variables. In *Proc. of the 31st FOCS*, pages 544–553, 1990.
2. Xavier Boyen. Reusable cryptographic fuzzy extractors. In *Proceedings of the 11th ACM conference on Computer and Communications Security*, pages 82–91. ACM Press, 2004.
3. Ee-Chien Chang, Vadym Fedyukovych, and Qiming Li. Secure sketch for multi-set difference. Cryptology ePrint Archive, Report 2006/090, 2006. http://eprint.iacr.org/.
4. Ee-Chien Chang, Ren Shen, and Francis Weijian Teo. Finding the original point set hidden among chaff. In *ASIACCS*, 2006. To appear.
5. T.C. Clancy, N. Kiyavash, and D.J. Lin. Secure smartcard-based fingerprint authentication. In *ACM Workshop on Biometric Methods and Applications*, 2003.
6. Michael D.Garris and R.Michael McCabe. Fingerprint minutiae from latent and matching tenprint images. *NIST Special Database 27*, 2000.
7. Yevgeniy Dodis, Leonid Reyzin, and Adam Smith. Fuzzy extractors: How to generate strong keys from biometrics and other noisy data. In *Eurocrypt'04*, volume 3027 of *LNCS*, pages 523–540. Springer-Verlag, 2004.
8. E.G. Coffman Jr., L. Flatto, and P. Jelenković. Interval packing: The vacant-interval distribution. *The Annals of Applied Probability*, 10(1):240–257, 2000.
9. Ari Juels and Madhu Sudan. A fuzzy vault scheme. In *IEEE Intl. Symp. on Information Theory*, 2002.
10. Ari Juels and Martin Wattenberg. A fuzzy commitment scheme. In *Proc. ACM Conf. on Computer and Communications Security*, pages 28–36, 1999.
11. D. Maltoni, D. Maio, A.K. Jain, and S. Prabhakar. *Handbook of Fingerprint Recognition*. Springer, 2003.
12. Yaron Minsky, Ari Trachtenberg, and Richard Zippel. Set reconciliation with nearly optimal communications complexity. In *ISIT*, 2001.
13. I. Palasti. On some random space filling problems. *Publ. Math. Inst. Hung. Acad. Sci.*, 5:353–359, 1960.
14. A. Rényi. On a one-dimensional problem concerning random space-filling. *Publ. Math. Inst. Hung. Acad. Sci.*, 3:109–127, 1958.
15. P. Tuyls and J. Goseling. Capacity and examples of template-protecting biometric authentication systems. In *ECCV Workshop BioAW*, pages 158–170, 2004.
16. Shenglin Yang and Ingrid Verbauwhede. Automatic secure fingerprint verification system based on fuzzy vault scheme. In *IEEE Intl. Conf. on Acoustics, Speech, and Signal Processing (ICASSP)*, pages 609–612, 2005.

Parallel and Concurrent Security of the HB and HB$^+$ Protocols

Jonathan Katz* and Ji Sun Shin**

Dept. of Computer Science, University of Maryland
{jkatz, sunny}@cs.umd.edu

Abstract. Juels and Weis (building on prior work of Hopper and Blum) propose and analyze two shared-key authentication protocols — HB and HB$^+$ — whose extremely low computational cost makes them attractive for low-cost devices such as radio-frequency identification (RFID) tags. Security of these protocols is based on the conjectured hardness of the "learning parity with noise" (LPN) problem: the HB protocol is proven secure against a passive (eavesdropping) adversary, while the HB$^+$protocol is proven secure against active attacks.

Juels and Weis prove security of these protocols only for the case of *sequential* executions, and explicitly leave open the question of whether security holds also in the case of *parallel* or *concurrent* executions. In addition to guaranteeing security against a stronger class of adversaries, a positive answer to this question would allow the HB$^+$ protocol to be parallelized, thereby substantially reducing its round complexity.

Adapting a recent result by Regev, we answer the aforementioned question in the affirmative and prove security of the HB and HB$^+$ protocols under parallel/concurrent executions. We also give what we believe to be substantially *simpler* security proofs for these protocols which are more *complete* in that they explicitly address the dependence of the soundness error on the number of iterations.

1 Introduction

Low-cost, severely resource-constrained devices such as radio-frequency identification (RFID) tags or sensor nodes demand extremely efficient algorithms and protocols. Securing such devices is a challenge since, in many cases, "traditional" cryptographic protocols are simply too computationally-intensive to be utilized. With this motivation in mind, Juels and Weis [20] — building upon work of Hopper and Blum [18, 19] — investigate two highly-efficient, shared-key (unidirectional) authentication protocols suitable for an RFID *tag* identifying itself to a tag *reader*. (We will sometimes refer to the tag as a *prover* and the tag reader as a *verifier*.) These protocols are extremely lightweight, requiring both parties to perform only a relatively small number of primitive bit-wise operations such

---

* This research was supported by NSF Trusted Computing grants #0310499 and #0310751, and NSF CAREER award #0447075.
** Supported by NSF Trusted Computing grant #0310499.

S. Vaudenay (Ed.): EUROCRYPT 2006, LNCS 4004, pp. 73–87, 2006.

as "XOR" and "AND," and can thus be implemented using fewer than the 5-10K gates required to implement even a block cipher such as DES or AES [20].

The two protocols studied by Juels and Weis are both proven secure via reduction to the "learning parity with noise" (LPN) problem [4, 5, 6, 9, 17, 21, 18, 19, 25]; a formal definition of this problem as well as evidence for its difficulty are reviewed in Section 2.1. The first protocol (called the HB protocol [18, 19]) is proven secure against a *passive* (eavesdropping) adversary, while the second (called HB$^+$) is proven secure against the stronger class of *active* adversaries. In each case, Juels and Weis focus on a single, "basic authentication step" of the protocol and prove that a computationally-bounded adversary cannot succeed in impersonating a tag in this case with probability noticeably better than 1/2; that is, a single iteration of the protocol has *soundness error* 1/2. The implicit assumption (though see below) is that repeating these "basic authentication steps" sufficiently-many times yields a protocol with negligible soundness error.

Difficulties and limitations. There are, however, some subtle limitations of the security proofs given by Juels and Weis. Most serious, perhaps, is a difficulty explicitly highlighted by Juels and Weis and regarded by them as a potential barrier to usage of the HB$^+$ protocol in practice [20, Section 6]: the proof of security for HB$^+$ requires that the adversary's interactions with the tag (i.e., when the adversary is impersonating a tag reader) be *sequential*. Besides leaving in question the security of HB$^+$ under *concurrent* executions, this also means that the HB$^+$ protocol itself (which, recall, consists of sufficiently-many repetitions of an underlying basic authentication step) requires very high round complexity since the multiple iterations of the basic authentication step cannot be *parallelized* but must instead be performed sequentially. The difficulty and importance of proving security of various identification protocols under concurrent or parallel composition is well-understood, and many results are known: for example, the (black-box) zero-knowledge property of an identification protocol is not preserved under parallel [14] or concurrent [8] composition (though it is preserved under sequential composition [16]), whereas witness indistinguishability *is* preserved in these cases [11]. Unfortunately, the HB$^+$ protocol is not known to satisfy either zero knowledge or witness indistinguishability and so such results are of no help here.

An additional difficulty, not explicitly mentioned in [20], is that it is unclear what is the exact relationship between the soundness error and the number of repetitions of the basic authentication step; this is true for both the HB and HB$^+$ protocols, regardless of whether the repetitions are carried out in parallel or sequentially.[1] This is related to the more general question of "when is solving multiple instances of a problem more difficult than solving a single instance?" (i.e., *hardness amplification*) which has been studied in many contexts [26, 15, 3, 13, 24, 7] and turns out to be surprisingly non-trivial to answer.

---

[1] Indeed, Juels and Weis only prove soundness 1/2 for a basic authentication step and never make any claims regarding the security of multiple iterations (for either HB or HB$^+$); this indicates that those authors also recognized the difficulty of characterizing the dependence of soundness on the number of iterations.

Unfortunately, there does not seem to be any prior work that applies in our setting. Specifically:

- For the HB and HB$^+$ protocols it is not possible to efficiently verify whether a given transcript is "successful" without possession of the secret key; thus, Yao's "XOR-lemma" [26, 15] and related techniques that require efficient verifiability do not apply.
- Work on hardness amplification for "weakly-verifiable puzzles" [7] does not apply either. Although the HB/HB$^+$ protocols can be viewed as efficiently-verifiable puzzles, hardness amplification in [7] is only proved for *completely independent* instances of the "puzzle." In particular, then, the work of [7] implies that running the basic authentication step of the HB protocol n times *using n independent keys* yields soundness (roughly) $1/2^n$, but says nothing about running n iterations using the *same* key (which is the case we are interested in).
- The HB/HB$^+$ protocols are *computationally*-sound only, and thus known results [13, Appendix C] [24] on soundness reduction for interactive proof systems (which apply only when soundness holds even against an all-powerful cheating prover) do not apply either.
- Bellare, et al. [3] study soundness reduction in computationally-sound protocols, and show a positive result [3, Sect. 4] for the case of protocols running in 3 rounds. Unfortunately, their result is specifically stated to apply *only* when the verifier does not hold a secret key (or, more generally, only when the verifier does not share state across different iterations). As in the case of weakly-verifiable puzzles, then, this result is of no help when the same secret key is used across all iterations.

An additional difficulty in our setting is that the verifier is supposed to accept even when some iterations have not been answered successfully; indeed, crucial to both the HB and HB$^+$ protocols is that the honest prover injects "noise" into its answers and so even the honest prover does not succeed with probability 1. This was not explicitly addressed in the security proofs of [20], either, and introduces additional complications.

1.1 Our Contributions

In this work we address the difficulties and open questions mentioned above, and show the following results: (1) the HB$^+$ protocol remains secure under arbitrary concurrent interactions of the adversary with the honest prover/tag, and so in particular the iterations of the HB$^+$ protocol can be parallelized; furthermore, (2) our security proofs explicitly incorporate the dependence of the soundness error on the number of iterations as well as the error introduced by the honest prover.

Besides the results themselves, we expect that the techniques and proofs we give here will be of independent interest for future work on cryptographic applications of the LPN problem. Our main technical tool is a result due to Regev [25] (see also [5]) showing that the hardness of the LPN problem implies the pseudorandomness of a certain distribution. Using this, we give proofs which we believe are substantially *simpler* than those given in [20], and also more *complete* (in that, in contrast to [20], they explicitly deal with the dependence of soundness on the number of iterations and also the issues arising due to non-perfect completeness).

1.2   Additional Discussion

The problem of secure authentication using a shared, secret key is by now well-understood, and many widely-known solutions based on, e.g., block ciphers are available. We stress that the aim of the line of research considered here, as in [20], is to develop protocols which are *exceptionally* efficient while still guaranteeing some useful level of (provable) security. The estimates from [20] are that 5,000–10,000+ gates are needed for block-cipher implementations, whereas a typical RFID tag may only have 2,000 gates that can be dedicated to security. Moore's Law will not necessarily help here, either: as pointed out in [20], there is intense pressure to keep prices for RFID tags low; as computational power per fixed unit of currency increases, the trend has been to reduce the cost of tags and thus expand their application domain rather than to increase their computational power while keeping costs fixed. In short, there seems to be "little effective change in tag resources for some time to come, and thus a pressing need for new lightweight primitives" [20].

Gilbert, et al. [12] have recently shown a man-in-the-middle attack on the HB$^+$ protocol. Although their attack would be debilitating if carried out successfully, the possibility of such an attack does not mean that it is now useless to explore the security of the HB/HB$^+$ protocols in weaker attack models! (Indeed, only recently have man-in-the-middle attacks on identification protocols been formally considered in general [2], yet certainly research in the area conducted up to that point is not valueless.) There will always be some tradeoff between efficiency and security, and our work can be viewed as mapping out where the HB/HB$^+$ protocols lie on this spectrum. Moreover, Juels and Weis [20, Appendix A] note that the man-in-the-middle attack of [12] does not apply in a *detection-based* system where numerous failed authentication attempts immediately raise an alarm. Furthermore, especially in the case of RFID (where communication is inherently short range), it appears much more difficult to mount a man-in-the-middle attack than an active attack.[2] The reader is referred to the work of Wool, et al. [22, 23], for an illuminating discussion on the feasibility of man-in-the-middle attacks in RFID systems.

2   Definitions and Preliminaries

We formally define the LPN problem and state and prove the main technical lemma on which we rely. We also define our notion(s) of security for identifica-

---

[2] Though there have been claims of being able to read some RFID tags over as much as 69 feet [1], the maximum distance from which many commonly-used cards can be read appears to be almost two orders of magnitude lower [22]. Note further that a man-in-the-middle attack requires the ability to *send* data to the tag (and reader).

tion; these are standard, but some complications arise due to the fact that the HB/HB$^+$ protocols do not have perfect completeness.

2.1 The LPN Problem

View k as a security parameter. If $\mathbf{s}, \mathbf{a}_1, \ldots, \mathbf{a}_\ell$ are binary vectors of length k, let $z_i = \langle \mathbf{s}, \mathbf{a}_i \rangle$ denote the dot product of $\mathbf{s}$ and $\mathbf{a}_i$ (modulo 2). Given the values $\mathbf{a}_1, z_1, \ldots, \mathbf{a}_\ell, z_\ell$ for randomly-chosen $\{\mathbf{a}_i\}$ and $\ell = O(k)$, it is possible to efficiently solve for $\mathbf{s}$ using standard linear-algebraic techniques. However, in the presence of *noise* where each z_i is flipped (independently) with probability ε, finding $\mathbf{s}$ becomes much more difficult. We refer to the problem of learning $\mathbf{s}$ in this latter case as *the LPN problem*.

For the formal definition, let Ber_ε be the Bernoulli distribution with parameter $\varepsilon \in (0, \frac{1}{2})$ (so if $\nu \sim \mathsf{Ber}_\varepsilon$ then $\Pr[\nu = 1] = \varepsilon$ and $\Pr[\nu = 0] = 1 - \varepsilon$), and let $A_{\mathbf{s},\varepsilon}$ be the distribution defined by:

$$\left\{ \mathbf{a} \leftarrow \{0,1\}^k; \nu \leftarrow \mathsf{Ber}_\varepsilon : (\mathbf{a}, \langle \mathbf{s}, \mathbf{a} \rangle \oplus \nu) \right\}.$$

Also let $A_{\mathbf{s},\varepsilon}$ denote an oracle which outputs (independent) samples according to this distribution. Algorithm M is said to (t, q, δ)-*solve the* LPN_ε *problem* if

$$\Pr\left[\mathbf{s} \leftarrow \{0,1\}^k : M^{A_{\mathbf{s},\varepsilon}}(1^k) = \mathbf{s} \right] \geq \delta,$$

and furthermore M runs in time at most t and makes at most q queries to its oracle.[3] In asymptotic terms, in the standard way, the LPN_ε problem is "hard" if every probabilistic polynomial-time algorithm solves the LPN_ε problem with only negligible probability (where the algorithm's running time and success probability are functions of k).

Note that ε is usually taken to be a fixed constant independent of k, as will be the case here. The value of ε to use depends on a number of tradeoffs and design decisions: although, roughly speaking, the LPN_ε problem becomes "harder" as ε increases, a larger value of ε also implies that the honest prover is rejected more often (as will become clear when we describe the HB/HB$^+$ protocols, below). In any case, our results are meaningful for all $\varepsilon \in (0, \frac{1}{4})$. For concreteness, the reader can think of $\varepsilon \approx \frac{1}{8}$.

The hardness of the LPN_ε problem (for constant $\varepsilon \in (0, \frac{1}{2})$) has been studied in many previous works. It can be formulated also as the problem of decoding a random linear code [4, 25], and is $\mathcal{NP}$-complete [4] as well as hard to approximate within a factor better than 2 (where the optimization problem is phrased as finding an $\mathbf{s}$ satisfying the most equations) [17]. These worst-case hardness results are complemented by numerous studies of the average-case hardness of the problem [5, 6, 9, 21, 18, 19, 25]. Most relevant for our purposes is that the current best-known algorithm for solving the LPN_ε problem [6] requires $t, q = 2^{\Theta(k/\log k)}$.

---

[3] Our formulation of the LPN problem follows, e.g., [25]; the formulation in, e.g., [20] allows M to output any $\mathbf{s}$ satisfying $\geq (1 - \varepsilon)$ fraction of the equations returned by $A_{\mathbf{s},\varepsilon}$. It is easy to see that for q large enough these formulations are essentially equivalent as with overwhelming probability there will be a unique such $\mathbf{s}$.

We refer the reader to [20, Appendix D] for more exact estimates of the running time of this algorithm, as well as suggested practical values for k.

2.2   A Technical Lemma

In this section we prove a key technical lemma: hardness of the LPN_ε problem implies "pseudorandomness" of $A_{\mathbf{s},\varepsilon}$. Specifically, let U_{k+1} denote the uniform distribution on $(k+1)$-bit strings. The following lemma shows that oracle access to $A_{\mathbf{s},\varepsilon}$ (for randomly-chosen $\mathbf{s}$) is indistinguishable from oracle access to U_{k+1}. A proof of the following is essentially in [25, Sect. 4], although we have fleshed out some of the details and worked out the concrete parameters of the reduction.

Lemma 1. *Say there exists an algorithm D making q oracle queries, running in time t, and such that*

$$\left| \Pr\left[\mathbf{s} \leftarrow \{0,1\}^k : D^{A_{\mathbf{s},\varepsilon}}(1^k) = 1 \right] - \Pr\left[D^{U_{k+1}}(1^k) = 1 \right] \right| \geq \delta.$$

Then there exists an algorithm M making $q' = O\left(q \cdot \delta^{-2} \log k\right)$ oracle queries, running in time $t' = O\left(t \cdot k\delta^{-2} \log k\right)$, and such that

$$\Pr\left[\mathbf{s} \leftarrow \{0,1\}^k : M^{A_{\mathbf{s},\varepsilon}}(1^k) = \mathbf{s} \right] \geq \delta/4.$$

(Various tradeoffs are possible between the number of queries/running time of M and its success probability in solving LPN_ε; see [25, Sect. 4]. We aimed for simplicity in the proof rather than trying to optimize parameters.)

Proof. Set $N = O\left(\delta^{-2} \log k\right)$. Algorithm $M^{A_{\mathbf{s},\varepsilon}}(1^k)$ proceeds as follows:

1. M chooses random coins ω for D and uses these for the remainder of its execution.
2. M runs $D^{U_{k+1}}(1^k; \omega)$ for a total of N times to obtain an estimate p for the probability that D outputs 1 in this case. (The probability here is over the responses from the oracle.)
3. M obtains $q \cdot N$ samples $\{(\mathbf{a}_{1,j}, z_{1,j})\}_{j=1}^q, \ldots, \{(\mathbf{a}_{N,j}, z_{N,j})\}_{j=1}^q$ from $A_{\mathbf{s},\varepsilon}$. Then for $i \in [k]$:
   (a) Run $D(1^k; \omega)$ for a total of N times, each time using a fresh set of samples $\{(\mathbf{a}_j, z_j)\}_{j=1}^q$ to answer the q oracle queries of D. Answer the jth oracle query of D in each iteration by choosing a random bit c_j and returning $(\mathbf{a}_j \oplus (c_j \cdot \mathbf{e}_i), z_j)$, where $\mathbf{e}_i$ is the vector with 1 at position i and 0s elsewhere. Obtain an estimate p_i for the probability that D outputs 1 in this case.
   (b) If $|p_i - p| \geq \delta/4$ set $s_i' = 0$; else set $s_i' = 1$.
4. Output $\mathbf{s}' = (s_1', \ldots, s_k')$.

Let us analyze the behavior of M. First note that, by standard averaging argument, with probability at least $\delta/2$ over choice of $\mathbf{s}$ and random coins ω it holds that

$$\left| \Pr\left[D^{A_{\mathbf{s},\varepsilon}}(1^k; \omega) = 1 \right] - \Pr\left[D^{U_{k+1}}(1^k; \omega) = 1 \right] \right| \geq \delta/2, \tag{1}$$

where the probabilities are taken over the answers D receives from its oracle. We restrict our attention to $\mathbf{s}, \omega$ for which Eq. (1) holds and show that in this case M outputs $\mathbf{s}' = \mathbf{s}$ with probability at least $1/2$. The theorem follows.

By our choice of N we have that

$$\left| \Pr \left[D^{U_{k+1}}(1^k; \omega) = 1 \right] - p \right| \leq \delta/16 \tag{2}$$

except with probability at most $O(1/k)$. Next focus on a particular iteration i of steps 3(a) and 3(b). Letting hyb_i denote the distribution of the answers returned to D in this iteration, we again have

$$\left| \Pr \left[D^{\mathsf{hyb}_i}(1^k; \omega) = 1 \right] - p_i \right| \leq \delta/16 \tag{3}$$

except with probability at most $O(1/k)$. Applying a union bound (and setting parameters appropriately) we see that with probability at least $1/2$ Eqs. (2) and (3) hold (the latter for all $i \in [k]$), and so we assume this to be the case for the rest of the proof.

We claim that if $s_i = 0$ then $\mathsf{hyb}_i = A_{\mathbf{s}, \varepsilon}$, while if $s_i = 1$ then $\mathsf{hyb}_i = U_{k+1}$. To see this note that when $s_i = 0$ the answer $(\mathbf{a}_j \oplus (c_j \cdot \mathbf{e}_i), z_j)$ returned to D is distributed exactly according to $A_{\mathbf{s}, \varepsilon}$ since $\langle \mathbf{s}, \mathbf{a}_j \oplus (c_j \cdot \mathbf{e}_i) \rangle = \langle \mathbf{s}, \mathbf{a}_j \rangle = z_j$. On the other hand, if $s_i = 1$ then $z_j = \langle \mathbf{s}, \mathbf{a}_j \rangle$ is independent of $\mathbf{a}_j \oplus (c_j \cdot \mathbf{e}_i)$ since c_j is random (and unknown to D).

It follows that if $s_i = 0$ then

$$\left| \Pr \left[D^{\mathsf{hyb}_i}(1^k; \omega) = 1 \right] - \Pr \left[D^{U_{k+1}}(1^k; \omega) = 1 \right] \right| \geq \delta/2$$

(by Eq. (1)), and so $|p_i - p| \geq \frac{\delta}{2} - 2 \cdot \frac{\delta}{16} = \frac{3\delta}{8}$ (by Eqs. (2) and (3)) and $s_i' = 0 = s_i$. When $s_i = 1$ then

$$\Pr \left[D^{\mathsf{hyb}_i}(1^k; \omega) = 1 \right] = \Pr \left[D^{U_{k+1}}(1^k; \omega) = 1 \right],$$

and so $|p_i - p| \leq 2 \cdot \frac{\delta}{16} = \frac{\delta}{8}$ (again using Eqs. (2) and (3)) and $s_i' = 1 = s_i$. Since this holds for all $i \in [k]$, we conclude that $\mathbf{s}' = \mathbf{s}$.

2.3   Overview of the HB/HB$^+$ Protocols, and Security Definitions

The HB and HB$^+$ protocols as analyzed here consist of n *parallel* iterations of a "basic authentication step." We describe the basic authentication step for the HB protocol, and defer a discussion of the HB$^+$ protocol to Section 3.2. In the HB protocol, a tag $\mathcal{T}$ and a reader $\mathcal{R}$ share a random secret key $\mathbf{s} \in \{0, 1\}^k$; a basic authentication step consists of the reader sending a random challenge $\mathbf{a} \in \{0, 1\}^k$ to the tag, which replies with $z = \langle \mathbf{s}, \mathbf{a} \rangle \oplus \nu$ for $\nu \sim \mathsf{Ber}_\varepsilon$. The reader can then verify whether the response z of the tag satisfies $z \overset{?}{=} \langle \mathbf{s}, \mathbf{a} \rangle$; we say the iteration is *successful* if this is the case. See Figure 1.

Even for an honest tag a basic iteration is unsuccessful with probability ε. For this reason, a reader accepts upon completion of all n iterations of the basic authentication step as long as $\approx \varepsilon \cdot n$ of these iterations were unsuccessful. More

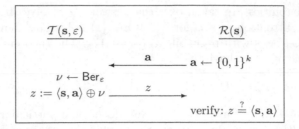

Fig. 1. The basic authentication step of the HB protocol

precisely, let l, u be such that $l \le \varepsilon \cdot n \le u$; then the reader accepts as long as the number of unsuccessful iterations lies in the range $[l, u]$. Since $\varepsilon \cdot n$ is the expected number of unsuccessful iterations for an honest tag, the completeness error ε_c (i.e., the probability that an honest tag is rejected) can be calculated via a Chernoff bound.[4] Overall, then, the entire HB protocol is parameterized by ε, l, u, and n.

Observe that by sending random answers in each of the n iterations, an adversary trying to impersonate a valid tag succeeds with probability

$$\delta^*_{\varepsilon,l,u,n} \stackrel{\text{def}}{=} 2^{-n} \cdot \sum_{i=l}^{u} \binom{n}{i};$$

that is, $\delta^*_{\varepsilon,l,u,n}$ is the *best* possible soundness error we can hope to achieve for the given setting of the parameters. Our definitions of security will be expressed in terms of the adversary's ability to do better than this. Looking at asymptotic security (taking k as a security parameter), note that for any constant $\varepsilon < 1/2$ it is easy to find functions l, u, n of k such that $n = O(k)$ and furthermore both the completeness error ε_c and the "best achievable" soundness error $\delta^*_{\varepsilon,l,u,n}$ are negligible.

Let $\mathcal{T}^{\mathsf{HB}}_{\mathbf{s},\varepsilon,n}$ denote the tag algorithm in the HB protocol when the tag holds secret key $\mathbf{s}$ (note that the tag algorithm is independent of l, u), and let $\mathcal{R}^{\mathsf{HB}}_{\mathbf{s},\varepsilon,l,u,n}$ similarly denote the algorithm run by the tag reader. We denote a complete execution of the HB protocol between a party $\hat{\mathcal{T}}$ and the reader $\mathcal{R}$ by $\left\langle \hat{\mathcal{T}}, \mathcal{R}^{\mathsf{HB}}_{\mathbf{s},\varepsilon,l,u,n} \right\rangle$ and say this equals 1 iff the reader accepts.

For a passive attack on the HB protocol, we imagine an adversary $\mathcal{A}$ running in two stages: in the first stage the adversary obtains q transcripts[5] of (honest) executions of the protocol by interacting with an oracle $\mathsf{trans}^{\mathsf{HB}}_{\mathbf{s},\varepsilon,n}$ (this models

---

[4] Note in particular that if u is set to *exactly* $\varepsilon \cdot n$ then the completeness error will be rather high. One can imagine changing the protocol so that the tag introduces *at most* $\varepsilon \cdot n$ errors; see Section 4 for discussion of this point.

[5] Following [18, 19, 20], a transcript comprises only the messages exchanged between the parties and does not include the reader's decision of whether or not to accept. If the adversary is given this additional information, the adversary's advantage may increase by (at most) an additive factor of $q \cdot \varepsilon_c$.

eavesdropping); in the second stage, the adversary interacts with the reader and tries to impersonate the tag. We define the adversary's advantage as

$$\mathsf{Adv}^{passive}_{\mathcal{A},\mathsf{HB}}(\varepsilon,\mathsf{l},\mathsf{u},n) \overset{\mathrm{def}}{=}$$
$$\Pr\left[\mathbf{s} \leftarrow \{0,1\}^k; \mathcal{A}^{\mathsf{trans}^{\mathsf{HB}}_{\mathbf{s},\varepsilon,n}}(1^k) : \left\langle \mathcal{A}, \mathcal{R}^{\mathsf{HB}}_{\mathbf{s},\varepsilon,\mathsf{l},\mathsf{u},n} \right\rangle = 1\right] - \delta^*_{\varepsilon,\mathsf{l},\mathsf{u},n}.$$

As we will describe in Section 3.2, the HB$^+$ protocol uses two keys $\mathbf{s}_1, \mathbf{s}_2$. We let $\mathcal{T}^{\mathsf{HB}^+}_{\mathbf{s}_1,\mathbf{s}_2,\varepsilon,n}$ denote the tag algorithm in this case, and let $\mathcal{R}^{\mathsf{HB}^+}_{\mathbf{s}_1,\mathbf{s}_2,\varepsilon,\mathsf{l},\mathsf{u},n}$ denote the algorithm run by the tag reader. For the case of an active attack on the HB$^+$ protocol, we again imagine an adversary running in two stages: in the first stage the adversary interacts at most q times with the honest tag algorithm (with concurrent executions allowed), while in the second stage the adversary interacts only with the reader.[6] The adversary's advantage in this case is

$$\mathsf{Adv}^{active}_{\mathcal{A},\mathsf{HB}^+}(\varepsilon,\mathsf{l},\mathsf{u},n) \overset{\mathrm{def}}{=}$$
$$\Pr\left[\mathbf{s}_1,\mathbf{s}_2 \leftarrow \{0,1\}^k; \mathcal{A}^{\mathcal{T}^{\mathsf{HB}^+}_{\mathbf{s}_1,\mathbf{s}_2,\varepsilon,n}}(1^k) : \left\langle \mathcal{A}, \mathcal{R}^{\mathsf{HB}^+}_{\mathbf{s}_1,\mathbf{s}_2,\varepsilon,\mathsf{l},\mathsf{u},n} \right\rangle = 1\right] - \delta^*_{\varepsilon,\mathsf{l},\mathsf{u},n}.$$

We remark that in both the HB and HB$^+$ protocols, the tag reader's actions are independent of the secret key(s) it holds except for its final decision whether or not to accept. So, allowing the adversary to interact with the reader multiple times (even concurrently) does not give the adversary much additional advantage (other than the fact that, as usual, the probability that the adversary succeeds in at least one impersonation attempt scales linearly with the number of attempts).

3 Proofs of Security for the HB and HB$^+$ Protocols

3.1 Security of the HB Protocol Against Passive Attacks

Recall from the previous section that we parameterize the HB protocol by ε (a measure of the noise introduced by the tag), l,u (which determine the completeness error ε_c as well as the best achievable soundness δ^*), and n (the number of iterations of the basic authentication step given in Figure 1). We stress that these n iterations are run in parallel, and so the entire protocol requires only two rounds.

The following result characterizes security of the HB protocol against passive attack. This can be compared to [20, Lemma 1], where Juels and Weis prove security for a single iteration of the HB protocol (i.e., they fix $n = 1$) and do not explicitly take the non-zero completeness error into account (this is taken into account in the following via the dependence on l,u).

---

[6] As we have already noted, this is the "classical" notion of security against active attacks which does not take into account man-in-the-middle attacks.

Theorem 1. *Say there exists an adversary $\mathcal{A}$ eavesdropping on q executions of the HB protocol, running in time t, and achieving $\mathsf{Adv}_{\mathcal{A},\mathsf{HB}}^{passive}(\varepsilon, \mathsf{l}, \mathsf{u}, n) \geq \delta$. Then there exists an algorithm D making $(q+1) \cdot n$ oracle queries, running in time $O(t)$, and such that*

$$\left| \Pr\left[\mathbf{s} \leftarrow \{0,1\}^k : D^{A_{\mathbf{s},\varepsilon}}(1^k) = 1 \right] - \Pr\left[D^{U_{k+1}}(1^k) = 1 \right] \right|$$

$$\geq \delta + \delta_{\varepsilon,\mathsf{l},\mathsf{u},n}^* - \varepsilon_c - 2^{-n} \cdot \sum_{i=0}^{2\,\mathsf{u}} \binom{n}{i}.$$

Asymptotically, for any $\varepsilon < \frac{1}{4}$ and $n = \Theta(k)$ all terms of the above expression (other than δ) are negligible for appropriate choice of l, u. We thus conclude that the HB protocol is secure (for $n = \Theta(k)$ and appropriate choice of l, u) assuming the hardness of the LPN_ε problem.

Proof. Algorithm D, given access to an oracle returning $(k+1)$-bit strings $(\mathbf{a}, z)$, proceeds as follows:

1. D runs the first phase of $\mathcal{A}$. Each time $\mathcal{A}$ requests to view a transcript of the protocol, D obtains n samples $\{(\mathbf{a}_i, z_i)\}_{i=1}^n$ from its oracle and returns these to $\mathcal{A}$.

2. When $\mathcal{A}$ is ready for the second phase, D again obtains n samples $\{(\bar{\mathbf{a}}_i, \bar{z}_i)\}_{i=1}^n$ from its oracle. D then sends the challenge $(\bar{\mathbf{a}}_1, \ldots, \bar{\mathbf{a}}_n)$ to $\mathcal{A}$ and receives in return a response $Z' = (z_1', \ldots, z_n')$.

3. D outputs 1 iff $\bar{Z} = (\bar{z}_1, \ldots, \bar{z}_n)$ and Z' differ in at most $2\mathsf{u}$ entries.

When D's oracle is U_{k+1}, it is clear that D outputs 1 with probability exactly $2^{-n} \cdot \sum_{i=0}^{2\mathsf{u}} \binom{n}{i}$ since $\bar{Z}$ is in this case uniformly distributed and independent of everything else. On the other hand, when D's oracle is $A_{\mathbf{s},\varepsilon}$ then the transcripts D provides to $\mathcal{A}$ during the first phase of $\mathcal{A}$'s execution are distributed identically to real transcripts in an execution of the HB protocol. Let $Z^* \stackrel{\text{def}}{=} (\langle \mathbf{s}, \bar{\mathbf{a}}_1 \rangle, \ldots, \langle \mathbf{s}, \bar{\mathbf{a}}_n \rangle)$ be the vector of "correct" answers to the challenge $(\bar{\mathbf{a}}_1, \ldots, \bar{\mathbf{a}}_n)$ sent by D in the second phase. Then with probability at least $\delta + \delta_{\varepsilon,\mathsf{l},\mathsf{u},n}^*$ it holds that Z' and Z^* differ in at most u entries (since $\mathcal{A}$ successfully impersonates the tag with this probability). Also, since $\bar{Z}$ is distributed exactly as the answers of an honest tag, $\bar{Z}$ and Z^* differ in at most u positions except with probability at most ε_c. It follows that with probability at least $\delta + \delta_{\varepsilon,\mathsf{l},\mathsf{u},n}^* - \varepsilon_c$ the vectors Z' and $\bar{Z}$ differ in at most $2\mathsf{u}$ entries, and so D outputs 1 with at least this probability.   □

The above result provides a useful security guarantee only when $\varepsilon < 1/4$, since when $\varepsilon \geq 1/4$ then $2\mathsf{u} \geq 2\varepsilon \cdot n \geq n/2$ and so $2^{-n} \cdot \sum_{i=0}^{2\mathsf{u}} \binom{n}{i} \geq 1/2$. We also note that the concrete security reduction obtained above leaves much to be desired, and in particular it is not clear whether useful levels of security are achieved for reasonably-efficient settings of the parameters. On the other hand, it is unclear what can be said about the tightness of the security reductions obtained by Juels and Weis [20] since they do not explicitly handle multiple iterations of the protocol nor do they consider the effect that the acceptance criteria (i.e., in terms of l, u) have on the soundness.

We believe that the security reduction can be improved by taking into account the distribution on $\bar{Z}$ when D's oracle is $A_{s,\varepsilon}$ (and modifying step 3 of D appropriately), as well as by focusing on protocols with perfect completeness. See Section 4 for some discussion of the latter possibility.

3.2 Security of the HB$^+$ Protocol Against Active Attacks

The HB protocol is insecure against an active attack, as an adversary can simply repeatedly query the tag with the same challenge vector $(\mathbf{a}_1, \dots, \mathbf{a}_n)$ and thereby determine with high probability the correct values of $\langle \mathbf{s}, \mathbf{a}_1 \rangle, \dots, \langle \mathbf{s}, \mathbf{a}_n \rangle$ (after which solving for $\mathbf{s}$ is easy). To combat such an attack, Juels and Weis propose to modify the HB protocol by having the tag and reader share *two* (independent) keys $\mathbf{s}_1, \mathbf{s}_2 \in \{0,1\}^k$. A basic authentication step now consists of three rounds: first the tag sends a random "blinding factor" $\mathbf{b} \in \{0,1\}^k$; the reader replies with a random challenge $\mathbf{a} \in \{0,1\}^k$ as before; and finally the tag replies with $z = \langle \mathbf{s}_1, \mathbf{b} \rangle \oplus \langle \mathbf{s}_2, \mathbf{a} \rangle \oplus \nu$ for $\nu \sim \mathsf{Ber}_\varepsilon$. As in the HB protocol, the tag reader can then verify whether the response z of the tag satisfies $z \stackrel{?}{=} \langle \mathbf{s}_1, \mathbf{b} \rangle \oplus \langle \mathbf{s}_2, \mathbf{a} \rangle$, and we again say the iteration is *successful* if this is the case. See Figure 2.

$$\mathcal{T}(\mathbf{s}_1, \mathbf{s}_2, \varepsilon) \qquad\qquad \mathcal{R}(\mathbf{s}_1, \mathbf{s}_2)$$

$$\mathbf{b} \leftarrow \{0,1\}^k \xrightarrow{\quad \mathbf{b} \quad}$$

$$\xleftarrow{\quad \mathbf{a} \quad} \mathbf{a} \leftarrow \{0,1\}^k$$

$$\nu \leftarrow \mathsf{Ber}_\varepsilon$$

$$z := \langle \mathbf{s}_1, \mathbf{b} \rangle \oplus \langle \mathbf{s}_2, \mathbf{a} \rangle \oplus \nu \xrightarrow{\quad z \quad}$$

$$\text{verify: } z \stackrel{?}{=} \langle \mathbf{s}_1, \mathbf{b} \rangle \oplus \langle \mathbf{s}_2, \mathbf{a} \rangle$$

Fig. 2. The basic authentication step of the HB$^+$ protocol

The actual HB$^+$ protocol consists of n parallel iterations of the basic authentication step (and so the entire protocol requires only three rounds). The protocol also depends upon parameters l, u as in the case of the HB protocol, and the values ε_c and $\delta^*_{\varepsilon,l,u,n}$ are defined exactly as there.

The following result characterizes security of the HB$^+$ protocol under *active* attacks. It can be compared to [20, Lemma 3], where Juels and Weis prove security for a single iteration of the HB$^+$ protocol (i.e., they fix $n = 1$). Their proof requires rewinding of the adversary $\mathcal{A}$ in order to simulate the first phase of $\mathcal{A}$, and therefore their proof does *not* extend to the case of parallel or concurrent executions of the basic authentication step.

We remark that by combining the proofs of Theorem 2 and Lemma 1 (i.e., reducing the HB$^+$ protocol directly to the LPN problem rather than relying on Lemma 1 as an intermediate step) we can improve the security reduction stated in the following theorem. By applying techniques from [25, Sect. 4], the parameters of the reduction can be improved further.

Theorem 2. *Say there exists an adversary $\mathcal{A}$ interacting with the tag in at most q executions of the HB^+ protocol (possibly concurrently), running in time t, and achieving $\mathsf{Adv}_{\mathcal{A},\mathsf{HB}^+}^{active}(\varepsilon, \mathsf{l}, \mathsf{u}, n) \geq \delta$. Then there exists an algorithm D making $q \cdot n$ oracle queries, running in time $O(t)$, and such that*

$$\left| \Pr\left[\mathbf{s} \leftarrow \{0,1\}^k : D^{A_{\mathbf{s},\varepsilon}}(1^k) = 1 \right] - \Pr\left[D^{U_{k+1}}(1^k) = 1 \right] \right|$$

$$\geq \left(\frac{\delta + \delta_{\varepsilon,\mathsf{l},\mathsf{u},n}^*}{2} \right)^3 - \frac{2^n}{2^k} - 2^{-n} \cdot \sum_{i=0}^{2\mathsf{u}} \binom{n}{i}.$$

Asymptotically, for any $\varepsilon < \frac{1}{4}$ and appropriate choice of $n, \mathsf{l}, \mathsf{u}$ the last two terms of the above expression (and also ε_c) are negligible. We thus conclude that the HB^+ protocol is secure (for appropriate choice of $n, \mathsf{l}, \mathsf{u}$) assuming the hardness of the LPN_ε problem.

Proof. Algorithm D, given access to an oracle returning $(k+1)$-bit strings $(\mathbf{b}, \bar{z})$, proceeds as follows:

1. D chooses $\mathbf{s}_2 \in \{0,1\}^k$ uniformly at random. Then, it runs the first phase of $\mathcal{A}$. To simulate a basic authentication step, D does the following: it obtains a sample $(\mathbf{b}, \bar{z})$ from its oracle and sends $\mathbf{b}$ as the initial message. $\mathcal{A}$ replies with a challenge $\mathbf{a}$, and then D responds with $z = \bar{z} \oplus \langle \mathbf{s}_2, \mathbf{a} \rangle$. Note that since D does not rewind $\mathcal{A}$ here, there is no difficulty in simulating parallel executions of the basic authentication step (i.e., as part of an execution of the "full" HB^+ protocol) or concurrent executions of the HB^+ protocol.

2. When $\mathcal{A}$ is ready for the second phase of the HB^+ protocol, $\mathcal{A}$ sends an initial message $\mathbf{b}_1, \ldots, \mathbf{b}_n$ (we now explicitly look at the actual HB^+ protocol rather than focusing on a single basic authentication step). In response, D chooses random $\mathbf{a}_1^1, \ldots, \mathbf{a}_n^1 \in \{0,1\}^k$, sends these challenges to $\mathcal{A}$, and records $\mathcal{A}$'s response $z_1^1, \ldots, z_n^1$. Then D rewinds $\mathcal{A}$, chooses random $\mathbf{a}_1^2, \ldots, \mathbf{a}_n^2 \in \{0,1\}^k$, sends these to $\mathcal{A}$, and records $\mathcal{A}$'s response $z_1^2, \ldots, z_n^2$.

3. Let $z_i^\oplus := z_i^1 \oplus z_i^2$ and set $Z^\oplus \overset{\text{def}}{=} (z_1^\oplus, \ldots, z_n^\oplus)$. Let $\hat{\mathbf{a}}_i = \mathbf{a}_i^1 \oplus \mathbf{a}_i^2$ and $\hat{z}_i = \langle \mathbf{s}_2, \hat{\mathbf{a}}_i \rangle$, and set $\hat{Z} \overset{\text{def}}{=} (\hat{z}_1, \ldots, \hat{z}_n)$. D outputs 1 iff $Z^\oplus$ and $\hat{Z}$ differ in at most $2\mathsf{u}$ entries.

Let us analyze the behavior of D:

Case 1: Say D's oracle is U_{k+1}. In step 1, above, since $\bar{z}$ is uniformly distributed and independent of everything else, the answers z that D returns to $\mathcal{A}$ are uniformly distributed and independent of everything else. It follows that $\mathcal{A}$'s view throughout the experiment is independent of the secret $\mathbf{s}_2$ chosen by D.

The $\{\hat{\mathbf{a}}_i\}_{i=1}^n$ are uniformly and independently distributed, and so except with probability $\frac{2^n}{2^k}$ they are linearly independent and non-zero (cf. the claim proved below). Assuming this to be the case, $\hat{Z}$ is uniformly distributed over $\{0,1\}^n$

from the point of view of $\mathcal{A}$. But then the probability that $Z^\oplus$ and $\hat{Z}$ differ in at most 2u entries is exactly $2^{-n} \cdot \sum_{i=0}^{2u} \binom{n}{i}$. We conclude that D outputs 1 in this case with probability at most $\frac{2^n}{2^k} + 2^{-n} \cdot \sum_{i=0}^{2u} \binom{n}{i}$.

Case 2: Say D's oracle is $A_{\mathbf{s}_1,\varepsilon}$ for randomly-chosen $\mathbf{s}_1$. In this case, D provides a perfect simulation for the first phase of $\mathcal{A}$. By a standard averaging argument, with probability at least $\hat{\delta} \stackrel{\text{def}}{=} \frac{\delta + \delta^*_{\varepsilon,l,u,n}}{2}$ over the randomness used in the first phase of $\mathcal{A}$ (which includes the keys $\mathbf{s}_1, \mathbf{s}_2$, the randomness of $\mathcal{A}$, and the randomness used in responding to $\mathcal{A}$'s queries) the probability (over random challenges $\mathbf{a}_1, \ldots, \mathbf{a}_n$ sent by the tag reader in the second phase) that $\mathcal{A}$ successfully impersonates the tag in the second phase is at least $\hat{\delta}$. Assume this is the case. Then the probability that $\mathcal{A}$ successfully responds to both sets of queries $\mathbf{a}_1^1, \ldots, \mathbf{a}_n^1$ and $\mathbf{a}_1^2, \ldots, \mathbf{a}_n^2$ is at least $\hat{\delta}^2$. But this means that $(z_1^1, \ldots, z_n^1)$ differs in at most u entries from the "correct" answer

$$\text{ans}^1 \stackrel{\text{def}}{=} \left(\langle \mathbf{s}_1, \mathbf{b}_1 \rangle \oplus \langle \mathbf{s}_2, \mathbf{a}_1^1 \rangle, \ldots, \langle \mathbf{s}_1, \mathbf{b}_n \rangle \oplus \langle \mathbf{s}_2, \mathbf{a}_n^1 \rangle \right)$$

and also $(z_1^2, \ldots, z_n^2)$ differs in at most u entries from the "correct" answer

$$\text{ans}^2 \stackrel{\text{def}}{=} \left(\langle \mathbf{s}_1, \mathbf{b}_1 \rangle \oplus \langle \mathbf{s}_2, \mathbf{a}_1^2 \rangle, \ldots, \langle \mathbf{s}_1, \mathbf{b}_n \rangle \oplus \langle \mathbf{s}_2, \mathbf{a}_n^2 \rangle \right).$$

But then $(z_1^1, \ldots, z_n^1) \oplus (z_1^2, \ldots, z_n^2) = Z^\oplus$ differs in at most 2u entries from

$$\begin{aligned} \text{ans}^1 \oplus \text{ans}^2 &= \left(\langle \mathbf{s}_2, \mathbf{a}_1^1 \rangle \oplus \langle \mathbf{s}_2, \mathbf{a}_1^2 \rangle, \ldots, \langle \mathbf{s}_2, \mathbf{a}_n^1 \rangle \oplus \langle \mathbf{s}_2, \mathbf{a}_n^2 \rangle \right) \\ &= \left(\langle \mathbf{s}_2, (\mathbf{a}_1^1 \oplus \mathbf{a}_1^2) \rangle, \ldots, \langle \mathbf{s}_2, (\mathbf{a}_n^1 \oplus \mathbf{a}_n^2) \rangle \right) = \hat{Z}. \end{aligned}$$

We conclude that D outputs 1 in this case with probability at least $\hat{\delta} \cdot \hat{\delta}^2$. This completes the proof of the theorem. $\square$

The following technical claim, used above, is quite straightforward:

Claim. Assume n vectors $\mathbf{a}_1, \ldots, \mathbf{a}_n$ are chosen uniformly at random from $\{0,1\}^k$. The probability that these vectors are not linearly independent is less than $\frac{2^n}{2^k}$.

Proof. Say event Bad_i occurs if $\mathbf{a}_i$ is linearly dependent on the previous $i-1$ vectors chosen (for the case $i=1$ this is the event $\mathbf{a}_1 = 0^k$). Since the subspace spanned by $i-1$ vectors has size at most 2^{i-1}, the probability of Bad_i is at most $\frac{2^{i-1}}{2^k}$. Applying a union bound, we have:

$$\Pr \left[\bigvee_{i=1}^n \mathsf{Bad}_i \right] \leq 2^{-k} \cdot \sum_{i=0}^{n-1} 2^i < \frac{2^n}{2^k},$$

yielding the claim. $\square$

A typical range of parameters might be $k \approx 200$ and $n \approx 40\text{--}50$, so the $\frac{2^n}{2^k}$ term above is truly inconsequential.

4   Conclusions and Open Questions

The main technical results of this paper are the first rigorous proofs of (1) security of the HB^+ protocol against active attacks, even under parallel and concurrent executions; and (2) "hardness amplification" for the HB and HB^+ protocols as the number of iterations of the basic authentication step increases. Our proofs are also the first to explicitly take into account the non-zero completeness error and the impact this has on the security of the protocol as a whole.

We believe our proofs are remarkably simple, and view this as an additional contribution of this work (rather than as a drawback!). Indeed, we expect there will be further applications of Lemma 1 to the analysis of other cryptographic constructions based on the LPN problem, and hope this paper inspires and aids others in exploring such applications.

It would be nice to improve the analysis (or propose new protocols) so as to obtain meaningful security guarantees even in the case $\frac{1}{4} \leq \varepsilon < \frac{1}{2}$. It would also be wonderful to improve the concrete security reductions obtained here, or to propose new protocols with tighter security reductions. (As we have mentioned, it is not clear whether the proofs provided here yield sufficiently-high security for practically-efficient settings of the parameters.) As one possible approach toward this goal, one can imagine changing the HB/HB^+ protocols so that the tag always introduces *at most* $\varepsilon \cdot n$ errors, rather than introducing errors in each of the n iterations with independent probability ε.[7] (A related idea, in a different context, was explored in [5]; their analysis does not seem to apply to our setting.) This would give protocols with perfect completeness, and would help improve the concrete security bounds as well since the upper bound u could be set to exactly $\varepsilon \cdot n$ and the "problem" mentioned in footnote 5 would also go away. On the other hand it is not clear what can be said of the hardness of the natural variant of the LPN problem such protocols would be based on.

It would also be very interesting to see an efficient protocol based on the LPN problem that is provably resistant to man-in-the-middle attacks.

References

1. Associated Press. "Geeks Flex Hacker Muscles at Defcon." Article appeared Aug. 2, 2005 on CNN.com.
2. M. Bellare, M. Fischlin, S. Goldwasser, and S. Micali. Identification Protocols Secure against Reset Attacks. *Adv. in Cryptology — Eurocrypt 2001*, LNCS vol. 2045, Springer-Verlag, pp. 495–511, 2001.
3. M. Bellare, R. Impagliazzo, and M. Naor. Does Parallel Repetition Lower the Error in Computationally-Sound Protocols? *38th IEEE Symposium on Foundations of Computer Science*, IEEE, pp. 374–383, 1997.
4. E.R. Berlekamp, R.J. McEliece, and H.C.A. van Tilborg. On the Inherent Intractability of Certain Coding Problems. *IEEE Trans. Info. Theory* 24: 384–386, 1978.

---

[7] Note that introducing *exactly* $\varepsilon \cdot n$ errors in the n iterations is insecure.

5. A. Blum, M. Furst, M. Kearns, and R. Lipton. Cryptographic Primitives Based on Hard Learning Problems. *Adv. in Cryptology — Crypto '93*, LNCS vol. 773, Springer-Verlag, pp. 278–291, 1994.
6. A. Blum, A. Kalai, and H. Wasserman. Noise-Tolerant Learning, the Parity Problem, and the Statistical Query Model. *J. ACM* 50(4): 506–519, 2003.
7. R. Canetti, S. Halevi, and M. Steiner. Hardness Amplification of Weakly Verifiable Puzzles. *2nd Theory of Cryptography Conference (TCC 2005)*, LNCS vol. 3378, Springer-Verlag, pp. 17–33, 2005.
8. R. Canetti, J. Kilian, E. Petrank, and A. Rosen. Black-Box Concurrent Zero-Knowledge Requires (Almost) Logarithmically Many Rounds. *SIAM J. Computing* 32(1): 1–47, 2002.
9. F. Chabaud. On the Security of Some Cryptosystems Based on Error-Correcting Codes. *Adv. in Cryptology — Eurocrypt '94*, LNCS vol. 950, Springer-Verlag, pp. 131–139, 1995.
10. W. Diffie and M. Hellman. New Directions in Cryptography. *IEEE Trans. Info. Theory* 22(6): 644–654 (1976).
11. U. Feige and A. Shamir. Witness Indistinguishability and Witness Hiding Protocols. *22nd ACM Symposium on Theory of Computing*, ACM, pp. 416–426, 1990.
12. H. Gilbert, M. Robshaw, and H. Silbert. An Active Attack against HB$^+$ — a Provably Secure Lightweight Authentication Protocol. Available at http://eprint.iacr.org/2005/237
13. O. Goldreich. *Modern Cryptography, Probabilistic Proofs, and Pseudorandomness.* Springer-Verlag, 1998.
14. O. Goldreich and H. Krawczyk. On the Composition of Zero-Knowledge Proof Systems. *SIAM J. Computing* 25(1): 169–192, 1996.
15. O. Goldreich, N. Nisan, and A. Wigderson. On Yao's XOR-Lemma. Available at http://eccc.uni-trier.de/eccc-reports/1995/TR95-050/
16. O. Goldreich and Y. Oren. Definitions and Properties of Zero-Knowledge Proof Systems. *J. Cryptology* 7(1): 1–32, 1994.
17. J. Håstad. Some Optimal Inapproximability Results. *J. ACM* 48(4): 798–859, 2001.
18. N. Hopper and M. Blum. A Secure Human-Computer Authentication Scheme. Technical Report CMU-CS-00-139, Carnegie Mellon University, 2000.
19. N. Hopper and M. Blum. Secure Human Identification Protocols. *Adv. in Cryptology — Asiacrypt 2001*, LNCS vol. 2248, pp. 52–66, 2001.
20. A. Juels and S. Weis. Authenticating Pervasive Devices with Human Protocols. *Adv. in Cryptology — Crypto 2005*, LNCS vol. 3621, Springer-Verlag, pp. 293–308, 2005. Updated version available at: http://www.rsasecurity.com/rsalabs/staff/bios/ajuels/publications/pdfs/lpn.pdf
21. M. Kearns. Efficient Noise-Tolerant Learning from Statistical Queries. *J. ACM* 45(6): 983–1006, 1998.
22. Z. Kfir and A. Wool. Picking Virtual Pockets using Relay Attacks on Contactless Smartcard Systems. Available at http://eprint.iacr.org/2005/052
23. I. Kirschenbaum and A. Wool. How to Build a Low-Cost, Extended-Range RFID Skimmer. Available at http://eprint.iacr.org/2006/054
24. R. Raz. A Parallel Repetition Theorem. *SIAM J. Computing* 27(3): 763–803, 1998.
25. O. Regev. On Lattices, Learning with Errors, Random Linear Codes, and Cryptography. *37th ACM Symposium on Theory of Computing*, ACM, pp. 84–93, 2005.
26. A. C.-C. Yao. Theory and Applications of Trapdoor Functions. *23rd IEEE Symposium on Foundations of Computer Science*, IEEE, pp. 80–91, 1982.

Polling with Physical Envelopes: A Rigorous Analysis of a Human-Centric Protocol[*]

Tal Moran[1] and Moni Naor[1,**]

Department of Computer Science and Applied Mathematics,
Weizmann Institute of Science, Rehovot, Israel

Abstract. We propose simple, realistic protocols for polling that allow the responder to plausibly repudiate his response, while at the same time allow accurate statistical analysis of poll results. The protocols use simple physical objects (envelopes or scratch-off cards) and can be performed without the aid of computers. One of the main innovations of this work is the use of techniques from theoretical cryptography to rigorously prove the security of a realistic, physical protocol. We show that, given a few properties of physical envelopes, the protocols are unconditionally secure in the universal composability framework.

1 Introduction

In the past few years, a lot of attention has been given to the design and analysis of electronic voting schemes. Constructing a protocol that meets all (or even most) of the criteria expected from a voting scheme is generally considered to be a tough problem. The complexity of current protocols (in terms of how difficult it is to describe the protocol to a layperson) reflects this fact. A slightly easier problem, which has not been investigated as extensively, is that of polling schemes.

Polling schemes are closely related to voting, but usually have slightly less exacting requirements. In a polling scheme the purpose of the pollster is to get a good statistical profile of the responses, however some degree of error is admissible. Unlike voting, absolute secrecy is generally not a requirement for polling, but some degree of response privacy is often necessary to ensure respondents' cooperation.

The issue of privacy arises because polls often contain questions whose answers may be incriminating or stigmatizing (e.g., questions on immigration status, drug use, religion or political beliefs). Even if promised that the results of the poll will be used anonymously, the accuracy of the poll is strongly linked to the trust responders place in the pollster. A useful rule of thumb for polling sensitive questions is "better privacy implies better data": the more respondents trust that their responses cannot be used against them, the likelier they are to

---

[*] This work was partially supported by the Minerva Foundation.
[**] Incumbent of the Judith Kleeman Professorial Chair.

S. Vaudenay (Ed.): EUROCRYPT 2006, LNCS 4004, pp. 88–108, 2006.

answer truthfully. Using polling techniques that clearly give privacy guarantees can significantly increase the accuracy of a poll.

A well-known method for use in these situations is the "randomized response technique" (RRT), introduced by Warner [25]. Roughly, Warner's idea was to tell responders to lie with some fixed, predetermined, probability (e.g., roll a die and lie whenever the die shows one or two). As the probability of a truthful result is known exactly, statistical analysis of the results is still possible[1], but an individual answer is always plausibly deniable (the respondent can always claim the die came up one).

Unfortunately, in some cases this method causes its own problems. In pre-election polls, for example, responders have a strong incentive to always tell the truth, ignoring the die (since the results of the polls are believed to affect the outcome of the elections). In this case, the statistical analysis will give the cheating responders more weight than the honest responders. Ambainis, Jakobsson and Lipmaa [1] proposed the "Cryptographic Randomized Response Technique" to deal with this problem. Their paper contains a number of different protocols that prevent malicious responders from biasing the results of the poll while preserving the deniability of the randomized response protocol. Unlike Warner's original RRT, however, the CRRT protocols are too complex to be implemented in practice without the aid of computers. Since the main problem with polling is the responders' lack of trust in the pollsters, this limitation makes the protocols of [1] unsuitable in most instances.

The problem of trust in complex protocols is not a new one, and actually exists on two levels. The first is that the protocol itself may be hard to understand, and its security may not be evident to the layman (even though it may be formally proved). The second is that the computers and operating system actually implementing the protocol may not be trusted (even though the protocol itself is). This problem is more acute than the first. Even for an expert, it is very difficult to verify that a computer implementation of a complex protocol is correct.

Ideally, we would like to design protocols that are simple enough to grasp intuitively and can also be implemented transparently (so that the user can follow the steps and verify that they are correct).

1.1   Our Results

In this paper we propose two very simple protocols for cryptographic randomized response polls, based on *tamper-evident seals* (introduced in a previous paper by the authors [18]). A tamper-evident seal is a cryptographic primitive that captures the properties of a sealed envelope: while the envelope is sealed, it is

---

[1]   For instance, suppose $p > \frac{1}{2}$ is the probability of a truthful response, n is the total number of responses, x is the number of responders who actually belong in the "yes" category and R is the random variable counting the number of "yes" responses. R is the sum of n independent indicator random varables, so R is a good estimation for $E(R) = px + (1-p)(n-x) = x(2p-1) + n(1-p)$. Therefore, given R, we can accurately estimate the actual number of "yes" responders: $x = \frac{E(R) - n(1-p)}{2p-1}$.

impossible to tell what's inside, but if the seal is broken the envelope cannot be resealed (so any tampering is evident). In fact, our CRRT protocols are meant to be implemented using physical envelopes (or scratch-off cards) rather than computers. Since the properties of physical envelopes are intuitively understood, even by a layman, it is easy to verify that the implementation is correct.

The second important contribution of this paper, differentiating it from previous works concerning human-implementable protocols, is that we give a formal definition and a rigorous proof of security for the protocols. The security is unconditional: it relies only on the physical tamper-evidence properties of the envelopes, not on any computational assumption. Furthermore, we show that the protocols are "universally composable" (as defined by Canetti [3]). This is a very strong notion of security that implies, via Canetti's Composition Theorem, that the security guarantees hold even under general concurrent composition

Our protocols implement a relaxed version of CRRT (called *weakly secure* in [1]). We also give an inefficient strong CRRT protocol (that requires a large number of rounds), and give impossibility results and lower bounds for strong CRRT protocols with a certain range of parameters (based on Cleve's lower bound for coin flipping [8]). These suggest that constructing a strong CRRT protocol using scratch-off cards may be difficult (or even impossible if we require a constant number of rounds).

1.2 Related Work

Randomized Response Technique. The randomized response technique for polling was first introduced in 1965 [25]. Since then many variations have been proposed (a survey can be found in [6]). Most of these are attempts to improve or change the statistical properties of the poll results (e.g., decreasing the variance), or changing the presentation of the protocol to emphasize the privacy guarantee (e.g., instead of lying, tell the responders to answer a completely unrelated question). A fairly recent example is the "Three Card Method" [14], developed for the United States Government Accountability Office (GAO) in order to estimate the size of the illegal resident population. None of these methods address the case where the responders maliciously attempt to bias the results.

To the best of our knowledge, the first polling protocol dealing explicitly with malicious bias was given by Kikuchi, Akiyama, Nakamura and Gobioff. [17], who proposed to use the protocol for voting (the protocol described is a randomized response technique, although the authors do not appear to have been aware of the previous research on the subject). Their protocol is still subject to malicious bias using a "premature halting" attack (this is equivalent to the attack on the RRT protocol in which the responder rolls a die but refuses to answer if result of the die is not to his liking). A more comprehensive treatment, as well as a formal definition of cryptographic randomized response, was given by Ambainis et al. [1]. In their paper, Ambainis et al. also give a protocol for *Strong* CRRT, in which the premature halting attack is impossible. In both the papers [17,1], the protocols are based on cryptographic assumptions and require computers to implement.

Independently of this work, Stamm and Jakobsson show how to implement the protocol of [1] using playing cards [24]. They consider this implementation only as a visualization tool. However, if we substitute envelopes for playing cards (and add a verification step), this protocol gives a Responder-Immune protocol (having some similarities to the one described in Section 3.2).

Deniable and Receipt-Free Protocols. The issues of deniability and coercion have been extensively studied in the literature (some of the early papers in this area are [2,22,4,5,15]). There are a number of different definitions of what it means for a protocol to be *deniable*. Common to all of them is that they protect against an adversary that attacks actively only *after* the protocol execution: in particular, this allows the parties to lie about their random coins. Receipt-Free protocols provide a stronger notion of security: they guarantee that even if a party is actively colluding with the adversary, the adversary should have no verifiable information about which input they used. Our notion of "plausible deniability" is weaker than both "traditional" deniability and receipt-freeness, in that we allow the adversary to gain some information about the input. However, as in receipt-freeness, we consider an adversary that is active before and during the protocol, not just afterwards.

Secure Protocols Using "Real" Objects. The idea of using real objects to provide security predates cryptography: people have been using seals, locks and envelopes for much of history. Using real objects to implement protocols that use them in non-obvious ways is a newer notion. Fagin, Naor and Winkler [16] propose protocols for comparing secret information that use various objects, from paper cups to the telephone system. In a more jocular tone, Naor, Naor and Reingold [19] propose a protocol that provides a "zero knowledge proof of knowledge" of the correct answer to the children's puzzle "Where's Waldo" using "low-tech devices" (e.g., a large newspaper and scissors). In all these works the security assumptions and definitions are informal or unstated. Crépeau and Kilian [10] show how to use a deck of cards to play "discreet" solitary games (these involve hiding information from yourself). Their model is formally defined, however it is not malicious; the solitary player is assumed to be honest but curious.

A related way of using real objects is as aids in performing a "standard" calculation. Examples in this category include Schneier's "Solitaire" cipher [23] (implemented using a pack of cards), and the "Visual Cryptography" of Naor and Shamir [21] (which uses the human visual system to perform some basic operations on images). The principles of Visual Cryptography form the basis for some more complex protocols, such as the "Visual Authentication" protocol of Naor and Pinkas [20], and Chaum's human verifiable voting system [7].

Tamper-Evident Seals. This work can be viewed as a continuation of a previous work by the authors on tamper-evident seals [18]. In [18], we studied the possibility of implementing basic cryptographic primitives using different variants of physical, tamper-evident seals. In the current work we focus on their use in realistic cryptographic applications, rather than theoretical constructs (for instance, there is a very sharp limit on the number of rounds and the number of envelopes that can be used in a protocol that we expect to be practical for humans). We limit ourselves to the "distinguishable envelope" (DE) model, as this model has a number of intuitive physical embodiments, while at the same time is powerful enough, in theory, to implement many useful protocols[2] (an informal description of this model is given in Section 2.3; for a formal definition see [18]).

Overview of Paper. In Section 2, we give formal definitions of the functionalities we would like to realize and the assumptions we make about the humans implementing the protocols. Section 3 gives an informal description of the CRRT protocols. In Section 4, we show how to amplify a weak CRRT protocol in order to construct a strong CRRT protocol, and give some impossibility results and lower bounds for strong CRRT protocols. Finally, a discussion and some open problems appear in Section 5.

The formal protocol specification and proof of security for our Pollster-Immune CRRT protocol appears in Appendix A. Due to space constraints, the complete specifications and formal proofs for the other protocols will appear only in the full version of this paper.

2   The Model

Ideal Functionalities. Many two-party functionalities are easy to implement using a trusted third party that follows pre-agreed rules. In proving that a two-party protocol is secure, we often want to say that it behaves "as if it were performed using the trusted third party". The "Universal Composability" framework, defined by Canetti [3], is a formalization of this idea. In the UC model, the trusted third party is called the *ideal functionality*. If every attack against the protocol can also be carried out against the ideal functionality, we say the protocol realizes the functionality. Canetti's Composition Theorem says that *any* protocol that is secure using the ideal functionality, will remain secure if we replace calls to the ideal functionality with executions of the protocol.

Defining the security guarantees of our protocols as ideal functionalities has an additional advantage as well: it is usually easier to understand what it means for a protocol to satisfy a definition in this form than a definition given as a list of properties. Below, we describe the properties we wish to have in a CRRT protocol, and give formal definitions in the form of ideal functionalities.

2.1   Cryptographic Randomized Response

A randomized response protocol involves two parties, a pollster and a responder. The responder has a secret input bit b (this is the true response to the poll question). In the ideal case, the pollster learns a bit c, which is equal to b with probability p (p is known to the pollster) and to $1 - b$ with probability $1 - p$.

---

[2] Although the "indistinguishable envelope model" (also defined in [18]) is stronger (e.g., oblivious transfer *is* possible in this model), it seems to be very hard to devise a secure, physical realization of this functionality.

Since p is known to the pollster, the distribution of responders' secret inputs can be easily estimated from the distribution of the pollster's outputs.

The essential property we require of a Randomized Response protocol is *plausible deniability*: A responder should be able to claim that, with reasonable probability, the bit learned by the pollster is not the secret bit b. This should be the case even if the pollster maliciously deviates from the protocol.

A *Cryptographic* Randomized Response protocol is a Randomized Response protocol that satisfies an additional requirement, *bounded bias*: The probability that $c = b$ must be at most p, even if the responder maliciously deviates from the protocol. The bounded bias requirement ensures that malicious responders cannot bias the results of the poll (other than by changing their own vote). Note that even in the ideal case, a responder can always choose any bias p' between p and $1 - p$, by randomly choosing whether to vote b or $1 - b$ (with the appropriate probability).

Strong p-CRRT. In a *strong* CRRT protocol, both the deniability and bounded bias requirements are satisfied. Formally, this functionality has a single command:

Vote b. The issuer of this command is the responder. On receiving this command the functionality tosses a weighted coin c, such that $c = 0$ with probability p. It then outputs $b \oplus c$ to the pollster and the adversary.

Unfortunately, we do not know how to construct a practical strong CRRT protocol that can be implemented by humans. In Section 4, we present evidence to suggest that finding such a protocol may be hard (although we do show an impractical strong CRRT protocol, that requires a large number of rounds). The protocols we propose satisfy relaxed conditions: The first protocol is immune to malicious pollsters (it is equivalent to strong CRRT if the pollster is honest), while the second is immune to malicious responders (it is equivalent to strong CRRT if the responder is honest).

Pollster-Immune p-CRRT (Adapted from Weak CRRT in [1]). This is a weakened version of CRRT, where a malicious pollster cannot learn more than an honest pollster about the responder's secret bit. A malicious responder can bias the result by deviating from the protocol (halting early). A cheating responder will be caught with fixed probability, however, so the pollster can accurately estimate the number of responders who are cheating (and thus bound the resulting bias). When the pollster catches the responder cheating, it outputs $\boxtimes$ instead of its usual output. Formally, the ideal functionality accepts the following commands:

Query. The issuer of this command is the pollster, the other party is the responder. The functionality ignores all commands until it receives this one. On receiving this command the functionality chooses a uniformly random bit r and a bit v, such that $v = 1$ with probability $2p - 1$. If the responder is corrupted, the functionality then sends both bits to the adversary.

Vote b. On receiving this command from the responder, the functionality checks
whether $v = 1$. If so, it outputs b to the pollster, otherwise it outputs r to
the pollster.

Halt. This command captures the responder's ability to cheat. On receiving
this command from a corrupt responder, the functionality outputs ⊠ to the
pollster and halts.

The functionality is slightly more complex (and a little weaker) than would
appear to be necessary, and this requires explanation. Ideally, the functionality
should function as follows: the responder casts her vote, and is notified of the
actual bit the pollster would receive. The responder then has the option to halt
(and prevent the pollster from learning the bit). Our protocol gives the corrupt
responder a little more power: the responder first learns whether the pollster will
receive the bit sent by the responder, or whether the pollster will receive a bit
fixed in advance (regardless of what the responder sends). The responder can
then plan her actions based on this information. The functionality we describe
is the one that is actually realized by our protocol (for $p = \frac{3}{4}$).

Responder-Immune p-CRRT. In this weakened version of CRRT, malicious
responders cannot bias the results more than honest responders, but a malicious
pollster can learn the responder's secret bit. In this case, however, the responder
will discover that the pollster is cheating. When the responder catches the pollster cheating, it outputs ⊠ to signify this. The functionality accepts the following
commands:

Vote b. The issuer of this command is the responder. On receiving this
command the functionality tosses a weighted coin c, such that $c = 0$ with
probability p. It then outputs $b \oplus c$ to the pollster and adversary.

Reveal. The command may only be sent by a corrupt pollster *after* the Vote
command was issued by the responder. On receiving this command, the
functionality outputs b to the adversary and ⊠ to the responder.

Test x. The command may only be sent by a corrupt pollster, after the Vote
command was issued by the responder. On receiving this command:
  - if $x = b$, then with prob. $\frac{1}{2}$ it outputs b to the adversary and ⊠ to the
    responder, and with prob. $\frac{1}{2}$ it outputs $\perp$ to the adversary (and nothing
    to the responder).
  - if $x = 1 - b$ the functionality outputs $\perp$ to the adversary (and nothing
    to the responder).

Ideally, we would like to realize responder-immune CRRT without the **Test**
command. Our protocol realizes this slightly weaker functionality (for $p = \frac{2}{3}$).
It may appear that a corrupt pollster can cheat without being detected using
the **Test** command. However, for any corrupt pollster strategy, if we condition
on the pollster's cheating remaining undetected, the pollster gains no additional
information about the responder's choice (since in that case the response to the
Test command is always $\perp$).

2.2 Modelling Humans

The protocols introduced in this paper are meant to be implemented by humans. To formally prove security properties of the protocols, it is important to make explicit the abilities and limitations we expect from humans.

Following Instructions. The most basic assumption we make about the parties participating in the protocol is that an honest party will be able to follow the instructions of the protocol correctly. While this requirement is clearly reasonable for computers, it may not be so easy to achieve with humans (e.g., one of the problems encountered with the original randomized response technique is that the responders sometimes had difficulty understanding what they were supposed to do). The ability to follow instructions depends on the complexity of the protocol (although this is a subjective measure, and hard to quantify). Our protocols are secure and correct only assuming the honest parties are actually following the protocol. Unfortunately, we do not know how to predict whether this assumption actually holds for a specific protocol without "real" experimental data.

Random Choice. Our protocols require the honest parties to make random choices. Choosing a truly random bit may be very difficult for a human (in fact, even physically tossing a coin has about 0.51 probability of landing on the side it started on [13]). For the purposes of our analysis, we assume that whenever we require a party to make a random choice it is uniformly random. In practice, a random choice may be implemented using simple physical means (e.g., flipping a coin or rolling a die). In practice, the slight bias introduced by physical coin flipping will not have a large effect on the correctness or privacy of our protocols.

Non-requirements. Unlike many protocols involving humans, we do not assume any additional capabilities beyond those described above. We don't require parties to forget information they have learned, or to perform actions obliviously (e.g., shuffle a deck without knowing what the permutation was). Of particular note, we don't require the parties to watch each other during the protocol: this means the protocols can be conducted by mail.

2.3 Distinguishable Envelopes

Our CRRT protocols require a physicial assumption: tamper-evident envelopes or scratch-off cards. Formally, we model these by an ideal functionality we call "Distinguishable Envelopes" (defined in [18]). Loosely speaking, a distinguishable envelope is an envelope in which a message can be sealed. Anyone can open the envelope (and read the message), but the broken seal will be evident to anyone looking at the envelope.

3 An Informal Presentation of the Protocols

It is tempting to try to base a CRRT protocol on oblivious transfer (OT), since if the responder does not learn what the pollster's result is, it may be hard to influence it (in fact, one of the protocols in [1] is based on OT). However, OT is impossible in the DE model [18]. As we show in Section 4.1, this proof implies that in any CRRT protocol using distinguishable envelopes, the responder *must* learn a lot about the pollster's result. In both our protocols, the responder gets complete information about the final result.

To make the presentation more concrete, suppose the poll question is "do you eat your veggies?". Clearly, no one would like to admit that they do not have a balanced diet. On the other hand, pressure groups such as the "People for the Ethical Treatment of Salad" have a political interest in biasing the results of the poll, making it a good candidate for CRRT.

3.1 Pollster-Immune CRRT

This protocol can be implemented with pre-printed scratch-off cards: The responder is given a scratch-off card with four scratchable "bubbles", arranged in two rows of two bubbles each. In each row, the word "Yes" is hidden under one bubble and the word "No" under the other (the responder doesn't know which is which). The responder scratches a random bubble in each row. Suppose the responder doesn't eat her veggies. If one of the rows (or both) show the word "No", she "wins" (and the pollster will count the response as expressing dislike of vegetables). If both bubbles show "Yes", she "loses" (and the pollster will count the response as expressing a taste for salad). In any case, before returning the card to the pollster, the responder "eliminates" the row that shows the unfavored answer by scratching the entire row (she picks one of the rows at random if both rows show the same answer) Thus, as long as the responder follows the protocol,

Fig. 1. Sample execution of pollster-immune protocol

the pollster receives a card that has one "eliminated" (entirely scratched) row and one row showing the result he will count. An example of protocol execution appears in Figure 1.

Security Intuition. Note that in exactly $\frac{3}{4}$ of the cases the counted result will match the responder's intended result. Moreover, without invalidating the entire card, the responder cannot succeed with higher probability. On the other hand, this provides the responder with plausible deniability: she can always claim both rows were "bad", and so the result didn't reflect her wishes. Because the pollster doesn't know which were the two bubbles that were scratched first, he cannot refute this claim. An important point is that plausible deniability is preserved even if the pollster attempts to cheat (this is what allows the responder to answer the poll accurately even when the pollster isn't trusted). Essentially, the only way the pollster can cheat without being unavoidably caught is to put the *same* answer under both bubbles in one of the rows. To get a feeling for why this doesn't help, write out the distribution of responses in all four cases (cheating/honest, Yes/No). It will be evident that the pollster does not get any additional information about the vote from cheating in this way.

On the other hand, the responder learns the result before the pollster, and can decide to quit if it's not to her liking (without revealing the result to the pollster). Since the pollster does not know the responder's outcome, this has the effect of biasing the result of the poll. However, by counting the number of prematurely halted protocol executions, the pollster can accurately estimate the *number* of cheating responders.

The formal protocol specification and proof appear in Appendix A.

Generalizing to Any Rational p. The protocol above realizes Pollster-Immune $\frac{3}{4}$-CRRT. In some cases we require a p-CRRT protocol for different values of p. In particular, if we need to repeat the poll, we need the basic protocol to have p closer to $\frac{1}{2}$ (in order to maintain the plausible deniability).

The following protocol will work for any rational $p = \frac{k}{n}$ (assume $k > \frac{1}{2}n$): As in the former protocol, the pollster generates two rows of bubbles. One row contains k "Yes" bubbles and $n - k$ "No" bubbles in random order (this row is the "Yes" row), and the other contains k "No" bubbles and $n - k$ "Yes" bubbles (this row is the "No" row). The rows are also in a random order. The responder's purpose is to find the row matching her choice. She begins by scratching a single bubble in each row. If both bubbles contain the same value, she "eliminates" a random row (by scratching it out completely). Otherwise, she "eliminates" the row that does not correspond to her choice. The pollster's output is the majority value in the row that was not eliminated. The probability that the pollster's output matches the responder's choice is exactly p.

Unfortunately, this protocol is completely secure only for a semi-honest pollster (one that correctly generates the scratch-off cards). A malicious pollster can cheat in two possible ways: he can replace one of the rows with an invalid row (one that does not contain exactly k "Yes" bubbles or exactly k "No" bubbles), or he can use two valid rows that have the same majority value (rather than opposite majority values). In both cases the pollster will gain additional information about the responder's choice. This means the protocol does not realize the ideal Pollster-Immune CRRT functionality.

If the pollster chooses to use an invalid row, he will be caught with probability at least $\frac{1}{2}(1-p)$ (since with this probability the responder will scratch identical bubbles in both rows, and choose to eliminate the invalid row). We can add "cheating detection" to the protocol to increase the probability of detecting this attack. In a protocol with cheating detection, the pollster gives the responder ℓ scratch-off cards rather than just one (each generated according to the basic protocol). The responder chooses one card to use as in the basic protocol. On each of the other cards, she scratches off a single row (chosen randomly), and verifies that it contains either exactly k "Yes" bubbles or exactly k "No" bubbles. She then returns all the cards to the pollster (this step is necessary to prevent the responder from increasing her chances by trying multiple cards until one gives the answer she wants). A pollster that cheats by using an invalid row will be caught with probability $1 - \frac{1}{\ell}$.

A malicious pollster can still cheat undetectably by using two valid rows with identical majorities. This gives only a small advantage, however, and in practice the protocol may still be useful when p is close to $\frac{1}{2}$.

3.2   Responder-Immune CRRT

The responder takes three envelopes (e.g., labelled "1", "2" and "3"), and places one card containing either "Yes" or "No" in each of the envelopes. If she would like to answer "No", she places a single "Yes" card in a random envelope, and one "No" card in each of the two remaining envelopes. She then seals the envelopes and gives them to the pollster (remembering which of the envelopes contained the "Yes" card).

The pollster chooses a random envelope and opens it, revealing the card to the responder. He then asks the responder to tell him which of the two remaining envelopes contains a card with the *opposite* answer. He opens that envelope as well. If the envelope does contain a card with the opposite answer, he records the answer on the first card as the response to the poll, and returns the third (unopened) envelope to the responder.

If both opened envelopes contain the same answer, it can only be because the responder cheated. In this case, the pollster opens the third envelope as well. If the third envelope contains the opposite answer, the pollster records the answer on the first card as the response to the poll. If, on the other hand, all three envelopes contain the same answer, the pollster rolls a die: A result of 1 to 4 (probability $\frac{2}{3}$) means he records the answer that appears in the envelopes, and a result of 5 or 6 means she records the opposite answer. An example of protocol execution (where both parties follow the protocol) appears in Figure 2.

Security Intuition. In this protocol, the responder gets her wish with probability at most $\frac{2}{3}$ no matter what she does. If she follows the protocol when putting the answers in the envelopes, the pollster will choose the envelope containing the other answer with probability $\frac{1}{3}$. If she tries to cheat by putting the same answer in all three envelopes, the pollster will roll a die and choose the opposite

Fig. 2. Sample execution of responder-immune protocol

answer with probability $\frac{1}{3}$. The pollster, on the other hand, can decide to open all three envelopes and thus discover the real answer favored by the responder. If he does this, however, the responder will see that the seal on the returned envelope was broken and know the pollster was cheating.

The pollster may also be able to cheat in an additional way: he can open two envelopes before telling the responder which envelope he opened, and hope that the responder will not require him to return an envelope that was already opened. This attack is what requires us to add the **Test** command to the functionality.

Implementation Notes. This protocol requires real envelopes (rather than scratch-off cards) to implement, since the responder must choose what to place in the envelopes (and we cannot assume the responder can create a scratch-off card). In general, tamper-evidence for envelopes may be hard to achieve (especially as the envelopes will most likely be provided by the pollster). In this protocol, however, the pollster's actions can be performed in full view of the responder, so any opening of the envelopes will be immediately evident. When this is the case, the responder can tell which envelope the pollster opened first, so the protocol actually realizes the stronger version of the Responder-Immune CRRT functionality (without the **Test** command).

If the penalty for a pollster caught cheating is large enough, the privacy guaranteed by this protocol, may be enough to convince responders to answer accurately in a real-world situation even with the weaker version of the functionality. This is because any pollster cheating that can possibly reveal additional information about the responder's choice carries with it a corresponding risk of detection.

Generalizing to Any Rational p. When the pollster's actions are performed in view of the responder (in particular, when the responder can see exactly which envelopes are opened by the pollster), this protocol has a straightforward generalization to any rational $p = \frac{k}{n}$, where $k > \frac{1}{2}n$: the responder uses n (rather than 3) envelopes, of which k contain her choice and $n - k$ contain its opposite. After the pollster chooses an envelope to open, the responder shows him $n - k$ envelopes that contain the opposite value.

Note that when this generalized protocol is performed by mail, it does *not* realize the ideal functionality defined in Section 2.1.

4   Strong CRRT Protocols

Ideally, we would like to have CRRT protocols that cannot be biased at all by malicious responders, while perfectly preserving the responder's deniability, even against malicious pollsters. Unfortunately, the protocols described in Section 3 do not quite achieve this. At the expense of increasing the number of rounds, we can get arbitrarily close to the Strong-CRRT functionality defined in Section 2.1.

Consider a protocol in which the pollster and responder perform the pollster-immune p-CRRT protocol r times, one after the other (with the responder using the same input each time). The pollster outputs the majority of the subprotocols' outputs. If the responder halts at any stage, the pollster uses uniformly random bits in place of the remaining outputs.

This protocol gives a corrupt responder at most $O(\frac{1}{\sqrt{r}})$ advantage over an honest responder. We give here only the intuition for why this is so: Clearly, if a corrupt responder wants to bias the result to some bit b, it is in her best interest to use b for all the inputs. Since the subprotocol securely realizes p-CRRT, the only additional advantage she can gain is by halting at some round i. However, halting affects the result only if the other $r - 1$ rounds were balanced (this is the only case in which the outcome of the ith round affects the majority). In the case where $p = \frac{1}{2}$, it is easy to see that the probability for this occurring is $O(\frac{1}{\sqrt{r}})$. However, the probability that $r - 1$ independent weighted coin flips are balanced is maximized when $p = \frac{1}{2}$. Thus, the additional advantage that can be gained by the adversary is at most $O(\frac{1}{\sqrt{r}})$.

The problem with the amplification protocol described above is that the probability that an honest responder will get the result she wants tends to 1 as the number of rounds grows, for any constant $p > \frac{1}{2}$. Therefore, to preserve plausible deniability we must use a p-CRRT protocol where p is very close to $\frac{1}{2}$, such as the protocol described in Section 3.1 that works for any rational p. This adds further complexity to the protocol (e.g., our generalized Pollster-Immune protocol requires $\Omega(\frac{1}{\epsilon})$ bubbles on the scratch-off card for $p = \frac{1}{2} + \epsilon$). This, this multi-round protocol is probably not feasible in practice.

4.1 Lower Bounds and Impossibility Results

In this section we attempt to show that constructing practical strong CRRT protocols is a difficult task. We do this by giving impossibility results and lower bounds for implementing subclasses of the strong CRRT functionality. We consider a generalization of the strong p-CRRT functionality defined in Section 2.1, which we call (p, q)-CRRT. The (p, q)-CRRT functionality can be described as follows:

Vote b. The issuer of this command is the responder. On receiving this command the functionality tosses a weighted coin c, such that $c = 0$ with probability p. It then outputs $b \oplus c$ to the pollster. The functionality supplies the responder with exactly enough additional information so that she can guess c with probability $q \geq p$.

In the definition of strong CRRT given in Section 2.1, we specify exactly how much information the pollster learns about the responder's choice, but leave completely undefined what a cheating responder can learn about the pollster's result. The (p, q)-CRRT functionality quantifies this information: in a (p, p)-CRRT, the responder does not gain any additional information (beyond her pre-existing knowledge that the pollster's result will equal her choice with probability p). In a $(p, 1)$-CRRT, the responder learns the pollster's result completely. We show that (p, p)-CRRT implies oblivious transfer (and is thus impossible in the DE model), while $(p, 1)$-CRRT implies strong coin-flipping (and thus we can lower-bound the number of rounds required for the protocol). For values of q close to p or close to 1, the same methods can still be used to show lower bounds.

(p, q)-CRRT When q Is Close to p. First, note that when $p = q$ we can view the (p, q)-CRRT functionality as a binary symmetric channel (BSC) with error probability $1 - p$. Crépeau and Kilian have shown that a protocol for Oblivious Transfer (OT) can be constructed based on any BSC [9]. However, it is impossible to implement OT in the Distinguishable Envelope (DE) model [18]. Therefore (p, p)-CRRT cannot be implemented in the DE model. It turns out that this is also true for any q close enough to p. This is because, essentially, the (p, q)-CRRT functionality is a $(1 - q, 1 - p)$-Passive Unfair Noisy Channel (PassiveUNC), as defined by Damgård, Kilian and Salvail [12]. A (γ, δ)-PassiveUNC is a BSC with error δ which provides the corrupt sender (or receiver) with additional information that brings his perceived error down to γ; (i.e., a corrupt sender can guess the bit received by the receiver with probability $1 - \gamma$, while an honest sender can guess this bit only with probability $1 - \delta$). For γ and δ that are close enough (the exact relation is rather complex), Damgård, Fehr, Morozov and Salvail [11] show that a (γ, δ)-PassiveUNC is sufficient to construct OT. For the same range of parameters, this implies that realizing (p, q)-CRRT is impossible in the DE model.

(p, q)-CRRT When q Is Close to 1. When $q = 1$, both the pollster and the responder learn the poll result together. A $(p, 1)$-CRRT can be used as a protocol for strongly fair coin flipping with bias $p - \frac{1}{2}$. In a strongly fair coin flipping protocol with bias ϵ, the bias of an honest party's output is at most ϵ regardless of the other party's actions — even if the other party aborts prematurely. If q is close to 1, we can still construct a coin flipping protocol, albeit without perfect consistency. The protocol works as before, except that the responder outputs his best guess for the pollster's output: both will output the same bit with probability q.

A result by Cleve [8] shows that even if all the adversary can do is halt prematurely (and must otherwise follow the protocol exactly), any r-round protocol in which honest parties agree on the output with probability $\frac{1}{2} + \epsilon$ can be biased by at least $\frac{\epsilon}{4r+1}$. Cleve's proof works by constructing $4r + 1$ adversaries, each of which corresponds to a particular round. An adversary corresponding to round i follows the protocol until it reaches round i. It then halts immediately, or after one extra round. The adversary's decision is based only on what the ourput of an honest player would be in the same situation, should the *other* party halt after this round. Cleve shows that the average bias achieved by these adversaries is $\frac{\epsilon}{4r+1}$, so at least one of them must achieve this bias. The same proof also works in the DE model, since all that is required is that the adversary be able to compute what it would output should the other player stop after it sends the messages (and envelopes) for the current round. This calculation may require a party to open some envelopes (the problem being that this might prevent the adversary from continuing to the next round). However, an honest player would be able to perform the calculation in the next round, after sending this round's envelopes, so it cannot require the adversary to open any envelopes that may be sent in the next round.

Cleve's lower bound shows that a (p, q)-CRRT protocol must have at least $\frac{q - \frac{1}{2}}{4(p - \frac{1}{2})} - \frac{1}{4}$ rounds. Since a protocol with a large number of rounds is impractical for humans to implement, this puts a lower bound on the bias p (finding a CRRT protocol with a small p is important if we want to be able to repeat the poll while still preserving plausible deniability).

This result also implies that it is impossible to construct a $(p, 1)$-CRRT protocol in which there is a clear separation between the responder's choice and the final output. That is, the following functionality, which we call *p-CRRT with confirmation*, is impossible to implement in the DE model:

Vote b. The issuer of this command is the responder. On receiving this command the functionality outputs "Ready?" to the pollster. When the pollster answers with "ok" the functionality tosses a weighted coin c, such that $c = 0$ with probability p. It then outputs $b \oplus c$ to the pollster and responder.

p-CRRT with confirmation is identical to $(p, 1)$-CRRT, except that the output isn't sent until the pollster is ready. The reason it is impossible to implement is that this functionality can be amplified by parallel repetition to give a strongly fair coin flipping protocol with arbitrarily small p. Since the amplification is in parallel, it does not increase the number of rounds required by the protocol, and thus contradicts Cleve's lower bound. Briefly, the amplified protocol works as follows: the responder chooses k inputs randomly, and sends each input to a separate (parallel) instance of p-CRRT with confirmation. The pollster waits until all the inputs have been sent (i.e., it receives the "Ready?" message from all the instances), then sends "ok" to all the instances. The final result will be the xor of the outputs of all the instances. Since the different instances act independently, the bias of the final result is exponentially small in k.

5   Discussion and Open Problems

Polling Protocols by Mail. The pollster-immune CRRT protocol requires only a single round; This makes it convenient to use in polls through the post (it only requires the poll to be sent to the responder, "filled out" and returned). The responder-immune protocol presents additional problems when used through the post. First, in this case the protocol realizes a slightly weaker functionality than in the face-to-face implementation. Second, it requires two rounds, and begins with the responder. This means, in effect, that it would require an extra half-round for the pollster to notify the responder about the existence of the poll. It would be interesting to find a one-round protocol for the responder-immune functionality as well. It may be useful, in this context, to differentiate between "information-only" communication (which can be conducted by phone or email), and transfer of physical objects such as envelopes (which require "real" mail).

Efficient Generalization to Arbitrary p. We describe efficient p-CRRT protocols for specific values of p: $p = \frac{3}{4}$ in the Pollster-Immune case, and $p = \frac{2}{3}$ in the Responder-Immune case. Our generalized protocols are not very efficient: for $p = \frac{1}{2} + \epsilon$ they require $\Omega(\frac{1}{\epsilon})$ envelopes. In a protocol meant to be implemented by humans, the efficiency of the protocol has great importance. It would be useful to find an efficient general protocol to approximate arbitrary values of p (e.g., logarithmic in the approximation error).

Side-Channel Attacks. The privacy of our protocols relies on the ability of the responder to secretly perform some actions. For instance, in the pollster-immune protocol we assume that the order in which the bubbles on the card were scratched remains secret. In practice, some implementations may be vulnerable to an attack on this assumption. For example, if the pollster uses a light-sensitive dye on the scratch-off cards that begins to darken when the coating is scratched off, he may be able to tell which of the bubbles was scratched first. Side-channel attacks are attacks on the model, not on the CRRT protocols themselves. As these attacks highlight, when implementing CRRT using a physical implementation of Distinguishable Envelopes, it is important to verify that this implementation actually does realize the required functionality.

Dealing with Human Limitations. Our protocols make two assumptions about the humans implementing them: that they can make random choices and that they can follow instructions. The former assumption can be relaxed: if the randomness "close to uniform" the security and privacy will suffer only slightly (furthermore, simple physical aids, such as coins or dice, make generating randomness much easier). The latter assumption is more critical; small deviations from the protocol can result in complete loss of privacy or security. Constructing protocols that are robust to human error could be very useful.

Practical Strong CRRT Protocols. As we discuss in Section 4.1, for a range of parameters p, q-CRRT is impossible, and for a different range of parameters it is impractical. For some very reasonable values, such as $\frac{3}{4}$-Strong CRRT, we can approximate the functionality using a large number of rounds, but do not know how to prove any lower bound on the number of rounds required. Closing this gap is an obvious open question. Alternatively, finding a physical model in which efficient Strong CRRT is possible is also an interesting direction.

Acknowledgements

We would like to thank Adi Shamir and Yossi Oren for pointing out possible side-channel attacks in the scratch-off card model, and the anonymous reviewers for many helpful comments.

References

1. A. Ambainis, M. Jakobsson, and H. Lipmaa. Cryptographic randomized response techniques. In *PKC '04*, volume 2947 of *LNCS*, pages 425–438, 2004.
2. J. Benaloh and D. Tuinstra. Receipt-free secret-ballot elections. In *STOC '94*, pages 544–553, 1994.
3. R. Canetti. Universally composable security: A new paradigm for cryptographic protocols. In *FOCS '01*, pages 136–145, 2001.
4. R. Canetti, C. Dwork, M. Naor, and R. Ostrovsky. Deniable encryption. In *CRYPTO '97*, volume 1294 of *LNCS*, pages 90–104, 1997.
5. R. Canetti and R. Gennaro. Incoercible multiparty computation. In *FOCS '96*, pages 504–513, 1996.
6. A. Chaudhuri and R. Mukerjee. *Randomized Response: Theory and Techniques*, volume 85. Marcel Dekker, 1988.
7. D. Chaum. E-voting: Secret-ballot receipts: True voter-verifiable elections. *IEEE Security & Privacy*, 2(1):38–47, Jan./Feb. 2004.
8. R. Cleve. Limits on the security of coin flips when half the processors are faulty. In *STOC '86*, pages 364–369, 1986.
9. C. Crépeau and J. Kilian. Achieving oblivious transfer using weakened security assumptions. In *FOCS '88*, pages 42–52, 1988.
10. C. Crépeau and J. Kilian. Discreet solitary games. In *CRYPTO '93*, volume 773 of *LNCS*, pages 319–330, 1994.
11. I. B. Damgård, S. Fehr, K. Morozov, and L. Salvail. Unfair noisy channels and oblivious transfer. In *TCC '04*, volume 2951 of *LNCS*, pages 355–373, 2004.
12. I. B. Damgård, J. Kilian, and L. Salvail. On the (im)possibility of basing oblivious transfer and bit commitment on weakened security assumptions. In *Eurocrypt '99*, volume 1592 of *LNCS*, pages 56–73, 1999.
13. P. Diaconis, S. Holmes, and R. Montgomery. Dynamical bias in the coin toss, 2004. http://www-stat.stanford.edu/~cgates/PERSI/papers/headswithJ.pdf.

14. J. A. Droitcour, E. M. Larson, and F. J. Scheuren. The three card method: Estimating sensitive survey items–with permanent anonymity of response. In *Proceedings of the American Statistical Association, Social Statistics Section [CD-ROM]*, 2001.
15. C. Dwork, M. Naor, and A. Sahai. Concurrent zero knowledge. In *STOC '98*, pages 409–418, New York, NY, USA, 1998. ACM Press.
16. R. Fagin, M. Naor, and P. Winkler. Comparing information without leaking it. *Commun. ACM*, 39(5):77–85, 1996.
17. H. Kikuchi, J. Akiyama, G. Nakamura, and H. Gobioff. Stochastic voting protocol to protect voters privacy. In *WIAPP '99*, pages 102–111, 1999.
18. T. Moran and M. Naor. Basing cryptographic protocols on tamper-evident seals. In *ICALP 2005*, volume 3580 of *LNCS*, pages 285–297, July 2005.
19. M. Naor, Y. Naor, and O. Reingold. Applied kid cryptography, Mar. 1999. http://www.wisdom.weizmann.ac.il/~naor/PAPERS/waldo.ps.
20. M. Naor and B. Pinkas. Visual authentication and identification. In *CRYPTO '97*, volume 1294 of *LNCS*, pages 322–336, 1997.
21. M. Naor and A. Shamir. Visual cryptography. In *Eurocrypt '94*, volume 950 of *LNCS*, pages 1–12, 1995.
22. K. Sako and J. Kilian. Receipt-free mix-type voting schemes. In *EUROCRYPT '95*, volume 921 of *LNCS*, pages 393–403, 1995.
23. B. Schneier. The solitaire encryption algorithm, 1999. http://www.schneier.com/solitaire.html.
24. S. Stamm and M. Jakobsson. Privacy-preserving polling using playing cards. Cryptology ePrint Archive, Report 2005/444, December 2005.
25. S. Warner. Randomized response: a survey technique for eliminating evasive answer bias. *Journal of the American Statistical Association*, pages 63–69, 1965.

A   A Pollster-Immune $\frac{3}{4}$-CRRT Protocol

A.1   Formal Specification

Let $\mathcal{P}$ be the pollster and $\mathcal{R}$ the responder. Denote $\mathcal{P}$'s random bits p_0, p_1 and $\mathcal{R}$'s random bits r_0, r_1, r_2.

1. To implement **Query:** $\mathcal{P}$ creates two pairs of envelopes, each pair containing a 0 and a 1. The first pair contains $(p_0, 1 - p_0)$ and the second $(p_1, 1 - p_1)$. $\mathcal{P}$ sends both pairs to $\mathcal{R}$.
2. To implement **Vote** b: $\mathcal{R}$ opens a random envelope from each pair (the index of the first envelope opened is given by r_0 and the second by r_1. Denote the values of the opened envelopes $x_0 = p_0 \oplus r_0$ and $x_1 = p_1 \oplus r_1$.
   (a) If $x_0 = x_1$ (i.e., both the opened values are equal), $\mathcal{R}$ chooses a random pair and opens the remaining envelope in that pair (the first pair if $r_2 = 0$ and the second if $r_2 = 1$).
   (b) If $x_0 \neq x_1$, $\mathcal{R}$ opens the remaining envelope in the pair whose open envelope is not equal to b.
   (c) In both cases, $\mathcal{R}$ verifies that the envelopes in the completely opened pair contain different values (i.e., that the pair is valid). If so, $\mathcal{R}$ then sends all four envelopes back to $\mathcal{P}$, otherwise $\mathcal{R}$ halts.

3. If $\mathcal{R}$ halted in the previous step, $\mathcal{P}$ outputs ⊠ and halts. Otherwise, $\mathcal{P}$ verifies that exactly three of the four envelopes received from $\mathcal{R}$ are open. If so, $\mathcal{P}$ outputs the contents of the open envelope in the pair that contains the sealed envelope. If not, $\mathcal{P}$ outputs ⊠.

A.2   Proof of Security

In this section we give the proof that the protocol securely realizes Pollster-Immune $\frac{3}{4}$-CRRT in the UC model. The proof follows the standard outline for a UC proof: we describe an ideal adversary, $\mathcal{I}$, that works in the ideal world by simulating a real adversary, $\mathcal{A}$ (given black-box access to $\mathcal{A}$), along with the envelope functionalities used to implement the protocol in the real world. We then show that no environment machine, $\mathcal{Z}$ (which is allowed to set the parties' inputs) can distinguish between the case that it is communicating with $\mathcal{A}$ in the real world, and the case where it is communicating with $\mathcal{I}$ in the ideal world (for a more in-depth explanation of the UC model, see [3]). We'll deal separately with the case when $\mathcal{A}$ corrupts $\mathcal{P}$ and when it corrupts $\mathcal{R}$ (since we assume the corruption occurs as a first step). The proof that the views of $\mathcal{Z}$ in the real and ideal worlds are identical is by exhaustive case analysis.

$\mathcal{A}$ Corrupts $\mathcal{P}$

1. $\mathcal{I}$ waits to receive c, the outcome of the poll from the ideal functionality. $\mathcal{I}$ now begins simulating $\mathcal{F}_{DE}$ and $\mathcal{R}$ (as if he were a real honest party). The simulation runs until $\mathcal{P}$ sends four envelopes as required by the protocol (up to this point $\mathcal{R}$ did not participate at all in the protocol).

2. If both pairs of envelopes are valid (contain a 0 and a 1), $\mathcal{I}$ chooses one of the pairs at random, and simulates opening the envelope in the pair that contains c and both envelopes in the other pair (there is an assignment to the random coins of $\mathcal{R}$ which would have this result in the real world). It then simulates the return of all four envelopes to $\mathcal{P}$.

3. If both pairs of envelopes are invalid, $\mathcal{I}$ simulates $\mathcal{R}$ halting (this would eventually happen in a real execution as well).

4. If exactly one pair of envelopes is invalid, denote the value in the invalid pair by z.
   (a) If $c = z$, $\mathcal{I}$ simulates opening both envelopes in the valid pair, and a random envelope in the invalid pair (depending on the random coins of $\mathcal{R}$, this is a possible result in the real world). It then simulates the return of all four envelopes to $\mathcal{P}$
   (b) If $c \neq z$, $\mathcal{I}$ simulates $\mathcal{R}$ halting (depending on the random coins of $\mathcal{R}$, this is also a possible result in the real world).

5. $\mathcal{I}$ continues the simulation until $\mathcal{A}$ halts.

Note that throughout the simulation, all simulated parties behave in a manner that is feasible in the real world as well. Thus, the only possible difference between the views of $\mathcal{Z}$ in the ideal and real worlds is the behavior of the simulated $\mathcal{R}$, which depends only on the contents of the four envelopes sent by $\mathcal{P}$ and the output of the ideal functionality (which in turn depends only on b). It is easy (albeit tedious) to go over all 32 combinations of envelopes and input, and verify that the distribution of $\mathcal{R}$'s output in both cases (the real and ideal worlds) are identical. We enumerate the basic cases below. All other cases are identical to one of the following by symmetry:

1. $\mathcal{A}$ sends two valid pairs of envelopes. Assume it sends $[(b, 1-b), (b, 1-b)]$ (the other combinations follow by symmetry). $\mathcal{I}$ returns the following distribution ("*" denotes a sealed envelope):
   (a) With probability $\frac{3}{4}$ ($c = b$) it selects uniformly from
   $$\{[(b, *), (b, 1 - b)], [(b, 1 - b), (b, *)]\}$$
   (b) With probability $\frac{1}{4}$ ($c \neq b$) it selects uniformly from
   $$\{[(*, 1 - b), (b, 1 - b)], [(b, 1 - b), (*, 1 - b)]\}$$
   In the real world, the order of envelopes opened by $\mathcal{R}$ would be distributed uniformly from one of the following sets (each with probability $\frac{1}{4}$):
   (a) $\{[(1, *), (3, 2)]\}$
   (b) $\{[(1, *), (2, 3)], [(1, 3), (2, *)]\}$
   (c) $\{[(3, 1), (2, *)]\}$
   (d) $\{[(3, 1), (*, 2)], [(*, 1), (3, 2)]\}$
   Note that the observed result is distributed identically in both cases.

2. $\mathcal{A}$ sends two invalid pairs of envelopes: in this case, in both the real and ideal worlds the adversary will see the responder halting with probability 1.

3. $\mathcal{A}$ sends one valid and one invalid pair of envelopes:
   (a) $\mathcal{A}$ sends $[(b, b), (b, 1 - b)]$ (the other case where the invalid pair matches b is symmetric). The distribution of the returned envelopes in the ideal world is:
      i. With probability $\frac{3}{4}$ ($c = b$) it selects uniformly from
      $$\{[(b, *), (b, 1 - b)], [(*, 1 - b), (b, 1 - b)]\}$$
      ii. With probability $\frac{1}{4}$ ($c \neq b$) it halts.
      In the real world, the order of envelopes opened by $\mathcal{R}$ would be distributed uniformly from one of the following sets (each with probability $\frac{1}{4}$); the sets marked with † lead to $\mathcal{R}$ halting:
      i. $\{[(1, *), (3, 2)]\}$
      ii. $\{[(1, *), (2, 3)], [(1, 3), (2, *)]^\dagger\}$
      iii. $\{[(*, 1), (2, 3)], [(3, 1), (2, *)]^\dagger\}$
      iv. $\{[(*, 1), (3, 2)]\}$
      Note that in both worlds $\mathcal{R}$ halts with probability $\frac{1}{4}$, and otherwise the returned envelopes are identically distributed.
   (b) $\mathcal{A}$ sends $[(1 - b, 1 - b), (b, 1 - b)]$ (the other case where the invalid pair matches $1 - b$ is symmetric). The distribution of the returned envelopes in the ideal world is:
      i. With probability $\frac{1}{4}$ ($c \neq b$) it selects uniformly from
      $$\{[(1 - b, *), (b, 1 - b)], [(*, 1 - b), (b, 1 - b)]\}$$
      ii. With probability $\frac{3}{4}$ ($c = b$) it halts.
      In the real world, the order of envelopes opened by $\mathcal{R}$ would be distributed uniformly from one of the following sets (each with probability $\frac{1}{4}$); the sets marked with † lead to $\mathcal{R}$ halting:

   i. $\{[(1, *), (3, 2)], [(1, 3), (*, 2)]^\dagger\}$
  ii. $\{[(1, 3), (2, *)]^\dagger\}$
 iii. $\{[(*, 1), (3, 2)], [(3, 1), (*, 2)]^\dagger\}$
 iv. $\{[(3, 1), (2, *)]^\dagger\}$

Note that in both worlds $\mathcal{R}$ halts with probability $\frac{3}{4}$, and otherwise the returned envelopes are identically distributed.

$\mathcal{A}$ Corrupts $\mathcal{R}$

1. $\mathcal{I}$ waits to receive v and r from the ideal functionality (in response to the **Query** command sent by the ideal $\mathcal{P}$).
2. $\mathcal{I}$ simulates $\mathcal{R}$ receiving four envelopes. The remainder of the simulation depends on the values of v and r:
   (a) If $v = 1$, $\mathcal{I}$ chooses a uniformly random bit t. The first envelope $\mathcal{R}$ opens in the first pair will have the value t, and the first envelope opened in the second pair will have the value $1 - t$. The values revealed in the remaining envelopes will always result in a valid pair.
   (b) If $v = 0$, The first envelope $\mathcal{R}$ opens in each pair will have the value r, and the remaining envelopes the value $1 - r$.
3. $\mathcal{I}$ continues the simulation until $\mathcal{R}$ sends all four envelopes back to $\mathcal{P}$. If $\mathcal{R}$ opened exactly three envelopes, $\mathcal{I}$ sends **Vote** b to the ideal functionality, where b is calculated as by the pollster in the protocol description. If $\mathcal{R}$ did not open exactly three envelopes, $\mathcal{I}$ sends the **Halt** command to the ideal functionality.

Note that throughout the simulation, all simulated parties behave in a manner that is feasible in the real world as well. Furthermore, the outputs of the ideal and simulated $\mathcal{P}$ are always identical. Thus, the only possible difference between the views of $\mathcal{Z}$ in the ideal and real worlds is the contents of the envelopes opened by $\mathcal{R}$. In the real world, the envelope contents are random. In the ideal world, v and r are i.i.d. uniform bits. Therefore the order in which the envelopes are opened does not matter; any envelope in the first pair is independent of any envelope in the second. Hence, the distributions in the ideal and real worlds are identical.

QUAD: A Practical Stream Cipher with Provable Security*

Côme Berbain[1], Henri Gilbert[1], and Jacques Patarin[2]

[1] France Telecom Research and Development,
38-40 rue du Général Leclerc, F-92794 Issy-les-Moulineaux, France
[2] Université de Versailles,
45 avenue des Etats-Unis, F-78035 Versailles cedex, France

Abstract. We introduce a practical stream cipher with provable security named QUAD. The cipher relies on the iteration of a multivariate quadratic system of m equations in $n < m$ unknowns over a finite field. The security of the keystream generation of QUAD is provably reducible to the conjectured intractability of the MQ problem, namely solving a multivariate system of quadratic equations. Our recommended version of QUAD uses a 80-bit key, 80-bit IV and an internal state of $n = 160$ bits. It outputs 160 keystream bits ($m = 320$) at each iteration until 2^{40} bits of keystream have been produced.

1 Introduction

Stream ciphers represent, together with block ciphers, one of the two main classes of symmetric encryption algorithms. Generally speaking stream ciphers seem to allow faster encryption and to require lower computing resources than block ciphers, and the fastest known stream ciphers (e.g. SEAL, RC4, SNOW 2.0, the Shrinking Generator) are indeed significantly faster in software than an efficient block cipher such as AES [27]. However, the design of secure stream ciphers is not currently as well understood as the design of secure block ciphers. The state of the art of the cryptanalysis of stream ciphers, e.g. LFSR based stream ciphers, has evolved significantly over the last ten years and many recent proposals still suffer from security weaknesses. This is illustrated by the fact that none of the candidate stream ciphers submitted to the call for cryptographic primitives of the European project NESSIE were retained since attacks more efficient than exhaustive search were found for all candidates during the evaluation period. This is also illustrated by the ongoing eSTREAM [11] call for stream ciphers proposals of the European project ECRYPT. Stream ciphers complying with two main profiles have been called for, namely stream ciphers allowing much faster software encryption than existing block ciphers (profile 1) and stream ciphers requiring much lower resources for hardware implementation than existing block

---

* The work described in this paper has been supported by the French Ministry of Research RNRT X-CRYPT project and by the European Commission through the IST Program under Contract IST-2002-507932 ECRYPT.

S. Vaudenay (Ed.): EUROCRYPT 2006, LNCS 4004, pp. 109–128, 2006.

ciphers (profile 2). However, more than one third of the 34 submitted stream ciphers, which cover these two profiles, have already been shown to be insecure.

Our aim is to propose a practical cipher with unusually strong security arguments. The novel stream cipher we propose was designed with another trade-off between security, speed and computing resources than reflected by the eS-TREAM profiles 1 and 2. We slightly relax the requirements on speed and computing resources, i.e. we only require a stream cipher that is sufficiently fast for most practical purposes. But we introduce an unusually strong security requirement for symmetric cryptography (which is out of reach of the current state of the art for block ciphers), namely that the security of the cipher be provably reducible to the conjectured intractability of a well-known and studied mathematical problem. The security of the novel stream cipher is provably reducible to the intractability of the MQ problem [15], which consists of finding a solution (if any) to a multivariate quadratic system of m quadratic equations in n variables over a finite field $GF(q)$, typically $GF(2)$. The MQ problem is conjectured to be difficult for suitably chosen values of n and m. In general the associated decision problem is known to be NP-complete even in the case where the considered field is $GF(2)$, and moreover no efficient algorithm to solve MQ with a significant success probability is known to exist for sufficiently large values of n (say $n > 100$) when the quadratic equations are randomly chosen. The implementation complexity of our stream cipher is reasonable and the encryption speed (4.6 Mbit/s for a software implementation in C on a standard PC), though lower than AES, is more than sufficient for many practical purposes.

Constructing a provably secure stream cipher is not a novel topic. However, designing a practical provably secure stream cipher is an open problem. Following seminal work by Shamir, Blum and Micali [4], Yao [31], Levin and Goldreich [25] in the 80's, considerable research effort has been dedicated to the construction of provably secure pseudo-random number generators (PRNG) that expand a short seed (e.g. a key) into a larger bit string. This can be used as the keystream for encryption purposes. Available security results typically state that if the iterated function underlying the construction of a number generator satisfies suitable one-wayness properties, then the generator is a secure PRNG, i.e. its L-bit output is computationally indistinguishable from the uniform distribution over $\{0, 1\}^L$. This research effort has led to remarkable generic results, e.g. the proof by Impagliazzo, Levin, Luby and Håstad [21] that a secure PRNG can be constructed based upon any one way function (OWF). It has also led to provably secure PRNG constructions based on the conjectured intractability of specific problems. The first provably secure PRNG was introduced by Blum and Micali [4] and relates the security of the PRNG to the one-wayness of exponentiation modulo a prime number. The provably secure PRNG proposed by L. Blum, M. Blum and M. Shub [3] exploits the conjectured intractability of quadratic residuosity modulo Blum integers. Alexi, Chor, Goldreich and Schnorr proposed a PRNG construction with security that relies upon the RSA assumption. Impagliazzo and Naor [24] and Fisher and Stern [13] proposed PRNG constructions respectively relying on the difficulty of the subset sum problem and of the syndrome decoding problem. Even in the case of specific constructions, current provably secure PRNGs are too inefficient to provide a practical stream cipher. This is due to the fact that the function iterated by the PRNG is usually too computationally expensive, and that only a restricted number of bits can be produced at each iteration (this number is generally at most proportional to the logarithm of the input length n of the iterated function). However some efforts have been made to improve the constructions. A first idea is to extract more than $\log n$ bits at each round. Constructions based on the discrete logarithm problem makes it possible to extract $n - \log(n)$ bits at each iteration instead of $\log n$. Despite this fact, the fastest generator based on discrete logarithm proposed by Gennaro [16] is still impractical: it requires 350 multiplications of 3000-bit numbers to extract 2775 bits. Another problem for which it is possible to extract more than $\log n$ bits is the syndrome decoding problem. A PRNG has been proposed by Fisher and Stern in [13] but the number of extracted bits, although higher than $\log n$, is still small for practical values of n. Another recently proposed idea is to replace a slow iterated function by some primitive which is much faster to compute. Håstad and Näslund proposed BMGL [30], a stream cipher with security that relies on the difficulty of extracting the key from one plaintext ciphertext couple in AES. Their practical construction consists of iterating AES and extracting $\log n$ bits at each round. This cipher is fast, especially compared to other provably secure ciphers, but its security relies only on the security of the AES and not on a simple and well-studied mathematical problem.

On the contrary, MQ is a simple and well-studied mathematical problem and the values of n for which the problem is difficult are small (around 100 bits), particularly when compared to discrete logarithm or factorisation, where at least 1024 bits are required. Furthermore a large number of bits (e.g. $\frac{n}{2}$) bits or even more can be produced at each iteration.

This paper is organized as follows. We first give some preliminary background on the status of the MQ problem and basic security definitions in a concrete (non asymptotic) security model. Then we describe the new construction and give a formal proof of security for the associated keystream generator. Finally we give the encryption speed of software implementations of our stream cipher.

2   Preliminaries

2.1   Multivariate Quadratic Systems

We consider a finite field $GF(q)$. A multivariate quadratic equation (or equivalently a multivariate quadratic form) in n variables over $GF(q)$ is a polynomial of degree at most 2 in $GF(q)[x_1, \ldots, x_n]$ which can be written as

$$Q(x) = \sum_{1 \leq i \leq j \leq n} \alpha_{i,j} x_i x_j + \sum_{1 \leq i \leq n} \beta_i x_i + \gamma$$

with all the coefficients $\alpha_{i,j}$, β_i, and γ in $GF(q)$. In the particular case $q = 2$, which will be considered in the sequel, the monomial forms $x_i x_i$ and x_i are equal.

It is easy to see that the set $\mathcal{Q}$ of multivariate quadratic forms in n variables is an N-dimensional vector space over $GF(q)$, where $N = \frac{n(n+3)}{2} + 1$ if $q \neq 2$ and $N = \frac{n(n+1)}{2} + 1$ if $q = 2$. A basis of this vector space is given by the $N - 1$ distinct monomial functions of degree 1 or 2 and the constant form 1. Any element of $\mathcal{Q}$ can be represented by the N-tuple of its $GF(q)$ coefficients in this basis. Throughout the rest of this paper, we mean by a randomly chosen quadratic form in n unknowns the quadratic form represented in the above basis by a uniformly and independently drawn N-tuple of $GF(q)$ coefficients.

A multivariate quadratic system S of m quadratic equations in n variables over $GF(q)$ is a set $(Q_1, \ldots Q_m)$ of m quadratic equations in n variables over $GF(q)$. In the sequel, we mean by a randomly chosen system of m quadratic form in n unknowns, n independently and randomly chosen quadratic forms. Such a system is represented by mN uniformly and independently drawn $GF(q)$ coefficients.

A quadratic form Q over n unknowns over $GF(2)$ is called non degenerate iff Q is not equivalent to a quadratic form in strictly fewer than n linear combinations of the n input variables. There exists a polynomial time algorithm to check whether a given quadratic form is non degenerate and more generally to compute the so-called rank of a quadratic form [26]. The number of solutions of the quadratic equation $Q = 0$ associated with a non degenerate quadratic form Q over n unknowns is either 2^{n-1} or $2^{n-1} + 2^{\frac{n-2}{2}}$ or $2^{n-1} - 2^{\frac{n-2}{2}}$ depending on the parity of n and the value of γ. Thus for sufficient large values of n, say $n > 100$, non degenerate quadratic forms are either perfectly balanced (odd n values) or have an undetectable bias (even n values).

2.2   Status of the MQ Problem

We define the problem of solving simultaneous **multivariate quadratic** equations (**MQ problem**) as follows: given a multivariate quadratic system of m quadratic equations over $GF(q)$ $S = (Q_1, \ldots, Q_m)$, find a value $x \in GF(q)^n$, if any, such that $Q_i(x) = 0$ for all $1 \leq i \leq m$.

Depending on the respective values of n and m, instances of MQ can be either easy or very difficult to solve. For $m = 1$ the number of solutions is known [26] and it is quite easy to find one solution. When m is significantly smaller than n, that is for an underdefined quadratic system, finding a solution is easy [6]. In the opposite situation of an overdefined system $(m > n)$ providing $N = \frac{n(n+1)}{2} + 1$ ($q = 2$ case) or $\frac{n(n+3)}{2} + 1$ ($q \neq 2$ case) linearly independent quadratic equations, or more generally when nearly N linearly independent quadratic equations are available, solving an MQ problem is easy by linearization. The total complexity is then only $O(n^6)$. However for general values of m and n the MQ problem is known to be NP-hard, even when restricted to quadratic equations over $GF(2)$ [15] [14] or over any finite field [28].

Moreover, what seems to make the MQ problem particularly well suited to cryptographic applications is that it is conjectured to be very difficult not only asymptotically and in worst case, but already for small suitably selected values of m and n and in terms of the average complexity of solving a random instance. The problem seems to be most difficult when m is close to n. For $m = n$ and $q = 2$ the complexity of the best known solving algorithms is $2^{n-O(\sqrt{(n)})}$ and thus rather close to the 2^n complexity of exhaustive search, and totally out of reach of existing computers for a random instance and n values larger than 100. Even when $q = 2$, $m = kn$ and $k > 1$ is small enough compared with $\frac{n}{2}$, the best known computer algebra algorithms such as XL [10] and improved variants of Buchbergers's Groebner basis computation algorithm such as Faugère's F4 and F5 algorithms [12] are exponential in n for a randomly chosen quadratic system. Much research has been dedicated in the past years to the above problem [9], [7]. Magali Bardet's PHD thesis [1] provides an accurate analysis of the complexity of the most efficient known Groebner basis computation algorithm for solving a random system of $m = kn$ equations in n unknowns. We will use some complexity estimates of [1] when discussing practical recommendations of the parameter values of our cipher.

Though we expect degenerate instances of the systems used in our construction leading to a weak stream cipher to be extremely unlikely, we suggest the following extra precaution when drawing these systems at random to provide some extra guaranties that some of the weakest instances are avoided: check that each quadratic equation is non degenerate or at least has a high rank value close to the one of a non degenerate form, and discard any quadratic equation which would not satisfy this condition. In order to discard a slightly larger subset of weak instance, one can also check that low weight linear combinations of the selected quadratic equations satisfy the above rank conditions. Also check that the obtained quadratic equations are linearly independent in $\mathcal{Q}$.

2.3   Basic Security Notions

All the security definitions used throughout this paper relate to the concrete (non asymptotic) security model. We are using the following basic security notions that we state here informally. Two probability distributions D_1 and D_2 over a finite set Ω are said to be **computationally distinguishable** with computing resources R and advantage ϵ if there exits a probabilistic testing algorithm A which on any input value $x \in \Omega$ outputs a binary answer "1" (accept) or "0" (reject) using computing resources at most R and satisfies

$$|Pr_{x \in D_1}(A(x) = 1) - Pr_{x \in D_2}(A(x) = 1)| \geq \epsilon.$$

Though this is not explicitly reflected in our notation, the above probabilities are not only taken over x values distributed according to D_1 or D_2, but also over the random choices of algorithm A. Algorithm A is called a distinguisher with advantage ϵ. If no such algorithm exists, then we say that D_1 and D_2 are computationally indistinguishable with advantage better than ϵ. When the computing resources R is not specified, we implicitly mean feasible computing resources (i.e. say less than 2^{80} simple operations).

Let n and L denote integers such that $L > n$. A n-bit to L-bit function G is said to be a **Pseudo Random Number Generator (PRNG)** if for a random n-bit input variable x selected according to the uniform law on $\{0,1\}^n$ the probability distribution of the random variable $G(x)$ is computationally indistinguishable from the uniform law over $\{0,1\}^L$.

3   QUAD: A New Stream Cipher

We now introduce the proposed stream cipher, named QUAD.

$S = (Q_1, \ldots, Q_{kn})$ denotes a multivariate quadratic system of kn randomly chosen equations in n variables over $GF(q)$, and S_0 and S_1 denote two (k times smaller) additional multivariate systems of n randomly chosen equations in n variables over $GF(q)$. S, S_0 and S_1 are fixed and publicly known. During the key and IV loading and the keystream generation, the internal register state is a $x = (x_1, \ldots, x_n)$ n-tuple of $GF(q)$ values.

3.1   Keystream Generation and Encryption

The keystream generation process simply consists in iterating the three following steps in order to produce $(k-1)n$ $GF(q)$ keystream values at each iteration.

- Compute the kn-tuple of $GF(q)$ values $S(x) = (Q_1(x), \ldots, Q_{kn}(x))$ where x is the current value of the internal state;
- Output the sequence $S_{out}(x) = (Q_{n+1}(x), \ldots, Q_{kn}(x))$ of $(k-1)n$ $GF(q)$ keystream values
- Update the internal state x with the sequence of n $GF(q)$ first generated values $S_{it}(x) = (Q_1(x), \ldots, Q_n(x))$

The maximal keystream sequence that may be generated with a single (key,iv) pair is L $GF(q)$ values. In order to encrypt a plaintext of length $l \leq L$ $GF(q)$ symbols, each of the first l $GF(q)$ values of the keystream sequence is added (using the $GF(q)$ addition) with the corresponding plaintext value.

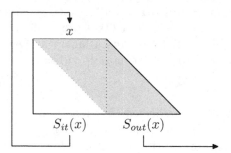

3.2   Key and IV Setup

Before generating any keystream we need to initialize the internal state x, with the key K and the initialization vector IV, which are respectively represented by a sequence of $GF(q)$ elements of length $|K|$ and a binary sequence of $\{0,1\}$ values of length $|IV|$. We assume for the time being, for simplicity of the subsequent proofs [1] that $|K|$ is chosen exactly equal to n.

The initialization is done as follows : we use two carefully randomly chosen multivariate quadratic systems S_0 and S_1 of n equations over n unknowns. We initially set the internal state value x to the n bit value K. Then for each of the $|IV|$ bits IV_1 to $IV_{|IV|}$ of the IV value the internal state x is updated as follows: if $IV_i = 0$, x is replaced by the $GF(q)^n$ value $S_0(x)$; if $IV_i = 1$, x is replaced by the $GF(q)^n$ value $S_1(x)$. These $|IV|$ steps provide a key and IV dependent internal state value x. We then clock the cipher $|IV|$ additional times as described in section 3.1, but without outputting the keystream in order to further transform the internal state value x, and then enter the keystream generation mode to produce the keystream.

4   Security

We now give a proof that for a randomly chosen multivariate quadratic system our PRNG is secure. For simplicity of the proof we will work over $GF(2)$. The proof can be divided in three parts, which can be informally outlined as follows.

In the first part (Theorem 1), we prove that if the L-bit keystream sequence associated with a known fixed or randomly chosen system S of $m = kn$ quadratic equations and an unknown randomly chosen initial internal state $x \in \{0,1\}^n$ is distinguishable from the L-bit output of a perfectly uniform generator, then for a known random quadratic system S of $m = kn$ equations and an unknown randomly chosen input value $x \in \{0,1\}^n$, $S(x)$ is distinguishable from a random kn bit word.

In the second part (Theorem 2), we prove that if for a known randomly chosen quadratic system S and an unknown randomly chosen x, $S(x)$ is distinguishable from a random kn bit word then, for any n-bit to 1-bit quadratic form R (in particular any linear form R), one has the property that for a randomly chosen n bit value x, $R(x)$ can be predicted better than at random given $S(x)$.

In the third part (Theorem 3), we prove that, for a known fixed or randomly chosen S and a randomly chosen linear form R, $R(x)$ can be predicted better than at random given $S(x)$, then with non negligible probability a preimage of $S(x)$ can be efficiently computed given $S(x)$. Thus S is not strongly one way. This part is essentially a proof of Goldreich-Levin's theorem [25], in which a fast Walsh transform computation is used to get a tighter reduction.

4.1   Distinguishing the Keystream Allows to Distinguish the Output of a Random Quadratic System

Theorem 1 states that if one can distinguish the keystream of the generator based on the iteration of a quadratic system S from a random L-bit sequence, then one

---

[1] Note however that we will consider later on, in section 4.5, an extended key loading method allowing to set the key length to values strictly lower than n, for instance to $|K| = \frac{n}{2}$ if one wishes the key length to reflect the complexity of the best known attack.

can distinguish the output of S from a random m-bit sequence. Though we consider a randomly chosen system S because we need distinguishing properties related to a random system for the sequel, the property we prove would also hold if we considered a fixed system S. Our proof is inspired by the proof given in [20] that a similar result holds for the generator based on iteration of any fixed n-bit to m-bit function, where $m > n$, but provides a tighter bound for the advantage.

Theorem 1. *Let* $L = \lambda(k-1)n$ *be the number of keystream bits produced in time* λT_S *using* λ *iterations of our construction. Suppose there is an algorithm A that distinguishes the L-bit keystream sequence associated with a known randomly chosen system S and an unknown randomly chosen initial internal state* $x \in \{0,1\}^n$ *from a random L-bit sequence in time T with advantage* ϵ. *Then there exists an algorithm B that for a randomly chosen S distinguishes* $S(x)$ *corresponding to an unknown random input x, from a random value of size kn in time* $T' = T + \lambda T_S$ *with advantage* $\frac{\epsilon}{\lambda}$.

Proof. We introduce the hybrid probability distributions $D^i(S)$ over $\{0,1\}^L$xi. For $0 \leq i \leq \lambda$ respectively associated with the random variables

$$t^i(S, x) = (r_1, r_2, \ldots, r_i, S_{out}(x), S_{out}(S_{it}(x)), \ldots, S_{out}(S_{it}^{\lambda-i-1}(x)))$$

where the r_j and x are random independent uniformly distributed values of $\{0,1\}^n$ and the notational conventions that $(r_1, r_2, \ldots, r_i)$ is the null string if $i = 0$ and that $(S_{out}(x), \ldots, S_{out}(S_{it}^{\lambda-i-1)}(x)))$ is the null string if $i = \lambda$. Consequently $D^0(S)$ is the distribution of the L-bit keystream and $D^\lambda(S)$ is the uniform distribution over $\{0,1\}^L$. We denote by $p^i(S)$ the probability that A accepts a random L-bit sequence distributed according to $D^i(S)$, and denote by p^i the average value of $p^i(S)$ over the $(k-1)n(n\frac{(n+1)}{2} + 1)$-dimensional vector space of quadratic systems S. We have supposed that algorithm A distinguishes between $D^0(S)$ and $D^\lambda(S)$ with advantage ϵ, in other words that $|p^0 - p^\lambda| \geq \epsilon$. Algorithm B works as follows : on input $(x_1, x_2) \in \{0,1\}^{kn}$ with $x_1 \in \{0,1\}^n$ and $x_2 \in \{0,1\}^{(k-1)n}$, it selects randomly an i such that $0 \leq i \leq \lambda - 1$ and constructs the L-bit vector

$$t(S, x_1, x_2) = (r_1, r_2, \ldots, r_i, x_2, S_{out}(x_1), S_{out}(S_{it}(x_1)), \ldots, S_{out}(S_{it}^{\lambda-i-2}(x_1))).$$

If (x_1, x_2) is distributed accordingly to the output distribution of S, i.e. $(x_1, x_2) = S(x) = (S_{it}(x), S_{out}(x))$ for a uniformly distributed value of x, then

$$t(S, x_1, x_2) = (r_1, r_2, \ldots, r_i, S_{out}(x), S_{out}(S_{it}(x)), \ldots, S_{out}(S_{it}^{\lambda-i-1}(x)))$$

is distributed according to $D^i(S)$. Now if (x_1, x_2) is distributed according to the uniform distribution, then

$$t(S, x_1, x_2) = (r_1, r_2, \ldots, r_i, x_2, S_{out}(x_1), S_{out}(S_{it}(x_1)), \ldots, S_{out}(S_{it}^{\lambda-i-2}(x_1))).$$

Thus $t(S, x_1, x_2)$ is distributed according to $D^{i+1}(S)$. In order to distinguish the output distribution of S from the uniform law, algorithm B calls algorithm A with inputs $(S, t(S, x_1, x_2))$ and returns the value returned by A. Thus

$$|Pr_{S,x}(B(S, S(x)) = 1) - Pr_{S,x_1,x_2}(B(S, (x_1, x_2)) = 1)|$$
$$= |\frac{1}{\lambda} \sum_{i=0}^{\lambda-1} p^i - \frac{1}{\lambda} \sum_{i=1}^{\lambda} p^i| = \frac{1}{\lambda}|p^0 - p^\lambda| \geq \frac{\epsilon}{\lambda}.$$

Thus B distinguishes the output distribution of S from the uniform distribution with probability at least $\frac{\epsilon}{\lambda}$ in time $T + \lambda T_S$.

4.2 Distinguishing the Output of a Random Quadratic System Allows to Predict Any Quadratic Equation

Now we prove that if there exists a distinguisher between $S(x)$ and a kn-bit random value such as the one considered in the above theorem, it can be converted into an algorithm that predicts the result of any quadratic polynomial (and in particular any linear polynomial).

Theorem 2. *Suppose there is an algorithm A that, given a randomly chosen known multivariate quadratic system S of kn equations in n unknowns, distinguishes $S(x)$, where x is an unknown random input value, from a random string of length kn with advantage at least ϵ and in time T. Then there is an algorithm B that, given a randomly chosen quadratic system S of kn equations in n unknowns, any n-bit to 1-bit quadratic form R, and $y = S(x)$ where x is a random input value, predicts $R(x)$ with success probability at least $\frac{1}{2} + \frac{\epsilon}{4}$ using at most $T' = T + 2T_S$ operations.*

Proof. We first show that there exists an algorithm A' which returns 1 on input $(S, S(x))$ with probability at least $\frac{1}{2} + \frac{\epsilon}{2}$ and returns 1 on input (S, u) for some random u with probability $\frac{1}{2}$: if the acceptance probability of A is larger (by at least ϵ) on an input $(S, S(x))$ than on a random input. Then it suffices to consider A' which on input (S, r) either returns $A(S, r)$ or draws a random value u and returns $1 - A(S, u)$ with probability $\frac{1}{2}$ for each case. In the opposite situation, it suffices to consider A' which on input (S, r) either returns $1 - A(S, r)$ or draws a random value and returns $A(S, u)$ with probability $\frac{1}{2}$ for each case.

Algorithm B works as follows. On input $S = (Q_1, \ldots Q_{kn})$, R and a kn-bit value y, B selects a random kn-bit vector $a = (a_1, \ldots, a_{kn})$ and a random bit b, which represents an hypothesis for $R(x)$. Then it computes for all i from 1 to kn the quadratic equation $P_i = Q_i + (a_i \cdot R)$. All the equations P_i form the quadratic system S'. Then B invokes the algorithm A' with input the new quadratic system S' and the value $y + (b \cdot a)$. Finally B returns what A' returns.

Now assume that $y = S(x)$ where x is an unknown random value. We have $\forall i, x, P_i(x) = Q_i(x) + (a_i \cdot R(x)) = y_i + (a_i \cdot R(x))$.

Suppose b is really equal to $R(x)$, then $S'(x) = y + (b \cdot a)$ so the distinguisher A' has been fed with the random quadratic system $S' = (P_1, \cdots, P_{kn})$ and $S'(x)$:

$$Pr_{S,x\in U_n}(B(S,S(x),R)=R(x))=Pr_{S',x\in U_n}(A'(S',S'(x))=1)\geq\frac{1}{2}+\frac{\epsilon}{2}.$$

On the contrary, suppose b is not equal to $R(x)$, then $S'(x)=y+((1+b)\cdot a)=(y+(b\cdot a))+a$. Thus there is an error of a on the value furnished to A' as compared with $S'(x)$. Because a is randomly chosen, we have:

$$Pr_{S,x\in U_n}(B(S,S(x),R)=R(x))=Pr_{S',x\in U_n}(A'(S',S'(x)+a)=0)$$
$$=Pr_{S',t\in U_{kn}}(A'(S',t)=0)=\frac{1}{2}$$

Thus we have:

$$Pr_{S,x\in U_n}(B(S,S(x),R)=R(x))\geq\frac{1}{2}\left(\left(\frac{1}{2}+\frac{\epsilon}{2}\right)+\frac{1}{2}\right)=\frac{1}{2}+\frac{\epsilon}{4}$$

The total running time of B is at most $T+2T_S$, since computing the kn P_i requires for each i to compute all the $\frac{n(n-1)}{2}$ monomials of Q_i and R, which does not cost more than two evaluations of the system for some entry.

4.3   A Linear Form Is a Hard Core Bit for Any One Way Function

Now we show that if for a fixed or random quadratic system S and more generally any fixed or random n-bit to m-bit function f there exists a predictor such as the one considered in the former theorem, i.e. a predictor allowing, given an n-bit to 1-bit linear form R, to predict $R(x)$ with a success probability (over all S and x values) strictly larger than $\frac{1}{2}$, then a preimage of $S(x)$ (resp. f(x)) can be efficiently computed, so that S (resp f) is not one way. This result is the Goldreich-Levin theorem [25] that we prove as to get a tight reduction. Before proving the theorem, which relates to the computation, given the image $S(x)$ or $f(x)$ for a random unknown value x and a random system S, of a list containing x, we first establish a lemma representing the technical core of the proof in which a fixed (unknown) value of x is considered. Our proofs are inspired by the simplified treatment of the original Goldreich-Levin proofs developed by Rackoff, Goldreich[18] and Bellare [2], and also by the proofs provided by Håstad and Näslund in their BMGL paper [30].

Lemma 1. *Let us denote by x a fixed unknown n-bit value and denote by f a fixed n-bit to m-bit function. Suppose there exists an algorithm B that given the value of $f(x)$ allows to predict the value of any linear equation R over n unknowns with probability $\frac{1}{2}+\epsilon$ over R, using at most T operations. Then there exists an algorithm C, which given $f(x)$ produces in time at most T' a list of at most $4n^2\epsilon^{-2}$ values such that the probability that x appears in this list is at least $1/2$.*

$$T'=\frac{2n^2}{\epsilon^2}\left(T+\log\left(\frac{2n}{\epsilon^2}\right)+2\right)+\frac{2n}{\epsilon^2}T_f$$

The proof of lemma 1 is given in the Appendix. Lemma 1 applies to a fixed x and a fixed system S (or a fixed n-bit to m-bit function f). However, the success probability of the predictor of Theorem 2 is taken over all (x, S) pairs for any linear form R. Consequently, we need a theorem allowing us to exploit the existence of such a predictor to show the applicability of the lemma to a non-negligible fraction of (x, S) pairs.

Theorem 3. *Suppose there is an algorithm B, that given a randomly chosen quadratic system S of m quadratic equations, a randomly chosen n-bit to 1-bit quadratic form R and the image $S(x)$ of a randomly chosen (unknown) n-bit value x, predicts the value of $R(x)$ with probability at least $\frac{1}{2} + \epsilon$ over all possible (x, S, R) triplets using T operations. Then there is an algorithm C, which given the image $S(x)$ of a randomly chosen (unknown) n-bit value x produces a preimage of $S(x)$ with probability at least $\epsilon/2$ (over all possible values of x and S) in time T'.*

$$T' = \frac{8n^2}{\epsilon^2}\left(T + \log\left(\frac{8n}{\epsilon^2}\right) + 2\right) + \frac{8n}{\epsilon^2}T_f$$

Proof. The assumption about algorithm B can be written as

$$Pr_{(x,S,R)\in\{0,1\}^{n+mN+n}}\left\{B(S, S(x), R) = R(x)\right\} \geq \frac{1}{2} + \epsilon.$$

It results that for a fraction at least ϵ of all the (x, S) pairs one has

$$Pr_{R\in\{0,1\}^n}\left\{B(S, S(x), R) = R(x)\right\} \geq \frac{1}{2} + \frac{\epsilon}{2}.$$

Otherwise, there would exist a fraction at least $1 - \epsilon$ of the (x, S) pairs which associated prediction probability over the R values would be strictly less than $\frac{1}{2} + \frac{\epsilon}{2}$, and therefore $Pr_{(x,S,R)\in\{0,1\}^{n+mN+n}}\left\{B(S, S(x), R) = R(x)\right\}$ would be upper bounded by $(1 - \epsilon)(\frac{1}{2} + \frac{\epsilon}{2}) + \epsilon = \frac{1}{2} + \epsilon - \epsilon^2$, which contradicts the assumption about Algorithm B.

Thus for a fraction at least ϵ of all the (x, S) pairs the conditions of lemma 1 are met and algorithm C of the lemma provides a preimage of $S(x)$ with probability at least $1/2$.

4.4 A Security Proof for the Proposed PRNG

Now it is easy to see that if we sequentially apply theorems 1, 2, and 3, we obtain the following reduction theorem, which states that if, for a random system and a random initial value, the L-bit keystream sequence was distinguishable from a random L-bit sequence then there would exist an efficient algorithm allowing to find a preimage of the image of a random n-bit input value by a random quadratic n-bit to m-bit system, which for suitably chosen values of n would contradict the assumptions made in Section 2 on the difficulty of solving MQ.

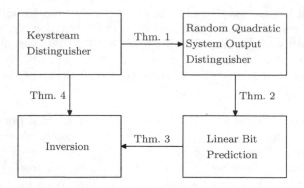

Theorem 4. *Let $L = \lambda(k-1)n$ be the number of keystream bits produced by in time λT_S using λ iterations of our construction. Suppose there exists an algorithm A that distinguishes the L-bit keystream sequence associated with a known randomly chosen system S and an unknown randomly chosen initial internal state $x \in \{0,1\}^n$ from a random L-bit sequence in time T with advantage ϵ. Then there exists an algorithm C, which given the image $S(x)$ of a randomly chosen (unknown) n-bit value x by a randomly chosen n-bit to m-bit quadratic system S produces a preimage of $S(x)$ with probability at least $\frac{\epsilon}{2^3\lambda}$ over all possible values of x and S in time upper bounded by T'.*

$$T' = \frac{2^7 n^2 \lambda^2}{\epsilon^2}\left(T + (\lambda+2)T_S + \log\left(\frac{2^7 n\lambda^2}{\epsilon^2}\right) + 2\right) + \frac{2^7 n\lambda^2}{\epsilon^2}T_S$$

Proof. Theorems 1 to 3 state that if an algorithm X exists, then another algorithm Y exists. In the case of Theorem 1, the resulting algorithm Y can be directly play the role of algorithm X in Theorem 2. In the case of Theorem 2, the resulting algorithm Y, named algorithm B, has the property

$$\forall R \in \{0,1\}^N Pr_{(x,S)\in\{0,1\}^{n+mN}}\{B(S,S(x),R) = R(x)\} \geq \frac{1}{2} + \frac{\epsilon}{4}$$

which implies

$$Pr_{(x,S,R)\in\{0,1\}^{n+mN+N}}\{B(S,S(x),R) = R(x)\} \geq \frac{1}{2} + \frac{\epsilon}{4}$$

Thus algorithm Y can play the role of algorithm X in Theorem 3, and if we compose the distinguishing probability and complexity expressions of the three concatenated theorems, we obtain the claimed distinguishing probability and complexity bounds.

Discussion. Theorem 4 above relates to the keystream generation part of QUAD, i.e. to the expansion of a randomly chosen initial state into the keystream and does not include the key and IV loading for deriving the initial state. Moreover it does not guarantee the strength of a particular instance of QUAD associated with a fixed system S but (informally) it shows that if MQ is intractable then most instances of QUAD are secure.

4.5   Specifying the Parameter Values for QUAD

We now propose concrete parameters n, k, L, $|K|$ and $|IV|$ for our construction. We restrict ourselves to the $GF(2)$ case. We want to ensure a security level of at least 2^{80}. More precisely we want Theorem 4 to ensure that if for a random system and a random initial internal state value at the beginning of the keystream generation there exists a testing algorithm that allows us to distinguish an L-bit keystream produced by QUAD from a uniformly drawn keystream sequence with an advantage of more than $\epsilon = \frac{1}{100}$ in time less than $T = 2^{80}$ this would imply the existence of an inversion algorithm of non negligible success probability $\epsilon' = \frac{\epsilon}{2^3 \lambda}$ allowing, given a random n-bit to kn-bit system of quadratic equations and the $S(x)$ image by S of a random input value x, to find a preimage by S of $S(x)$ in time T' lower by a factor of more than ϵ' than the best known inversion algorithms for the MQ problem, and thus result in the existence of a large set of weak instances of MQ.

Depending on the intended application of the stream cipher, the maximum keystream length L can vary from a few hundreds bits for a mobile phone application to up to 2^{40} bits. Consequently the allowed parameter values for n and k will also vary, since it is much more demanding to get a security argument for $L = 2^{40}$ bits than for $L = 1000$ bits. We will however retain the latter value $L = 2^{40}$ for a first estimate of the corresponding required value of n.

In her thesis, Magali Bardet [1] shows that the best Groebner basis algorithm to solve a system of kn equations in k unknowns has (in the case of a regular system) a complexity of $T(k, n) = \left(\binom{n+1}{D}\right)^{2.37}$, where D is close to $\left(-k + \frac{1}{2} + \frac{1}{2}\sqrt{2k^2 - 10k - 1 + 2(k+2)\sqrt{k(k+2)}}\right) n$. To obtain a contradiction, we need to have T' lower than $\epsilon'T(k, n)$. For $k = 2$ and with the previous values of $L = 2^{40}$, $T = 2^{80}$ and $\epsilon = \frac{1}{100}$, we get $\epsilon' = 2^{-42}$ and we need to have n greater than 350. For $n = 256$ and $k = 2$, we only get a contradiction if we produce less than $L = 2^{22} = 4$ Mbits of keystream for each key and IV pair.

Practical Values. For practical use of QUAD we recommend an internal state length of $n = 160$ bits and an expansion factor k of 2 and a maximum keystream length $L = 2^{40}$. We further recommend an IV length $|IV|$ of 80 bits. For such n, k and L values, we do not get a contradiction as for the former parameter values. However our proof reduction is not optimal, and we expect that these parameter values suffice to provide the desired security level of about 2^{80}.

If instead of the n-bit key length assumed (for simplicity of the security arguments) in sections 2 and 3, a keylength $|K|$ strictly lower than n is preferred in order for $|K|$ to better reflect the expected security level, we suggest the following extension of the key loading method described in section 3: periodically repeat the $|K|$ bits of K to get an expanded key of length n, and apply the key and IV procedure of section 3 to this expanded key. We suggest, if this extended key loading method option is retained, to select a key length $|K| = 80$. Though the shorter key option weakens the security arguments of section 4 and can thus be considered less conservative than the full length $n = 160$-bit key, we are not aware of any major security weakness resulting from this option.

An indication of the advantages of the use of the MQ problem for constructing a provably secure stream cipher, in terms of the required internal state size, is given by a comparison with the fastest known provably secure stream cipher, namely a discrete log based construction proposed by Gennaro in [16] with internal state length $n = 3000$ bits (to be compared with the $n = 350$ and 256 internal state lengths derived above) and which produces 2775 bits per iteration and applies 335 modular multiplications of 3000-bit numbers at each iteration. Moreover the security argument of [16] does not assume the existence of a keystream sequence distinguishing algorithm in time $T = 2^{80}$ to get a contradiction, but only a distinguishing algorithm in time $T = 3.5 \cdot 10^{10} \simeq 2^{35}$. Another advantage of MQ is that MQ is NP-hard, whereas the Discrete Logarithm Problem is only in NP $\cap$ co-NP. Moreover the best known algorithm to solve the Discrete Logarithm problem are subexponential, while for MQ, those algorithm are exponential.

5   Cryptanalysis

In this section, we consider various attacks and verify whether they are applicable to our construction. We focus on security aspects not covered by the proof of security of the former section, e.g. the protection against resynchronization attacks provided by the key and IV loading mechanism.

Resistance Against Algebraic Attacks. QUAD was designed to resist algebraic attack techniques. As a matter of fact, the key and IV loading and keystream generation mechanisms of QUAD are based upon the iteration of quadratic systems whose associated equations are conjectured to be computationally impossible to solve [2]. In more details, recovering the initial state x of the keystream generator from the whole keystream is more difficult than recovering x from $S(x)$, i.e. solving an intractable quadratic system of kn equations. As for the key and IV loading mechanism, it is possible to express any keystream block, as a set of $(k-1)n$ algebraic equations on the $|K| \geq 80$ key bits. However since the key and IV setup consists of $2|K|$ rounds of a quadratic function, this set consists of $(k-1)n$ equations of degree $|K|$ or nearly $|K|$ on the $|K|$ key bits. It is quite natural to conjecture that such a system is highly intractable.

Correlation Attacks and Distinguishing Attacks. We expect QUAD to be immune to such attacks except for extremely unlikely degenerate instances of the quadratic system S, for example if one of the n-bit to 1-bit quadratic forms of S_{out} or a linear combination of these $(k-1)n$ quadratic forms has an exceptionally low rank and therefore (for even values of n) a detectable bias.

---

[2] except for a small fraction of degenerate instances of S, S_0 and S_1 whose occurrence is extremely unlikely if these systems are selected as described in section 4.5.

Time-Memory-Data Tradeoffs and other Generic Attacks. The internal state of our construction has a size n of at least 160 bits in order to resist against generic time-data tradeoff, which have a complexity of $2^{\frac{n}{2}}$.

Since QUAD is based upon the iteration of the quadratic system S_{it}, the keystream sequences it produces are ultimately periodic. Moreover, since S_{it} is not one to one, the order of magnitude of the period can be expected to be $2^{\frac{n}{2}}$ $(k-1)n$-bit keystream blocks. One of the consequences of specifying a maximal keystream length $L << 2^{\frac{n}{2}}$ (a typical order of magnitude is $L = 2^{40}$) is that the detection of short cycles is extremely unlikely.

Guess and Determine Attacks. The analysis of attacks of this type allows us to fix an upper bound on k. Let us assume that an adversary is able to guess p bits of the internal state. Then this adversary gets a system of $(k-1)n$ equations in the $(n-p)$ remaining internal state variables. If the number of monomials generated by these $n-p$ variables $n_p = \frac{1}{2}(n-p)(n-p+1)$ is close to $(k-1)n$, the adversary can linearize the system and recover the internal state. Solving $n_p = (k-1)n$ gives us a number $p_0 = n + \frac{1-\sqrt{1+8n(k-1)}}{2}$ such that for $p \geq p_0$ the linearization is possible. The complexity C of the resulting "attack" is about $2^{p_0}((k-1)n)^\omega$, where ω is between 2 and 3. If C is lower than $2^{|K|}$, then the attack is better than exhaustive search. Consequently, k has to be chosen such that C be larger than $2^{|K|}$. For instance for $n = 160$ and $|K| = 80$, $k < 21$ implies that $p_0 > 80$, and therefore $C >> 2^{80}$. More conservative (i.e. lower) values of k than the one given by this simple bound are of course recommended.

Unsurprisingly, the attack would become more efficient for unlikely degenerate instances of S, for instance if several quadratic forms of S could be all expressed as quadratic functions of substantially less than n linear combinations of the n state variables.

Resistance to Resynchronization Attacks with Chosen IVs. Our proof does not cover the Key and IV sctup but only the keystream generation. They provide a strong argument towards the conjecture that the keystream sequence resulting from any single known or chosen IV value cannot be distinguished from a random sequence, but do not provide guarantees regarding the independence of the sequences resulting from several chosen IVs and the resistance of QUAD against resynchronization attacks. However the following informal argument indicates that the key and IV setup construction of QUAD prevents such resynchronization attacks, or more generally any detectable statistical bias on the joint distribution of the keystream sequences resulting from the same key and several chosen IVs. Let us consider any t-tuple $(IV^1, \cdots, IV^t)$ of t distinct IV values and one randomly chosen n-bit initial state value before IV loading x. By applying the security proofs of section 4 to the $S = (S_0, S_1)$ system of $2n$ quadratic equations, the n-bit to $2n$-bit mapping S_0, S_1 is a strong pseudorandom generator. However, the key and IV loading consists of applying a tree-based construction proposed by Goldreich, Goldwasser and Micali [19] to this generator, so that we can expect the distribution of the $(x^1, \cdots, x^t)$ t-tuple of internal state values resulting from the loading of x and IV^1 to IV^t to be indistinguishable from a t-tuple of random independent values. Moreover, the subsequent runnup rounds during which the keystream generator is run without outputting keystream bits provide an extra security margin, since only high degree functions of x^1 to x^t are available to an adversary instead of quadratic functions. If instead of the proposed key and IV setup the key and IV values the IV had been loaded into the initial state and an insufficient number of quadratic mappings had been applied to the initial state before activating the keystream generation, then chosen-IV attacks exploiting the higher degree differential properties of low degree functions could have been mounted.

Dual Ciphers. Because of the structure of the QUAD equations, it is easy to find dual ciphers of QUAD, i.e. simple (e.g. linear) transformations f and g of the key K and the keystream as to ensure that for each triplet of quadratic systems (S, S_0, S_1) there exist quadratic systems (S', S_0', S_1') such that for any key K and any IV value IV, the keystream associated with $(f(K), IV, S', S_0', S_1')$ is the image by g of the keystream associated with $(f(K), IV, S, S_0, S_1)$. We do not expect this property to represent a security threat for QUAD.

6   Performance

In this Section we give performance results for our recommended version of QUAD, which has 160 bits of internal state, an expansion factor of 2 and a 80-bit key and IV length. On a Pentium IV clocked at 2.5GHz with 512 kByte of cache and using the Intel compiler, our recommended version of QUAD reaches a speed of 4347 cycles/byte (4.6 Mbit/s). On a Pentium 4 with 1MByte of cache, the same version reaches a speed of 2915 cycles/byte (5.7 Mbit/s). This cache effect is due to the fact that the quadratic system used contains more than 4 millions of binary coefficients, which requires around 1MByte to store. A version of QUAD running on an Opteron clocked at 2.1 GHz with a 64-bit architecture reaches the speed of 2176 cycles/byte (quite close from 1MByte/s). An optimised version of Blum Blum Shub's generator with an internal state of 1024 bits, which is far from the number of bits of the internal state required for proven security, reaches 30374 cycles/byte. In his paper[16], Gennaro claimed his discrete logarithm based generator to be twice faster for these parameters. We can therefore assume that this generator runs at about 15000 cycles/byte. Though QUAD is significantly slower than AES, which runs at 25 cycles/byte, it is much more efficient than other provably secure pseudo random generator. Moreover, implementations of QUAD with quadratic system over larger fields (e.g. $GF(16)$ or $GF(256)$) are much faster and even reach 106 cycles/byte.

7   Conclusion

In this paper we introduced QUAD, a novel synchronous stream cipher based on MQ with a security proof in the concrete security model. Eventhough this construction relies on a mathematical problem and has a proof of security, its internal state is of small size n and it extracts a small multiple of n bits at each round. A software implementation of our recommended version of QUAD reaches a speed of 4.6 Mb/s on a standard PC. This makes QUAD of great interest for applications where security is the main concern. We do not preclude that it might be possible to derive tighter bounds in some parts of the proof, which would allow us to further reduce the internal state size and increase the number of extracted bits.

We would like to thank Matt Robshaw and Olivier Billet for helpful comments.

References

1. Magali Bardet. *Étude des systèmes algébriques surdéterminés. Applications aux codes correcteurs et à la cryptographie.* PhD thesis, Université Paris VI, 2004.
2. Mihir Bellare. The Goldreich-Levin Theorem. http://www-cse.ucsd.edu/users/mihir/courses.html, 1999.
3. Lenore Blum, Manuel Blum, and Mike Shub. A simple unpredictable pseudo-random number generator. *SIAM J. Comput.*, 15(2):364–383, 1986.
4. Manuel Blum and Silvio Micali. How to generate cryptographically strong sequences of pseudo-random bits. *SIAM J. Comput.*, 13(4):850–864, 1984.
5. Don Coppersmith, Shai Halevi, and Charanjit S. Jutla. Cryptanalysis of stream ciphers with linear masking. In Moti Yung, editor, *Advances in Cryptology – CRYPTO 2002*, volume 2442 of *Lecture Notes in Computer Science*, pages 515–532. Springer-Verlag, 2002.
6. Nicolas Courtois, Louis Goubin, Willi Meier, and Jean-Daniel Tacier. Solving underdefined systems of multivariate quadratic equations. In *Public Key Cryptography*, pages 211–227, 2002.
7. Nicolas Courtois, Alexander Klimov, Jacques Patarin, and Adi Shamir. Efficient algorithms for solving overdefined systems of multivariate polynomial equations. In Bart Preneel, editor, *Advances in Cryptology – EUROCRYPT 2000*, volume 1807 of *Lecture Notes in Computer Science*, pages 392–407. Springer-Verlag, 2000.
8. Nicolas Courtois and Willi Meier. Algebraic attacks on stream ciphers with linear feedback. In Eli Biham, editor, *Advances in Cryptology – EUROCRYPT 2003*, volume 2656 of *Lecture Notes in Computer Science*, pages 345–359. Springer-Verlag, 2003.
9. Nicolas Courtois and Jacques Patarin. About the XL Algorithm over $GF(2)$. In Marc Joye, editor, *Topics in Cryptology – CT-RSA 2003*, volume 2612 of *Lecture Notes in Computer Science*, pages 141–157. Springer-Verlag, 2003.
10. Claus Diem. The XL-Algorithm and a Conjecture from Commutative Algebra. In Pil Joong Lee, editor, *Advances in Cryptology – ASIACRYPT 2004*, volume 3329 of *Lecture Notes in Computer Science*, pages 323–337. Springer-Verlag, 2004.
11. ECRYPT. eSTREAM: ECRYPT Stream Cipher Project, IST-2002-507932. Available at http://www.ecrypt.eu.org/stream/, Accessed September 29, 2005, 2005.
12. Jean-Charles Faugère, Hideki Imai, Mitsuru Kawazoe, Makoto Sugita, and Gwénolé Ars. Comparison Between XL and Grbner Basis Algorithms. In Pil Joong Lee, editor, *Advances in Cryptology – ASIACRYPT 2004*, volume 3329 of *Lecture Notes in Computer Science*, pages 338–353. Springer-Verlag, 2004.
13. Jean-Bernard Fischer and Jacques Stern. An efficient pseudo-random generator provably as secure as syndrome decoding. In *EUROCRYPT*, pages 245–255, 1996.

14. Aviezri S. Fraenkel and Yaacov Yesha. Complexity of solving algebraic equations. *Inf. Process. Lett.*, 10(4/5):178–179, 1980.
15. Michael R. Garey and David S. Johnson. *Computers and Intractability: A Guide to the Theory of NP-Completeness*, chapter 7.2 Algebraic Equations over $GF(2)$. W H Freeman & Co, 1979.
16. Rosario Gennaro. An improved pseudo-random generator based on discrete log. In *CRYPTO*, pages 469–481, 2000.
17. Oded Goldreich. Three xor-lemmas an exposition. Technical report, Weizmann Instritute of Science, Revohot, Israel, 1995.
18. Oded Goldreich. *Fondationsof Cryptography*, volume 1. Cambridge University Press, 2001.
19. Oded Goldreich, Shafi Goldwasser, and Silvio Micali. How to construct random functions. *J. ACM*, 33(4):792–807, 1986.
20. Shafi Goldwasser and Mihir Bellare. Lecture notes on cryptography. Available at `http://www-cse.ucsd.edu/users/mihir/courses.html`, 2001.
21. Johan Håstad, Russell Impagliazzo, Leonid A. Levin, and Michael Luby. A pseudo-random generator from any one-way function. *SIAM J. Comput.*, 28(4):1364–1396, 1999.
22. Russel Impagliazzo, Leonid A. Levin, and Michael Luby. Pseudo-random generation from one-way functions. In D.S.Johnson, editor, *21th ACM Symposium on Theory of Computing – STOC '89*, pages 12–24. ACM Press, 1989.
23. Russel Impagliazzo and Moni Naor. Efficient cryptographic schemes provably as secure as subset sum. *Journal of Cryptology*, 9(4):199–216, 1996.
24. Russell Impagliazzo and Moni Naor. Efficient cryptographic schemes provably as secure as subset sum. *J. Cryptology*, 9(4):199–216, 1996.
25. Leonid A. Levin and Oded Goldreich. A hard-core predicate for all one-way functions. In D. S. Johnson, editor, *21th ACM Symposium on Theory of Computing – STOC '89*, pages 25–32. ACM Press, 1989.
26. Rudolf Lidl and Haradl Niederreiter. *Finite Fields*. Cambride University Press, 1997.
27. National Institute of Standards and Technology. FIPS-197: Advanced Encryption Standard, November 2001. Available at `http://csrc.nist.gov/publications/fips/`.
28. Jacques Patarin and Louis Goubin. Asymmetric cryptography with s-boxes. In *ICICS*, pages 369–380, 1997.
29. Jacques Patarin and Louis Goubin. Asymmetric cryptography with s-boxes. In *ICICS*, pages 369–380, 1997.
30. Johan Håstad and Mats Näslund. Bmgl: Synchronous key-stream henerator with provable security. submitted to Nessie Project, 2000.
31. Andrew Yao. Theory and applications of trapdoor function. In *Foundations of Cryptography FOCS 1982*, 1982.

Appendix. Proof of Lemma 1

We denote by L_i, $1 \leq i \leq n$ the n-bit to 1-bit linear forms defined by $L_i(x)=x_i$, where x is represented by the binary string $x_1 x_2 \cdots x_n$. The idea of the proof is to call algorithm B sufficiently many times to recover all the $x_i = L_i(x)$ one by one. To do so, we introduce a parameter t, whose order of magnitude is $\log n$ which will be specified later. We use t randomly chosen n-bit to 1-bit linear forms

$R_1, \ldots, R_t$ to randomize our requests to algorithm B. For each $L_i(x)$ we want to retrieve, we call algorithm B 2^t times, using the 2^t linear combinations $\bigoplus_j \alpha_j R_j$ of the R_k forms in order to randomize L_i. Suppose we know the t values for $R_j(x)$, then for any α we can also compute the value of $\bigoplus_j \alpha_j R_j(x)$ and add this value to $B(\bigoplus_j \alpha_j R_j \oplus L_i, f(x))$. We denote

$$C(i, \alpha) = B(\bigoplus_j \alpha_j R_j \oplus L_i, f(x)) \oplus \bigoplus_j \alpha_j R_j(x)$$

If we make a correct assumption on the t values $R_1(x)$ to $R_t(x)$ and if B returned the right value of $(\bigoplus_j \alpha_j R_j \oplus L_j)(x)$, then we have

$$C(i, \alpha) = (\bigoplus_j \alpha_j R_j \oplus L_i)(x) \oplus \bigoplus_j \alpha_j R_j(x)$$

$$= L_i(x) \oplus \bigoplus_j \alpha_j R_j(x) \oplus \bigoplus_j \alpha_j R_j(x) = L_i(x).$$

For all the possible α values, we collect the vote $C(i, \alpha)$ for the value of $L_i(x)$. Since algorithm B is supposed to answer correctly most of the time, taking the majority of the votes $C(i, \alpha)$ will provide us with the value of $L_i(x)$ with a high probability if we assume that 2^t requests are enough. The counterpart of this technique is that we have to guess the real values of $R_j(x)$ for all j but since t is of logarithmic size this is achievable.

We now give a more formal proof with a small difference: we use fast Walsh transform computations to simultaneously compute the 2^t results of the votes on the $C(i, \alpha)$ values for all the 2^t possible t-tuples of assumptions $R_j(x)$, $1 \leq j \leq t$, instead of computing them independently.

Before we give the proof, we need to recall some results on the Walsh transform. Given a real function of t binary variables $g(x_1, \ldots, x_t)$, the Walsh transform of g is the real function of t binary variables $G = W(g)$ defined by

$$G(u_1, \ldots, u_t) = \sum_{x_1, \ldots, x_t \in \{0,1\}^t} f(x_1, \ldots, x_t)(-1)^{u_1 x_1 + \ldots + u_t x_t}$$

It is known that the time needed to compute the Walsh transform of a function of t binary variables is $t \cdot 2^t$.

Proof. The algorithm C works as follows : first it randomly selects t elements $R_1, \ldots, R_t$ of the n-dimensional vector space over $GF(2)$ of the n-bit to 1-bit linear forms.

Then for each $i = 1, \ldots, n$ it executes the following process: for all the 2^t possible $\alpha = (\alpha_1, \ldots \alpha_t)$ t-tuples $\in \{0, 1\}^t$ store $(-1)^{B(\bigoplus_j \alpha_j r_j \oplus L_i, f(x))}$ in a table of size 2^t, say $(c_0, \ldots c_{2^t-1})$ (thus the coefficient associated with α is $c_{\sum_{j=0}^{t-1} \alpha_j \cdot 2^{j-1}}$). Then it applies the Walsh transform to this table (which represents a function of α. This gives 2^t numbers $(\beta_0^i, \ldots, \beta_{2^t-1}^i)$ such that

$$\beta_k^i = \sum_\alpha (-1)^{B(\bigoplus_j \alpha_j R_j \oplus L_i, f(x))}(-1)^{<k, \alpha>}$$

$$= |\{\alpha | C(i, \alpha) = 0\}| - |\{\alpha | C(i, \alpha) = 1\}|$$

β_k^i is the difference of the number of 0 votes and 1 for $L_i(x)$ corresponding to the assumption that $R_j(x) = k_j$ for all j comprised between 1 and t. Consequently if β_k^i is positive, then C sets bit i of the n-bit candidate value C_k associated with the assumption k to $C_k^i = 0$, otherwise this bit is set to $C_k^i = 1$.

After this process has been completed for all the n values of i, one is left with a list of 2^t n-bit candidate values for x corresponding to each of the 2^t assumptions for $R_1(x)$ to $R_t(x)$. For each candidate value C_k, algorithm C then computes $f(C_k)$ and compares it to $f(x)$. If a match occurs, C keeps C_k in the list of at most 2^t candidate values for x it outputs, otherwise C_k is discarded from the list.

The total running time of algorithm C is $n2^t(T + t + 2) + 2^t T_f$ where T_f is the time needed to compute $f(y)$ for an n-bit value y.

Let us now upper bound the probability that algorithm C fails to select x in the list of pre-images of $f(x)$ it produces. Over the 2^t assumptions for $R_1(x)$ to $R_t(x)$, only the correct one is to be considered. The failure probability of C is upper bounded by the sum of the n probabilities p_i that the vote for $L_i(x)$ is incorrect and we have:

$$p_i = Pr\left\{|\{\alpha|C(i,\alpha) = L_i(x)\}| < \frac{2^t}{2}\right\}$$

$|\{\alpha|C(i,\alpha) = L_i(x)\}|$ is the sum of the 2^t pairwise independent 0-1 variables $C(i,\alpha) \oplus L_i(x) \oplus 1$ of average value $\mu_\alpha \geq \frac{1}{2} + \frac{\epsilon}{2}$ and variance $v_\alpha = \frac{1}{4} - \frac{\epsilon^2}{4}$. Thus p_i has average value $\mu = 2^t\left(\frac{1}{2} + \frac{\epsilon}{2}\right)$ and variance $\sigma^2 = 2^t\left(\frac{1}{4} - \frac{\epsilon^2}{4}\right)$. By applying Chebyshev's inequality, we have

$$p_i = Pr\left\{\sum_\alpha C(i,\alpha) \oplus L_i(x) \oplus 1 < \frac{2^t}{2}\right\}$$

$$= Pr\left\{\sum_\alpha C(i,\alpha) \oplus L_i(x) \oplus 1 - \mu < -\frac{2^t\epsilon}{2}\right\}$$

$$\leq Pr\left\{\left|\sum_\alpha C(i,\alpha) \oplus L_i(x) \oplus 1 - \mu\right| > \frac{2^t\epsilon}{2}\right\} \leq \frac{\sigma^2}{(2^t\frac{\epsilon}{2})^2} \leq \frac{1}{2^t\epsilon^2}$$

Thus the failure probability of C is upper bounded by $\frac{n}{2^t\epsilon^2}$. If we want to have a probability of success for algorithm C higher than $\frac{1}{2}$, then we have to choose t such that $2^t = \frac{n}{\epsilon^2}$. Finally the total complexity of algorithm C is given by

$$\frac{2n^2}{\epsilon^2}\left(T + \log(\frac{2n}{\epsilon^2}) + 2\right) + \frac{2n}{\epsilon^2}T_f$$

How to Strengthen Pseudo-random Generators by Using Compression*

Aline Gouget** and Hervé Sibert

France Telecom Research and Development,
42 rue des Coutures, BP6243, F-14066 Caen Cedex 4, France
{aline.gouget, herve.sibert}@francetelecom.com

Abstract. Sequence compression is one of the most promising tools for strengthening pseudo-random generators used in stream ciphers. Indeed, adding compression components can thwart algebraic attacks aimed at LFSR-based stream ciphers. Among such components are the Shrinking Generator and the Self-Shrinking Generator, as well as recent variations on Bit-Search-based decimation. We propose a general model for compression used to strengthen pseudo-random sequences. We show that there is a unique (up to length-preserving permutations) construction that reaches an optimal trade-off between output rate and security against several attacks.

1 Introduction

The huge amount of work impulsed by the ECRYPT call for stream ciphers [5] shows how much progress has been made in stream ciphers analysis in the recent years. While researchers in the area are still willing to design new proposals with innovative, yet not always secure, ideas. If cryptanalysis seems to put the fate of stream ciphers at stake, this is also the consequence of a lack of theoretical security results for stream ciphers and pseudo-random generators.

Compression of sequences can strengthen pseudo-random generators used in stream ciphers. In particular, adding compression components can thwart algebraic attacks aimed at LFSR-based stream ciphers [1, 4]. Such components include decimation components such as the Shrinking Generator [3] and the Self-Shrinking Generator [15]. Decimation has come back in focus recently with the Bit-Search Generator [9] and subsequent variations on it [10].

Compression mechanisms may suffer from timing attacks [12] since the speed of the output is variable in a manner that depends on the generator's state. Thus, LFSR-based ciphers involving a decimation mechanism may be easily breakable in case of leakage of the number of times LFSRs are clocked for each output. However, such side channel attacks are usually alleviated by buffering the output, as described for instance in [14]; these issues are not discussed in this paper.

---

* Work partially supported by the French Ministry of Research RNRT X-CRYPT Project and by the European Commission under contract IST-2002-507932 via the ECRYPT Network of Excellence.

** Current e-mail address: aline.gouget@gemplus.com

S. Vaudenay (Ed.): EUROCRYPT 2006, LNCS 4004, pp. 129–146, 2006.

Our main purpose is to propose a general model for compression used in the generation of pseudo-random sequences, in order to build compression components upon theoretical results. In Section 2, we detail related work on the subject including the Shrinking Generator and the Bit-Search generator variation used in the DECIM proposal to the ECRYPT stream cipher project. In Section 3, we construct our framework for compression components using prefix codes dedicated to pseudo-random generation. In Section 4, we focus on the case when the compression output is 0 or 1. We show that there is then a unique (up to length-preserving permutations) construction that reaches an optimal trade-off between output rate and security against several attacks, including entropy-based reconstruction, linear equations retrieval, and FBDD attacks. In Section 5, we apply our results to the Self-Shrinking Generator and Bit-Search based decimation.

2 Related Work

Generation of pseudo-random sequences using *compression techniques* relies on the use of a *compression function*. A *compression function* is a function that compresses m-bit inputs (m is not necessarily a fixed value) to n-bit outputs, where $m \geq n$. The properties required for such functions depend on the application context. For instance, one-wayness is required for cryptographic hash functions, whereas compression functions for data compression must not be one-way. The properties of a compression function to be used to *shrink* pseudo-random sequences are yet to be defined.

Decimation components are a particular case of compression components. The Shrinking Generator (SG) [3] compresses two sources of pseudo-random bits to create a third source of potentially better quality than the original sources; the term quality stands for the difficulty of predicting the pseudo-random sequence. Similarly, the Self-Shrinking Generator (SSG) [15], the Bit-Search Generator (BSG) [9] and its variants such as the ABSG [10] all compress a single source of pseudo-random bits in order to produce a second source of potentially better quality. The ABSG is used in the DECIM stream cipher proposed to the ECRYPT stream cipher project. The general running of DECIM is to produce a pseudo-random bit sequence from an LFSR filtered by a Boolean function which is next compressed by the ABSG.

The *output rate* is usually considered to compare the efficiency of compression components. The BSG and the ABSG have the advantage over the SG and the SSG that they operate at a rate $1/3$ instead of $1/4$ (i.e. producing n bits in the output requires on average $3n$ bits of the input sequence instead of $4n$ bits).

Security criteria are crucial for cryptographic compression components. Since many stream ciphers are LFSR-based, most theoretical results on compression components concern the period or the linear complexity of the sequences obtained by applying these components on the output of a maximal length LFSR. First, algebraic results show that regular decimation is not suitable [16]. Then, several attacks on stream cipher based on a compression component are known. The first type of attack focuses on the properties of the compression function when assuming that the input sequence is uniformly chosen. For instance, FBDD-attacks, proposed by Krause [13], rely on properties of the compression function in the context of LFSR-based generators. The attacks given in [10, 11] use the most probable case (when it exists) in order to reconstruct the input sequence in the context of LFSR-based generators. A second type of attack exploits more information on LFSR-based generators. For instance, the attack on the SG given in [3] exploits the knowledge of the feedback polynomials, and the attack on the SSG given in [6, 7] applies only for particular feedback polynomials.

3   A Compression Model for Pseudo-random Generation

One usually expects data compression techniques to transform an input sequence into a very short output sequence while keeping the ability to recover the input from the output, which means no information on the input shall be lost.

In the context of pseudo-random generation, the purpose is different. We focus on the use of the compressed output as the keystream used to cipher a message in a stream cipher. The input sequence s is supposed to be the pseudo-random output of a public mechanism with secret parameters (e.g., the output of an LFSR initialized with a secret key and an initialization vector). This mechanism may have weaknesses with respect to attacks aiming at correlation or algebraic properties of its output. Our aim is to delete from s enough information to prevent such attacks that may apply to s, by hiding algebraic properties of the input sequence. At the same time, our output should not be too short compared with its input, so that it can be used for the same applications as s.

Thus, our aim is opposite to usual data compression: we expect the compression algorithm to process the input into an output sequence which delivers as little information on the input as possible, while remaining as long as possible.

In the sequel, we call *random input sequences* those sequences that follow the uniform distribution of binary words: each word w is a prefix of a random input sequence with probability $1/2^{|w|}$, and all words are assumed to be independent.

3.1   Prefix Codes and Binary Trees

A *binary code* is a subset of words of $\{0, 1\}^+$. The language $\mathcal{C}^*$ of a binary code $\mathcal{C}$ is the set of all binary words that are concatenation of words in $\mathcal{C}$. A code $\mathcal{C}$ is a *prefix code* if no codeword has a strict prefix in $\mathcal{C}$. Notice that, in this case, the words of $\mathcal{C}^*$ parse into codewords in a unique manner. A code is *maximal prefix* when no other prefix code properly contains it. A code $\mathcal{C}$ is *right complete* if every word w can be completed into a word $v = ww'$ that belongs to $\mathcal{C}^*$ or, equivalently, if every word w with no prefix in $\mathcal{C}$ has a multiple $v = ww'$ in $\mathcal{C}$.

Proposition 1. *A code is maximal prefix if, and only if, it is prefix and right complete.*

Proof. Suppose $\mathcal{C}$ is maximal prefix. Let w be a non-empty word which has no prefix in $\mathcal{C}$. As $\mathcal{C}$ is maximal prefix, $\mathcal{C} \cup \{w\}$ is not a prefix code, so w has a right multiple in $\mathcal{C}$. Hence, $\mathcal{C}$ is right complete.

Conversely, let C be prefix and right complete, and C' be a prefix code that contains C. Let $w \in C'$. As C is right complete, w has a right multiple w' in C^*. Let then m be the smallest prefix of w' in C. As C' is prefix, this implies $m = w$, so we have $w \in C$, and consequently $C' = C$. Therefore, C is maximal prefix.   □

Throughout the paper, we will see that all suitable codes for our constructions are maximal prefix codes. There is a natural bijection between binary prefix codes and binary trees called *coding trees*, in which a node either is a leaf, or it has two children. This bijection links the words of the code and the leaves of the tree. Thus, we often use the equivalence between binary prefix codes and binary trees in the sequel. An example of a coding tree is given in Figure 1.

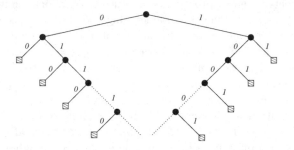

Fig. 1. ABSG code tree

3.2   General Framework

We consider an infinite input sequence of bits $\mathbf{s} = (s_i)_{i \geq 0}$, a binary prefix code C and a mapping $f : C \to \{0, 1\}^*$, called the *compression function*. We call $f(C)$ the *output set*. The sequence $\mathbf{s}$ parses into a sequence of codewords $\mathbf{w} = (w_i)_{i \geq 0}$, each w_i being the unique codeword such that $w_0 \ldots w_i$ is a prefix of $\mathbf{s}$ that belongs to C^*. Each w_i is then mapped by f to its image in $f(C)$. The output sequence is $(f(w_i))_{i \geq 0}$, seen as a bit sequence. We denote this output sequence by

$$\mathbf{y} = Enc_{C,f}(\mathbf{s}).$$

The framework extends to finite input sequences, by parsing the input sequence in the same way, until the remainder has no prefix in C.

Definition 1. *The* output rate *of the pair* (C, f), *denoted by* $Rate(C, f)$, *is the average number of output bits generated by one bit of a random input sequence.*

Obviously, not all binary codes and functions are suitable for this framework. For instance, choosing $C = \{00\}$ does not enable to process a sequence containing ones. As for the function, choosing the projection onto the empty word ε produces an empty output sequence. In order to apply the framework to every possible input sequence, it is then necessary to determine what the requirements on the following components are:

1. the choice of C must enable the parsing of every random input sequence,
2. the choice of f must be such that, for uniformly distributed input sequences, the corresponding output sequences also follow the uniform distribution.

3.3   Requirements on C

First, there are some straight requirements on C. In our framework, we consider prefix codes only. Indeed, if C contained two distinct words w and w' with w a prefix of w', then w' would never appear in the decomposition $\mathbf{w}$ of $\mathbf{s}$. Therefore, we may delete from C all the codewords that already have a prefix in C with no loss of generality, thus transforming C into a binary prefix code. Next, we want every random input sequence to be processable. This implies that C is right complete. Overall, in order to effectively process any random input, we introduce the following definition:

Definition 2. *A binary code C is* suitable *if it is prefix and if the expected length $E(C)$ of an element of C in the decomposition of a random input sequence is finite.*

Proposition 2. *For a suitable code C, the following equality holds:*

$$\sum_{w \in C} \frac{1}{2^{|w|}} = 1.$$

Proof. Let us consider the binary tree corresponding to C. We denote by L_n and N_n respectively the number of leaves and nodes of depth n. Then, we have $L_0 = 0$, $N_0 = 1$, and for every $n \geq 1$, the relation $L_n + N_n = 2N_{n-1}$ holds. Let $S_n = \sum_{0 \leq k \leq n} \frac{L_n}{2^n}$. Then, we have $\frac{L_n}{2^n} = \frac{N_{n-1}}{2^{n-1}} - \frac{N_n}{2^n}$, which gives $S_n = N_0 - \frac{N_n}{2^n} = 1 - \frac{N_n}{2^n}$, so we only have to prove $N_n = o(2^n)$.

Now, N_n is the number of nodes of depth n, and a random input sequence begins with n bits corresponding to such a node with probability $\frac{N_n}{2^n}$. For each one of these nodes, the first word of the input sequence recognized as a word of C has length at least n. Thus, these nodes contribute at least $n\frac{N_n}{2^n}$ to $E(C)$. As $E(C)$ is finite, this implies that $\frac{N_n}{2^n}$ tends to 0 when n tends to ∞. $\qquad\square$

Therefore, in the case of a suitable code, $E(C)$ is equal to the mean length of the words of C for the uniform distribution on the alphabet $\{0, 1\}$, so we have:

Proposition 3. *Let C be a suitable code. Then, we have the equality*

$$E(C) = \sum_{w \in C} \frac{|w|}{2^{|w|}}.$$

Remark 1. If a prefix code C satisfies the equality in Proposition 2 (for binary codes, otherwise 2 is replaced by the size of the alphabet), then it is maximal prefix, and the equivalence holds when C is finite (see for instance [2]). Thus, suitable codes are maximal prefix, and the converse is true for finite codes, with $E(C)$ then being given in Proposition 3. However, being maximal prefix may not be sufficient for $E(C)$ to converge when C is infinite, as Example 1 will show.

Example 1. Let us consider the code $\mathcal{C}$ defined iteratively as follows: for every n starting from $n = 0$, $\mathcal{C}$ contains all the words $w1^{2^n}$, where w is a word of length 2^n with no prefix already in $\mathcal{C}$. The defined code is prefix, and every word w with no prefix in $\mathcal{C}$ with $2^{n-1} < |w| \leq 2^n$ can be completed into a word of $\mathcal{C}$ by concatenating enough 1's to reach length 2^{n+1}, so $\mathcal{C}$ is right complete. Therefore, $\mathcal{C}$ is maximal prefix. However, the number of words of length 2^{n+1} in $\mathcal{C}$ being at most 2^n, we get

$$\sum_{w \in \mathcal{C}} \frac{1}{2^{|w|}} \leq \sum_{n \geq 0} \frac{1}{2^{2^n}},$$

this last sum being strictly less than 1. Thus, a random binary sequence may never fall into $\mathcal{C}$ with non-zero probability. Hence, $E(\mathcal{C})$ is infinite, and it is no longer equal to the mean length of code words, which, here, is finite. The code $\mathcal{C}$ is an example of maximal prefix code which is not suitable.

3.4   Requirements on f

As for $\mathcal{C}$, there are also several immediate requirements on $f(\mathcal{C})$. However, they are more practical than theoretical: at first glance, $f(\mathcal{C})$ may be any set of binary words, including ε. Now, it must obviously contain at least two non-empty words, one beginning with 0, and the other with 1, in order to make it possible for the output to look random for random inputs. Moreover, it must be possible to construct every binary sequence with the elements of $f(\mathcal{C})$.

In order to be able to process every random input sequence, we introduce the following definition, which corresponds to the requirement of Definition 2:

Definition 3. *Suppose $\mathcal{C}$ is a suitable code. Let f be a compression function $f : \mathcal{C} \rightarrow \{0,1\}^*$. We say that the pair $(\mathcal{C}, f)$ is a proper encoder if the expected length $E(f(\mathcal{C}))$ of the image by f of an element of $\mathcal{C}$ in the decomposition of a randomly chosen input sequence is finite and nonzero.*

As we review the properties of the output sequences with respect to uniformly distributed input sequences, we have:

Proposition 4. *For a proper encoder $(\mathcal{C}, f)$, the expected length of the image by f of an element of $\mathcal{C}$ in the decomposition of a randomly chosen input sequence, denoted by $E(f(\mathcal{C}))$, is given by*

$$E(f(\mathcal{C})) = \sum_{w \in \mathcal{C}} \frac{|f(w)|}{2^{|w|}}.$$

Definitions 2 and 3 ensure the finiteness of $E(\mathcal{C})$ and $E(f(\mathcal{C}))$, so we get:

Proposition 5. *The* output rate *of a proper encoder $(\mathcal{C}, f)$ is given by*

$$Rate(\mathcal{C}, f) = \frac{E(f(\mathcal{C}))}{E(\mathcal{C})}.$$

Now, we are going to determine an optimal choice for the output set $f(C)$ against reconstruction of the input. To every word in the output corresponds a set of preimages in C. Knowing an output word thus reduces the possible choices of preimages to one particular set. We will show that, in order to minimize the information rate, the set $f(C)$ should be as small as possible.

In order to ensure that the distribution of the output sequences satisfies randomness properties such as those described in [8], each bit of the output sequence must have equal probability to be 0 or 1. Therefore, we need, for every $n \geq 1$:

$$\sum_{w\in C,|f(w)|\geq n, f(w)_n=0} \frac{1}{2^{|w|}} = \sum_{w\in C,|f(w)|\geq n, f(w)_n=1} \frac{1}{2^{|w|}},$$

where $f(w)_n$ is the n-th bit of the word $f(w)$.

The prefix code output case. First, we consider the case where $f(C)$ is a prefix code. If it contains two elements, the only possible choice such that the probability distribution of the output for random inputs is that of a random sequence is $f(C) = \{0,1\}$. In this case, 0 and 1 must have probability $\frac{1}{2}$ to appear in the output sequence for a random input sequence.

Suppose now that $f(C)$ has more than 2 elements. We want to prove that, given a random input sequence, knowing the output sequence, we can retrieve more information on the first element of C than in the case $f(C) = \{0,1\}$.

Proposition 6. *Let (C, f) be a proper encoder, and, for $x \in f(C)$, let $P(x) = \sum_{w\in f^{-1}(x)} \frac{1}{2^{|w|}}$. Then, for a random input sequence $\mathbf{s}$, each word of the decomposition of $\mathbf{s}$ over C has average length $E(C)$, and it is known with an average entropy*

$$E(C) + \sum_{x\in f(C)} P(x) \log P(x).$$

Proof. For $x \in f(C)$, let us denote by C_x the preimage of x in C. Then, the probability that the first element of C recognized in a random input sequence is mapped by f to x is $P(x) = \sum_{w\in C_x} \frac{1}{2^{|w|}}$. Similarly, the expected length of an element in the preimage of x is $E(C_x) = \frac{1}{P(x)}\sum_{w\in C_x} \frac{|w|}{2^{|w|}}$. At last, we compute the entropy on the elements in C_x:

$$H(C_x) = -\sum_{w\in C_x} \frac{1}{P(x)2^{|w|}} \log \frac{1}{P(x)2^{|w|}} = \sum_{w\in C_x} \frac{1}{P(x)2^{|w|}} (\log P(x) + |w|)$$

$$= \frac{1}{P(x)} \sum_{w\in C_x} \frac{|w|}{2^{|w|}} + \frac{\log P(x)}{P(x)} \sum_{w\in C_x} \frac{1}{2^{|w|}} = E(C_x) + \log P(x).$$

The average number of bits retrieved is therefore $\sum_{x\in f(C)} P(x)E(C_x) = E(C)$ for a random input sequence, so it does not depend on $f(C)$. The average entropy is

$$\sum_{x\in f(C)} P(x)(E(C_x) + \log P(x)) = E(C) + \sum_{x\in f(C)} P(x) \log P(x),$$

with $\sum_{x\in f(C)} P(x) = 1$. □

It is always possible, given a suitable code $\mathcal{C}$, to divide $\mathcal{C}$ into two equiprobable subsets (the probabilities of leaves in the tree being of the form $\frac{1}{2^n}$ with $n \geq 1$, and their sum being 1). Thus, for every suitable code, there exists a mapping $f : \mathcal{C} \to \{0,1\}$ such that 0 and 1 are output with probability $\frac{1}{2}$.

Therefore, in order to maximize the entropy for a given suitable code $\mathcal{C}$, the value of $|\sum_{x \in f(\mathcal{C})} P(x) \log P(x)|$ should be as small as possible, which implies $\#(f(\mathcal{C})) = 2$. Therefore, the optimal choice of the output set is $f(\mathcal{C}) = \{0,1\}$, with 0 and 1 having probability $\frac{1}{2}$ to be output for a random input sequence.

The non-prefix output case. We now consider the case where $f(\mathcal{C})$ does not contain the empty word ε, but $f(\mathcal{C})$ is not a prefix code. Let $\mathcal{C}(\mathbf{y})$ be the set of words of $\mathcal{C}$ such that, for every $w \in \mathcal{C}(\mathbf{y})$, the sequence $\mathbf{y}$ begins with w. Then, the probability that $\mathbf{s}$ begins with w depends on $\mathbf{y}$.

Example 2. Suppose $f(\mathcal{C}) = \{0, 01, 10, 11\}$ with $P_0 = P_{11} = \frac{1}{3}$ and $P_{01} = P_{10} = \frac{1}{6}$. Then, the first word of the finite output sequence 010 corresponds to a pair of words (w, w') of $\mathcal{C}$, with either $f(w) = 0$ and $f(w') = 10$, or $f(w) = 01$ and $f(w') = 0$. As we have $P_0 P_{10} = P_{01} P_0$, the probabilities that $f(w) = 0$ and $f(w) = 01$ are equal whereas $P_0 > P_{01}$.

Thus, it is no longer possible to determine with certainty each word $(f(w_i))$ in the image of the input sequence. However, a path similar to that of Section 3.4 can be followed. The corresponding Proposition 12 and its proof are provided in Appendix A. They lead to the same conclusion as Section 3.4, namely that the optimal choice of the output set is $f(\mathcal{C}) = \{0,1\}$ (thus being prefix), with 0 and 1 having probability $\frac{1}{2}$ to appear in the output sequence for a random input sequence. This case is discussed in Section 4.

General case. We now suppose that ε can belong to the output set $f(\mathcal{C})$.

Proposition 7. *Let $(\mathcal{C}, f)$ be a proper encoder such that $f(\mathcal{C})$ contains ε. Then, there exists a proper encoder $(\mathcal{C}', f')$ such that $f'(\mathcal{C}')$ does not contain ε and that, for every infinite binary sequence $\mathbf{s}$, we have*

$$Enc_{\mathcal{C}, f}(\mathbf{s}) = Enc_{\mathcal{C}', f'}(\mathbf{s}).$$

Moreover, defining $P_\varepsilon = \sum_{w \in f^{-1}(\varepsilon)} \frac{1}{2^{|w|}}$, we have

$$E(\mathcal{C}') = \frac{1}{1 - P_\varepsilon} E(\mathcal{C}), \quad and \quad E(f'(\mathcal{C}')) = \frac{1}{1 - P_\varepsilon} E(f(\mathcal{C})).$$

Proof. Denote by $\mathcal{C}_\varepsilon$ the set of preimages of ε, and by $\mathcal{C}_{\bar{\varepsilon}}$ the complement of $\mathcal{C}_\varepsilon$ in $\mathcal{C}$. Let $\mathcal{C}'$ be the binary code defined by $\mathcal{C}' = \mathcal{C}_\varepsilon^* \mathcal{C}_{\bar{\varepsilon}}$, that is, the set of binary words that parse into a sequence of words of $\mathcal{C}_\varepsilon$, followed by a word of $\mathcal{C}_{\bar{\varepsilon}}$. Consider the function f' that maps each element ww' of $\mathcal{C}'$, with $w \in \mathcal{C}_\varepsilon^*$, and $w' \in \mathcal{C}_{\bar{\varepsilon}}$, to $f(w')$. As the decomposition is unique, f' is well-defined. Moreover, for every input sequence $\mathbf{s}$, the equality $Enc_{\mathcal{C}, f}(\mathbf{s}) = Enc_{\mathcal{C}', f'}(\mathbf{s})$ is obviously satisfied. At last, we have $f(\mathcal{C}') = f(\mathcal{C}) \backslash \{\varepsilon\}$, so the image of f' does not contain ε.

There remains to show that the new pair $(\mathcal{C}', f')$ is also a proper encoder. First, $\mathcal{C}'$ is also a prefix code because of unicity of the decomposition over $\mathcal{C}$. Next, as the length of ε is 0, we have

$$E(f'(\mathcal{C}')) = \sum_{v \in \mathcal{C}_\varepsilon^*, w \in \mathcal{C}_\varepsilon} \frac{|f(w)|}{2^{|v|+|w|}} = \sum_{n \geq 0} \left(\sum_{v \in \mathcal{C}_\varepsilon} \frac{1}{2^{|v|}} \right)^n \times \sum_{w \in \mathcal{C}_\varepsilon} \frac{|f(w)|}{2^{|w|}} = \frac{E(f(\mathcal{C}))}{1 - P_\varepsilon}.$$

As the two encoders $(\mathcal{C}, f)$ and $(\mathcal{C}', f')$ are equivalent, they have the same output rate, which yields the same relation between $E(\mathcal{C}')$ and $E(\mathcal{C})$. Hence, $(\mathcal{C}', f')$ is a proper encoder. $\square$

Proposition 7 shows that we can suppose without loss of generality that $f(\mathcal{C})$ does not contain ε. Therefore, the optimal choice for $f(\mathcal{C})$ is

$$f(\mathcal{C}) = \{0, 1\}.$$

4   The $\{0, 1\}$-Case

In this section, we focus on the optimal choice of the proper encoder $(\mathcal{C}, f)$ when $f(\mathcal{C}) = \{0, 1\}$, with 0 and 1 equiprobable relatively to the uniform distribution over the input sequence. We first give the results that arise from Section 3 in this case, and we study the security of the framework against well-known attacks: exhaustive reconstruction, most probable case reconstruction, equations retrieval and FBDD attacks. Then, using these security results, we deduce the optimal choice for $(\mathcal{C}, f)$ against these attacks.

4.1   Parameters of the $\{0, 1\}$-Case

Firstly, we give some general properties of the framework in the $\{0, 1\}$ case. We denote by $\mathcal{C}_0$ and $\mathcal{C}_1$ the two sets of preimages of respectively 0 and 1 by f. We also define $\mathcal{C}_b^n = \{w \in \mathcal{C}_b, |w| = n\}$ and $D_b^n = \#(\mathcal{C}_b^n)$.

Proposition 8. *Let $(\mathcal{C}, f)$ be a proper encoder with $f(\mathcal{C}) = \{0, 1\}$. Then, for a random input sequence $\mathbf{s}$, the average length and entropy of each word of the decomposition of $\mathbf{s}$ over $\mathcal{C}$ are respectively $E(\mathcal{C})$ and $E(\mathcal{C}) - 1$.*

This result comes from Proposition 6 when applied to the case $f(\mathcal{C}) = \{0, 1\}$, with 0 and 1 being equiprobable. This equiprobability also implies:

Proposition 9. *Given a bit b of the output sequence, a word $w \in \mathcal{C}_b$ is the preimage of b with probability $\frac{1}{2^{|w|-1}}$.*

Proof. Each word w of $\mathcal{C}$ appears in the input sequence with probability $\frac{1}{2^{|w|}}$, and the probability that w belongs to $\mathcal{C}_b$ is $\frac{1}{2}$, which gives the result. $\square$

4.2   Security Analysis

This section is dedicated to the general analysis of the security provided by the compression component. We also focus on the case when the input sequence is the output of a maximal length LFSR.

Exhaustive reconstruction. Exhaustive reconstruction consists in reconstructing consecutive bits of the input sequence from the output sequence starting from a fixed point in the output sequence. When a bit b appears in the output, the expected length and the entropy on the preimage of b in the input sequence are respectively equal to

$$E_b = \sum_{w \in \mathcal{C}_b} \frac{|w|}{2^{|w|-1}} \text{ and } H_b = - \sum_{w \in \mathcal{C}_b} \frac{1}{2^{|w|-1}} \log_2\left(\frac{1}{2^{|w|-1}}\right).$$

Developing H_b gives

$$H_b = \sum_{w \in \mathcal{C}_b} \frac{|w|-1}{2^{|w|-1}} = E_b - 1.$$

Therefore, for a bit b in the output, one can deduce E_b bits in the input, with entropy $E_b - 1$.

Suppose that the input sequence is given by a LFSR of length L with a public feedback polynomial and with the secret key as its initial state. Let $E = \frac{E_0 + E_1}{2}$. It is therefore possible to retrieve the complete state of the LFSR with an attack of average complexity $\mathcal{O}(2^{\frac{E-1}{E}L})$, requiring $\mathcal{O}(\frac{L}{E})$ consecutive output bits.

Moreover, when $E_0 \neq E_1$ holds, the complexity of the attack can be reduced by seeking for a sequence where mostly bits b appear, with b such that $E_b < E_{\bar{b}}$. This yields an attack with better complexity, but requiring the knowledge of more output bits. The general running of this attack consists in taking a window of consecutive bits in the keystream sequence where most bits are b. The difficulty when mounting this attack is to determine the better trade-off between the length of the window and the required number of bits b in this window in order to retrieve L equations involving consecutive bits of the input sequence. Such an attack is described in [10] in the case of the BSG decimation algorithm.

Reconstruction based on the most probable case. Another reconstruction attack consists in betting each time that the preimage of a bit b is (one of) the most probable. Consequently, for each bit b, we set $\ell_b = \min\{|w|, w \in \mathcal{C}_b\}$, and $\mathcal{C}_b^{\text{short}} = \{w \in \mathcal{C}_b, |w| = \ell_b\}$. Contrary to the previous attack, we cannot choose the point from which consecutive input bits will be effectively reconstructed.

For a bit b in the output, the preimage of b is $w \in \mathcal{C}_b^{\text{short}}$ with probability $1/2^{\ell_b-1}$. Thus, we recover ℓ_b bits of the input with probability $1/2^{\ell_b-1}$.

Suppose now that the input sequence is given by a LFSR of length L. Let $\ell = \frac{\ell_0 + \ell_1}{2}$. It is then possible to retrieve the complete state of the LFSR with an attack of average complexity $\mathcal{O}(2^{\frac{\ell-1}{\ell}L})$, requiring $\mathcal{O}(2^{\frac{\ell-1}{\ell}L})$ output bits (namely, enough for the bet to succeed). In the case where not all the preimages of b have the same length, we have $\ell_b < E_b$, so the complexity of this attack is less than that of exhaustive reconstruction.

Like in exhaustive reconstruction, when $\ell_0 \neq \ell_1$ holds, the attack complexity can be reduced by seeking sequences where most bits are b, such that $\ell_b < \ell_{\bar{b}}$.

Equations retrieval. In some cases, and in particular when the input sequence is given by a maximum-length LFSR, it is sufficient to retrieve linear equations on bits that are not consecutive in the input sequence.

However, it is not necessarily easier to retrieve bits that are apart in the input sequence, because the compression process creates entropy on the length of the preimages of words in the output sequence. Thus, retrieving bits that are apart means that we are able to control the length of the gaps between the bits retrieved in the input sequence.

For a bit b in the output, the preimage of b has length n with probability $\frac{D_b^n}{2^{n-1}}$, where D_b^n is the number of preimages of b of length n. Now, if the preimage of b has length n, then we can derive a number ϕ_b^n of linear equations on the input bits satisfying

$$\max\left(0, n - (D_b^n - 1)\right) \le \phi_b^n \le n - \lceil \log(D_b^n) \rceil.$$

Therefore, we can retrieve at least $n + 1 - D_b^n$ equations with probability $\frac{D_b^n}{2^{n-1}}$. For a bit b in the output, the average number of retrieved linear equations is thus

$$\bar{\phi}_b = \sum_{n \ge 1} \frac{D_b^n \phi_b^n}{2^{n-1}},$$

the entropy on the length of the preimage of b being

$$H_b^{\text{length}} = -\sum_{n \ge 1} \frac{D_b^n}{2^{n-1}} \log\left(\frac{D_b^n}{2^{n-1}}\right).$$

In the best case (which can always be achieved by properly choosing $\mathcal{C}$ and f), where ϕ_b^n is the least possible, we obtain:

Proposition 10. *Consider a proper encoder $(\mathcal{C}, f)$ such that $f(\mathcal{C}) = \{0, 1\}$, with 0 and 1 having the same probability for random input sequences. Let $\bar{\phi}_b$ and H_b^{length} be the average number of retrieved linear equations for a bit b and the associated entropy on the length of the preimage of b. Then, we have*

$$\bar{\phi}_b = E_b - \delta_b^\phi, \ \text{with} \ \delta_b^\phi = \sum_{n \ge 1} \frac{D_b^n}{2^{n-1}} \min(n, D_b^n - 1),$$

and

$$H_b^{length} = E_b - 1 - \delta_b^H, \ \text{with} \ \delta_b^H = \sum_{n \ge 1} \frac{D_b^n}{2^{n-1}} \log D_b^n.$$

Moreover, δ_b^ϕ and δ_b^ϕ are both positive, and they satisfy $\delta_b^\phi \ge \delta_b^H$.

Proof. The formulas for δ_b^ϕ and δ_b^H both follow from straight computation. Now, we always have $D_b^n \le 2^n$, so $\log D_b^n$ is always at most n. Moreover, for every integer $x > 1$, we have $x - 1 \ge \log x$. So, for every n such that $D_b^n \ne 0$, we have $\min(n, D_b^n - 1) \ge \log D_b^n$. $\qquad \square$

These results link the complexity of equations retrieval attacks with exhaustive reconstruction by way of E_b. As a consequence of this proposition, when 0 and 1 have the same number of preimages of each given length, retrieving L equations has complexity at least $\mathcal{O}(2^{\frac{E-1-\delta^\phi}{E-\delta^\phi}L})$, while exhaustive reconstruction of L bits has complexity $\mathcal{O}(2^{\frac{E-1}{E}L})$. Thus, for $\delta^\phi = 0$, equations retrieval is not more effective than exhaustive reconstruction. This happens only when each bit has at most one preimage of each length.

Suppose now the input sequence is given by a LFSR of length L. It is therefore possible to retrieve L linear equations on the input bits of the LFSR with an attack of average complexity

$$\mathcal{O}(2^{(\frac{H_0^{\text{length}}}{\phi_0}+\frac{H_1^{\text{length}}}{\phi_1})\frac{L}{2}}),$$

requiring $\mathcal{O}(L)$ consecutive output bits.

Like in the previous attacks, when $\frac{\bar{\phi_0}}{H_0^{\text{length}}} \neq \frac{\bar{\phi_1}}{H_1^{\text{length}}}$ holds, the complexity of the attack can be reduced by seeking for a sequence where mostly bits 0 or 1 appear (depending on the inequality direction). The attack thus obtained has better complexity, but requires the knowledge of more output bits.

Example 3. We consider the ABSG code tree. For every length $n \geq 2$, there is exactly one preimage of 0 and one preimage of 1 of length n. We obtain

$$\bar{\phi_b} = \sum_{n \geq 2} \frac{n}{2^{n-1}} = 3 = E_b, \text{ and } H_b^{\text{length}} = \sum_{n \geq 2} \frac{n-1}{2^{n-1}} = \bar{\phi_b} = H_b.$$

The equations retrieval attack is thus as difficult as exhaustive reconstruction for the ABSG.

FBDD attacks. Krause [13] introduced the FBDD-attack (standing for Free Binary Decision Diagram) which is a cryptanalysis method for LFSR-based generators, i.e., a generator LG that, for each initial state $x \in \{0,1\}^n$, outputs a linear bitstream $LG(x)$, and a compression function which compresses the linear bitstream. The cryptanalysis method relies on two assumptions called the *FBDD Assumption* and the *Pseudo-randomness Assumption* (see [13] for details).

The cost of the cryptanalysis depends on two properties of the compression function that are a parameter γ linked to the maximal length of the sequence output by the compression function when applied on all sequences of length m, and some parameter α (see [13] and some details in [10]); the two parameters α and γ are reals between 0 and 1. Then, the time and space complexity of the FBDD-attack is $L^{\mathcal{O}(1)}2^{\frac{1-\alpha}{1+\alpha}L}$ and it requires $\lceil \gamma\alpha^{-1}L \rceil$ consecutive bits of the keystream in order to compute L consecutive bits of the input sequence.

When the probability that the image of a randomly chosen finite input sequence is a prefix of a given output sequence varies according to the output sequence, it is not clear whether the original FBDD-attack may be improved to be more efficient.

4.3   Optimal Choices

In this part, we construct an optimal proper encoder in light of the attacks considered previously.

Requirements based on security analysis. In order to thwart attacks based on asymmetry between the preimage of 0 and that of 1, each output bit must have the same number of preimages of a given length.

Next, in order to maximize the complexity of most probable case attacks while keeping a good output rate, the length of the shortest word in $\mathcal{C}$ should be as close as possible to the average length of the words in $\mathcal{C}$.

Example 3 shows that the ABSG compression mechanism is optimal regarding equations retrieval attacks, meaning that it is not easier to retrieve equations than to reconstruct consecutive bits of the input sequence.

In the general case, equations retrieval attacks can have a better complexity than exhaustive reconstruction. However, as shown in Proposition 10, in order to lessen their efficiency, each bit should have at most one preimage of each length.

Construction of an optimal framework. For an output rate at least $\frac{1}{2}$, the number of choices for the proper encoder are finite, because of symmetry requirements, and the output rate is either equal to 1 or $\frac{1}{2}$ exactly. For $Rate = 1$, there are two proper encoders, with $\mathcal{C} = \{0, 1\}$, which is insecure. For $Rate = \frac{1}{2}$, one can construct 6 proper encoders. The suitable code is $\mathcal{C} = \{00, 01, 10, 11\}$, and the function f is such that 0 and 1 have two preimages each. For each choice, as the length of the preimages is constant, we can apply the equations retrieval attack and solve the corresponding system. The complexity is then $\mathcal{O}(L)$.

Let then h be the minimal depth of leaves in the tree. As each output bit must have the same number of preimages of a given length, the number of preimages of 0 and 1 of depth h in $\mathcal{C}$ is the same. Then, the complexity of reconstruction using the most probable case is $O(\frac{h-1}{h}L)$. In order to maximize the output rate, we have to choose $h = 2$, and no level in the tree should have only internal nodes. This implies that, at every depth more than 2, the tree must have exactly 2 leaves, until the last level with depth d, where it has 4 leaves. We denote by T_2^d the set of code trees of depth d, and exactly 2 leaves of depth $2, 3, \ldots, d-1$ (hence 4 leaves of depth d for $d < \infty$). The ABSG code tree belongs to T_2^∞. In order to obtain proper encoders using these codes, one only has to use functions f such that the number of preimages of 0 and 1 of each depth in $\mathcal{C}$ is the same.

This optimal code can be adapted for smaller output rates, beginning at depth $h > 2$. This makes most probable case attacks more complex, though another way of complexifying them is to act on the input sequence using, for instance, a longer LFSR. The tree considered then has exactly 2^{h-1} leaves of depth $h, \ldots, d-1$ and maximal depth d (reached by exactly 2^h leaves when d is finite). Notice that the trees T_h^d can be constructed by putting 2^{h-2} trees of T_2^{d+2-h} at depth $h - 2$ in a tree with all internal nodes until depth $h - 2$.

However, equations retrieval attacks are more efficient for $h > 2$:

Proposition 11. *Consider a proper encoder $(\mathcal{C}, f)$ such that the code tree of $\mathcal{C}$ is a T_h^d tree, and that 0 and 1 have the same number of preimages of each given* *length. Suppose also that $\mathcal{C}$ is such that the number of equations linking the preimages of b of length n is the least possible, namely $\phi_b^n = \max(0, n+1 - D_b^n)$. Then, we have:*

1. *for every $h \geq 2$, the entropy on the length of the preimage of a given output bit b is $H^{length} = 2 - 2^{h+1-d}$,*
2. *for $2 \leq h \leq 4$, we have $\bar{\phi}_b = h + 2 - 2^{h-2} - 2^{h-d} - 2^{2h-d-2}$, which is equal to 3 for $h = 3$ and $d = \infty$.*

Proof. These are the results of straight, yet tedious computations.   □

As a consequence of these results, and namely of the entropy remaining less than 2, the complexity of equations retrieval attacks does not grow fast when the output rate decreases. Therefore, the optimal framework against these attacks is reached when the code tree belongs to T_2^∞. However, the attack complexity remains at least $\mathcal{O}(2^{\frac{L}{2}})$ for trees in T_2^d with $d > 2$.

Definition 4. *We say that a proper encoder $(\mathcal{C}, f)$ is an optimal encoder if the associated code tree belongs to T_2^∞, and if 0 and 1 have exactly one preimage by f of length ℓ, for $\ell \geq 2$.*

In Table 1, we provide the characteristics of proper encoders constructed on the basis of general T_h^d trees as defined in Proposition 11. We also provide a comparison with the SSG. We left aside polynomial terms in the computational complexity. One should also note than most probable case attacks require much known keystream, whereas the other attacks considered require only a number of bits linear in L, where L is the number of bits we want to retrieve. The results for $FBDD$ attacks are taken from [13, 10] for the SSG and ABSG. Moreover, the complexity of FBDD attacks is the same for all optimal encoders, including the ABSG. We see that equations retrieval attacks are more powerful against T_2^d trees than exhaustive reconstruction, which is why we did not consider them as optimal. However, they may be easier to protect against timing attacks than optimal encoders, because the length of their codewords is bounded.

5   Applications

5.1   Bit-Search-Based Generators

In [9], the BSG algorithm was proposed, and was presented together with the ABSG, which was then described in [10]. Both share the same code tree presented in Figure 1, which belongs to T_2^∞, and thus fits in our framework. The corresponding code is $\mathcal{C} = \{01^k0, 10^k1, k \geq 0\}$.

In the case of the BSG, the compression function f_{BSG} maps codewords of length 2 to 0, and the other codewords to 1. Therefore, it is not an optimal encoder. This asymmetry resulted in several attacks [10, 11]. For instance, the equations retrieval attack takes advantage of it and it is especially efficient against the BSG, with complexity $\mathcal{O}(2^{\frac{1}{3}L})$.

Table 1. Characteristics and attack exponent against T_h^d trees filtering LFSRs

	Output rate	Exhaustive reconstruction	Most probable case	Equations retrieval	FBDD attacks
T_h^d	$\frac{1}{h+1-2^{h-d}}$	$\frac{h-2^{h-d}}{h+1-2^{h-d}}L$	$\frac{h-1}{h}L$	see Prop.11	n/a
T_h^∞	$\frac{1}{h+1}$	$\frac{h}{h+1}L$	$\frac{h-1}{h}L$	see Prop.11	n/a
T_2^d	$\frac{1}{3-2^{2-d}}$	$\frac{2-2^{2-d}}{3-2^{2-d}}L$	$\frac{1}{2}L$	$\frac{2-2^{3-d}}{3-2^{3-d}}L$	n/a
T_2^∞ (ABSG)	$\frac{1}{3}$	$\frac{2}{3}L$	$\frac{1}{2}L$	$\frac{2}{3}L$	$\simeq 0.532L$
T_3^∞	$\frac{1}{4}$	$\frac{3}{4}L$	$\frac{2}{3}L$	$\frac{2}{3}L$	$\simeq 0.615L$
SSG	$\frac{1}{4}$	$\frac{3}{4}L$	$\frac{1}{2}L$	$\frac{2}{3}L$(see Section 5.2)	$\simeq 0.656L$

In the case of the ABSG, the compression function f_{ABSG} maps codewords to their second bit, so it is an optimal encoder. Therefore, the ABSG is optimal against the attacks we described. Their complexity is given in table 1.

5.2 Self-shrinking Generator

Let us set $\mathcal{C} = \{00, 01, 10, 11\}$, and define $f : \mathcal{C} \to \{0, 1, \varepsilon\}$ by : $f(00) = f(01) = \varepsilon$, $f(10) = 0$ and $f(11) = 1$. The Self-Shrinking Generator is exactly the scheme corresponding to the pair $(\mathcal{C}, f)$ in our framework.

The pair $(\mathcal{C}, f)$ is a proper encoder, but it contains ε. Following the transformation described in Proposition 7, we set $\mathcal{C}' = \{(0\{0, 1\})^*1\{0, 1\}\}$, and we define $f' : \mathcal{C}' \to \{0, 1\}$ by $f(w) = b$ for $w \in \{(0\{0, 1\})^*1b\}$.

The pair $(\mathcal{C}', f')$ is a proper encoder that has an optimal output set and satisfies the symmetry requirement: at every level of the corresponding tree, described in Figure 2, there are exactly as many preimages of 0 and 1.

The SSG is neither an optimal encoder, nor is it optimal among proper encoders having the same output rate. This comes from the fact that one out of two levels in the tree is empty. Let us compare the corresponding scheme to the optimal choice for the same output rate ($\frac{1}{4}$ for the SSG), whose code tree is a T_3^∞ tree. For both schemes, the complexity of the exhaustive reconstruction attack is the same, namely $\mathcal{O}(2^{\frac{3}{4}L})$. However, the complexity of the most probable case

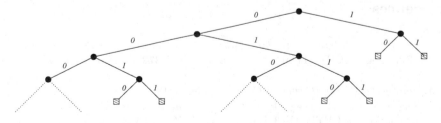

Fig. 2. SSG code tree

attack against the SSG is $\mathcal{O}(2^{\frac{L}{2}})$, requiring $\mathcal{O}(2^{\frac{L}{2}})$ bits of the output. For the T_3^∞ choice, this attack has complexity $\mathcal{O}(2^{\frac{2}{3}L})$, and requires $\mathcal{O}(2^{\frac{2}{3}L})$ output bits. Therefore, the SSG is not optimal against most probable case attacks.

Moreover, for each output bit, the input has length $2n$ with probability $\frac{1}{2^n}$, in which case one can recover $n+1$ equations. This yields that 3 equations are known on average, with an entropy of 2. Therefore, the equations retrieval attack has complexity $\mathcal{O}(2^{\frac{2}{3}L})$, which is the same as T_3^∞, but also as T_2^∞ (ABSG). As this attack requires a number of bits linear in L, it is as practical as exhaustive reconstruction. Notice that the equations retrieval attack against the SSG has almost the same complexity as the FBDD attack of Krause [13], while not requiring a large amount of memory.

Therefore, an optimal encoder such as the ABSG is as secure against the attacks considered in this paper as the Self-Shrinking Generator (apart from FBDD-attacks), while providing a better output rate ($\frac{1}{3}$ instead of $\frac{1}{4}$).

6   Conclusion and Further Work

In this paper, we have extensively studied how to compress efficiently and securely the output of pseudo-random generators. It turns out that the ABSG, which was introduced in [9, 10], and is part of the DECIM proposal to the ECRYPT stream cipher project [5], has the optimal properties against several well-known attacks. But it is also possible to design several other optimal encoders with the same properties, using code trees taken from the T_2^∞ infinite family. At last, we have also shown compression components based on these trees are almost as secure as the Self-Shrinking Generator [15], while providing an output rate of $\frac{1}{3}$ instead of $\frac{1}{4}$. We consider two main directions for research in this area. First, one could use another generator to choose the compression function at each iteration, while keeping the same code tree. The idea is thus to generalize this framework by using other pseudo-random generators to control compression. This should provide us with comparisons with the Shrinking Generator [3]. Second, if the compression function and the code are chosen properly, a compression component may also erase the bias of a pseudo-random generator that does not produce every bit sequence with equal probability. It then seems possible to construct a general design for bias-erasing compression.

References

1. F. Armknecht, M. Krause, *Algebraic Attacks on Combiners with Memory*, Advances in Cryptology – CRYPTO'03 Proceedings, LNCS **2729**, Springer-Verlag, (2003), 162–176.
2. J. Berstel, D. Perrin, *Theory of Codes*, Academic Press, (1985).
3. D. Coppersmith, H. Krawczyk, Y. Mansour, *The Shrinking Generator*, Advances in Cryptology – CRYPTO'93 Proceedings, LNCS **773**, Springer-Verlag, (1993), 22–39.

4. N. Courtois, W. Meier, *Algebraic Attacks on Stream Ciphers with Linear Feedback* Advances in Cryptology – EUROCRYPTO'03 Proceedings, LNCS **2656**, Springer-Verlag, (2003), 345–359.

5. eStream, Stream cipher project of the European Network of Excellence in Cryptology ECRYPT, http://www.ecrypt.eu.org/stream/.

6. P. Ekdahl, T. Johansson, W. Meier, *Predicting the Shrinking Generator with Fixed Connections*, Advances in Cryptology – EUROCRYPT 2003 Proceedings, LNCS **2656**, Springer-Verlag, E. Biham, ed., (2003), 330–344.

7. P. Ekdahl, T. Johansson, W. Meier, *A note on the Self-Shrinking Generator*, In *Proc. of International Symposium on Information Theory*, page 166, IEEE, (2003).

8. S. Golomb, *Shift Register Sequences*, Revised Edition, Aegean Park Press, (1982).

9. A. Gouget, H. Sibert, *The Bit-Search Generator*, In *The State of the Art of Stream Ciphers: Workshop Record, Brugge, Belgium, October 2004*, pages 60–68, (2004).

10. A. Gouget, H. Sibert, C. Berbain, N. Courtois, N. Debraize and C. Mitchell, *Analysis of the Bit-Search Generator and sequence compression techniques*, Proceedings of FSE'05, LNCS **3557**, Springer-Verlag, (2005).

11. M. Hell, T. Johansson, *Some attacks on the Bit-Search Generator* Proceedings of FSE'05, LNCS **3557**, Springer-Verlag, (2005).

12. P. Kocher, *Timings attacks on implementations of Diffie–Hellman, RSA, DSS and other systems*, Proceedings of Crypto 1996, LNCS **1109**, Springer-Verlag, (1996).

13. M. Krause. *BDD-based Cryptanalysis of Keystream Generators*, In EURO-CRYPT 2002, pp. 222-237, LNCS 2332, Springer, (2002).

14. I. Kessler, H. Krawczyk, *Minimum Buffer Length and Clock Rate for the Shrinking Generator Cryptosystem*, IBM Research Report, RC 19938 (88322), (1995).

15. W. Meier, O. Staffelbach, *The Self-Shrinking Generator*, Advances in Cryptology – EUROCRYPT'94 Proceedings, LNCS **950**, Springer-Verlag, (1994), 205–214.

16. R. A. Rueppel, *Analysis and Design of Stream Ciphers*, Springer-Verlag, (1986).

A   Choice of the Output Set: Non-prefix Case

We consider the case where $f(\mathcal{C})$ is not a prefix code and does not contain the empty word ε. Thus, it is no longer possible to determine with certainty each word $(f(w_i))$ of the image of the input sequence. For statistical reasons, $f(\mathcal{C})$ contains at least one word beginning with 0, and one beginning with 1. Moreover, as it is not prefix, it also contains two words beginning with the same bit. Therefore, $f(\mathcal{C})$ contains at least three elements.

Proposition 12. *Let $(\mathcal{C}, f)$ be a proper encoder such that $f(\mathcal{C})$ is a non-prefix set that does not contain the empty word ε. Then, the average expected length of the first word of the decomposition of the input sequence over $\mathcal{C}$ is $E(\mathcal{C})$ when the input sequence is chosen uniformly. This word is known with average entropy $E(\mathcal{C}) + \Delta(\mathcal{C})$, with $\Delta(\mathcal{C}) < -1$.*

Proof. For $x \in f(\mathcal{C})$, we denote by $\mathcal{C}_x$ the set of preimages of x in $\mathcal{C}$, and we define $P(x) = \sum_{w \in \mathcal{C}_x} \frac{1}{2^{|w|}}$. Let $\mathbf{y}$ be the output sequence corresponding to a randomly chosen input sequence $\mathbf{s}$. Let $\mathcal{C}(\mathbf{y})$ be the set of words of $\mathcal{C}$ such that, for every $w \in \mathcal{C}(\mathbf{y})$, the sequence $\mathbf{y}$ begins with w.

Let $P_{\mathbf{y}}(x)$ denote the probability that the image by f of the first element of $\mathcal{C}$ recognized in $\mathbf{s}$ is x, given $\mathbf{y}$. Then, we have $\sum_{x \in \mathcal{C}(\mathbf{y})} P_{\mathbf{y}}(x) = 1.$

Now, each element in $\mathcal{C}_x$ has probability $\frac{1}{P(x)2^{|w|}}$ to be the preimage of x. Thus, the average length of the first element of $\mathcal{C}$ recognized in $\mathbf{s}$, given $\mathbf{y}$, is

$$E_{\mathbf{y}}(\mathcal{C}) = \sum_{x \in \mathcal{C}(\mathbf{y})} P_{\mathbf{y}}(x) \sum_{w \in \mathcal{C}_x} \frac{|w|}{P(x)2^{|w|}}.$$

Output sequences are chosen following the uniform distribution on input sequences, so the average length of the first element of $\mathcal{C}$ recognized in a random input sequence knowing the output is $E(\mathcal{C}) = \sum_{w \in \mathcal{C}} \frac{|w|}{2^{|w|}}$. Hence, the average value of $E_{\mathbf{y}}$ for random input sequences is $E(\mathcal{C})$, which is the first result.

Next, the entropy on the first element of $\mathcal{C}$ recognized in $\mathbf{s}$, given $\mathbf{y}$, is:

$$H_{\mathbf{y}}(\mathcal{C}) = - \sum_{x \in \mathcal{C}(\mathbf{y})} \sum_{w \in \mathcal{C}_x} P_{\mathbf{y}}(x) \frac{1}{P(x)2^{|w|}} \log \frac{P_{\mathbf{y}}(x)}{P(x)2^{|w|}}$$

$$= \sum_{x \in \mathcal{C}(\mathbf{y})} \sum_{w \in \mathcal{C}_x} \frac{P_{\mathbf{y}}(x)}{P(x)2^{|w|}} (|w| + \log P(x) - \log P_{\mathbf{y}}(x))$$

$$= E_{\mathbf{y}}(\mathcal{C}) + \sum_{x \in \mathcal{C}(\mathbf{y})} P_{\mathbf{y}}(x)(\log P(x) - \log P_{\mathbf{y}}(x)).$$

The average value of $H_{\mathbf{y}}$ for uniformly chosen input sequences is thus the sum of $E(\mathcal{C})$ and of the average value of $\Delta_{\mathbf{y}}(\mathcal{C}) = \sum_{x \in \mathcal{C}(\mathbf{y})} P_{\mathbf{y}}(x)(\log P(x) - \log P_{\mathbf{y}}(x)).$ for random input sequences. Let b be the first bit of $\mathbf{y}$. Then, $\mathcal{C}(\mathbf{y})$ is included in the subset $\mathcal{C}(b*)$ of $\mathcal{C}$ consisting of those words that are mapped by f to a word whose first bit is b. For statistical reasons, we have $\sum_{x \in \mathcal{C}(b*)} \frac{1}{2^{|x|}} = \frac{1}{2}$, which yields

$$\sum_{x \in \mathcal{C}(\mathbf{y})} \frac{1}{2^{|x|}} \leq \frac{1}{2} \quad . \tag{1}$$

Moreover, as there are at least 3 elements in $f(\mathcal{C})$, there are some output sequences $\mathbf{y}$ such that the inequality in Equation (1) is strict.

Using equality $\sum_{x \in \mathcal{C}(\mathbf{y})} P_{\mathbf{y}}(x) = 1$ and inequality (1), we obtain $\Delta_{\mathbf{y}} \leq -1$ for every output $\mathbf{y}$, the inequality being strict when that of (1) is. Therefore, the average value of $\Delta_{\mathbf{y}}$ for all random input sequences is strictly less than -1. $\quad\square$

Hence, the optimal choice of the output set, even if non-prefix output sets are considered, is still $f(\mathcal{C}) = \{0, 1\}$, with 0 and 1 having probability $\frac{1}{2}$ to appear in the output sequence for a random input sequence.

Efficient Computation of Algebraic Immunity for Algebraic and Fast Algebraic Attacks

Frederik Armknecht[1], Claude Carlet[2], Philippe Gaborit[3], Simon Künzli[4],
Willi Meier[4], and Olivier Ruatta[3]

[1] Universität Mannheim, 68131 Mannheim, Germany
armknecht@th.informatik.uni-mannheim.de
[2] INRIA, Projet CODES, BP 105, 78153 Le Chesnay Cedex;
also with Univ. of Paris 8, France
claude.carlet@inria.fr
[3] Université de Limoges, 87060 Limoges, France
{gaborit, olivier.ruatta}@unilim.fr
[4] FH Nordwestschweiz, 5210 Windisch, Switzerland
{simon.kuenzli, willi.meier}@fhnw.ch

Abstract. In this paper we propose several efficient algorithms for assessing the resistance of Boolean functions against algebraic and fast algebraic attacks when implemented in LFSR-based stream ciphers. An algorithm is described which permits to compute the algebraic immunity d of a Boolean function with n variables in $\mathcal{O}(D^2)$ operations, for $D \approx \binom{n}{d}$, rather than in $\mathcal{O}(D^3)$ operations necessary in all previous algorithms. Our algorithm is based on multivariate polynomial interpolation. For assessing the vulnerability of arbitrary Boolean functions with respect to fast algebraic attacks, an efficient generic algorithm is presented that is not based on interpolation. This algorithm is demonstrated to be particularly efficient for symmetric Boolean functions. As an application it is shown that large classes of symmetric functions are very vulnerable to fast algebraic attacks despite their proven resistance against conventional algebraic attacks.

Keywords: Algebraic Attacks, Algebraic Degree, Boolean Functions, Fast Algebraic Attacks, Stream Ciphers, Symmetric Functions.

1 Introduction

Many keystream generators consist of combining several linear feedback shift registers (LFSRs) and possibly some additional memory. One example is the E_0 keystream generator which is part of the Bluetooth standard. LFSRs are very efficient in hardware and can be designed such that the produced bitstream has maximum period and good statistical properties. Various approaches to the cryptanalysis of LFSR-based stream ciphers were discussed in literature (e.g., time-memory-tradeoff, fast correlation attacks or BDD-based attacks). For some keystream generators, algebraic attacks and fast algebraic attacks outmatched all previously known attacks [3, 12, 13].

S. Vaudenay (Ed.): EUROCRYPT 2006, LNCS 4004, pp. 147–164, 2006.

For LFSR-based filter or combining generators, their security mainly relies on a nonlinear Boolean output function f filtering the contents of one LFSR or combining the outputs of several ones. The present paper studies the resistance of this kind of stream ciphers to (fast) algebraic attacks.

In view of algebraic attacks, the notion of algebraic immunity (or annihilator immunity) has been introduced (the algebraic immunity $\mathcal{AI}$ of a Boolean function f is the minimum value of d such that f or $f + 1$ admits a function g of degree d such that $fg = 0$). The construction of Boolean functions for LFSR-based stream ciphers with large algebraic immunity achieved much attention recently, [5, 6, 7, 15, 17]. However, many of these functions do not allow for other good cryptographic properties like large non-linearity or large orders of resiliency, and as will be shown later, have undesirable properties with regard to fast algebraic attacks. It seems therefore relevant to be able to efficiently determine the immunity of existing and newly constructed Boolean functions against algebraic and fast algebraic attacks.

Until now, the best algorithms known for computing the algebraic immunity d of a function with n variables work roughly in $\mathcal{O}(D^3)$ operations, where $D \approx \binom{n}{d}$. This is impractical for functions with 20 or more variables. In this paper, we give an algorithm which computes the $\mathcal{AI}$ of a function in $\mathcal{O}(D^2)$ operations. The algorithm is based on multivariate polynomial interpolation, and it is applied to two particular families of Boolean functions: the inverse functions and the Kasami power functions. The quadratic nature of the algorithm is experimentally verified, and for the first time, the $\mathcal{AI}$ of a function with 20 variables is computed to be $\mathcal{AI} = 9$.

Resistance against fast algebraic attacks is not fully covered by algebraic immunity, as has been demonstrated, e.g., by a fast algebraic attack on the eSTREAM candidate SFINKS, [11]. For determining immunity against fast algebraic attacks, we give a new algorithm that is based on methods different from interpolation, and that for general Boolean functions allows to efficiently assess immunity against fast algebraic attacks. The complexity of our second algorithm is in $\mathcal{O}(DE^2)$, where $E \approx \binom{n}{e}$ and e in many cases of interest is much smaller than d. This compares favorably with the known algorithms, which are in $\mathcal{O}(D^3)$. The algorithm is applied to several of the above mentioned classes of Boolean functions with optimal algebraic immunity, including symmetric Boolean functions, like the majority functions. Symmetric functions are attractive as the hardware complexity grows only linearly with the number of input variables. However, it is shown in this paper that the specific structure of these functions can be exploited in a much refined algorithm for determining resistance against algebraic attacks that is particularly efficient. It is concluded that large classes of symmetric functions are very vulnerable to fast algebraic attacks despite their optimal algebraic immunity. A symmetric function would not be implemented by itself but rather in combination with other nonlinear components in stream ciphers. It seems nevertheless essential to know the basic cryptographic properties of each component used.

The paper is organized as follows. In Section 2, the basics of algebraic and fast algebraic attacks are described. Section 3 derives an algorithm for efficient computation of the algebraic immunity as well as a modified algorithm to determine all minimal degree annihilators. In Section 4, an algorithm for efficient computation of immunity against fast algebraic attacks is presented. In Section 5, the algorithm is adapted and improved for symmetric functions, and it is proven that the class of majority functions which have maximum $\mathcal{AI}$ is very vulnerable to fast algebraic attacks. We finally conclude in Section 6.

2   Algebraic Attacks and Fast Algebraic Attacks

2.1   Algebraic Attacks

For an LFSR L with N entries filtered by a Boolean function f with n variables, algebraic attacks consist of two steps [12]:

- **First step.** Finding functions g of low degree d such that $fg = 0$ or $(f + 1)g = 0$. Until this paper, the complexity of this step was roughly in D^3, for $D := \sum_{i=0}^{d} \binom{n}{i}$ (which is about $\binom{n}{d}$ for $d < n/2$) and where 3 is taken for the exponent of the matrix inversion.
- **Second step.** Solving a nonlinear system of multivariate equations $g(L^i (x_1, \ldots, x_N)) = 0$ for adequate i, induced by the functions g of the annihilator sets $\mathrm{Ann}(f)$ and $\mathrm{Ann}(f + 1)$. Usually this system is solved by linearization with a complexity of D_N^3 for $D_N := \sum_{i=0}^{d} \binom{N}{i}$. The number of required bits of keystream is proportional to D_N, whereas this value can be reduced if several annihilators of f and/or $f \oplus 1$ with minimum degree are known. Alternatively, this system can be solved by Gröbner basis, but then the complexity of solving is difficult to evaluate, see [4, 19].

The lowest degree of the function $g \neq 0$ for which $fg = 0$ or $(f+1)g = 0$ is called the algebraic (or annihilator) immunity $\mathcal{AI}$ of f. In [12] it has been shown that for any function f with n-bit input vector, functions $g \neq 0$ and h exist, with $fg = h$ such that e and d are at most $\lceil n/2 \rceil$. This implies that $\mathcal{AI}(f) \leq \lceil n/2 \rceil$.

2.2   Fast Algebraic Attacks

Fast algebraic attacks were introduced by Courtois in [13]. They were confirmed and improved later by Armknecht in [3] and Hawkes and Rose in [20]. A prior aim of fast algebraic attacks is to find a relation $fg = h$ with $e := \deg g$ small and $d := \deg h$ larger. In classical algebraic attacks, the degree d of h would necessarily lead to considering a number of unknowns of the order of D_N. In fast algebraic attacks, one considers that the sequence of the functions $h(L^i(x_1, \cdots, x_N))$ can be obtained as an LFSR with linear complexity D_N. One uses then the Berlekamp-Massey algorithm to eliminate all monomials of degree superior to e in the equations, such that eventually one only needs to solve a system in $E_N := \sum_{i=0}^{e} \binom{N}{i}$ unknowns. The complexity of fast algebraic attacks can be summarized in these four steps:

- **Relation search step.** One searches for functions g and h of low degrees such that $fg = h$. For g and h of degrees e and d respectively, with associated values $D := \sum_{i=0}^{d} \binom{n}{i}$ and $E := \sum_{i=0}^{e} \binom{n}{i}$, such g and h can be found when they exist by solving a linear system with $D + E$ equations, and with complexity $\mathcal{O}((D + E)^3)$. Usually one considers $e < d$.
- **Pre-computation step.** In this step, one searches for particular linear relations which permit to eliminate monomials with degree greater than e in the equations. This step needs a sequence of $2D_N$ bits of stream and has a complexity of $\mathcal{O}(D_N \log^2(D_N))$, see [20].
- **Substitution step.** At this step, one eliminates the monomials of degrees greater than e. This step has a natural complexity in $\mathcal{O}(E_N^2 D_N)$ but using discrete Fourier transform, it is claimed in [20] that a complexity $\mathcal{O}(E_N D_N \log(D_N))$ can be obtained.
- **Solving step.** One solves the system with E_N linear equations in $\mathcal{O}(E_N^3)$.

Notice that, for arbitrary non-zero functions f, g, h, the relation $fg = h$ implies $fh = h$, thus we have $d \geq \mathcal{AI}(f)$ and we can restrict to values e with $e \leq d$. Fast algebraic attacks are always more efficient than conventional algebraic attacks if $d = \mathcal{AI}(f)$ and $e < d - 1$. In case that e turns out to be large for this d, it is of interest to determine the minimum e where d is slightly larger than $\mathcal{AI}(f)$.

3   Efficient Computation of the Algebraic Immunity

In this section, we present an algorithm which computes the algebraic immunity $\mathcal{AI}$ of a Boolean function in $\mathcal{O}(D^2)$ operations. In particular, the algorithm returns a non-zero annihilator of minimum degree d, without necessitating a prior guess of d. The algorithm is based on the notion of multivariate polynomial interpolation, it generalizes the classical incremental Newton interpolation algorithm to the multivariate case. We also explain how to modify the algorithm to return the set of *all* non-zero annihilators with minimum degree. Eventually we give experimental results of our algorithm.

3.1   Multivariate Lagrange Interpolation

Before stating what is the multivariate Lagrange interpolation problem when it is specified to binary polynomials, we need to introduce some notation. We denote by $\mathbb{F}$ the finite field $GF(2)$ and by $\mathbb{F}^k$ the vector space of dimension k over $\mathbb{F}$. Consider $x := x_1, \ldots, x_k$ a set of k binary variables, $\alpha := (\alpha_1, \ldots, \alpha_k) \in \mathbb{F}^k$ a multi-index, $z := (z_1, \ldots, z_k)$ an element of $\mathbb{F}^k$. We denote $x^\alpha := x_1^{\alpha_1} \cdots x_k^{\alpha_k}$ and $z^\alpha := z_1^{\alpha_1} \cdots z_k^{\alpha_k}$. Let $E := \{\alpha_1, \ldots, \alpha_D\} \subseteq \mathbb{F}^k$ be a set of multi-indices, then we denote by $x^E := \{x^{\alpha_1}, \ldots, x^{\alpha_D}\}$ the set of associated monomials. We identify the ring of boolean functions in n variables with $\mathbb{F}[x]/\langle x_i^2 - x_i, i = 1, \ldots, n \rangle$, the quotient ring of the ring of polynomials with coefficients in $\mathbb{F}$ by the ideal generated by the relations $x_i^2 - x_i$, $i \in \{1, \ldots, n\}$. We will use, explicitly or not, several times this identification. In our framework, the multivariate Lagrange problem can be stated as follows:

Problem 1. Let $E := \{\alpha_1, \ldots, \alpha_D\} \subseteq \mathbb{F}^n$, $\mathcal{Z} := \{z_1, \ldots, z_D\} \subseteq \mathbb{F}^n$ and $\bar{v} := (v_1, \ldots, v_D) \in \mathbb{F}^D$. Does there exist a polynomial $g \in \mathbb{F}[x_1, \ldots, x_n]$ whose monomial support is included in x^E and such that $g(z_i) = v_i$, $\forall i \in \{1, \ldots, D\}$?

Remark 1. The general multivariate Lagrange interpolation problem has been addressed in [23], but the proposed algorithm has cubic complexity (on the number of monomials). We will present an algorithm with a quadratic complexity over $\mathbb{F}$ instead.

An answer to Problem 1 in terms of existence and uniqueness is presented by means of the following definition:

Definition 1. *Let* $\mathcal{Z} := \{z_1, \ldots, z_D\} \subseteq \mathbb{F}^n$ *and* $E := \{\alpha_1, \ldots, \alpha_D\} \subseteq \mathbb{F}^n$, *we define the Vandermonde matrix as*

$$V_{\mathcal{Z},E} := \begin{pmatrix} z_1^{\alpha_1} & \cdots & z_1^{\alpha_D} \\ \vdots & \ddots & \vdots \\ z_D^{\alpha_1} & \cdots & z_D^{\alpha_D} \end{pmatrix}, \tag{1}$$

and we define the Vandermonde determinant to be $v_{\mathcal{Z},E} := \det(V_{\mathcal{Z},E})$.

Proposition 1. *There exists an unique solution* $g \in \mathbb{F}[x]$ *to Problem 1 if* $v_{\mathcal{Z},E} \neq 0$. *Furthermore, the solution* g *is given by* $g(x) = \bigoplus_{j=1}^{D} g_{\alpha_j} x^{\alpha_j}$, *where the vector* $\bar{g} := (g_{\alpha_1}, \ldots, g_{\alpha_D})^t$ *is the only solution of the system*

$$V_{\mathcal{Z},E} \, \bar{g} = \bar{v} . \tag{2}$$

Remark 2. Given the set $\mathcal{Z} := \{z_1, \ldots, z_D\} \subseteq \mathbb{F}^n$, the existence of a set $E := \{\alpha_1, \ldots, \alpha_D\} \subseteq \mathbb{F}^n$ such that $v_{\mathcal{Z},E} \neq 0$ is ensured since it is enough to take for E the set of monomials which are not in the monomial ideal generated by the leading monomials of a Gröbner basis of the ideal of the polynomials vanishing at each point of $\mathcal{Z}$.

With the following proposition, the minimum annihilator problem can be reduced to a multivariate Lagrange interpolation problem:

Proposition 2. *Let* f *be a Boolean function,* $\mathcal{Z} := f^{-1}(1)$ *and* E *such that* x^E *is the complementary of the monomial ideal generated by the leading monomials of a Gröbner basis for a graduated order of the ideal of the polynomials vanishing at each point of* $\mathcal{Z}$. *Then, if* $\beta \notin E$ *is of minimum weight, the function* R_β *defined below is a minimum-degree annihilator of* f,

$$R_\beta := \det \begin{pmatrix} x^\beta & x^{\alpha_1} & \cdots & x^{\alpha_D} \\ z_1^\beta & z_1^{\alpha_1} & \cdots & z_1^{\alpha_D} \\ \vdots & \vdots & \ddots & \vdots \\ z_D^\beta & z_D^{\alpha_1} & \cdots & z_D^{\alpha_D} \end{pmatrix} .$$

Furthermore, $R_\beta = x^\beta \oplus g$ *where* g *is the solution of Problem 1 with* $\bar{v} = (z_1^\beta, z_2^\beta, \ldots, z_D^\beta)$.

Proof. The function $R_\beta(x)$ is an annihilator of f, as for an argument $x \in \mathcal{Z}$ the above matrix becomes singular. In addition, R_β has minimum degree because E is the complementary of the monomial ideal generated by the leading terms of a Gröbner basis for a graduated monomial order. The relation $R_\beta = x^\beta \oplus g$ is obtained by developing the determinant defining R_β with respect to the first row, and by considering g obtained with Cramer's rule in Eq. 2.   □

3.2   General Description of the Algorithm

The general idea of the algorithm is to apply Prop. 2 incrementally with a linear complexity at each step. Let us introduce some more notation: E^d is the set of all α of weight equal to d. Then $E^{\leq d} := E^0 \cup \ldots \cup E^d$ (ordered by increasing weight) and $E_i := \{\alpha_1, \ldots, \alpha_i\}$, which are the first i elements of $E^{\leq d}$. Let $\mathcal{Z} := f^{-1}(1) \subseteq \mathbb{F}^n$ (with arbitrary ordering) and $\mathcal{Z}_i := \{z_1, \ldots, z_i\}$. We assume $v_{\mathcal{Z}_i, E_i} \neq 0$ for all $i \in \{1, \ldots, |\mathcal{Z}|\}$, this condition[1] is sufficient to apply Prop. 2 on the sets $\mathcal{Z}_i$, E_i.

Then the algorithm works as follows: apply Prop. 2 for an intermediate set of points $\mathcal{Z}_i$ and an associated set of exponents E_i, with $\beta = \alpha_{i+1}$. A particular solution $g_i = R_\beta \oplus x^\beta$ is an intermediate annihilator of f on the set $\mathcal{Z}_i$. If one can verify that g_i is also an annihilator of f on the global set $\mathcal{Z}$, then a minimum degree annihilator of f is found. Otherwise, one considers a new point z_{i+1} and $\mathcal{Z}_{i+1}$ with associated set of exponents E_{i+1}, until an annihilator of f on $\mathcal{Z}$ is found.

Remark 3. Notice that the original interpolation problem with $\bar{v} = 0$ is turned into a sequence of interpolation problems with (in general) non-zero $\bar{v}$, depending on the exponent α_{i+1} used at each step. In particular, the fact $\bar{v} = 0$ on $f^{-1}(1)$ is used implicitly in the computation of the ordered set E associated to $f^{-1}(1)$.

For each intermediate step, the updating procedure can be done in linear time, resulting in an overall complexity of $\mathcal{O}(D^2)$ rather than $\mathcal{O}(D^3)$. In fact, this is a multivariate generalization of the Newton interpolation scheme: recall that a Newton basis for the polynomial interpolation problem allows to introduce interpolation nodes one by one (without the requirement to recalculate previous coefficients). In addition, the Newton basis leads to a triangular Vandermonde matrix, which can be solved in quadratic time (on the number of interpolation nodes).

3.3   Computing a Minimum Degree Annihilator

Define $V_i := V_{\mathcal{Z}_i, E_i}$, and consider an LU-decomposition of V_i, i.e. $V_i = L_i U_i$, where U_i is triangular superior and L_i is triangular inferior. Then, the system $V_i \bar{g}_i = \bar{v}_i$ with $\bar{v}_i := (v_1, \ldots, v_i)$ is equivalent to $U_i \bar{g}_i = L_i^{-1} \bar{v}_i$, and the solution $\bar{g}_i$ can be found by solving two triangular systems (i.e computing the inverses of U_i and L_i). If the polynomial associated to $\bar{g}_i$ is not an annihilator of f, then we

---

[1] Such kind of ordered sets of points and exponents always exists and can be computed incrementally in quadratic time (see [23]), so we do not lose any generality.

solve the system for V_{i+1} and $\bar{v}_{i+1} = (\bar{v}_i, v_{i+1},)^t$. However, instead of computing a complete LU-decomposition of V_{i+1}, we write

$$V_{i+1} = \begin{pmatrix} V_i & C_{i+1} \\ R_i & z_{i+1}^{\alpha_{i+1}} \end{pmatrix} = \begin{pmatrix} L_i & 0 \\ 0 & 1 \end{pmatrix} \begin{pmatrix} U_i & L_i^{-1}C_{i+1} \\ R_i & z_{i+1}^{\alpha_{i+1}} \end{pmatrix}$$

with $C_{i+1} := (z_1^{\alpha_{i+1}}, \ldots, z_i^{\alpha_{i+1}})^t$ and $R_i := (z_{i+1}^{\alpha_1}, \ldots, z_{i+1}^{\alpha_i})$. Consequently, knowledge of an LU-decomposition of V_i yields an almost LU-decomposition of V_{i+1} (with the exception of R_i). This is a basic fact usually exploited to design efficient LU-factorization algorithms.

In our framework, one can avoid a direct computation of L_i as follows. Denote $X_i := (x_1, \ldots, x_i)^t$, where the elements x_j are considered as indeterminate, and denote $P_i(x_1, \ldots, x_i) := L_i^{-1}X_i$. Then we have

$$V_{i+1} = \begin{pmatrix} L_i & 0 \\ 0 & 1 \end{pmatrix} \begin{pmatrix} U_i & P_i(z_1^{\alpha_{i+1}}, \ldots, z_i^{\alpha_{i+1}}) \\ z_{i+1}^{\alpha_1} \cdots z_{i+1}^{\alpha_i} & z_{i+1}^{\alpha_{i+1}} \end{pmatrix} \tag{3}$$

$$\bar{v}_{i+1} = \begin{pmatrix} L_i & 0 \\ 0 & 1 \end{pmatrix} \begin{pmatrix} P_i(v_1, \ldots, v_i) \\ v_{i+1} \end{pmatrix} . \tag{4}$$

Consequently, the system of equations $V_{i+1}\bar{g}_{i+1} = \bar{v}_{i+1}$ is equivalent to

$$\begin{pmatrix} U_i & P_i(z_1^{\alpha_{i+1}}, \ldots, z_i^{\alpha_{i+1}}) \\ z_{i+1}^{\alpha_1} \cdots z_{i+1}^{\alpha_i} & z_{i+1}^{\alpha_{i+1}} \end{pmatrix} \bar{g}_{i+1} = \begin{pmatrix} P_i(v_1, \ldots, v_i) \\ v_{i+1} \end{pmatrix} . \tag{5}$$

Triangulation of the left matrix is an easy task since U_i is triangular (it is achieved by elimination of the i first entries of the last row by row operations). The same operations are carried out on the right matrix. This yields U_{i+1} and $P_{i+1}(x_1, \ldots, x_{i+1})$, and the system $U_{i+1}\bar{g}_{i+1} = P_{i+1}(v_1, \ldots, v_{i+1})$ can be solved. If the polynomial associated to $\bar{g}_{i+1}$ is not an annihilator of f (i.e. if $\exists z \in \mathcal{Z}$ such that $g(z) \neq 0$), the subsequent elements in $E^{\leq d}$ and $\mathcal{Z}$ are added and so on. Practically, the only points introducing new constraints on the annihilator are those for which the input polynomial does not vanish already. The method terminates because the degree of the annihilator is bounded. Denote by t be the number of iterations of Alg. 1.

As an input of the algorithm, we do not take a monomial expansion of f, but the vector of its evaluation at points z_i. In the case $t = |\mathcal{Z}|$, this vector can be computed with asymptotically $\mathcal{O}(t \log(t))$ operations using a method based on fast Fourier transform, and more easily in $\mathcal{O}(tN_m)$ operations over the ground field (where N_m is the number of monomials in the algebraic normal form of f) by simply adding the evaluation of each monomial at the t points. The algorithm incrementally computes the values of the annihilator at every point and lifts them in the monomial basis in order to compute the power expansion. Let us discuss the most costly operations at the ith step of the algorithm:

- The triangulation in step 4 requires i arithmetic operations. As U_i is already a upper triangular matrix, we only need to eliminate the first $i - 1$ entries in the last row, and update the entry in the bottom right corner. This is done by replacing $z_i^{\alpha_1}, \ldots, z_i^{\alpha_{i-1}}$ by 0 and $z_i^{\alpha_i}$ by $z_i^{\alpha_i} - \sum_{j=1}^{i-1} z_i^{\alpha_1} \cdot P_{i,j}$ where $(P_{i,1}, \ldots, P_{i,i-1}) = P(z_1^{\alpha_i}, \ldots, z_{i-1}^{\alpha_i})$.

Algorithm 1. Computation of an annihilator of minimum degree

Input: f, $\mathcal{Z} := f^{-1}(1)$, $E^{\leqslant \lceil n/2 \rceil}$.
Output: An annihilator of f of minimum degree.
1: Initialization: $U_1 \leftarrow (z_1^{\alpha_1})$, $v_1 \leftarrow f(z_1) \oplus 1$, $\bar{g} \leftarrow 1$, $P \leftarrow (x_1)$, $i \leftarrow 1$.
2: **while** the polynomial associated to $\bar{g}$ is not an annihilator of f **do**
3:     $i \leftarrow i + 1$.
4:     $\begin{pmatrix} U_i & P(z_1^{\alpha_i}, \ldots, z_{i-1}^{\alpha_i}) \\ z_i^{\alpha_1} \ldots z_i^{\alpha_{i-1}} & z_i^{\alpha_i} \end{pmatrix} \overset{\text{row op.}}{\longmapsto} \begin{pmatrix} U_i & P(z_1^{\alpha_i}, \ldots, z_i^{\alpha_i}) \\ 0 \ldots 0 & \end{pmatrix} =: U_{i+1}$.
5:     Use the same row operations from $\left(P(z_1^{\alpha_i}, \ldots, z_{i-1}^{\alpha_i}), z_i^{\alpha_i} \right) \mapsto P(z_1^{\alpha_i}, \ldots, z_i^{\alpha_i})$ to
       perform the update $(P(v_1, \ldots, v_{i-1}), v_i) \mapsto P(v_1, \ldots, v_i)$.
6:     Solve $U_i \bar{g}_i = P(v_1, \ldots, v_i)$ with $\bar{g}_i = (g_1, \ldots, g_i)$.
7: **end while**
8: Output $g(x) := \bigoplus_{j=1}^{i} g_j x^{\alpha_j}$.

- The updating process of P requires i arithmetic operations.
- Solving the system in step 6 basically requires i^2 arithmetic operations. However, this is also feasible with i arithmetic operations by the following remark, allowing to correct g_i in order to compute g_{i+1}:

$$U_{i+1}\, g_{i+1} = \begin{pmatrix} U_i & P_i(z_1^{\alpha_{i+1}}, \ldots, z_i^{\alpha_{i+1}}) \\ 0 & * \end{pmatrix} \begin{pmatrix} g_i' \\ * \end{pmatrix} = \begin{pmatrix} P_i(v_1, \ldots, v_i) \\ v_{i+1} \end{pmatrix}.$$

We do not introduce any new complex computation to check whether g is an annihilator of f. Namely, we compute the values of g at points which are not introduced yet. This can be done by updating a vector storing the evaluations of g at each point considered so far, where a new step leads to a linear number of operations (corresponding to the number of coordinates). Again, the overal cost of this computation is quadratic on the number of points.

The arithmetic complexity $AC(N)$ of the proposed algorithm is given by $AC(N) = AC(N-1) + \text{const} \cdot i + \mathcal{O}(D)$. An simple computation shows that $AC(N) = \mathcal{O}(t^2 + tD)$. Since t is the number of monomials occurring in a minimum degree annihilator of f, t has the same order of magnitude as D. This is summarized in the following proposition:

Proposition 3. *The arithmetic complexity of Alg. 1 to compute the minimum degree d of an annihilator of f is $\mathcal{O}(D^2)$.*

In order to obtain the quadratic behavior, it is necessary to handle memory allocation with care (in particular, management of the extension operations on the matrix are delicate, and a bad memory allocation leads to an implementation cubic in space and time). We finally remark that the above method can also be used to construct functions with high algebraic immunity.

3.4   Computing All Minimum Degree Annihilators

In this section, we explain how to modify Alg. 1 to compute all minimum-degree annihilators g of a polynomial f. Notice that $\text{Ann}(f, d) := \{g \in \langle x^{E^{\leq d}} \rangle \subseteq$

$\mathbb{F}[x] \| fg = 0\}$ is a vector space. Consequently, we only have to compute a basis of $\text{Ann}(f, d)$, and this for the minimum value of d. The idea of the method proposed here is to run Alg. 1 until we find the first annihilator together with d. Then, the algorithm searches for further annihilators, considering only exponents in $E^{\leq d}$. In addition, if α_i is the exponent lastly introduced and resulting in an annihilator, we can execute a further search without α_i (this can be implemented by backtracking the last update). The reason is that if g and $g' \neq g$ are both annihilators which contain x^{α_i}, then one can construct another annihilator $g \oplus g'$ which is independent of x^{α_i}. Hence, the new algorithm can still be run incrementally, and it terminates after introduction of α^D. As the number of steps required to find the first annihilator is of the same order of magnitude as D, the asymptotic performance of the new algorithm does not increase. This is resumed in the following proposition:

Proposition 4. *The above modifications of Alg. 1 allow to compute the minimum degree d of an annihilator of f, and a basis of $\text{Ann}(f, d)$, using $\mathcal{O}(D^2)$ arithmetic operations.*

3.5   Experimental Results

In this section, we apply Alg. 1 to two particular families of Boolean power functions: the inverse functions and the Kasami type functions (see [10]). We verified that an implementation of the algorithm in C code followed the announced quadratic time complexity on the number of variables.

The inverse function is of particular interest, since this function is used with $n = 8$ variables in the S-box of **AES**, and almost directly as a filtering function in **SFINKS** [6]. For different values of n, Tab. 1 lists the power exponent of the function f (which is equal to -1 here), its weight, its algebraic degree, its nonlinearity and its algebraic immunity.

The Kasami functions in n variables have exponents of the form $2^{2k} - 2^k + 1$ with $\gcd(k, n) = 1$ and $k \leq n/2$. These functions are of interest since we can see

Table 1. Computation of the weight, degree, nonlinearity and algebraic immunity for the inverse function and some Kasami power functions for $12 \leq n \leq 20$

n	|exp.|	weight	deg.	nonlin.	$\mathcal{AI}$	exp.	|weight|	deg.	nonlin.	$\mathcal{AI}$
		Inverse function					Kasami power functions			
12	-1	2048	11	1984	5	993	2048	11	1984	5
13	-1	4096	12	4006	6	993	4096	6	$2^{12} - 2^6$	6
14	-1	8192	13	8064	6	4033	8192	6	$2^{13} - 2^7$	6
15	-1	2^{14}	14	16204	6	4033	2^{14}	7	$2^{14} - 2^8$	7
16	-1	2^{15}	15	$2^{15} - 2^8$	6	$2^{14} - 2^7 + 1$	2^{15}	15	$2^{15} - 2^7$	7
17	-1	2^{16}	16	65174	7	$2^{14} - 2^7 + 1$	2^{16}	8	$2^{16} - 2^8$	8
18	-1	2^{17}	17	$2^{17} - 2^9$	7	$2^{16} - 2^8 + 1$	2^{17}	9	$2^{17} - 2^9$	8
19	-1	2^{18}	18	261420	7	$2^{16} - 2^8 + 1$	2^{18}	9	$2^{18} - 2^9$	9
20	-1	2^{19}	19	$2^{19} - 2^{10}$	7	$2^{18} - 2^9 + 1$	2^{19}	19	$2^{19} - 2^9$	9

that, for the number n of variables which is currently usual in cryptography, they have a high algebraic immunity.[2] We consider several Kasami type exponents (where $\gcd(k, n)$ may be different from 1), see Tab. 1. For $n = 12, 16, 20$, we converted non-balanced functions to balanced ones by flipping the first entries in the truth tables. For the first time, we accomplish computation of the $\mathcal{AI}$ of a function with 20 variables, $\mathcal{AI} = 9$ and good nonlinearity.

4   Efficient Computation of Immunity Against Fast Algebraic Attacks

Let us first introduce some notation for this section. Any Boolean function f with an n-bit input vector $x := (x_1, \ldots, x_n)$ can be characterized by its truth table $T(f) := (f(0), \ldots, f(2^n - 1)) \in \mathbb{F}^{2^n}$ or by its algebraic normal form $f(x) = \bigoplus_\alpha f_\alpha x^\alpha$, with coefficients $f_\alpha \in \mathbb{F}$, multi-indices $\alpha \in \mathbb{F}^n$ (which can also be identified by their integers) and the abbreviation $x^\alpha := x_1^{\alpha_1} \cdots x_n^{\alpha_n}$. Consequently, we define the coefficient vector of f by $C(f) := (f_0, \ldots, f_{2^n-1}) \in \mathbb{F}^{2^n}$.

Given a Boolean function f with n input variables, the goal is to decide whether g of degree e and h of degree d exist, such that $fg = h$. The known function f is represented preferably by the truth table $T(f)$, which allows to efficiently access the required elements, and the unknown functions g and h are represented by the coefficient vectors $C(g)$ and $C(h)$, which leads to the simple side conditions $g_\beta = 0$ for $|\beta| > e$ and $h_\gamma = 0$ for $|\gamma| > d$. In order to decide if g and h exist, one has to set up a number of linear equations in g_β and h_γ. Such equations are obtained, e.g., by evaluation of $f(z) \cdot \bigoplus_\beta g_\beta z^\beta = \bigoplus_\gamma h_\gamma z^\gamma$ for some values of z. There are $D + E$ variables, so one requires at least the same number of equations. The resulting system of equations can be solved by Gaussian elimination with time complexity $\mathcal{O}((D+E)^3) = \mathcal{O}(D^3)$. If any $D+E$ equations are linearly independent, then no nontrivial g and h of corresponding degree exist. Otherwise, one may try to verify a nontrivial solution. Certainly, there are more sophisticated algorithms, namely we are able to express a single coefficient h_γ as a linear combination of coefficients g_β. If these relations hold for any value of γ, one may choose γ with $|\gamma| > d$ such that $h_\gamma = 0$, in order to obtain relations in g_β only. Consequently, equations for coefficients of g can be completely separated from equations for coefficients of h. As there are only E variables g_β, one requires at least E equations, and the system of equations can be solved in $\mathcal{O}(E^3)$. Depending on the parameters n, d, e and on the structure of f, there are different strategies how to efficiently set up equations.

4.1   Setting Up Equations

In this section, we consider the product $fg = h$ where f, g and h are arbitrary Boolean functions in n variables. Here are some additional notational conventions: For $\alpha, \beta, \gamma \in \mathbb{F}^n$, let $\alpha \subseteq \beta$ be an abbreviation for $\mathrm{supp}(\alpha) \subseteq \mathrm{supp}(\beta)$,

---

[2] However, it is shown in [24] that Kasami functions have bad algebraic immunity when n is very large.

where $\text{supp}(\alpha) := \{i | \alpha_i = 1\}$, and let $\alpha \vee \beta := (\alpha_1 \vee \beta_1, \dots, \alpha_n \vee \beta_n)$. For $B, C \in \mathbb{F}^{2^n}$, we define the scalar product $B \cdot C := \bigoplus_{k=0}^{2^n-1} [B]_k \cdot [C]_k$. All expressions are modulo 2 here. With the following theorem, we are able to express a single coefficient h_γ as a linear combination of coefficients g_β, where the linear combination is computed either with $T(f)$ or with $C(f)$.

Theorem 1. *Let* $f(x) = \bigoplus_\alpha f_\alpha x^\alpha$ *and* $g(x) = \bigoplus_\beta g_\beta x^\beta$. *Set* $h(x) = \bigoplus_\gamma h_\gamma x^\gamma := f(x) \cdot g(x)$. *With* $A_{i,j} \in \mathbb{F}$ *and* $B_{i,j} \in \mathbb{F}^{2^n}$, *we have for each* γ

$$h_\gamma = \bigoplus_\beta \binom{\gamma}{\beta} A_{\gamma,\beta} \cdot g_\beta \tag{6}$$

$$A_{i,j} := B_{i,j} \cdot T(f) = B_{i,i-j} \cdot C(f) \tag{7}$$

$$[B_{i,j}]_k := \binom{i}{k} \cdot \binom{k}{j}. \tag{8}$$

Proof. The binary Moebius transform relates the ANF of a Boolean function with the corresponding truth table, namely considering Lucas' theorem $f(k) = \bigoplus_\alpha \binom{k}{\alpha} f_\alpha = \bigoplus_{\alpha \subseteq k} f_\alpha$ and $f_k = \bigoplus_\alpha \binom{k}{\alpha} f(\alpha) = \bigoplus_{\alpha \subseteq k} f(\alpha)$. We obtain the relation $h_\gamma = \bigoplus_{\alpha \subseteq \gamma} f(\alpha) g(\alpha)$. With $g(\alpha) = \bigoplus_{\beta \subseteq \alpha} g_\beta$, this becomes $h_\gamma = \bigoplus_{\alpha \subseteq \gamma} \bigoplus_{\beta \subseteq \alpha} g_\beta f(\alpha)$. Rearranging the coefficients, we finally have the product $h_\gamma = \bigoplus_{\beta \subseteq \gamma} g_\beta \bigoplus_{\beta \subseteq \alpha \subseteq \gamma} f(\alpha) = \bigoplus_\beta \binom{\gamma}{\beta} g_\beta B_{\gamma,\beta} \cdot T(f)$. In order to prove the second relation, we multiply the ANF of both functions and obtain $h_\gamma = \bigoplus_{\alpha \vee \beta = \gamma} f_\alpha g_\beta$. This binary sum can then be partitioned according to $h_\gamma = \bigoplus_{\beta \subseteq \gamma} g_\beta \bigoplus_{\alpha \subseteq \gamma; \alpha \vee \beta = \gamma} f_\alpha$. With Lucas' theorem again, we have the relation $h_\gamma = \bigoplus_{\beta \subseteq \gamma} g_\beta \bigoplus_{\gamma - \beta \subseteq \alpha \subseteq \gamma} f_\alpha = \bigoplus_\beta \binom{\gamma}{\beta} g_\beta B_{\gamma,\gamma-\beta} \cdot C(f)$. $\square$

4.2   Determining the Existence of Solutions

We propose an efficient algorithm to determine the existence of g and h with corresponding degrees, see Alg. 2. The algorithm is based on the equation $h_\gamma = \bigoplus_{\beta \subseteq \gamma} g_\beta \bigoplus_{\beta \subseteq \alpha \subseteq \gamma} f(\alpha)$, which is a variant of Th. 1.

Let us discuss the complexity of Alg. 2. Initialization of G takes at most $\mathcal{O}(E^2)$ time and memory, and $\mathcal{I}$ can be constructed in $\mathcal{O}(E)$ time. Iteration initiates by choosing a fixed γ of weight $d + 1$, this step will be repeated E times to set up the same number of equations. Notice that the set $\{\gamma : |\gamma| = d + 1\}$ is sufficient to choose E different values of γ, as $E < \binom{n}{d+1}$ in the case of $e \ll d$ and $d \approx n/2$ (which is the typical scope of fast algebraic attacks). Thereafter, one chooses a fixed β of weight b. This step will be repeated for all $\binom{d+1}{b}$ elements of weight b, and for all $b = 0, \dots, e$. Given this choice of γ and β, we find $|\mathcal{A}| = 2^{d+1-b}$, which corresponds to the number of operations to compute A. Overall complexity of the iteration becomes $E \sum_{b=0}^{e} \binom{d+1}{b} 2^{d+1-b} < E(e+1) \binom{d+1}{e} 2^{d+1} < DE^2$, where the last inequality holds in the specified range of parameters. Time complexity of the final step of Alg. 2 is $\mathcal{O}(E^3)$. The dominating term, and hence complexity of Alg. 2 corresponds to $\mathcal{O}(DE^2)$. Compared to the complexity $\mathcal{O}(D^3)$ of Alg. 2 in [21], Alg. 2 is very efficient for g of low degree.

Algorithm 2. Determine the existence of g and h for any f

Input: A Boolean function f with n input variables and two integers $0 \le e \le \mathcal{AI}(f)$
and $\mathcal{AI}(f) \le d \le n$.
Output: Determine if g of degree at most e and h of degree at most d exist such that
$fg = h$.
1: Initialize an $E \times E$ matrix G, and let each entry be zero.
2: Compute an ordered set $\mathcal{I} \leftarrow \{\beta : |\beta| \le e\}$.
3: **for** i from 1 to E **do**
4:     Choose a random γ with $|\gamma| = d + 1$.
5:     Determine the set $\mathcal{B} \leftarrow \{\beta : \beta \subseteq \gamma, |\beta| \le e\}$.
6:     **for all** β in $\mathcal{B}$ **do**
7:         Determine the set $\mathcal{A} \leftarrow \{\alpha : \beta \subseteq \alpha \subseteq \gamma\}$.
8:         Compute $A \leftarrow \bigoplus_{\mathcal{A}} f(\alpha)$.
9:         Let the entry of G in row i and column β (in respect to $\mathcal{I}$) be 1 if $A = 1$.
10:     **end for**
11: **end for**
12: Solve the linear system of equations, and output **no** g **and** h **of corresponding**
degree if there is only a trivial solution.

4.3   Experimental Results

In [15], a class of (non-symmetric) Boolean functions f with maximum algebraic
immunity is presented; these functions will be referred here as DGM functions.
Application of Alg. 2 on their examples for $n = 5, 6, 7, 8, 9, 10$ reveals that h
and g exist with $d = \mathcal{AI}(f) = \lceil n/2 \rceil$ and $e = 1$. We point out that this is
the most efficient situation for a fast algebraic attack. Explicit functions g with
corresponding degree are also obtained by Alg. 2, see Tab. 2 (where dim denotes
the dimension of the solution space for g of degree e). A formal expansion of
$f(x) \cdot g(x)$ was performed to verify the results. A reaction on this attack is
presented in [16].

Table 2. Degrees of the functions h and g for DGM functions f with n input variables

n	deg f	deg h	deg g	g	dim
5	4	3	1	$1 + x_4$	4
6	4	3	1	$1 + x_6$	4
7	5	4	1	$1 + x_4 + x_5$	1
8	5	4	1	$1 + x_5 + x_6$	1
9	8	5	1	$x_4 + x_5 + x_6 + x_7$	1
10	8	6	1	$x_5 + x_6 + x_7 + x_8$	1

5   Efficient Computation of Immunity for Symmetric Functions

Consider the case that $f(x)$ is a symmetric Boolean function. This means that
$f(x) = f(x_1, \ldots, x_n)$ is invariant under changing the variables x_i. Therefore,

we have $f(y) = f(y')$ if $|y| = |y'|$ and we can identify f with its (abbreviated) truth table $T^s(f) := (f^s(0), \ldots, f^s(n)) \in \mathbb{F}^{n+1}$ where $f^s(i) := f(y)$ for a y with $|y| = i$. Let $\sigma_i(x) := \bigoplus_{|\alpha|=i} x^\alpha$ denote the elementary symmetric polynomial of degree i. Then, each symmetric function f can be expressed by $f(x) = \bigoplus f_i^s \sigma_i(x)$ with $f_i^s \in \mathbb{F}$. Similarly to the non-symmetric case, f can be identified with its (abbreviated) coefficient vector $C^s(f) := (f_0^s, \ldots, f_n^s) \in \mathbb{F}^{n+1}$.

In this section, we present a general analysis of the resulting system of equations for symmetric functions and propose a generic and a specific algorithm in order to determine the existence of g and h of low degrees.

5.1   Setting Up Equations

One can derive a much simpler relation for the coefficients h_γ in the case of symmetric functions f.

Corollary 1. *Let* $f(x) = \bigoplus_{i=0}^n f_i^s \sigma_i(x)$ *be a symmetric function and* $g(x) = \bigoplus_\beta g_\beta x^\beta$. *Set* $h(x) = \bigoplus_\gamma h_\gamma x^\gamma := f(x) \cdot g(x)$. *Then, with* $A_{i,j}^s \in \mathbb{F}$ *and* $B_{i,j}^s \in \mathbb{F}^{n+1}$, *we have for each* γ

$$h_\gamma = \bigoplus_\beta \binom{\gamma}{\beta} A_{|\gamma|,|\beta|}^s \cdot g_\beta \tag{9}$$

$$A_{i,j}^s := B_{i,j}^s \cdot T^s(f) = B_{i,i-j}^s \cdot C^s(f) \tag{10}$$

$$\left[B_{i,j}^s\right]_k := \binom{i-j}{i-k}. \tag{11}$$

Proof. Notice that Th. 1 holds for any function f, including symmetric functions. Computation of $A_{\gamma,\beta} = B_{\gamma,\beta} \cdot T(f)$ for symmetric functions may be simplified by collecting all terms of the truth table with the same weight. Therefore, let $i := |\gamma|$ and $j := |\beta|$ and define $[B_{i,j}^s]_k := \bigoplus_{|\alpha|=k} [B_{\gamma,\beta}]_\alpha$, such that $A_{\gamma,\beta} = A_{i,j}^s := B_{i,j}^s \cdot T^s(f)$. For $j \leq i$ we have $\bigoplus_{|\alpha|=k} \binom{\gamma}{\alpha}\binom{\alpha}{\beta} = \bigoplus_{|\alpha|=k; \beta \subseteq \alpha \subseteq \gamma} 1$. Counting the number of choices of the k elements of the support of α, we find that the above sum equals $\binom{i-j}{k-j}$. The proof of $A_{i,j}^s = B_{i,i-j}^s \cdot C^s(f)$ is similar. $\square$

5.2   Determining the Existence of Solutions

Given a symmetric function f, the existence of g and h with corresponding degrees can be determined by an adapted version of Alg. 2 (which will be referred as Alg. 2^s): step 7 is omitted, and step 8 is replaced by $A \leftarrow A_{i,j}^s$. The discussion of this slightly modified algorithm is similar to Sect. 4.2. However, computation of $A_{i,j}^s$ requires only $n + 1$ evaluations of the function f, which can be neglected in terms of complexity. Consequently, time complexity to set up equations is only about $\mathcal{O}(E^2)$, and overall complexity of Alg. 2^s becomes $\mathcal{O}(E^3)$.

Next, we will derive a method of very low (polynomial) complexity to determine the existence of g and h of low degree for a symmetric function f, but with the price that the method uses only sufficient conditions (i.e. some solutions may be lost). More precisely, we constrict ourselves to homogeneous functions g of degree e (i.e. g contains monomials of degree e only), and Eq. 9 becomes $h_\gamma = A^s_{|\gamma|,e} \bigoplus_{|\beta|=e} \binom{\gamma}{\beta} g_\beta$. Remember that $h_\gamma = 0$ for $|\gamma| > d$, so the homogeneous function g is determined by the corresponding system of equations for all γ with $|\gamma| = d + 1, \ldots, n$. In this system, the coefficient $A^s_{|\gamma|,e}$ is constant for $\binom{n}{|\gamma|}$ equations. If $A^s_{|\gamma|,e} = 0$, then all these equations are linearly dependent (i.e. of type $0 = 0$). On the other hand, if $A^s_{|\gamma|,e} = 1$, then a number of $\binom{n}{|\gamma|}$ additional equations is possibly linearly independent. Consequently, if the sum of all possibly linearly independent equations for $|\gamma| = d + 1, \ldots, n$ is smaller than the number of variables $\binom{n}{e}$, then nontrivial homogeneous functions g exist. This sufficient criterion is formalized by

$$\sum_{i=d+1}^{n} A^s_{i,e} \cdot \binom{n}{i} < \binom{n}{e}. \tag{12}$$

Given some degree e, the goal is to find the minimum value of d such that Eq. 12 holds. This can be done incrementally, starting from $d = n$. We formalized Alg. 3 of polynomial complexity $\mathcal{O}(n^3)$. This algorithm turned out to be very powerful (but not necessarily optimal) in practice, see Sect. 5.4 for some experimental results.

Algorithm 3. Determine the degrees of g and h for symmetric f

---

Input: A symmetric Boolean function f with n input variables.
Output: Degrees of specific homogeneous functions g and h such that $fg = h$.
1: **for** e from 0 to $\lceil n/2 \rceil$ **do**
2:    Let $d \leftarrow n$, number of equations $\leftarrow 0$, number of variables $\leftarrow \binom{n}{e}$.
3:    **while** number of equations < number of variables **and** $d + 1 > 0$ **do**
4:       Compute $A \leftarrow A^s_{d,e}$.
5:       Add $A \cdot \binom{n}{d}$ to the number of equations.
6:       $d \leftarrow d - 1$.
7:    **end while**
8:    Output $\deg g = e$ and $\deg h = d + 1$.
9: **end for**

For a specified class of symmetric Boolean functions f, it is desirable to prove some general statements concerning the degrees of g and h for any number of input variables n. In the next section, we apply technique based on Alg. 3 in order to prove a theorem for the class of majority functions.

5.3 Fast Algebraic Attacks on the Majority Function

We denote by f the symmetric Boolean majority function with $n \geq 2$ input variables, defined by $f^s(i) := 0$ if $i \leq \lfloor n/2 \rfloor$ and $f^s(i) := 1$ otherwise. For example, $T^s(f) := (0, 0, 1)$ for $n = 2$, and $T^s(f) := (0, 0, 1, 1)$ for $n = 3$. The algebraic degree of this function is $2^{\lfloor \log_2 n \rfloor}$. In [7] and [17], it could be proven independently that f has maximum algebraic immunity[3]. However, in the following theorem, we disclose the properties of f (and related functions) with respect to fast algebraic attacks.

Theorem 2. *Let f be the majority function with any $n \geq 2$ input variables. Then there exist Boolean functions g and h such that $fg = h$, where $d := \deg h = \lfloor n/2 \rfloor + 1$ and $e := \deg g = d - 2^j$, and where $j \in \mathbb{N}^0$ is maximum so that $e > 0$.*

Proof. According to Eq. 9 for symmetric functions, we set up a system of equations in the coefficients of g only. The coefficients $A_{i,j}^s$ of Eq. 10 have a simple form in the case of the majority function, namely $A_{i,j}^s = \bigoplus_{k \geq d} \binom{i-j}{k-j} = \bigoplus_{k \geq d} \binom{i-j-1}{k-j-1} + \bigoplus_{k \geq d} \binom{i-j-1}{k-j} = \binom{i-j-1}{d-j-1} + 2 \bigoplus_{k \geq d} \binom{i-j-1}{k-j} = \binom{i-j-1}{d-j-1}$ for $i > d$. Additionally, we assume that g is homogeneous of degree $e := d - 2^j$ where j is chosen maximum such that $e \geq 1$. According to Lucas' theorem, we find $A_{d+i,e}^s = 0$ for $1 \leq i < d - e$. Consequently, only equations with $|\gamma| = 2d - e, \ldots, n$ may impose conditions on the coefficients g_β. As we can show that $\sum_{i=0}^{e-1} \binom{n}{i} < \binom{n}{e}$, the sufficient criterion (12) is satisfied, and nontrivial solutions exist. □

Algebraic and fast algebraic attacks are invariant with regard to binary affine transformations in the input variables. Consequently, Th. 2 is valid for all Boolean functions which are derived from the majority function by means of affine transformations. We notice that such a class of functions was proposed in a recent paper, discussing design principles of stream ciphers [5,6].

5.4 Experimental Results

Application of Alg. 2^s reveals that Th. 2 is optimal for the majority function where $d = \lfloor n/2 \rfloor + 1$ (verification for $n = 5, 6, \ldots, 16$). An explicit homogeneous function g can be constructed according to $g(x) = \prod_{i=1}^{e}(x_{2i-1} + x_{2i})$. We verified that Alg. 3 can discover the solutions of Th. 2.

In [7], a large pool of symmetric Boolean functions with maximum algebraic immunity is presented (defined for n even). One of these functions is the majority function, whereas the other functions are nonlinear transformations of the majority function. Application of Alg. 3 brings out that Th. 2 is valid for all functions f (verification for $n = 6, 8, \ldots, 16$). For some functions f, Alg. 3 finds better solutions than predicted by Th. 2 (e.g. for $T^s(f) := (0, 0, 0, 1, 1, 0, 1)$ where $d = 3$ and $e = 1$), which means that Th. 2 is not optimal for all symmetric functions. All solutions found by Alg. 3 can be constructed according to the above equation. Furthermore, Alg. 2^s finds a few solutions which are (possibly) better than predicted by Alg. 3 (e.g. for $T^s(f) := (0, 0, 0, 1, 1, 1, 0)$ where $d = 3$ and $e = 2$), which means that Alg. 3 is not optimal for all symmetric functions.

---

[3] Notice that for n odd, it is verified in [17] up to $n = 11$ that the majority function is the only symmetric Boolean function with maximum $\mathcal{AI}$.

6   Conclusions

In this paper, several efficient algorithms have been derived for assessing resistance of LFSR-based stream ciphers against conventional as well as fast algebraic attacks. This resistance is directly linked to the Boolean output function used. In many recent proposals, the number of inputs for this function is about 20 or larger. For such input sizes, verification of immunity against (fast) algebraic attacks by existing algorithms is infeasible. Due to improved efficiency of our algorithms, provable resistance of these stream ciphers against conventional and fast algebraic attacks has become amenable. Our algorithms have been applied to various classes of Boolean functions. In one direction the algebraic immunity of two families of Boolean power functions, the inverse functions and Kasami type functions, have been determined. For the first time, the algebraic immunity $\mathcal{AI}$ of a highly nonlinear function with 20 variables is computed to be as large as $\mathcal{AI} = 9$. In another direction, our algorithms have been applied to demonstrate that large classes of Boolean functions with optimal algebraic immunity are very vulnerable to fast algebraic attacks. This applies in particular to classes of symmetric functions including the majority functions.

Acknowledgments

The first author has been supported by grant Kr 1521/7-2 of the DFG (German Research Foundation). The fourth and fifth author are supported in part by grant 5005-67322 of NCCR-MICS (a center of the Swiss National Science Foundation). The fifth author also receives partial funding through GEBERT RÜF STIFTUNG. We would like to thank Subhamoy Maitra for valuable discussions.

References

1. F. Armknecht, and G. Ars. Introducing a New Variant of Fast Algebraic Attacks and Minimizing Their Successive Data Complexity. In *Progress in Cryptology - Mycrypt 2005*, LNCS 3715, pages 16–32. Springer Verlag, 2005.
2. F. Armknecht. Algebraic Attacks and Annihilators. In *WEWoRC 2005*, volume P-74 of *LNI*, pages 13–21. Gesellschaft für Informatik, 2005.
3. F. Armknecht. Improving Fast Algebraic Attacks. In *Fast Software Encryption 2004*, LNCS 3017, pages 65–82. Springer Verlag, 2004.
4. G. Ars. Application des Bases de Gröbner à la Cryptographie. Thèse de l'Université de Rennes, 2005.
5. A. Braeken, and J. Lano. On the (Im)Possibility of Practical and Secure Nonlinear Filters and Combiners. In *Selected Areas in Cryptography - SAC 2005*, LNCS 3897, pages 159–174. Springer Verlag, 2006.
6. A. Braeken, J. Lano, N. Mentens, B. Preneel, and I. Verbauwhede. SFINKS: A Synchronous Stream Cipher for Restricted Hardware Environments. In *eS-TREAM, ECRYPT Stream Cipher Project*, Report 2005/026. Available at http://www.ecrypt.eu.org/stream.

7. A. Braeken, and B. Preneel. On the Algebraic Immunity of Symmetric Boolean Functions. In *Progress in Cryptology - INDOCRYPT 2005*, LNCS 3797, pages 35–48. Springer Verlag, 2005.
8. P. Camion, C. Carlet, P. Charpin, and N. Sendrier. On Correlation-Immune Functions. In *Advances in Cryptology - CRYPTO 1991*, LNCS 576, pages 86–100. Springer Verlag, 1991.
9. A. Canteaut, and M. Videau. Symmetric Boolean Functions. In *IEEE Transactions on Information Theory*, volume 51/8, pages 2791–2811, 2005.
10. C. Carlet, and P. Gaborit. On the Construction of Boolean Functions with a Good Algebraic Immunity. In *Boolean Functions: Cryptography and Applications - BFCA*, 2005.
11. N. Courtois. Cryptanalysis of SFINKS. To appear in *Information Security and Cryptology - ICISC*, 2005.
12. N. Courtois, and W. Meier. Algebraic Attacks on Stream Ciphers with Linear Feedback. In *Advances in Cryptology - EUROCRYPT 2003*, LNCS 2656, pages 345–359. Springer Verlag, 2003.
13. N. Courtois. Fast Algebraic Attacks on Stream Ciphers with Linear Feedback. In *Advances in Cryptology - CRYPTO 2003*, LNCS 2729, pages 176–194. Springer Verlag, 2003.
14. N. Courtois, and J. Pieprzyk. Cryptanalysis of Block Ciphers with Overdefined Systems of Equations. In *Advances in Cryptology - ASIACRYPT 2002*, LNCS 2501, pages 267–287. Springer Verlag, 2002.
15. D. K. Dalai, K. C. Gupta, and S. Maitra. Cryptographically Significant Boolean Functions: Construction and Analysis in Terms of Algebraic Immunity. In *Fast Software Encryption 2005*, LNCS 3557, pages 98–111. Springer Verlag, 2005.
16. D. K. Dalai, K. C. Gupta, and S. Maitra. Notion of Algebraic Immunity and its Evaluation related to Fast Algebraic Attacks. In *Second International Workshop on Boolean Function Cryptography and Applications*, 2006.
17. D. K. Dalai, S. Maitra, and S. Sarkar. Basic Theory in Construction of Boolean Functions with Maximum Possible Annihilator Immunity. To appear in *Design, Codes and Cryptography*. Springer Verlag, 2006.
18. N. J. Fine. Binomial Coefficients Modulo a Prime. In *The American Mathematical Monthly*, volume 54, pages 589–592, 1947.
19. J.-C. Faugère, and G. Ars. An Algebraic Cryptanalysis of Nonlinear Filter Generators using Gröbner bases. In *Rapport de Recherche INRIA*, volume 4739, 2003.
20. P. Hawkes, and G. G. Rose. Rewriting Variables: The Complexity of Fast Algebraic Attacks on Stream Ciphers. In *Advances in Cryptology - CRYPTO 2004*, LNCS 3152, pages 390–406. Springer Verlag, 2004.
21. W. Meier, E. Pasalic, and C. Carlet. Algebraic Attacks and Decomposition of Boolean Functions. In *Advances in Cryptology - EUROCRYPT 2004*, LNCS 3027, pages 474–491. Springer Verlag, 2004.
22. W. Meier, and O. Staffelbach. Nonlinearity Criteria for Cryptographic Functions. In *Advances in Cryptology - EUROCRYPT 1989*, LNCS 434, pages 549–562. Springer Verlag, 1990.
23. B. Mourrain, and O. Ruatta. Relations Between Roots and Coefficients, Interpolation and Application to System Solving. In *J. Symb. Comput.*, volume 33/5, pages 679–699, 2002.
24. Y. Nawaz, G. Gong, and K. Gupta. Upper Bounds on Algebraic Immunity of Power Functions. To appear in *Fast Software Encryption 2006*. Springer Verlag, 2006.

25. P. J. Olver. On Multivariate Interpolation. In *Stud. Appl. Math.*, volume 116, pages 201–240, 2006.
26. T. Siegenthaler. Correlation-Immunity of Nonlinear Combining Functions for Cryptographic Applications. In *IEEE Transactions on Information Theory*, volume 30/5, pages 776–780, 1984.
27. T. Siegenthaler. Decrypting a Class of Stream Ciphers Using Ciphertext Only. In *IEEE Transactions on Computer*, volume 34/1, pages 81–85, 1985.

VSH, an Efficient and Provable Collision-Resistant Hash Function

Scott Contini[1], Arjen K. Lenstra[2], and Ron Steinfeld[1]

[1] Department of Computing, Macquarie University, NSW 2109, Australia
[2] EPFL IC LACAL, INJ 330, Station 14, 1015-Lausanne, Switzerland

Abstract. We introduce VSH, *very smooth hash*, a new S-bit hash function that is provably collision-resistant assuming the hardness of finding nontrivial modular square roots of very smooth numbers modulo an S-bit composite. By very smooth, we mean that the smoothness bound is some fixed polynomial function of S. We argue that finding collisions for VSH has the same asymptotic complexity as factoring using the Number Field Sieve factoring algorithm, i.e., subexponential in S.

VSH is theoretically pleasing because it requires just a single multiplication modulo the S-bit composite per $\Omega(S)$ message-bits (as opposed to $O(\log S)$ message-bits for previous provably secure hashes). It is relatively practical. A preliminary implementation on a 1GHz Pentium III processor that achieves collision resistance at least equivalent to the difficulty of factoring a 1024-bit RSA modulus, runs at 1.1 MegaByte per second, with a moderate slowdown to 0.7MB/s for 2048-bit RSA security.

VSH can be used to build a fast, provably secure randomised trapdoor hash function, which can be applied to speed up provably secure signature schemes (such as Cramer-Shoup) and designated-verifier signatures.

Keywords: hashing, provable reducibility, integer factoring.

1 Introduction

Current collision-resistant hash algorithms that have provable security reductions are too inefficient to be used in practice. One example [17, 20] that is provably reducible from integer factorisation is of the form $x^m \bmod n$ where m is the message, n a supposedly hard to factor composite, and x is some prespecified base value. A collision $x^m \equiv x^{m'} \bmod n$ reveals a multiple $m - m'$ of the order of x (which in itself divides $\phi(n)$). Such information can be used to factor n in polynomial time assuming certain properties of x.

The above algorithm is quite inefficient because it requires on average 1.5 multiplications modulo n per message-bit. Improved provable algorithms exist [7] which require a multiplication per $O(\log \log n)$ message-bits, but beyond that it seems that so far all attempts to gain efficiency came at the cost of losing provability (see also [1]). We propose a hash algorithm that uses a single multiplication per $\Omega(\log n)$ message-bits. It uses RSA-type arithmetic, obviating the need for completely separate hash function code such as SHA-1. Our algorithm may therefore be useful in embedded environments where code space is limited.

S. Vaudenay (Ed.): EUROCRYPT 2006, LNCS 4004, pp. 165–182, 2006.

We say that an integer is *very smooth* if its prime factors are bounded by $(\log n)^c$ for a fixed constant c. We use VSH, for *very smooth hash*, to refer to our new hash because finding a collision (i.e., strong collision resistance) for VSH is provably as difficult as finding a nontrivial modular square root of a very smooth number modulo n. We show that the latter problem, which we call VSSR, is connected to integer factorisation, and that it is reasonable to believe that VSSR is hard as well (until quantum computers are built). We emphasize that VSH is 'only' collision-resistant and not suitable as a substitute for a random oracle.

Given the factorisation of the VSH-modulus, collisions can be created (cf. trapdoor hashes in [20]). Therefore, for wide-spread application of a single VSH-modulus one has to rely on a trusted party to generate the modulus (and not to create collisions). Or one could use [2] to generate a modulus with knowledge of its factorisation shared among a group of authorities. For a one time computation the overhead may be acceptable. If each party would have it own VSH-modulus, the repudiation concerns are the same as those concerning regular RSA.

On the positive side, we show how VSH can be used to build a provably secure randomised trapdoor hash function which requires only about 4 modular multiplications to evaluate on fixed-length messages of length $k < \log_2 n$ bits (compared to the fastest construction in [20], which requires about k multiplications). Randomised trapdoor hash functions are used in signature schemes to achieve provable security against adaptive chosen message attack [20], and in designated-verifier signature schemes to achieve privacy [11, 21]. Our function can replace the trapdoor function used in the Cramer-Shoup signature scheme [6], maintaining its provable security while speeding up verification time by about 50%.

We also present a variant of VSH using a prime modulus p (with no trapdoor), which has about the same efficiency and is provably collision-resistant assuming the hardness of finding discrete logarithms of very smooth numbers modulo p.

Related Work. Previous hash functions with collision resistance provably related to factoring have lower efficiency than VSH. The $x^m \bmod n$ function mentioned above appeared in [17, 20]. A collision-resistant hash function based on a claw free permutation pair (where claw finding is provably as hard as factoring an RSA modulus) was proposed in [9]—this function requires 1 squaring per bit processed. In [7] the construction is generalised to use families of $r \geq 2$ claw free permutations, such that $\log_2(r)$ bits can be processed per permutation evaluation. Two factoring based constructions are presented, which require 2 multiplications per permutation evaluation. In the first construction the modulus n has $1 + \log_2(r)$ prime factors and thus becomes impractical already for small $\log_2(r)$. The second one uses a regular RSA modulus, but requires publishing r random quadratic residues modulo n. This becomes prohibitive too for relatively small $\log_2(r)$; as a result the construction requires a multiplication modulo an S-bit RSA modulus n per $O(\log S)$ message-bits while consuming polynomial space $(r = O(poly(S)))$. The constructions in [1] are more efficient but are only provably collision-resistant assuming an underlying hash function is modeled as a random oracle (we make no such assumption).

Section 2 introduces VSSR. VSH and its variants are presented in Section 3. Section 4 describes a VSH-based randomised trapdoor hash function which speeds up the Cramer-Shoup signature scheme. Section 5 concludes with implementation results.

2   Security Definitions

Notation. Throughout this paper, let $c > 0$ be a fixed constant and let n be a hard to factor S-bit composite for an integer $S > 0$. The ring of integers modulo n is denoted $\mathbf{Z}_n$, and its elements are represented by $\{0, 1, \ldots, n-1\}$ or $\{-n+1, -n+2, \ldots, 0\}$. It will be clear from the context which representation is being used. The ith prime is denoted p_i: $p_1 = 2$, $p_2 = 3$, $\ldots$, and $p_0 = -1$. An integer is p_k-smooth if all its prime factors are $\leq p_k$. We use

$$L[n, \alpha] = e^{(\alpha + o(1))(\log n)^{1/3}(\log \log n)^{2/3}}$$

for constant $\alpha > 0$ and $n \to \infty$, and where the logarithms are natural.

Definition 1. *An integer b is a very smooth quadratic residue modulo n if the largest prime in b's factorisation is at most $(\log n)^c$ and there exists an integer x such that $b \equiv x^2 \bmod n$. The integer x is said to be a modular square root of b.*

Definition 2. *An integer x is said to be a* trivial *modular square root of an integer b if $b = x^2$, i.e. b is a perfect square and x is the integer square root of b.*

Trivial modular square roots have no relation to the modulus n. Such identities are easy to create, and therefore they are not allowed in the security reduction. A sufficient condition for a very smooth integer b representing a quadratic residue not to have a trivial modular square root is having some prime p such that p divides b but p^2 does not. Another sufficient condition is that b is negative. Our new hardness assumption is that it is difficult to find a nontrivial modular square root of a very smooth quadratic residue modulo n. Before formulating our assumption, we give some relevant background on integer factorisation.

Background. General purpose integer factoring algorithms are used for the security evaluation of RSA, since they do not take advantage of properties of the factors. They all work by constructing nontrivial congruent squares modulo n since such squares can be used to factor n: if $x, y \in \mathbf{Z}$ are such that $x^2 \equiv y^2 \bmod n$ and $x \not\equiv \pm y \bmod n$, then $\gcd(x \pm y, n)$ are proper factors of n. To construct such x, y a common strategy uses so-called *relations*. An example of a relation would be an identity of the form

$$v^2 \equiv \prod_{0 \leq i \leq u} p_i^{e_i(v)} \bmod n,$$

where u is some fixed integer, $v \in \mathbf{Z}$, and $(e_i(v))_{i=0}^u$ is a $(u + 1)$-dimensional integer vector. Given $u + 1 + t$ relations, at least t linearly independent dependencies modulo 2 among the $u + 1 + t$ vectors $(e_i(v))_{i=0}^u$ can be found using linear algebra. Each such dependency corresponds to a product of v^2-values that equals a product modulo n of p_i's with all even exponents, and thus a solution to $x^2 \equiv y^2 \bmod n$. If $x \not\equiv \pm y \bmod n$, then it leads to a proper factor of n. A relation with all even exponents $e_i(v)$ leads to a pair x, y right away, which has, in our experience with practical factoring algorithms, never happened unless n is very small. It may safely be assumed that for each relation found at least one of the $e_i(v)$'s is odd—actually most that are non-zero will be equal to 1.

For any u, relations are easily computed if n's factorisation is known, since square roots modulo primes can be computed efficiently and the results can be assembled via the Chinese Remainder Theorem. If the factorisation is unknown, however, relations in practical factoring algorithms are found by a deterministic process that depends on the factoring algorithm used. It is sufficiently unpredictable that the resulting x, y may be assumed to be random solutions to $x^2 \equiv y^2 \bmod n$, implying that the condition $x \not\equiv \pm y \bmod n$ holds for at least half of the dependencies. Despite the lack of a rigorous proof, this heuristic argument has not failed us yet. A few dependencies usually suffice to factor n.

The expected relation collection runtime is proportional to the product of u (approximately the number of relations one needs) and the inverse of the smoothness probability of the numbers that one hopes to be p_u-smooth, since this probability is indicative for the efficiency of the collection process. For the fastest factoring algorithms published so far, the Number Field Sieve (NFS, cf. [13, 5]), the overall expected runtime (including the linear algebra) is minimised—based on loose heuristic grounds—when, asymptotically for $n \to \infty$, u behaves as $L[n, 0.96...]$. For this u, the running time is $L[n, 1.923...]$, i.e., the square of u.

With the current state of the art of integer factorisation, one cannot expect that, for any value of u, a relation can be found faster than $L[n, 1.923...]/u$ on average, asymptotically for $n \to \infty$. For u-values much smaller than the optimum, the actual time to find a relation will be considerably larger (cf. remark below and [14]). For $u \approx (\log n)^c$, it is conservatively estimated that finding a relation requires runtime at least

$$\frac{L[n, 1.923...]}{(\log n)^c} = L[n, 1.923...],$$

asymptotically for $n \to \infty$, because the denominator gets absorbed in the numerator's $o(1)$. This observation that finding relations for very small u (i.e., u's that are bounded by a polynomial function of $\log n$) can be expected to be asymptotically as hard as factoring n, is the basis for our new hardness assumption.

Before formulating it, we present two ways to use the hardness estimate $L[n, 1.923...]/u$ for small u in practice. One way is to use the asymptotics and assume that finding a relation is as hard as factoring n. A more conservative approach incorporates the division by u in the estimate. In theory this is a futile exercise because, as argued, a polynomially bounded u disappears in the $o(1)$ for $n \to \infty$. In practice, however, n does not go to infinity but actual values have to be dealt with. If n' is a hard to factor integer for which $\log n$ and $\log n'$ are relatively close, then it is widely accepted that the ratio of the NFS-factoring runtimes for n and n' approximates $L[n, 1.923...]/L[n', 1.923...]$ where the $o(1)$'s are dropped. To assess the hardness estimate $L[n, 1.923...]/u$ for very small u, one therefore finds the least integer S' for which, after dropping the $o(1)$'s,

$$L[2^{S'}, 1.923...] \geq \frac{L[n, 1.923...]}{u}, \tag{1}$$

and assumes that finding a relation for this n and u may be expected to be (at least) as hard as NFS-factoring a hard to factor S'-bit integer. Note that S' will be less than S, the length of n. Examples of matching S, S', u values are given in Section 5.

This factoring background provides the proper context for our new problem and its hardness assumption.

Definition 3. *(VSSR: Very Smooth number nontrivial modular Square Root)* *Let n be the product of two unknown primes of approximately the same size and let $k \leq (\log n)^c$. VSSR is the following problem: Given n, find $x \in Z_n^*$ such that $x^2 \equiv \prod_{i=0}^{k} p_i^{e_i} \bmod n$ and at least one of $e_0, \ldots, e_k$ is odd.*

VSSR Assumption. The VSSR assumption is that there is no probabilistic polynomial (in $\log n$) time algorithm which solves VSSR with non-negligible probability (the probability is taken over the random choice of the factors of n and the random coins of the algorithm).

One can contrive moduli where VSSR is not difficult, such as if n is very close to a perfect square. However, such examples occur with exponentially small probability assuming the factors of n are chosen randomly, as required. According to proper security definitions [18], these examples do not even qualify as weak keys since the time-to-first-solution is slower than factoring, and therefore are not worthy of further consideration.

The VSSR Assumption is rather weak and useless in practice since it does not tell us for what size moduli VSSR would be sufficiently hard. This is similar to the situation in integer factorisation where the hardness assumption does not suffice to select secure modulus sizes. We therefore make an additional, stronger assumption that links the hardness of VSSR to the current state of the art in factoring. It is based on the conservative estimate for the difficulty of finding a relation for very small u given above.

Computational VSSR Assumption. The computational VSSR assumption is that solving VSSR is as hard as factoring a hard to factor S'-bit modulus, where S' is the least positive integer for which equation (1) holds (where, as in (1), the $o(1)$'s in the $L[...]$'s are dropped).

Remark. For existing factoring algorithms, the relation collection runtime increases sharply for smoothness bounds that are too low, almost disastrously so if the bound is taken as absurdly low as in VSSR (cf. [14]). Therefore, the Computational VSSR Assumption is certainly overly conservative. Just assuming—as suggested above—that solving VSSR is as hard as factoring n may be more accurate. Nevertheless, the runtime estimates for our new hash function will be based on the overly conservative Computational VSSR Assumption.

Although our analysis is based on the *average* runtime to find a relation using the NFS, it is very conservative (i.e., leads to a large n) compared to a more direct analysis involving the relevant smoothness probability of squares modulo n. The latter would lead to a hardness estimate for finding even a single very smooth relation that is more similar to the runtime of the Quadratic Sieve integer factorisation algorithm, and thereby to much smaller 'secure' modulus sizes (obviously, unless n's factorisation is known or n has a special form which it will not have when properly chosen). Thus, we feel more comfortable using our NFS-based approach.

3   Very Smooth Hash Algorithm

The basic version of VSH follows below. More efficient variants of VSH are discussed later in this section.

VSH Algorithm. Let k, the block length, be the largest integer such that $\prod_{i=1}^{k} p_i < n$. Let m be an ℓ-bit message to be hashed, consisting of bits $m_1, \ldots, m_\ell$, and assume $\ell < 2^k$. To compute the hash of m perform steps 1 through 5:

1. Let $x_0 = 1$.
2. Let $\mathcal{L} = \lceil \frac{\ell}{k} \rceil$ (the number of blocks). Let $m_i = 0$ for $\ell < i \leq \mathcal{L}k$ (padding).
3. Let $\ell = \sum_{i=1}^{k} \ell_i 2^{i-1}$ with $\ell_i \in \{0, 1\}$ be the binary representation of the message length ℓ and define $m_{\mathcal{L}k+i} = \ell_i$ for $1 \leq i \leq k$.
4. For $j = 0, 1, \ldots, \mathcal{L}$ in succession compute

$$x_{j+1} = x_j^2 \times \prod_{i=1}^{k} p_i^{m_{j \cdot k+i}} \bmod n.$$

5. Return $x_{\mathcal{L}+1}$.

Message Length. The message length does not need to be known in advance, which is useful for applications involving streaming data. In an earlier version which appeared on eprint [4], the message length was prepended, which may prove inconvenient and also required usage of p_{k+1}. If one uses the common method of appending a single 1 bit prior to zero-padding the final block, collisions can easily be created for the above version of VSH.

Compression Function H. VSH applies the compression function $H(x, m)$: $\mathbf{Z}_n^* \times \{0, 1\}^k \to \mathbf{Z}_n^*$ with $H(x, m) = x^2 \prod_{i=1}^{k} p_i^{m_i} \bmod n$, and applies a variant of the Merkle-Damgård transformation [15, 8] to extend H to arbitrarily long inputs. We comment on why this works in Section 3.1.

1024-Bit n. For 1024-bit n, the value for k would be 131. The requirement $\ell < 2^k$ is therefore not a problem in any real application, and most of the bits ℓ_i will be zero. The Computational VSSR Assumption with $S = 1024$ and $k = u = 131$ leads to $S' = 840$. The security level obtained by VSH using

1024-bit n is therefore at least the security level obtained by 840-bit RSA and, given recent hash developments, by SHA-1.

Efficiency. Because $\prod_{1 \le i \le K} p_i$ is asymptotically proportional to $e^{(1+o(1))K \log K}$, for $K \to \infty$, the k used in the basic version of VSH is proportional to $\frac{\log n}{\log \log n}$. It follows that the product $\prod_{i=1}^{k} p_i^{m_{j \cdot k+i}}$ can be computed in time $O((\log n)^2)$ using straightforward multiplication without modular reduction. Therefore the cost of each iteration is less than the cost of 3 modular multiplications. Since k bits are processed per iteration, the basic version of VSH requires a single modular multiplication per $\Omega(\frac{\log n}{\log \log n})$ message-bits, with a small constant in the Ω.

Creating Collisions. With $e_i = \sum_{j=0}^{\mathcal{L}} m_{j \cdot k+i} 2^{\mathcal{L}-j}$ for $1 \le i \le k$, the value calculated by the VSH algorithm equals the multi-exponentiation $\prod_{i=1}^{k} p_i^{e_i} \bmod n$. Given $\phi(n)$ and assuming large enough $\mathcal{L}$, collisions can be generated by replacing e_i by $e_i + t_i \phi(n)$ for any set of i's with $1 \le i \le k$ and positive integers t_i (see also VSH-DL below). Thus, parties that know n's factorisation can create collisions at will. But collisions of this sort immediately reveal $\phi(n)$ and thereby n's factorisation. Creating collisions that cannot immediately be used to factor n is a harder problem, involving discrete logarithms of very smooth numbers.

To avoid repudiation concerns if VSH would be used 'globally' with the same modulus it would be advisable to generate n using the method from [2]. On the other hand, it is conceivable—and may be desirable—to expand PKIs to allow one to choose one's own hash function, rather than using a 'fixed target' for all. In this setting, one cannot allow the owner of a VSH-modulus to claim he did not sign something by displaying a collision. Especially taking into consideration that the only easy way the user can create a collision would also reveal the factorisation of n, this would be analogous to somebody using RSA who anonymously posts the factorisation of their modulus in order to fraudulently claim that he did not sign something. Thus, in such a situation the VSH-modulus should be considered compromised and the user's certificate should be revoked.

Short Message Inversion. The VSH algorithm described above allows easy inversion of short and some sparse messages since there may be no wrap-around modulo n. The attacker first guesses the length, divides the hash modulo n by the corresponding $\prod_{i=1}^{k} p_i^{m_{\mathcal{L} \cdot k+i}}$, and checks if the resulting value is very smooth. This type of invertibility may be undesirable for some applications, but others require just collision resistance (cf. below). See [16] for a related application.

A solution to this invertibility problem that does not affect our proof of security (cf. below) is to square the final output enough times to ensure wrap-around (no more than $\log_2 \log_2 n$ times). Other, more efficient solutions may be possible. Note that for all hash functions, the hash of extremely sparse or short messages can always be 'inverted' by trial and error.

Undesirable Properties. It is easy to find messages for which the hashes h and h' satisfy $h = 2h'$. Our solution to the invertibility problem addresses this issue as well. Other similar possibly undesirable properties can be constructed. We again emphasize that VSH is not intended to model a random oracle, and therefore cannot be blindly substituted as is into constructions that depend upon them (such as RSA signatures and some MAC constructions). We remind the reader that random oracles do not exist in the real world, and therefore relying on them too much is not recommended. On the other hand, entirely provable solutions do exist which require only collision resistance: for example, see Section 4.

Having stressed upfront in the last three remarks the disadvantages of VSH, we turn to its most attractive property, namely its provable collision resistance.

3.1   Security Proof for VSH

We prove that VSH is (strongly) collision-resistant. Using proper security notions [19], (strong) collision resistance also implies second preimage resistance.

Theorem 1. *Finding a collision in VSH is as hard as solving VSSR (i.e., VSH is collision-resistant under the assumptions from Section 2).*

Proof. We show that different colliding messages m and m' lead to a solution of VSSR. Let $x'_{...}$ denote the $x_{...}$ values in the VSH algorithm applied to m' and let $\ell, \mathcal{L}$ and $\ell', \mathcal{L}'$ be the bitlengths and number of blocks of m and m', respectively. Since m and m' collide, $m \neq m'$ and $x_{\mathcal{L}+1} = x'_{\mathcal{L}'+1}$.

First consider the case of $\ell = \ell'$. Let $m[j]$ denote m's jth k-bit block, $m[j] = (m_{j \cdot k+i})_{i=1}^{k}$, and let $t \leq \mathcal{L}$ be the largest index such that $(x_t, m[t]) \neq (x'_t, m'[t])$ but $(x_j, m[j]) = (x'_j, m'[j])$ for $t < j \leq \mathcal{L} + 1$. Then,

$$(x_t)^2 \times \prod_{i=1}^{k} p_i^{m_{t \cdot k+i}} \equiv (x'_t)^2 \times \prod_{i=1}^{k} p_i^{m'_{t \cdot k+i}} \bmod n \ . \tag{2}$$

Let $\Delta = \{i : m_{t \cdot k+i} \neq m'_{t \cdot k+i}, 1 \leq i \leq k\}$ and $\Delta_{10} = \{i \in \{1, \ldots, k\} : m_{t \cdot k+i} = 1 \text{ and } m'_{t \cdot k+i} = 0\}$. Because all factors in Equation (2) are invertible modulo n, it is equivalent to

$$\left[(x_t/x'_t) \times \prod_{i \in \Delta_{10}} p_i \right]^2 \equiv \prod_{i \in \Delta} p_i \bmod n \ . \tag{3}$$

If $\Delta \neq \emptyset$, Equation (3) solves VSSR. If $\Delta = \emptyset$, then $(x_t)^2 \equiv (x'_t)^2 \bmod n$ and $t \geq 1$ (since $m \neq m'$ and using the definition of t). With $x_t \not\equiv \pm x'_t \bmod n$ VSSR can be solved by factoring n. If $x_t \equiv \pm x'_t \bmod n$ then $x_t \equiv -x'_t \bmod n$, since $\Delta = \emptyset$ implies (by definition of t) that $x_t \neq x'_t$. But $x_t \equiv -x'_t \bmod n$ leads to $(x_{t-1}/x'_{t-1})^2$ being congruent to -1 times a very smooth number and thus solves VSSR.

Now consider the case $\ell \neq \ell'$. Since $x_{\mathcal{L}+1} = x'_{\mathcal{L}'+1}$, we have $(x_{\mathcal{L}}/x'_{\mathcal{L}'})^2 \equiv \prod_{i=1}^{k} p_i^{\ell'_i - \ell_i} \bmod n$. Since $|\ell'_i - \ell_i| = 1$ for at least one i, VSSR is solved using a transformation as in Equation (3).   □

Why Merkle-Damgård Works. VSH applies a variant of the Merkle-Damgård transformation [15, 8] to hash arbitrary length messages using the compression function $H : \mathbf{Z}_n^* \times \{0,1\}^k \to \mathbf{Z}_n^*$. The proof in [8] shows that a sufficient condition for a hash function to be collision-resistant is that its compression function H is collision-resistant, i.e. it is hard to find any $(x,m) \neq (x',m')$ with $H(x,m) = H(x',m')$. However, our compression function $H(x,m) = x^2 \prod_{i=1}^k p_i^{m_i} \bmod n$ is not strictly collision-resistant ($H(-x \bmod n, m) = H(x,m)$), and yet we proved that H is still sufficiently strong to make VSH collision-resistant. Therefore, one may ask whether we can strengthen the result in [8] to state explicitly the security properties of a compression function (which are weaker than full collision resistance) that our compression function satisfies and that are still sufficient in general to make the resulting hash function collision-resistant. Indeed, these conditions can be readily generalised from our proof of Theorem 1, so we only state them here:

(1) Collision Resistance in Second input: It is hard to find $(x,m), (x',m') \in \mathbf{Z}_n^* \times \{0,1\}^k$ with $m \neq m'$ such that $H(x,m) = H(x',m')$.
(2) Preimage Resistance for a collision in first input: It is hard to find $(x,m) \neq (x',m') \in \mathbf{Z}_n^* \times \{0,1\}^k$ and $m^* \in \{0,1\}^k$ such that $H(y,m^*) = H(y',m^*)$, where $y = H(x,m)$, $y' = H(x',m')$ and $y \neq y'$.

The VSH compression function H satisfies these properties, under the VSSR Assumption.

3.2 Example: A Related Algorithm That Can Be Broken

To emphasize the importance of the nontrivialness, consider a hash function that works similarly to VSH, except breaks the message into blocks r_1, r_2, ... of $K > 1$ bits and uses the compression function $x_{j+1} = x_j^2 \times 2^{r_j+1} \bmod n$. Because $K > 1$ collisions can simply be created. For example, for any e with $0 < e < 2^{K-1}$ the message blocks $r_1 = e$ and $r_2 = 2e$ collide with $r_1' = 2e$ and $r_2' = 0$. The colliding values are $(2^e)^2 2^{2e}$ and $(2^{2e})^2 2^0$, but this does not lead to a solution of VSSR or a chance to factor n. Such trivial relations are useless, and the security of this hash algorithm is not based on a hard problem. The fix is to use the costlier compression function $x_{j+1} = x_j^{2^K} \times 2^{r_j+1}$, but that results in the same function $x^m \bmod n$ from [17, 20].

3.3 Combining VSH and RSA

Since the output length of VSH is the length of a secure RSA modulus (thus 1024–2048 bits), VSH seems quite suitable in practice for constructing 'hash-then-sign' RSA signatures for arbitrarily long messages. However, such a signature scheme must be designed carefully to ensure its security. To illustrate a naive insecure scheme, let (n,e) be the signer's public RSA key, where the modulus n is used for both signing and hashing. The signing function $\sigma : \{0,1\}^* \to \mathbf{Z}_n$ is $\sigma(m) = VSH_n(m)^{1/e} \bmod n$, where $VSH_n : \{0,1\}^* \to \mathbf{Z}_n$ is VSH with modulus n. For a k-bit message $m = (m_1, \ldots, m_k) \in \{0,1\}^k$, the signature is thus $\sigma(m) = (\kappa \prod_{i=1}^k p_i^{2m_i})^{1/e} \bmod n$, for a κ that is the same for all k-bit messages.

This scheme is insecure under the following chosen message attack. After obtaining signatures on three k-bit messages: $s_0 = \sigma((0,0,0,\ldots,0)) = \kappa^{1/e} \bmod n$, $s_1 = \sigma((1,0,0,\ldots,0)) = (\kappa p_1^2)^{1/e} \bmod n$, and $s_2 = \sigma((0,1,0,\ldots,0)) = (\kappa p_2^2)^{1/e} \bmod n$, the attacker easily computes the signature $\frac{s_1 s_2}{s_0} \bmod n$ on the new k-bit forgery message $(1,1,0,\ldots,0)$. It is easy to see that $k+1$ signatures on $k+1$ properly chosen messages suffice to sign any k-bit message.

To avoid such attacks, we suggest a more theoretically sound design approach for using VSH with 'hash-then-sign' RSA signatures that does not rely on any property of VSH beyond the collision resistance which it was designed to achieve:

Step 1. Let $\bar{n}$ be an $(S+1)$-bit RSA modulus, with $\bar{n}$ and the S-bit VSH modulus n chosen independently at random. So, $\bar{n} > 2^S$. Specify a one-to-one one-way encoding function $f : \{0,1\}^S \to \{0,1\}^S$, and define the short-message (S-bit) RSA signature scheme with signing function $\sigma_{\bar{n}}(m) = (f(m))^{1/e} \bmod \bar{n}$. The function f is chosen such that the short-message scheme with signing function $\sigma_{\bar{n}}$ is existentially unforgeable under chosen message attack. In the standard model no provable techniques are known to find f, but since f is one-to-one, there are no collision resistance issues to consider when designing f.

Step 2. With $(\bar{n}, n, e)$ as the signer's public key, the signature scheme for signing arbitrary length messages is now constructed with signing function $\sigma_{\bar{n},n}(m) = \sigma_{\bar{n}}(VSH_n(m))$. It is easy to prove that the scheme with signing function $\sigma_{\bar{n},n}$ is existentially unforgeable under chosen message attack, assuming that the scheme with signing function $\sigma_{\bar{n}}$ is and that VSH_n is collision-resistant. We emphasize that the proof no longer holds if $\bar{n} = n$: in order to make the proof work in that case, one needs the stronger assumption that VSH_n is collision-resistant even given access to a signing oracle σ_n. However, it is worth remarking that if the function f is modeled as a random oracle, then the proof of security works (under the RSA and VSSR assumptions) even with a shared modulus ($\bar{n} = n$).

3.4   Variants of VSH

Cubing Instead of Squaring. Let $H' : \mathbf{Z}_n^* \times \{0,1\}^k \to \mathbf{Z}_n^*$ with $H'(x,m) = x^3 \prod_{i=1}^{k} p_i^{m_i} \bmod n$ be a compression function that replaces the squaring in H by a cubing. If $\gcd(3, \phi(n)) = 1$ then thanks to the injectivity of the RSA cubing map modulo n, the function H' is collision-resistant, assuming the difficulty of computing a modular cube root of a very smooth cube-free integer of the form $\prod_{i=1}^{k} p_i^{e_i} \neq 1$, where $e_i \in \{0,1,2\}$ for all i. This problem is related to RSA inversion, and is also conjectured to be hard. Although H' requires about 4 modular multiplications per k message-bits (compared to 3 for H), it has the interesting property that H' itself is collision-resistant, while this is not quite the case for H (because $x^2 \prod_i p_i^{m_i} \equiv (-x)^2 \prod_i p_i^{m_i} \bmod n$).

Increasing the Number of Small Primes. A speed-up is obtained by allowing the use of larger k than the largest one for which $\prod_{i=1}^{k} p_i < n$. This does not affect the proof of security and reduction to VSSR, as long as k is still polynomially bounded in $\log n$. The Computational VSSR Assumption implies that a larger modulus n has to be used to maintain the same level of security.

Furthermore, the intermediate products in Step 4 of the VSH algorithm may get larger than n and may thus have to be reduced modulo n every so often. Nevertheless, the resulting smaller $\mathcal{L}$ may outweigh these disadvantages.

Precomputing Products of Primes. An implementation speed-up may be obtained by precomputing products of primes. Let $b > 1$ be a small integer, and assume that $k = \bar{k}b$ for some integer $\bar{k}$. For $i = 1, 2, \ldots, \bar{k}$ compute the 2^b products over all subsets of the set of b primes $\{p_{(i-1)b+1}, p_{(i-1)b+2}, \ldots, p_{ib}\}$, resulting per i in 2^b moderately sized values $v_{i,t}$ for $0 \le t < 2^b$. The k message-bits per iteration of VSH are now split into $\bar{k}$ chunks $m[0], m[1], \ldots, m[\bar{k} - 1]$ of b bits each, interpreted as non-negative integers $< 2^b$. The usual product is then calculated as $\prod_{i=1}^{\bar{k}} v_{i,m[i-1]}$. This has no effect on the number of iterations or the modulus size to be used to achieve a certain level of security.

Fast VSH. Redefining the above $v_{i,t}$ as $p_{(i-1)2^b+t+1}$ and using $i = 1, 2, \ldots, k$ instead of $i = 1, 2, \ldots, \bar{k}$, the block length increases from k to bk, the number of iterations is reduced from $\lceil \frac{\ell}{k} \rceil$ to $\lceil \frac{\ell}{bk} \rceil$, and the calculation in Step 4 of the VSH algorithm becomes

$$x_{j+1} = x_j^2 \times \prod_{i=1}^{k} p_{(i-1)2^b+m[jbk+i-1]+1} \bmod n,$$

where $m[r]$ is the rth b-bit chunk of the message, with $0 \le m[r] < 2^b$. Because the number of small primes increases from k to $k2^b$, a larger modulus would, conservatively, have to be used to maintain the same level of security. But this change does not affect the proof of security and, as shown in the analysis below and the runtime examples in the final section, it is clearly advantageous.

Analysis of Fast VSH. Since $p_{(i-1)2^b+m[jbk+i-1]+1} \le p_{i2^b}$, each intermediate product in the compression function for Fast VSH will be less than n if $\prod_{i=1}^{k} p_{i2^b} < n$. If k is maximal such that $\prod_{i=1}^{(k+1)2^b} p_i \le (2n)^{2^b}$, then

$$\prod_{i=1}^{(k+1)2^b} p_i = \prod_{t=1}^{2^b} \prod_{i=0}^{k} p_{i2^b+t} \le (2n)^{2^b},$$

so that $\prod_{i=0}^{k} p_{i2^b+1} \le 2n$. With $p_{i2^b} < p_{i2^b+1}$ it follows that $\prod_{i=1}^{k} p_{i2^b} < n$. Thus, for $(k+1)2^b$ proportional to $\frac{2^b \log(2n)}{\log(2^b \log(2n))}$ and k to $\frac{\log(2n)}{\log(2^b \log(2n))} - 1$, the cost of Fast VSH is one modular multiplication per $\Omega(bk)$ message-bits, with bk proportional to $\frac{b \log(2n)}{\log(2^b \log(2n))} - b$. Selecting 2^b as any fixed positive power of $\log(2n)$, it follows that bk is proportional to $\log n$ and thus that Fast VSH requires a single modular multiplication per $\Omega(\log n)$ message-bits. It also follows that the number of small primes $k2^b$ is polynomially bounded in $\log n$ so that, with S' the overly conservative RSA security level obtained according to the Computational

VSSR Assumption, Fast VSH requires a single modular multiplication per $\Omega(S')$ message-bits.

Zero Chunks in Fast VSH. A negligible speed-up and tiny saving in the number of primes can be obtained in Fast VSH if for a particular b-bit pattern (such as all zeros) no prime is multiplied in (as was the case in basic VSH).

Fast VSH with Increased Block Length. Fast VSH can be used in a straightforward fashion with a larger block length than suggested by the above analysis. If, for instance, the number of small primes is taken almost w times larger, for some integer $w > 1$, the small prime product can be split into w factors each less than n. Per iteration this results in a single modular squaring, $w - 1$ modular multiplications plus the time to build the w products. The best value for w is best determined experimentally, and will depend on various processor characteristics (such as cache size to hold a potentially rather large table of primes).

Generating Collisions. For all variants given above knowledge of $\phi(n)$ can be used to generate collisions, though displaying such a collision is not in the user's interest since it would give out a break to the user's hash function (i.e. it would be similar to someone giving out the factorisation of their RSA modulus).

VSH-DL, a Discrete Logarithm Variant. We present a discrete logarithm (DL) variant of VSH that has no trapdoor. Its security depends on the following problem and its hardness assumption.

Definition 4. *(VSDL: Very Smooth number Discrete Log) Let p, q be primes with $p = 2q + 1$ and let $k \leq (\log p)^c$. VSDL is the following problem: given p, find integers $e_1, e_2, \ldots, e_k$ such that $2^{e_1} \equiv \prod_{i=2}^{k} p_i^{e_i} \bmod p$ with $|e_i| < q$ for $i = 1, 2, \ldots, k$, and at least one of $e_1, e_2, \ldots, e_k$ is non-zero.*

VSDL Assumption. The VSDL assumption is that there is no probabilistic polynomial (in $\log p$) time algorithm which solves VSDL with non-negligible probability (the probability is taken over the random choice of the prime p and the random coins of the algorithm).

A solution to a VSDL instance produces the base 2 DL modulo p of a very smooth number (the requirements on the exponents e_i avoids trivial solutions in which all exponents are zero modulo q). Given k random VSDL solutions, the base 2 DL of nearly all primes $p_1, \ldots, p_k$ can be solved with high probability by linear algebra modulo q. Although computing the DLs of a polynomial number of small primes is an impressive feat, it does not help to solve arbitrary DL problems. To solve the DL of an arbitrary group element with respect to some generator one could include both generator and element among the p_i, but there is no guarantee that solutions to VSDL contain the appropriate elements. Nevertheless, there is a strong connection between the hardness of VSDL and the hardness of computing DLs modulo p, which is reminiscent of, but seems to be somewhat weaker than, the connection between VSSR and factorisation. See also [3]. As was the case for VSSR, moduli for which VSDL is not difficult are easily constructed and not worthy of further consideration.

Let p be an S-bit prime of the form $2q + 1$ for prime q, let k be a fixed integer length (number of small primes, typically $k \approx S/\log S$), and let $\mathcal{L} \leq S - 2$. We define a VSH-DL compression function $H_{DL} : \{0,1\}^{\mathcal{L}k} \rightarrow \{0,1\}^S$, where m is an $\mathcal{L}k$-bit message consisting of bits $m_1, m_2 \ldots, m_{\mathcal{L}k}$:

- Set $x_0 = 1$. For $j = 0, 1, \ldots, \mathcal{L}-1$, compute $x_{j+1} = x_j^2 \times \prod_{i=1}^{k} p_i^{m_{j \cdot k+i}} \bmod p$.
- Return $H_{DL}(m) = x_{\mathcal{L}}$ interpreted as a value in $\{0,1\}^S$.

If $e_i = \sum_{j=0}^{\mathcal{L}-1} m_{j \cdot k+i} 2^{\mathcal{L}-j-1}$ for $1 \leq i \leq k$, then $H_{DL}(m) = \prod_{i=1}^{k} p_i^{e_i} \bmod p$. A collision $m, m' \in \{0,1\}^{\mathcal{L}k}$ with $m \neq m'$ therefore implies that $\prod_{i=1}^{k} p_i^{e_i} \equiv \prod_{i=1}^{k} p_i^{e_i'} \bmod p$, where $e_i' = \sum_{j=0}^{\mathcal{L}-1} m_{j \cdot k+i}' 2^{\mathcal{L}-j-1}$ and m' consists of the bits $m_1', \ldots, m_{\mathcal{L}k}'$. Rearranging this congruence, a solution $2^{e_1-e_1'} \equiv \prod_{i=2}^{k} p_i^{e_i'-e_i} \bmod p$ to VSDL follows, because $|e_i' - e_i| < 2^{\mathcal{L}} \leq 2^{S-2} \leq q$ for all i and $e_i' - e_i \neq 0$ for some i since $m \neq m'$. Hence the compression function H_{DL} is collision-resistant under the VSDL assumption. We remark that VSH-DL can be viewed as a (more efficient) special case of the collision-resistant function in [3], which uses random group elements in place of the small primes p_i.

The compression function H_{DL} uses the same iteration as the basic VSH algorithm. Hence, for the same modulus length S and number of primes k it has the same throughput efficiency of a single modular multiplication per about $\frac{k}{3}$ message-bits. By applying the Merkle-Damgård transformation [15,8], H_{DL} can be used to hash messages of arbitrary length in blocks of $\mathcal{L}k - S$ message-bits per evaluation of H_{DL}. This leads to a reduction in throughput by a factor of $\frac{\mathcal{L}k-S}{\mathcal{L}k}$ (since only $\mathcal{L}k - S$ of the $\mathcal{L}k$ bits processed in each H_{DL} evaluation are new message-bits) relative to factoring based VSH. However, for long messages, this throughput reduction factor can be made close to 1 by choosing a sufficiently large block length $\mathcal{L}k$; indeed, the construction allows block lengths up to $\mathcal{L}k = k(S-2)$, and for this choice the throughput reduction factor is $1 - \frac{S}{k(S-2)} \approx 1 - \frac{1}{k} \approx 1$.

Reducing the Length. A possible drawback of VSH is its relatively large output length. We are investigating length-reduction possibilities by combining VSH-DL with elliptic curve, trace, or torus-based methods [10,12,22].

4   VSH Randomised Trapdoor Hash and Applications

Let $\mathcal{M}, \mathcal{R}, \mathcal{H}$ be a message, randomiser, and hash space, respectively. A *randomised trapdoor hash function* [20] $F_{pk} : \mathcal{M} \times \mathcal{R} \rightarrow \mathcal{H}$ is a collision-resistant function that can be efficiently evaluated using a public key pk, but for which certain randomly behaving collisions can be found given a secret trapdoor key sk:

Collision Resistance in Message Input. Given pk, it is hard to find $m, m' \in \mathcal{M}$ and $r, r' \in \mathcal{R}$ for which $m \neq m'$ and $F_{pk}(m, r) = F_{pk}(m', r')$.

Random Trapdoor Collisions. There exists an efficient algorithm that given trapdoor (sk, pk), $m, m' \in \mathcal{M}$ with $m \neq m'$, and $r \in \mathcal{R}$, finds a randomiser

$r' \in \mathcal{R}$ such that $F_{pk}(m,r) = F_{pk}(m',r')$. Furthermore, if r is chosen uniformly from $\mathcal{R}$ then r' is uniformly distributed in $\mathcal{R}$.

Randomised trapdoor hash functions have applications in provably strengthening the security of signature schemes [20], and constructing designated-verifier proofs/signature schemes [11, 21]. The factorisation trapdoor of VSH suggests that it can be used to build such a function. Here we describe a provably secure randomised trapdoor hash family which preserves the efficiency of VSH.

Key Generation: Choose two $S/2$-bit random primes p, q with $p \equiv q \equiv 3 \bmod 4$ and S-bit product n. The public key is n with trapdoor key $sk = (p,q)$. Let k be as in the basic VSH algorithm, $\mathcal{M} = \cup_{\ell=0}^{2^k-1}\{0,1\}^\ell$, and $\mathcal{R} = \mathbf{Z}_n^*$.

Hash Function: Let $m \in \mathcal{M}$ of length $\ell < 2^k$ and $r \in \mathcal{R}$. Calculate the basic VSH of m with x_0 replaced by r to compute $x_{\mathcal{L}+1}$ and output $F_n(m,r) = x_{\mathcal{L}+1}^2 \bmod n$.

Theorem 2. *The above construction satisfies the security requirements for randomised trapdoor hash functions, under the VSSR assumption.*

Proof. Collision Resistance in Message Input: The proof follows the same lines as the proof of Theorem 1 since the value of x_0 and the squaring at the end do not affect the security reduction.

Random Trapdoor Collisions: Let $m, m' \in \mathcal{M}$ with $m \neq m'$ and $r \in \mathcal{R}$. Because $F_n(m,r) = (r^{2^{\mathcal{L}+1}} \prod_{i=1}^{k} p_i^{e_i})^2 \bmod n$, where $m_{\mathcal{L}k+i} = \ell_i$ and $e_i = \sum_{j=0}^{\mathcal{L}} m_{j \cdot k+i} 2^{\mathcal{L}-j}$ for $1 \leq i \leq k$, finding $r' \in \mathcal{R}$ with $F_n(m,r) = F_n(m',r')$ amounts to finding r' such that

$$(r')^{2^{\mathcal{L}'+2}} \equiv r^{2^{\mathcal{L}+2}} \cdot (\prod_{i=1}^{k} p_i^{e_i - e_i'})^2 \bmod n$$

(where $m'_{\mathcal{L}'k+i} = \ell_i'$ and $e_i' = \sum_{j=0}^{\mathcal{L}'} m'_{j \cdot k+i} 2^{\mathcal{L}'-j}$ for $1 \leq i \leq k$), i.e., finding an $(\mathcal{L}'+2)$nd square root modulo n of the right hand side g of the equation for $(r')^{2^{\mathcal{L}'+2}}$. Given the trapdoor key (p,q) this is achieved as follows.

Let $QR_n = \{y \in \mathbf{Z}_n^* : (\frac{y}{p}) = (\frac{y}{q}) = 1\}$ denote the subgroup of quadratic residues of $\mathbf{Z}_n^*$. The choice $p \equiv q \equiv 3 \bmod 4$ implies that -1 is a quadratic non-residue in $\mathbf{Z}_p^*$ and $\mathbf{Z}_q^*$, so for each element of QR_n exactly one of its 4 square roots in $\mathbf{Z}_n^*$ belongs to QR_n. Hence the squaring map on QR_n permutes QR_n and given (p,q) it can be efficiently inverted by computing the proper square roots modulo p and q and combining them by Chinese remaindering. Since $g \in QR_n$, its $(\mathcal{L}'+1)$st square root $d \in QR_n$ can thus be computed, and r' is then chosen uniformly at random among the 4 square roots in $\mathbf{Z}_n^*$ of d.

If r is uniformly distributed in $\mathbf{Z}_n^*$, then (since each element of QR_n has 4 square roots in $\mathbf{Z}_n^*$) the value $r^2 \bmod n$ is uniformly distributed in QR_n. The squaring map on QR_n permutes QR_n, so that g and d are also uniformly distributed in QR_n. It follows that r' is uniformly distributed in $\mathbf{Z}_n^*$.   □

Efficiency. For short fixed-length messages with $\ell \leq k$ (i.e., 1 block), the message length can be omitted, so that $F_n(m,r) = (r^2 \prod_{i=1}^{k} p_i^{m_i})^2 \bmod n$. Evaluation requires only about 4 compared to at least k modular multiplications required by the trapdoor functions in [20]. On the other hand, the trapdoor collision-finding algorithm for F_n is not very fast, requiring a square root modulo n per message block. This is not a major issue because in many applications of randomised hash functions, the collision-finding algorithm is only used in the *security proof* of a signature scheme rather than in the scheme itself. However, it reduces the efficiency of the reduction and thus requires slightly increased security parameters.

'Inversion' Trapdoor Property. It follows from the proof of Theorem 2 that F_n also satisfies the 'inversion' trapdoor property [20]. This is stronger than the trapdoor collision property, and can be used to upgrade a signature scheme's resistance against random message attacks to chosen message attacks: Given the trapdoor key, a random element $d \in QR_n$ in the range of F_n and an $m \in \mathcal{M}$, it is easy to find a randomiser $r \in \mathbf{Z}_n^*$ such that $F_n(m, r) = d$ and r is uniformly distributed in $\mathbf{Z}_n^*$ when d is uniformly distributed in QR_n.

Application. As an example application, we mention the Cramer-Shoup (CS) signature scheme [6], which to our knowledge is the most efficient factoring-based signature scheme provably secure in the standard model (under the strong-RSA assumption). The CS scheme makes use of an RSA-based randomised trapdoor hash function to achieve security against adaptive message attacks. Using F_n instead cuts the signing and verification costs by about a double exponentiation each, while preserving the proven security. The modified CS scheme is as follows:
Key Generation: Choose two safe random $\approx S/2$-bit primes $\bar{p}, \bar{q}$ and two random $\approx S/2$ bit primes p, q with $p \equiv q \equiv 3 \bmod 4$ that result in S-bit moduli $\bar{n} = \bar{p}\bar{q}$ and $n = pq$, and choose $x, z \in QR_{\bar{n}}$ at random. Let $h : \{0, 1\}^S \to \{0, 1\}^\ell$ be a collision-resistant hash function for a security parameter ℓ for which an ℓ-bit (traditional) hash and S-bit RSA offer comparable security (typically $\ell = 160$ when $S = 1024$). The public key is $(x, z, n, \bar{n}, h)$ and the secret key is $(\bar{p}, \bar{q})$.
Signing: To sign $m \in \{0, 1\}^*$, choose a random $(\ell + 1)$-bit prime e and a random $r \in \mathbf{Z}_n^*$ and compute $y = (x \cdot z^{h(F_n(m,r))})^{1/e} \bmod \bar{n}$. The signature is (e, y, r).
Verifying: To verify message/signature pair $(m, (e, y, r))$, check that e is an odd $(\ell + 1)$-bit integer and that $y^e z^{-h(F_n(m,r))} \equiv x \bmod \bar{n}$.

The cost of verification in the original CS scheme is about two double exponentiations with ℓ-bit exponents. The modified scheme requires approximately one such double exponentiation, so a saving in verification time of about 50% can be expected. The relative saving in signing time is smaller. However, the length of the public key is larger than in the original scheme by typically 25%.

Because VSH's output length S is typically much larger than ℓ, VSH cannot be used for the ℓ-bit collision-resistant hash function h above. To avoid the need for an ad-hoc ℓ-bit hash function, h may be dropped and e chosen as an $(S + 1)$-bit prime, making the scheme much less efficient. The variant below eliminates the need for h and maintains almost the computational efficiency of the scheme above, but has a larger public key and requires some precomputation.

Key Generation: Let $\bar{p}, \bar{q}, p, q, \bar{n}, n$ be as above, let $s = \lceil \frac{S}{\ell} \rceil$ and randomly choose $x, z_1, \ldots, z_s \in QR_{\bar{n}}$. The public key is $(x, z_1, \ldots, z_s, n, \bar{n})$ with secret key $(\bar{p}, \bar{q})$.

Signing: To sign $m \in \{0, 1\}^*$, choose a random $(\ell + 1)$-bit prime e and a random $r \in \mathbf{Z}_n^*$, and compute $F_n(m, r)$. Interpret $F_n(m, r)$ as a value in $\{0, 1\}^{s \cdot \ell}$ (possibly after padding) consisting of s consecutive ℓ-bit blocks $F_{n,1}(m, r), \ldots, F_{n,s}(m, r)$ and compute $y = (x \cdot \prod_{u=1}^s z_u^{F_{n,u}(m,r)})^{1/e} \bmod \bar{n}$. The signature is (e, y, r).

Verifying: To verify message/signature pair $(m, (e, y, r))$, check that e is an odd $(\ell + 1)$-bit integer and that $y^e \prod_{u=1}^s z_u^{-F_{n,u}(m,r)} \equiv x \bmod \bar{n}$.

For typical parameter values such as $S = 1024$, $\ell = 171$, $s = 6$, the $2^s = 64$ subset products modulo $\bar{n}$ of the z_u's may be precomputed. Using multi-exponentiation, that would make the above scheme about as efficient as the previous variant. It can be proved (cf. [4]) that the above CS signature variant is secure assuming the strong-RSA and VSSR assumptions. Thus we have obtained an efficient signature scheme proven secure without ad-hoc assumptions. This is unlike the original CS scheme, which relied on a collision resistance or universal one-wayness assumption regarding a 160-bit hash function—as far as we are aware, the only practical provably secure design for such a function is an inefficient discrete log based construction using an elliptic curve defined over a 160-bit order finite field. A disadvantage of our variant is that its public key is typically 9 kbits, which is about 3 times more than in the original CS scheme.

5   Efficiency of VSH in Practice

Let the cost of a multiplication modulo n be $O((\log n)^{1+\epsilon})$ operations, where $\epsilon = 1$ if ordinary multiplication is used, and where $\epsilon > 0$ can be made arbitrarily small if fast multiplication methods are used. Asymptotically the cost of the basic VSH algorithm is $O(\frac{(\log n)^{1+\epsilon}}{k}) = O((\log n)^\epsilon \log \log n)$ operations per message-bit. Given n's factorisation one can do better for long messages by reducing the k exponents of the p_i's modulo $\phi(n)$. Asymptotically, Fast VSH costs $O((\log n)^\epsilon)$ operations per message-bit. It is faster in practice too, cf. below.

The table below lists VSH runtimes obtained using a gmp-based implementation on a 1GHz Pentium III. The two security levels conservatively correspond to 1024-bit and 2048-bit RSA (based on the Computational VSSR Assumption, where an S-bit VSH-modulus leads to a lower RSA security level S' depending on the number of small primes). In the 2nd and 6th rows basic VSH is used with more small primes, in the 3rd and 7th rows extended with precomputed prime products and message processing $b = 8$ bits at a time. Fast VSH also processed $b = 8$ message-bits at a time. With $S' = 1024$ and $S = 1516$ (i.e., at least 1024-bit RSA security, at the cost of a 1516-bit VSH-modulus) Fast VSH is about 25 times slower than Wei Dai's SHA-1 benchmark [23]. Better throughput will be obtained under the more aggressive assumption that VSH with an S-bit modulus achieves S-bit RSA security. A similarly more favorable comparison will be obtained when using VSH with parameters matching the actual SHA-1 security level; at the time of writing that is 63 bits, but as it is a moving target

S'	Method	# small primes	S	b	# products	Megabyte/second
1024	Basic VSH	152	1234	1	n/a	0.355
		1024	1318	1	n/a	0.419
				8	128 * 256	0.486
	Fast VSH	$2^{16} = 65536$	1516	8	n/a	1.135
2048	Basic VSH	272	2398	1	n/a	0.216
		1024	2486	1	n/a	0.270
				8	128 * 256	0.303
	Fast VSH	$2^{18} = 262144$	2874	8	n/a	0.705

we prefer not to specify matching VSH parameters. In any case, the slowdown is a small price for avoiding heuristically collision-resistant hashes. Nevertheless, except for its lack of other nice properties, VSH has been criticised for being too slow. We consider the prospects of faster VSH software more realistic than a proof that SHA-2 offers any security at all.

Acknowledgements. We gratefully acknowledge inspiring discussions with Igor Shparlinski and Eran Tromer, and we thank Yvo Desmedt, Josef Pieprzyk, Benne de Weger, and the anonymous Eurocrypt'06 reviewers for their insightful comments. This article was written while the second author was employed by Lucent Technologies' Bell Laboratories and was affiliated to the Technische Universiteit Eindhoven.

References

1. M. Bellare and D. Micciancio. A new paradigm for collision-free hashing: incrementality at reduced cost. In EUROCRYPT 97, volume 1233 of *LNCS*, page 163–192, Berlin, 1997, Springer-Verlag.
2. D. Boneh and M. Franklin. Efficient generation of shared RSA keys. In CRYPTO 97, volume 1294 of *LNCS*, page 425–439, Berlin, 1997, Springer-Verlag.
3. D. Chaum, E. van Heijst, and B. Pfitzmann. Cryptographically strong undeniable signatures, unconditionally secure for the signer. In CRYPTO 91, volume 576 of *LNCS*, page 470–484, Berlin, 1991, Springer-Verlag.
4. S. Contini, A.K. Lenstra, and R. Steinfeld. VSH, an efficient and provable collision resistant hash function. Report 2005/193, Cryptology ePrint Archive, 2005. `eprint.iacr.org/2005/193/`.
5. R. Crandall and C. Pomerance. *Prime Numbers: a Computational Perspective*, New York, 2001, Springer-Verlag.
6. R. Cramer and V. Shoup. Signature schemes based on the strong RSA assumption. In volume 3 of *ACM Transactions on Information and System Security (ACM TISSEC)*, page 161–185, 2000.
7. I. Damgård. Collision-free hash functions and public key signature schemes. In EUROCRYPT 87, volume 304 of *LNCS*, page 203–216, Berlin, 1987, Springer-Verlag.
8. I. Damgård. A design principle for hash functions. In CRYPTO 89, volume 435 of *LNCS*, page 416–427, Berlin, 1989, Springer-Verlag.
9. S. Goldwasser, S. Micali, and R. Rivest. A digital signature scheme secure against adaptively chosen message attacks. *SIAM J. on Comp.*, 17(2):281–308, 1988.
10. S. Hankerson, A. Menezes, S. Vanstone. *Guide to Elliptic Curve Cryptography*, New York, 2004, Springer-Verlag.

11. M. Jakobsson, K. Sako, and R. Impagliazzo. Designated verifier proofs and their applications. In EUROCRYPT 96, volume 1070 of *LNCS*, page 143–154, Berlin, 1996, Springer-Verlag.
12. A.K. Lenstra and E.R. Verheul. The XTR public key system. In CRYPTO 2000, volume 1880 of *LNCS*, page 1–19, Berlin, 2000, Springer-Verlag.
13. A.K. Lenstra and H.W. Lenstra Jr. *The Development of the Number Field Sieve*, Berlin, 1993, Springer-Verlag.
14. A.K. Lenstra, E. Tromer, A. Shamir, W. Kortsmit, B. Dodson, J. Hughes, and P. Leyland, Factoring estimates for a 1024-bit RSA modulus. In Chi Sung Laih, editor, ASIACRYPT 2003, volume 2894 of *LNCS*, page 55–74, Berlin, 2003, Springer-Verlag.
15. R. Merkle. One way hash functions and DES. In CRYPTO 89, volume 435 of *LNCS*, page 428–446, Berlin, 1989, Springer-Verlag.
16. D. Naccache and J. Stern A new public-key cryptosystem. In Walter Fumy, editor, EUROCRYPT 97, volume 1233 of *LNCS*, page 27–36, Berlin, 1997, Springer-Verlag.
17. D. Pointcheval. The composite discrete logarithm and secure authentication. In PKC 2000, volume 1751 of *LNCS*, page 113–128, Berlin, 2000, Springer-Verlag.
18. R.L. Rivest and R.D. Silverman. Are 'strong' primes needed for RSA. Report 2001/007, Cryptology ePrint Archive, 2001. `eprint.iacr.org/2001/007/`.
19. P. Rogaway and T. Shrimpton. Cryptographic hash-function basics: definitions, implications, and separations for preimage resistance, second-preimage resistance, and collision resistance. In B. Roy and W. Meier, editors, FSE 2004, volume 3017 of *LNCS*, page 371–388, Berlin, 2004, Springer-Verlag.
20. A. Shamir and Y. Tauman. Improved online/offline signature schemes. In CRYPTO 2001, volume 2139 of *LNCS*, page 355–367, Berlin, 2001, Springer-Verlag.
21. R. Steinfeld, H. Wang, and J. Pieprzyk. Efficient extension of standard Schnorr/RSA signatures into universal designated-verifier signatures. In PKC 2004, volume 2947 of *LNCS*, page 86–100, Berlin, 2004, Springer-Verlag.
22. K. Rubin and A. Silverberg. Torus-based cryptography. In CRYPTO 2003, volume 2729 of *LNCS*, page 349–365, Berlin, 2003, Springer-Verlag.
23. Wei Dai. *Crypto++ 5.2.1 Benchmarks*. `www.eskimo.com/~weidai/benchmarks.html`.

Herding Hash Functions and the Nostradamus Attack

John Kelsey[1] and Tadayoshi Kohno[2]

[1] National Institute of Standards and Technology
john.kelsey@nist.gov
[2] CSE Department, UC San Diego
tkohno@cs.ucsd.edu

Abstract. In this paper, we develop a new attack on Damgård-Merkle hash functions, called the *herding attack*, in which an attacker who can find many collisions on the hash function by brute force can first provide the hash of a message, and later "herd" any given starting part of a message to that hash value by the choice of an appropriate suffix. We focus on a property which hash functions should have–Chosen Target Forced Prefix (CTFP) preimage resistance–and show the distinction between Damgård-Merkle construction hashes and random oracles with respect to this property. We describe a number of ways that violation of this property can be used in arguably practical attacks on real-world applications of hash functions. An important lesson from these results is that hash functions susceptible to collision-finding attacks, especially brute-force collision-finding attacks, cannot in general be used to prove knowledge of a secret value.

1 Introduction

Cryptographic hash functions are usually assumed to have three properties: Collision resistance, preimage resistance, and second preimage resistance. And yet many additional properties, related to the above in unclear ways, are also required of hash function in practical applications. For example, hash functions are sometimes used in "commitment" schemes, to prove prior knowledge of some information, priority on an invention, etc. When the information takes on more than a small number of possible values, there does not appear to be an obvious way to extend a collision finding attack to break the commitment scheme; therefore, collision resistance does not seem to be necessary to use the hash function in this way. This appears fortunate in light of the many recent attacks on collision resistance of existing hash functions[2, 3, 13, 19, 21, 22, 23, 24] and the widespread use of hash functions short enough to fall to brute-force collision attacks[20].

We show that the natural intuition above is incorrect. Namely, we uncover (what we believe to be) subtle ways of exploiting the iterative property of Damgård-Merkle[6, 16] hash functions to extend certain classes of collision-finding attacks against the compression function to attack commitment schemes and other uses of hash function that do not initially appear to be related to collision resistance.

S. Vaudenay (Ed.): EUROCRYPT 2006, LNCS 4004, pp. 183–200, 2006.

1.1   Example: Proving Prior Knowledge with a Hash Function

Consider the following example. One day in early 2006, the following ad appears in the *New York Times*:

> I, Nostradamus, hereby provide the MD5 hash H of many important predictions about the future, including most importantly, the closing prices of all stocks in the S&P500 as of the last business day of 2006.

A few weeks after the close of business in 2006, Nostradamus publishes a message. Its first few blocks contain the precise closing prices of the S&P500 stocks. It then continues with many rambling and vague pronouncements and prophecies which haven't come true yet. The whole message hashes to H.

The main question we address in this paper is whether this should be taken as evidence that Nostradamus really knew the closing prices of the S&P500 many months in advance. MD5 has been the subject of collision attacks, and indeed is susceptible to brute force collision attacks, but there are no known preimage attacks. And yet, it seems that a preimage attack on MD5 would be necessary to allow Nostradamus to first commit to a hash, and then produce a message which so precisely describes the future after the fact.

1.2   Chosen Target Forced Prefix (CTFP) Preimage Resistance

The first question to address when considering the situation outlined above is to ask exactly what property of a hash function would have to be violated by Nostradamus in order to falsely "prove" prior knowledge of these closing prices. The property is not directly one of the commonly discussed properties of hash functions (collision resistance[1], preimage resistance, and second preimage resistance). Instead, we need an atypical property, which we will call "chosen target forced prefix" (CTFP) preimage resistance[2].

In order to falsely prove his knowledge of the closing prices of the S&P500, Nostradamus would first have to choose a target hash value, H. He then would have to wait until the closing values of the S&P500 stocks for 2006 were available. Finally, he would have to find some way to form a message that started with a description of those closing values, P, and ended up with the originally committed-to hash H.

Following this example, we can define CTFP preimage resistance as follows: In the first phase of his attack Nostradamus performs some precomputation and then outputs an n-bit hash value H; H is his "chosen target". The challenger then selects some prefix P and supplies it to Nostradamus; P is the "forced

---

[1] Collision resistance would preclude the attack, but does not appear to be necessary for the attack to fail.

[2] We are indebted to Dan Brown for pointing out a previous use of the same idea: In one of three independent proofs of the security of Pinstov-Vanstone signatures, the same property with a different name, "target value resistance," was used. See [4], in which it was conjectured that SHA1 had this property; our result shows that it does not if one can find collisions starting from two arbitrary IVs.

prefix." In our informal security definition we place no restriction on how the challenger picks P, but for simplicity we may assume that the challenger picks P uniformly at random from some large but finite set of strings. In the second phase of his attack, Nostradamus computes and outputs some string S. Nostradamus compromises the CTFP preimage resistance of the hash function if $\mathsf{hash}(P\|S) = H$. If we model the hash function as a random oracle [1], then unless Nostradamus is lucky and guesses P in the first phase of his attack, we would expect him to have to try $O(2^n)$ values for S in the second phase before finding one such that $\mathsf{hash}(P\|S) = H$. Consequently, it might seem reasonable to expect that Nostradamus would have to perform $O(2^n)$ hash function computations to compromise the CTFP preimage resistance of a real hash function. (While one could consider a more formal definition of CTFP for hash function families, and consider the relationship between CTFP-resistance and other security goals, we do not do so here but instead focus on our attacks.)

As described in detail below, the ability to violate the CTFP preimage resistance property allows an attacker to carry out a number of surprising attacks on applications of a hash function. Almost any use of a hash function to prove knowledge of some information can be attacked by someone who can violate this property. Many applications of hashing for signatures or for fingerprinting some information which are not vulnerable to attack by straightforward collision-finding techniques are broken by an attacker who can violate CTFP preimage resistance.

Further, when the CTFP definition is relaxed somewhat (for example, by allowing Nostradamus some prior limited knowledge or control over the format of P, giving him prior knowledge of the full (large) set of possible P strings that might be presented, or allowing him to use any of a large number of encodings of P with the same meaning), the attacks become still cheaper and more practical.

1.3 Herding Attacks

The major result of this paper is as follows: For Damgård-Merkle[6, 16] construction hash functions, CTFP preimage resistance can always be violated by repeated application of brute-force collision-finding attacks. More efficient collision-finding algorithms for the hash function being attacked may be used to make the attack more efficient, if the details of the collision-finding algorithms support this. An attack that violates this property effectively "herds" a given prefix to the desired hash value; we thus call any such attack violating the CTFP preimage resistance property a "herding attack."

The herding attack shows that the CTFP preimage resistance of a hash function like MD5 or SHA1 is ultimately limited by the collision resistance of the hash function. At a high level, and in its basic variant, the attack is parameterized by some positive integer k, e.g., $k = 50$, and by the output size n of the hash function. In the first phase of a herding attack, the attacker, Alice, repeatedly applies a collision-finding attack against a hash function to build a *diamond structure*, which is a data structure reminiscent of a binary tree. With high probability it takes at most $2^{k/2+n/2+2}$ applications of the hash compression function

Table 1. Herding with Short Suffixes

output size	example	diamond width(k)	suffix length (blocks)	work
128	MD5	41	48	2^{87}
160	SHA1	52	59	2^{108}
192	Tiger	63	70	2^{129}
256	SHA256	84	92	2^{172}
512	Whirlpool	169	178	2^{343}
n		$(n-5)/3$	$k + \lg(k) + 1$	2^{n-k}

(and possibly fewer, depending on details of more efficient collision-finding attacks[3]) to create a diamond structure with $2^{k+1} - 2$ intermediate hash states, of which 2^k are used in the basic form of the attack. In the second phase of the attack, Alice exhaustively searches for a string S' such that $P\|S'$ collides with one of the diamond structure's intermediate states; this step requires trying $O(2^{n-k})$ possibilities for S'. Having found such a string S', Alice can construct a sequence of message blocks Q from the diamond structure, and thus build a suffix $S = S'\|Q$ such that $\mathsf{hash}(P\|S) = H$; this step requires a negligible amount of work, and the resulting suffix S will be $k+1$-blocks long. We stress that Alice can have significant control over the contents of S, which means that S may not be "random looking" but may instead contain structured data suitable for the application that Alice is trying to attack. Table 1 present some parameters for a version of our attack.

1.4 Practical Impact

Our techniques for carrying out herding attacks have much in common with the long message second preimage attacks of [12]. However, those attacks required implausibly long messages, and so probably could never be applied in practice. By contrast, our herding attacks require quite short suffixes, and appear to be practical in many situations. Similarly, many recent cryptanalytic results on hash functions, such as [22, 23], require very careful control over the format of the messages to be attacked. This is not generally true of our herding attacks, though more efficient variants that make use of cryptanalytic results on the underlying hash functions will naturally have to follow the same restrictions as those attacks.

Near the end of this paper, we describe a number of ways in which our herding attacks and variations on them can be exploited. In developing the herding

---

[3] The collision finding attacks needed for constructing the diamond structure are somewhat different than those in recent results on MD5, SHA0, and SHA1[22, 23]. We are uncertain whether these attacks can be adapted to the requirements of constructing the diamond structure, though it seems plausible that it might work. For the diamond structure we need collisions between two messages starting with different IVs.

attack, we also describe a new method of building multicollisions for Damgård-Merkle hash functions which we believe to be of independent interest, and which may be useful in many other hash function attacks.

1.5   Related Work

The herding attack is closely related to the long message second preimage attacks in [8] and [12], and is ultimately built upon the multicollision-finding technique of [10]. Our technique for herding is related to the result of Lai and Massey [14] showing a meet-in-the-middle second preimage attack when pseudopreimages can be found cheaper than exhaustive search; in our attack, instead of finding pseudopreimages, we construct a message by repeated collision searches, and then do a meet-in-the-middle type attack to find a large set of possible second preimages on our own chosen message. Our results complement Coron, Dodis, Malinaud, and Puniya's work[5], which does not present attacks like the ones we present, but which shows that iterative hash functions like MD5 and SHA1 are not random oracles, even when their compression functions are. Variants of our attacks works against Coron, et al's fixes but do not violate their provable security bounds.

More broadly, our result re-enforces the lessons that might sensibly be taken from [7, 10, 11, 12, 15] on the many ways in which seemingly impractical hash function collisions may be applied in practice. The security properties of Damgård-Merkle hash functions against attackers who can find collisions are currently not well understood.

2   The Diamond Structure: A Building Block for Herding

In this section we introduce the *diamond structure*. This is a structure of messages constructed to produce a large multicollision of a quite different format than that of Joux[10]. Our multicollision is more expensive, and the same length. For example, a 2^k diamond-structure multicollision costs about $2^{n/2+k/2+2}$ work, relative to Joux' $k \times 2^{n/2}$ work. There are two reasons why the diamond structure lets an attacker do things which are not possible with only a Joux multicollision:

1. The diamond structure allows 2^k choices for the first block of a 2^k multicollision, whereas Joux multicollisions involve a sequence of pairs of choices for each part of the message.
2. The diamond structure contains $2^{k+1} - 2$ intermediate hash values, making the herding attack possible with short suffixes.

A diamond structure is essentially a Merkle tree built by brute force.

Figure 1 describes the basic idea, where edges represent messages and values like $h[i, j]$ represent intermediate hash states. In the diagram, the attacker starts with eight different first message blocks, each leading to a different hash value; he then searches for collisions between pairs of these hash values, yielding four resulting intermediate hash values (at the cost of about $8 \times 2^{n/2}$ work using a

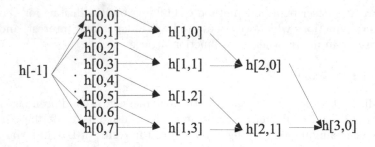

Fig. 1. The Basic Diamond Structure

naive algorithm). He repeats the process with the four remaining values, then the two remaining ones. The result is a diamond structure which is 2^k states wide, and contains $2^{k+1} - 1$ states total.

Producing a Suffix from an Intermediate Hash Value. Consider any of the starting hash values. A suffix which maps that hash value to the final hash H is constructed by walking down the tree from the leaves to the root, appending the message blocks from each edge in the tree to produce a suffix.

Consider any intermediate hash value. Similarly, walking from that node down to the root of the tree yields a suffix which maps the intermediate hash value to the final hash H. Subsequently we discuss how to augment the suffix if the hash function includes the length of the message in its last block.

Building the Structure. Building the structure is more efficient than a naive approach suggests. Instead of fixing the position of each node within the tree and then searching for collisions, the attacker dynamically builds the tree structure during the collision search. To map 2^k hash values down to 2^{k-1}, she generates about $2^{n/2+1/2-k/2}$ candidate message blocks from each starting hash value in a single level of the structure, and then finds collisions between the different starting values dynamically. The total work done to reduce 2^k hash values to 2^{k-1} is about $2^{n/2+k/2+1/2}$, and thus the work done to construct a full diamond structure with 2^k hash values at its widest point is about $2^{n/2+k/2+2}$.

The work done to build the diamond structure is based on how many messages must be tried from each of 2^k starting values, before each has collided with at least one other value. Intuitively, we can make the following argument, which matches experimental data for small parameters: When we try $2^{n/2+k/2+1/2}$ messages spread out from 2^k starting hash values (lines), we get $2^{n/2+k/2+1/2-k}$ messages per line, and thus between any pair of these starting hash values, we expect about $(2^{n/2+k/2+1/2-k})^2 \times 2^{-n} = 2^{n+k+1-2k-n} = 2^{-k+1}$ collisions. We thus expect about $2^{-k+k+1} = 2^1 = 2$ other hash values to collide with any given starting hash value.

If this search is done on a single processor, then each time a pair of lines collide, no further searching is done from those lines. There may be cases where two pairs of lines collide on the same hash value. This very slightly decreases the number of reachable hash values, but the expected number of these is extremely small. For example, in a 2^{55} diamond structure, there are about 2^{56} intermediate hashes which are the results of these collision searches. For a 160-bit hash, we thus expect roughly 2^{-49} such collisions, so we can ignore the effect of them on our result.

Parallelizeability. It is easy to adapt the parallel collision search algorithm of [20] to the construction of a diamond structure. The result of each iteration of the search algorithm yields both a seed for the next message block to try, and also a choice of which of the 2^k starting chaining values will be used.

Employing Cryptanalytic Attacks. The above discussion has focused on brute-force search as a way to build the diamond structure. An alternative is to use some cryptanalytic results on the hash function. Whether this will work depends on details of the cryptanalysis:

1. A collision-finding algorithm which produces a pair of messages from the same initial value is not useful in constructing the diamond structure. Similarly, an algorithm that can find collisions only from initial chaining values with a single difference is not useful.
2. An algorithm which works for any known IV difference can be directly applied to build the diamond structure, though one must fix the positions of the nodes within the diamond structure in advance. If the work to find a collision pair is 2^w, then this algorithm should be used to reduce 2^k lines of hash values to 2^{k-1} lines so long as $w + k - 1 < n/2 + k/2 + 1/2$.
3. An algorithm which works for a subset 2^{-p} of all pairs of IVs can be used to construct the diamond structure if the pairs can be recognized efficiently. This is done by inserting one extra message block at each layer of the diamond structure, and using this to force selected pairs of lines to initial values from which the collision-search algorithm will work. The work necessary to find one collision between lines is now $2^{p/2+1} + 2^w$. This algorithm should be used to reduce 2^k lines to 2^{k+1} so long as $\lg(2^{p/2+1} + 2^w) + k - 1 < n/2 + k/2 + 1/2$.

Expandable Messages. Using the notation from [12], an (a, b)-expandable message is a set of messages of varying lengths, between a and b inclusive, all of which yield the same intermediate hash. Expandable messages may be found from any initial hash value using the techniques found in [12], and more efficiently found for some hash functions, including MD5 an SHA1, using techniques from [8]; in the latter case, the cost is around twice that of a brute-force collision finding attack.

If all $2^{k+1} - 2$ intermediate hash values from the diamond structure are used in the later steps of herding, then a $(1, k+1)$-expandable message must be produced at the end of the diamond structure, to ensure that the final herded message is always a fixed length. This is necessary since we assume that the length of the message will be included in the last block. If only the widest layer of 2^k hash values is used, no expandable message is required.

Precomputation of the Prefix. If the full set of prefixes are known and small enough, the diamond structure can be computed from their resulting intermediate hashes. This follows from the fact that the starting hash values are arbitrary. This is discussed at more depth in Sections 3.1 and 4.

Variant: The Elongated Diamond Structure. Using ideas from [12], long messages offer a naive way to mount the attack; the diamond structure offers much shorter suffixes. However, the attacker can build a diamond structure with many intermediate hashes more cheaply than above, if she is willing to tolerate unreasonably long messages.

The widest layer of the diamond structure is chosen, with 2^k hash values. Then, the attacker computes 2^r message blocks for each of the 2^k hash values, thus producing a total of 2^{k+r} reachable intermediate states. He then constructs the collision tree as described above.

The total work done to build a 2^r-long elongated diamond structure with 2^k values at its widest point is about $2^{r+k}+2^{k/2+n/2+2}$; this structure contains 2^{k+r} intermediate hash values, and yields suffixes of about 2^{r-1} message blocks on average. In general, for reasonable suffix lengths, the elongated diamond structure has only a small advantage over regular diamond structures. An elongated diamond structure must have an $(r, 2^r + r)$-expandable message appended to its end, to ensure that the final herded messages are always the same length, and so always have the same final hash value.

It is possible to parallelize much of the production of an elongated diamond structure. If the width is 2^k hash values at the beginning, then the construction of the structure can be parallelized up to 2^k ways.

3   How to Herd a Hash Function

The herding attack allows an attacker to commit to the hash of a message she doesn't yet fully know, at the cost of a large computation. This attack is closely related to the long message second-preimage attacks of [8, 12] and the multicollision-finding techniques of [10].

At a high level, the attack works as follows:

1. *Build the Diamond Structure:* Alice produces a search structure which contains many intermediate hash values. From any of these intermediate hash values, a message can be produced which will lead to the same final hash H. Alice may commit to H at this point.
2. *Determine the Prefix:* Later, Alice gains knowledge of P.
3. *Find a Linking Message:* Alice now searches for a single-block which, if appended to P, would yield an intermediate hash value which appears in her search structure.
4. *Producing the Message:* Finally, Alice produces a sequence of message blocks from her structure to link this intermediate hash value back to the previously sent H.

At the end of this process, Alice has first committed to a hash H, then decided what message she will provide which hashes to H and which begins with the prefix P.

Building the Diamond Structure. This step is described in Section 2.

Finding a Linking Message. Once a diamond structure is constructed and its hash H is committed to, the attacker learns the prefix P. She must then find a linking message–a message which allows her to link the prefix P into the diamond structure. See Figure 2. When there are 2^k intermediate hash values in the diamond structure, the attacker expects to try about 2^{n-k} trial messages in order to find a linking message.

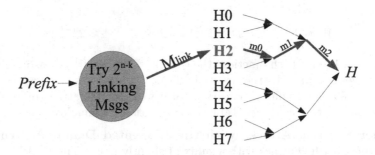

Fig. 2. Finding a Linking Message and Producing the Suffix

The starting chaining values for the diamond structure can be chosen arbitrarily. This makes it easy to parallelize the search for linking messages when herding a prefix into the first (widest) layer of the diamond structure. For example, the starting chaining values may be chosen to have their low 64 bits all zeros[18]; then each processor searching for a linking message need only check the list of starting hash values about once per 2^{64} trials.

Producing the Message. Once a linking message from P, M_{link}, is found, the suffix is produced as described above–basically, the attacker walks up the tree from the linked-to hash value to the root, producing another message block on each step. See Figure 2. If all $2^{k+1} - 2$ intermediate hash values from the diamond structure are used when finding M_{link}, then the pre-determined expandable message must be appended to the end of the suffix.

3.1 Work Done for Herding Attacks

A maximally short suffix for the herding attack is found by producing a 2^k hash value wide diamond structure, and only searching for linking messages to the outermost (widest) level of hash values in the diamond structure, so that no expandable message is needed. In this case, the length of the suffix is $k + 1$ message blocks, and the work done for the herding attack is approximately

$$2^{n-k} + 2^{n/2+k/2+2} . \tag{1}$$

Searching for linking messages to all $2^{k+1}-2$ intermediate hashes in the structure requires adding an additional $\lg(k) + 1$ message blocks for a $(\lg(k), k + lg(k))$-expandable message, and decreases the work required to

$$2^{n-k-1} + 2^{n/2+k/2+2} + k \times 2^{n/2+1} , \tag{2}$$

the $k \times 2^{n/2+1}$ term arising from the search for an expandable message[12].

The cheapest herding attack with a reasonably short suffixes can be determined by setting the work done for constructing the diamond structure and finding the linking message equal. We thus get a diamond structure of width 2^k, suffix length L, and total work W, where:

$$k = \frac{n-5}{3} \tag{3}$$

$$L = \lg(k) + k + 1 \tag{4}$$

$$W = 2^{n-k-1} + 2^{n/2+k/2+2} + k \times 2^{n/2+1} \approx 2^{n-k} . \tag{5}$$

Thus, using a 160-bit hash function, the cheapest attack with a reasonably short suffix involves a diamond structure with about 2^{52} messages at its widest point, producing a 59-block suffix, and with a total work for the attack of about 2^{108} compression function calls. See Table 1 for additional examples.

Work for Herding Attacks with the Elongated Diamond Structure.
The cheapest herding attack with a suffix of slightly more than 2^r blocks can be determined by once again setting the work done for constructing the diamond structure and finding the linking message equal, so long as $k + r < k/2 + n/2$. We thus get an elongated diamond structure of width 2^k, suffix length L, and total work W, where:

$$k = \frac{n-2r-3}{3} \tag{6}$$

$$L = \lg(k + 2^r) + k + 1 + 2^r \tag{7}$$

$$W = 2^{n-k-r} + 2^{n/2+k/2+2} + k \times 2^{n/2+1} + 2^{k+r} \approx 2^{n-k-r+1} . \tag{8}$$

Thus, with a 160-bit hash function and a 2^{55} block suffix (about as long as is allowed for SHA1 or RIPEMD-160), an attacker would end up doing about 2^{90} work total to herd any prefix into the previously published hash value.

Work for Herding from Precomputed Prefixes. If the set of possible prefixes contains 2^k possible messages, the diamond structure can be built from the resulting 2^k intermediate hashes. In this case, there is no search for a linking message, and the total work for the attack is done in building the diamond structure.

3.2   Making Messages Meaningful

These attacks all involve producing a suffix to some forced prefix, which forces the complete message to have a specific hash value H. In order to use herding in a real deception, however, the attacker probably cannot just append a bunch of random blocks to the end of her predictions or other messages. Instead, she needs to produce a suffix which is at least somewhat meaningful or plausible. There are a number of tricks for doing this.

Using Yuval's Trick. Using Yuval's clever trick[25], the attacker can prepare a basic long document appropriate to her intended deception, and produce many independent variation points in the document. This allows the use of meaningful-looking messages for most contexts. For example, each message block in layer i of the diamond structure could be a variation on the same theme, using about $n/2$ possible variation points. In practice, this likely will make the suffix longer, since it is hard to put 80 variation points in a 64-character message. However, this has almost no effect on the herding attack. If the attacker needs ten message blocks (640 characters) for each collision, her suffixes will be ten times longer, but no harder to find. The algorithm for finding them works the same way.

The contents of these suffixes must be pretty general. The natural way to handle this in most applications of herding is to write some common text discussing how the results are supposed to have been obtained ("I consulted my crystal ball, and spent many hours poring over the manuscripts of the ancient prophets...."). These can then be varied at many different points, independently, to yield many possible bitstrings all having the same meaning.

Committing to Meaning, Not Bits. For many of the attacks for which herding is useful, the goal is to falsely commit to some actual meaning, not necessarily some specific message string. For example, an attacker trying to prove her ability to predict the stock market is not really forced to use any fixed format for the contents of her stock market predictions, so long as anyone reading them will unambiguously be able to tell whether she got her predictions right.

This provides a great deal of extra flexibility for the attacker in using Yuval's trick, and also in arranging the different parts of the message to be committed to, in order to maximize her convenience.

4   Exploiting Prior Knowledge of the Prefix Space

As suggested in Sections 2 and 3.1, the attack becomes much more efficient if the prefix can be precomputed. In fact, it is often possible to precompute the message piecemeal in ways that leave a huge number of possible prefixes available, without requiring a huge amount of work.

Just as with the full herding attack, the precomputed version would not be useful against a random oracle–we make use of the iterative structure of existing hash functions to make the attack work.

Precomputing All Possible Prefixes. In the herding attack, the attacker may reasonably expect to produce a diamond structure with 2^{50} or more possible hash values. For a great many possible applications of the herding attack, this may be more than the possible number of prefix messages. The attacker may now take advantage of an interesting feature of the diamond structure: There is no restriction on the choice of starting hash values for the structure.

Let 2^k, the width of the diamond structure, be the number of possible prefix messages that the attacker may need to herd to her fixed hash value. (If there are fewer prefix messages, the attacker appends one block to all the possible prefix messages, and varies that block to produce a set of prefix messages that is exactly the right size.) She computes the intermediate hash after processing each prefix message, and uses these intermediate hashes as the starting hash values for the diamond structure.

The initial work to construct the diamond structure in this way is the same as for the more general herding attack. However, the attacker now has the ability to immediately produce a message which starts with any possible prefix with the desired hash value. That is, she need not do a second expensive computation to herd the prefix she is given.

The attacker who has a larger set of possible prefixes than this is not lost; she may precompute the hashes of the most likely 2^k prefixes. Then, if any of those prefixes is presented to her, she can herd it immediately; otherwise, she must do the large computation, or simply allow her prediction or other deception to fail with some probability.

Using Joux Multicollisions. Joux multicollisions are not sufficient for the general herding attack. However, when the set of possible messages to be committed to is of the right form and can be precomputed, Joux multicollisions can be used to mount a weaker form of the herding attack.

Consider the case where the attacker wishes to commit to a sequence of "yes" or "no" predictions, without knowing which she will need to reveal later. An example of this would be a list of famous people who will or will not marry during the year. In the precomputation phase of the attack, the attacker determines a list of famous people and the order in which she will predict whether they will marry. Following the Joux multicollision technique, she produces a list of about $2^{n/2}$ variations on a "Yes, this person will marry this year" prediction and about $2^{n/2}$ variations on a "No, this person will not marry this year" prediction. Each prediction is independent; the attacker finds a colliding yes/no prediction for the first famous person, then for the second, and so on. See Figure 3. When finished, she publishes her list of famous people and the hash of her predictions for the future. At the end of the year, she "reveals" her predictions, choosing for each pair of colliding blocks the one that reflects what did happen that year.

Fig. 3. Using Joux Multicollisions to Predict Who Will Get Married

This variant of the attack is much cheaper than those based on the diamond structure, but is also much less flexible. It can use existing cryptanalytic techniques on SHA1 and MD5 since, at each stage, the attacker is looking for two messages that collide starting from the same IV; of course, the use of existing cryptanalytic techniques might influence the structure of the attacker's yes/no predictions. Precomputations of enormous sets of prefixes become possible using this technique. Most importantly, it can be combined with the diamond structure and variations of the Joux multicollision to provide even more flexibility to the attacker, as we discuss below.

Combining Precomputations and Joux Multicollisions. In some cases, some large part of the information to be committed to will fit cleanly into the Joux multicollision structure, but other parts will not. For example, consider a prediction of the course and outcome of a national election in the United States[4]. Before the election is run, the attacker produces a set of 32 prefixes which describe the course of the election in broad terms, e.g., "Smith won a decisive victory," "Jones narrowly carried the critical swing states and won," etc. After this, each state's outcome is listed, e.g., "Alabama went for Smith, Alaska went for Jones, ..." The first part of the message is a precomputed diamond structure; the second part is a Joux multicollision allowing 2^{50} different outcomes.

Applying the Joux Multicollision Idea to Diamond Structures. An even more powerful way to structure these predictions is to concatenate precomputed diamond structures in a kind of super-Joux collision.

Consider the above description, but now suppose we wanted to specify one of 32 possible descriptions of how the election went in each state, e.g., "In Alabama, Smith won a resounding victory," or "In Maryland, Jones narrowly won after a series of vicious attack ads."

The attacker can string together 51 diamond structures total, one to describe the whole election, one for each state. This allows the attacker to "commit" to a prediction with 2^{255} possible values (requiring $2^{127.5+n/2+2}$ work with an n-bit hash function using a straightforward precomputed diamond structure), while doing much less work ($51 \times 2^{2.5+n/2+2}$). The attacker also gains enormous flexibility by being able to avoid the strict format of the Joux multicollisions.

5   Applying the Attacks: Herding for Fun and Prophets

In this section, we describe how the herding attack can be used in many different contexts to do (what we believe to be) surprising things.

Predicting the Future: The Nostradamus Attack. The "Nostradamus attack" is the use of herding to commit to the hash of a message that the attacker doesn't even know. This destroys the ability to use hashes, for which collisions can be found, to prove prior knowledge of any information.

The Nostradamus attack is carried out in order to convince people that the attacker can tell the future. This could be based on some claimed psychic power, but also on some claimed improved understanding in science or economics, allowing detailed prediction of the weather, elections, markets, etc. This can also be used to "prove" access to some inside information, as with some attacker attempting to convince a reporter or intelligence agent that she has inside access to a terrorist cell or secretive government agency.

---

[4] The only detail about US politics needed to understand this example is that all elections ultimately produce exactly one victor.

At a very general level, this attack works as follows:

1. The attacker presents the victim with a hash H, along with a claim about the kind of information this represents. She promises to produce the message that yields the hash after the events predicted have occurred.
2. The attacker waits for the events to unfold, just as the victim does.
3. The attacker herds a description of the events as they did unfold into her hash output, and provides the resulting message to the victim, thus "proving" her prior knowledge.

There are many variations on this theme; the predictions can be fully precomputed, completely unpredictable until they come to pass, or some mix of the two.

Committing to an Ordering. The techniques for many of the variants of the Nostradamus attack follow from the discussions in Sections 3 and 4. Here we suggest another possibility, which uses what we call a "hash router;" see Figure 4. Alice decides to prove (perhaps in a gambling context) that she can predict the outcome of a race with 32 entrants. She commits to a sequence of 32 hash outputs, $H_{0,1,...,31}$. After the race is over, she produces 32 strings, $S_{0,1,...,31}$ such that S_i describes the entrant in the race who finished in ith place, and $H_i = \mathsf{hash}(S_i)$.

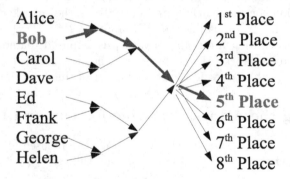

Fig. 4. Committing to an Ordering Using a "Hash Router"

Alice builds a precomputed diamond structure starting from the names of the 32 entrants. When the diamond structure yields a final hash H, she produces 32 new message strings (probably simply strings like "1st place", "2nd place", etc.), and processes them from H to get 32 different hash outputs. She commits to these hash outputs. When the time comes to reveal her choices, she produces 32 strings which commit her to the correct ordering of entrants in the race. Note that Alice can route any of her starting precomputed prefixes to any of the hash outputs.

Retroactive Collisions. Under normal circumstances, someone creating a hash collision must broadly know to what he is committing. While some clever attacks have gotten around this by using some bits of the two colliding messages to change the meaning of later parts of a message[7, 9], these attacks are easy to detect by looking at the underlying data.

The herding attack may be used to "backdate" a collision. That is, the attacker sets up a collision today, and commits to its hash and perhaps one message with that hash. Later, she decides what document she wishes collided with the one she committed to, and so she herds that document to the same hash. The property of the hash function being violated is identical as in the case of "proving" prior knowledge, but the applications are quite different.

Stealing Credit for Inventions. The attacker can use the same idea to claim to be a brilliant inventor, while actually stealing other peoples' work. He submits hashes to a digital timestamping service periodically. After he sees some new invention he wants to claim, he herds a description of the invention to some old hash value.

To save the attacker from building multiple diamond structures, the attacker could construct a single diamond structure, and append a single message block which would vary for each submission.

Tweaking a Signed Document. Consider the case where Alice has a very reasonable document which she has signed, making some sensible predictions about the future or statements of fact or terms of agreement. She wants to make sure she can later "tweak" this document in some ways. Herding will permit this:

1. Using the precomputed variant with Joux multicollisions, she can produce two alternatives for each paragraph or section of the document.
2. Using the precomputed diamond with Joux multicollisions, she can produce many variations for some sections, and pairs of variations for others. She chooses one to produce initially, but can change to another without changing the hash.
3. Using the full herding attack, she can produce one "herded" document. Any variation in the "prefix" part of the document she wishes to make later can be made by carrying out another herding attack.

This attack can be used to tweak messages, contracts, news stories, signed/hashed software, etc.

Random Number Fixing. Alice and Bob want to agree on a shared random sequence for some game. Alice sends $\mathsf{hash}(X_1)$, then Bob responds with X_2. Finally, Alice reveals X_1, and Alice and Bob each derive random bits by combining X_1 and X_2 in some way. The herding attacks and its variations can be used to allow Alice to exert substantial control over the resulting random bit sequence. If the full herding attack isn't practical in this scenario, Alice can at least use the Joux multicollision variant to allow herself two choices per agreed random number, where Bob has no choice.

6   Finding Multiblock Fixed Points

Attacks on commitment schemes are not the only applications of the diamond structure and herding attack ideas. We can also find short cycles in hash functions. This is done in a simple way: We first construct a diamond structure, where each of the starting hash values in the structure are found by generating a random message block, and computing the compression function result of that message block from the hash function's initial value. If the diamond structure is 2^k wide, we then compute 2^{n-k} trial message blocks from the end of the diamond structure. We expect an intermediate collision, which yields a k-block fixed point for the hash algorithm.

This can be extended; with 2^{n-k+r} work, we expect about 2^r different k-block fixed points, all reachable from a legitimate message. These can be concatenated together; we can choose which of the 2^r k-block chunks of message we wish to append to the message next, without reference to previous choices. Further, any message can be "herded" to this set of fixed points with about 2^{n-k} work and k appended blocks. For completeness, we recall that [17] show how to find single-block fixed points in Davies-Meyer constructions and [12] show how to find single-block fixed points in Snefru.

7   Conclusions

In this paper, we have defined a property of a hash function, Chosen Target Forced Prefix (CTFP) preimage resistance, which is both surprisingly important for real-world applications of hash functions, and also surprisingly dependent on collision resistance of the hash function. We have described a variation on the Joux multicollision technique for building tree-like structures of multicollisions called "diamond structures," and enumerated a number of techniques made possible by these structures. We have described a number of arguably practical attacks which use these techniques.

At a very basic level, we believe that the most important lesson the reader can take from this paper is that using iterated hash functions whose collision resistance has been violated is very difficult, even when the relevant security property does not appear to depend on collision resistance.

A great deal of research remains to be done in this area. The diamond structure seems likely to us to be about as useful in developing new attacks as the Joux multicollision result, and we hope to see others building on the work in this paper by finding other surprising things to do to iterated hash functions using herding attacks and the diamond structure. Additionally, there may be many other surprising ways in which iterated hash functions built on the Damgård-Merkle construction may be attacked when the attacker can find intermediate collisions.

Acknowledgments

The authors wish to thank Dan Brown, Morris Dworkin, Niels Ferguson, Hal Finney, Stuart Haber, Ulrich Kuehn, Bart Preneel, Christian Rechberger, Bruce Schneier, the many participants of the NIST hash workshop, and the anonymous referees for helpful comments and discussions on the subject of this paper.

T. Kohno was supported by NSF CCR-0208842, NSF ANR-0129617, and NSF CCR-0093337. Part of this research was performed while T. Kohno was visiting the University of California at Berkeley.

References

1. M. Bellare and P. Rogaway. Random oracles are practical: A paradigm for designing efficient protocols. In *ACM CCS 93*, pages 62–73. ACM Press, 1993.
2. E. Biham and R. Chen. Near-collisions of SHA-0. In M. Franklin, editor, *CRYPTO 2004*, volume 3152 of *LNCS*, pages 290–305. Springer-Verlag, Berlin, Germany, 2004.
3. E. Biham, R. Chen, A. Joux, P. Carribault, C. Lemuet, and W. Jalby. Collisions of SHA-0 and Reduced SHA-1. In R. Cramer, editor, *EUROCRYPT 2005*, volume 3494 of *LNCS*. Springer-Verlag, Berlin, Germany, 2005.
4. D. R. Brown and D. B. Johnson. Hash functions based on block ciphers. In D. Naccache, editor, *CT-RSA 2001*, volume 2020 of *LNCS*. Springer-Verlag, Berlin, Germany, 2001.
5. J.-S. Coron, Y. Dodis, C. Malinaud, and P. Puniya. Merkle-Damgård revisited: How to construct a hash function. In V. Shoup, editor, *CRYPTO 2005*, volume 3621 of *LNCS*. Springer-Verlag, Berlin, Germany, 2005.
6. I. Damgård. A design principle for hash functions. In G. Brassard, editor, *CRYPTO'89*, volume 435 of *LNCS*, pages 416–427. Springer-Verlag, Berlin, Germany, 1989.
7. M. Daum and S. Lucks. Attacking hash functions by poisoned messages: The story of Alice and her boss, 2005. http://www.cits.rub.de/MD5Collisions.
8. R. D. Dean. *Formal Aspects of Mobile Code Security*. PhD thesis, Princeton University, Jan. 1999.
9. M. Gebhardt, G. Illies, and W. Schindler. A note on practical value of single hash collisions for special file formats. NIST Cryptographic Hash Workshop, 2005. No published proceedings, available online at http://www.csrc.nist.gov/pki/ HashWorkshop/2005/Oct31_Presentations/Illies_NIST_05.pdf.
10. A. Joux. Multicollisions in iterated hash functions. Application to cascaded constructions. In M. Franklin, editor, *CRYPTO 2004*, volume 3152 of *LNCS*, pages 306–316. Springer-Verlag, Berlin, Germany, 2004.
11. D. Kaminsky. MD5 to be considered harmful someday. Cryptology ePrint Archive, Report 2004/357, 2004. http://eprint.iacr.org/.
12. J. Kelsey and B. Schneier. Second preimages on n-bit hash functions for much less than 2^n work. In R. Cramer, editor, *EUROCRYPT 2005*, volume 3494 of *LNCS*, pages 474–490. Springer-Verlag, Berlin, Germany, 2005.
13. V. Klima. Finding MD5 collisions on a notebook PC using multi-message modifications. Cryptology ePrint Archive, Report 2005/102, 2005. http:// eprint.iacr.org/.
14. X. Lai and J. L. Massey. Hash functions based on block ciphers. In R. A. Rueppel, editor, *EUROCRYPT'92*, volume 658 of *LNCS*. Springer-Verlag, Berlin, Germany, 1992.
15. A. Lenstra, X. Wang, and B. de Weger. Colliding X.509 certificates. Cryptology ePrint Archive, Report 2005/067, 2005. http://eprint.iacr.org/.
16. R. C. Merkle. One way hash functions and DES. In G. Brassard, editor, *CRYPTO' 89*, volume 435 of *LNCS*, pages 428–446. Springer-Verlag, Berlin, Germany, 1989.

17. S. Miyaguchi, K. Ohta, and M. Iwata. Confirmation that some hash functions are not collision free. In I. Damgård, editor, *EUROCRYPT'90*, volume 473 of *LNCS*. Springer-Verlag, Berlin, Germany, May 1990.
18. B. Preneel, 2005. Personal communication.
19. V. Rijmen and E. Oswald. Update on SHA-1. In A. Menezes, editor, *CT-RSA 2005*, volume 3376 of *LNCS*, pages 58–71. Springer-Verlag, Berlin, Germany, 2005.
20. P. van Oorschot and M. Wiener. Parallel collision search with cryptanalytic applications. *Journal of Cryptology*, 12(1):1–28, 1999.
21. X. Wang, X. Lai, D. Feng, H. Chen, and X. Yu. Cryptanalysis of the hash functions MD4 and RIPEMD. In R. Cramer, editor, *EUROCRYPT 2005*, volume 3494 of *LNCS*. Springer-Verlag, Berlin, Germany, 2005.
22. X. Wang, Y. L. Yin, and H. Yu. Finding collisions in the full SHA-1. In V. Shoup, editor, *CRYPTO 2005*, volume 3621 of *LNCS*. Springer-Verlag, Berlin, Germany, 2005.
23. X. Wang and H. Yu. How to break MD5 and other hash functions. In R. Cramer, editor, *EUROCRYPT 2005*, volume 3494 of *LNCS*, pages 19–35. Springer-Verlag, Berlin, Germany, 2005.
24. X. Wang, H. Yu, and Y. L. Yin. Efficient collision search attacks on SHA-0. In V. Shoup, editor, *CRYPTO 2005*, volume 3621 of *LNCS*. Springer-Verlag, Berlin, Germany, 2005.
25. G. Yuval. How to swindle Rabin. *Cryptologia*, 3(3):187–189, 1979.

Optimal Reductions Between Oblivious Transfers Using Interactive Hashing

Claude Crépeau* and George Savvides*

McGill University, Montéral, QC, Canada
{crepeau, gsavvi1}@cs.mcgill.ca

Abstract. We present an asymptotically optimal reduction of one-out-of-two String Oblivious Transfer to one-out-of-two Bit Oblivious Transfer using Interactive Hashing in conjunction with Privacy Amplification. Interactive Hashing is used in an innovative way to test the receiver's adherence to the protocol. We show that $(1 + \epsilon)k$ uses of Bit OT suffice to implement String OT for k-bit strings. Our protocol represents a two-fold improvement over the best constructions in the literature and is asymptotically optimal. We then show that our construction can also accommodate weaker versions of Bit OT, thereby obtaining a significantly lower expansion factor compared to previous constructions. Besides increasing efficiency, our constructions allow the use of any 2-universal family of Hash Functions for performing Privacy Amplification. Of independent interest, our reduction illustrates the power of Interactive Hashing as an ingredient in the design of cryptographic protocols.

Keywords: interactive hashing, oblivious transfer, privacy amplification.

1 Introduction

The notion of Oblivious Transfer was originally introduced by Rabin [12]. However, a variant of OT was first invented by Wiesner [14] but his work was only published post-facto. Its application to multi-party computation was shown by Even, Goldreich and Lempel in [8]. One-out-of-two String Oblivious Transfer, denoted $\binom{2}{1}$–String OT^k, is a primitive that allows a sender Alice to send one of two k-bit strings, a_0, a_1 to a receiver Bob who receives a_c for a choice bit $c \in \{0, 1\}$. It is assumed that the joint probability distribution $P_{a_0 a_1 c}$ from which the inputs are generated is known to both parties. The primitive offers the following security guarantees to an honest party facing a dishonest party:

- (Dishonest) Alice does not learn any extra information about Bob's choice c beyond what can be inferred from her inputs a_0, a_1 under distribution $P_{a_0 a_1 c}$.

- (Dishonest) Bob can learn information about only one of a_0, a_1. This excludes any joint information about the two strings except what can be inferred from Bob's input, (legitimate) output, and $P_{a_0 a_1 c}$.

---

* Supported in part by NSERC, MITACS, and CIAR.

S. Vaudenay (Ed.): EUROCRYPT 2006, LNCS 4004, pp. 201–221, 2006.

One-out-of-two Bit Oblivious Transfer, denoted $\binom{2}{1}$–Bit OT or simply Bit OT, is a simpler primitive which can be viewed as a special case of $\binom{2}{1}$–String OT^k with $k = 1$. Its apparent simplicity belies its surprising power as a cryptographic primitive: it is by itself sufficient to securely implement any two-party computation [9]. It is therefore not surprising that $\binom{2}{1}$–String OT^k can, in principle at least, be reduced to Bit OT. However, as such generic reductions are typically inefficient and impractical, many attempts at finding direct and efficient reductions have been made in the past. Besides increasing efficiency an orthogonal goal of some of these reductions has been to reduce $\binom{2}{1}$–String OT^k to weaker variants of Bit OT such as XOR OT, Generalized OT and Universal OT.

Contributions of This Paper. The original motivation behind our work was to highlight the potential of Interactive Hashing [11, 10] as an ingredient in the design of cryptographic protocols. This paper shows how in the context of reductions between Oblivious Transfers, Interactive Hashing (both its round-unbounded and constant-round version[6]) can be used for the selection of a small subset of positions to be subsequently used for tests. This selection is sufficiently random to thwart any dishonest receiver's attempts at cheating as well as sufficiently under the honest receiver's control to protect his privacy.

We show how such tests can be embedded in the reduction of String OT to Bit OT and weaker variants given by Brassard, Crépeau and Wolf [3]. The tests ensure that the receiver cannot deviate from the protocol more than an arbitrarily small fraction of the time, leading to two important improvements over the original reduction:

1. The *expansion factor* n/k (namely, the ratio of Bit OT uses to string length) is significantly reduced. Specifically:
   - In the case of Bit OT and XOR OT it decreases from $2 + \epsilon$ to $1 + \epsilon'$. This is in fact asymptotically optimal as the receiver has n bits of entropy after the n executions of Bit OT. For a formal proof that any reduction of $\binom{2}{1}$–String OT^k requires at least k executions of Bit OT, see [7].
   - In the case of Generalized OT it decreases from 4.8188 to $1 + \epsilon''$, which is again optimal.
   - In the case of Universal OT it is reduced by a factor of at least $8 \ln 2 = 5.545$ (its exact value is a function of the channel's characteristics).

2. The construction is more general as it allows any 2-Universal Family of Hash Functions to be used for Privacy Amplification.

2   Oblivious Transfer Variants and Their Specifications

2.1 $\binom{2}{1}$–$\mathbf{ROT}^k$ and Its Equivalence to $\binom{2}{1}$–String $\mathbf{OT}^k$

$\binom{2}{1}$–ROT^k is a randomized variant of $\binom{2}{1}$–String OT^k where Alice sends to Bob two independently chosen random strings $r_0, r_1 \in_R \{0,1\}^k$, of which Bob learns r_c for $c \in_R \{0,1\}$.

Security Requirements. Let R_0, R_1 be two independent random variables uniformly distributed in $\{0,1\}^k$ corresponding to the strings sent by Alice. Let C be a binary random variable uniformly distributed in $\{0,1\}$ corresponding to Bob's choice. The security requirements for $\binom{2}{1}$–ROT^k are captured by the following information-theoretic conditions:

1. (Dishonest) Alice does not gain any information about C during the protocol. In other words $\mathbf{H}(C) = 1$.

2. (Dishonest) Bob obtains information about only one of the two random strings during the protocol. Formally, at the end of each run of the protocol, there exists some $d \in \{0,1\}$ such that $\mathbf{H}(R_d \mid R_{\bar{d}}) = k$.

Equivalence to $\binom{2}{1}$–String OT^k. It is easy to see that $\binom{2}{1}$–ROT^k reduces to $\binom{2}{1}$–String OT^k. Conversely, as Protocol 1 shows, it is also possible to reduce $\binom{2}{1}$–String OT^k to $\binom{2}{1}$–ROT^k in a straightforward way. As $\binom{2}{1}$–ROT^k and $\binom{2}{1}$–String OT^k are equivalent, in this paper we will focus on reductions of $\binom{2}{1}$–ROT^k to Bit OT. This choice is motivated by the fact that the randomized nature of $\binom{2}{1}$–ROT^k and the independence of the two parties' inputs yield simpler constructions with easier to prove security.

Protocol 1. Reducing $\binom{2}{1}$–String OT^k to $\binom{2}{1}$–ROT^k

---

Let the inputs to $\binom{2}{1}$–String OT^k be $a_0, a_1 \in \{0,1\}^k$ for Alice and $c \in \{0,1\}$ for Bob.

1. Alice uses $\binom{2}{1}$–ROT^k to send $r_0, r_1 \in_R \{0,1\}^k$ to Bob, who receives $r_{c'}$ for some randomly chosen $c' \in \{0,1\}$.

2. Bob sends $d = c \oplus c'$ to Alice.

3. Alice sets $e_0 = a_0 \oplus r_d$ and $e_1 = a_1 \oplus r_{\bar{d}}$ and sends e_0, e_1 to Bob.

4. Bob decodes $a_c = e_c \oplus r_{c'}$.

Note that Step 1 of Protocol 1 can be performed before the two parties' inputs to $\binom{2}{1}$–String OT^k have been determined and its results stored for later. In Step 2 Bob sends to Alice a "flip bit" d which effectively allows him to invert the order in which Alice's strings are encrypted and thus to eventually learn the string a_c of his choice regardless of his initial random choice of c' in Step 1.

2.2 Weaker Variants of Bit OT

By relaxing the security guarantees against a dishonest receiver (Bob) we obtain weaker variants of Bit OT, as described below. In all cases b_0, b_1 denote Alice's input bits. Whatever extra choices may be available to Bob, he can always act honestly and request b_c for a choice $c \in \{0,1\}$. As in 'regular' Bit OT, dishonest Alice never obtains information about Bob's choice.

XOR OT (XOT). Bob can choose to learn one of $b_0, b_1, b_\oplus$ where $b_\oplus \overset{\text{def}}{=} b_0 \oplus b_1$.

Generalized OT (GOT). Bob can choose to learn $f(b_0, b_1)$ where f is any of the 16 possible one-bit functions of b_0, b_1.

Universal OT (UOT). Bob can choose to learn $\Omega(b_0, b_1)$ where Ω is any arbitrary discrete memoryless channel whose input is a pair of bits and whose output satisfies the following constraint: let $B_0, B_1 \in \{0, 1\}$ be uniformly distributed random variables and let $\alpha \le 1$ be a constant. Then,

$$\mathbf{H}(B_0, B_1 \mid \Omega(B_0, B_1)) \ge \alpha.$$

Note that we disallow $\alpha > 1$ as the channel would not allow Bob to act honestly.

3   Tools and Mathematical Background

3.1   Encoding of Subsets as Bit Strings

Let x be a small constant. In our protocols we will need to encode subsets of xn elements out of a total of n as bit strings. Let $K = \binom{n}{xn}$ be the number of such subsets. There exists a simple and efficiently computable bijection between the K subsets and the integers $0, \ldots, K - 1$, providing an encoding scheme with output length $m = \lceil \log(K) \rceil \le n\mathbf{H}(x)$. See [5] (Section 3.1) for details on its implementation. Note that in this encoding scheme, the bit strings in $\{0, 1\}^m$ that correspond to valid encodings, namely the binary representations of numbers $0, \ldots, K - 1$, could potentially make up only slightly more than half of all strings. In order to avoid having to deal with invalid encodings, we will consider any string $w \in \{0, 1\}^m$ to encode the same subset as $w \pmod{K}$. Thus in our modified encoding scheme each string in $\{0, 1\}^m$ is a valid encoding of some subset, while to each of the K subsets correspond either 1 or 2 bit strings in $\{0, 1\}^m$. This imbalance[1] in the number of encodings per subset turns out to be of little importance in our scenario thanks to Lemma 1 below.

Lemma 1. *Assume the modified encoding of Section 3.1 mapping subsets to bit strings in $\{0, 1\}^m$. If the fraction of subsets possessing a certain property is f, then the fraction f' of bit strings in $\{0, 1\}^m$ that map to subsets possessing that property satisfies $f' \le 2f$.*

Proof. Let P be the set containing all subsets possessing the property, and let Q be its complement. Then $f = \frac{|P|}{|P|+|Q|}$. The maximum fraction of strings in $\{0, 1\}^m$ mapping to subsets in P occurs when all subsets in P have two encodings each, while all subsets in Q have only one. Consequently, $f' \le \frac{2|P|}{2|P|+|Q|} \le \frac{2|P|}{|P|+|Q|} = 2f$ $\square$

---

[1] Note that this imbalance could be further reduced, if necessary, at the cost of a slight increase in the encoding length. Let $M \ge m$ and let every $w \in \{0, 1\}^M$ map to the same subset as $w \pmod{K}$. Then each of the K subsets will have at least $\lfloor \frac{2^M}{K} \rfloor$ and at most $\lceil \frac{2^M}{K} \rceil$ different encodings.

3.2   Interactive Hashing

Interactive Hashing is a primitive (first appearing in [11, 10] in the context of perfectly hiding commitments) that allows a sender to send an m–bit string s to a receiver, who receives both s and another, effectively random string in $\{0, 1\}^m$. The security properties of this primitive that are relevant to our setting are:

1. *The receiver cannot tell which of the two output strings was the original input.* Let the two output strings be s_0, s_1 (labeled according to lexicographic order). Then if both strings were apriori equally likely to have been the sender's input s, then they are aposteriori equally likely as well.
2. *When both participants are honest, the input is equally likely to be paired with any of the other strings.* Let s be the sender's input and let s' be the second output of Interactive Hashing. Then provided that both participants follow the protocol, s' will be uniformly distributed among all $2^m - 1$ strings different from s.
3. *The sender cannot force both outputs to have a rare property.* Let G be a subset of $\{0, 1\}^m$ such that $\frac{|G|}{2^m}$ is exponentially small in m. Then the probability that a dishonest sender will succeed in having both outputs s_0, s_1 be in G is also exponentially small in m.

Implementation of Interactive Hashing. In our reductions we will use Protocol 2 to implement Interactive Hashing. All operations below take place in $\mathcal{F}_2$.

Protocol 2. Interactive Hashing

Let s be a m-bit string that the sender wishes to send to the receiver.

1. The receiver chooses a $(m - 1) \times m$ matrix Q of rank $m - 1$. Let q_i be the i-th query, consisting of the i-th row of Q.
2. The receiver sends query q_1 to the sender. The sender responds with $c_1 = q_1 \cdot s$ where $\cdot$ denotes the dot product.
3. For $2 \leq i \leq m - 1$ do:
   (a) Upon receiving c_{i-1} the receiver sends query q_i to the sender.
   (b) The sender responds with $c_i = q_i \cdot s$
4. Both parties compute the two solutions to the resulting system of $m - 1$ equations and m unknowns and label them s_0, s_1 according to lexicographic order.

Security of Protocol 2. The properties of the linear system resulting from the interaction between the two parties easily establish that the first security requirement is met: that the receiver cannot guess which of the two output strings was the sender's original input to the protocol. Let V be the receiver's (marginal) view at the end of the protocol and let s_0, s_1 be the corresponding output strings. Note that V would be identical whether the sender's input was s_0 or s_1 as the responses obtained after each challenge would be the same in both cases. Consequently, if before the protocol begins the sender is equally likely to have chosen s_0 and s_1 as input — both with some small probability α — then at the end of the protocol each of these two strings has equal probability $1/2$ of having been the original input string given V. We remark that a dishonest receiver would gain nothing by selecting a matrix Q in a non-random fashion or with rank less than $t - 1$.

As for the second property, let s be the sender's input and let s' be the second output of Interactive Hashing. We first note that since the linear system has two distinct solutions, it is always the case that $s' \neq s$. To see that s' is uniformly distributed among all strings in $\{0,1\}^m \setminus s$, it suffices to observe that Q is randomly chosen among all rank $m - 1$ matrices and that the number of such Q's satisfying $Q(s) = Q(s') \Leftrightarrow Q(s - s') = 0$ is the same for any $s' \neq s$.

Concerning the third security requirement, it can be shown (see [5], Lemma 6) that if G is an exponentially small (in m) subset of $\{0,1\}^m$, then whatever dishonest strategy the receiver might use with the aim of forcing both outputs s_0 and s_1 to be strings from G, he will only succeed in doing so with exponentially small probability. We remark that more recent, unpublished results by the second author of this paper establish a tight upper bound of $15.682 \cdot |G|/2^m$ for this probability and that this upper bound remains valid for all ratios $|G|/2^m$.

More Efficient Implementations of Interactive Hashing. A constant-round Interactive Hashing protocol appears in [6]. The construction capitalizes on results from pseudorandomness, in particular efficient implementations of *almost t-wise independent permutations*, to significantly reduce the amount of interaction necessary. Specifically, it is shown that 4 rounds are sufficient for inputs of any size, in contrast to Protocol 2 that requires $m - 1$ rounds for inputs of size m. The main disadvantages of this constant-round implementation are its much greater complexity as well as the fact that some parameters in the construction require prior knowledge of an upper bound on G. As our only efficiency concern in this paper is the number of Bit OT executions, we will not deal with this alternative construction any further even though the authors believe that it would be a suitable replacement to Protocol 2, at least in the context of our reductions.

3.3   Tail Bounds

Markov's Inequality. Let X be a random variable assuming only positive values and let $\mu = \mathbf{E}[X]$. Then $\mathbf{Pr}[X \geq t] \leq \frac{\mu}{t}$.

Chernoff Bounds. Let $B(n,p)$ be the binomial distribution with parameters n, p and mean $\mu = np$. We will use the following versions of the Chernoff bound for $0 < \delta \leq 1$:

$$\mathbf{Pr}[B(n,p) \leq (1 - \delta)\mu] \leq e^{-\delta^2 \mu/2} \tag{1}$$

$$\mathbf{Pr}[B(n,p) \geq (1 + \delta)\mu] \leq e^{-\delta^2 \mu/3} \tag{2}$$

From (1) we can also deduce the following inequality

$$\mathbf{Pr}[B(n,p) \leq \mu - \Delta n] \leq e^{-\Delta^2 n/2} \tag{3}$$

3.4   Error Probability and Its Concentration on an Erasure Event

Fano's Lemma (Adapted from [3]). Let X be a random variable with range $\mathcal{X}$ and let Y be another, related random variable. Let p_e be the (average) error probability of correctly guessing the value of X with any strategy given the outcome of Y and let $h(p) \overset{\text{def}}{=} -p \log p - (1-p) \log(1-p)$. Then p_e satisfies:

$$h(p_e) + p_e \cdot \log_2(|\mathcal{X}| - 1) \geq \mathbf{H}\left(X \mid Y\right) \tag{4}$$

Specifying an Erasure Event Δ. Let X be a *binary* random variable and let p_e be the error probability of guessing X correctly using an optimal strategy (in other words, p_e is the minimum average error probability). Let $p \leq p_e$. For a specific guessing strategy with average guessing error at most $1/2$, let E be an indicator random variable corresponding to the event of guessing the value of X incorrectly. Note that $\mathbf{Pr}\left[\bar{E}\right] \geq \mathbf{Pr}\left[E\right] \geq p_e \geq p$. Define Δ to be another indicator random variable such that

$$\mathbf{Pr}\left[\Delta \mid E\right] = \frac{p}{\mathbf{Pr}\left[E\right]} \qquad\qquad \mathbf{Pr}\left[\Delta \mid \bar{E}\right] = \frac{p}{\mathbf{Pr}\left[\bar{E}\right]} \tag{5}$$

It follows that $\mathbf{Pr}\left[\Delta\right] = 2p$ and that $\mathbf{Pr}\left[E \mid \Delta\right] = \mathbf{Pr}\left[\bar{E} \mid \Delta\right] = \frac{1}{2}$. Suppose that the value of Δ is provided as side information by an oracle. Then with probability $2p$ we have $\Delta = 1$ in which case X is totally unknown We will refer to this event as an *erasure* of X. This leads to the following lemma:

Lemma 2. *Let X be a binary random variable and let p_e be the error probability when guessing X. Then X can be* erased *with probability $2p \leq 2p_e$.*

3.5   Privacy Amplification

Privacy Amplification [2] is a technique that allows a partially known string R to be shrunk into a shorter but almost uniformly distributed string r that can be used effectively as a one-time pad in cryptographic applications. For our needs we will use a simplified version of the Generalized Privacy Amplification Theorem [1] (also covered in [2]) which assumes that there are always u or more unknown physical bits about R (as opposed to general bounds on R's entropy).

Theorem 1. *Let R be a random variable uniformly distributed in $\{0,1\}^n$. Let V be a random variable corresponding to Bob's knowledge of R and suppose that any value $V = v$ provides no information about u or more physical bits of R. Let s be a security parameter and let $k = u - s$. Let $\mathcal{H}$ be a 2-Universal Family of Hash functions mapping $\{0,1\}^n$ to $\{0,1\}^k$ and let H be uniformly distributed in $\mathcal{H}$. Let $r = H(R)$ Then the following holds:*

$$\mathbf{H}\left(r \mid VH\right) \geq k - \log\left(1 + 2^{k-u}\right) \geq k - \frac{2^{k-u}}{\ln 2} = k - \frac{2^{-s}}{\ln 2} \tag{6}$$

It follows from Equation (6) that $I(r; VH) \leq 2^{-s}/\ln 2$. From Markov's inequality it follows that the probability that Bob has more than $2^{-s/2}$ bits of information about r is no larger than $2^{-s/2}/\ln 2$.

4   Previous Work

All reductions of $\binom{2}{1}$–ROT^k to Bit OT fall within two major categories: reductions based on Self-Intersecting Codes (Section 4.1) and reductions based on Privacy Amplification (Section 4.2).

4.1   Reductions Based on Self-intersecting Codes

These reductions use a special class of error-correcting codes called "self-intersecting codes" encoding k-bit input strings into n-bit codewords. They have the extra property that any two non-zero codewords c_0, c_1 must have a position i such that $c_{0i} \neq 0 \neq c_{1i}$. Consult [4] for more details.

Advantages and Disadvantages. The main advantage of this approach is that the self-intersecting code can be chosen ahead of time and embedded once and for all in the protocol. One of its main disadvantages is the rather large expansion factor n/k, theoretically lower-bounded by 3.5277 [13] and in practice roughly 4.8188. Another important limitation is that this approach does not lend itself to generalizations to weaker forms of Bit OT, such as XOT, GOT and UOT.

4.2   Reductions Based on Privacy Amplification

In Protocol 3 we introduce the construction of [3] upon which our own construction (Protocol 4) builds and expands.

Protocol 3. Reducing $\binom{2}{1}$–ROT^k to Bit OT

---

1. Alice selects $R_0, R_1 \in_R \{0,1\}^n$. Bob selects $c \in_R \{0,1\}$.

2. Alice sends R_0, R_1 to Bob using n executions of Bit OT, where the i-th round contains bits R_0^i, R_1^i. Bob receives R_c.

3. Let $k = n/2 - s$ where s is a security parameter. Alice randomly chooses two $k \times n$ binary matrices M_0, M_1 of rank k and sets $r_0 = M_0 \cdot R_0$ and $r_1 = M_1 \cdot R_1$.

4. Alice sends M_0, M_1 to Bob, who sets $r_c = M_c \cdot R_c$.

It is easy to see that Protocol 3 always succeeds in achieving $\binom{2}{1}$–ROT^k when both parties are honest. The properties of Bit OT guarantee that (dishonest) Alice cannot obtain any information on Bob's choice bit c at Step 2. On the other hand, at the end of Step 2 (dishonest) Bob is guaranteed to be missing at least $n/2$ bits of R_d for some $d \in \{0,1\}$. This is exploited at Step 3 by performing Privacy Amplification with output length $k = n/2 - s$. Specifically, the 2-universal family of Hash Functions used in Protocol 3 guarantees that r_d is uniformly distributed in $\{0,1\}^k$ and independent of $r_{\bar{d}}$ except with probability exponentially small in s. It is shown in [3] that using this family of hash functions this property can be maintained even if Bit OT is replaced with weaker variants such as XOR OT, Generalized OT and Universal OT — albeit at the cost of further reducing the size of k.

Advantages and Disadvantages. Besides its apparent simplicity and straight-forward implementation, the reduction of Protocol 3 has two main advantages over reductions based on Self-Intersecting Codes: Using n executions of Bit OT one can achieve $\binom{2}{1}$–ROT^k for k slightly less than $n/2$, leading to an expansion factor of $2 + \epsilon$. Consequently, it achieves a lower expansion factor than any reduction based in Self-Intersecting Codes. Using the 2-universal family of Hash Functions defined at Step 3, the reduction works without any modification when Bit OT is replaced with XOT and requires only a decrease in the size of k to work with GOT and UOT.

The construction suffers from two disadvantages: The proof of security relies heavily on the properties of matrices in $\mathcal{F}_2$ used for Privacy Amplification in Step 3. A general result for any universal class of hash functions was left as an open problem. In every run of the protocol a new set of matrices M_0, M_1 must be selected and transmitted, thereby increasing the amount of randomness needed as well as the communication complexity by $\Theta(n^2)$ bits.

5   The New Reduction of $\binom{2}{1}$–ROT^k to Bit OT

Notation and Conventions. In our reduction, two randomly chosen strings $T_0, T_1 \in_R \{0,1\}^n$ are transmitted pairwise using n executions of Bit OT. We denote by t_0^i, t_1^i the bits at position i of T_0, T_1, respectively. Let I be the set of all n positions. For a subset $s \subseteq I$ let $T(s)$ be the substring of T consisting of the bits at all positions $i \in s$ in increasing order of position. Note that $T(I) = T$. Subsets of I of cardinality xn will be mapped to bit strings of length $m = \lceil \log \left(\binom{n}{xn} \right) \rceil$ using the encoding/decoding scheme of Section 3.1.

Intuition Behind Protocol 4. At Step 1, the two parties agree on the value of x which will determine the proportion of bits sacrificed for tests.

At Step 2 Alice selects the two random n-bit strings to be transmitted to Bob using n executions of Bit OT.

At Step 3 Bob randomly chooses his choice bit $c \in \{0,1\}$. He also selects a small subset $s \in I$ of cardinality xn. This selection is made by first choosing an encoding w uniformly at random among $\{0,1\}^m$ and then mapping it to the corresponding subset s. This guarantees that on one hand, s is sufficiently random and on the other hand, that every string in $\{0,1\}^m$ is equally likely to be Bob's initial choice. The latter fact will be crucial in preventing Alice from guessing Bob's choice bit in later steps.

At Step 4 Alice transmits T_0, T_1 using n executions of Bit OT. Bob selects to learn t_c^i at all positions except at the few positions in s where his choice is reversed. As a result he knows most bits of T_c and only xn bits of $T_{\bar{c}}$. See Fig. 1.

The goal of the protocol at Step 5 is to select a second, effectively random subset. Bob starts by sending w to Alice using Interactive Hashing, the output of which will be w_0, w_1. As from Alice's point of view both strings are equally likely to have been Bob's original choice at Step 3, Property 1 of Interactive Hashing (Section 3.2) guarantees to Bob that Alice cannot guess the value of b such that $w_b = w$. At the same time Property 3 of Interactive Hashing provides

Protocol 4. New reduction of $\binom{2}{1}$–ROT^k to Bit OT using Interactive Hashing

1. Alice and Bob select x to be a (very small) positive constant less than 1.

2. Alice chooses two random strings $T_0, T_1 \in_R \{0,1\}^n$.

3. Bob chooses a random $c \in_R \{0,1\}$. Let $m = \lceil \log\left(\binom{n}{xn}\right) \rceil$. Bob selects $w \in_R \{0,1\}^m$ uniformly at random and decodes w into a subset $s \subset I$ of cardinality xn according to the encoding/decoding scheme of Section 3.1.

4. Alice transmits T_0, T_1 to Bob using n executions of Bit OT, with round i containing bits t_0^i, t_1^i. Bob chooses to learn t_c^i if $i \notin s$ and $t_{\bar{c}}^i$ if $i \in s$.

5. Bob sends w to Alice using Interactive Hashing (Protocol 2). Alice and Bob compute the two output strings, labeled w_0 , w_1 according to lexicographic order, as well as the corresponding subsets $s_0, s_1 \subset I$. Bob computes $b \in \{0,1\}$ s.t. $w_b = w$.

6. Alice checks that $|s_0 \cap s_1| \le 2 \cdot x^2 n$ and aborts otherwise.

7. Both parties compute $s_0' = s_0 \setminus (s_0 \cap s_1)$ and $s_1' = s_1 \setminus (s_0 \cap s_1)$.

8. Bob announces $a = b \oplus c$ to Alice. He also announces $T_0(s_{1-a}')$ and $T_1(s_a')$.

9. Alice checks that the strings announced by Bob are consistent with a and contain no errors. Otherwise she aborts the protocol.

10. Alice and Bob discard the Bit OT's at positions $s_0 \cup s_1$ and concentrate on the remaining positions in $J = I \setminus (s_0 \cup s_1)$. Let $j = |J|$ and $R_0 = T_0(J), R_1 = T_1(J)$.

11. Alice chooses two functions h_0, h_1 randomly and independently from a 2-universal family of hash functions with input length j and output length $k = j - 6xn \ge n - 8xn$. She sets $r_0 = h_0(R_0)$ and $r_1 = h_1(R_1)$. She sends h_0, h_1 to Bob.

12. Bob sets $r_c = h_c(R_c)$.

Alice with the guarantee that the choice of one of w_0, w_1 was effectively random and beyond Bob's control. We will see that this implies that the corresponding subset is also random enough to ensure that a cheating Bob will fail the tests at Step 9 except with negligible probability.

At Step 6 Alice makes sure that the intersection of s_0, s_1 is not too large as this would interfere with the proof of security against a dishonest Bob.

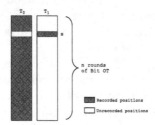

Fig. 1. During the n Bit OT executions Bob chooses t_c^i at positions $i \in I \setminus s$, and $t_{\bar{c}}^i$ at positions $i \in s$. In the Figure, $c = 0$ so in the end Bob knows $T_0(I \setminus s)$ and $T_1(s)$. Note that while $s \subset I$ is shown here as a contiguous block, in reality the positions it represents occur throughout the n executions.

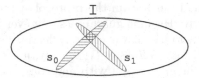

Fig. 2. Honest Bob sends his subset s to Alice through Interactive Hashing. With overwhelming probability this procedure produces two outputs s_0, s_1 of which one is s and the other is effectively randomly chosen. Alice does not know which of the two was Bob's original choice. The intersection of s_0, s_1 is later excluded to form s_0', s_1'.

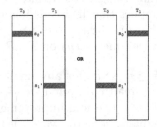

Fig. 3. After establishing sets s_0', s_1', Alice expects Bob to announce either $T_0(s_0')$ and $T_1(s_1')$ or $T_0(s_1')$ and $T_1(s_0')$ depending on the value of a. If Bob's choice was $c = 0$ as in Figure 1 and $s = s_0$ after Interactive Hashing, then he would choose the latter option.

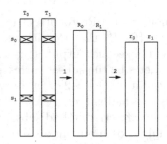

Fig. 4. After Bob has passed the tests, both players ignore the Bit OT executions at positions $s_0 \cup s_1$ and form strings R_0, R_1 from the remaining bits. Then independent applications of Privacy Amplification on R_0, R_1 produce $r_0, r_1 \in \{0,1\}^k$.

At Step 7 the two parties exclude the bits in this intersection from the tests that will follow since Bob cannot be expected to know both $T_0(s_0 \cap s_1)$ and $T_1(s_0 \cap s_1)$. What remains of s_0, s_1 is denoted s_0', s_1'.

At Step 8 Bob effectively announces $T_c(s_{\bar{b}}')$ and $T_{\bar{c}}(s_b')$ in both cases. Note that the only information related to c which is implied by the announced bits is the value of a, which is already made available to Alice at the beginning of the step. Alice can correctly guess $c = a \oplus b$ if and only if she can correctly guess b.

At Step 9 Alice checks that the strings were announced correctly and are consistent with the value of a — see Fig. 3. If that is the case then Alice is convinced that Bob has not deviated much from the protocol at Step 4. In a nutshell the idea here is that Interactive Hashing guarantees that even if Bob behaves dishonestly, without loss of generality s_1 was chosen effectively at random. Therefore, if Bob can announce all bits in $T_0(s_0'), T_1(s_1')$, say, it must have been the case that he knew most bits in T_1 to begin with and consequently few bits in T_0. In fact, we prove that if (dishonest) Bob learns more than $5xn$ bits of both T_0 and T_1 during Step 4 then he gets caught with overwhelming probability.

In Step 10 the two players discard the Bit OT executions at positions $s_0 \cup s_1$ that were used for tests and concentrate on the remaining j executions. Note that $j \geq n - 2xn$. As Bob passed the tests of Step 9, Alice is convinced that there is a $d \in \{0, 1\}$ such that Bob knows at most $5xn$ bits in T_d and thus at most $5xn$ bits in R_d. This implies that he is missing at least $j - 5xn$ bits of R_d.

In Step 11 she thus sets $k = (j - 5xn) - xn \geq n - 8xn$ and performs Privacy Amplification (with security parameter xn) on R_0, R_1 to get r_0, r_1. See Fig. 4.

Gains in Efficiency. As $k \geq n - 8xn$ for any small constant x, the expansion factor n/k is $1 + \epsilon$ for some small constant $\epsilon = \frac{8x}{1-8x}$. This is asymptotically optimal (see [7]) and represents a two-fold improvement over the corresponding reduction in [3] where the expansion factor was at least $2 + \epsilon'$.

5.1   Proof of Security and Practicality

Theorem 2. *The probability of failure of Protocol 4 with honest participants is exponentially small in n.*

Proof. If both parties are honest then Protocol 4 can only fail at Step 6. We will show that for any (fixed) $w \in \{0, 1\}^m$ that Bob inputs to Interactive Hashing at Step 5, the probability that the second output w' is such that $|s \cap s'| > 2 \cdot x^2 n$ is exponentially small in n. Let s be the subset corresponding to Bob's choice of w. We will call a subset s' *bad* if $|s \cap s'| > 2 \cdot x^2 n$. Likewise, we will call a string $w' \in \{0, 1\}^m$ *bad* if it maps to a bad subset.

We start by showing that the fraction of bad subsets is exponentially small in n. Suppose $s' \subset I$ is randomly chosen among all subsets of cardinality xn. One way to choose s' is by sequentially selecting xn positions uniformly at random without repetition among all n positions in I. The probability q_i that the i–th position thus chosen happens to collide with one of the xn positions in s satisfies

$$q_i < \frac{xn}{n - xn} = \frac{x}{1 - x}$$

As a thought experiment, suppose that one were to choose xn positions independently at random, so that each position collides with an element of s with probability exactly $q = \frac{x}{1-x}$. This artificial way of choosing xn positions can only increase the probability of ending up with more than $2x^2 n$ collisions. We can use the Chernoff bound (2) to upper bound this (larger) probability. Assuming $x < 1/2$ and setting $\delta = 1 - 2x$ we get

$$\mathbf{Pr}\left[B(xn, \frac{x}{1-x}) > 2x^2 n \right] \leq \epsilon'$$

where $\epsilon' = e^{-\frac{(1-2x)^2 x^2}{3(1-x)} n}$. This in turn guarantees that when s' is selected in the appropriate way, the event $|s \cap s'| > 2 \cdot x^2 n$ occurs with probability $\epsilon < \epsilon'$. In other words, the fraction of bad subsets is upper bounded by $\epsilon < \epsilon'$.

By Lemma 1, the fraction of bad strings in $\{0,1\}^m$ is at most 2ϵ. As w itself is bad, it follows that among all $2^m - 1$ strings other than w the fraction of bad strings is no larger than 2ϵ. Since by Property 2 of Interactive Hashing, w is paired to some uniformly chosen $w' \neq w$, the probability that the protocol aborts at Step 6 is upper bounded by 2ϵ which is exponentially small in n. □

Theorem 3. *Alice learns nothing about (honest) Bob's choice bit c.*

Proof. During Bob's interaction with Alice, his choice bit c comes into play only during the Bit OT executions of Step 4 and later at Step 8 when Bob announces $a = b \oplus c$. As Bit OT is secure by assumption, Alice cannot obtain any information about c in Step 4. As for Step 8, since (honest) Bob chooses w uniformly at random in $\{0,1\}^m$, both w_0 and w_1 are apriori equally likely choices. By Property 1 of Interactive Hashing (see Section 3.2), the aposteriori probabilities of w_0, w_1 having been Bob's input are then equal as well. Consequently, Alice cannot guess b with probability higher than $1/2$ and the same holds for $c = a \oplus b$. □

Security Against a Dishonest Bob. The proof of security against a dishonest Bob is considerably more involved. The main idea is that if Bob deviates from the protocol more than a small fraction of the time then he gets caught by the end of Step 9 with overwhelming probability. If, on the other hand, he deviates only a small fraction of the time, then Privacy Amplification effectively destroys any illegal information he may have obtained. We start with some definitions and lemmas that will help to prove the main theorem (Theorem 4) of this section.

Definition 1. *For a bit string σ, define $u_p(\sigma)$ to be the number of bits in σ that can be guessed correctly with probability at most $p < 1$. These bits will be referred to as* unknown *bits.*

Definition 2. *Let $s \subset I$. Assuming Definition 1, we call s good for T_c if $u_p(T_c(s)) \leq 3x^2 n$. Otherwise, we call s bad for T_c. We say that s is good for either T_0 or T_1 if at least one of $u_p(T_0(s)), u_p(T_1(s))$ is at most $3x^2 n$.*

Definition 3. *Let w be a string in $\{0,1\}^m$. We call w good for T_c if the subset s it encodes is good for T_c according to Definition 2. Otherwise, w is bad for T_c.*

Lemma 3. *Let $u_p(T_c) \geq 5xn$. Then among all subsets $s \subset I$ of cardinality xn the fraction of good subsets for T_c is less than $e^{-x^2 n/8}$.*

Proof. We will use the Probabilistic Method to show that the probability that a randomly chosen subset s is good for T_c is less than $e^{-x^2 n/8}$. One way of choosing s would be to sequentially choose xn positions in I at random and without replacement. Note that regardless of previous choices, for all $1 \leq i \leq xn$ the probability q_i of position i being chosen among the $u_p(T_c)$ positions of unknown bits always satisfies

$$q_i > \frac{u_p(T_c) - xn}{|I|} \geq \frac{5xn - xn}{n} = 4x$$

This implies that the probability of choosing a good subset for T_c would be greater if we were to choose the xn positions independently at random so that each position corresponds to an unknown bit with probability $q = 4x$. In this artificial case the distribution of the number of unknown bits is binomial with parameters xn, $4x$ and mean $\mu = 4x^2n$. Applying the Chernoff bound (Equation 1) with $\delta = 1/4$ we get

$$\mathbf{Pr}\left[B(xn, 4x) \leq 3x^2n\right] \leq e^{-x^2n/8}$$

We conclude that a subset s chosen randomly in the appropriate way has probability smaller than $e^{-x^2n/8}$ of being good for T_c, which establishes the claim.   □

Lemma 4. *Let both $u_p(T_0), u_p(T_1) \geq 5xn$. Then the fraction of strings in $\{0, 1\}^m$ that are good for either T_0 or T_1 is no larger than $4 \cdot e^{-x^2n/8}$.*

Proof. It follows from Lemma 3 and the Union Bound that the proportion of good subsets for either T_0 or T_1 is no larger than $2 \cdot e^{-x^2n/8}$. Lemma 1 in turn guarantees that the fraction of strings in $\{0, 1\}^m$ that are good for either T_0 or T_1 in $\{0, 1\}^m$ is at most $4 \cdot e^{-x^2n/8}$.   □

Lemma 5. *Let both $u_p(T_0), u_p(T_1) \geq 5xn$. Then the probability that (dishonest) Bob will clear Step 9 is exponentially small in n.*

Proof. By Lemma 4, the proportion of good strings in $\{0, 1\}^m$ for either T_0 or T_1 is at most $4 \cdot e^{-x^2n/8}$. By Property 3 of Interactive Hashing, the probability that both w_0, w_1 will be good at Step 5 of the protocol is at most ϵ_1 which is exponentially small in m (and hence in n). Consequently, with probability at least $1 - \epsilon_1$, at least one of the two bit strings (without loss of generality, w_1) is bad for *both* T_0 and T_1. In other words, w_1 corresponds to a subset s_1 with both $u_p(T_0(s_1)), u_p(T_1(s_1)) \geq 3x^2n$. Moreover, as Alice did not abort at Step 6 it must be the case that $|s_0 \cap s_1| \leq 2x^2n$. It follows that both $u_p(T_0(s_1')), u_p(T_1(s_1')) \geq 3x^2n - 2x^2n = x^2n$. Therefore, however Bob decides to respond in Step 8, he must correctly guess the value of at least x^2n unknown bits in one of T_0, T_1. As the bits were independently chosen, the probability of guessing them is $\epsilon_2 \leq p^{x^2n}$.

Bob will clear Step 9 only if he got two good strings from Interactive Hashing or got at least one bad string and then correctly guessed all the relevant bits. This probability is upper bounded by $\epsilon_1 + \epsilon_2$ (exponentially small in n).   □

Theorem 4. *The probability of (dishonest) Bob successfully cheating in Protocol 4 is exponentially small in n.*

Proof. Let $v_0, v_1 \subseteq I$ be the positions where (dishonest) Bob requested t_0^i, t_1^i respectively during Step 4. Note that $v_0 \cap v_1 = \emptyset$. We distinguish two cases: (Case 1 and Case 2 taken together establish the claim.)

Case 1: Both $|v_0|, |v_1| \leq n - 5xn$
In this case $u_{1/2}(T_0), u_{1/2}(T_1) \geq 5xn$, so by Lemma 5 (dishonest) Bob will fail to clear Step 9 except with exponentially (in n) small probability.

Case 2: One of $|v_0|, |v_1|$ is greater than $n - 5xn$
Without loss of generality, let $|v_0| > n - 5xn$. Then Bob knows less than $5xn$ bits about T_1, and consequently, less than $5xn$ bits about $R_1 = T_1(J)$. Note that as T_0, T_1 are independently chosen, even if an oracle were to subsequently provide all the bits of T_0 (or R_0 , or r_0), Bob would obtain no new information about R_1. As $u_{1/2}(R_1) \geq j - 5xn$, Privacy Amplification with output length $k = (j - 5xn) - xn$ destroys all but an exponentially (in n) small amount of information about r_1, with probability exponentially close to 1.   □

6   Extension to Weaker Variants of Bit OT

We demonstrate that Protocol 4 can accommodate weaker versions of Bit OT. Specifically, it requires no modification at all if Bit OT is replaced with XOT, while a virtually imperceptible decrease in the output length k guarantees its security with GOT. Decreasing k even further allows us to prove the Protocol's security when Bob has access to UOT with $\alpha \leq 1$. As in all three cases honest Bob's choices during Step 4 are identical to the case of Bit OT and remain equally well hidden from Alice's view, the proofs of Theorems 2 and 3 (establishing the Protocol's practicality and security against dishonest Alice) carry over verbatim to the new settings.

On the other hand, arguing that the Protocol remains secure against dishonest Bob is more involved and requires a separate analysis in each case. The basic idea, however, is the same as in the case of Bit OT and consists in showing that if Bob has deviated 'significantly' from the protocol then he gets caught with overwhelming probability, and if he has not, then Privacy Amplification effectively eliminates any illegal information he may have accumulated.

6.1   Security Against a Dishonest Bob Using XOT

Theorem 5. *The probability of (dishonest) Bob successfully cheating in Protocol 4 is exponentially small in n even if the Bit OT protocol is replaced with XOT.*

Proof. Let $v_0, v_1, v_\oplus \subseteq I$ denote the sets of positions i where (dishonest) Bob requested $t_0^i, t_1^i, t_\oplus^i = t_0^i \oplus t_1^i$ respectively during Step 4. As in the proof of Theorem 3, we distinguish two cases, in both of which the probability of cheating is exponentially small in n, as desired.

Case 1: One of $|v_0|, |v_1|$ is greater than $n - 5xn$
Without loss of generality, let $|v_0| > n - 5xn$. Then $|v_1 \cup v_\oplus| < 5xn$. Consequently, Bob knows less than $5xn$ bits about R_1 even if he is provided with all the bits of T_0 by an oracle after Step 4. We note in passing that such oracle information can only be helpful for the positions in $v_\oplus$. Since $u_{1/2}(R_1) > j - 5xn$, Privacy Amplification with output length $k = (j - 5xn) - xn$ would destroy all but an exponentially (in n) small amount of information about r_1, with probability exponentially close to 1.

Case 2: Both $|v_0|, |v_1| \le n - 5xn$.

This implies that both $|v_1 \cup v_\oplus|$ and $|v_0 \cup v_\oplus|$ are at least $5xn$ and consequently, $u_{1/2}(T_0), u_{1/2}(T_1) \ge 5xn$. By Lemma 5, Bob will fail to clear Step 9 except with exponentially (in n) small probability. □

Gains in Efficiency. The expansion factor is identical to the case of Bit OT (and optimal). Compared to the reduction in [3], ours is again twice as efficient.

6.2 Security Against a Dishonest Bob Using GOT

In the case of Generalized OT, during round i of Step 4 dishonest Bob can choose to obtain $f(t_0^i, t_1^i)$ for any of the 16 functions $f : \{0,1\}^2 \mapsto \{0,1\}$. Without loss of generality, we will assume that Bob never requests the two constant functions as this would provide him with no information. It is not difficult to see that in our context the information content of each of the remaining 14 functions is equivalent to that of one of the four functions $f_0, f_1, f_\oplus, f_{\mathrm{AND}}$ defined in Equation (7) below. We will thus assume that Bob always requests the output of one of these functions. In keeping with the notation of previous sections we let $v_0, v_1, v_\oplus, v_{\mathrm{AND}} \subseteq I$ be the positions where Bob requested $f_0, f_1, f_\oplus, f_{\mathrm{AND}}$ respectively.

$$f_0(t_0, t_1) = t_0, \quad f_1(t_0, t_1) = t_1, \quad f_\oplus(t_0, t_1) = t_0 \oplus t_1, \quad f_{\mathrm{AND}}(t_0, t_1) = t_0 \wedge t_1 \quad (7)$$

A Necessary Modification to Protocol 4. Our proof of security requires that k be slightly shorter than in the case of Bit OT and XOR OT, that is $k = (j - 8xn) - xn \ge n - 11xn$.

The security analysis of the protocol in this setting is somewhat more complicated compared to the case of Bit OT and XOT. This is due to the fact that requesting f_{AND} may or may not result in loss of information about (t_0, t_1): with probability $1/4$ the output of f_{AND} is 1 and so Bob learns both bits while with complementary probability $3/4$ the output is 0 in which case the input bits were $(0,0), (0,1), (1,0)$, all with equal probability. Note that in this latter case both t_0, t_1 are unknown as each can be guessed correctly with probability at most $2/3$.

Complications Arising from Adaptive Strategies. If dishonest Bob's requests could be assumed to be fixed ahead of time, our analysis would be quite straightforward since we could claim that among all requests in v_{AND}, with high probability a fraction $3/4 - \epsilon$ would produce an output of 0 and thus both t_0, t_1 would be added to the set of unknown bits in T_0, T_1. Our task is complicated by the fact that Bob obtains the output of the function he requested immediately after each round and can thus adapt his future strategy to past results. For example, Bob may be very risk-averse and start by asking for f_{AND} in the first round. If he is lucky and the output is 1, he asks for f_{AND} again, until he gets unlucky in which case he starts behaving honestly. This strategy makes it almost impossible to catch Bob cheating while it allows Bob to learn both r_0, r_1 with some nonzero — but admittedly quite small— probability. This example illustrates that we cannot assume that $|v_{\mathrm{AND}}|$ is known ahead of time and remains independent of results obtained during the n executions of Step 4.

Dealing with Adaptive Strategies. In order to prove the security of the protocol for any conceivable strategy that dishonest Bob might use, we start by observing that at the end of Step 4 one of the following two cases always holds:

Case 1: One of $|v_0|, |v_1| > n - 8xn$,   **Case 2:** Both $|v_0|, |v_1| \leq n - 8xn$

Note that these two cases refer only to the types of requests issued by Bob during Step 4 and do not depend in any way on the results obtained along the way. Given any (adaptive) strategy S for Bob, one can construct the following two strategies: Strategy S_1 begins by making the same choices as S but ensures that eventually the condition in Case 1 will be met: it "applies the brakes" just before this constraint becomes impossible to meet in the future and makes its own choices from that point on in order to meet its goal. Similarly, Strategy S_2 initially copies the choices of S but if necessary, stops following them to ensure that the condition of Case 2 is met. Let $\delta, \delta_1, \delta_2$ be the probabilities of successfully cheating using Strategies S, S_1, S_2, respectively. We will argue that $\delta \leq \delta_1 + \delta_2$. To see this, imagine three parallel universes in which Bob is interacting with Alice using strategies S, S_1, S_2 respectively. Recall that by the end of Step 4 the universe of Strategy S is identical either to the Universe of Strategy S_1 or to the Universe of Strategy S_2 (one of the two never had to "apply the brakes"). Therefore, Strategy S succeeds only if one of S_1, S_2 succeeds and so $\delta \leq \delta_1 + \delta_2$.

It remains to prove that both δ_1, δ_2 are exponentially small in n. To do this, we let Σ_1, Σ_2 be *any* adaptive strategies ensuring that the conditions of Case 1 and Case 2, respectively, are met. We will show that for any such strategies (thus, for S_0, S_1 as well), the probabilities of success Δ_1, Δ_2 are exponentially small in n, and therefore so is δ (since $\delta \leq \delta_1 + \delta_2 \leq \Delta_1 + \Delta_2$).

Theorem 6. *The probability of (dishonest) Bob cheating in (modified) Protocol 4 is exponentially small in n even if Bit OT is replaced with GOT.*

Proof. We will prove that Δ_1, Δ_2 are both exponentially small in n.

Without loss of generality, let $|v_0| > n - 8xn$ at the end of Step 4. Then Bob knows at most $8xn$ bits about T_1, even if he is provided with all the bits of T_0 by an oracle. Consequently, $u_{1/2}(R_1) > j - 8xn$ and therefore using Privacy Amplification with output length $k = (j - 8xn) - xn \geq n - 11xn$ will result in Bob having only an exponentially small amount of information about r_1 (even given r_0), except with an exponentially small probability Δ_1.

As for Strategies Σ_2, we start by showing that $\mathbf{Pr}\left[u_{2/3}(T_1) \leq 5xn\right]$ is small. Since any such strategy guarantees that $|v_1| \leq n - 8xn$, it follows that $|v_0 \cup v_\oplus \cup v_{AND}| \geq 8xn$. Given this constraint, the probability that $u_{2/3}(T_1) \leq 5xn$ is maximized if $|v_{AND}| = 8xn, |v_0| = |v_\oplus| = 0$. This is because each request in v_0 and $v_\oplus$ results with certainty in the corresponding bit in T_1 being unknown, while a request in v_{AND} produces an unknown bit in T_1 with probability $3/4$ (moreover, in this case the unknown bit can be guessed correctly with probability $2/3$ instead of $1/2$). Using the Chernoff bound (Equation 1) with $(n, p, \delta) \mapsto (8xn, 3/4, 1/6)$ gives

$$\mathbf{Pr}\left[u_{2/3}(T_1) \leq 5xn\right] \leq \mathbf{Pr}\left[B(8xn, \frac{3}{4}) \leq 5xn\right] \leq e^{-xn/12}$$

and similarly for $u_{2/3}(T_0)$. By the Union Bound, both $u_{2/3}(T_0), u_{2/3}(T_1) \geq 5xn$ except with probability at most $2 \cdot e^{-xn/12}$. In this case, Lemma 5 guarantees that Bob will manage to clear Step 9 with some probability ϵ exponentially small in n. We conclude that using any Strategy Σ_2, Bob can successfully cheat with probability $\Delta_2 \leq 2 \cdot e^{-xn/12} + \epsilon$ which is exponentially small in n.

Probability of Successfully Cheating Using Any Adaptive Strategy S.
As argued above, for any adaptive strategy S, the probability δ of cheating is upper bounded by $\delta_1 + \delta_2 \leq \Delta_1 + \Delta_2$ and hence exponentially small in n.   □

Gains in Efficiency. As $k \geq n - 11xn$ for any small constant x, the expansion factor n/k is $1 + \epsilon'$ for some (related) small constant ϵ'. It is only slightly larger than the expansion factor in the case of Bit OT and XOR OT and remains asymptotically optimal. This represents an increase in efficiency by a factor of about 4.8188 over the corresponding reduction in [3].

6.3   Security Against a Dishonest Bob Using Universal OT

In this case, in each round of Bit OT at Step 4 dishonest Bob can choose to obtain the output of any discrete, memoryless channel subject to the following constraint: let B_0, B_1 be independent, uniformly distributed random variables corresponding to Alice's inputs to Bit OT and let $\Omega = \Omega(B_0, B_1)$ be the channel's output to Bob. Then for some constant $\alpha \leq 1$ the following holds:

$$\mathbf{H}((B_0, B_1) \mid \Omega) \geq \alpha \tag{8}$$

Note that we require α to be at most 1 since otherwise, the channel would disallow honest behavior as well. Let ϵ to be any (very small) positive constant strictly less than $1/2$. We can then partition all possible channels satisfying the constraint of Equation 8 into the following three categories.

Ω_0: All channels satisfying $\mathbf{H}(B_0 \mid \Omega) < \epsilon\alpha$ and $\mathbf{H}(B_1 \mid B_0\Omega) > (1 - \epsilon)\alpha$.

Ω_1: All channels satisfying $\mathbf{H}(B_1 \mid \Omega) < \epsilon\alpha$ and $\mathbf{H}(B_0 \mid B_0\Omega) > (1 - \epsilon)\alpha$.

Ω_b: All channels satisfying $\mathbf{H}(B_0 \mid \Omega), \mathbf{H}(B_1 \mid \Omega) \geq \epsilon\alpha$.

Let $\rho(\alpha)$ be the unique solution $x \in [0, 1/2]$ to the equation $h(x) = \alpha$. Let $p_0 = p_1 = \rho((1 - \epsilon)\alpha)$ and $p_b = \rho(\epsilon\alpha)$. Then from Fano's inequality and Lemma 2 (Section 3.4) we can assert the following:

– p_0 is a lower bound on the error probability when guessing the value of B_1 after using a channel of type Ω_0 and this is true even if the value of B_0 is known with certainty. There thus exists an indicator random variable Δ_0 (provided as side information by an oracle) which leads to an erasure of B_1 with probability $2p_0$. Note: when there is no erasure ($\Delta_0 = 0$) it is not necessarily the case the corresponding bit is known with certainty.

- Likewise, p_1 lower bounds the error probability when guessing B_0 given the value of B_1 and the output of a channel of type Ω_1. This implies the existence of side information in the form of an indicator random variable Δ_1 that leads to an erasure of B_0 with probability $2p_1 = 2p_0$.

- When using a channel of type Ω_b, the probability of guessing B_0 incorrectly given the channel's output is at least p_b, and the same holds when guessing the value of B_1. Thus, there exists an indicator random variable Δ_b^0 (resp. Δ_b^1) which, if provided by an oracle, would lead to an erasure of B_0 (resp. B_1) with probability $2p_b$. Note that this statement is true only if the oracle provides *one* of Δ_b^0, Δ_b^1 each time. To see why this is so, suppose both were provided at the same time. Since Δ_b^0 along with Ω might contain more information about B_1 than was available in Ω alone, one can no longer assume that the event $\Delta_b^1 = 1$ would necessarily correspond to an erasure of B_1.

In order to simplify our analysis we will assume that after each round of UOT in Step 4, an oracle supplies Bob with the following side information, depending on the type of channel that Bob used:

Ω_0: The exact value of B_0, as well as the value of Δ_0. Note that this leads to B_1 being erased with probability $2p_0$.

Ω_1: The exact value of B_1, as well as the value of Δ_1. Note that this leads to B_0 being erased with probability $2p_1 = 2p_0$.

Ω_b: One of Δ_b^0, Δ_b^1, chosen at random with equal probability. Note that this leads to each of B_0, B_1 being erased with probability p_b in each round (not independently, though: B_0 and B_1 cannot be erased at the same time).

Another Modification to Protocol 4. For any very small positive constant ϵ, let $p_b \overset{\text{def}}{=} \rho(\epsilon\alpha)$ and $p_0 \overset{\text{def}}{=} \rho((1-\epsilon)\alpha)$. Our proof of security will require that we reduce k even further at step 11, by setting $k = 2p_0(j - 8p_b n) \geq 2p_0 n - 9p_0 p_b n$. For convenience, we will also set $x = p_b^2$ in Step 1.

Theorem 7. *The probability of dishonest Bob successfully cheating in (modified) Protocol 4 is exponentially small in n even if the Bit OT protocol is replaced with UOT satisfying the constraint of Equation (8).*

Proof. Let $v_0, v_1, v_b \subseteq I$ be the positions in Step 4 where Bob selected a channel of type $\Omega_0, \Omega_1, \Omega_b$, respectively. Then, at the end of Step 4 one of the following two cases always holds:

Case 1: One of $|v_0|, |v_1| > n - 6p_b n$

Case 2: Both $|v_0|, |v_1| \leq n - 6p_b n$

We proceed as in the proof of security for GOT in Section 6.2.

Without loss of generality, let $|v_0| > n - 6p_b n$ at the end of Step 4. This implies that at least $j - 6p_b n$ of the bits of R_1 were received over a channel of type Ω_0. Let μ_1 be the expected number of erasures in R_1, resulting from the side information Δ_0 provided by the oracle in each round. Then $\mu_1 \geq 2p_0 (j - 6p_b n)$. From Equation (3) we deduce that with probability exponentially close to 1 there will be at least $2p_0 (j - 7p_b n)$ erasures, in which case $u_{1/2} (R_1) \geq 2p_0 (j - 7p_b n)$.

Applying Privacy Amplification with output length $k = 2p_0 (j - 8p_b n)$ will thus produce an almost-uniformly distributed k-bit string r_1 (independent of r_0), except with exponentially (in n) small probability. Note that as $p_b^3 < 1/2$ and $j \geq n - 2x^2 n = n - 2p_b^4 n$, the output size k satisfies $k = 2p_0 (j - 8p_b n) \geq 2p_0 (n - 2p_b^4 n - 8p_b n) \geq 2p_0 (n - 9p_b n) = 2p_0 n - 9p_0 p_b n$.

The probability of any strategy Σ_1 successfully cheating is at most equal to the probability that there are too few erasures to begin with plus the probability that Privacy Amplification failed to produce an almost-uniformly distributed string. Our choices guarantee that this probability is exponentially small in n.

We show that with near certainty both $u_{1/2} (T_0)$ and $u_{1/2} (T_1)$ are at least $5xn$, which by Lemma 5 guarantees that Bob will fail to clear Step 9 with probability exponentially close to 1. We start by upper bounding the probability that $u_{1/2} (T_1) \leq 5xn$. Since $|v_1| \leq n - 6p_b n$, there are at least $6p_b n$ bits that were either sent over a channel of type Ω_0 or Ω_b. We will assume that exactly $6p_b n$ bits were sent over a channel of type Ω_b, as this choice minimizes the expected number of erasures in T_1 given our constraints, and hence maximizes the probability that $u_{1/2} (T_1) \leq 5xn$. Note that the expected number of erasures of B_1 in this case is $p_b \cdot 6p_b n = 6p_b^2 n = 6xn$. By the Chernoff bound

$$\mathbf{Pr} \left[u_{1/2} (T_1) \leq 5xn \right] \leq \mathbf{Pr} \left[B(6p_b n, p_b) \leq 5p_b^2 n \right] \leq \lambda$$

where λ is exponentially small in n.

The same argument applies to $u_{1/2} (T_0)$. Therefore, both $u_{1/2} (T_0), u_{1/2} (T_1) \geq 5xn$ except with probability at most 2λ. Then Lemma 5 guarantees that Bob will fail to clear Step 9 with probability $1 - \epsilon'$ for some ϵ' exponentially small in n. We conclude that using any Strategy Σ_2, Bob can successfully cheat with probability at most $2\lambda + \epsilon'$ which is exponentially small in n.

Probability of Cheating Using Any Adaptive Strategy S. As argued in Section 6.2, the probability of successful cheating for any adaptive strategy S is upper bounded by the sum of the probabilities of success of any strategies Σ_1, Σ_2. We have shown that both of these are exponentially small.   □

Gains in Efficiency. In both our reduction and that of [3], the expansion factor is a function of α. In our case $k \geq 2p_0 n - 9p_0 p_b n$. Since $p_b = \rho(\epsilon\alpha), p_0 = \rho((1-\epsilon)\alpha)$, for $\epsilon \to 0$ we get $p_0 \to \rho(\alpha), p_b \to 0$ and therefore $k \approx 2\rho(\alpha)n$, which translates to an expansion factor of $\frac{1}{2\rho(\alpha)} + \epsilon'$. The corresponding expansion factor in [3] is at least $\frac{4\ln 2}{p_e}$ where p_e is the unique solution in $(0, 1/2]$ to the equation $h(p_e) + p_e \log_2 3 = \alpha$. It is easy to verify by means of a graph that for all $0 \leq \alpha \leq 1$, we have $\rho(\alpha) > p_e$. Consequently, our expansion factor is always at least $8 \ln 2 = 5.545$ times smaller than the one in [3]. It is noteworthy that in the special case where $\alpha = 1$ we have $\rho(\alpha) = 1/2$ and therefore the expansion factor is $1 + \epsilon'$, which is optimal. Proving optimality for other values of α is left as an open problem.

7   Conclusions, and Open Problems

We have demonstrated how the properties of Interactive Hashing can be exploited to increase the efficiency and generality of existing String OT reductions. Specifically, we have shown that our reductions are optimal in the case of Bit OT, XOT and GOT, as well as for the special case of UOT where $\alpha = 1$. We conclude by listing some problems that our current work leaves open. (1) Modify Protocol 4 so that it never aborts when both participants are honest. This will require proving that Interactive Hashing would not allow a dishonest Bob to obtain strings w_0, w_1 such that the corresponding subsets s_0, s_1 have a large intersection. (2) Prove that our reduction is optimal for all α in the case of UOT, or modify it accordingly to achieve optimality. (3) Replace the Interactive Hashing Protocol (Protocol 2) with an appropriately adapted implementation of the constant round Protocol of [6] and prove that the ensuing reduction (Protocol 4) remains secure. (4) Further explore the potential of Interactive Hashing as an ingredient in cryptographic protocols design.

References

1. C. H. Bennett, G. Brassard, C. Crépeau, and U. Maurer. Generalized privacy amplification. *IEEE Trans. on Info. Theory*, 41(6):1915–1923, November 1995.
2. C. H. Bennett, G. Brassard, and J.-M. Robert. Privacy amplification by public discussion. *SIAM J. Comput.*, 17(2):210–229, 1988.
3. G. Brassard, C. Crépeau, and S. Wolf. Oblivious transfers and privacy amplification. *IEEE Transaction on Information Theory*, 16(4):219–237, 2003.
4. G. Brassard, C. Crépeau, and M. Santha. Oblivious transfers and intersecting codes. *IEEETIT: IEEE Transactions on Information Theory*, 42, 1996.
5. C. Cachin, C. Crépeau, and J. Marcil. Oblivious transfer with a memory-bounded receiver. In *IEEE Symposium on Foundations of Computer Science*, 1998.
6. Y. Zong Ding, D. Harnik, A. Rosen, and R. Shaltiel. Constant-round oblivious transfer in the bounded storage model. In *TCC*, pages 446–472, 2004.
7. Y. Dodis and S. Micali. Lower bounds for oblivious transfer reductions. *Lecture Notes in Computer Science*, 1592, 1999.
8. S. Even, O. Goldreich, and A. Lempel. A randomized protocol for signing contracts. *Commun. ACM*, 28(6):637–647, 1985.
9. Joe Kilian. Founding crytpography on oblivious transfer. In *STOC '88: Proceedings of the twentieth annual ACM symposium on Theory of computing*, pages 20–31, New York, NY, USA, 1988. ACM Press.
10. M. Naor, R. Ostrovsky, R. Venkatesan, and M. Yung. Perfect zero-knowledge arguments for NP using any one-way permutation. *Journal of Cryptology*, 11(2), 1998.
11. R. Ostrovsky, R. Venkatesan, and M. Yung. Fair games against an all-powerful adversary. *AMS DIMACS Series in Discrete Mathematics and Theoretical Computer Science*, 13, 1993.
12. M. O. Rabin. How to exchange secrets by oblivious transfer. Technical Memo TR–81, Aiken Computation Laboratory, Harvard University, 1981.
13. D.R. Stinson. Some results on nonlinear zigzag functions. *Journal of Combinatorial Mathematics and Combinatorial Computing*, 29:127–138, 1999.
14. S. Wiesner. Conjugate coding. *SIGACT News*, 15(1):78–88, 1983.

Oblivious Transfer Is Symmetric

Stefan Wolf and Jürg Wullschleger

Computer Science Department, ETH Zürich, Switzerland
{wolf, wjuerg}@inf.ethz.ch

Abstract. We show that oblivious transfer of bits from A to B can be obtained from a single instance of the same primitive from B to A. Our reduction is perfect and shows that oblivious transfer is in fact a symmetric functionality. This solves an open problem posed by Crépeau and Sántha in 1991.

1 Introduction

Modern cryptography is an increasingly broad discipline and deals with many subjects besides the classical tasks of encryption or authentication. An example is *multi-party computation*, where two or more parties, mutually distrusting each other, want to collaborate in a secure way in order to achieve a common goal, for instance, to carry out an electronic election. An example of a specific multi-party computation is *secure function evaluation*, where every party holds an input to a function, and the output should be computed in a way such that no party has to reveal unnecessary information about her input.

A primitive of particular importance in the context of two- and multi-party computation is *oblivious transfer*. In classical *Rabin oblivious transfer* [19] or *Rabin OT* for short, one of the parties—*the sender*—sends a bit b which reaches *the receiver* with probability $1/2$; the sender hereby remains ignorant of about whether the message has arrived or not. In other words, Rabin OT is nothing else than a binary erasure channel. Another variant of oblivious transfer is *chosen one-out-of-two oblivious transfer*—$\binom{2}{1}$–OT for short—, where the sender sends two bits b_0 and b_1 and the receiver's input is a choice bit c; the latter then learns b_c but gets no information about the other bit b_{1-c}. Chosen one-out-of-two oblivious transfer can be generalized to a primitive where the sender sends n messages, k of which the receiver can choose to read: *chosen k-out-of-n l-bit string oblivious transfer* or $\binom{n}{k}$–OT^l. One reason for the importance of oblivious transfer is its *universality*, i.e., it allows, in principle, for carrying out *any* two-party computation [14].

Besides *computational* cryptographic security, which is based on the assumed hardness of certain computational problems and a limitation on the adversary's computing power, there also exists *unconditional* security, which is based on the fact that the *information* the potential adversary obtains is limited. This latter type of security withstands attacks even by a computationally unlimited adversary; clearly, it is, *a priori*, more desirable to realize cryptographic primitives in such an unconditionally secure way. Unfortunately, oblivious transfer is impossible to achieve in an unconditionally secure way from scratch, i.e., between

S. Vaudenay (Ed.): EUROCRYPT 2006, LNCS 4004, pp. 222–232, 2006.

parties connected by a noiseless channel; in fact, not even if this is a *quantum* channel over which the parties can exchange not only "classical" bits but quantum states [15]. However, if some additional weak and realistic primitives are available such as noisy channels and noisy correlations, then unconditional security *can* be often achieved [7], [6], [8], [13], [22], [23].

Another way of realizing unconditionally secure oblivious transfer is from (a weaker form of) oblivious transfer itself: All the variants of oblivious transfer have been shown equivalent to different extents. For instance, $\binom{2}{1}$−OT can be reduced to m realizations of Rabin OT as long as a failure probability of 2^{-m} can be accepted [5]. On the other hand, $\binom{2}{1}$−OT^l can be reduced to $\Theta(l)$ realizations of $\binom{2}{1}$−OT—with or without failure probability, where the reduction can be made more efficient in terms of the hidden constant if a small probability of failure can be accepted [4]. In [2], a protocol was presented that reduces $\binom{2}{1}$−OT to a $\binom{2}{1}$−OT being available at an earlier point in time. This means that $\binom{2}{1}$−OT can be *precomputed* (or *stored* and used at any time later).

In [18] and [9], methods were proposed for obtaining $\binom{2}{1}$−OT from A to B from n instances of $\binom{2}{1}$−OT from B to A, where a failure probability of $2^{-\Theta(n)}$ has to be tolerated. The protocol of [18] is based on the realization of so-called "XOT" (i.e., the receiver can also choose to receive the XOR of the two bits sent) from two realizations of $\binom{2}{1}$−TO—the *reversed* version of $\binom{2}{1}$−OT. Note, however, that the resulting reduction of $\binom{2}{1}$−OT to $\binom{2}{1}$−TO of [18] also requires $\Theta(\log(1/\varepsilon))$ realizations of $\binom{2}{1}$−TO if ε is the tolerated failure probability.

1.1   Our Contribution

In [9], Crépeau and Sántha raised the question of whether it is possible to implement oblivious transfer in one direction using fewer instances of oblivious transfer in the other. In this paper, we answer this question with *yes* by presenting a protocol that needs *one* instance of oblivious transfer, *one* bit of communication and *one* bit of additional (local) randomness. All these parameters are optimal. Our reduction is very simple; in other words, the reversed version of oblivious transfer is basically just another way of *looking* at it. The symmetry is already there, *oblivious transfer is symmetric*.

Our reduction can be used to transform *any* protocol for $\binom{2}{1}$−OT—offering either computational or information-theoretic security for A and B, respectively—into a protocol for oblivious transfer from B to A having exactly the same security both for A and B as the original protocol; no additional failure can occur.

1.2   Outline

In Section 3, we first present protocols from [2] that allow $\binom{2}{1}$−OT to be "stored", i.e., to transform oblivious transfer into an *oblivious key*. Then, we will show that such an oblivious key can very easily be reversed—by a simple XOR executed by both players on their local data. It follows that oblivious transfer can be reversed equally easily. In Section 4 we present an even simpler protocol for reversing oblivious transfer and prove its security.

2   Definitions and Security

We define $\binom{2}{1}$–OT as a *black-box* (see Figure 1).

Definition 1. By $\binom{2}{1}$–OT or *chosen one-out-of-two oblivious transfer* we denote the following primitive between a sender A and a receiver B. A has two inputs b_0 and b_1 and no output, and B has input c and output y such that $y = b_c$.

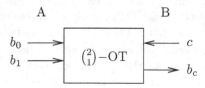

Fig. 1. Chosen one-out-of-two oblivious transfer

For the *reversed* version of $\binom{2}{1}$–OT, where B is the sender and A is the receiver, we will write $\binom{2}{1}$–TO.

This black-box model of oblivious transfer is called the *ideal model*, and it is how the world is supposed to be: The players have no other way of accessing the box than by the defined inputs and outputs: cheating is impossible. However, in reality such a perfect box does normally not exist. It must be *simulated* by a *protocol*. In this *real model*, the players can cheat in principle by not following the rules. A protocol is called a *secure implementation* of oblivious transfer if an adversary cannot do anything in the real model that he could not just as well have done in the ideal model. Thus, it must be shown that for any adversary in the real model, there exists an equivalent adversary in the ideal model: he gets the same information and the honest player obtains the same outputs as in the real model; the resulting views are *indistinguishable*.

We follow the formalism of [16] and [1] (see also [11]) to define when a protocol perfectly securely evaluates a function $f : \mathcal{X} \times \mathcal{Y} \to \mathcal{U} \times \mathcal{V}$. A *protocol* is a pair of algorithms $A = (A_1, A_2)$ that can interact by two-way message exchange. A pair $(\overline{A_1}, \overline{A_2})$ of algorithms is *admissible* for protocol A if at least one of the parties is honest, i.e., if $\overline{A_1} = A_1$ or $\overline{A_2} = A_2$ holds. (Note that in the case where both parties are cheaters, no security is required.) By z, we denote some additional auxiliary input that can potentially be used by both parties. For instance, z could include information about previous executions of the protocol. Note, however, that an honest party never makes use of z.

The Ideal Model. In the *ideal* model, the two parties can make use of a trusted party to calculate the function. The algorithms $\overline{B_1}$ and $\overline{B_2}$ of the protocol $\overline{B} = (\overline{B_1}, \overline{B_2})$ receive the inputs x and y, respectively, and the auxiliary input z. They send values x' and y' to the trusted party, who sends them back the values u' and v'—satisfying $(u', v') = f(x', y')$. Finally, $\overline{B_1}$ and $\overline{B_2}$ output the values u and v. The two *honest* algorithms B_1 and B_2 always send $x' = x$ and $y' = y$ to the trusted party, and always output $u = u'$ and $v = v'$. Now, if $\overline{B} = (\overline{B_1}, \overline{B_2})$ is an admissible pair of algorithms for protocol $B = (B_1, B_2)$, the *joint execution of f under $\overline{B}$ in the ideal model*,

$$\mathbf{ideal}_{f,\overline{B}(z)}(x, y) ,$$

is the resulting output pair, given the inputs x and y and the auxiliary input z.

The Real Model. In the *real* model, the parties have to compute f by a protocol $\Pi = (A_1, A_2)$ without the help of a trusted party. Let $\overline{A} = (\overline{A_1}, \overline{A_2})$ be an admissible pair for A. Then the *joint execution of Π under $\overline{A}$ in the real model*,

$$\mathbf{real}_{\Pi,\overline{A}(z)}(x, y) ,$$

is the resulting output pair, given the inputs x and y and the auxiliary input z.

Perfect Security: "Real = Ideal". A protocol Π computes a function f *perfectly securely* if, intuitively speaking, every "real" cheater has an equally powerful counterpart in the ideal model. Definition 2 also applies to *reduction* protocols from one functionality to another; here, the algorithms are allowed to call an oracle which perfectly implements the given functionality.

Definition 2. A protocol Π *computes f perfectly securely* if for every admissible $\overline{A} = (\overline{A_1}, \overline{A_2})$ there exists an admissible $\overline{B} = (\overline{B_1}, \overline{B_2})$—as efficient as $\overline{A}^1$ and with identical set of honest players—such that for all $x \in \mathcal{X}$, $x \in \mathcal{Y}$, and $z \in \mathcal{Z}$,

$$\mathbf{real}_{\Pi,\overline{A}(z)}(x, y) \equiv \mathbf{ideal}_{f,\overline{B}(z)}(x, y)$$

holds, where $\equiv$ means that the distributions are identical.

3 Storing and Reversing Oblivious Transfer

Oblivious transfer protocols rely either on tools borrowed from public-key cryptography [19], [10] or on additional assumptions [3], [6], [20], [4], [23]. In the first case, we have to deal with relatively slow algorithms which may be the bottleneck of the protocol execution. In the second case, one depends on these additional assumptions being present at the time of the execution of the protocol. In both cases it is, therefore, desirable to carry out as much of the computation as possible *in advance*, and to make the actual execution of oblivious transfer as fast and simple as possible, based on this pre-computation. Actually, almost the *entire* computation can be done beforehand: Protocols 1 and 2, proposed in [2], show how $\binom{2}{1}$−OT can be transformed into a so-called *oblivious key*, and *vice versa*.

An *oblivious key* is, intuitively speaking, the distribution that arises when A and B choose their inputs at random and execute $\binom{2}{1}$−OT.

Definition 3. By an *oblivious key*, $\binom{2}{1}$−OK, we denote the primitive where a sample of two random variables $U = (X_0, X_1)$ and $V = (C, Y)$ is given to A and B, respectively, where X_0, X_1, and C are independently and uniformly distributed bits, and where $Y = X_C$ holds.

---

[1] The running time of $\overline{B}$ must by polynomial in the running time of $\overline{A}$.

Note that $\binom{2}{1}$–OK is a *key* for oblivious transfer in very much the same sense as a shared secret bit is an encryption key in the one-time pad.

Protocol 1. $\binom{2}{1}$–OK from $\binom{2}{1}$–OT

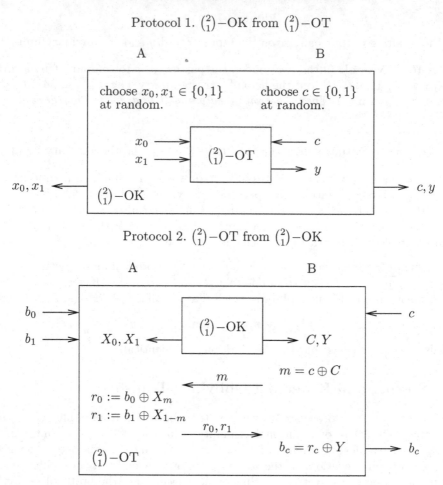

Protocol 2. $\binom{2}{1}$–OT from $\binom{2}{1}$–OK

The proofs that Protocols 1 and 2 are perfect single-copy reductions between the primitives $\binom{2}{1}$–OT and $\binom{2}{1}$–OK are given in [2] (and straight-forward). Note that both Protocols 1 and 2 work in the *honest-but-curious* model, whereas their combination is even perfectly secure in the *malicious* model.

The distribution of $\binom{2}{1}$–OK is given and illustrated on the left hand side of Figure 2.

$$P_{UV}((x_0, x_1), (c, y)) = \begin{cases} 1/8 & \text{if } y = x_c \\ 0 & \text{otherwise .} \end{cases}$$

When the symbols of U and V are renamed in a suitable way, the distribution corresponds to the one arising when Shannon's so-called *"noisy-typewriter chan-nel"* [21] is used with random input (see on the right hand side of Figure 2). Obviously, this distribution is *symmetric*. On the other hand, $\binom{2}{1}$–OK

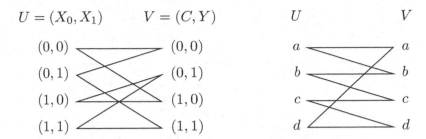

Fig. 2. Left hand side: The distribution of $\binom{2}{1}$–OK. (Each edge is a possible combination with probability 1/8.) Right hand side: The distribution arising from the "noisy-typewriter channel." Obviously, the two distributions are equivalent.

is equivalent to $\binom{2}{1}$–OT, which is, hence, symmetric as well: *A single* instance of $\binom{2}{1}$–TO allows for generating a realization of $\binom{2}{1}$–OT. The reduction is not only single-copy but also *perfect*, i.e., unconditionally secure without any error. This solves an open problem posed in [9] in a very simple way. Lemma 1 shows how the values in the distribution of a $\binom{2}{1}$–OK must be renamed in order for the oblivious key to be reversed.

Lemma 1. *Let X_0, X_1, C, and Y be binary random variables and let $(U, V) = ((X_0, X_1), (C, Y))$ be a $\binom{2}{1}$–OK. Then $((\overline{X_0}, \overline{X_1}), (\overline{C}, \overline{Y})) := ((Y, C \oplus Y), (X_0 \oplus X_1, X_0))$ is a $\binom{2}{1}$–OK as well.*

Proof. $\overline{Y} = X_0 = X_C \oplus C(X_0 \oplus X_1) = Y \oplus C(X_0 \oplus X_1) = \overline{X_0} \oplus (\overline{X_0} \oplus \overline{X_1})\overline{C} = \overline{X_{\overline{C}}}.$

A formal proof of the security of this transformation is omitted here. Intuitively, the privacy of both players is preserved since the ignorance of one player about the XOR of X_0 and X_1 is transformed into the ignorance of C, and *vice versa*.

4   Optimally Reversing Oblivious Transfer

The protocol outlined in the end of Section 3 requires *three* bits of additional communication. We present an even simpler protocol, Protocol 3, using only *one* bit of additional communication from A to B; this is optimal.

Theorem 1. *Protocol 3 perfectly securely reduces $\binom{2}{1}$–OT to one realization of $\binom{2}{1}$–TO.*

Proof. Let first both parties be honest, i.e., $\overline{A} = (A_1, A_2)$ in Protocol 3. Then we have, for all $(b_0, b_1) \in \{0, 1\}^2$, $c \in \{0, 1\}$, and $z \in \mathcal{Z}$,

$$\mathbf{real}_{3, \overline{A}(z)}((b_0, b_1), c) = (\bot, r \oplus (b_0 \oplus a))$$
$$= (\bot, b_0 \oplus (b_0 \oplus b_1)c)$$
$$= (\bot, b_c)$$
$$= \mathbf{ideal}_{\binom{2}{1}-\mathrm{OT}, B(z)}((b_0, b_1), c) \ .$$

Protocol 3. $\binom{2}{1}$–OT from $\binom{2}{1}$–TO

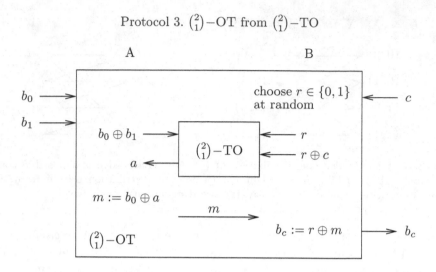

Let now the first party be honest, i.e., $\overline{A} = (A_1, \overline{A_2})$. In the real model, $\overline{A_2}$ receives (c, z) and sends $(a_0, a_1) = a(c, z)$ to $\binom{2}{1}$–TO. Then he receives $m = b_0 \oplus a_{b_0 \oplus b_1}$, and outputs $v(c, z, a_0, a_1, m)$. Let the adversary $\overline{B_2}$ in the ideal model be defined as follows: On inputs (c, z), he sends (c, z) to $\overline{A_2}$, and gets $(\overline{a_0}, \overline{a_1}) = a(c, z)$ back. He sends $\overline{c} := \overline{a_0} \oplus \overline{a_1}$ to $\binom{2}{1}$–OT and gets $b_{\overline{c}}$ back. Then he sends $\overline{m} := b_{\overline{c}} \oplus \overline{a_0}$ to $\overline{A_2}$, gets $\overline{v} = v(c, z, \overline{a_0}, \overline{a_1}, \overline{m})$ back and outputs $\overline{v}$.

Since

$$\overline{m} = \overline{a_0} \oplus b_{\overline{c}} = \overline{a_0} \oplus b_{\overline{a_0} \oplus \overline{a_1}} = b_0 \oplus \overline{a_0} \oplus (b_0 \oplus b_1)(\overline{a_0} \oplus \overline{a_1}) = b_0 \oplus \overline{a_{b_0 \oplus b_1}},$$

we have, for all $(b_0, b_1) \in \{0, 1\}^2$, $c \in \{0, 1\}$, and $z \in \mathcal{Z}$,

$$
\begin{aligned}
\mathbf{real}_{3, \overline{A}(z)}((b_0, b_1), c) &= (\bot, v(c, z, a_0, a_1, m)) \\
&= (\bot, v(c, z, a_0, a_1, b_0 \oplus a_{b_0 \oplus b_1})) \\
&\equiv (\bot, v(c, z, \overline{a_0}, \overline{a_1}, b_0 \oplus \overline{a_{b_0 \oplus b_1}})) \\
&= (\bot, v(c, z, \overline{a_0}, \overline{a_1}, \overline{m})) \\
&= \mathbf{ideal}_{\binom{2}{1} - \mathrm{OT}, (B_1, \overline{B_2})(z)}((b_0, b_1), c) .
\end{aligned}
$$

Assume now that the second party is honest, i.e., $\overline{A} = (\overline{A_1}, A_2)$. In the real model, $\overline{A_1}$ receives $((b_0, b_1), z)$ and sends $d = d((b_0, b_1), z)$ to $\binom{2}{1}$–TO, which returns $l = r \oplus dc$. Then, he sends $m = m((b_0, b_1), z, l)$ to A_2 and outputs $u((b_0, b_1), z, d, l, m)$. Let the adversary $\overline{B_1}$ in the ideal model be defined as follows: On inputs $((b_0, b_1), z)$, $\overline{B_1}$ sends $((b_0, b_1), z)$ to $\overline{A_1}$ and gets $\overline{d} = d((b_0, b_1), z)$ back. He chooses $\overline{l}$ uniformly at random and sends it to $\overline{A_1}$, who sends $\overline{m} = m((b_0, b_1), z, \overline{l})$ and $\overline{u} = u((b_0, b_1), z, \overline{d}, \overline{l}, \overline{m})$ back. He sends $(\overline{l} \oplus \overline{m}, \overline{l} \oplus \overline{m} \oplus d)$ to $\binom{2}{1}$–OT and outputs $\overline{u}$.

The honest player will output $\overline{l} \oplus \overline{m} \oplus c\overline{d}$. Since $l = r \oplus dc$ and r is uniform and independent of everything else, l is uniform and independent as well, which means that it has the same joint distribution as $\bar{l}$ with everything else. Therefore we have, for all $(b_0, b_1) \in \{0,1\}^2$, $c \in \{0,1\}$, and $z \in \mathcal{Z}$,

$$
\begin{aligned}
\mathbf{real}_{3,\overline{A}(z)}((b_0, b_1), c) &= (u((b_0, b_1), z, d, l, m), r \oplus m) \\
&= (u((b_0, b_1), z, d, l, m, l \oplus dc \oplus m) \\
&\equiv (u((b_0, b_1), z, \overline{d}, \overline{l}, \overline{m}, \overline{l} \oplus \overline{d}c \oplus \overline{m}) \\
&= \mathbf{ideal}_{\binom{2}{1}-\mathrm{OT}, \overline{B}(z)}((b_0, b_1), c) .
\end{aligned}
$$

Obviously, the simulated adversary is as efficient as the real adversary.   □

Our protocol is optimal: First of all, since it is impossible to construct unconditionally secure oblivious transfer from scratch, using *a single* instance of $\binom{2}{1}-\mathrm{TO}$ is optimal. Since $\binom{2}{1}-\mathrm{TO}$ does not allow any communication from Bob to Alice, but $\binom{2}{1}-\mathrm{OT}$ does allow one bit of communication, any protocol must communicate at least one bit. Furthermore, there cannot exist a protocol where Bob does not use any randomness, because then his inputs to $\binom{2}{1}-\mathrm{TO}$ would be deterministic functions of c. These functions could not both be constant, since then the output of $\binom{2}{1}-\mathrm{TO}$ would not depend on c and be useless, and therefore no oblivious transfer would be possible. But if the functions are not constant, A is able to obtain information about c.

5   Oblivious Linear-Function Evaluation

In contrast to $\binom{2}{1}-\mathrm{OT}$, all the other forms of oblivious transfer cannot be reversed without loss, i.e., in the perfect single-copy sense of Sections 3 and 4. This can easily be seen from the *monotones*, defined in [24]: A primitive can only be reversed without loss if

$$
H(Y \seardown X | X) = H(X \searchdown Y | Y),
$$

and $\binom{2}{1}-\mathrm{OT}$ is the only example of $\binom{n}{k}-\mathrm{OT}^l$ having this property.

In this section, we present another natural generalization of oblivious transfer to strings that *can* be reversed perfectly: *oblivious linear-function evaluation over $GF(q)$* or $GF(q)-\mathrm{OLFE}$ for short. Roughly speaking, the sender's input is a linear function $f : x \mapsto y = a_0 + a_1 x$, where $a_0, a_1, x, y \in GF(q)$, and the receiver's input is an argument $x \in GF(q)$ for which he then learns the evaluation of the function, $y = f(x)$ (see Figure 3). $GF(q)-\mathrm{OLFE}$ is a special case of oblivious polynomial evaluation [17]. It can easily be verified that $GF(2)-\mathrm{OLFE}$ is equivalent to $\binom{2}{1}-\mathrm{OT}$. Furthermore, [20] shows that with one instance of $GF(q)-\mathrm{OLFE}$ a very simple commitment scheme can be implemented, which allows to commit to a value $x \in GF(q)$. The scheme is perfectly hiding and $1/q$-binding.

The protocols of Sections 3 and 4 generalize to $GF(q)-\mathrm{OLFE}$ in a straightforward way: $GF(q)-\mathrm{OLFE}$ is, as oblivious transfer, equivalent to a non-interactive key—and can, therefore, be stored in the same sense. Moreover, this key is, as $\binom{2}{1}-\mathrm{OK}$, symmetric. Hence, $GF(q)-\mathrm{OLFE}$ from A to B can be reduced to

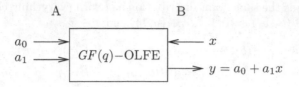

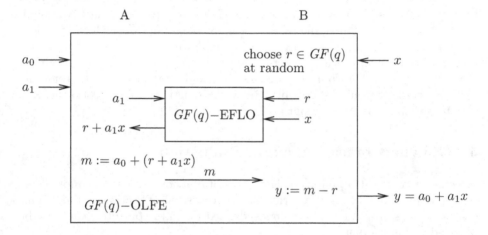

Fig. 3. Oblivious linear-function evaluation over $GF(q)$

$GF(q)$–OLFE from B to A—$GF(q)$–EFLO for short—in a perfect and single-copy sense. Protocol 4 is, in addition, optimal with respect to the required communication.

Protocol 4: $GF(q)$–OLFE from $GF(q)$–EFLO

6   Concluding Remarks

We have shown that chosen one-out-of-two bit oblivious transfer can be optimally reversed *very* easily. Furthermore, we have have presented a more general primitive with the same property: oblivious linear-function evaluation.

Acknowledgments

This work was carried out while both authors were with Université de Montréal, Canada. This research was supported by Canada's NSERC, Québec's FQRNT, and Switzerland's SNF.

References

1. D. Beaver, Foundations of Secure Interactive Computing, *Advances in Cryptology* *—Proceedings of CRYPTO '91*, LNCS, Vol. 576, pp. 377–391, Springer-Verlag, 1992.
2. D. Beaver, Precomputing oblivious transfer, *Advances in Cryptology—Proceedings of CRYPTO '95*, LNCS, Vol. 963, pp. 97–109, Springer-Verlag, 1995.

3. C. H. Bennett, G. Brassard, C. Crépeau, and H. Skubiszewska, Practical quantum oblivious transfer, *Advances in Cryptology—Proceedings of EUROCRYPT '91*, LNCS, Vol. 576, pp. 351–366, Springer-Verlag, 1992.
4. G. Brassard, C. Crépeau, and S. Wolf, Oblivious transfers and privacy amplification, *Journal of Cryptology*, Vol. 16, No. 4, pp. 219–237, 2003.
5. C. Crépeau, *Correct and private reductions among oblivious transfers*, Ph. D. Thesis, Massachusetts Institute of Technology, 1990.
6. C. Crépeau, Efficient cryptographic protocols based on noisy channels, *Advances in Cryptology—Proceedings of CRYPTO '97*, LNCS, Vol. 1233, pp. 306–317, Springer-Verlag, 1997.
7. C. Crépeau and J. Kilian, Achieving oblivious transfer using weakened security assumptions, *Proceedings of the 28th Symposium on Foundations of Computer Science (FOCS '88)*, pp. 42–52, IEEE, 1988.
8. C. Crépeau, K. Morozov, and S. Wolf, Efficient unconditional oblivious transfer from almost any noisy channel, *Proceedings of Fourth Conference on Security in Communication Networks (SCN) '04*, LNCS, Springer-Verlag, 2004.
9. C. Crépeau and M. Sántha, On the reversibility of oblivious transfer, *Advances in Cryptology—Proceedings of EUROCRYPT '91*, LNCS, Vol. 547, pp. 106–113, Springer-Verlag, 1991.
10. S. Even, O. Goldreich, and A. Lempel, A randomized protocol for signing contracts, *Communications of the ACM*, Vol. 28, No. 6, pp. 637–647, 1985.
11. O. Goldreich, *Foundations of Cryptography, Volume II: Basic Applications*, Cambridge University Press, 2004.
12. H. Imai, J. Müller-Quade, A. Nascimento, and A. Winter, Rates for bit commitment and coin tossing from noisy correlation, *Proceedings of the IEEE International Symposium on Information Theory (ISIT) '04*, IEEE, 2004.
13. H. Imai, A. Nascimento, and A. Winter, Oblivious transfer from any genuine noise, *unpublished manuscript*, 2004.
14. J. Kilian, Founding cryptography on oblivious transfer, *Proceedings of the Twentieth Annual ACM Symposium on Theory of Computing (STOC '88)*, pp. 20–31, 1988.
15. D. Mayers, Unconditionally secure quantum bit commitment is impossible, *Phys. Rev. Lett.*, Vol. 78, pp. 3414–3417, 1997.
16. S. Micali and P. Rogaway, Secure computation, *Advances in Cryptology—Proceedings of CRYPTO '91*, LNCS, Vol. 576, pp. 392–404, Springer-Verlag, 1992.
17. M. Naor and B. Pinkas. Oblivious transfer and polynomial evaluation, *Proceedings of the Thirty-First Annual ACM Symposium on Theory of Computing (STOC '99)*, pp. 245-354, 1999.
18. R. Ostrovsky, R. Venkatesan, and M. Yung, Fair games against an all-powerful adversary, *AMS DIMACS Series in Discrete Mathematics and Theoretical Computer Science*, Vol. 13, pp. 155–169, 1990.
19. M. Rabin, How to exchange secrets by oblivious transfer, *Technical Report TR-81, Harvard Aiken Computation Laboratory*, 1981.
20. R. L. Rivest, Unconditionally secure commitment and oblivious transfer schemes using private channels and a trusted initializer, *unpublished manuscript*, 1999.
21. C. E. Shannon, A mathematical theory of communication, *Bell System Technical Journal*, Vol. 27, pp. 379–423, 623–656, 1948.
22. A. Winter, A. Nascimento, and H. Imai, Commitment capacity of discrete memoryless channels, *Cryptography and Coding*, LNCS, Vol. 2898, pp. 35–51, Springer-Verlag, 2003.

23. S. Wolf and J. Wullschleger, Zero-error information and applications in cryptography, *Information Theory Workshop (ITW) '04*, IEEE, 2004.
24. S. Wolf and J. Wullschleger. New monotones and lower bounds in unconditional two-party computation. In *Advances in Cryptology—Proceedings of CRYPTO '05*, LNCS, Vol. 3621, pp. 467–477, Springer-Verlag, 2005.
25. A. D. Wyner, The wire-tap channel, *Bell System Technical Journal*, Vol. 54, No. 8, pp. 1355–1387, 1975.
26. R. W. Yeung, A new outlook on Shannon's information measures, *IEEE Transactions on Information Theory*, Vol. 37, No. 3, pp. 466–474, 1991.

Symplectic Lattice Reduction and NTRU

Nicolas Gama[1], Nick Howgrave-Graham[2], and Phong Q. Nguyen[3]

[1] École normale supérieure, DI, 45 rue d'Ulm, 75005 Paris, France
Nicolas.Gama@ens.fr
[2] NTRU Cryptosystems, Burlington, MA, USA
nhowgravegraham@ntru.com
[3] CNRS/École normale supérieure, DI, 45 rue d'Ulm, 75005 Paris, France
http://www.di.ens.fr/~pnguyen

Abstract. NTRU is a very efficient public-key cryptosystem based on polynomial arithmetic. Its security is related to the hardness of lattice problems in a very special class of lattices. This article is motivated by an interesting peculiar property of NTRU lattices. Namely, we show that NTRU lattices are proportional to the so-called symplectic lattices. This suggests to try to adapt the classical reduction theory to symplectic lattices, from both a mathematical and an algorithmic point of view. As a first step, we show that orthogonalization techniques (Cholesky, Gram-Schmidt, QR factorization, etc.) which are at the heart of all reduction algorithms known, are all compatible with symplecticity, and that they can be significantly sped up for symplectic matrices. Surprisingly, by doing so, we also discover a new integer Gram-Schmidt algorithm, which is faster than the usual algorithm for all matrices. Finally, we study symplectic variants of the celebrated LLL reduction algorithm, and obtain interesting speed ups.

1 Introduction

The NTRU cryptosystem [12] is one of the fastest public-key cryptosystems known, offering both encryption (under the name NTRUENCRYPT) and digital signatures (under the name NTRUSIGN [11]). Besides efficiency, another interesting feature of NTRU compared to traditional public-key cryptosystems based on factoring or discrete logarithm is its potential resistance to quantum computers: no efficient quantum algorithm is known for NP-hard lattice problems. The security and insecurity of NTRU primitives has been a popular research topic in the past 10 years, and NTRU is now being considered by the *IEEE P1363.1* standards [16].

The security of NTRU is based on the hardness of two famous lattice problems, namely the shortest and closest vector problems (see for instance the survey [23]), in a very particular class of lattices called convolution modular lattices by [20]. More precisely, it was noticed by the authors of NTRU and by Coppersmith and Shamir [6] that ideal lattice reduction algorithms could heuristically recover NTRU's secret key from the public key. This does not necessarily imply that NTRU is insecure, since there is a theoretical and experimental gap between

S. Vaudenay (Ed.): EUROCRYPT 2006, LNCS 4004, pp. 233–253, 2006.

existing reduction algorithms (such as LLL [18] or its block improvements by Schnorr [25]) and ideal lattice reduction (which is assumed to be solving NP-hard lattice problems), while NTRU is so far the only lattice-based cryptosystem known that can cope with high dimensions without sacrificing performances. Nor does it mean that the security of NTRU primitives is strictly equivalent to the hardness of lattice problems. In fact, the main attacks on NTRU primitives have bypassed the hard lattice problems: this was notably the case for the decryption failure attacks [15] on NTRUEncrypt, the attacks [7,8] on the ancestor NSS [13] of NTRUSign [11], as well as the recent attack [21] on NTRUSign [11] without perturbation. Almost ten years after the introduction of NTRU [12], no significant weakness on NTRU lattices has been found, despite the very particular shape of NTRU lattice bases: both the public and secret NTRU bases are $2N \times 2N$ matrices formed by four blocks of $N \times N$ circulant matrices. It is this compact representation that makes NTRU much more efficient than other lattice-based or knapsack-based schemes (see the survey [23]). A fundamental open question is whether this particular shape makes NTRU lattices easier to reduce or not.

Our Results. We propose to exploit the structure of NTRU lattices in lattice reduction algorithms. As a starting point, we observe a peculiar property of NTRU lattices: we show that NTRU lattices are proportional to the so-called *symplectic* lattices (see the survey [2]). As their name suggests, symplectic lattices are related to the classical symplectic group [28]: a lattice is said to be symplectic if it has at least one basis whose Gram matrix is symplectic, which can only occur in even dimension. Such lattices are *isodual*: there exists an isometry of the lattice onto its dual. Interestingly, most of the well-known lattices in low even dimension are proportional to symplectic lattices, *e.g.* the roots lattices $\mathbb{A}_2$, $\mathbb{D}_4$ and $\mathbb{E}_8$, the Barnes lattice P_6, the Coxeter-Todd lattice K_{12}, the Barnes-Wall lattice BW_{16} and the Leech lattice Λ_{24} (see the bible of lattices [5]). Besides, there is a one-to-one correspondence between symplectic lattices and principally polarized complex Abelian varieties, among which Jacobians form an interesting case (see [3]). This has motivated the study of symplectic lattices in geometry of numbers.

However, to our knowledge, symplectic lattices have never been studied in reduction theory. The long-term goal of this paper is to explore the novel concept of symplectic lattice reduction in which the classical reduction theory is adapted to symplectic lattices, from both a mathematical and an algorithmic point of view in order to speed up reduction algorithms. As a first step, we show that the Gram-Schmidt orthogonalization process – which is at the heart of all lattice reduction algorithms known – preserves symplecticity, and that is made possible by a slight yet essential change on the classical definition of a symplectic matrix, which is fortunately compatible with the standard theory of the symplectic group. We then exploit this property to speed up its computation for symplectic lattices. In doing so, we actually develop a new and faster method to compute integral Gram-Schmidt, which is applicable to all matrices, and not just symplectic matrices. The method is based on duality: it is faster than the classical method, because it significantly reduces the number of long-integer divisions. When applied to symplectic matrices, a further speed up is possible thanks to the links between symplecticity and duality: in practice, the method then becomes roughly 30 times faster than the classical GS method, which is roughly the time it would take on a matrix of halved dimension. Finally, we study symplectic versions of the celebrated LLL lattice basis reduction algorithm [18] and obtain a speedup of 6 for NTRU lattices of standard size. We restrict to the so-called integral version of LLL to facilitate comparisons: it might be difficult to compare two floating-point variants with different stability properties. We leave the cases of floating-point variants [22] and improved reduction algorithms [25] to future work, but the present work seems to suggest that reduction algorithms might be optimized to NTRU lattices in such a way that a $2n$-dimensional NTRU lattice would not take more time to reduce than an αn-dimensional lattice for some $\alpha < 2$. This is the case for Gram-Schmidt orthogonalization and LLL.

Related Work. Incidentally, the compatibility of the symplectic group with respect to standard matrix factorizations has recently been studied in [19]: however, because they rely on the classical definition of a symplectic matrix, they fail to obtain compatibility with Gram-Schmidt orthogonalization or the QR decomposition.

Road Map. The paper is organized as follows. In Section 2, we provide necessary background on lattice reduction and the symplectic group. In Section 3, we explain the relationship between NTRU lattices and symplecticity. In Section 4, we show that the Gram-Schmidt orthogonalization process central to all lattice reduction algorithms known is fully compatible with symplecticity. In Section 5, we present a new integral Gram-Schmidt algorithm, which leads to significant speed-ups for symplectic matrices. The final section 6 deals with symplectic variants of integral LLL.

2   Background

Let $\|.\|$ and $\langle.,.\rangle$ be the Euclidean norm and inner product of $\mathbb{R}^n$. Vectors will be written in bold, and we will use row-representation for matrices. The notations $\mathcal{M}_n(\mathbb{R})$ represents the $n \times n$-dimensional matrices over $\mathbb{R}$, and $GL_n(\mathbb{R})$ the n-dimensional invertible matrices of $\mathcal{M}_n(\mathbb{R})$. For a matrix M whose name is a capital letter, we will usually denote its coefficients by $m_{i,j}$: if the name is a Greek letter like μ, we will keep the same symbol for both the matrix and its coefficients. The matrix norm $|M|$ represents the maximum of the Euclidean norms of the rows of M. The notation $\lceil x \rfloor$ denotes a closest integer to x.

2.1   Lattices

We refer to the survey [23] for a bibliography on lattices. In this paper, by the term lattice, we mean a discrete subgroup of $\mathbb{R}^n$. The simplest lattice is $\mathbb{Z}^n$. It turns out that in any lattice L, not just $\mathbb{Z}^n$, there must exist linearly independent vectors $\mathbf{b}_1, \ldots, \mathbf{b}_d \in L$ such that:

$$L = \left\{ \sum_{i=1}^{d} n_i \mathbf{b}_i \mid n_i \in \mathbb{Z} \right\}.$$

Any such d-tuple of vectors $\mathbf{b}_1, \ldots, \mathbf{b}_d$ is called a *basis* of L: a lattice can be represented by a basis, that is, a row matrix. Two lattice bases are related to one another by some matrix in $GL_d(\mathbb{Z})$. The *dimension* of a lattice L is the dimension d of the linear span of L. Let $[\mathbf{b}_1, \ldots, \mathbf{b}_d]$ be vectors: the lattice is full-rank if $d = n$, which is the usual case. We denote by $G(\mathbf{b}_1, \ldots, \mathbf{b}_d)$ their *Gram matrix*, that is the $d \times d$ symmetric matrix $(\langle \mathbf{b}_i, \mathbf{b}_j \rangle)_{1 \le i,j \le d}$ formed by all the inner products. The *volume* vol(L) (or *determinant*) of the lattice L is the square root of the determinant of the Gram matrix of any basis of L: here, the Gram matrix is symmetric definite positive. The dual lattice $L^\times$ of a lattice L is:

$$L^\times = \{ \mathbf{v} \in \text{span}L, \ \forall \mathbf{u} \in L, \ \langle \mathbf{u}, \mathbf{v} \rangle \in \mathbb{Z} \}.$$

They have the same dimension and their volumes satisfy vol(L) $\cdot$ vol($L^\times$) = 1. If $B = [\mathbf{b}_1, ..., \mathbf{b}_d]$ is a basis of L, and $\delta_{i,j}$ the Kronecker symbol, then the dual family $B^\times = [\mathbf{b}_1^\times, ..., \mathbf{b}_d^\times]$ with $\mathbf{b}_i^\times \in \text{span}(L)$ satisfying $\langle \mathbf{b}_i^\times, \mathbf{b}_j \rangle = \delta_{i,j}$ is a basis of $L^\times$ called the dual basis of B. The Gram matrices of B and $B^\times$ are inverse of each other, and when L is a full rank lattice, $B^\times = B^{-t}$.

2.2   The Symplectic Group

The symplectic group is one of the classical groups [28], whose name is due to Weyl. Given four matrices $A, B, C, D \in \mathcal{M}_n(\mathbb{R})$ we denote by $Q[A, B, C, D]$ the $(2n) \times (2n)$ matrix with A, B, C, D as its quadrants:

$$Q[A, B, C, D] = \begin{pmatrix} A & B \\ C & D \end{pmatrix}.$$

Symplectic matrices are matrices preserving a nondegenerate antisymmetric bilinear form. Let σ be an isometry of $\mathbb{R}^n$. Then $\sigma^2 = -1$ if and only if there exists an orthonormal basis of $\mathbb{R}^n$ over which the matrix of σ is $J_{2n} = Q[0, I_n, -I_n, 0]$, where I_n is the $n \times n$ identity matrix. Usually, a matrix $M \in \mathcal{M}_{2n}(\mathbb{R})$ is said to be *symplectic* if and only if

$$M^t J_{2n} M = J_{2n}. \tag{1}$$

where M^t is the transpose of M. This is equivalent to M being invertible and having inverse equal to:

$$M^{-1} = -J_{2n} M^t J_{2n}. \tag{2}$$

The set of such matrices is denoted by $Sp(2n, \mathbb{R})$, which is a subgroup of the special linear group $SL_{2n}(\mathbb{R})$: a symplectic matrix has determinant +1. A matrix is symplectic if and only if its transpose is symplectic. The matrix $Q[A, B, C, D]$ is symplectic if and only if $AD^t - BC^t = I_n$ and both the matrices AB^t and CD^t are symmetric. It follows that a triangular matrix $Q[A, 0, C, D]$ may only be symplectic if A and D are diagonal, which is too restrictive to make the symplectic group fully compatible with standard matrix factorizations involving triangular matrices.

To fix this, we consider a variant of the usual symplectic group obtained by equation (1) with another matrix J_{2n}. Fortunately, this is allowed by the theory, as while as J_{2n} is a nonsingular, skew-symmetric matrix. From now on, we thus let $J_{2n} = Q[0, R_n, -R_n, 0]$ where R_n is the reversed identity matrix: the identity where the rows (or the columns) are in reverse order, that is, the (i, j)-th coefficient is the Kronecker symbol $\delta_{i,n+1-j}$. This new matrix J_{2n} still satisfies $J_{2n}^2 = -I_{2n}$, and is therefore compatible with symplecticity. From now on, by a symplectic matrix, we will mean a matrix satisfying equation (1) or (2) with this choice of J_{2n}, and this will be our symplectic group $Sp(2n, \mathbb{R})$. Now, a matrix $Q[A, B, C, D]$ is symplectic if and only if the following conditions hold:

$$BA^s = AB^s, \ DC^s = CD^s, AD^s - BC^s = R_n \qquad (3)$$

where $M^s = R_n M^t R_n$ for any $M \in \mathcal{M}_n(\mathbb{R})$, which corresponds to reflecting the entries in the off-diagonal $m_{i,j} \leftrightarrow m_{n+1-j,n+1-i}$. The matrix $R_n M$ reverses the rows of M, while $M R_n$ reverses the columns. In other words, compared to the usual definition of symplectic matrices, we have replaced the transpose operation M^t and I_n by respectively the reflection M^s and R_n. This will be the general rule to switch from the usual symplectic group to our variant. In fact, it can be checked that the reflection $M \mapsto M^s$ is an involution of $Sp(2n, \mathbb{R})$: M^s is symplectic (though R_n is not symplectic), and $(M^s)^s = M$. Now a triangular matrix $Q[A, 0, C, D]$ may be symplectic without requiring A and D to be diagonal. Naturally, M^{-s} will mean the inverse of M^s.

To conclude this subsection, let us give a few examples of symplectic matrices with our own definition of $Sp(2n, \mathbb{R})$, which will be very useful in the rest of the paper:

- Any element of $SL_2(\mathbb{R})$, that is, any 2x2 matrix with determinant 1.
- A diagonal matrix of $\mathcal{M}_{2n}(\mathbb{R})$ with coefficients $d_1, ..., d_{2n}$ is symplectic if and only if $d_i = 1/d_{2n+1-i}$ for all i.
- Any $\begin{bmatrix} A & 0 & B \\ 0 & M & 0 \\ C & 0 & D \end{bmatrix}$ including $\begin{bmatrix} I & 0 & 0 \\ 0 & M & 0 \\ 0 & 0 & I \end{bmatrix}$ where $\begin{cases} M \in Sp(2n, \mathbb{R}) \\ Q[A, B, C, D] \in Sp(2m, \mathbb{R}) \end{cases}$.
- $Q[U, 0, 0, U^{-s}]$ for any invertible matrix $U \in GL_n(\mathbb{R})$.
- $Q[I_n, 0, A, I_n]$ for any $A \in \mathcal{M}_n(\mathbb{R})$ such that $A = A^s$, that is, A is *reversed-symmetric*.

The symplecticity can be checked by equations (1), (2) or (3). In particular, these equations prove the following elementary lemma, which gives the structure of symplectic triangular matrices.

Lemma 1. *A lower-triangular $2n$-dimensional matrix L can always be decomposed as follows:*

$$L = \begin{bmatrix} \alpha & 0 & 0 \\ \mathbf{u}^t & M & 0 \\ \beta & \mathbf{v} & \gamma \end{bmatrix} \quad where \quad \begin{cases} \alpha, \beta, \gamma \in \mathbb{R} \\ \mathbf{u}, \mathbf{v} \in \mathbb{R}^{2n-2} \\ M \in \mathcal{M}_{2n-2}(\mathbb{R}) \ is \ triangular \end{cases}.$$

Then the matrix L is symplectic if and only if M is symplectic (and triangular), $\gamma = \frac{1}{\alpha}$ and $\mathbf{u} = -\alpha \mathbf{v} J_{2n-2} M^t$.

2.3 Symplectic Lattices

A lattice L is said to be *isodual* if there exists an isometry σ of L onto its dual (see the survey [2]). One particular case of isodualities is when $\sigma^2 = -1$, in which case the lattice is called "*symplectic*". Then there exists an orthogonal basis of $\mathrm{span}(L)$ over which the matrix of σ: is J_{2n} there is at least a basis of L whose Gram matrix is symplectic.

A symplectic lattice has volume equal to 1. In this paper, we will say that an integer full-rank lattice $L \in \mathbb{Z}^{2n}$ is *q-symplectic* if the lattice $L/\sqrt{q}$ is symplectic where $q \in \mathbb{N}^*$. Its volume is equal to q^n. Our q-symplectic lattices seem to be a particular case of the modular lattices introduced by Quebbemann [24], which are connected to modular forms.

2.4 Orthogonalization

Cholesky. Let $G \in \mathcal{M}_n(\mathbb{R})$ be symmetric definite positive. There exists a unique lower triangular matrix $L \in \mathcal{M}_n(\mathbb{R})$ with strictly positive diagonal such that $G = LL^t$. The matrix L is the *Cholesky factorization* of G, and its Gram matrix is G.

The $\mu D \mu^t$ Factorization. This factorization is the analogue of the so-called "LDL decomposition" in [9, Chapter 4.1]. Let $G \in \mathcal{M}_n(\mathbb{R})$ be symmetric definite. There exists a unique lower triangular matrix $\mu \in \mathcal{M}_n(\mathbb{R})$ with unit diagonal and a unique diagonal matrix $D \in \mathcal{M}_n(\mathbb{R})$ such that $G = \mu D \mu^t$. The pair (μ, D) is the *$\mu D \mu^t$ factorization* of G. When G is positive definite, then D is positive diagonal, and the relation between the $\mu D \mu^t$ and Cholesky factorizations of G is $L = \mu \sqrt{D}$.

QR or LQ. Let $M \in GL_n(\mathbb{R})$. There exists a unique pair $(Q, R) \in \mathcal{M}_n(\mathbb{R})^2$ such that $M = QR$, where Q is unitary and R is upper triangular with strictly positive diagonal. This is the standard QR factorization. Since we deal with row matrices, we prefer to have a lower triangular matrix, which can easily be achieved by transposition. It follows that there also exists a unique pair $(L, Q) \in \mathcal{M}_n(\mathbb{R})^2$ such that $M = LQ$, where Q is unitary and L is lower triangular with strictly positive diagonal. Note that L is the Cholesky factorization of the Gram matrix MM^t of M.

Gram-Schmidt. Let $[\mathbf{b}_1, \ldots, \mathbf{b}_d]$ be linearly independent vectors represented by the $d \times n$ matrix B. Their *Gram-Schmidt orthogonalization* (GSO) is the orthogonal family $[\mathbf{b}_1^*, \ldots, \mathbf{b}_d^*]$ defined recursively as follows: $\mathbf{b}_1^* = \mathbf{b}_1$ and $\mathbf{b}_i^*$ is the component of $\mathbf{b}_i$ orthogonal to the subspace spanned by $\mathbf{b}_1, \ldots, \mathbf{b}_{i-1}$. We have $\mathbf{b}_i^* = \mathbf{b}_i - \sum_{j=1}^{i-1} \mu_{i,j} \mathbf{b}_j^*$ where $\mu_{i,j} = \langle \mathbf{b}_i, \mathbf{b}_j^* \rangle / \|\mathbf{b}_j^*\|^2$ for all $i < j$. We let $\mu \in \mathcal{M}_d(\mathbb{R})$ be the lower triangular matrix whose coefficients are $\mu_{i,j}$ above the diagonal, and 1 on the diagonal. If B^* is the $d \times n$ row matrix representing $[\mathbf{b}_1^*, \ldots, \mathbf{b}_d^*]$, then $B = \mu B^*$. If we let G be the Gram matrix BB^t of B, then

μ is exactly the matrix from the $\mu D \mu^t$ decomposition of G, and its Cholesky factorization $L = (\ell_{i,j})$ is related to the GSO by: $\ell_{i,j} = \mu_{i,j} \|\mathbf{b}_j^*\|$ for $i < j$. The matrices L and B have the same Gram matrix, so the GSO can be viewed as a trigonalization of the lattice Λ spanned by B. Note that $\mathrm{vol}(\Lambda) = \prod_{i=1}^{d} \|\mathbf{b}_i^*\|$.

Integral Gram-Schmidt. In practice, we are interested in the case where the $\mathbf{b}_i$'s are in $\mathbb{Z}^n$. Then the $\mathbf{b}_i^*$'s and the $\mu_{i,j}$'s are in general rational. To avoid rational arithmetic, it is customary to use the following integral quantities (as in [27] and in the complexity analysis of [18]): for all $1 \leq i \leq d$, let: $\lambda_{i,i} = \prod_{j=1}^{i} \|\mathbf{b}_j^*\|^2 = \mathrm{vol}(\mathbf{b}_1, \ldots, \mathbf{b}_i)^2 \in \mathbb{Z}$. Then let $\lambda_{i,j} = \mu_{i,j} \lambda_{j,j}$ for all $j < i$, so that $\mu_{i,j} = \frac{\lambda_{i,j}}{\lambda_{j,j}}$. It is known that $\lambda_{i,j} \in \mathbb{Z}$. When using the GSO for lattice reduction, one does not need to compute the $\mathbf{b}_i^*$'s themselves: one only needs to compute the $\mu_{i,j}$'s and the $\|\mathbf{b}_i^*\|^2$. Since $\|\mathbf{b}_i^*\|^2 = \lambda_{i,i}/\lambda_{i-1,i-1}$ (if we let $\lambda_{0,0} = 1$), it follows that it suffices to compute the integral matrix $\lambda = (\lambda_{i,j})_{1 \leq i,j \leq d}$ for lattice reduction purposes. This is done by Algorithm 1, whose running time is $O(nd^4 \log^2 |B|)$ where $|B|$ is an upper bound of the $\|\mathbf{b}_i\|$'s.

Algorithm 1. Standard GS

Input: A set of d linearly independent vectors $[\mathbf{b_1}, ..., \mathbf{b_d}]$ of $\mathbb{Z}^n$
Output: The λ matrix of the GSO of $[\mathbf{b_1}, ..., \mathbf{b_d}]$.
1: **for** $i = 1$ to d **do**
2:     $\lambda_{i,1} \leftarrow \langle \mathbf{b}_i, \mathbf{b}_1 \rangle$
3:     **for** $j = 2$ to i **do**
4:         $S = \lambda_{i,1} \lambda_{j,1}$
5:         **for** $k = 2$ to $j - 1$ **do**
6:             $S \leftarrow (\lambda_{k,k} S + \lambda_{j,k} \lambda_{i,k})/\lambda_{k-1,k-1}$
7:         **end for**
8:         $\lambda_{i,j} \leftarrow \langle \mathbf{b}_i, \mathbf{b}_j \rangle \lambda_{j-1,j-1} - S$
9:     **end for**
10: **end for**

2.5   LLL Reduction

Size Reduction. A basis $[\mathbf{b}_1, \ldots, \mathbf{b}_d]$ is *size-reduced* with factor $\eta \geq 1/2$ if its GSO family satisfies $|\mu_{i,j}| \leq \eta$ for all $j < i$. An individual vector $\mathbf{b}_i$ is size-reduced if $|\mu_{i,j}| \leq \eta$ for all $j < i$. Size reduction usually refers to $\eta = 1/2$, and is typically achieved by successively size-reducing individual vectors.

LLL Reduction. A basis $[\mathbf{b}_1, \ldots, \mathbf{b}_d]$ is LLL-reduced [18] with factor (δ, η) for $1/4 < \delta \leq 1$ and $1/2 \leq \eta < \sqrt{\delta}$ if the basis is size-reduced with factor η and if its GSO satisfies the $(d - 1)$ Lovász conditions $(\delta - \mu_{i,i-1}^2) \|\mathbf{b}_{i-1}^*\|^2 \leq \|\mathbf{b}_i^*\|^2$, which means that the GSO vectors never drop too much. Such bases have several useful properties (see [4, 18]), notably the following one: the first basis vector is relatively short, namely

$$\|\mathbf{b}_1\| \leq \beta^{(d-1)/4} \mathrm{vol}(L)^{1/d}, \text{ where } \beta = 1/(\delta - \eta^2).$$

LLL-reduction usually refers to the factor $(3/4, 1/2)$ because this was the choice considered in the original paper by Lenstra, Lenstra and Lovász [18]. But the closer δ and η are respectively to 1 and $1/2$, the more reduced the basis is. The classical LLL algorithm obtains in polynomial time a basis reduced with factor $(\delta, 1/2)$ where δ can be arbitrarily close to 1. Reduction with a factor $(1, 1/2)$ is closely related to a reduction notion introduced by Hermite [10].

The LLL Algorithm. The basic LLL algorithm [18] computes an LLL-reduced basis in an iterative fashion: there is an index κ such that at any stage of the algorithm, the truncated basis $[\mathbf{b}_1, \ldots, \mathbf{b}_{\kappa-1}]$ is LLL-reduced. At each loop iteration, κ is either incremented or decremented: the loop stops when κ eventually reaches the value $d + 1$, in which case the entire basis $[\mathbf{b}_1, \ldots, \mathbf{b}_d]$ is already LLL-reduced.

LLL uses two kinds of operations: swaps of consecutive vectors and Babai's nearest plane algorithm [1], which performs at most d translations of the form $\mathbf{b}_\kappa \leftarrow \mathbf{b}_\kappa - m\mathbf{b}_i$, where m is some integer and $i < \kappa$. Swaps are used to achieve Lovász conditions, while Babai's algorithm is used to size-reduce vectors.

If L is a full-rank lattice of dimension n and $|B|$ is an upper bound on the $\|\mathbf{b}_i\|$'s, then the complexity of the LLL algorithm (using integral Gram-Schmidt) without fast integer arithmetic is $O(n^6 \log^3 |B|)$. The recent L^2 algorithm [22] (based on floating-point Gram-Schmidt) by Nguyen and Stehlé achieves a factor of (δ, ν) arbitrarily close to $(1, 1/2)$ in faster polynomial time: the complexity is $O(n^5(n + \log |B|) \log |B|)$ which is essentially $O(n^5 \log^2 |B|)$ for large entries. This is the fastest LLL-type reduction algorithm known.

3   NTRU Lattices

The NTRU [12] cryptosystem has many variants. To simplify our exposition, we focus on the usual version, but our results apply to all known variants of NTRU.

Let n be a prime number about several hundreds (e.g.251), and q be a small power of two (e.g.128 or 256). Let $\mathcal{R}$ be the ring $\mathbb{Z}[X]/(X^n - 1)$ whose multiplication is denoted by $*$. The NTRU secret key is a pair of polynomials $(f, g) \in \mathcal{R}^2$ with tiny coefficients compared to q, say 0 and 1. The polynomial f is chosen to be invertible modulo q, so that the polynomial $h = g/f \bmod q$ is well-defined in $\mathcal{R}$. The NTRU public key is the polynomial $h \in \mathcal{R}$ with coefficients modulo q. Its fundamental property is: $f * h \equiv g \bmod q$ in $\mathcal{R}$.

In order to fit multiplicative properties of polynomials of $\mathcal{R}$, we use circulant matrices. The application φ that maps a polynomial in $\mathcal{R}$ to its circulant matrix in $\mathcal{M}_n(\mathbb{Z})$ is defined by:

$$\varphi\left(\sum_{i=0}^{n-1} h_i X^i\right) = \begin{bmatrix} h_0 & h_1 & \cdots & h_{n-1} \\ h_{n-1} & h_0 & \cdots & h_{n-2} \\ \vdots & \ddots & \ddots & \vdots \\ h_1 & \cdots & h_{n-1} & h_0 \end{bmatrix}$$

1. This application is a ring morphism.
2. Circulant matrices are reversed-symmetric: $\varphi(a)^s = \varphi(a)$ for any $a \in \mathcal{R}$.

There is a natural lattice Λ in $\mathbb{Z}^{2n}$ corresponding to the set of pairs of polynomials $(u, v) \in \mathcal{R}^2$ such that $v * h \equiv u \bmod q$ (see [6, 23]). This lattice can be defined by the following basis, which is public since it can be derived from the public key:

$$B = Q\left[\varphi(q), \varphi(0), \varphi(h), \varphi(1)\right].$$

This basis is in fact the Hermite normal form of Λ. It follows that the dimension of Λ is $2n$ and its volume is q^n. Notice that $B/\sqrt{q}$ is symplectic by equation (1), and therefore the public basis B and the NTRU lattice Λ are q-symplectic.

Because of the fundamental property of the public key h, there is a special lattice vector in Λ corresponding to (g, f), which is heuristically the shortest lattice vector. All the vectors corresponding to the rotations $(g * X^k, f * X^k)$ also belong to Λ. In fact, in NTRUSIGN [11], the pair (f, g) is selected in such a way that there exists another pair of polynomials $(F, G) \in \mathcal{R}^2$ such that $f * G - g * F = q$ in $\mathcal{R}$. It follows that the following matrix is a secret basis of Λ:

$$C = Q\left[\varphi(g), \varphi(f), \varphi(G), \varphi(F)\right].$$

This is the basis used to sign messages in NTRUSIGN.

Hence, if a, b, c, d are polynomials in $\mathcal{R}$, the matrix $M = Q[\varphi(a), \varphi(b), \varphi(c), \varphi(d)]$ satisfies:

$$-MJ_{2n}M^t J_{2n} = Q[\varphi(a * d - b * c), 0, 0, \varphi(a * d - b * c)].$$

In particular, the secret basis satisfies: $-CJ_{2n}C^t J_{2n} = qI_{2n}$, which proves that C is a q-symplectic matrix like B, only with smaller coefficients. Hence, the unimodular transformation that transforms the public basis B into the secret basis C is symplectic. One may wonder if it is possible to design special (possibly faster) lattice reduction algorithms for NTRU lattices, which would restrict their elementary row transformations to the symplectic subgroup of $GL_{2n}(\mathbb{Z})$. This motivates the study of symplectic lattice reduction.

4   Symplectic Orthogonalization

All lattice reduction algorithms known are based on Gram-Schmidt orthogonalization, which we recalled in Section 2. In this section, we show that Cholesky factorization, LQ decomposition and Gram-Schmidt orthogonalization are compatible with symplecticity. The key result of this section is the following symplectic analogue of the so-called LDL^t factorization of a symmetric matrix:

Theorem 1 (Symplectic $\mu D\mu^t$). *Let G be a symmetric matrix in $Sp(2n, \mathbb{R})$. There exists a lower-triangular matrix $\mu \in Sp(2n, \mathbb{R})$ whose diagonal is 1, and a diagonal matrix $D \in Sp(2n, \mathbb{R})$ such that, $G = \mu D\mu^t$. And the pair (μ, D) is unique.*

Before proving this theorem, let us give three important consequences on Cholesky factorization, LQ factorization and Gram-Schmidt orthogonalization:

Theorem 2 (Symplectic Cholesky). *If $G \in S_p(2n, \mathbb{R})$ is a symmetric positive definite matrix, then its Cholesky factorization is symplectic.*

Proof. Apply Theorem 1 to G, then μ is lower-triangular symplectic with only 1 on the diagonal. Since G is positive definite, the diagonal matrix $D = \mu^{-1}G\mu^{-t}$ is positive definite. But D is also symplectic, so its coefficients satisfy $d_{i,i} = 1/d_{2n+1-i,2n+1-i}$ (see the end of Section 2.2). For these reasons, the square root C of D (with $c_{i,i} = \sqrt{d_{i,i}}$) is also symplectic. It is clear that $L = \mu C$ is symplectic and satisfies $G = LL^t$. Since the Cholesky factorisation of G is unique, it must be L and it is therefore symplectic.   □

Theorem 3 (Symplectic LQ). *If B is symplectic, then its LQ decomposition is such that both L and Q are symplectic.*

Proof. L is the Cholesky factorization of the matrix BB^t, which is symplectic, so the previous theorem shows that L is symplectic. Then $Q = L^{-1}B$ is also symplectic, because $Sp(2n, \mathbb{R})$ is a group.   □

Theorem 4 (Symplectic Gram-Schmidt). *If B is symplectic, then the μ matrix of its Gram-Schmidt orthogonalisation is also symplectic.*

Proof. Apply Theorem 1 to $G = BB^t$, then μB represents an orthogonal basis, because its Gram matrix is diagonal.   □

Thus, the isometry σ represented by J_{2n} that sends the symplectic basis onto its dual basis is also an isometry between each part of the GSO of the symplectic basis and its dual basis:

Corollary 1. *Let $[\mathbf{b}_1, ..., \mathbf{b}_{2n}]$ be a symplectic basis of a $2n$-dimensional lattice, then the GSO satisfy for all $i \leq n$, $\mathbf{b}_{2n+1-i}^* = \frac{1}{\|\mathbf{b}_i^*\|^2}\mathbf{b}_i^* J$ and $\mathbf{b}_i^* = \frac{1}{\|\mathbf{b}_{2n+1-i}^*\|^2}\mathbf{b}_{2n+1-i}^* J$.*

Proof. Consider the LQ factorization of $[\mathbf{b}_1, ..., \mathbf{b}_{2n}]$. The unitary matrix Q is symplectic, therefore equation (2) implies that $Q = -JQJ$. Hence, a unitary symplectic matrix always has the form:

$$\begin{pmatrix} C & D \\ -R_n D R_n & R_n C R_n \end{pmatrix} = \begin{pmatrix} A \\ R_n A J_{2n} \end{pmatrix}.$$

This proves that the directions of the $\mathbf{b}_i^*$ in this corollary are correct. Their norm are the diagonal coefficients of L, and this matrix is lower-triangular symplectic, so Lemma 1 implies that $\|\mathbf{b}_{2n+1-i}^*\| = 1/\|\mathbf{b}_i^*\|$.   □

We now prove Theorem 1 by induction over n. There is nothing to prove for $n = 0$. Thus, assume that the result holds for $n - 1$ and let $G = (g_{i,j})$ be a symmetric matrix in $Sp(2n, \mathbb{R})$. The main idea is to reduce the first column G

with a symplectic transformation, then verify that it automatically reduces the last row, and finally use the induction hypothesis to reduce the remaining $2n - 2$ dimensional block in the center. The symplectic transformation has the form:

$$P = \begin{bmatrix} 1 & 0 & 0 \\ (\alpha_2, ..., \alpha_{2n-1})^t & I_{2n-2} & 0 \\ \alpha_{2n} & (\alpha_2, ..., \alpha_{2n-1})J_{2n-2} & 1 \end{bmatrix} \text{ where } \alpha_2, ..., \alpha_{2n-1} \in \mathbb{R}.$$

This is a symplectic matrix because of Lemma 1. Apply the transformation with $\alpha_i = -\frac{g_{i,1}}{g_{1,1}}$. Then PGP^t has the following shape:

$$PGP^t = \begin{bmatrix} g_{1,1} & 0 & \gamma \\ 0 & S & \mathbf{u}^T \\ \gamma & \mathbf{u} & \beta \end{bmatrix},$$

where S is a $(2n - 2) \times (2n - 2)$ symmetric matrix, $\mathbf{u}$ is a $(2n - 2)$ dimensional row vector, and $\beta \in \mathbb{R}$ and $\gamma = 0$. The coefficient γ in the bottom left corner of PGP^t is equal to zero, because α_{2n} satisfies $\alpha_{2n}g_{1,1}+\alpha_{2n-1}g_{2,1}+...+\alpha_{n+1}g_{n,1}-\alpha_n g_{n+1,1} - ... - \alpha_2 g_{2n-1,1} + g_{2n,1} = 0$.

Since PGP^t is symplectic, the image by J_{2n} of the first row $\mathbf{r}_1$ of PGP^t has the form $\mathbf{e}_{2n} = (0, ..., 0, g_{1,1})$ and its jth row $\mathbf{r}_j$ satisfies $\langle \mathbf{e}_{2n}, \mathbf{r}_j \rangle = \delta_{2n,j}$ for all $j \geq 2$ (where δ is the Kronecker symbol): in other words, $\mathbf{u} = 0$ and $\beta = 1/g_{11}$.

$$PGP^t = \begin{bmatrix} g_{1,1} & 0 & 0 \\ 0 & S & 0 \\ 0 & 0 & \frac{1}{g_{1,1}} \end{bmatrix}.$$

As a result, S is symmetric positive definite and symplectic. The induction hypothesis implies the existence of a pair (μ_S, D_S) such that $S = \mu_S D_s \mu_S^t$, and we can extend μ_S to a lower-triangular matrix $U \in Sp(2n, \mathbb{R})$ using the third property at the end of Section 2.2:

$$U = \begin{bmatrix} 1 & 0 & 0 \\ 0 & \mu_S & 0 \\ 0 & 0 & 1 \end{bmatrix}.$$

Hence, the product $\mu = UP^{-1}$ is a lower-triangular symplectic matrix whose diagonal is 1, and $D = \mu^{-1}G\mu^{-t} \in Sp(2n, \mathbb{R})$ is diagonal. This concludes the proof of Theorem 1 by induction.

5   Speeding-Up Gram-Schmidt by Duality

The standard integral Gram Schmidt algorithm we recalled in Section 2 is based on a formula which computes $\mu_{i,j}$ from the $\mu_{i,k}$'s $(k < j)$ on the same row. This leads to many integer divisions for each coefficient as in the innerloop rows 5-7 of Algorithm 1.

5.1   The General Case

We now show that most of these divisions can be avoided. Consider a basis $B = [\mathbf{b}_1, ..., \mathbf{b}_d]$ and its Gram matrix $G = BB^t$. We know that if μ is the Gram-Schmidt orthogonalization of B, then $G = \mu D \mu^t$ where D is a positive diagonal matrix. If we rewrite the previous equation as $\mu = G\mu^{-t}D^{-t}$, it appears that for any integer $k < d$, if we know the $k \times k$ topleft triangle of μ and D, we can compute the $k \times k$ topleft triangle of $\mu^{-t}D^{-t}$ and the first full k columns of μ. The matrix μ^{-t} is just a rotation of μ^{-s}, which is the Gram-Schmidt orthogonalization of the dual basis of $(\mathbf{b}_d, ..., \mathbf{b}_1)$. At the end, we get not only the GSO of B, but also the one of its reverse dual basis: this method which we call "Dual Gram-Schmidt", is surprisingly faster than the classical one despite computing more information.

Theorem 5. *Let $G \in \mathcal{M}_n(\mathbb{Z})$ be a symmetric (positive) definite matrix, and μ the $\mu D \mu^t$ factorization of G. As in Section 2, we define $\lambda_{0,0} = 1$ and $\lambda_{k,k} = \det G_k$ where G_k is the $k \times k$ topleft block of G. Let $\lambda = \mu \cdot \mathrm{diag}(\lambda_{1,1}, ..., \lambda_{n,n})$ and $U = \mu^{-t} \cdot \mathrm{diag}(\lambda_{0,0}, ..., \lambda_{n-1,n-1})$. Then the following three relations hold:*

$$\lambda \in \mathcal{M}_n(\mathbb{Z}), \tag{4}$$

$$U \in \mathcal{M}_n(\mathbb{Z}), \tag{5}$$

$$\lambda = GU. \tag{6}$$

Proof. From $G = \mu D \mu^t$, we know that $\mu^{-1}G$ is uppertriangular: for all i and t with $i > t$, then $\sum_{j=1}^i \mu_{i,j}^{-1} g_{j,t} = 0$. If we call G' the $(i-1) \times (i-1)$ topleft block of G, $\mathbf{v} = (\mu_{i,1}^{-1}, ..., \mu_{i,i-1}^{-1})$ and $\mathbf{g}' = (g_{i,1}, ..., g_{i,i-1})$, then the previous equation is equivalent to $\mathbf{g}' = -\mathbf{v}G'$. By Cramer's rule, we deduce the following relation for all $j < i$, which proves relation (5):

$$u_{j,i} = \det G' \cdot \mathbf{v}_j = (-1)^{i-j} \det \begin{pmatrix} g_{1,1} & \cdots & g_{1,j-1} & g_{1,j+1} & \cdots & g_{1,i} \\ \vdots & \ddots & \vdots & \vdots & \ddots & \vdots \\ g_{i-1,1} & \cdots & g_{i-1,j-1} & g_{i-1,j+1} & \cdots & g_{i-1,i} \end{pmatrix}.$$

The relation (6) is obtained by multiplying $\mu = G\mu^{-t}D^{-t}$ by $\mathrm{diag}(\lambda_{1,1}, ..., \lambda_{n,n})$. It also implies that $\lambda_{p,i} = \sum_{k=1}^i g_{p,k}u_{k,i}$ is the last row development of an integer determinant:

$$\lambda_{p,i} = \det \begin{pmatrix} g_{1,1} & \cdots & g_{1,i} \\ \vdots & \ddots & \vdots \\ g_{i-1,1} & \cdots & g_{i-1,i} \\ g_{p,1} & \cdots & g_{p,i} \end{pmatrix},$$

which concludes the proof.   □

Note that these determinants prove that $\lambda_{p,i} \leq |G|^i$ and $U_{j,i} \leq |G|^{i-1}$.

   We derive from $\mu^{-t}\mu^t = \mathrm{Id}$ a column formula to compute the matrix U defined in the previous theorem:

$$(\mu^{-t})_{i,j} = -\mu_{i,j}^t - \sum_{k=j+1}^{i-1} \mu_{i,k}^t (\mu^{-t})_{k,j}. \tag{7}$$

From the definition of U, we know that $(\mu^{-t})_{i,j} = \frac{U_{i,j}}{\lambda_{j-1,j-1}}$. Replacing $\mu_{i,j}$ by $\frac{\lambda_{i,j}}{\lambda_{j,j}}$, we may rewrite the formula as: $-U_{i,j} = \left(\sum_{k=j+1}^{i-1} \lambda_{k,i} U_{k,j} \right) / \lambda_{i,i}$. Hence if we know the $i \times (i-1)$ top-left triangle of λ, we can compute the ith column of U using this formula (from the diagonal to the top), and then the ith column of λ using relation (6) of Theorem 5. It is not necessary to keep the ith column of U after that.

We deduce Algorithm 2 to compute the GSO of a basis B. The correctness of this algorithm is a consequence of the results described in this section. If we look at the number of operations requested, there are $i^2/2$ multiplications and one division in the innerloop lines 4-6, and $i(n + 1 - i)$ small multiplications between the input Gram matrix and U in the innerloop lines 7-9. This gives a total of $n^3/6$ large multiplications and $n^2/2$ divisions. In the Standard GS algorithm, there was as many multiplications, but $\Theta(n^3)$ divisions.

Algorithm 2. Dual Gram Schmidt

Input: A basis $B = (\mathbf{b}_1, ..., \mathbf{b}_d)$ or its Gram matrix G
Output: The GSO decomposition λ of B
1: **for** $i = 1$ to d **do**
2:     $U_i \leftarrow \lambda_{i-1,i-1}$
3:     $U_{i-1} \leftarrow -\lambda_{i,i-1}$
4:     **for** $j = i - 2$ downto 1 **do**
5:         compute $U_j = -(\sum_{k=j+1}^{i} \lambda_{k,j} U_k)/\lambda_{j,j}$
6:     **end for**
7:     **for** $j = i$ to d **do**
8:         compute $\lambda_{j,i} = \sum_{k=1}^{i} \langle \mathbf{b}_j, \mathbf{b_k} \rangle U_k$
9:     **end for**
10: **end for**

5.2   The Symplectic Case

We derive an algorithm specialized to q-symplectic bases to compute the λ matrix of the GSO, and we show why it is faster than the Dual Gram-Schmidt procedure applied to symplectic bases (see Algorithm 1 of Section 2). Let B be a q-symplectic basis. We know that L in the LQ decomposition of B is q-symplectic and the μ matrix corresponding to the GSO of B is symplectic. We will also use the integer dual matrix U we introduced in Theorem 5. Let us denote the quadrants of μ, and λ by:

$$\mu = \begin{pmatrix} \mu_a & 0 \\ \mu_\gamma & \mu_\delta \end{pmatrix}, \lambda = \begin{pmatrix} \lambda_a & 0 \\ \lambda_\gamma & \lambda_\delta \end{pmatrix} \quad \text{and} \quad U = \begin{pmatrix} U_\alpha & 0 \\ U_\gamma & U_\delta \end{pmatrix}.$$

Because of Theorem 4, we know that $\mu_\delta = \mu_\alpha^{-s}$. Together with Corollary 1, we have $U_\alpha = R_n \lambda_\delta R_n$ for symplectic matrices and the following relation for q-symplectic matrices: $U_\alpha \cdot \mathrm{diag}(q^2, q^4, ..., q^{2n}) = R_n \lambda_\delta R_n$. For this reason, it is only necessary to run DualGS up to the middle of the matrix, and fill the columns of λ_δ using those of U_α. Given as input a symplectic basis $B = (\mathbf{b}_1, ..., \mathbf{b}_{2n})$, Algorithm 3 computes the λ matrix of the GSO of B in time $O(n^5 \log^2 B)$ using standard arithmetic.

Algorithm 3. Symplectic Gram-Schmidt

Input: A q-symplectic basis $B = [\mathbf{b}_1, ..., \mathbf{b}_{2n}]$
Output: The λ matrix of the GSO of B.
1: precompute: q^{2i} for $i = 1$ to n
2: **for** $i = 1$ to n **do**
3:     $U_i \leftarrow \lambda_{i-1,i-1}$ (for all k, U_k represents $U_{k,i}$)
4:     $U_{i-1} \leftarrow -\lambda_{i,i-1}$
5:     **for** $j = i - 2$ downto 1 **do**
6:         $U_j \leftarrow -(\sum_{k=j+1}^{i} \lambda_{k,j} U_k)/\lambda_{j,j}$
7:     **end for**
8:     **for** $j = 1$ to i **do**
9:         $\lambda_{2n+1-j, 2n+1-i} \leftarrow q^{2(n+1-i)} \cdot U_j$
10:    **end for**
11:    **for** $j = i$ to $2n$ **do**
12:        compute $\lambda_{j,i} = \sum_{k=1}^{i} G_{j,k} U_k$
13:    **end for**
14: **end for**

5.3   Experiments

We performed tests on randomly generated bases, whose coefficients are uniformly distributed 128-bit integers (see Table 1). On these random matrices, the speedup is rather moderate, but it will be much more significant when considering symplectic matrices.

We also performed tests on secret bases of NTRUSign as described in Section 3 (see Table 2). Roughly speaking, Algorithm 2 is at least 3 times as fast as Standard GS, and the specialized algorithm is from 10 to 30 times as fast as Standard GS. We give in function of the dimension of the input matrix, the running time in seconds to compute the GSO for each algorithm on 2Ghz Opteron

Table 1. Timing of Gram-Schmidt algorithms on random matrices

n	StandardGS in seconds	DualGS in seconds	**DualGS speedup**
100	37.3	25.1	1.49
200	881	579	1.52
300	5613	3644	1.54

Table 2. Timing of Gram-Schmidt algorithms on NTRUSign bases

$2n$	Standard GS	DualGS	SympGS	speedup DualGS	speedup SympGS
502	179	122	8.7	1.46	20.5
802	1822	1254	67	1.45	27.1
1214	12103	8515	390	1.42	31.0

processors with a 32-bit version of NTL 5.4. Note that the speed-up of 31 in Symplectic GS seems to indicate that the cost of computing the GSO of a $2n$-dimensional symplectic basis is roughly the one of computing the GSO of a standard n-dimensional standard matrix.

6 Symplectic LLL

When applied to a symplectic basis, the standard LLL algorithm will likely not preserve symplecticity, because its elementary operations are not necessarily symplectic. In this section, we show how one can slightly modify the notion of size-reduction used in LLL to make it compatible with symplecticity, and we then deduce a symplectic version of LLL. We do not know if every symplectic lattice contains an LLL-reduced basis which is also symplectic. But we show how to obtain efficiently a symplectic basis which is *effectively LLL-reduced* (as introduced in [14]). Such bases satisfy the most important properties of LLL-reduced bases, those which are sufficient to guarantee the shortness of the first basis vector.

6.1 Symplectic Size-Reduction

The first condition in LLL-reduction is size-reduction, which we recalled in Section 2. Unfortunately, size-reduction transformations are not necessarily symplectic. However, we show that it is still possible to force half of the coefficients of μ to be very small using symplectic transformations, and at the same time, bound the size of the remaining coefficients.

We say that a matrix $B \in \mathcal{M}_{2n}(\mathbb{R})$ is *semi-size reduced* if its GSO satisfies: for all $j \leq n$, for all $i \in [j + 1, 2n + 1 - j]$, $|\mu_{i,j}| \leq \frac{1}{2}$.

Theorem 6. *If $B \in Sp(2n, \mathbb{R})$ is semi-size reduced, then its GSO μ is bounded by $\|\mu\|_{\infty} \leq n \cdot (\frac{3}{2})^n$.*

Proof. For the block μ_δ , Equation (7) gives for $i \geq n + 1$ and $j \geq n + 1$, $|\mu_{i,j}| \leq \frac{1}{2} + \frac{1}{2}\sum_{k=j+1}^{i-1} |\mu_{i,k}|$, which is bounded by a geometric sequence of ratio $\frac{3}{2}$. Hence, the bottom diagonal block is bounded by $|\mu_{i,j}| \leq \frac{1}{2}(\frac{3}{2})^{i-j-1}$, and this bound can be reached for $\mu_{i\in[1,n],j\in[1,i-1]} = -1/2$.

For the bound on block μ_γ , apply Equation (1) to μ^s in order to get a column formula. This gives for $i \geq n + 1$ and $j \geq 2n - i$:

$$\mu^s_{n+1-j,n+1-i} = \mu^s_{i,j} + \sum_{k=j+1}^{n} \mu^s_{i,k}\mu^s_{2n+1-j,2n+1-k}$$

$$- \sum_{k=n+1}^{i-1} \mu^s_{i,k}\mu^s_{2n+1-j,2n+1-k}.$$

After reindexing the matrix and applying the triangular inequality to this sum, we obtain

$$|\mu_{i,j}| \le \frac{1}{2} + \frac{2n+1-2j}{4} + \frac{1}{2}\sum_{k=2n+2-j}^{i-1} |\mu_{k,j}|.$$

It is still bounded by a geometric sequence of ratio $\frac{3}{2}$, but the initial term $\mu_{i+1,i}$ is less than $\frac{2n+3-2j}{4} \le n$. Thus $|\mu_{i,j}| \le n \cdot (\frac{3}{2})^{i-2n+j-2}$.   □

6.2 Lovász Conditions and Effective Reduction

A basis satisfying all Lovász conditions and $|\mu_{i,i-1}| \le 1/2$ is referred to as *effectively LLL-reduced* in [14]. Such bases have the same provable bounds on the size of $\mathbf{b}_1$ (which is typically the most interesting vector in a basis) as LLL-reduced bases. Besides, it is easy to derive an LLL-reduced basis from an effectively LLL-reduced basis, using just size reductions (no swaps). In general the reason for weakly reducing the other $\mu_{i,j}$ for $1 \le j < i-1$ is to prevent coefficient explosion in the explicit $\mathbf{b}_i$, but there are many other strategies for this that don't require as strict a condition as $|\mu_{i,j}| \le 1/2, 1 \le j < i-1$ (see [17, 26]). It is not difficult to see that this notion of *"effective LLL-reduction"* can be reached by symplectic transformations.

Lemma 2. *A symplectic $2n$-dimensional basis B is effectively-reduced if and only if its first $n+1$ vectors are effectively LLL-reduced.*

Proof. Let μ be the GSO matrix of B, for $i \le n$, since μ is symplectic, we know that $\mu_{2n+2+i,2n+1-i} = \mu_{i,i-1}$. Using Corollary 1, the Lovasz condition for the ith index is equivalent to $\delta \frac{1}{\|\mathbf{b}^*_{2n+2-i}\|^2} \le \frac{1}{\|\mathbf{b}^*_{2n+1-i}\|^2} - \mu_{2n+2+i,2n+1-i}\frac{1}{\|\mathbf{b}^*_{2n+2-i}\|^2}$, which is precisely the Lovasz condition for the $2n+2-i^{\text{th}}$ index.   □

This means that for all $i \le n$, every operation made on the rows i and $i-1$ that reduces B can be blindly applied on the rows $2n-i+2$ and $2n-i+1$ without knowing the GSO of the second block. A symplectic basis is said to be *symplectic-LLL reduced* if it is both *effectively LLL-reduced* and *semi-size-reduced*.

6.3 A Symplectic-LLL Algorithm

It is easy to find polynomial algorithms for symplectic-LLL reduction, but the difficulty is to make them run faster than LLL in practice. Here, we present an algorithm which reaches symplectic-LLL reduction with an experimental running-time 6 times smaller than LLL on NTRU public bases of standard size (but the speed up may be larger for higher-dimensional lattices).

Symplectic LLL is an iterative algorithm that reduces a symplectic lattice L from its center. It takes as input the integer GSO λ of a symplectic lattice and outputs the GSO of a symplectic-LLL reduced basis and the unimodular transformation that achieves the reduction. More precisely, it only keeps one half of the GSO of symplectic matrices, since the other half can be easily deduced with (1) or Lemma 1. Here, we chose to keep the left triangle $\lambda' = \lambda_{i,j}, 1 \leq j \leq n, j \leq i \leq 2n+1-j$. During the algorithm, every elementary operation (swap or a linear combination of rows) is deduced from λ', and λ' is updated incrementally like in the standard integer LLL (see [18, 4]). As a result, symplecticLLL can generate the complete sequence of elementary operations of the reduction without knowing the basis. Unfortunately, having only the GSO of the LLL reduced basis is not sufficient to compute its coefficients, so every operation that occur in symplectic LLL algorithm is in fact performed on a third part matrix U. If U is initially equal to the input basis (resp. the identity matrix), then it is transformed into the LLL-reduced basis (resp. the unitary transformation).

In this paragraph, we explain the principles of SymplecticLLL on the projected lattice vectors, but in practice, all operations are done on the GSO λ' (see Algorithm 4 for details). Let $C_k = [\pi_{n+1-k}(\mathbf{b}_{n+1-k}), ..., \pi_{n+1-k}(\mathbf{b}_{n+k-1})]$ where $1 \leq k \leq n$. The $2k$-dimensional lattice $L(C_k)$ is symplectic, and its GSO matrix μ is the $2k \times 2k$ block located in the center of the GSO of the basis. When the algorithm begins, the counter k is set to 1. At the start of each loop iteration, C_{k-1} is already symplectic-LLL-reduced (there is no condition if $k = 1$). If $k = 1$ then the projected lattice C_1 is Lagrange-reduced and the counter k is set to 2. If the determinant of the transformation computed by Lagrange reduction is -1, we negate $\mathbf{b}_{n+1}$ to preserve the symplecticity of the basis. In the general case, C_k is semi-size-reduced, which means that $\lambda_{i,n+1-k}$ is made lower than $\frac{1}{2}\lambda_{n+1-k,n+1-k}$ with symplectic combinations of rows for $i = n + 2 - k$ to $n + k$. If the pair $(n + 1 - k, n + 2 - k)$ does not satisfy Lovász condition (by symplecticity neither does the pair $(n + k - 1, n + k)$), then the pairs of consecutive vectors $(\mathbf{b}_{n-k+1}, \mathbf{b}_{n-k+2})$ and $(\mathbf{b}_{k-1}, \mathbf{b}_k)$ are swapped and the counter k is decremented, otherwise k is incremented. The loop goes on until k eventually reaches the value $n + 1$.

Experiments show that this basic symplecticLLL algorithm is already as fast as LLL in dimension 200, and becomes faster in higher dimension. The quality of the output basis is similar to the one of StandardLLL. The drop of the GSO obtained with symplecticLLL is in general smoother (better) than with standardLLL, because both the basis and its dual are reduced in the same time (see Figure 1). Note also that the curve of $\log \|\mathbf{b}_i^*\|$ obtained after symplecticLLL is symmetric because of Corollary 1. We now describe optimizations.

6.4   Optimizations

The following two optimizations do not modify the output of the algorithm, but considerably speed up the algorithm in practice:

Early Reduction. Let C_i be the $2i$-dimensional central projection of the input basis for $2 \leq i \leq n$. Suppose that Algorithm 4 found the unimodular matrix

---

Algorithm 4. symplectic LLL

Input: A GSO matrix λ of a q-symplectic basis (at least the left triangle λ')
Output: The GSO λ' of the reduced basis, and the unitary transformation U
1: $k \leftarrow 1, U = I_{2n}$ (or $U = U_{\text{init}}$ initially given by the user)
2: **while** $k \leq n$ **do**
3:    **if** $k = 1$ **then**
4:      compute $\lambda_{n+1,n+1} = q^2 \lambda'_{n-1,n-1}$
5:      find the 2×2 unimodular transformation P that Lagrange-reduces the GSO of C_1
6:      ensure that the determinant of P is not -1, negate one row of P if necessary
7:      Apply P on the two middle rows of λ' and U, and update $\lambda'_{n,n}, \lambda'_{n+1,n}$
8:      $k \leftarrow 2$
9:    **end if**
10:   **for** $i = n + 2 - k$ to $n + k$ **do**
11:     $r \leftarrow \lfloor \lambda_{i,n+1-k}/\lambda_{n+1-k,n+1-k} \rceil$
12:     $\mathbf{u}_i \leftarrow \mathbf{u}_i - r\,\mathbf{u}_{n+1-k}$ and $\lambda'_i \leftarrow \lambda'_i - r\,\lambda'_{n+1-k}$
13:     $\mathbf{u}_{n+k} \leftarrow \mathbf{u}_{n+k} + r\,\mathbf{u}_{2n+1-i}$ and $\lambda'_{n+k} \leftarrow \lambda'_{n+k} + r\,\lambda'_{2n+1-i}$ if $i \leq n$
14:     $\mathbf{u}_{n+k} \leftarrow \mathbf{u}_{n+k} - r\,\mathbf{u}_{2n+1-i}$ and $\lambda'_{n+k} \leftarrow \lambda'_{n+k} - r\,\lambda'_{2n+1-i}$ if $n+1 \leq i \leq n+k-1$
15:   **end for**
16:   **if** Lovász does not hold for the pair $(n - k + 1, n - k + 2)$ **then**
17:     compute $\lambda_{n+k,n-k+2}$ using Lemma 1
18:     swap $\mathbf{u}_{n+k} \leftrightarrow \mathbf{u}_{n+k-1}$ and $\lambda'_{n+k,j} \leftrightarrow \lambda'_{n+k-1,j}$ for $1 \leq j \leq n - k$
19:     swap $\mathbf{u}_{n-k+2} \leftrightarrow \mathbf{u}_{n-k+1}$ and $\lambda'_{n-k+2,j} \leftrightarrow \lambda'_{n-k+2,j}$ for $1 \leq j \leq n - k$
20:     update $\lambda'_{n-k+2,n-k+1}$ and $\lambda'_{i,n-k+1}, \lambda'_{i,n-k+2}$ for $n - k + 3 \leq i \leq n + k$ using the same swap formula as standard LLL
21:     $k \leftarrow k - 1$
22:   **else**
23:     $k \leftarrow k + 1$
24:   **end if**
25: **end while**

$U_p \in Sp(2p, \mathbb{Z})$ such that $U_p C_p$ is symplecticLLL reduced and the reduced GSO λ'_p. If we want to reduce the initial C_{p+1} using Algorithm 4, we know that when the counter k reaches $p+1$ for the first time, the current state U_{p+1} and λ'_{p+1} is:

$$U_{p+1} = \begin{pmatrix} 1 & 0 & 0 \\ 0 & U_p & 0 \\ 0 & 0 & 1 \end{pmatrix} \quad \text{and} \quad \lambda'_{p+1} = \begin{pmatrix} \lambda_{n-p,n-p} & 0 \\ U_p \lambda_{.,n-p} & \lambda'_p \\ \lambda_{n+p+1,n-p} & 0 \end{pmatrix}.$$

So we can launch Algorithm 4 on λ'_{p+1} with $U_{\text{init}} = U_{p+1}$ to finish the reduction. Using this simple trick for $p = 2$ to n, the first transformations of Algorithm 4 apply to lower dimensional matrices. On NTRU matrices, the execution is almost two times faster than the basic symplecticLLL. In the Standard LLL algorithm, the analogue is to update only the first p rows of the GSO, where p is the maximum index of vectors already modified by the main loop of LLL since the beginning of the execution.

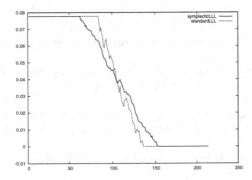

Fig. 1. Quality of the input basis ($\log \|\mathbf{b}_i^*\|$ as a function of i)

Table 3. Experimental results (on 2GHz Opteron with 32-bit NTL 5.4)

n	q	Standard LLL *in seconds*	SympLLL Early red. *in seconds*	SympLLL Integer triang. optim. *in seconds*	speedup **Early reduction**	speedup **integer triang.**
half of dim.	*max. coefs*					
40	64	3.09	2.27	1.98	1.36	1.56
83	64	26.89	6.62	4.46	4.06	6.02
107	64	44.7	6.13	4.51	7.29	9.91
167	128	410.8	98.86	65.40	4.15	6.28
253	128	2028	553	294	3.66	6.89
317	128	3688	1131	519	3.26	7.10

Integer Triangular Matrices. This last optimization only works on matrices for which every $\|\mathbf{b}_k^*\|^2$ is an integer (at the beginning). It is the case of all NTRU public key matrices, and all integer triangular matrices. The key result is that in the previous algorithm, each λ_p' is initially divisible by $D_{n-p} = \prod_{i=1}^{n-p} \|\mathbf{b}_i^*\|^2$. The only improvement is to use reduced GSO λ_p'/D_{n-p} instead of λ_p' in the previous algorithm. Then the first transformations of Algorithm 4 apply to matrices of lower dimension, but also with smaller coefficients. On NTRU matrices, the execution becomes almost 4 times faster than the basic symplecticLLL.

Acknowledgements. Part of this work, as well as a visit of the second author to the ENS, were supported by the Commission of the European Communities through the IST program under contract IST-2002-507932 ECRYPT. We would like to thank Joe Silverman, Mike Szydlo and William Whyte for useful conversations.

References

1. L. Babai. On Lovász lattice reduction and the nearest lattice point problem. *Combinatorica*, 6:1–13, 1986.
2. A.-M. Bergé. Symplectic lattices. In *Quadratic forms and their applications (Dublin, 1999)*, volume 272 of *Contemp. Math.*, pages 9–22. Amer. Math. Soc., Providence, RI, 2000.
3. P. Buser and P. Sarnak. On the period matrix of a Riemann surface of large genus. *Invent. Math.*, 117(1):27–56, 1994. With an appendix by J. H. Conway and N. J. A. Sloane.
4. H. Cohen. *A Course in Computational Algebraic Number Theory.* Springer-Verlag, 1995. Second edition.
5. J. Conway and N. Sloane. *Sphere Packings, Lattices and Groups.* Springer-Verlag, 1998. Third edition.
6. D. Coppersmith and A. Shamir. Lattice attacks on NTRU. In *Proc. of Eurocrypt '97*, volume 1233 of *LNCS*. IACR, Springer-Verlag, 1997.
7. C. Gentry, J. Jonsson, J. Stern, and M. Szydlo. Cryptanalysis of the NTRU signature scheme (NSS) from Eurocrypt 2001. In *Proc. of Asiacrypt '01*, volume 2248 of *LNCS*. Springer-Verlag, 2001.
8. C. Gentry and M. Szydlo. Cryptanalysis of the revised NTRU signature scheme. In *Proc. of Eurocrypt '02*, volume 2332 of *LNCS*. Springer-Verlag, 2002.
9. G. H. Golub and Charles F. Van Loan. *Matrix Computations.* The John Hopkins University Press, third edition, 1996.
10. C. Hermite. Extraits de lettres de M. Hermite à M. Jacobi sur différents objets de la théorie des nombres, deuxième lettre. *J. Reine Angew. Math.*, 40:279–290, 1850. Also available in the first volume of Hermite's complete works, published by Gauthier-Villars.
11. J. Hoffstein, N. A. Howgrave-Graham, J. Pipher, J. H. Silverman, and W. Whyte. NTRUSIGN: Digital signatures using the NTRU lattice. In *Proc. of CT-RSA*, volume 2612 of *LNCS*. Springer-Verlag, 2003.
12. J. Hoffstein, J. Pipher, and J. Silverman. NTRU: a ring based public key cryptosystem. In *Proc. of ANTS III*, volume 1423 of *LNCS*, pages 267–288. Springer-Verlag, 1998. First presented at the rump session of Crypto '96.
13. J. Hoffstein, J. Pipher, and J. H. Silverman. NSS: An NTRU lattice-based signature scheme. In *Proc. of Eurocrypt '01*, volume 2045 of *LNCS*. Springer-Verlag, 2001.
14. N. Howgrave-Graham. Finding small roots of univariate modular equations revisited. In *Cryptography and coding (Cirencester, 1997)*, volume 1355 of *Lecture Notes in Comput. Sci.*, pages 131–142. Springer, Berlin, 1997.
15. N. A. Howgrave-Graham, P. Q. Nguyen, D. Pointcheval, J. Proos., J. H. Silverman, A. Singer, and W. Whyte. The impact of decryption failures on the security of NTRU encryption. In *Proc. of the 23rd Cryptology Conference (Crypto '03)*, volume 2729 of *LNCS*, pages 226–246. IACR, Springer-Verlag, 2003.
16. IEEE P1363.1 Public-Key Cryptographic Techniques Based on Hard Problems over Lattices, June 2003. IEEE., Available from http://grouper.ieee.org/groups/1363/lattPK/index.html.
17. B. A. LaMacchia. Basis reduction algorithms and subset sum problems. Technical Report AITR-1283, 1991.
18. A. K. Lenstra, H. W. Lenstra, Jr., and L. Lovász. Factoring polynomials with rational coefficients. *Mathematische Ann.*, 261:513–534, 1982.

19. D. S. Mackey, N. Mackey, and F. Tisseur. Structured factorizations in scalar product spaces. *SIAM J. of Matrix Analysis and Appl.*, 2005. To appear.
20. A. May and J. H. Silverman. Dimension reduction methods for convolution modular lattices. In *Proc. of CALC '01*, volume 2146 of *LNCS*. Springer-Verlag, 2001.
21. P. Q. Nguyen and O. Regev. Learning a parallelepiped: cryptanalysis of GGH and NTRU signatures. In *Proc. of Eurocrypt '06, LNCS*. Springer-Verlag, 2006.
22. P. Q. Nguyen and D. Stehlé. Floating-point LLL revisited. In *Proc. of Eurocrypt '05*, volume 3494 of *LNCS*, pages 215–233. IACR, Springer-Verlag, 2005.
23. P. Q. Nguyen and J. Stern. The two faces of lattices in cryptology. In *Proc. of CALC '01*, volume 2146 of *LNCS*. Springer-Verlag, 2001.
24. H.-G. Quebbemann. Modular lattices in Euclidean spaces. *J. Number Theory*, 54(2):190–202, 1995.
25. C. P. Schnorr. A hierarchy of polynomial lattice basis reduction algorithms. *Theoretical Computer Science*, 53:201–224, 1987.
26. M. Seysen. Simultaneous reduction of a lattice basis and its reciprocal basis. *Combinatorica*, 13(3):363–376, 1993.
27. B. M. M. de Weger. Solving exponential Diophantine equations using lattice basis reduction algorithms. *J. Number Theory*, 26(3):325–367, 1987.
28. H. Weyl. *The classical groups*. Princeton Landmarks in Mathematics. Princeton University Press, 1997. Their invariants and representations, Fifteenth printing, Princeton Paperbacks.

The Function Field Sieve in the
Medium Prime Case

Antoine Joux[1,3] and Reynald Lercier[1,2]

[1] DGA
[2] CELAR, Route de Laillé, 35170 Bruz, France
Reynald.Lercier@m4x.org
[3] Université de Versailles St-Quentin-en-Yvelines, PRISM
45, avenue des Etats-Unis, 78035 Versailles Cedex, France
Antoine.Joux@m4x.org

Abstract. In this paper, we study the application of the function field sieve algorithm for computing discrete logarithms over finite fields of the form $\mathbb{F}_{q^n}$ when q is a medium-sized prime power. This approach is an alternative to a recent paper of Granger and Vercauteren for computing discrete logarithms in tori, using efficient torus representations. We show that when q is not too large, a very efficient $L(1/3)$ variation of the function field sieve can be used. Surprisingly, using this algorithm, discrete logarithms computations over some of these fields are even easier than computations in the prime field and characteristic two field cases. We also show that this new algorithm has security implications on some existing cryptosystems, such as torus based cryptography in T_{30}, short signature schemes in characteristic 3 and cryptosystems based on super-singular abelian varieties. On the other hand, cryptosystems involving larger basefields and smaller extension degrees, typically of degree at most 6, such as LUC, XTR or T_6 torus cryptography, are not affected.

1 Introduction

Computing discrete logarithms is, with integer factorization, one of the two number-theoretical hard problems upon which public-key cryptography is usually based. Two kind of groups are often considered, elliptic curves and multiplicative groups of finite fields. The latter case is further partitioned into several sub-cases, prime fields $\mathbb{F}_p$, characteristic two fields $\mathbb{F}_{2^n}$, where n is usually prime and extensions of medium-sized fields $\mathbb{F}_{q^n}$, where q is a medium-sized prime power[1] p^k. Until recently, the last case was rarely considered in cryptography. However, two recent developments make use of such fields, pairing-based cryptography [13, 5, 6, 7] and torus-based cryptography [19, 8, 21, 27]. For this reason, practical evaluation of the hardness of discrete logarithms in such fields is becoming an important issue. Recently, an approach based on rational torus representation was proposed by Granger and Vercauteren [12], it was applied in [22].

---

[1] Remark that we use the notation $\mathbb{F}_{q^n}$ to emphasize the fact that for composite extension degrees, viewing the field as $\mathbb{F}_{p^{kn}}$ is not necessarily optimal.

S. Vaudenay (Ed.): EUROCRYPT 2006, LNCS 4004, pp. 254–270, 2006.
© International Association for Cryptologic Research 2006

In this paper, we revisit a much older approach, the function field sieve. This algorithm was originally introduced by Adleman [3] as an extension of Coppersmith's algorithm [9]. Its complexity was subsequently improved by Adleman and Huang in [4]. This algorithm is known to be efficient when the base field is fixed and the extension degree grows. Moreover, it was shown to be practical and applied to characteristic 2 in [14]. Later on, it was also used in characteristic 3 in [11]. However, when both p and the total extension degree nk grow, the reference is the approach of Adleman and Demarrais in [2, 1], which makes use of a variation of Coppersmith's algorithm, involving function fields, when $p \leq nk$. As soon as this bound is exceeded they use a different algorithm based on number fields. This approach gives an $L(1/2)$ complexity for medium-sized base fields. In this paper, we describe a new variation of the function field sieve which is dedicated to medium-sized values of q and allows for fast computation of discrete logarithms in $\mathbb{F}_{q^n}$, even when q is much larger than n. For such fields, we show that our approach is faster, both from a theoretical complexity viewpoint with an $L(1/3)$ complexity and as a practical tool. More precisely, this variation of the function field sieve is applicable with $L(1/3)$ complexity whenever $\log q$ remains smaller than $O(\sqrt{n} \log n)$.

The paper is organized as follows, in section 2 we describe the function field sieve variation we are considering, in section 3 we show that the asymptotic complexity is the same as the complexity of the function field sieve with small base fields, in section 4 we describe real sized experimentations with this algorithm, finally in section 5 we discuss the impact of our algorithm on the security of some cryptosystems.

2 A Medium Sized Variation on the Function Field Sieve

The function field sieve algorithm for computing discrete logarithms over $\mathbb{F}_{p^n}$ is quite similar to the number field sieve for computing discrete logarithms over $\mathbb{F}_p$ (see [10, 28]). Both algorithms consider multiplicative identities using smooth objects over well-chosen smoothness bases. With the number field sieve, the objects are numbers in number fields and the smoothness bases contain ideals of small norm. With the function field sieve, the objects are polynomials in function fields and the smoothness bases contain ideals whose norms are polynomials of small degree. The complexity of such algorithms, is usually expressed using a notation, initially introduced for fast integer factorization algorithms [20]. This now classical notation is defined as follows:

$$L_Q(\alpha, c) = \exp((c + o(1))(\log Q)^\alpha (\log \log Q)^{1-\alpha}).$$

For the two extreme cases, prime fields $\mathbb{F}_p$ and extension fields $\mathbb{F}_{p^n}$ with fixed characteristic p, the number field sieve and the function field sieve respectively yield $L(1/3, (64/9)^{1/3})$ and $L(1/3, (32/9)^{1/3})$ algorithms. In the intermediate cases, the best available complexity is $L(1/2)$ as described by Adleman and Demarrais in [1, 2]. We would like to further remark, that using the function field sieve with fixed p, we have a smaller constant in the $L(1/3)$ expression than with the number field sieve. This is due to the fact that $\mathbb{F}_{p^n}$ has a large number of different representations, one for each irreducible polynomial of degree n over $\mathbb{F}_p$. This was discovered in [4] and a practical variant was presented in [14]. Surprisingly, even with medium-sized base fields, a similar construction that makes use of well chosen representations is possible, as shown below. The most important question is how to choose a good smoothness basis. With a medium sized base field $\mathbb{F}_{q^n}$, when q has just the right size, this is in fact very simple. It suffices to choose as the smoothness bases the sets of ideals whose norms are degree one polynomials, no more, no less. When $\log q$ and $\sqrt{n} \log n$ are correctly balanced, this choice yields a very efficient algorithm with complexity $L(1/3, 3^{1/3})$. In section 2.2, we discuss different choices of smoothness bases that should be used instead of this simple choice, when the balance between q and n varies.

More precisely, let q be the cardinality of the base field and n the degree of the extension. In order to define $\mathbb{F}_{q^n}$, we proceed as follows. First, choose a minimal pair (d_1, d_2), with $d_2 = d_1$ or $d_1 + 1$, and with $d_1 d_2 \geq n$. Then, find two polynomials f_1 and f_2, in two unknowns, X and t, of the form:

$$f_1(X, t) = X - g_1(t), \quad f_2(X, t) = g_2(X) + t,$$

where g_1 and g_2 are univariate polynomials of degree d_1 and d_2, such that, $g_2(g_1(t)) + t$ has an irreducible factor $F(t)$ of degree n over $\mathbb{F}_q$. We claim that such polynomials are easy to find (see section 4 for examples). We use $F(t)$ as our definition polynomial for $\mathbb{F}_{q^n}$. Clearly, f_1 and f_2 have a common root $X = g_1(t)$ in $\mathbb{F}_{q^n}$. As a consequence, f_1 and f_2 define good function fields for the function field sieve algorithm. Using standard vocabulary, we say that f_1 defines the linear side of the sieve.

The next step of the algorithm is to send objects of the form $a(t)X - b(t)$ in the two function fields. At this point, we slightly differ from standard practice and consider only a subset of such objects, by fixing $a(t) = wt + 1$ and choosing $b(t) = ut + v$, where u, v and w are elements of the base field $\mathbb{F}_q$. As usual, we then compute the norm of $a(t)X - b(t)$ in the two function fields. This restriction on $a(t)$ comes from the fact that, since we are working with polynomials, all factorizations are defined up to a constant in the base field. This choice of $a(t)$ avoids multiple sieving of the same objects. Note that from a practical point of view, when q is large enough, it is even better to reduce the sieving space and fix $a(t) = 1$ only. Then, on the linear side, we find $b(t) - g_1(t)$ a degree d_1 polynomial. On the other side, we find $g_2(b(t)) + t$ a degree d_2 polynomial. This contrasts with the general case, where the respective degrees are $d_1 + 1$ and $d_2 + 1$. It is a well-known fact that among polynomials of degree d over $\mathbb{F}_q$, the proportion of degree d polynomials having d roots quickly tends towards $1/d!$ as q grows. We say that $b(t)$ generates a relation when both sides completely split into degree 1 factors. Using the traditional heuristic and assuming that the sieving process generates random looking polynomials, this occurs with a probability which is very close to either $1/(d_1! \cdot d_2!)$ or $1/((d_1 + 1)! \cdot (d_2 + 1)!)$. It remains to see whether we obtain enough relations. On the linear side, our chosen smoothness basis contains the q possible unitary polynomials of degree 1, namely the polynomials $t + u$, with u in $\mathbb{F}_q$. On the other side, due to our particular choice of f_2, the smoothness basis also contains q elements, which are ideals of norm $t + g_2(u)$, with u in $\mathbb{F}_q$. As a consequence, we need $2q$ equations. Since we are sieving over either q^2 or q^3 elements, this particular choice works when either $q \geq 2\, d_1! \cdot d_2!$ with reduced sieving space or $q^2 \geq 2\, (d_1 + 1)! \cdot (d_2 + 1)!$ with full sieving space.

After generating the multiplicative identities as above, we transform them into linear equations involving logarithms of polynomials on the linear side and "logarithms of ideals" on the other side[2]. The resulting system of equations is then solved using a sparse linear algebra algorithm such as Lanczos or Wiedeman [18, 23, 30, 17]. This linear algebra step is performed modulo $(q^n - 1)/(q - 1)$. Indeed, the multiplicative identities are defined up to a multiplicative constant in $\mathbb{F}_q$ and the logarithms are computed in the quotient group of $\mathbb{F}_{q^n}^*$ by $\mathbb{F}_q^*$. It is interesting to note that due to the very specific form of the equations we use, with exactly d_1 (or $d_1 + 1$) unknowns (potentially counting multiplicities) on the left-hand side and d_2 (or $d_2 + 1$) unknowns on the right-hand side, our system does not have full rank over the rationals. There is a "parasitic" solution with all the left-hand side unknowns set to d_2 and all right-hand side unknowns set to d_1. This means that after the linear algebra, the resulting solution does not contain pure discrete logarithms, the result is masked by some additive constant. However, by considering fractions such as $(t + u)/(t + v)$, the contribution of this constant can be cancelled. Moreover, if we can find even a single equation with a different structure, the masking constant can easily be found. The simplest way to proceed is to find a linear polynomial which completely splits in the function field defined by f_2. This yields a specific kind of equation[3] which nicely breaks the above symmetry and allows us to find and remove the unwanted constant. An example of this technique is given in section 4.

2.1 Individual Discrete Logarithms

Once the two steps described above, sieving and linear algebra, have been performed, we obtained the logarithms of the elements of the smoothness bases. This is well and good, but does not fully solve the discrete logarithm problem. An additional step is required to compute the logarithms of large elements in the finite field. We propose a classical approach based on "special-q" descent, which is similar to the approach proposed in [9, 14] for the case of logarithms over an extension of a small base field. The idea is the following. Given an element y in the finite field, whose logarithm is wanted, we first build many elements of the form $y^i \cdot t^j$. Each of these elements can be represented as a polynomial in t of degree n. Alternatively, using continued fractions, we can also find representations

---

[2] This notion of logarithms of ideals is described and used in [15, 14]. With the specific choice of f_2 we have given, there is a simpler description, because the function field is principal and all ideals can be represented by a single element in the finite field.

[3] These equations are often used in function field sieve algorithms and are called systematic equations.

by rational fractions, whose numerators and denominators have degrees near $n/2$. From an asymptotic viewpoint, both approaches are equivalent. However, in practice, the latter is more efficient. Once we obtain such a representation, we test whether it can be factored in polynomials of degree $\mu\sqrt{n}$ for a constant μ to be determined in the sequel. After testing sufficiently many representations, we find an adequate one and are left with the problem of computing logarithms of polynomials of degree at most $\mu\sqrt{n}$. Let $\mathfrak{q}$ be such a low degree polynomial. We can now find its logarithm by sieving again on elements of the form $a(t)X - b(t)$, where $a(t)$ and $b(t)$ are polynomials of degree at most $\mu\sqrt{n}$ chosen to ensure that $\mathfrak{q}$ divides the linear side (in the function field defined by f_1) of the resulting equations. After finding an element $a(t)X - b(t)$ that factor in both function fields into polynomials of degree smaller than the degree of $\mathfrak{q}$, we iterate the descent down to degree one, where all logarithms are known. This descent alternates between special-$\mathfrak{q}$ on the linear and the high degree function fields. Once the descent reaches degree one, we backtrack and compute the logarithms of each special-$\mathfrak{q}$ and finally the logarithms of $y^i \cdot t^j$ and y. If the special-$\mathfrak{q}$ values occurring at the first level are small enough, then the total degree of the objects to be factored in the next levels are strictly smaller and the bottleneck of this step is the search for a good representation of $y^i \cdot t^j$. In fact, this can be ensured by choosing μ such that $\mu\sqrt{n} \cdot (d_2+1) + d_1 < n$. We show that in the complexity analysis of section 3 and prove that choosing a value of μ between $1/2$ and 1 ensures a good behavior of the individual logarithm phase.

2.2   Extension to Smaller Base Fields

From a practical point of view, the above case is probably the most interesting. However, it is nice to know whether the approach can be extended to different choices of q and n. We now briefly describe a family of algorithm which neatly cover all the cases where q is smaller than above. Each algorithm depends on a main parameter D, which bounds the degree of norms of elements in the smoothness bases. The previous algorithm corresponds to $D = 1$. The general case of D is very similar to the restricted case $D = 1$. We construct the function fields in the same manner as above, only changing the choices of d_1 and d_2. More precisely, we take $d_1 \approx \sqrt{Dn}$ and $d_2 \approx \sqrt{n/D}$. We sieve over $a(t)X - b(t)$, where $a(t)$ and $b(t)$ are also degree D polynomials and $a(t)$ is unitary. The total size of the sieving space is q^{2D+1}. On the linear side, we need to factor a polynomial of degree at most $d_1 + D$ over the smoothness basis. On the other side, we need to factor a polynomial of degree at most $d_2D + 1$. Dismissing constants and low order terms, both degrees are near $\sqrt{nD}$. In this context, we need to know the asymptotic smoothness probability of a degree n polynomial into factors of degree at most m. This problem has been widely studied and very precise estimates are given in [24]. However, results are usually given for fixed q, when both n and m grow. Here, m is fixed and both q and n grow. Yet, the logarithm of the probability of smoothness is still equivalent to $n/m \log(m/n)$. Moreover, in order to prove our complexity result, a lower bound on the probability is sufficient. For the sake of completeness, we prove this lower bound in appendix A.

For simplicity of exposition, in the rest of this section, we work with equivalent expressions, however, the argument can easily be rewritten to accommodate a lower bound only.

From these estimates, we deduce that the logarithm of the probability of smoothness on each side is approximately $-\sqrt{n/D}\log\sqrt{n/D}$. Adding the two, we obtain a total logarithm of heuristic probability of $-\sqrt{n/D}\log(n/D)$. Moreover, the total size of the two smoothness bases is about $2q^D$. As with the case $D = 1$, we should make sure that we obtain enough equations, this approximately requires:

$$(D + 1)\log(q) \geq \sqrt{n/D}\log(n/D).$$

With this algorithm, the individual logarithms phase remains almost identical. The only needed change is to use polynomials of degree $\mu\sqrt{Dn}$ to represent the element under consideration. We analyze the heuristic complexity of this extension in section 3 and show that, as in the case $D = 1$, the right choice is to take μ between $1/2$ and 1.

2.3   Practical Improvements

Large primes variation. In the asymptotic analysis below (section 3), we observe that when q decreases below some point, we need to increase the parameter D to successfully compute discrete logarithms. However, this change of algorithm greatly increases the overall complexity. This happens when the sieving space is not large enough to get enough equations. However, right at the boundary, the number of missing equations is quite small. In that case, it is a good idea to use a large prime variation by allowing a small number of higher degree polynomials in each decomposition when splitting polynomials over the smoothness bases. This does not improve the asymptotic complexity, however, in practice, it can make the difference between a feasible and an infeasible computation. We do not further discuss this idea, which is classical in implementations of number field and function field sieves.

Use of Galois Action. In some specific cases, it is possible to use additional structure of the field $\mathbb{F}_{q^n}$ to improve the practicality of our algorithm. The basic idea is to use the Galois group in order to reduce the size of the smoothness bases. Assume that there exists an element ϕ of the Galois group which acts on both smoothness bases by sending any element to a conjugate also belonging to the smoothness basis. If we further express ϕ as a Frobenius power, we can reduce the number of unknowns in the linear algebra by a factor which is equal to the order of the action of ϕ on $\mathbb{F}_{q^n}$. We can also speed up the sieving process by the same factor, since less equations are needed. However, we should take care and avoid sieving on values for $a(t)X - b(t)$ yielding conjugate equations.

To make this idea more precise, let us discuss the specific case of $\mathbb{F}_{2^{nk}}$, where $q = 2^k$ and n and k are coprime. In that case, we can view $\mathbb{F}_{2^{nk}}$ as a tower of extensions or alternatively as a compositum. With the latter representation, we independently define $\mathbb{F}_{2^n}$ and $\mathbb{F}_{2^k}$ and put the two representations together to get $\mathbb{F}_{q^n}$. This means that f_1 and f_2 can both have their coefficients in $\mathbb{F}_2$. In that case, we take for ϕ the n-th Frobenius power, i.e., the mapping which sends x to x^{2^n} in $\mathbb{F}_{q^n}$. Clearly, if t is a root of the irreducible polynomial $F(t)$ defined by f_1 and f_2, it is an element of $\mathbb{F}_{2^n}$ and thus fixed by ϕ. However, the action of ϕ on $a(t)$ and $b(t)$ is not trivial. Indeed, assume that we work with parameter $D = 1$, then $b(t) = ut + v$ with u and v in $\mathbb{F}_{2^k}$. Unless u and v are both in $\mathbb{F}_2$ the image of $b(t)$ by ϕ is a different polynomial. Repeating the application of ϕ, we find yet another polynomial, and so on ... Since k and n are coprime, the order of the action of ϕ on $\mathbb{F}_{2^k}$ is k. As a consequence, the sieving process can be sped up by a factor[4] of k. Moreover, choose $t + u$ an element of the smoothness basis on the linear side. Clearly, $(t + u)^{2^n} = t + \phi(u)$ is another element of the same smoothness basis and the logarithms of the two elements say l_u and $l_{\phi(u)}$ are related by $l_{\phi(u)} = 2^n l_u$. This implies that the number of unknowns on the linear side can be divided by k. A similar argument also applies on the other side. As a consequence, we gain a speed-up by k on the sieving process and a speed-up by k^2 on the linear algebra.

Clearly, the same construction works for any small characteristic. Use of Galois action to speed-up the computation is also possible in other cases. In particular, for all fields of the form $\mathbb{F}_{q^2}$, it is possible to gain a constant speed-up of two. In some cases, it is also possible to have a larger speed-up. However, the details are much more technical and in particular may require to construct f_1 and f_2 in a different manner than the construction given at the beginning of the present section. We only illustrate this by giving an example in section 4.

3   Asymptotic Heuristic Complexity

In this section, given the respective values of q and n, we give the asymptotic complexity of our algorithm both for $D = 1$, for other fixed values of D and, finally, in the general case. It is convenient to let Q denote q^n and to assume, when the parameter D is fixed, that there exists a parameter α such that:

$$n = \frac{1}{\alpha} \cdot \left(\frac{\log Q}{\log \log Q} \right)^{2/3}, \qquad q = \exp\left(\alpha \cdot \sqrt[3]{\log Q \cdot \log^2 \log Q} \right).$$

Using this notation, we can analyze the complexity of each algorithm in the family, determined by the parameter D. Since there are two main phases, sieving and linear algebra, the total complexity expressed by $L(1/3, c)$ is determined by the maximum of the complexities of each phase. Let $L(1/3, c_1)$ be the complexity of the sieving and $L(1/3, c_2)$ be the complexity of linear algebra. Then, recalling from the analysis of section 2 that the smoothness basis has $O(q^D)$ elements and that the logarithm of the heuristic probability of finding a relation is $-\sqrt{n/D} \log(n/D)$, we find:

$$c_1 = \frac{2}{3\sqrt{\alpha D}} + \alpha D \quad \text{and} \quad c_2 = 2\alpha D.$$

---

[4] Disregarding the rare cases where both $a(t)$ and $b(t)$ have all their coefficients in $\mathbb{F}_2$.

Moreover, we need to check that we obtain enough equations. We recall that this approximately requires:

$$(D+1)\log(q) \geq \sqrt{n/D}\log(n/D) \quad \text{or} \quad (D+1)\alpha \geq \frac{2}{3\sqrt{\alpha D}}.$$

Whenever this condition is satisfied, we say that the algorithm with parameter D is applicable. Putting all these conditions together, we find that for each value of α we should use the lowest possible parameter D yielding an applicable algorithm. Moreover, in the range of applicability the complexity of each algorithm decreases with α. The optimal case for each algorithm happens when:

$$(D+1)\alpha = \frac{2}{3\sqrt{\alpha D}}.$$

Just below this threshold, we need to use the next algorithm in the family and the complexity jumps up to $L(1/3, c_2(D+1)) = L(1/3, 2\alpha(D+1))$. Thus at each threshold, a discontinuity occurs in the complexity . The largest such gap is between $D = 1$ and $D = 2$, at $\alpha = 3^{-2/3}$. On both sides of the gap, the respective complexities are $L(1/3, \sqrt[3]{3})$ for $D = 1$ and $L(1/3, \sqrt[3]{64/9})$ for $D = 2$. All the other gaps are smaller and the gap size decreases with D and tends to 0 as D grows. Moreover, the complexity tends to $L(1/3, \sqrt[3]{32/9})$. Thus, up to a single exception, the complexity of our family of algorithms is at worst the complexity of the number field sieve, is at best even better than the complexity of the function field sieve with fixed prime and tends to this latter complexity when D grows. The exception happens when α becomes too large even for $D = 1$, more precisely when $\alpha > \sqrt[3]{8/9}$. Indeed, in that case, the complexity $L(1/3, 2\alpha)$ is larger than the complexity of the number field sieve. We summarize this complexity analysis in figure 1. The three horizontal lines on this graph represents the constants $\sqrt[3]{3}$, $\sqrt[3]{32/9}$ and $\sqrt[3]{64/9}$.

General Values of q. Another interesting question is to study the complexity of the algorithm when q grows more slowly than $L_Q(1/3, \epsilon)$ for all ϵ. In that case, we no longer use a fixed parameter D, but let it grow slowly with Q. More precisely, we choose for D the nearest integer to the solution d of the following equation:

$$q^d = L_Q(1/3, \sqrt[3]{4/9}).$$

This choice yields complexity $L(1/3, \sqrt[3]{32/9})$ in all the considered cases. Note that this includes the usual function field sieve, for fixed q, as a special case. We also remark, that as announced in introduction the $L(1/3)$ boundary on q corresponds to the range where $\log q$ remains smaller than $O(\sqrt{n}\log n)$.

Individual Logarithms. Concerning individual logarithms, we should choose a constant μ, both small enough to guarantee that the degrees of polynomials occurring during the descent are strictly decreasing and large enough to ensure that the initial good representation is found efficiently. Let $m = \mu\sqrt{Dn}$ be the maximal degree of the polynomials appearing in a good representation. Once

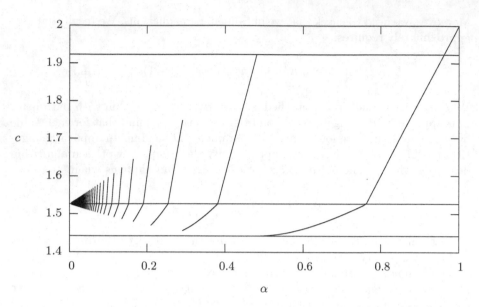

Fig. 1. Complexity $L_Q(1/3, c)$ as a function of $q = L_Q(1/3, \alpha)$

again, we need to use the fact that the logarithm of the probability of smoothness is still equivalent to $n/m \log(m/n)$, even when m, n and q all grow. Moreover, as before, the argument could be rewritten using only the lower bound given in A.

Replacing n by its expression in term of Q we find that the probability is equivalent to

$$1/L_Q\left(1/3, \frac{1}{3\mu\sqrt{\alpha D}}\right).$$

We would like to ensure that the constant in this expression is smaller than the constant in the complexity of the main phase of the algorithm. This implies:

$$\frac{1}{3\mu\sqrt{\alpha D}} < \max(c_1, c_2), \text{ with } c_1 = \frac{2}{3\sqrt{\alpha D}} + \alpha D \text{ and } c_2 = 2\alpha D.$$

It clearly suffices to have: $\frac{1}{3\mu\sqrt{\alpha D}} < \frac{2}{3\sqrt{\alpha D}}$, where the right hand side is the first summand in c_1. This is true whenever $\mu > 1/2$.

Moreover, we need to make sure the special-q descent involves polynomials of decreasing degree. Since the degrees of $a(t)$ and $b(t)$ during the descent are at most the degree of the special-q itself, substituting in f_1 and f_2, we require: $(d_2\mu\sqrt{Dn} + 1) + (d_1 + \mu\sqrt{Dn}) < n$. Replacing d_1 and d_2 by their values and disregarding low order terms, we get: $\mu n < n$. This can be ensured by choosing $\mu < 1$. As a consequence, we can choose any value of μ in the range $]1/2; 1[$.

Finally, we need to check that at each step of the special-q descent, there are sufficiently many pairs $(a(t), b(t))$ to obtain at least one relation. Potentially, we might expect to encounter problems for a special-q of degree two, when trying to relate it to polynomials of degree one. In that case, the natural choice would be to select linear polynomials for $a(t)$ and $b(t)$. Since $a(t)$ has the restricted form $wt + 1$, there are only $q^3/q^2 = q$ possible pairs involving the special-q value. As a consequence, with such a choice for $a(t)$ and $b(t)$, we cannot guarantee that a relation can be found for this special-q value. Thus, we need to use polynomials of degree two for $a(t)$ and $b(t)$. Of course, this lowers the smoothness probability which becomes:

$$\frac{1}{((d_1 + 2)! \cdot (2d_2 + 1)!)}$$

instead of $1/((d_1+1)! \cdot (d_2+1)!)$ in the main phase. However, since a single relation is needed, we keep the good asymptotic complexity. Indeed, in the least favorable case with respect to this issue, which happens to be the extreme case of the basic $(D = 1)$ algorithm, the main phase probability is almost equal to $1/q^2$ and the main sieving costs q^3. Using the same parameters, the smoothness probability of the individual logarithm phase is asymptocally equivalent to $1/q^3$. Thus, in this worst case, the individual logarithm phase has the same asymptotic cost as the main phase. In all other cases, the main phase dominates the complexity.

4   Numerical Examples

4.1   Basic Example

Our first example is the computation of discrete logarithms over $\mathbb{F}_{65537^{25}}$. The cardinality Q of this field is a number of about 400 bits or 120 decimal digits. It can be factored as:

$$Q = 65536 \cdot 3571 \cdot 37693451 \cdot 137055701 \cdot 10853705894563968937051 \cdot P_{247}$$

Since the largest prime factor has 247 bits, Pollard's rho [25] is not practical for this example. As far as we know, this sets a new record for the computation over medium characteristic fields.

We first choose our function fields, fixing the two definition polynomials f_1 and f_2 as follows:

$$f_1(X,t) = X - t^5 - t - 3, \qquad f_2(X,t) = X^5 + X + 1 + t.$$

Taking the resultant of f_1 and f_2, thus eliminating X, we find an irreducible polynomial $F(t)$ over $\mathbb{F}_{65537}$. We let α denote a root of $F(t)$ in the extension field. We also let β denote $\alpha^5 + \alpha + 3$.

Once this is done, we start the sieving process, using the reduced sieving space $X - (a\,t + b)$, with a and b in $\mathbb{F}_{65537}$. When we find a good pair (a, b) we obtain an equality between smooth objects. Indeed, the two function fields we are using are principal, thus whenever both norms are smooth, we can write an explicit identity between generators. For example, replacing X by $-2\,t + 20496$ in f_1 and f_2 yields smooth polynomials. Writing down explicit generators for the corresponding ideals, this yields the following equality:

$$(\alpha + 2445) \cdot (\alpha + 9593) \cdot (\alpha + 31166) \cdot (\alpha + 39260) \cdot (\alpha + 48610) =$$
$$\lambda(\beta + 43449) \cdot (\beta + 18727) \cdot (\beta + 17129) \cdot (\beta + 1946) \cdot (\beta + 49823),$$

where $\lambda = -2$ is an element of $\mathbb{F}_{65537}$.

The sieving process itself is extremely fast, we give in appendix B the source C code of the program we used. This program finds all good (a, b) pairs in two minutes on a Pentium laptop at 1.6 GHz. Once the sieving is complete, each good pair yields a linear equation between 5 logarithms of elements $\alpha + u$ and 5 logarithms of $\beta + v$. We converted the output of the C program into linear equations using a short interpreted PARI/GP script. This conversion took an additional two minutes. It was as long as the sieve itself, however, it did not seem necessary to write a faster program for this task.

Solving the linear algebra system was the bottleneck of the algorithm. After some structured gaussian elimination, we had to solve a sparse system of 79 466 equations in 78 465 unknowns and 3.8 million entries. This was done using the Lanczos algorithm. In order to avoid divisions by non-invertible elements, we worked modulo $q_0 = Q/(65536 \cdot 3571)$. This took a little more than two days on the same laptop. The resulting solution gave logarithms, up to an additive constant. As explained in section 2, we determined the constant using the following systematic equation:

$$\beta \cdot (\beta + 16) \cdot (\beta - 16) \cdot (\beta + 4096) \cdot (\beta - 4096) = -(\alpha + 1).$$

After removing this additive constant and renormalizing the result, we had all the logarithms of elements $\alpha + u$ and $\beta + v$ modulo q_0. For example,

$$l = 9580541088009323484229889821453339382943430459454536234824$$
$$8403754835240173532297063343231849297238533209444439485,$$
$$m = 4649571275692520918560124050338108397005057301288170051718$$
$$5566862384316422897306135296316764963935552585468877691$$

are the respective logarithms modulo q_0 of $\alpha + 1$ and β in base α. This can be checked by testing that $(\alpha + 1)^{3571 \cdot l} / \alpha^{3571}$ belongs to $\mathbb{F}_{65537}$, and similarly for $\beta^{3571 \cdot m} / \alpha^{3571}$.

The final step was to choose a random looking element of $\mathbb{F}_{65537^{25}}$ and to compute its complete logarithm. Since, α itself does not generate the full multiplicative group, we decided to express the logarithm in basis 3α, which is a generator. We took as challenge the element:

$$\lambda = \sum_{i=0}^{24} (\lfloor \pi \cdot 65537^{i+1} \rfloor \bmod 65537) \alpha^i = 41667\alpha^{24} + \cdots + 9279.$$

After finding a good representation of λ using polynomials of degree at most 3 and completing the special-$\mathfrak{q}$ descent, we added the contribution of the logarithm modulo the powers of 2 and modulo 3571. Finally, we concluded that the logarithm of λ in basis 3α is:

$$4053736945052440744587988507271545773377910517074639935754736$$
$$3481852609028577772820085371649268383536448936947412841146999.$$

4.2   Galois Action Example

We consider here a discrete logarithm challenge that is defined in $\mathbb{F}_{p^{30}}$ where $p = 370801$: such a finite field has got a 556-bit cardinality and it contains a 114-bit multiplicative subgroup. A smaller extension $\mathbb{F}_{p^{18}}$ has been recently performed by Vercauteren and Lercier [22] at the expense of one week over a network of 10 AMD's Athlon(TM) XP 2000+ for the sieving step and 12 hours for the linear algebra step, using the algorithm of [12]. To solve our $T_{30}(\mathbb{F}_p)$ challenge, we first experimented with the algorithm defined in section 2. It turns out that, with $f_1(X, t) = X - (t^6 + t + 30)$ and $f_2(X, t) = X^5 + X + 1 + t$, a three hours computation for the sieving step and a two days computation for the linear step[5] would have been necessary on a 1.15 GHz 16-processors HP AlphaServer GS1280.

Thus, it was preferable to make use of the Galois action idea and define

$$f_1(X, t) = X - t^5 \text{ and } f_2(X, t) = X^6 + X - 17 - t^5.$$

This yields a definition polynomial for $\mathbb{F}_{p^{30}}$ equal to $F(t) = t^{30} - 17$, the Galois group of which is generated by $\phi : t \mapsto t^p = 172960 \times t$. With such a choice, f_1 and f_2 have a common root $X = t^5$, which is fixed by ϕ^6 and thus lies in the subfield $\mathbb{F}_{p^6}$. As a result, the conjugates by ϕ^6 of places (in both algebraic function fields) in the smoothness basis reduced modulo p are still elements of the smoothness basis. Since discrete logarithms of conjugates differ from each other by a power of p^6, we clearly divide by *five* the size of the smoothness basis: only 74161 places in the linear side[6] and only 74114 places in the other side.

With a sieving program similar to the one given in appendix B, we found 329082 useful divisors of functions $X - (at + b)$ with a and b in $\mathbb{F}_p$, in 45 minutes. The supports of these divisors contain only degree one places and we restricted the values of a to avoid conjugate equations. Since, the reduction modulo p of these degree one places is equal to a suitable power of p^6 of one element of the smoothness basis, we clearly have enough equations.

We skipped the structured Gaussian elimination step, since at this time our code is not able to handle matrices with so many large coefficients. Of course, we had to modify our implementation of the Lanczos algorithm to handle this case. Finally, we were able to solve this sparse system of 150270 equations in 148270 variables (with 11 entries by row equal to powers p^{6i}, $i = 0, \ldots 4$) at the cost of a 10 hours computation on 8 processors of a 1.15 GHz HP AlphaServer GS1280. We worked modulo

$$q_0 = 129717983265199170691 \times 3780896193379818021601 \times$$
$$270849696832313136083187915736989 01.$$

---

[5] This matrix is twice as big as the one used by Vercauteren and Lercier in $\mathbb{F}_{p^{18}}$. It is also twice as heavy.

[6] In truth, only 12361 places in the linear side are really necessary because conjugates by ϕ itself are again elements of the smoothness basis. Due to the additional coding work that would have been required, we did not take advantage of this speed-up in our experiment.

Let us note that the kernel of this matrix has got only one vector (its coefficients are not all equal to one and thus, we do not have any "parasitic" solution). After this step, using the Galois action of ϕ^6, we have the logarithms, modulo q_0, of elements $t + u \in \mathbb{F}_{p^{30}}$ for any $u \in \mathbb{F}_p$.

In the final step, we took as challenge the element

$$\lambda = \sum_{i=0}^{29} (\lfloor \pi \times p^{i+1} \rfloor \bmod p) t^i = 162147t^{29} + \cdots + 52502.$$

We first write this element as a product of elements of degree at most four and using a special-$\mathfrak{q}$ descent, we finally found that the logarithm of λ in basis $t - 6$ is:

834934758318669039584738321669880646445961989720307919277

236643257447878787655408750007604393413253988463644325187

4051550980392237533812685076653542562214928407573371226.

5   Security Implications

In this section, we discuss the applicability of our variation of the function field sieve to cryptosystems that make use of extension fields. First of all, we remark that for some systems, our approach is slower than generic algorithms and does not improve upon known attacks. Let us start by giving examples of such cryptosystems which are immune to our attack. The relevant property is that the systems make use of extensions of quite small degree over prime fields. Typically, the security of systems which use extension degree 6 over prime fields are not affected by this algorithm. In particular, this includes LUC [19], XTR [8, 21], CEILIDH [27], some pairing-based schemes as the complex multiplication variation of the short signature scheme of [6, 7] and also torus-based cryptography in T_6. When the extension degree is larger than that, it is important to reassess the security on a case by case basis. In the rest of this section, we do so for torus-based cryptography in T_{30}, the short signature scheme of [6, 7] in characteristic three and some of the supersingular abelian varieties proposed in [26].

For the case T_{30}, the base field is quite large and the bottleneck of the algorithm is the linear algebra whose complexity is $(d_1 + d_2)p^2$ additions modulo some factor of $p^{30} - 1$. In typical instantiation the relevant factor is a prime q_0 between 160 and 256 bits, according to the expected security level. We should compare our algorithm to a generic algorithm such as Pollard's rho [25], whose complexity is $\sqrt{q_0}$ operations in the finite field. Since additions modulo q_0 are less expensive than operations in the finite field and since $d_1 + d_2$ is small, it seems fair to proceed by comparing p^2 with $\sqrt{q_0}$. We conclude that for 80-bit security, it is necessary to choose for p a prime of 40 bits or more. For 32-bit primes and 160-bit subgroup, as proposed in [29], the expected security level is not reached and the effective security level is around 2^{64}. On the other hand, the security of the 64-bit primes examples with 200-bit subgroups proposed in the same paper is unaffected.

The short signature scheme of [6, 7] can be instantiated in two different ways. Either by using complex multiplication technique to build elliptic curves over $\mathbb{F}_p$ with a pairing that outputs numbers in $\mathbb{F}_{p^6}$. Or by using special supersingular curves over $\mathbb{F}_{3^\ell}$ with a pairing having values in $\mathbb{F}_{3^{6\ell}}$. Note that since the journal version [7], the characteristic three instantiation is no longer recommended. As said above, our algorithm does not change the security of the first case. In the second case, we can restate the problem as discrete logarithm in $\mathbb{F}_{q^\ell}$, where $q = 3^6 = 729$. From a practical point of view, this opens the possibility to use our algorithm with a parameter D equal to 2 or 3. With luck, and depending on the exact value of ℓ, we may fall in a zone where our algorithm is more efficient than the regular function field sieve in characteristic 3. Let us consider some of the usual possibilities. The easiest case is $\ell = 121$, since the extension field can even be viewed as a degree 33 extension of $\mathbb{F}_{3^{22}}$, which can be adressed with parameter $D = 1$, yielding a complexity near 2^{70}, which might be improved using Galois action. The expected Pollard rho complexity is 2^{78}. Similarly for $\ell = 97$ using $D = 2$, we find a complexity around 2^{71} instead of the expected 2^{76} and for $\ell = 149$, using $D = 3$ we find a complexity around 2^{105} instead of 2^{110}.

In fact, when looking at fields of characteristic three, our attack applies even better with the proposal of [26] which is to work with supersingular abelian varieties. Indeed, in the most extreme case, this proposal relies on the security of discrete logarithms in $3^{30\ell}$, which due to large choice of possible subfields is extremely likely to fall in a good case of our algorithm. In the same paper, the use of fields of the form $2^{12\ell}$ is also considered. Our algorithm can again be used here, especially when ℓ is composite (the cases $\ell = 121$ and $\ell = 87$ for example).

6   Conclusion

In this paper, we have presented a new variation of the function field sieve algorithm, which unexpectedly applies to finite field of the form $\mathbb{F}_{q^n}$ when both q and n are of medium sized. This allows us to compute discrete logarithms in $\mathbb{F}_{q^n}$ faster than for discrete logarithms problems in field of a comparable size of the form $\mathbb{F}_p$ (with p prime) or even $\mathbb{F}_{2^n}$ (with n prime). This shows that despite former belief, discrete logarithms in some fields $\mathbb{F}_{q^n}$ are easier than in $\mathbb{F}_p$ or $\mathbb{F}_{2^n}$. As a consequence, we show that the security of some recent cryptosystems needs to be reassessed to account for this fact. We leave as an open question the problem of finding an efficient $L(1/3)$ algorithm for solving discrete logarithms in $\mathbb{F}_{q^n}$ when q is larger than $L_{q^n}(1/3)$. Up to q of the form $L(1/2)$, our algorithm with parameter $D = 1$ is the fastest known technique and has complexity q^2, beyond that one should turn to the number field based algorithm described in [1, 2] with complexity $L(1/2)$.

Last Minute News. A recent preprint [16] describes a generalization of the number field sieve that is applicable to finite finite fields of size $Q = q^n$, whenever q grows faster than $L_Q(1/3)$. Put together with the present paper, this gives asymptotic complexity $L_Q(1/3)$ for discrete logarithms in all finite fields.

References

1. L. Adleman and J. DeMarrais. A subexponential algorithm for discrete logarithms over all finite fields. In D. Stinson, editor, *Proceedings of CRYPTO'93*, volume 773 of *Lecture Notes in Comput. Sci.*, pages 147–158. Springer, 1993.
2. L. Adleman and J. DeMarrais. A subexponential algorithm for discrete logarithms over all finite fields. *Math. Comp.*, 61(203):1–15, 2003.
3. L. M. Adleman. The function field sieve. In *Algorithmic Number Theory, Proceedings of the ANTS-I conference*, volume 877 of *Lecture Notes in Comput. Sci.*, pages 108–121, 1994.
4. L. M. Adleman and M. A. Huang. Function field sieve method for discrete logarithms over finite fields. In *Information and Computation*, volume 151, pages 5–16. Academic Press, 1999.
5. D. Boneh and M. Franklin. Identity based encryption from the Weil pairing. In *Crypto '2001*, volume 2139 of *Lecture Notes in Computer Science*, pages 213–229, 2001.
6. D. Boneh, B. Lynn, and H. Shacham. Short signatures from the Weil pairing. In C. Boyd, editor, *Proceedings of ASIACRYPT'2001*, volume 2248 of *Lecture Notes in Comput. Sci.*, pages 514–532. Springer, 2001.
7. D. Boneh, B. Lynn, and H. Shacham. Short signatures from the Weil pairing. *J. of Cryptology*, 17(4):297–319, 2004.
8. A.E. Brouwer, R. Pellikaan, and E.R. Verheul. Doing More with Fewer Bits. In *Advances in Cryptology — ASIACRYPT '99*, volume 1716 of *Lecture Notes in Computer Science*, pages 321–332. Springer, 1999.
9. D. Coppersmith. Fast evaluation of logarithms in fields of characteristic two. *IEEE transactions on information theory*, IT-30(4):587–594, July 1984.
10. D. Gordon. Discrete logarithms in GF(p) using the number field sieve. *SIAM J. Discrete Math*, 6:124–138, 1993.
11. R. Granger, A. Holt, D. Page, N. Smart, and F. Vercauteren. Function field sieve in characteristic three. In D. Buell, editor, *Algorithmic Number Theory, Proceedings of the ANTS-VI conference*, volume 3076 of *Lecture Notes in Comput. Sci.*, pages 223–234. Springer, 2004.
12. R. Granger and F. Vercauteren. On the discrete logarithm problem on algebraic tori. In V. Shoup, editor, *Proceedings of CRYPTO'2005*, volume 3621 of *Lecture Notes in Comput. Sci.*, pages 66–85. Springer, 2005.
13. A. Joux. A one round protocol for tripartite diffie-hellman. In *Fourth Algorithmic Number Theory Symposium*, volume 1838 of *Lecture Notes in Computer Science*, pages 385–394, 2000.
14. A. Joux and R. Lercier. The function field sieve is quite special. In C. Fieker and D. Kohel, editors, *Algorithmic Number Theory, Proceedings of the ANTS-V conference*, volume 2369 of *Lecture Notes in Comput. Sci.*, pages 431–445. Springer, 2002.
15. A. Joux and R. Lercier. Improvements to the general number field sieve for discrete logarithms in prime fields. A comparison with the gaussian integer method. *Math. Comp.*, 72:953–967, 2003.
16. A. Joux, R. Lercier, N. Smart, and F. Vercauteren. The number field sieve in the medium prime case. Preprint.
17. B.A. LaMacchia and A.M. Odlyzko. Solving Large Sparse Linear Systems Over Finite Fields. In *Advances in Cryptology — CRYPTO '90*, volume 537 of *Lecture Notes in Computer Science*, pages 109–133. Springer-Verlag, 1991.
18. C. Lanczos. Solutions of systems of linear equations by minimized iterations. In *J. Res. Nat.*, volume 49, pages 33–53. Bureau of Standards, 1952.

19. M.J.J. Lennon and P.J. Smith. LUC: A New Public Key System. In *IFIP TC11 Ninth International Conference on Information Security IFIP/Sec*, pages 103–117, 1993.
20. A. K. Lenstra and H. W. Lenstra, Jr., editors. *The development of the number field sieve*, volume 1554 of *Lecture Notes in Mathematics*. Springer–Verlag, 1993.
21. A.K. Lenstra and E.R. Verheul. The XTR Public Key System. In *Advances in Cryptology — CRYPTO 2000*, volume 1880 of *Lecture Notes in Computer Science*, pages 1–19. Springer, 2000.
22. R. Lercier and F. Vercauteren. Discrete logarithms in $\mathbb{F}_{p^{18}}$ - 101 digits. NM-BRTHRY mailing list, June 2005.
23. A. M. Odlyzko. Discrete logarithms in finite fields and their cryptographic significance. In T. Beth, N. Cot, and I. Ingemarsson, editors, *Advances in Cryptology — EUROCRYPT '84*, volume 209 of *Lecture Notes in Computer Science*, pages 224–314. Springer–Verlag, 1985.
24. D. Panario, X. Gourdon, and P. Flajolet. An analytic approach to smooth polynomials over finite fields. In J. Buhler, editor, *Algorithmic Number Theory, Proceedings of the ANTS-III conference*, volume 1423, pages 226–236. Springer, 1998.
25. J. M. Pollard. Monte Carlo methods for index computation (mod p). *Math. Comp.*, 32:918–924, 1978.
26. K. Rubin and A. Silverberg. Supersingular abelian varieties in cryptology. In M. Yung, editor, *Proceedings of CRYPTO'2002*, volume 2442 of *Lecture Notes in Comput. Sci.*, pages 336–353. Springer, 2002.
27. K. Rubin and A. Silverberg. Torus-Based Cryptography. In *Advances in Cryptology — CRYPTO 2003*, volume 2442 of *Lecture Notes in Computer Science*, pages 349–365. Springer, 2003.
28. O. Schirokauer. Discrete logarithms and local units. *Phil. Trans. R. Soc. Lond. A 345*, pages 409–423, 1993.
29. M. van Dijk, R. Granger, D. Page, K. Rubin, A. Silverberg, M. Stam, and D. Woodruff. Practical cryptography in high dimensional tori. In R. Cramer, editor, *Proceedings of EUROCRYPT'2005*, volume 3494 of *Lecture Notes in Comput. Sci.*, pages 234–250. Springer, 2005.
30. D.H. Wiedemann. Solving Sparse Linear Equations Over Finite Fields. *IEEE Trans. Information Theory*, 32:54–62, 1986.

A   Lower Bound on the Smoothness Probability

In this appendix, we prove the lower bound of the probability of smoothness of polynomials of degree n over the basis of monic irreducible polynomials of degree at most m. As usual, it suffices to work with unitary polynomials of degree n and we denote by $N_q(n, m)$ the number of m-smooth unitary polynomials. Before giving our lower bound on $N_q(n, m)$, we recall that the number of monic irreducible polynomials of degree t is:

$$I_q(t) = \frac{1}{t} \sum_{d|t} \mu(t/d)q^t \geq \frac{1}{t}\left(q^t - \lceil \log_2 t \rceil q^{t/2}\right),$$

where μ denotes the Möbius function. We first show the expected lower bound when n is a multiple of m, $n = \ell m$. In that case, the number of smooth polynomials is greater than the number of possible products of ℓ distinct polynomials of degree m. This number is: $\frac{1}{\ell!} \prod_{i=0}^{\ell-1} I_q(m) - i$. Replacing I_q by its values, letting ℓ and q grow and dividing by q^n to get a probability, we obtain a lower bound of: $\frac{1}{\ell!(m+\epsilon)^\ell}$ for any value of $\epsilon > 0$. Taking the logarithm we find $\ell(\log \ell + m + \epsilon)$ which is asymptotically equivalent to $\ell \log \ell$ as expected.

In the general case, we write $n = \ell m + r$ with $r < m$ and proceed similarly with a product of one irreducible of degree r and ℓ distinct irreducibles of degree m. The lower bounds immediately follows.

B   Listing of Sieving C Code for 65537^{25}

```
#include <stdio.h>
#include <stdlib.h>
#define PRIME 65537
int RootTab[2*PRIME]; int AlphaTab[2*PRIME]; char Count[PRIME];
AddSieveElement(int root, int alpha) { static int count=0;
  RootTab[count]=root; AlphaTab[count]=alpha; count++;
}
InitLinearSide() { /* Polynomial X-(t^5+t+3) */
  int alpha,root; long long tmp;
  for (alpha=0;alpha<PRIME;alpha++) {
    tmp=alpha; tmp*=tmp; tmp%=PRIME; tmp*=tmp; tmp%=PRIME;
    tmp*=alpha; tmp%=PRIME; tmp+=alpha+3; tmp%=PRIME;
    root=tmp; AddSieveElement(root,alpha);
}}
InitOtherSide() { /* Polynomial X^5+X+1+t */
  int alpha,root; long long tmp;
  for (root=0;root<PRIME;root++) {
    tmp=root; tmp*=tmp; tmp%=PRIME; tmp*=tmp; tmp%=PRIME;
    tmp*=root; tmp%=PRIME; tmp+=root+1; tmp%=PRIME;
    alpha=tmp; if (alpha) alpha=PRIME-alpha;
    AddSieveElement(root,alpha);
}}
FindMultiCollisions(int line) { int i,root;
  for (i=0;i<PRIME;i++) Count[i]=0;
  for (i=0;i<2*PRIME;i++) {root=RootTab[i]; Count[root]++;}
  for (i=0;i<PRIME;i++)
    if (Count[i]>=9) printf("b(t)=%d*t+%d;\n",-line,i);
}
UpdateTables() { int i,root;
  for (i=0;i<2*PRIME;i++) { root=RootTab[i]+AlphaTab[i];
    if (root>=PRIME) root-=PRIME;
    RootTab[i]=root;
}}
main() { int line; InitLinearSide(); InitOtherSide();
  for(line=0;line<PRIME;line++) {
    FindMultiCollisions(line); UpdateTables();
}}
```

Learning a Parallelepiped:
Cryptanalysis of GGH and NTRU Signatures

Phong Q. Nguyen[1],* and Oded Regev[2],**

[1] CNRS & École normale supérieure, DI, 45 rue d'Ulm, 75005 Paris, France
http://www.di.ens.fr/~pnguyen/
[2] Department of Computer Science, Tel-Aviv University, Tel-Aviv 69978, Israel
http://www.cs.tau.ac.il/~odedr/

Abstract. Lattice-based signature schemes following the Goldreich-Goldwasser-Halevi (GGH) design have the unusual property that each signature leaks information on the signer's secret key, but this does not necessarily imply that such schemes are insecure. At Eurocrypt '03, Szydlo proposed a potential attack by showing that the leakage reduces the key-recovery problem to that of distinguishing integral quadratic forms. He proposed a heuristic method to solve the latter problem, but it was unclear whether his method could attack real-life parameters of GGH and NTRUSIGN. Here, we propose an alternative method to attack signature schemes à la GGH, by studying the following learning problem: given many random points uniformly distributed over an unknown n-dimensional parallelepiped, recover the parallelepiped or an approximation thereof. We transform this problem into a multivariate optimization problem that can be solved by a gradient descent. Our approach is very effective in practice: we present the first succesful key-recovery experiments on NTRUSIGN-251 without perturbation, as proposed in half of the parameter choices in NTRU standards under consideration by IEEE P1363.1. Experimentally, 90,000 signatures are sufficient to recover the NTRUSIGN-251 secret key. We are also able to recover the secret key in the signature analogue of all the GGH encryption challenges, using a number of signatures which is roughly quadratic in the lattice dimension.

1 Introduction

Inspired by the seminal work of Ajtai [1], Goldreich, Goldwasser and Halevi (GGH) proposed at Crypto '97 [9] a lattice analogue of the coding-theory-based public-key cryptosystem of McEliece [19]. The security of GGH is related to the hardness of approximating the closest vector problem (CVP) in a lattice. The GGH article [9] focused on encryption, and five encryption challenges were

* Part of this work is supported by the Commission of the European Communities through the IST program under contract IST-2002-507932 ECRYPT, and by the French government through the X-Crypt RNRT project.
** Supported by an Alon Fellowship, by the Binational Science Foundation, by the Israel Science Foundation, and by the EU Integrated Project QAP.

S. Vaudenay (Ed.): EUROCRYPT 2006, LNCS 4004, pp. 271–288, 2006.

Fig. 1. The Hidden Parallelepiped Problem in dimension two

issued on the Internet [8]. Two years later, Nguyen [22] found a flaw in the original GGH encryption scheme, which allowed to solve four out of the five GGH challenges, and obtain partial information on the last one. Although GGH might still be secure with an appropriate choice of the parameters, its efficiency compared to traditional public-key cryptosystems is perhaps debatable: it seems that a very high lattice dimension is required, while the keysize grows roughly quadratically in the dimension (even when using the improvement suggested by Micciancio [20]). The only lattice-based scheme known that can cope with very high dimension is NTRU [15] (see the survey [23]), which can be viewed as a very special instantiation of GGH with a "compact" lattice and different encryption/decryption procedures (see [20, 21]).

In [9], Goldreich *et al.* described how the underlying principle of their encryption scheme could also provide a signature scheme. The resulting GGH signature scheme did not attract much interest in the research literature until the company NTRU CRYPTOSYSTEMS proposed a relatively efficient signature scheme called NTRUSIGN [11], based exactly on the GGH design but using the compact NTRU lattices. NTRUSIGN had a predecessor NSS [14] less connected to the GGH design, and which was broken in [6, 7]. Gentry and Szydlo [7] observed that the GGH signature scheme has an unusual property (compared to traditional signature schemes): each signature released leaks information on the secret key, and once sufficiently many signatures have been obtained, a certain Gram matrix related to the secret key can be approximated. The fact that GGH signatures are not zero-knowledge can be explained intuitively as follows: for a given message, many valid signatures are possible, and the one selected by the secret key says something about the secret key itself.

This information leakage does not necessarily prove that such schemes are insecure. Szydlo [25] proposed a potential attack on GGH based on this leakage (provided that the exact Gram matrix could be obtained), by reducing the key-recovery problem to that of distinguishing integral quadratic forms. It is however unknown if the latter problem is easy or not, although Szydlo proposed a heuristic method based on existing lattice reduction algorithms applied to quadratic forms. As a result, it was unclear if Szydlo's approach could actually work on real-life instantiations of GGH and NTRUSIGN. The paper [12] claims that, for NTRUSIGN without perturbation, significant information about the secret key is leaked after 10,000 signatures. However, it does not identify any attack that would require less than 100 million signatures (see [11, Sect. 4.5] and [12, Sect. 7.2 and App. C]).

OUR RESULTS. In this article, we present a new key-recovery attack on lattice-based signature schemes following the GGH design, including NTRUSIGN. The basic observation is that a list of known pairs $(message, signature)$ gives rise to the following learning problem, which we call the hidden parallelepiped problem (HPP): given many random points uniformly distributed over an unknown n-dimensional parallelepiped, recover the parallelepiped or an approximation thereof. We transform the HPP into a multivariate optimization problem based on the fourth moment (also known as *kurtosis*) of one-dimensional projections. This problem can be solved by a gradient descent. Our approach is very effective in practice: we present the first succesful key-recovery experiments on NTRUSIGN-251 without perturbation, as proposed in half of the parameter choices in the NTRU standards [4] being considered by IEEE P1363.1 [18]; the number of required signatures can be as low as 90,000, but the true figure might even be lower. We have also been able to recover the secret key in the signature analogue of all five GGH encryption challenges, using a number of signatures which is roughly quadratic in the lattice dimension. When the number of signatures is sufficiently high, the running time of the attack is only a fraction of the time required to generate all the signatures.

RELATED WORK. Interestingly, it turns out that the HPP (as well as related problems) have already been looked at by people dealing with what is known as *Independent Component Analysis* (ICA) (see, e.g., the book by Hyvärinen *et al.* [16]). ICA is a statistical method whose goal is to find directions of independent components, which in our case translates to the n vectors that define the parallelepiped. It has many applications in statistics, signal processing, and neural network research. To the best of our knowledge, this is the first time ICA is used in cryptanalysis.

There are several known algorithms for ICA, and most are based on a gradient method such as the one we use in our algorithm. Our algorithm is closest in nature to the FastICA algorithm proposed in [17], who also considered the fourth moment as a goal function. We are not aware of any rigorous analysis of these algorithms; the proofs we have seen often ignore the effect of errors in approximations. Finally, we remark that the ICA literature offers other, more general, goal functions that are supposed to offer better robustness against noise etc. We have not tried to experiment with these other functions, since the fourth moment seems sufficient for our purposes.

Another closely related result is that by Frieze *et al.* [5], who proposed a polynomial-time algorithm to solve the HPP (and generalizations thereof). Technically, their algorithm is slightly different from those present in the ICA literature as it involves the Hessian, in addition to the usual gradient method. They also claim to have a fully rigorous analysis of their algorithm, taking into account the effect of errors in approximations. Unfortunately, most of the analysis is missing from the preliminary version, and to the best of our knowledge, a full version of the paper has never appeared.

Open Problems. It would be interesting to study natural countermeasures against our attack, such as:

- Perturbation techniques (as suggested by [12, 4, 13]), where the hidden parallelepiped is replaced by a more complicated set. For instance, the second half of parameter choices in NTRU standards [4] involves exactly a single perturbation. In this case, the attacker now has to solve a hidden parallelepiped problem for which the parallelepiped is replaced by the Minkowski sum of two hidden parallelepipeds: the lattice spanned by one of the parallepipeds is public, but not the other one.
- Using secret bases with much larger entries.

However, such countermeasures have an impact on the effiency of the signature scheme.

ROAD MAP. The paper is organized as follows. In Section 2, we provide notation and necessary background on lattices, GGH and NTRUSIGN. In Section 3, we introduce the hidden parallelepiped problem, and explain its relationship to GGH-type signature schemes. In Section 4, we present a method to solve the hidden parallelepiped problem. In Section 5, we present experimental results obtained with the attack on real-life instantiations of GGH and NTRUSIGN. In Section 6, we provide a theoretical analysis of the main parts of our attack.

2   Background and Notation

Vectors of $\mathbb{R}^n$ will be row vectors denoted by bold lowercase letters such as $\mathbf{b}$, and we will use row representation for matrices. For any ring $\mathcal{R}$, $\mathcal{M}_n(\mathcal{R})$ will denote the ring of $n \times n$ matrices with entries in $\mathcal{R}$. The group of $n \times n$ invertible matrices with real coefficients will be denoted by $GL_n(\mathbb{R})$ and $O_n(\mathbb{R})$ will denote the subgroup of orthogonal matrices. The transpose of a matrix M will be denoted by M^t, so M^{-t} will mean the inverse of the transpose. The notation $\lceil x \rfloor$ denotes a closest integer to x. Naturally, $\lceil \mathbf{b} \rfloor$ will denote the operation applied to all the coordinates of $\mathbf{b}$. If X is a random variable, we will denote by $\text{Exp}[X]$ its expectation. The gradient of a function f from $\mathbb{R}^n$ to $\mathbb{R}$ will be denoted by $\nabla f = (\frac{\partial f}{\partial x_1}, \ldots, \frac{\partial f}{\partial x_n})$.

2.1   Lattices

Let $\| \cdot \|$ and $\langle \cdot, \cdot \rangle$ be the Euclidean norm and inner product of $\mathbb{R}^n$. We refer to the survey [23] for a bibliography on lattices. In this paper, by the term lattice, we mean a full-rank discrete subgroup of $\mathbb{R}^n$. The simplest lattice is $\mathbb{Z}^n$. It turns out that in any lattice L, not just $\mathbb{Z}^n$, there must exist linearly independent vectors $\mathbf{b}_1, \ldots, \mathbf{b}_n \in L$ such that:

$$L = \left\{ \sum_{i=1}^{n} n_i \mathbf{b}_i \mid n_i \in \mathbb{Z} \right\}.$$

Any such n-tuple of vectors $[\mathbf{b}_1, \ldots, \mathbf{b}_n]$ is called a basis of L: an n-dimensional lattice can be represented by a basis, that is, a matrix of $GL_n(\mathbb{R})$. Reciprocally, any matrix $B \in GL_n(\mathbb{R})$ spans a lattice: the set of all integer linear combinations of its rows, that is, $\mathbf{m}B$ where $\mathbf{m} \in \mathbb{Z}^n$. The *closest vector problem* (CVP) is the following: given a basis of $L \subseteq \mathbb{Z}^n$ and a target $\mathbf{t} \in \mathbb{Q}^n$, find a lattice vector $\mathbf{v} \in L$ minimizing the distance $\|\mathbf{v} - \mathbf{t}\|$. If we denote by d that minimal distance, then approximating CVP to a factor k means finding $\mathbf{v} \in L$ such that $\|\mathbf{v} - \mathbf{t}\| \leq kd$. A measurable part D of $\mathbb{R}^n$ is said to be a *fundamental domain* of a lattice $L \subseteq \mathbb{R}^n$ if the sets $\mathbf{b} + D$, where $\mathbf{b}$ runs over L, cover $\mathbb{R}^n$ and have pairwise disjoint interiors. If B is a basis of L, then the parallelepiped $\mathcal{P}_{1/2}(B) = \{\mathbf{x}B, \mathbf{x} \in [-1/2, 1/2]^n\}$ is a fundamental domain of L. All fundamental domains of L have the same Lebesgue measure: the volume $\mathrm{vol}(L)$ of the lattice L.

2.2 The GGH Signature Scheme

The GGH scheme [9] works with a lattice L in $\mathbb{Z}^n$. The secret key is a non-singular matrix $R \in \mathcal{M}_n(\mathbb{Z})$, with very short row vectors (their entries are polynomial in n). Two distributions for the generation of R were suggested in [9]:

- The square distribution (called "random lattice" in [9]), where R is uniformly distributed over $\{-\ell, \ldots, \ell\}^{n \times n}$ for some integer bound ℓ. Goldreich *et al.* [9] suggested $\ell = 4$ because the value of ℓ had almost no effect on the quality of the bases in their experiments.
- The hypercubic distribution (called "almost rectangular lattice" in [9]), where R is the sum of kI_n and a noise with square distribution for some ℓ. The GGH challenges [8] used such a distribution, with parameters $k = 4\lceil \sqrt{n} + 1 \rfloor + 1$ and a noise with square distribution $\{-4, \ldots, +3\}^{n \times n}$. Micciancio [20] noticed that this distribution has the weakness that it discloses the rough directions of the secret vectors.

The lattice L is the lattice in $\mathbb{Z}^n$ spanned by the rows of R: the knowledge of R enables the signer to approximate CVP rather well in L. The basis R is then transformed to a non-reduced basis B, which will be public. In the original scheme [9], B is the multiplication of R by sufficiently many small unimodular matrices. Micciancio [20] suggested to use the Hermite normal form (HNF) of L instead. As shown in [20], the HNF gives an attacker the least advantage (in a certain precise sense) and it is therefore a good choice for the public basis. The messages are hashed onto a "large enough" subset of $\mathbb{Z}^n$, for instance a large hypercube. Let $\mathbf{m} \in \mathbb{Z}^n$ be the hash of the message to be signed. The signer applies Babai's round-off CVP approximation algorithm [3] to get a lattice vector close to $\mathbf{m}$:

$$\mathbf{s} = \lfloor \mathbf{m}R^{-1} \rceil R,$$

so that $\mathbf{s} - \mathbf{m} \in \mathcal{P}_{1/2}(R) = \{\mathbf{x}R, \mathbf{x} \in [-1/2, 1/2]^n\}$. Of course, any other CVP approximation algorithm could alternatively be applied, for instance Babai's nearest plane CVP approximation algorithm [3]. To verify the signature $\mathbf{s}$ of $\mathbf{m}$, one would first check that $\mathbf{s} \in L$ using the public basis B, and compute the distance $\|\mathbf{s} - \mathbf{m}\|$ to check that it is sufficiently small.

2.3   NTRUSign

NTRUSIGN [11] is a special instantiation of GGH with the compact lattices from the NTRU encryption scheme [15], which we briefly recall: we refer to [11, 4] for more details. In the NTRU standards [4] being considered by IEEE P1363.1 [18], one selects $n = 251$ and $q = 128$. Let $\mathcal{R}$ be the ring $\mathbb{Z}[X]/(X^n - 1)$ whose multiplication is denoted by $*$. Using resultants, one computes a quadruplet $(f, g, F, G) \in \mathcal{R}^4$ such that $f * G - g * F = q$ in $\mathcal{R}$ and f is invertible mod q, where f and g have 0–1 coefficients (with a prescribed number of 1), while F and G have slightly larger coefficients, yet much smaller than q. This quadruplet is the NTRU secret key. Then the secret basis is the following $(2n) \times (2n)$ matrix:

$$
R = \begin{bmatrix}
f_0 & f_1 & \cdots & f_{n-1} & g_0 & g_1 & \cdots & g_{n-1} \\
f_{n-1} & f_0 & \cdots & f_{n-2} & g_{n-1} & g_0 & \cdots & g_{n-2} \\
\vdots & & \ddots & \vdots & \vdots & & \ddots & \vdots \\
f_1 & \cdots & f_{n-1} & f_0 & g_1 & \cdots & g_{n-1} & g_0 \\
F_0 & F_1 & \cdots & F_{n-1} & G_0 & G_1 & \cdots & G_{n-1} \\
F_{n-1} & F_0 & \cdots & F_{n-2} & G_{n-1} & G_0 & \cdots & G_{n-2} \\
\vdots & & \ddots & \vdots & \vdots & & \ddots & \vdots \\
F_1 & \cdots & F_{n-1} & F_0 & G_1 & \cdots & G_{n-1} & G_0
\end{bmatrix},
$$

where f_i denotes the coefficient of X^i of the polynomial f. Due to the special structure of R, it turns out that a single row of R is sufficient to recover the whole secret key. Because f is chosen invertible mod q, the polynomial $h = g/f \,(\mathrm{mod}\, q)$ is well-defined in $\mathcal{R}$: this is the NTRU public key. Its fundamental property is that $f * h \equiv g \,(\mathrm{mod}\, q)$ in $\mathcal{R}$. The polynomial h defines a natural public basis of L, which we omit (see [11]).

The messages are assumed to be hashed in $\{0, \ldots, q-1\}^{2n}$. Let $\mathbf{m} \in \{0, \ldots, q-1\}^{2n}$ be such a hash. We write $\mathbf{m} = (\mathbf{m}_1, \mathbf{m}_2)$ with $\mathbf{m}_i \in \{0, \ldots, q - 1\}^n$. It is shown in [11] that the vector $(\mathbf{s}, \mathbf{t}) \in \mathbb{Z}^{2n}$ which we would obtain by applying Babai's round-off CVP approximation algorithm to $\mathbf{m}$ using the secret basis R can be alternatively computed using convolution products involving $\mathbf{m}_1$, $\mathbf{m}_2$ and the NTRU secret key (f, g, F, G). In practice, the signature is simply $\mathbf{s}$ and not $(\mathbf{s}, \mathbf{t})$, as $\mathbf{t}$ can be recovered from $\mathbf{s}$ thanks to $\mathbf{h}$. Besides, $\mathbf{s}$ might be further reduced mod q, but its initial value can still be recovered because it is such that $\mathbf{s} - \mathbf{m}_1$ ranges over a small interval (this is the same trick used in NTRU decryption). This gives rise for standard parameter choices to a signature length of $251 \times 7 = 1757$ bits. While this signature length is much smaller than other lattice-based signature schemes such as GGH, it is still significantly larger than more traditional signature schemes such as DSA.

This is the basic NTRUSIGN scheme [11]. In order to strengthen the security of NTRUSIGN, perturbation techniques have been proposed in [12, 4, 13]. Roughly speaking, such techniques perturb the hashed message $\mathbf{m}$ before signing with the NTRU secret basis. However, it is worth noting that there is no perturbation in half of the parameter choices recommended in NTRU standards [4] under consideration by IEEE P1363.1. Namely, this is the case for the parameter choices ees251sp2, ees251sp3, ees251sp4 and ees251sp5 in [4]. For the other half, only a single perturbation is recommended. But NTRU has stated that the parameter sets presented in [13] are intended to supersede these parameter sets.

3   The Hidden Parallelepiped Problem

Consider the signature generation in the GGH scheme described in Section 2. Let $R \in \mathcal{M}_n(\mathbb{Z})$ be the secret basis used to approximate CVP in the lattice L. Let $\mathbf{m} \in \mathbb{Z}^n$ be the message digest. Babai's round-off CVP approximation algorithm [3] computes the signature $\mathbf{s} = \lfloor \mathbf{m}R^{-1} \rceil R$, so that $\mathbf{s} - \mathbf{m}$ belongs to the parallelepiped $\mathcal{P}_{1/2}(R) = \{\mathbf{x}R, \mathbf{x} \in [-1/2, 1/2]^n\}$, which is a fundamental domain of L. In other words, the signature generation is simply a reduction of the message $\mathbf{m}$ modulo the parallelepiped spanned by the secret basis R. If we were using Babai's nearest plane CVP approximation algorithm [3], we would have another fundamental parallelepiped (spanned by the Gram-Schmidt vectors of the secret basis) instead: we will not further discuss this case in this paper, since it does not create any significant difference and since this is not the procedure chosen in NTRUSIGN.

GGH [9] suggested to hash messages into a set much bigger than the fundamental domain of L. This is for instance the case in NTRUSIGN where the cardinality of $\{0, \ldots, q-1\}^{2n}$ is much bigger than the lattice volume q^n. Whatever the distribution of the message digest $\mathbf{m}$ might be, it would be reasonable to assume that the distribution $\mathbf{s} - \mathbf{m}$ is uniform (or very close to uniform) in the secret parallelepiped $\mathcal{P}_{1/2}(R)$. This is because the parallelepiped is a fundamental domain, and we would expect the output distribution of the hash function to be random in a natural sense. In other words, it seems reasonable to make the following assumption:

Assumption 1 (The Uniformity Assumption). *Let R be the secret basis of the lattice $L \subseteq \mathbb{Z}^n$. When the GGH scheme signs polynomially many "randomly chosen" message digests $\mathbf{m}_1, \ldots, \mathbf{m}_k \in \mathbb{Z}^n$ using Babai's round-off algorithm, the signatures $\mathbf{s}_1, \ldots, \mathbf{s}_k$ are such that the vectors $\mathbf{s}_i - \mathbf{m}_i$ are independent and uniformly distributed over $\mathcal{P}_{1/2}(R) = \{\mathbf{x}R, \mathbf{x} \in [-1/2, 1/2]^n\}$.*

Note that this is only an idealized assumption: in practice, the signatures and the message digests are integer vectors, so the distribution of $\mathbf{s}_i - \mathbf{m}_i$ is discrete rather than continuous, but this should not be a problem if the lattice volume is sufficiently large, as is the case in NTRUSIGN. Similar assumptions have been used in previous attacks [7, 25] on lattice-based signature schemes. We emphasize that all our experiments on NTRUSIGN do not use this assumption and work with real-life signatures.

We thus arrive at the following geometric learning problem (see Fig.1):

Problem 2 (The Hidden Parallelepiped Problem or HPP). *Let $V = [\mathbf{v}_1, \ldots, \mathbf{v}_n] \in GL_n(\mathbb{R})$. Define the parallelepiped spanned by V as $\mathcal{P}(V) = \{\sum_{i=1}^n x_i \mathbf{v}_i, \ x_i \in [-1, 1]\}$. Denote by $U(\mathcal{P})$ the uniform distribution on a parallelepiped $\mathcal{P}$. Given $\mathrm{poly}(n)$ samples from $U(\mathcal{P}(V))$, find a good approximation of the rows of $\pm V$.*

Algorithm 1. Solving the Hidden Parallelepiped Problem

Input: A polynomial number of samples uniformly distributed over a parallelepiped
$\mathcal{P}(V)$.

Output: Approximations of rows of $\pm V$.

1: Compute an approximation G of the Gram matrix $V^t V$ of V^t (see Section 4.1).
2: Compute the Cholesky factor L of G^{-1}, so that $G^{-1} = LL^t$.
3: Multiply the samples of $\mathcal{P}(V)$ by L to the right to obtain samples of $\mathcal{P}(C)$ where
   $C = VL$.
4: Compute approximations of rows of $\pm C$ by Algorithm 2 from Section 4.3.
5: Multiply each approximation by L^{-1} to the right to derive an approximation of a
   row of $\pm V$.

In the definition of the HPP, we chose $[-1, 1]$ rather than $[-1/2, 1/2]$ like in Assumption 1 to simplify subsequent calculations. Clearly, if one could solve the HPP, then one would be able to approximate the secret basis in GGH by collecting random pairs (*message, signature*). If the approximation was sufficiently good, one would recover the secret vectors simply by rounding the coordinates to their closest integer. If simple rounding failed, one could apply approximate CVP algorithms to try to recover the secret lattice vectors, as one knows a lattice basis from the GGH public key. The experiments of [22] on the GGH-challenges [8] show that in practice, one does not need to be extremely close to the lattice to recover the closest lattice vector, even in high dimension. But only experiments can tell if the approximation will be sufficiently good for existing lattice reduction algorithms, if simple rounding failed.

4   Learning a Parallelepiped

In this section, we propose a method to solve the Hidden Parallelepiped Problem (HPP), based on the following steps. First, we approximate the covariance matrix of the given distribution. This covariance matrix is essentially $V^t V$ (where V defines the given parallelepiped). We then exploit this approximation in order to transform our hidden parallelepiped $\mathcal{P}(V)$ into a unit hypercube: in other words, we reduce the HPP to the case where the hidden parallelepiped is a hypercube. Finally, we show how hypercubic instances of the HPP are related to a multivariate optimization problem based on the fourth moment, which we solve by a gradient descent.

We remark that the idea of approximating the covariance matrix was already present in the work of Gentry and Szydlo [25, 7]; however, after this basic step, our strategy differs completely from theirs. We now describe our algorithm in more detail.

4.1   The Gram Matrix/Covariance Leakage

It was first observed by Gentry and Szydlo [7, 25] that GGH signatures leak an approximation of the Gram matrix of the transpose of the secret basis. Here, we simply translate this observation to the HPP setting:

Lemma 1 (Gram Leakage). *Let* $V \in GL_n(\mathbb{R})$. *Let* $\mathbf{v}$ *be chosen uniformly at random over the parallelepiped* $\mathcal{P}(V)$. *Then:*

$$\text{Exp}[\mathbf{v}^t\mathbf{v}] = V^tV/3.$$

Proof. We can write $\mathbf{v} = \mathbf{x}V$ where $\mathbf{x}$ has uniform distribution over $[-1,1]^n$. Hence,

$$\mathbf{v}^t\mathbf{v} = V^t\mathbf{x}^t\mathbf{x}V.$$

An elementary computation shows that $\text{Exp}[\mathbf{x}^t\mathbf{x}] = I_n/3$ where I_n is the $n \times n$ identity matrix, and the lemma follows. $\square$

Hence, by multiplying by 3 the average of $\mathbf{v}^t\mathbf{v}$ over all the samples $\mathbf{v}$ of the hidden parallelepiped $\mathcal{P}(V)$, we obtain an approximation of V^tV, which is the Gram matrix of V^t. Note that $V^tV/3$ is simply the covariance matrix of $U(\mathcal{P}(V))$.

4.2   Morphing a Parallelepiped into a Hypercube

The second stage is explained by the following result:

Lemma 2 (Hypercube Transformation). *Let* $V \in GL_n(\mathbb{R})$. *Denote by* $G \in GL_n(\mathbb{R})$ *the symmetric positive definite matrix* V^tV. *Denote by* $L \in GL_n(\mathbb{R})$ *the Cholesky factor[1] of* G^{-1}, *that is,* L *is the unique lower-triangular matrix such that* $G^{-1} = LL^t$. *Then the matrix* $C = VL \in GL_n(\mathbb{R})$ *satisfies the following:*

1. *The rows of* C *are unit vectors which are pairwise orthogonal. In other words, C is an orthogonal matrix in* $O_n(\mathbb{R})$ *and* $\mathcal{P}(C)$ *is a unit hypercube.*
2. *If* $\mathbf{v}$ *is chosen uniformly at random over the parallelepiped* $\mathcal{P}(V)$, *then* $\mathbf{c} = \mathbf{v}L$ *is uniformly distributed over the hypercube* $\mathcal{P}(C)$.

Proof. The Gram matrix $G = V^tV$ is clearly symmetric positive definite. Then $G^{-1} = V^{-1}V^{-t}$ is also symmetric positive definite: it has a Cholesky factorization $G^{-1} = LL^t$ where L is lower-triangular matrix. Hence, $V^{-1}V^{-t} = LL^t$. Let $C = VL \in GL_n(\mathbb{R})$. Then:

$$CC^t = VLL^tV^t = VV^{-1}V^{-t}V^t = I.$$

For the second claim, let $\mathbf{v}$ have the uniform distribution $U(\mathcal{P}(V))$, then $\mathbf{v} = \mathbf{x}V$ where $\mathbf{x}$ has uniform distribution over $[-1,1]^n$. It follows that $\mathbf{v}L = \mathbf{x}VL = \mathbf{x}C$ has uniform distribution over $\mathcal{P}(C)$. $\square$

Lemma 2 says that by applying the transform L, we can map our parallelepiped samples uniformly distributed over $\mathcal{P}(V)$ into hypercube samples uniformly distributed over $\mathcal{P}(C)$. If we could approximate the rows of $\pm C$, we could also approximate the rows of $\pm V$ thanks to L^{-1}. In other words, we have reduced the Hidden Parallelepiped Problem into what one might call the Hidden Hypercube Problem (see Fig. 2). From an implementation point of view, we note that

---

[1] Instead of the Cholesky factor, one can take any matrix L such that $G^{-1} = LL^t$. We work with Cholesky factorization as this turns out to be more convenient in our experiments.

Fig. 2. The Hidden Hypercube Problem in dimension two

the Cholesky factorization (required for obtaining L) can easily be computed by a process close to the Gram-Schmidt orthogonalization process (see [10]). Lemma 2 assumes that we know the Gram matrix $G = V^t V$ exactly. If we only have an approximation of the Gram matrix G, then C will only be close to some orthogonal matrix in $O_n(\mathbb{R})$: the Gram matrix CC^t of C will be close to the identity matrix, and the images of our parallelepiped samples will have a distribution close to the uniform distribution of some unit hypercube.

4.3   Learning a Hypercube

For any $V = [\mathbf{v}_1, \ldots, \mathbf{v}_n] \in GL_n(\mathbb{R})$ and any integer $k \geq 1$, we define the k-th moment over a vector $\mathbf{w} \in \mathbb{R}^n$ as:

$$\mathrm{mom}_{V,k}(\mathbf{w}) = \mathrm{Exp}[\langle \mathbf{u}, \mathbf{w} \rangle^k],$$

where $\mathbf{u}$ is uniformly distributed over the parallelepiped $\mathcal{P}(V)$. Clearly, $\mathrm{mom}_{V,k}(\mathbf{w})$ can be approximated thanks to the samples of $\mathcal{P}(V)$. We stress that our moments are different from the moments previously considered in [11, 13]: our moments are functions, rather than fixed values. We are interested in the second and fourth moments. A straightforward calculation shows that for any $\mathbf{w} \in \mathbb{R}^n$, they are given by

$$\mathrm{mom}_{V,2}(\mathbf{w}) = \frac{1}{3} \sum_{i=1}^{n} \langle \mathbf{v}_i, \mathbf{w} \rangle^2$$

$$\mathrm{mom}_{V,4}(\mathbf{w}) = \frac{1}{5} \sum_{i=1}^{n} \langle \mathbf{v}_i, \mathbf{w} \rangle^4 + \frac{1}{3} \sum_{i \neq j} \langle \mathbf{v}_i, \mathbf{w} \rangle^2 \langle \mathbf{v}_j, \mathbf{w} \rangle^2$$

Note that the second moment is related to the Gram matrix/covariance mentioned in Section 4.1. When $V \in O_n(\mathbb{R})$, the second moment becomes $\|\mathbf{w}\|^2/3$ while the fourth moment becomes

$$\mathrm{mom}_{V,4}(\mathbf{w}) = \frac{1}{3}\|\mathbf{w}\|^4 - \frac{2}{15} \sum_{i=1}^{n} \langle \mathbf{v}_i, \mathbf{w} \rangle^4.$$

The gradient of the latter is therefore

$$\nabla \mathrm{mom}_{V,4}(\mathbf{w}) = \sum_{i=1}^{n} \left(\frac{4}{3} \left(\sum_{j=1}^{n} \langle \mathbf{v}_j, \mathbf{w} \rangle^2 \right) \langle \mathbf{v}_i, \mathbf{w} \rangle - \frac{8}{15} \langle \mathbf{v}_i, \mathbf{w} \rangle^3 \right) \mathbf{v}_i.$$

For $\mathbf{w}$ on the unit sphere the second moment is constantly $1/3$, and

$$\text{mom}_{V,4}(\mathbf{w}) = \frac{1}{3} - \frac{2}{15}\sum_{i=1}^{n}\langle\mathbf{v}_i, \mathbf{w}\rangle^4$$

$$\nabla\text{mom}_{V,4}(\mathbf{w}) = \frac{4}{3}\mathbf{w} - \frac{8}{15}\sum_{i=1}^{n}\langle\mathbf{v}_i, \mathbf{w}\rangle^3\mathbf{v}_i. \tag{1}$$

Lemma 3. *Let $V = [\mathbf{v}_1, \ldots, \mathbf{v}_n] \in O_n(\mathbb{R})$. Then the global minimum of $\text{mom}_{V,4}(\mathbf{w})$ over the unit sphere of $\mathbb{R}^n$ is $1/5$ and this minimum is obtained at $\pm\mathbf{v}_1, \ldots, \pm\mathbf{v}_n$. There are no other local minima.*

Proof. The method of Lagrange multipliers shows that for $\mathbf{w}$ to be an extremum point of $\text{mom}_{V,4}$ on the unit sphere, it must be proportional to $\nabla\text{mom}_{V,4}(\mathbf{w})$. By writing $\mathbf{w} = \sum_{i=1}^{n}\langle\mathbf{v}_i, \mathbf{w}\rangle\mathbf{v}_i$ and using Eq. (1), we see that there must exist some α such that $\langle\mathbf{v}_i, \mathbf{w}\rangle^3 = \alpha\langle\mathbf{v}_i, \mathbf{w}\rangle$ for $i = 1, \ldots, n$. In other words, each $\langle\mathbf{v}_i, \mathbf{w}\rangle$ is either zero or $\pm\sqrt{\alpha}$. It is easy to check that among all such points, only $\pm\mathbf{v}_1, \ldots, \pm\mathbf{v}_n$ form local minima. $\qquad\square$

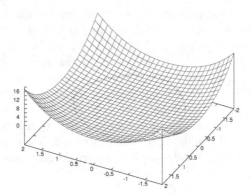

Fig. 3. The fourth moment for $n = 2$. The dotted line shows the restriction to the unit circle.

In other words, the hidden hypercube problem can be reduced to a minimization problem of the fourth moment over the unit sphere. A classical technique to solve such minimization problems is the gradient descent described in Algorithm 2. The gradient descent typically depends on a parameter δ, which has to be carefully chosen. Since we want to minimize the function here, we go in the opposite direction of the gradient. To approximate the gradient in Step 2 of Algorithm 2, we notice that

$$\nabla\text{mom}_{V,4}(\mathbf{w}) = \text{Exp}[\nabla(\langle\mathbf{u}, \mathbf{w}\rangle^4)] = 4\text{Exp}[\langle\mathbf{u}, \mathbf{w}\rangle^3\mathbf{u}].$$

This allows to approximate the gradient $\nabla\text{mom}_{V,4}(\mathbf{w})$ using averages over samples, like for the fourth moment itself.

Algorithm 2. Solving the Hidden Hypercube Problem by Gradient Descent

---

Parameters: A descent parameter δ.

Input: A polynomial number of samples uniformly distributed over a unit hypercube $\mathcal{P}(V)$.

Output: An approximation of some row of $\pm V$.

    1: Let $\mathbf{w}$ be chosen uniformly at random from the unit sphere of $\mathbb{R}^n$.

    2: Compute an approximation $\mathbf{g}$ of the gradient $\nabla \mathrm{mom}_4(\mathbf{w})$ (see Section 4.3).

    3: Let $\mathbf{w}_{new} = \mathbf{w} - \delta\mathbf{g}$.

    4: Divide $\mathbf{w}_{new}$ by its Euclidean norm $\|\mathbf{w}_{new}\|$.

    if $\mathrm{mom}_{V,4}(\mathbf{w}_{new}) \geq \mathrm{mom}_{V,4}(\mathbf{w})$ where the moments are approximated by sampling

    then

        return the vector $\mathbf{w}$.

    else

        Replace $\mathbf{w}$ by $\mathbf{w}_{new}$ and go back to Step 2.

    end if

---

5   Experimental Results

As usual in cryptanalysis, perhaps the most important question is whether or not the attack works in practice. We therefore implemented the attack in C++ and ran it on a 2GHz PC/Opteron. The critical parts of the code were written in plain C++ using double arithmetic, while the rest used Shoup's NTL library version 5.4 [24]. Based on early experiments, we chose $\delta = 0.7$ in the gradient descent (Algorithm 2), for all the experiments mentioned here. The choice of δ has a big impact on the behaviour of the gradient descent. We stress that our choices of the parameters may not be optimal, so the experimental results should be taken with caution. When doing several descents in a row, it is useful to relax the halting condition 5 in Algorithm 2 to abort descents which seem to make very little progress.

5.1   NTRUSign

We applied Algorithm 1 to real-life parameters of NTRUSign. More precisely, we ran the attack on NTRUSIGN-251 without perturbation, corresponding to the parameter choices ees251sp2, ees251sp3, ees251sp4 and ees251sp5 in the NTRU standards [4] under consideration by IEEE P1363.1 [18]. This corresponds to a lattice dimension of 502. We did not rely on the uniformity assumption: we generated genuine NTRUSIGN signatures of messages generated uniformly at random over $\{0, \ldots, q - 1\}^n$. The results of the experiments are summarized in Figure 4. For each given number of signatures, we generated a new set of signatures, and applied Algorithm 1: from the set of samples, we derived an approximation of the Gram matrix, and used it to transform the parallelepiped into a hypercube, and finally, we ran a series of random descents, starting with random points. The curve shows the average number of random descents needed to recover the secret key, based on the number of signatures: the average was

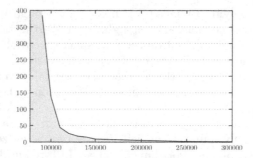

Fig. 4. Experiments on NTRUSIGN-251 without perturbation. The curve shows the average number of random descents required to recover the secret key, depending on the number of signatures, which is in the range 80,000–300,000.

computed using roughly a thousand descents. A success is counted when the simple rounding of the approximation obtained discloses *exactly* one of the vectors of the secret basis (which is sufficient to recover the whole secret basis in the case of NTRUSIGN): no additional approximate CVP stage is required here. Typically, a single random descent does not take much time: for instance, a usual descent for 150,000 signatures takes roughly ten minutes. When successful, a gradient descent may take as little as a few seconds. The minimal number of signatures to make the attack successful in our experiments was 90,000, in which case the required number of random descents to run was about 400. With 80,000 signatures, we tried 5,000 descents without any success, but maybe a substantially larger (yet realistic) number of descents would disclose the secret key: our experiments should not be considered as optimal. The curve given in Fig. 4 may vary a little bit, depending on the secret basis: for instance, for the basis used in the experiments of Fig. 4, the average number of random descents was 15 with 140,000 signatures, but it was 23 for another basis generated with the same NTRU parameters. It seems that the exact geometry of the secret basis has an influence, as will be seen in the analysis of Section 6. It is now clear that perturbation techniques are really mandatory for the security of NTRUSIGN, though it is currently unknown if such techniques are sufficient to prevent this kind of attacks.

5.2   The GGH Challenges

We also did a few experiments on the GGH-challenges [8], which range from dimension 200 to 400. Because there is actually no GGH signature challenge, we simply generated secret bases like in the GGH encryption challenges. This time, we relied on the uniformity assumption: we created samples uniformly distributed over the secret parallelepiped, and tried to recover the secret basis. Because the secret vectors are significantly longer than in the NTRUSIGN case, we never recovered directly the secret vector by simple rounding when the descent was successful, but the approximation obtained was sufficiently good to disclose the secret vector after applying Babai's CVP nearest plane algorithm [3], provided that the number of samples was large enough. When starting the descent, rather than starting with a random point on the unit sphere, we took advantage of the fact that we knew the rough directions of the secret vectors, due to the hyper-cubic distribution. As a result, the gradient descent takes very few iterations compared to the general case. For instance, in dimension 400, with 360,000 signatures, the gradient descent required only 6 iterations. The difference between the rounded vector found and the closest lattice vector was a 400-dimensional integer vector with only four ± 1 coefficients, the rest being zero. By applying Babai's CVP nearest plane algorithm [3] on an LLL-reduced basis (obtained by LLL reduction of the public HNF basis), we obtained immediately an exact vector of the secret basis: no strong reduction algorithm was needed (note that the error vector was much smaller than in the experiments of [22]). From our limited experiments ranging from dimension 200 to 400, it seemed that the number of required signatures was roughly quadratic in the dimension, if one wished for a very good success rate. To thwart such attacks, it seems that the GGH scheme would need to use secret bases with much bigger entries, unlike in the GGH challenges. This would certainly impact the efficiency of the signature scheme, notably the size of the signature, which is already not negligible even for NTRUSIGN.

6   Theoretical Analysis

Our goal in this section is to give a rigorous theoretical justification to the success of the attack. We will not try to give a tight estimate on the performance of the attack. Instead, we will show that given a large enough polynomial number of samples, Algorithm 1 succeeds in finding a good approximation to a row of V with some constant probability. Let us remark that it is possible that a rigorous analysis already exists in the ICA literature, although we were unable to find any (an analysis under some simplifying assumptions can be found in [17]). Also, Frieze et al. [5] sketch a rigorous analysis of a similar algorithm.

In order to approximate the covariance matrix, the fourth moment, and its gradient, our attack computes averages over samples. Mainly because the samples are independent (and identically distributed), we can use known bounds on large deviations such as the Chernoff bound (see, e.g., [2]) to obtain that with extremely high probability the approximations are very close to the true values. In our analysis below we omit the explicit calculations, as these are relatively standard.

6.1   Analysis of Algorithm 2

We start by analyzing Algorithm 2. For simplicity, we consider only the case in which the descent parameter δ equals $3/4$. A similar analysis holds for $0 < \delta < 3/4$. Another simplifying assumption we make is that instead of the stopping rule in Step 5 we simply repeat the descent step some small number r of times (which will be specified later).

For now, let us assume that the matrix V is an orthogonal matrix, so our samples are drawn from a unit hypercube $\mathcal{P}(V)$. We will later show that the actual matrix V, as obtained from Algorithm 1, is very close to orthogonal, and that this approximation does not affect the success of Algorithm 2.

Let us first analyze the behavior of Algorithm 2 under the assumption that all gradients are computed exactly without any error. Let $\mathbf{v}_1, \ldots, \mathbf{v}_n$ be n orthonormal vectors that define the parallelepiped $\mathcal{P}(V)$, and write any $\mathbf{w} \in \mathbb{R}^n$ as $\mathbf{w} = \sum_{i=1}^n w_i \mathbf{v}_i$. Then, using Eq. (1), we see that for $\mathbf{w}$ on the unit sphere,

$$\nabla \mathrm{mom}_{V,4}(\mathbf{w}) = \frac{4}{3}\mathbf{w} - \frac{8}{15}\sum_{i=1}^n w_i^3 \mathbf{v}_i.$$

Since we took $\delta = 3/4$, Step 3 in Algorithm 2 performs

$$\mathbf{w}_{new} = \frac{2}{5}\sum_{i=1}^n w_i^3 \mathbf{v}_i.$$

The vector is then normalized in Step 4. So we see that each step in the gradient descent takes a vector $(w_1, \ldots, w_n)$ to the vector $\alpha \cdot (w_1^3, \ldots, w_n^3)$ for some normalization factor α (where both vectors are written in the $\mathbf{v}_i$ basis). Hence, after r iterations, a vector $(w_1, \ldots, w_n)$ is transformed to the vector

$$\alpha \cdot (w_1^{3^r}, \ldots, w_n^{3^r})$$

for some normalization factor α.

Recall now that the original vector $(w_1, \ldots, w_n)$ is chosen uniformly from the unit sphere. It can be shown that with some constant probability, one of its coordinates is greater in absolute value than all other coordinates by a factor of at least $1 + \Omega(1/\log n)$ (first prove this for a vector distributed according to the standard multivariate Gaussian distribution, and then note that by normalizing we obtain a uniform vector from the unit sphere). For such a vector, after only $r = O(\log \log n)$ iterations, this gap is amplified to more than, say, $n^{\log n}$, which means that we have one coordinate very close to ± 1 and all others are at most $n^{-\log n}$ in absolute value. This establishes that if all gradients are known precisely, Algorithm 2 succeeds with some constant probability.

To complete the analysis of Algorithm 2, we now argue that it succeeds with good probability even in the presence of noise in the approximation of the gradients. First, it can be shown that for any $c > 0$, given a large enough polynomial number of samples, with very high probability all our gradient approximations are accurate to within an additive error of n^{-c} in the ℓ_2 norm (we have r such approximations during the course of the algorithm). This follows by a standard application of the Chernoff bound followed by a union bound. Now let $\mathbf{w} = (w_1, \ldots, w_n)$ be a unit vector in which one coordinate, say the jth, is greater in absolute value than all other coordinates by some factor $\eta \geq 1 + \Omega(1/\log n)$. Since $\mathbf{w}$ is a unit vector, this in particular means that $w_j > 1/\sqrt{n}$. As we saw before, in the vector $\tilde{\mathbf{w}}_{new} := \mathbf{w} - \delta\nabla \mathrm{mom}_4(\mathbf{w})$, this factor increases to η^3. We also have that $\tilde{\mathbf{w}}_{new,j} = \frac{2}{5}w_j^3 > \frac{2}{5}n^{-1.5} > n^{-2}$. By our assumption on the approximation $\mathbf{g}$, we have that for each i, $|\tilde{\mathbf{w}}_{new,i} - \mathbf{w}_{new,i}| \le n^{-c}$. So for any $k \ne j$,

$$\frac{|\mathbf{w}_{new,k}|}{|\mathbf{w}_{new,j}|} \le \frac{|\tilde{\mathbf{w}}_{new,k}| + n^{-c}}{|\tilde{\mathbf{w}}_{new,j}| - n^{-c}} \le \frac{|\tilde{\mathbf{w}}_{new,k}| + n^{-c}}{|\tilde{\mathbf{w}}_{new,j}|(1 - n^{-c+2})} \le (1 - n^{-c+2})^{-1}\eta^{-3} + 2n^{-c+2}.$$

So we see that a gap of η turns into a gap of $((1 - n^{-c+2})^{-1}\eta^{-3} + 2n^{-c+2})^{-1}$. A straightforward calculation shows that after $O(\log \log n)$ steps, this gap becomes $\Omega(n^{c-2})$. Hence, by choosing a large enough c, we can make the output vector very close to one of the $\pm\mathbf{v}_i$s. This completes our analysis of Algorithm 2.

6.2   Analysis of Algorithm 1

We now complete the analysis of the attack by analyzing Algorithm 1. Recall that a sample $\mathbf{v}$ from $\mathcal{P}(V)$ can be written as $\mathbf{x}V$ where $\mathbf{x}$ is chosen uniformly from $[-1, 1]^n$. So let $\mathbf{v}_i = \mathbf{x}_iV$ for $i = 1, \ldots, N$ be the input samples. Then our approximation G to the Gram matrix V^tV is given by $G = V^t\tilde{I}V$ where $\tilde{I} = \frac{3}{N}\sum\mathbf{x}_i^t\mathbf{x}_i$. We claim that with high probability, $\tilde{I}$ is very close to the identity matrix. Indeed, for $\mathbf{x}$ chosen randomly from $[-1, 1]^n$, each diagonal entry of $\mathbf{x}^t\mathbf{x}$ has expectation $1/3$ and each off-diagonal entry has expectation 0. Moreover, these entries take values in $[-1, 1]$. By the Chernoff bound we obtain that for any approximation parameter $c > 0$, if we choose, say, $N = n^{2c+1}$ then with very high probability each entry in $\tilde{I} - I$ is at most n^{-c} in absolute value. In particular, this implies that all eigenvalues of the symmetric matrix $\tilde{I}$ are in the range $1 \pm n^{-c+1}$.

Recall that we define L to be the Cholesky factor of $G^{-1} = V^{-1}\tilde{I}^{-1}V^{-t}$ and that $C = VL$. Now $CC^t = VLL^tV^t = \tilde{I}^{-1}$, which implies that C is close to an orthogonal matrix. Let us make this precise. Consider the singular value decomposition of C, given by $C = U_1DU_2$ where U_1, U_2 are orthogonal matrices and D is diagonal. Then $CC^t = U_1D^2U_1^t$ and hence $D^2 = U_1^t\tilde{I}^{-1}U_1$. From this it follows that the diagonal of D consists of the square roots of the reciprocals of the eigenvalues of $\tilde{I}$, which in particular means that all values on the diagonal of D are also in the range $1 \pm n^{-c+1}$.

Consider the orthogonal matrix $\tilde{C} = U_1U_2$. We will show below that with some constant probability, Step 4 yields a good approximation of a row of $\pm\tilde{C}$, call it $\tilde{\mathbf{c}}$. The output of Algorithm 1 is therefore

$$\tilde{\mathbf{c}}L^{-1} = \tilde{\mathbf{c}}C^{-1}V = (\tilde{\mathbf{c}}\tilde{C}^{-1})(\tilde{C}C^{-1})V = (\tilde{\mathbf{c}}\tilde{C}^{-1})(U_1D^{-1}U_1^t)V.$$

As we have seen before, all eigenvalues of $U_1D^{-1}U_1^t$ are close to 1, and therefore the above is a good approximation to a row of $\pm V$, given that $\tilde{\mathbf{c}}$ is a good approximation to a row of $\pm\tilde{C}$.

To complete the analysis, we will show that for large enough c, samples from $\mathcal{P}(C)$ 'look like' samples from $\mathcal{P}(\tilde{C})$. More precisely, assume that c is chosen such that the number of samples required by Algorithm 2 is less than, say, n^{c-4}. Then, it follows from Lemma 4 below that the statistical distance between a set of n^{c-4} samples from $\mathcal{P}(C)$ and a set of n^{c-4} samples from $\mathcal{P}(\tilde{C})$ is at most $O(n^{-1})$.

By our analysis of Algorithm 2, we know that when given samples from $\mathcal{P}(\tilde{C})$, it outputs an approximation of a row of $\pm\tilde{C}$ with some constant probability. Hence, when given samples from $\mathcal{P}(C)$, it must still output an approximation of a row of $\pm\tilde{C}$ with a probability that is smaller by at most $O(n^{-1})$ and in particular, constant.

Lemma 4. *The statistical distance between the uniform distribution on $\mathcal{P}(C)$ and that on $\mathcal{P}(\tilde{C})$ is at most $O(n^{-c+3})$.*

Proof. We first show that the parallelepiped $\mathcal{P}(C)$ is almost contained and almost contains the cube $\mathcal{P}(\tilde{C})$:

$$(1 - n^{-c+2})\mathcal{P}(\tilde{C}) \subseteq \mathcal{P}(C) \subseteq (1 + n^{-c+2})\mathcal{P}(\tilde{C}).$$

To show this, take any vector $\mathbf{y} \in [-1, 1]^n$. The second containment is equivalent to showing that all the coordinates of $\mathbf{y}U_1 DU_1^t$ are at most $1 + n^{-c+2}$ in absolute value. Indeed, by the triangle inequality,

$$\|\mathbf{y}U_1 DU_1^t\|_\infty \leq \|\mathbf{y}\|_\infty + \|\mathbf{y}U_1(D-I)U_1^t\|_\infty \leq 1 + \|\mathbf{y}U_1(D-I)U_1^t\|_2$$
$$\leq 1 + n^{-c+1}\sqrt{n} < 1 + n^{-c+2}.$$

The first containment is proved similarly. On the other hand, the ratio of volumes between the two cubes is $((1 + n^{-c+2})/(1 - n^{-c+2}))^n = 1 + O(n^{-c+3})$. From this it follows that the statistical distance between the uniform distribution on $\mathcal{P}(C)$ and that on $\mathcal{P}(\tilde{C})$ is at most $O(n^{-c+3})$. $\qquad\square$

References

1. M. Ajtai. Generating hard instances of lattice problems. In *Proc. of 28th STOC*, pages 99–108. ACM, 1996.
2. N. Alon and J. H. Spencer. *The probabilistic method.* Wiley-Interscience [John Wiley & Sons], New York, second edition, 2000.
3. L. Babai. On Lovász lattice reduction and the nearest lattice point problem. *Combinatorica*, 6:1–13, 1986.
4. Consortium for Efficient Embedded Security. Efficient embedded security standards #1: Implementation aspects of NTRUEncrypt and NTRUSign. Version 2.0 available available at [18], June 2003.
5. A. Frieze, M. Jerrum, and R. Kannan. Learning linear transformations. In *37th Annual Symposium on Foundations of Computer Science (Burlington, VT, 1996)*, pages 359–368. IEEE Comput. Soc. Press, Los Alamitos, CA, 1996.
6. C. Gentry, J. Jonsson, J. Stern, and M. Szydlo. Cryptanalysis of the NTRU signature scheme (NSS) from Eurocrypt 2001. In *Proc. of Asiacrypt '01*, volume 2248 of *LNCS*. Springer-Verlag, 2001.
7. C. Gentry and M. Szydlo. Cryptanalysis of the revised NTRU signature scheme. In *Proc. of Eurocrypt '02*, volume 2332 of *LNCS*. Springer-Verlag, 2002.
8. O. Goldreich, S. Goldwasser, and S. Halevi. Challenges for the GGH cryptosystem. Available at `http://theory.lcs.mit.edu/~shaih/challenge.html`.

9. O. Goldreich, S. Goldwasser, and S. Halevi. Public-key cryptosystems from lattice reduction problems. In *Proc. of Crypto '97*, volume 1294 of *LNCS*, pages 112–131. IACR, Springer-Verlag, 1997. Full version vailable at ECCC as TR96-056.
10. G. Golub and C. van Loan. *Matrix Computations*. Johns Hopkins Univ. Press, 1996.
11. J. Hoffstein, N. A. Howgrave Graham, J. Pipher, J. H. Silverman, and W. Whyte. NTRUSIGN: Digital signatures using the NTRU lattice. Full version of [12]. Draft of April 2, 2002, available on NTRU's website.
12. J. Hoffstein, N. A. Howgrave Graham, J. Pipher, J. H. Silverman, and W. Whyte. NTRUSIGN: Digital signatures using the NTRU lattice. In *Proc. of CT-RSA*, volume 2612 of *LNCS*, pages 122–140. Springer-Verlag, 2003.
13. J. Hoffstein, N. A. Howgrave Graham, J. Pipher, J. H. Silverman, and W. Whyte. Performances improvements and a baseline parameter generation algorithm for NTRUsign. In *Proc. of Workshop on Mathematical Problems and Techniques in Cryptology*, pages 99–126. CRM, 2005.
14. J. Hoffstein, J. Pipher, and J. H. Silverman. NSS: An NTRU lattice-based signature scheme. In *Proc. of Eurocrypt '01*, volume 2045 of *LNCS*. Springer-Verlag, 2001.
15. J. Hoffstein, J. Pipher, and J.H. Silverman. NTRU: a ring based public key cryptosystem. In *Proc. of ANTS III*, volume 1423 of *LNCS*, pages 267–288. Springer-Verlag, 1998. First presented at the rump session of Crypto '96.
16. A. Hyvärinen, J. Karhunen, and E. Oja. *Independent Component Analysis*. John Wiley & Sons, 2001.
17. A. Hyvärinen and E. Oja. A fast fixed-point algorithm for independent component analysis. *Neural Computation*, 9(7):1483–1492, 1997.
18. IEEE P1363.1. Public-key cryptographic techniques based on hard problems over lattices. http://grouper.ieee.org/groups/1363/lattPK/, June 2003.
19. R.J. McEliece. A public-key cryptosystem based on algebraic number theory. Technical report, Jet Propulsion Laboratory, 1978. DSN Progress Report 42-44.
20. D. Micciancio. Improving lattice-based cryptosystems using the Hermite normal form. In *Proc. of CALC '01*, volume 2146 of *LNCS*. Springer-Verlag, 2001.
21. D. Micciancio and S. Goldwasser. *Complexity of Lattice Problems: A Cryptographic Perspective*, volume 671 of *The Kluwer International Series in Engineering and Computer Science*. Kluwer Academic Publishers, Boston, Massachusetts, 2002.
22. P. Q. Nguyen. Cryptanalysis of the Goldreich-Goldwasser-Halevi cryptosystem from Crypto '97. In *Proc. of Crypto '99*, volume 1666 of *LNCS*, pages 288–304. IACR, Springer-Verlag, 1999.
23. P. Q. Nguyen and J. Stern. The two faces of lattices in cryptology. In *Proc. of CALC '01*, volume 2146 of *LNCS*. Springer-Verlag, 2001.
24. V. Shoup. NTL: A library for doing number theory. Available at http://www.shoup.net/ntl/.
25. M. Szydlo. Hypercubic lattice reduction and analysis of GGH and NTRU signatures. In *Proc. of Eurocrypt '03*, volume 2656 of *LNCS*. Springer-Verlag, 2003.

The Cramer-Shoup Encryption Scheme Is Plaintext Aware in the Standard Model

Alexander W. Dent

Royal Holloway, University of London,
Egham, Surrey, TW20 0EX, U.K.
a.dent@rhul.ac.uk

Abstract. In this paper we examine the notion of plaintext awareness as it applies to hybrid encryption schemes. We apply this theory to the Cramer-Shoup hybrid scheme acting on fixed length messages and deduce that the Cramer-Shoup scheme is plaintext-aware in the standard model. This answers a previously open conjecture of Bellare and Palacio on the existence of fully plaintext-aware encryption schemes.

1 Introduction

Plaintext awareness is a simple concept with a difficult explanation. An encryption scheme is plaintext aware if it is practically impossible for any entity to produce a ciphertext without knowing the associated message. This effectively renders a decryption oracle useless to an attacker, as any ciphertext submitted for decryption must either be invalid or the attacker must already know the decryption of that ciphertext and so does not gain any information by querying the oracle. Thus a scheme that is plaintext aware and semantically secure should be secure against adaptive attacks.

There are two problems with this simplistic approach. Firstly, if we wish to achieve the IND-CCA2 definition of security for an encryption scheme, then we have to be careful about how we define plaintext awareness, because, in this model, the attacker is always given one ciphertext for which he does not know the corresponding decryption (the challenge ciphertext). It is usually comparatively simple to achieve plaintext awareness when you do not have to consider the attacker as able to get hold of ciphertexts for which he does not know the corresponding decryption. We will follow the notation of Bellare and Palacio [4] and term this PA1 plaintext-awareness. A scheme that is IND-CPA and PA1 plaintext aware is only IND-CCA1 secure [4]. It is a lot harder to prove plaintext-awareness in full generality, when the attacker has access to an oracle that will return ciphertexts for which the attacker does not know the corresponding decryption, especially if the attacker has some measure of control over the probability distribution that the oracle uses to select the messages that it encrypts. This is termed PA2 plaintext awareness.

The second problem is that it is difficult to formally define plaintext awareness. The obvious way to define it is to say that for every attacker $\mathcal{A}$ that outputs a

S. Vaudenay (Ed.): EUROCRYPT 2006, LNCS 4004, pp. 289–307, 2006.

challenge ciphertext C, there exists a plaintext extractor $\mathcal{A}^*$ for $\mathcal{A}$ that outputs the decryption of C when given C as input. However, any encryption scheme that satisfies this definition of plaintext awareness in the standard model must necessarily fail to be IND-CPA secure. Hence, such a definition is not useful. For a satisfactory definition of plaintext awareness to be proposed, it is imperative that the plaintext extractor $\mathcal{A}^*$ be given some extra information about the actions that the attacker $\mathcal{A}$ took in order to compute the challenge ciphertext.

The original definition of plaintext awareness [2] was only given in the random oracle model and the plaintext extractor was given access to the oracle queries that the attacker made when constructing ciphertexts. This definition works well, but can only prove the security of a scheme in the random oracle model. Recently, a definition of plaintext awareness has been given in the standard model [4], where the plaintext extractor is also given access to the random coins that the attacker used in constructing the challenge ciphertext; thus the plaintext extractor can examine every action that the attacker took in its execution. Unfortunately, Bellare and Palacio were unable to prove that any scheme met their strongest (PA2) definition of plaintext awareness, although they suggested that the Cramer-Shoup scheme [5] was a very likely candidate.

This paper proves that the Cramer-Shoup scheme is plaintext aware in the standard model, thus proving the conjecture of Bellare and Palacio. The proof uses two new techniques: *encryption simulation* and *PA1+ plaintext awareness*. An encryption scheme that is simulatable is necessarily IND-CCA2 secure, and so the concept has limited use. However, the concept of PA1+ plaintext awareness may have further scope. The proof is obtained under several computational assumptions, including the controversial Diffie-Hellman Knowledge (DHK) assumption. We also assume the existence of groups on which the DDH problem is hard and the existence of suitably secure hash functions.

2   Preliminaries

2.1   Asymmetric Encryption Schemes

We briefly recap the notion of an asymmetric cipher and of a KEM-DEM hybrid cipher [5]. We will assume that the reader is familiar with the general theory of hybrid ciphers and will concentrate on introducing notation that will be used in this paper. An asymmetric encryption scheme is a triple of algorithms:

1. A probabilistic polynomial-time *key generation algorithm*, $\mathcal{G}$, which takes as input a security parameter 1^k and outputs a public/private key pair (pk, sk). The public key defines the *message space* $\mathcal{M}$, which is the set of all possible messages that can be submitted to the encryption algorithm, and the *ciphertext space* $\mathcal{C}$, which is the set of all possible ciphertexts that can be submitted to the decryption algorithm (and may be larger than the range of the encryption algorithm).
2. A (possibly) probabilistic polynomial-time *encryption algorithm*, $\mathcal{E}$, which takes as input a message $m \in \mathcal{M}$ and a public key pk, and outputs a ciphertext $C \in \mathcal{C}$. We will denote this as $C = \mathcal{E}(pk, m)$.

3. A deterministic polynomial-time *decryption algorithm*, $\mathcal{D}$, which takes as input a ciphertext $C \in \mathcal{C}$ and a secret key sk, and outputs either a message $m \in \mathcal{M}$ or the error symbol $\perp$. We denote this as $m = \mathcal{D}(sk, C)$.

The accepted notion of security for an asymmetric encryption scheme is assessed via the following game played between a two-stage attacker $\mathcal{A} = (\mathcal{A}_1, \mathcal{A}_2)$ and a hypothetical challenger:

1. The challenger generates a valid public/private key pair (pk, sk) by running $\mathcal{G}(1^k)$.
2. The attacker runs $\mathcal{A}_1$ on the input pk. It terminates by outputting two equal-length messages m_0 and m_1, as well as some state information *state*. During its execution $\mathcal{A}_1$ may query a decryption oracle that, when given $C \in \mathcal{C}$ will return $\mathcal{D}(sk, C)$.
3. The challenger picks a bit $b \in \{0, 1\}$ uniformly at random and computes the challenge ciphertext $C^* = \mathcal{E}(pk, m_b)$.
4. The attacker runs $\mathcal{A}_2$ on C^* and *state*. It terminates by outputting a guess b' for b. Again, during its execution, $\mathcal{A}_2$ may query a decryption oracle, subject to the restriction that it may not query the oracle on the input C^*.

The attacker wins the game if $b = b'$. The attacker's advantage is defined to be:

$$|Pr[b = b'] - 1/2| . \tag{1}$$

Definition 1. *If, for all polynomial-time attackers $\mathcal{A}$, the advantage that $\mathcal{A}$ has in winning the above game for an encryption scheme $(\mathcal{G}, \mathcal{E}, \mathcal{D})$ is negligible as a function of the security parameter k, then that encryption scheme is said to be IND-CCA2 secure.*

For more information on the basic security models for an asymmetric encryption scheme, the reader is referred to [2].

A hybrid cipher is an asymmetric cipher which uses a keyed symmetric algorithm, such as an encryption algorithm or a MAC, as a subroutine. Most hybrid ciphers can be presented as the combination of an asymmetric key encapsulation method (KEM) and a symmetric data encapsulation method (DEM). A KEM is a triple of algorithms consisting of:

1. A probabilistic, polynomial-time key generation algorithm, *Gen*, which takes as input a security parameter 1^k and outputs a public/private key pair (pk, sk).
2. A probabilistic, polynomial-time encapsulation algorithm, *Encap*, which takes as input a public key pk, and outputs a key K and an encapsulation of that key C. We denote this as $(C, K) = Encap(pk)$.
3. A deterministic, polynomial-time decapsulation algorithm, *Decap*, which takes as inputs the private key sk and an encapsulation C, and outputs a symmetric key K or the error symbol $\perp$. We denote this as $K = Decap(sk, C)$.

A DEM is a pair of algorithms consisting of:

1. A deterministic, polynomial-time encryption algorithm, ENC, which takes as input a message $m \in \{0,1\}^*$ of any length and a symmetric key K of some pre-determined length. It outputs an encryption $C = \text{ENC}_K(m)$.
2. A deterministic, polynomial-time decryption algorithm, DEC, which takes as input an encryption $C \in \{0,1\}^*$ and a symmetric key K of some pre-determined length, and outputs either a message $m \in \{0,1\}^*$ or the error symbol $\perp$.

A KEM and a DEM can be composed in the obvious way in create a hybrid encryption algorithm. The greatest advantage of designing a hybrid encryption scheme in terms of KEMs and DEMs is that Cramer and Shoup [5] were able to propose independent security criteria for the KEM and the DEM that guarantee that a secure KEM and a secure DEM combine to give a secure (IND-CCA2) encryption scheme. However, since our focus is on plaintext awareness, we will not need to discuss these security notions here.

2.2   Plaintext Awareness

We use the notions and notations given by Bellare and Palacio [4]. The notion of plaintext awareness in the standard model states that an encryption scheme $(\mathcal{G}, \mathcal{E}, \mathcal{D})$ is plaintext aware in the standard model if, for all ciphertext creators (attackers) $\mathcal{A}$, there exists a plaintext extractor $\mathcal{A}^*$ which takes as input the random coins of $\mathcal{A}$ and can answer the decryption queries of $\mathcal{A}$ in a manner that $\mathcal{A}$ cannot distinguish from a real decryption oracle. In order that $\mathcal{A}$ can be given access to ciphertexts for which it does not know the corresponding decryption, $\mathcal{A}$ will be allowed to query a plaintext creation oracle $\mathcal{P}$ with some query information aux. The plaintext creation oracle will pick a message at random (possibly from a distribution partially defined by aux) and returns the encryption of that message to the attacker[1]. Note that both $\mathcal{A}^*$ and $\mathcal{P}$ retain their state and their ability to access the same random tape between invocations.

We will assume that all the algorithms described are polynomial-time, probabilistic, state-based Turing machines, and that the random coins of the Turing machine $\mathcal{A}$ are denoted $R[\mathcal{A}]$. Plaintext awareness is formally defined using two games. First we define the **REAL** game:

1. The challenger generates a random key pair $(pk, sk) = \mathcal{G}(1^k)$ and creates an empty list of ciphertexts CLIST.
2. The attacker executes $\mathcal{A}$ on pk.
   - If the attacker queries the encryption oracle with query information aux, then the challenger generates a random message $m = \mathcal{P}(aux)$ and computes its encryption $C = \mathcal{E}(pk, m)$. It adds C to CLIST and returns C to the attacker.

---

[1] Technically, the plaintext creator will only generate a random message, and it will be left to the challenge to compute the encryption of that message. However, since the ciphertext creator and the plaintext extractor receive exactly the same inputs regardless of whether the challenger or the plaintext creator encrypts the message, we do not distinguish between the two cases.

– If the attacker queries the decryption oracle with a ciphertext C, then the decryption oracle returns $\mathcal{D}(sk, C)$. The attacker may not query the decryption oracle with any ciphertext appearing on CLIST.

The attacker terminates by outputting a bitstring x.

The **FAKE** game is defined as:

1. The challenger generates a random key pair $(pk, sk) = \mathcal{G}(1^k)$ and creates an empty list of ciphertexts CLIST.
2. The attacker executes $\mathcal{A}$ on pk.
   – If the attacker queries the encryption oracle with query information aux, then the challenger generates a random message $m = \mathcal{P}(aux)$ and computes its encryption $C = \mathcal{E}(pk, m)$. It adds C to CLIST and returns C to the attacker.
   – If the attacker queries the decryption oracle with a ciphertext C, then the decryption oracle returns $\mathcal{A}^*(C, pk, R[\mathcal{A}], \text{CLIST})$. The attacker may not query the decryption oracle with any ciphertext appearing on CLIST.

The attacker terminates by outputting a bitstring x.

Definition 2 (Plaintext awareness). *An asymmetric encryption scheme is said to be plaintext aware (PA2) if for all ciphertext creators $\mathcal{A}$, there exists a plaintext extractor $\mathcal{A}^*$ such that for all plaintext creators $\mathcal{P}$ and polynomial time algorithms Dist the advantage*

$$|Pr[Dist(x) = 1|\mathcal{A} \text{ plays } \textbf{REAL}] - |Pr[Dist(x) = 1|\mathcal{A} \text{ plays } \textbf{FAKE}]| \quad (2)$$

*that Dist has in distinguishing whether $\mathcal{A}$ interacts with the **REAL** game or the **FAKE** game is negligible as a function of the security parameter.*

An asymmetric encryption scheme is said to be PA1 if for all ciphertext creators $\mathcal{A}$ that make no encryption oracle queries, there exists a plaintext extractor $\mathcal{A}^*$ such that for all polynomial time algorithms Dist the advantage

$$|Pr[Dist(x) = 1|\mathcal{A} \text{ plays } \textbf{REAL}] - |Pr[Dist(x) = 1|\mathcal{A} \text{ plays } \textbf{FAKE}]| \quad (3)$$

is negligible as a function of the security parameter.

3 Simulatable Encryption Schemes

The aim of this paper is to show that the Cramer-Shoup scheme is plaintext-aware. In order to do this we take advantage of a very useful property that it possess: when instantiated with a suitable DEM, no attacker can distinguish valid ciphertexts from completely random bit strings. By this we mean that there exists a function f, which is in some sense invertible, that takes random bits as input and outputs bit strings that look like ciphertexts to an attacker. These bit strings are very unlikely to actually *be* valid ciphertexts (as we believe that the Cramer-Shoup scheme is plaintext aware) but no attacker can distinguish them from valid ciphertexts. We call this *encryption simulation*. For a simulatable encryption scheme, an attacker's ability to get hold of new ciphertexts in the PA2 model is roughly equivalent to an ability to get hold of blocks of random data. A scheme that remains plaintext-aware even when the attacker can get hold new fixed-length random strings on demand is said to be PA1+ plaintext aware. This notion is stronger than PA1, but conceptually weaker than PA2 plaintext awareness. In this section we will formally define simulation and PA1+ plaintext awareness, and show that any scheme that is both PA1+ and simulatable is PA2.

3.1   Simulatable Encryption

We will wish to work with encryption schemes that are simulatable, by which we mean that there exists a Turing machine f which take a string of random bits as input and produces an output that cannot be distinguished from real ciphertexts. The difference between f and the real encryption function is that f must be in some sense invertible. We envisage f taking long strings of random bits as input and producing a shorter output, and so we insist on the existence of a Turing machine f^{-1} which acts as a perfect inverse for f when used on the right, i.e.

$$f(f^{-1}(C)) = C \text{ for all } C \in \mathcal{C}. \tag{4}$$

However, since f^{-1} cannot act as a perfect inverse for f when used on the left, we merely require that $f^{-1}(f(r))$ looks like a randomly generated bit string, i.e. it is computationally infeasible to tell the difference between a random string r of the appropriate length and $f^{-1}(f(r))$. Hence, f^{-1} must be a probabilistic polynomial-time Turing machine; while, for technical reasons, f must be a deterministic polynomial-time Turing machine.

Definition 3 (Simulatable Encryption Scheme). *An asymmetric encryption scheme $(\mathcal{G}, \mathcal{E}, \mathcal{D})$ is simulatable if there exist two polynomial-time Turing machines (f, f^{-1}) such that:*

- *f is a deterministic Turing machine that takes the public key pk and an element $r \in \{0,1\}^l$ as input, and outputs elements of $\mathcal{C}$. For simplicity's sake, we shall often represent f as a function from $\{0,1\}^l$ to $\mathcal{C}$ and suppress the public key input.*
- *f^{-1} is a probabilistic Turing machine that takes the public key pk and an element $C \in \mathcal{C}$ as input, and outputs elements of $\{0,1\}^l$. Again, we will often represent f^{-1} as a function from $\mathcal{C}$ to $\{0,1\}^l$ and suppress the public key input.*
- *$f(f^{-1}(C)) = C$ for all $C \in \mathcal{C}$.*
- *There exists no polynomial-time attacker $\mathcal{A}$ that has a non-negligible advantage in winning the following game:*
  1. *The challenger generates a key pair $(pk, sk) = \mathcal{G}(1^k)$ and randomly chooses a bit $b \in \{0,1\}$.*
  2. *The attacker executes $\mathcal{A}$ on the input pk. The attacker has access to an oracle $\mathcal{O}_f$ that takes no input, generates a random element $r \in \{0,1\}^l$, and returns r if $b = 0$ and $f^{-1}(f(r))$ if $b = 1$. The attacker terminates by outputting a guess b' for b.*

The attacker wins if $b = b'$ and its advantage is defined in the usual way.

- *There exists no polynomial-time attacker $\mathcal{A}$ that has a non-negligible advantage in winning the following game:*
  1. *The challenger generates a key pair $(pk, sk) = \mathcal{G}(1^k)$, an empty list* CLIST, *and a bit b chosen randomly from $\{0, 1\}$.*
  2. *The attacker executes $\mathcal{A}$ on the input pk. The attacker has access to two oracles:*
     * *An encryption oracle that takes a message $m \in \mathcal{M}$ as input and returns an encryption C. If $b = 0$, then the oracle returns $C = \mathcal{E}(pk, m)$. If $b = 1$, then the oracle returns $C = f(r)$, for some randomly chosen $r \in \{0, 1\}^l$. In either case C is added to* CLIST.
     * *A decryption oracle that takes an encryption $C \in \mathcal{C}$ as input and returns $\mathcal{D}(sk, C)$. The attacker may not query the decryption oracle on any $C \in$ CLIST.*

     The attacker terminates by outputting a guess b' for b.

  The attacker wins if $b = b'$ and its advantage is defined in the usual way.

At this stage, and for technical reasons that will become apparent in the next section, we will restrict ourselves to encryption schemes that have fixed-length ciphertext spaces, i.e. the ciphertext space $\mathcal{C} = \{0, 1\}^n$ for some n. Normally, the simplest way of producing a cipher with fixed-length ciphertexts is to restrict the message space to fixed-length messages.

Theorem 1. *If $(\mathcal{G}, \mathcal{E}, \mathcal{D})$ is a simulatable encryption scheme then it is IND-CCA2 secure.*

Sketch Proof. Let $\mathcal{A}$ be an IND-CCA2 attacker for the scheme, and let **Game 1** be the game in which $\mathcal{A}$ interacts with the IND-CCA2 game properly. Let **Game 2** be similar to Game 1 except that the challenge ciphertext is computed by applying f to a randomly generated string $r \in \{0, 1\}^l$, rather than using the proper encryption algorithm. Let W_i be the event that $\mathcal{A}$ wins Game i.

Consider the following algorithm $\mathcal{B}$ against the simulatability of the encryption scheme:

1. The challenger generates a key pair $(pk, sk) = \mathcal{G}(1^k)$, an empty list CLIST, and a bit b chosen randomly from $\{0, 1\}$.
2. $\mathcal{B}$ executes $\mathcal{A}_1$ on the input pk. If $\mathcal{A}_1$ makes a decryption oracle query, then this is passed directly to $\mathcal{B}$'s decryption oracle and the result returned to $\mathcal{A}_1$. $\mathcal{A}_1$ terminates by outputting two equal-length messages m_0 and m_1, and some state information *state*.
3. $\mathcal{B}$ randomly chooses a bit $d \in \{0, 1\}$ and queries its encryption oracle with the message m_d. $\mathcal{B}$ receives back a ciphertext C^*.
4. $\mathcal{B}$ executes $\mathcal{A}_2$ on the input $(C^*, state)$. If $\mathcal{A}_2$ makes a decryption oracle query, then this is passed directly to $\mathcal{B}$'s decryption oracle and the result returned to $\mathcal{A}_2$. Note that $\mathcal{A}_2$ will never force $\mathcal{B}$ to make a decryption oracle query on $C^* \in$ CLIST due to the nature of the IND-CCA2 game. $\mathcal{A}_2$ terminates by outputting a guess d' for d.
5. If $d = d'$, then $\mathcal{B}$ outputs 1. Otherwise $\mathcal{B}$ outputs 0.

If $b = 0$ then $\mathcal{B}$ perfectly simulates Game 1 for $\mathcal{A}$. If $b = 1$ then $\mathcal{B}$ perfectly simulates Game 2 for $\mathcal{A}$. In both cases $\mathcal{B}$ outputs 1 if and only if $\mathcal{A}$ wins. It is well known that we may express $\mathcal{B}$'s advantage as:

$$\frac{1}{2}|Pr[\mathcal{B} \text{ outputs } 1|b = 0] - Pr[\mathcal{B} \text{ outputs } 1|b = 1]|. \tag{5}$$

However,

$$|Pr[\mathcal{B} \text{ outputs } 1|b = 0] - Pr[\mathcal{B} \text{ outputs } 1|b = 1]| = |Pr[W_0] - Pr[W_1]|. \tag{6}$$

Hence, $|Pr[W_1] - Pr[W_2]|$ is negligible, as the encryption algorithm is simulatable. In Game 2, though, the challenge ciphertext is completely independent of the messages supplied by the attacker. Therefore, $Pr[W_2] = 1/2$ and $(\mathcal{G}, \mathcal{E}, \mathcal{D})$ is IND-CCA2 secure.   □

Therefore, in some sense, the notion of encryption simulation is less useful than one might hope. It should be easier to prove that a scheme is IND-CCA2 secure, than to show that it is simulatable; and if we can show that a scheme is simulatable, then there is no need to consider whether it is plaintext aware, as we have already shown that it is IND-CCA2. However, since our goal in this paper is to show that PA2 schemes can exist, this notion will prove useful.

3.2   PA1+ Plaintext Awareness

For a simulatable encryption algorithm, a ciphertext creator's ability to get hold of new, randomly generated ciphertexts C (that are the encryption of messages drawn from some distribution) is roughly equivalent to being able to get hold of randomly generated strings $r = f^{-1}(C) \in \{0,1\}^l$. We define the PA1+ model as the extension of the PA1 model in which a ciphertext creator has access to an oracle which provides it with randomly generated bit strings of length l, and show that, for a simulatable encryption algorithm, this is enough to imply that the scheme is PA2 plaintext-aware.

We define the PA1+ model using the **REAL** and **FAKE** games as before. For an attacker $\mathcal{A}$ the **REAL** game works as follows:

1. The challenger generates a random key pair $(pk, sk) = \mathcal{G}(1^k)$.
2. The attacker executes $\mathcal{A}$ on pk. The attacker has access to a decryption oracle and to a randomness oracle.
   - If the attacker queries the randomness oracle, then the challenger generates a random strong $r \in \{0,1\}^l$, and returns r to the attacker.
   - If the attacker queries the decryption oracle with a ciphertext C, then the decryption oracle returns $\mathcal{D}(sk, C)$.
   The attacker terminates by outputting a bitstring x.

The **FAKE** game is defined in the obvious way:

1. The challenger generates a random key pair $(pk, sk) = \mathcal{G}(1^k)$ and creates an (empty) list RLIST of the random blocks that the attacker has been given.

2. The attacker executes A on pk. The attacker has access to a decryption oracle and to a randomness oracle.
   - If the attacker queries the randomness oracle, then the challenger generates a random strong $r \in \{0,1\}^l$, adds r to RLIST and returns r to the attacker.
   - If the attacker queries the decryption oracle with a ciphertext C, then the decryption oracle returns $A^*(C, pk, R[A], \text{RLIST})$.
   The attacker terminates by outputting a bitstring x.

Definition 4 (PA1+ Plaintext Awareness). *An asymmetric encryption scheme is said to be PA1+ plaintext aware if for all polynomial-time ciphertext creators A, there exists a polynomial-time plaintext extractor A^* such that for all polynomial-time distinguishing algorithms Dist the advantage*

$$|Pr[Dist(x) = 1|A \text{ plays } \boldsymbol{REAL}] - |Pr[Dist(x) = 1|A \text{ plays } \boldsymbol{FAKE}]| \quad (7)$$

that Dist has in distinguishing whether A interacts with the $\boldsymbol{REAL}$ game or the $\boldsymbol{FAKE}$ game is negligible as a function of the security parameter.

Intuitively, the difference between PA1 and PA1+ is in the ability for the ciphertext creator to act in a manner that is unpredictable by the plaintext extractor after the plaintext extractor has returned a message. For a scheme that is PA1, the plaintext extractor, when attempting to provide some sort of decryption of a ciphertext, knows exactly what the ciphertext creator is going to do with the ciphertext (as it has access to the ciphertext creator's random tape). Hence, the plaintext creator can tailor its response to make sure that that particular execution of the ciphertext creator cannot differentiate between the plaintext extractor's response and the response of a real decryption oracle. However, a PA1+ ciphertext creator has the ability to acquire random bits that could affect its execution *after* it has received the plaintext extractor's response, and so the plaintext extractor cannot tailor its response in the same way.

Theorem 2. *Let $(\mathcal{G}, \mathcal{E}, \mathcal{D})$ be a simulatable encryption algorithm. If $(\mathcal{G}, \mathcal{E}, \mathcal{D})$ is PA1+ then it is PA2.*

Proof. This proof works in several stages. We wish to show that for any PA2 ciphertext creator for the encryption scheme A, there exists a plaintext extractor A^*. First we show that any PA2 ciphertext creator A for the encryption scheme can be used to create a PA1+ ciphertext creator $\bar{A}$. Since the encryption scheme is PA1+ plaintext aware, there exist a plaintext extractor $\bar{A}^*$ for $\bar{A}$. We then show that we can use the plaintext extractor $\bar{A}^*$ for $\bar{A}$ to build a plaintext extractor A^* for A. We will use this technique liberally throughout this paper.

Let A be any PA2 ciphertext creator and let $\bar{A}$ be the PA1+ ciphertext creator that runs as follows.

1. Execute A.
   - If A makes a decryption oracle query, then $\bar{A}$ passes this query directly on to its own decryption oracle.

- If $\mathcal{A}$ makes an encryption oracle query (with query information aux), then $\bar{\mathcal{A}}$ queries its randomness oracle, receives back an l-bit block of randomness r, and returns $f(r)$ to $\mathcal{A}$.
2. $\mathcal{A}$ terminates by outputting a bitstring x. Output x.

Let $W_{0,Dist}$ be the event that $Dist(x) = 1$ when $\mathcal{A}$ interacts with the PA2 model and a real decryption oracle. Let $W_{1,Dist}$ be the event that $Dist(x) = 1$ when $\bar{\mathcal{A}}$ interacts with the PA1+ model and a real decryption oracle. It is clear that any non-negligible difference between $Pr[W_{0,Dist}]$ and $Pr[W_{1,Dist}]$ can be used to create an algorithm that can distinguish between ciphertexts and simulated ciphertexts, contravening Definition 3. Thus,

$$|Pr[W_{0,Dist}] - Pr[W_{1,Dist}]|$$

is negligible as a function of the security parameter.

Since $\bar{\mathcal{A}}$ is PA1+ ciphertext creator, there exists a plaintext extractor $\bar{\mathcal{A}}^*$ for $\bar{\mathcal{A}}$. Let $W_{2,Dist}$ be the event that $Dist(x) = 1$ when $\bar{\mathcal{A}}$ interacts with the PA1+ model and $\bar{\mathcal{A}}^*$ is used to simulate the decryption oracle. Since $\bar{\mathcal{A}}^*$ is a successful plaintext extractor for $\bar{\mathcal{A}}$, we have that

$$|Pr[W_{1,Dist}] - Pr[W_{2,Dist}]|$$

is negligible as a function of the security parameter.

We now alter slightly the way that the randomness oracle works. Instead of randomly generated a block of randomness r and returning this to $\bar{\mathcal{A}}$, consider an oracle that randomly generates a block of randomness $r \in \{0,1\}^l$ and returns $f^{-1}(f(r))$ to the ciphertext creator. Let $W_{3,Dist}$ be the event that $Dist(x) = 1$ when the randomness oracle behaves in this way. Clearly, any significant difference between $Pr[W_{2,Dist}]$ and $Pr[W_{3,Dist}]$ can be used to create an algorithm that can distinguish between random blocks r and $f^{-1}(f(r))$, thus contravening the properties of f given in Definition 3. Hence,

$$|Pr[W_{2,Dist}] - Pr[W_{3,Dist}]|$$

is negligible as a function of the security parameter.

If we examine the architecture now, we notice that RLIST contains elements of the form $f^{-1}(f(r))$, and $\mathcal{A}$ (being run as a subroutine of $\bar{\mathcal{A}}$) is given elements of the form $f(f^{-1}(f(r))) = f(r)$. Consider now a situation where

- the randomness oracle returns $f(r)$ instead of $f^{-1}(f(r))$
- to the ciphertext creator $\mathcal{A}$ (instead of $\bar{\mathcal{A}}$),
- and decryption queries are answered using a plaintext extractor $\mathcal{A}^*$. $\mathcal{A}^*$ works by executing $\bar{\mathcal{A}}^*$ on the input $(pk, C, R[\mathcal{A}], \text{RLIST})$, where C is the ciphertext to be decrypted and RLIST is the list of l-bit random blocks given by taking the responses C' returned the randomness oracle and computing $f^{-1}(C')$.

Let $W_{4,Dist}$ be the event that $Dist(x) = 1$ in this model. Clearly, the functionality of this model is identical to the previous model. Hence,

$$Pr[W_{3,Dist}] = Pr[W_{4,Dist}].$$

We may now consider the model in which the randomness oracle reverts to being an encryption oracle. I.e. instead of returning $f(r)$ for some randomly chosen l-bit block r, it returns the encryption $\mathcal{E}(m, pk)$ for message $m = \mathcal{P}(aux)$. Let $W_{5, Dist(x)}$ be the event that $Dist(x) = 1$ in this model. As before, if there is any significant difference between $Pr[W_{4, Dist(x)}]$ and $Pr[W_{5, Dist(x)}]$, then we may build an algorithm that distinguishes between ciphertexts and simulated ciphertexts, contravening Definition 3. Therefore,

$$|Pr[W_{4, Dist(x)}] - Pr[W_{5, Dist(x)}]|$$

is negligible. However, this means that

$$|Pr[W_{0, Dist(x)}] - Pr[W_{5, Dist(x)}]|$$

is negligible as a function of the security parameter, and so that $\mathcal{A}$ has a successful plaintext extractor $\mathcal{A}^*$. Therefore, $(\mathcal{G}, \mathcal{E}, \mathcal{D})$ is PA2 plaintext aware.   □

3.3 PA1+ Plaintext-Aware KEMs

It will be convenient for us to work with the hybrid version of the Cramer-Shoup encryption scheme. In this section we will show that a KEM-DEM scheme composed of a PA1+ KEM and an arbitrary DEM is PA1+.

We start by defining what we mean by a PA1+ KEM. The PA1+ model for a KEM is the obvious extension of the PA1+ model for an encryption scheme. Formally, we define the **REAL** game as:

1. The challenger generates a random key pair $(pk, sk) = Gen(1^k)$.
2. The attacker executes $\mathcal{A}$ on pk.
   - If the attacker queries the randomness oracle, then the oracle generates a fixed-length random string $r \in \{0, 1\}^l$ uniformly at random and returns r to the attacker.
   - If the attacker queries the decapsulation oracle with a ciphertext C, then the decapsulation oracle returns $Decap(sk, C)$.

   The attacker terminates by outputting a bitstring x.

The **FAKE** game is defined as follows:

1. The challenger generates a random key pair $(pk, sk) = Gen(1^k)$.
2. The attacker executes $\mathcal{A}$ on pk.
   - If the attacker queries the randomness oracle, then the oracle generates a fixed-length random string $r \in \{0, 1\}^l$ uniformly at random, adds r to RLIST and returns r to the attacker.
   - If the attacker queries the decapsulation oracle with a ciphertext C, then the decapsulation oracle returns $\mathcal{A}^*(C, pk, R[\mathcal{A}], \text{RLIST})$.

   The attacker terminates by outputting a bitstring x.

Definition 5. *A KEM is said to be PA1+ if, for all ciphertext creators $\mathcal{A}$, there exists a plaintext extractor $\mathcal{A}^*$ such that for all polynomial time distinguishers Dist the advantage*

$$|Pr[Dist(x) = 1|\mathcal{A} \text{ plays } \textbf{REAL}] - |Pr[Dist(x) = 1|\mathcal{A} \text{ plays } \textbf{FAKE}]| \quad (8)$$

*that Dist has in distinguishing whether $\mathcal{A}$ interacts with the **REAL** game or the **FAKE** game is negligible as a function of the security parameter.*

Theorem 3. *A hybrid encryption scheme composed of a PA1+ KEM and an arbitrary DEM is PA1+.*

Proof. We show that any ciphertext creator $\mathcal{A}$ for the encryption scheme can be used to create a ciphertext creator $\bar{\mathcal{A}}$ for the KEM. Since the KEM is plaintext aware, there exists a plaintext extractor $\bar{\mathcal{A}}^*$ for $\bar{\mathcal{A}}$. We then use $\bar{\mathcal{A}}^*$ to construct a plaintext extractor $\mathcal{A}^*$ for $\mathcal{A}$.

Let $\mathcal{A}$ be a ciphertext creator for the hybrid encryption scheme. We define the ciphertext creator $\bar{\mathcal{A}}$ for the KEM as the algorithm that executes $\mathcal{A}$. If $\mathcal{A}$ queries the decryption oracle with a ciphertext (C_1, C_2), then $\bar{\mathcal{A}}$ queries the decapsulation oracle with encapsulation C_1. If the oracle returns $\bot$ then $\bar{\mathcal{A}}$ returns $\bot$ to $\mathcal{A}$. Otherwise the oracle returns a key K and $\bar{\mathcal{A}}$ returns $\text{Dec}_K(C_2)$ to $\mathcal{A}$. Any queries that $\mathcal{A}$ makes to the randomness oracle are passed directly on to $\bar{\mathcal{A}}$'s randomness oracle, and the results returned to $\mathcal{A}$.

Since $\bar{\mathcal{A}}$ is a valid ciphertext creator for the KEM, there exists a plaintext extractor $\bar{\mathcal{A}}^*$. We define a plaintext extractor $\mathcal{A}^*$ for $\mathcal{A}$ as follows. On the submission of a ciphertext (C_1, C_2), $\mathcal{A}^*$ executes $\bar{\mathcal{A}}^*$ on C_1. If $\bar{\mathcal{A}}^*$ returns $\bot$, then $\mathcal{A}^*$ returns $\bot$ to $\mathcal{A}$. Otherwise $\bar{\mathcal{A}}^*$ returns a key K, and $\bar{\mathcal{A}}^*$ returns $\text{Dec}_K(C_2)$. It is easy to see that the system in which $\mathcal{A}$ interacts with its decryption oracle (in the **REAL** or **FAKE** game) is the same as $\bar{\mathcal{A}}$ interacting with its decryption oracle in the same game. Hence, the outputs of $\mathcal{A}$ must be indistinguishable regardless of the game which $\mathcal{A}$ is playing. □

4   The Cramer-Shoup Scheme

In this section we will show that the Cramer-Shoup scheme, when applied to fixed length messages, is fully plaintext aware (PA2). This will prove a conjecture of Bellare and Palacio [4] by showing PA2 schemes can exist in the standard model. For our purposes, the Cramer-Shoup scheme will consist of the Cramer-Shoup KEM and an Encrypt-then-MAC DEM using a suitably secure encryption algorithm and MAC algorithm. Note that this is slightly different to the Cramer-Shoup scheme proven PA1 plaintext aware by Bellare and Palacio [4], but that similar techniques could have been used to prove that this scheme is PA1. We will define the Cramer-Shoup KEM as working over an arbitrary group G: this will make it easier to separate the properties required from the scheme from those that are required from the group.

Definition 6 (Cramer-Shoup KEM). *The Cramer-Shoup KEM is defined by the following three algorithms:*

- *The key generation algorithm which runs as follows:*
  1. *Generate a cyclic group G of order q and a generator g for G.*
  2. *Randomly select $w \in \mathbb{Z}_q^*$ and set $W = g^w$.*
  3. *Randomly select elements x, y and z from $\mathbb{Z}_q$, and set $X = g^x$, $Y = g^y$, and $Z = g^z$.*
  4. *The public key consists of (g, q, W, X, Y, Z). The private key consists of (g, q, w, x, y, z). Note that both the encapsulation and decapsulation algorithms also make use of a hash function $Hash : G \times G \rightarrow \mathbb{Z}_q$ and a key derivation function $KDF : G \times G \rightarrow \{0,1\}^n$, where n is the (fixed) length of the required symmetric key.*
- *The encapsulation algorithm which runs as follows:*
  1. *Randomly select $u \in \mathbb{Z}_q$ and set $A = g^u$, $\hat{A} = W^u$ and $B = Z^u$.*
  2. *Set $K = KDF(A, B)$.*
  3. *Set $v = Hash(A, \hat{A})$.*
  4. *Set $D = X^u Y^{uv}$.*
  5. *Output the key K and the encapsulation $(A, \hat{A}, D)$.*
- *The decapsulation algorithm which runs as follows:*
  1. *Set $v = Hash(A, \hat{A})$.*
  2. *Check that $D = A^{x+yv}$ and that $\hat{A} = A^w$. If not, output $\bot$ and halt.*
  3. *Otherwise, set $B = A^z$.*
  4. *Output $K = KDF(A, B)$.*

4.1   Cramer-Shoup Is Simulatable

In order to show that the Cramer-Shoup scheme is PA2, we need to show two separate things: that it is PA1+ and that it is simulatable. In this section we will show that the Cramer-Shoup scheme is simulatable. In order to do this we have to show that we can find Turing machines f and f^{-1} that satisfy Definition 3. It is enough to show that there exists Turing machines (Kf, Kf^{-1}) and (Df, Df^{-1}) that simulate the KEM and DEM respectively. The function Df must accurately simulate the encryption of a fixed-length message by the DEM under a random key. The function Kf should produce encapsulations for which it is impossible to distinguish a correct encapsulation pair (C, K) from a simulated encapsulation pair $(Kf(r), K')$, where r is a randomly generated bitstring of length l and K' is a randomly generated symmetric key of the appropriate length. Formal treatments are given in the full version of the paper.

We construct our DEM from a suitably secure block cipher running in counter mode and from the EMAC MAC algorithm. Details of both of these schemes can be found in, for example, [7].

Theorem 4. *An Encrypt-then-MAC DEM composed of the counter mode encryption scheme and the EMAC MAC algorithm is simulatable if the underlying block cipher is indistinguishable from random.*

Sketch Proof. First, we note that the decryption oracle to which the attacker has access is of no use due to the unforgeability of the MAC. Hence, we remove it.

The result then follows from the indistinguishability of the MAC code [11] and the indistinguishability of the counter mode encryption [1].   □

We will now show that the Cramer-Shoup KEM is simulatable providing that it is instantiated on a group that is simulatable. Again, we only provide a loose description of a simulatable group here, leaving the formal description to the full version of this paper. A group is simulatable if there exists Turing machines (Gf, Gf^{-1}) analogous to those in Definition 3 for which it is impossible to distinguish between a randomly chosen group element $h \in G$ and a simulated group element $Gf(r)$, where r is chosen randomly from the set $\{0, 1\}^l$.

Theorem 5. *The Cramer-Shoup KEM is simulatable if it is instantiated on a simulatable group G on which the DDH problem is hard, and under the assumptions that the hash function Hash is target collision resistant and that the key derivation function KDF is unpredictable with random inputs.*

These assumptions are formally defined as follows. The notation is taken from the Cramer and Shoup paper [5].

Definition 7 (DDH). *For any polynomial-time algorithm $\mathcal{A}$ that outputs a single bit, we define AvdDDH to be*

$$|Pr[\mathcal{A}(p, q, g, g^x, g^y, g^{xy}) = 1 | x, y \text{ chosen randomly from } \mathbb{Z}_q]$$
$$-Pr[\mathcal{A}(p, q, g, g^x, g^y, g^z) = 1 | x, y, z \text{ chosen randomly from } \mathbb{Z}_q]| \quad (9)$$

The DDH assumption is that, for all polynomial-time algorithms $\mathcal{A}$, AvdDDH is negligible as a function of the security parameter.

Definition 8 (TCR). *Let Hash be the hash function used within the Cramer-Shoup scheme. For any polynomial-time algorithm $\mathcal{A}$, we define AvdTCR to be*

$$Pr[\mathcal{A}(\phi^*) \neq \phi^* \wedge Hash(\mathcal{A}(\phi^*)) = Hash(\phi^*)$$
$$|\phi^* \text{ chosen randomly from } \langle g \rangle \times \langle g \rangle] \quad (10)$$

The target collision resistance (TCR) assumption is that, for all polynomial-time algorithms $\mathcal{A}$, AvdTCR is negligible as a function of the security parameter.

Definition 9 (KDF). *Let KDF be the key derivation function used within the Cramer-Shoup scheme and l be the length of symmetric keys that the scheme is required to produce. Let E_1 be the event that A and B are chosen randomly from $\langle g \rangle$ and E_2 be the event that A is chosen randomly from $\langle g \rangle$ and K is chosen randomly from $\{0, 1\}^n$. For any polynomial time algorithm $\mathcal{A}$ that outputs a single bit, we define AvdDist(KDF) to be*

$$|Pr[\mathcal{A}(p, q, g, A, KDF(A, B)) = 1|E_1] - Pr[\mathcal{A}(p, q, g, A, K) = 1|E_2]| \quad (11)$$

The distribution assumption for KDF is that, for all polynomial-time algorithms $\mathcal{A}$, AvdDist(KDF) is negligible as a function of the security parameter.

Proof of Theorem 5. Let $\mathcal{A}$ be any attacker that is attempting to distinguish a real encapsulation pair (C, K) from a simulated encapsulation $(f(r), K')$ where r is a randomly generated bitstring of length l and K' is a randomly chosen symmetric key of the appropriate length. We will assume $\mathcal{A}$ makes at most q_E encapsulation oracle queries and q_D decapsulation oracle queries. Let **Game 1** be the game in which interacts with correct encryption and decryption oracles. Let **Game 2** be the game in which, for its first query to the encapsulation oracle, the attacker is interacting with the following algorithm rather than the true encapsulation algorithm:

1. Randomly select $u \in \mathbb{Z}_q$ and set $A = g^u$.
2. Randomly select $\hat{u} \in \mathbb{Z}_q \setminus \{u\}$ and set $\hat{A} = g^{\hat{u}}$.
3. Randomly select $K \in \{0, 1\}^n$.
4. Set $v = Hash(A, \hat{A})$ and $D = X^u Y^{uv}$.
5. Output the encapsulation $(A, \hat{A}, D)$ and the symmetric key K.

Let W_i be the event that the attacker $\mathcal{A}$ wins Game i. We use a result of Cramer-Shoup [5] to take us most of the way towards our goal.

Lemma 1 (Cramer-Shoup).

$$|Pr[W_1] - Pr[W_2]| \le AdvDDH + AdvTCR + AdvDist(KDF) + (q_E + 3)/q \quad (12)$$

Let **Game 3** be the game in which $\hat{A}$ is computed as follows:

2. Randomly select $\hat{u} \in \mathbb{Z}_q$ and set $\hat{A} = g^{\hat{u}}$.

Clearly the two games are identical unless $\hat{u} = u$, hence:

$$|Pr[W_2] - Pr[W_3]| \le 1/q. \quad (13)$$

Let **Game 4** be the game in which D is computed as follows:

4. Randomly select $r' \in \mathbb{Z}_q$ and set $D = g^{r'} Y^{uv}$.

Clearly, any difference in behaviour of the attacker between Game 3 and Game 4 means that he has distinguished between the Diffie-Hellman triple (A, X, X^u) and $(A, X, g^{r'})$. [Note that the proof makes use of the fact that we may compute Y^{uv} as A^{vy} in the case that we know y but do not know the discrete logarithm of A.] Hence,

$$|Pr[W_3] - Pr[W_4]| \le AdvDDH. \quad (14)$$

Let **Game 5** be the game in which D is computed as follows:

4. Randomly select $r' \in \mathbb{Z}_q$ and set $D = g^{r'}$.

This difference is pure conceptual, and so $Pr[W_4] = Pr[W_5]$. However, now each of the elements of the ciphertext, and the symmetric key, are randomly generated from their appropriate ranges. At this stage, and merely through altering the way we respond to the first encryption oracle query, we have

$$|Pr[W_1] - Pr[W_5]| \le 2 \cdot AdvDDH + AdvTCR + AdvDist(KDF) + (q_E + 4)/q. \quad (15)$$

Let **Game 6** be the game in which each of the encapsulation oracle queries is answered using the algorithm in Game 5, and not just the first one. By repeated application of the previous results we have that:

$$|Pr[W_1] - Pr[W_6]| \leq q_E\{2 \cdot AdvDDH + AdvTCR + AdvDist(KDF)) + (q_E + 4)/q\}.$$
(16)

Lastly, suppose the group G can be simulated by the pair of Turing machines (Gf, Gf^{-1}), and let **Game 7** be the game in which the encapsulation oracle computes the ciphertexts as follows.

1. Randomly select $r_1 \in \{0,1\}^l$ and set $A = Gf(r_1)$.
2. Randomly select $r_2 \in \{0,1\}^l$ and set $\hat{A} = Gf(r_2)$.
3. Randomly select $r_3 \in \{0,1\}^l$ and set $D = Gf(r_3)$.
4. Randomly select $K \in \{0,1\}^n$.
5. Output the encapsulation $(A, \hat{A}, D)$ and the symmetric key K.

Since the group is simulatable, the difference between success probabilities when the encapsulation is provided as in Game 6 is negligible. However this means that the difference between $Pr[W_1]$ and $Pr[W_7]$ is negligible, and so the KEM is simulatable.   □

Lastly, we need to show that simulatable groups exist. The obvious method to attempt to simulate a cyclic group G of order q with generator g is to define

$$f : \{0,1\}^l \to G \quad \text{by setting} \quad f(r) = g^r$$
(17)

where $l \gg q$. This provides a perfectly adequate definition of f, but leaves us know way of computing a machine f^{-1} (without solving the discrete logarithm problem in G). We are therefore required to use sneakier techniques.

Theorem 6. *If q and p are primes such that $p = 2q + 1$, and G is the subgroup of $\mathbb{Z}_p^*$ of order q, then G is simulatable.*

Sketch Proof. To show that G is simulatable, we are required to find Turing machines Gf and Gf^{-1} that are analogous to those given in Definition 3 but for which the output of Gf is a group element. Let k be an integer much larger than $\log_2(q)$ and let α be an integer. We consider a map $Gf : \{0,1\}^{\alpha k} \to G$ as follows. First, split the input r into α k-bit substrings $r = r_1 || r_2 || \ldots || r_\alpha$. Next, consider r_1 as an integer modulo p and test whether it is in G. If so, output $r_1 \bmod p$; otherwise consider the next substring of r in the same way. Since the distribution of $r_i \bmod p$ is almost uniform, we have that the probability that this algorithm fails to return a random element of G is approximately $1/2^\alpha$.

The inverse machine Gf^{-1} works similarly. Given a group element g, first chooses a random bit $b \in \{0,1\}$. If $b = 0$ then construct a random string r_1 of length k such that $r_1 \bmod p \equiv g$, append random data to r_1 so that it is αk-bits long and output the result. If b=1, then construct a random element r_1 such that $r_1 \bmod p \notin G$ and choose a new random bit for the block r_2. This process continues until either we choose a bit $b = 0$ or we have constructed α blocks of data (at which point the algorithm fails). Again, the probability that this happens is approximately $1/2^\alpha$.   □

4.2   Cramer-Shoup Is PA1+

Now we are only require to show that the Cramer-Shoup scheme is PA1+ to complete our proof that it is PA2. In this section we show that Cramer-Shoup is PA1+ on a simulatable group under the DHK assumption.

The DHK assumption states that any attacker given a random element W in a group generated by g, can only compute a Diffie-Hellman triple (W, g^u, W^u) if they know u.

Definition 10 (DHK). *Let G be a cyclic group G of order q and a generator g for G. The DHK assumption for G is that for any polynomial-time algorithm $\mathcal{A}$ there exists a polynomial-time extractor $\mathcal{A}^*$ such that the probability that $\mathcal{A}$ wins the following game is negligible.*

1. *The challenger randomly chooses an element $W \in G$.*
2. *The attacker executes $\mathcal{A}$ on the input W. The attacker has access to an oracle which, when given a triple $(W, A, \hat{A}) \in G^3$, executes $\mathcal{A}^*(W, A, \hat{A}, R[\mathcal{A}])$ and returns the result.*

The attacker wins the game if it submits a triple of the form (W, g^u, W^u) to the oracle and the oracle fails to return u. The challenger wins the game if $\mathcal{A}$ terminates without this event occurring.

The DHK assumption is certainly a very strong one. It was essentially introduced by Damgård in 1991 [6] and has been used in a number of applications [3,4,8,9]. However, it is unclear if the assumption holds true or not. Opponents of the assumption point out that it is not falsifiable (and so demonstrations that it is false must be complex) [10] and that variants of the assumption have been proven false [3]. Nevertheless, it is used to prove that a version of the Cramer-Shoup scheme is PA1 [4] and so we consider it a reasonable assumption under which to prove that the Cramer-Shoup scheme is PA2. The question of whether plaintext awareness can be demonstrated under weaker assumptions is a major open problem.

Theorem 7. *The Cramer-Shoup KEM is PA1+ in a simulatable group under the DHK assumption*

Sketch Proof Let $\mathcal{A}$ be any PA1+ ciphertext creator. We use the assumption that we can find algorithms that solve the DHK problem to build a plaintext extractor $\mathcal{A}^*$ for $\mathcal{A}$.

Consider the following plaintext extractor $\mathcal{A}^*$ for $\mathcal{A}$ that makes use of a DHK oracle. When it is first invoked, $\mathcal{A}^*$ receives the public key (W, X, Y, Z) and the random coins $R[\mathcal{A}]$ of $\mathcal{A}$. It first simulates the random coins of an attacker that only received W from the challenger and computed X, Y and Z. This is necessary because the DHK assumption is only valid when the challenger gives the attacker a single group element W. The simulated random coins string is given by:

$$R = Gf^{-1}(X) \| Gf^{-1}(Y) \| Gf^{-1}(Z) \| R[\mathcal{A}] \tag{18}$$

where Gf^{-1} is the inverse function associated with the simulatable group. If $\mathcal{A}$ makes a decryption oracle query on the ciphertext $(A, \hat{A}, D)$ then $\mathcal{A}^*$ proceeds as follows:

1. Query the DHK oracle with the triple $(W, A, \hat{A})$ and the coins (R, RLIST). The oracle will return a value $u \in \mathbb{Z}_q$ or the error symbol $\perp$. If the oracle returns $\perp$, then return $\perp$ and terminate.
2. Set $v = Hash(A, \hat{A})$.
3. Check that $A = g^u$, $\hat{A} = W^u$ and $D = X^u Y^{uv}$. If not, return $\perp$.
4. Set $B = Z^u$.
5. Set $K = KDF(A, B)$.
6. Return K.

It is clear that $\mathcal{A}^*$ correctly simulates the decapsulation algorithm providing that it obtains correct solutions to the DHK problem from the DHK oracle. The DHK assumption states that there exists an algorithm $\mathcal{A}'$ that can answer the queries of the DHK oracle given the randomness that $\mathcal{A}$ used in creating these queries. It is important to note that because the DHK oracle must give back answers which are completely correct, *and not answers that are merely indistinguishable from correct by* $\mathcal{A}$, it is sufficient to give $\mathcal{A}'$ access to the random coins that $\mathcal{A}$ used in creating its challenge. In other words, it is sufficient for $\mathcal{A}'$ to take as input the random coins R and all the random blocks RLIST that have been received by $\mathcal{A}$ up to the point at which the DHK oracle query was made. Hence, by the DHK assumption, there exists an algorithm $\mathcal{A}'$ that correctly responds to the DHK oracles queries, and so there exists a plaintext extractor $\mathcal{A}^*$ for $\mathcal{A}$. Hence, the Cramer-Shoup KEM is PA1+. □

5   Conclusion

We have shown that the Cramer-Shoup scheme is PA2 plaintext aware and therefore demonstrated the existence of fully plaintext aware encryption algorithms. However, in order to do this, we have had to use results which demonstrate that the Cramer-Shoup scheme is IND-CCA2 secure already. Therefore, if the primary goal of plaintext awareness is to make proving the security of an encryption scheme easier, then the results of this paper are of little use. We present these results not as a practical tool, but as a proof that PA2 plaintext aware schemes can be shown to exist.

Acknowledgements

The author would like to thank Martijn Stam for his detailed and insightful comments on the several drafts of this paper. Thanks should also be given to both Nigel Smart and the anonymous referees for their helpful comments. The author gratefully acknowledges the financial support of the EPSRC.

References

1. M. Bellare, A. Desai, E. Jokipii, and P. Rogaway. A concrete security treatment of symmetric encryption. In *Proceedings of the 38th Symposium on Foundations of Computer Science*, IEEE, 1997.
2. M. Bellare, A. Desai, D. Pointcheval, and P. Rogaway. Relations among notions of security for public-key encryption schemes. In H. Krawczyk, editor, *Advances in Cryptology – Crypto '98*, volume 1462 of *Lecture Notes in Computer Science*, pages 26–45. Springer-Verlag, 1998.
3. M. Bellare and A. Palacio. The knowledge-of-exponent assumptions and 3-round zero-knowledge protocols. In M. Franklin, editor, *Advances in Cryptology – Crypto 2004*, volume 3152 of *Lecture Notes in Computer Science*, pages 273–289. Springer-Verlag, 2004.
4. M. Bellare and A. Palacio. Towards plaintext-aware public-key encryption without random oracles. In P. J. Lee, editor, *Advances in Cryptology – Asiacrypt 2004*, volume 3329 of *Lecture Notes in Computer Science*, pages 48–62. Springer-Verlag, 2004.
5. R. Cramer and V. Shoup. Design and analysis of practical public-key encryption schemes secure against adaptive chosen ciphertext attack. *SIAM Journal on Computing*, 33(1):167–226, 2004.
6. I. B. Damård. Towards practical public key systems secure against chosen ciphertext attacks. In J. Feigenbaum, editor, *Advances in Cryptology – Crypto '91*, volume 576 of *Lecture Notes in Computer Science*, pages 445–456. Springer-Verlarg, 1991.
7. A. W. Dent and C. J. Mitchell. *User's Guide to Cryptography and Standards*. Artech House, 2005.
8. S. Hada and T. Tanaka. On the existence of 3-round zero-knowledge protocols. In H. Krawcyzk, editor, *Advances in Cryptology – Crypto '98*, volume 1462 of *Lecture Notes in Computer Science*, pages 408–423. Springer-Verlag, 1998.
9. H. Krawczyk. HMQV: A high-performance secure Diffie-Hellman protocol. In V. Shoup, editor, *Advances in Cryptology – Crypto 2005*, volume 3621 of *Lecture Notes in Computer Science*, pages 546–566. Springer-Verlag, 2005.
10. M. Naor. On cryptographic assumptions and challenges. In D. Boneh, editor, *Advances in Cryptology – Crypto 2003*, volume 2729 of *Lecture Notes in Computer Science*, pages 96–109. Springer-Verlag, 2003.
11. E. Petrank and C. Rackoff. CBC MAC for real-time data sources. *Journal of Cryptography*, 13(3):315–339, 2000.

Private Circuits II:
Keeping Secrets in Tamperable Circuits

Yuval Ishai*, Manoj Prabhakaran**, Amit Sahai***, and David Wagner†

Abstract. Motivated by the problem of protecting cryptographic hardware, we continue the investigation of *private circuits* initiated in [16]. In this work, our aim is to construct circuits that should protect the secrecy of their internal state against an adversary who may modify the values of an *unbounded* number of wires, *anywhere in the circuit*. In contrast, all previous works on protecting cryptographic hardware relied on an assumption that some portion of the circuit must remain *completely free* from tampering.

We obtain the first feasibility results for such private circuits. Our main result is an efficient transformation of a circuit C, realizing an arbitrary (reactive) functionality, into a private circuit C' realizing the same functionality. The transformed circuit can successfully detect any serious tampering and erase all data in the memory. In terms of the information available to the adversary, even in the presence of an unbounded number of adaptive wire faults, the circuit C' emulates a black-box access to C.

1 Introduction

Can you keep a secret when your brain is being tampered with? In this paper we study the seemingly paradoxical problem of constructing a circuit such that *all parts* of the circuit are open to tampering at the level of logic gates and wires, and yet the circuit can maintain the secrecy of contents of memory. We construct *private circuits* which, even as they are being tampered with, can detect such tampering and, if necessary, "self-destruct" to prevent leaking their secrets. We consider security against a powerful inquisitor who may adaptively query the circuit while tampering with an arbitrary subset of wires within the circuit, *including the part of the circuit that is designed to detect tampering.*

The above question is motivated by the goal of designing secure cryptographic hardware. While the traditional focus of cryptography is on analyzing *algorithms*, in recent years there have been growing concerns about *physical* attacks that

---

* Technion. This research was supported by grant 2004361 from the United States-Israel Binational Science Foundation (BSF) and grant 36/03 from the Israel Science Foundation.

** U.I.U.C.

*** U.C.L.A. This research was supported by grants from the NSF ITR and Cybertrust programs, a grant from BSF, an Alfred P. Sloan Foundation Fellowship, and a generous equipment grant from Intel.

† U.C. Berkeley.

S. Vaudenay (Ed.): EUROCRYPT 2006, LNCS 4004, pp. 308–327, 2006.

exploit the *implementations* (rather than the functionality) of cryptographic primitives. For instance, it is in some cases possible to learn the secret key of an encryption scheme by measuring the power consumed during an encryption operation or the time it takes for the operation to complete [19, 20]. Other types of physical attacks rely on inducing faults [6, 5, 20], electromagnetic radiation [28, 11, 29], magnetic fields [27], cache hit ratios [18, 24], probing wires using a metal needle [1], and others [17, 31, 32, 30, 2, 30]. In general, attacks of this type have proven to be a significant threat to the practical security of embedded cryptographic devices.

One possible approach for defeating the above type of attacks is by designing specific hardware countermeasures, such as adding large capacitors to hide the power consumption. Many such countermeasures have been proposed in the literature. An inherent limitation of these approaches is that each such countermeasure must be specially tailored for the set of *specific physical attacks* it is intended to defeat. For example, one might design physical protection against attacks based on electro-magnetic radiation, but still be vulnerable to attacks based on physical probes.

A different approach is to tackle the problem at the *logical* level, namely by designing algorithms that, when implemented, will be robust against a wide class of physical attacks. Here, we would want to classify attacks not based on the physical mechanism of the attack, but rather on the *logical effect of the attack* – for instance, can we defend against *all* physical attacks that toggle the value on a wire? Several ad-hoc approaches have been suggested (e.g., [10, 21, 15]) with some subsequently broken [7, 9]. Recently, a more general and theoretically sound study of physical security has been initiated in [16, 22, 12] (see Section 1.4 for an account of this related work).

The current paper continues this line of work, but departs from all previous work in the following fundamental way. All types of attacks that were previously considered from a theoretical perspective are either (1) in some sense *spatially limited*, and in particular cannot be applied to the entire circuitry on which they are mounted [12]; or (2) deal with observation rather than faults [16, 22]. The question that motivates our work is the intriguing possibility of offering protection even against adversaries that can tamper with the *entire* circuit. This goal might sound too ambitious. For instance, the adversary can easily modify the *functionality* of the circuit by simply destroying it completely. However, this does not rule out the possibility of preventing the adversary from learning the *secret information*, say a cryptographic key, stored in the circuit. Once the device is already in the hands of the adversary, secrecy is the primary relevant concern.

The above question is captured by our notion of a *private circuit*, which we also call a *self-destructing circuit*. Informally, such a circuit should carry out some specified functionality (say, encryption) while protecting the secrecy of its internal state (a key) even against an *unbounded* number of adversarial faults. A natural way for achieving this goal is to build a tamper detection mechanism which can detect faults and trigger a "self-destruction" mechanism to erase all internal state. (This is akin to a prisoner of war taking a suicide pill.) The central problem with implementing this approach in our setting is that *such a tamper detection circuitry as well as the self-destruction mechanism itself can be attacked and disabled by the adversary.* Thus, it is tempting to conjecture that such self-destructing circuits simply cannot exist.

In this paper, we obtain the first positive results establishing the feasibility of private circuits in the presence of adversarial faults that can affect *any* wire inside the circuit. Before describing our results, we give some further motivating discussion, and a more detailed account of the types of circuits and the fault model we consider.

1.1   Discussion of Difficulties

We briefly discuss some natural ideas to prevent loss of privacy due to faults and why they don't appear to work, as well as some inherent limitations to our model.

Natural Approaches. First, one can consider standard techniques for fault-tolerant circuits based on error-correcting codes or redundancy (see [26] and references therein). However, such approaches are limited to tolerating only a *bounded* number of total adversarial faults, whereas we are interested in the case where the adversary can induce, over time, an *unbounded* number of faults, eventually even faulting every wire in the circuit!

Next, one may think of using signature schemes or related techniques, which would work as follows at a high level: hard-wire into the circuit a signature on the circuit, and then verify the correctness of the signature before executing the original functionality, otherwise cause a "self-destruct" (*c.f.* [12]). In our context, this fails for a simple reason: the adversary can fault the wire in the circuit that contains the "Correct"/"Incorrect" output of the signature verification algorithm, so that this wire always reads "Correct", regardless of whether the signature verification succeeded or not.

Similarly, one may think of directly applying multi-party computing techniques [14, 4, 8] providing security against mobile Byzantine faults [23]. However, here we cannot rely on an "honest majority", since there is an unbounded number of faults, and every part of the circuit is susceptible to attacks. In protocols for multi-party computation with no honest majority, each party executes a large set of instructions, which invariably includes a verification step. In our model, the adversary can fault just this verification portion of each party's computation in order to fool the party into thinking that the verification always succeeds. Thus, whatever approach we take, we must somehow prevent this kind of attack.

Another idea that seems to immediately help is to use randomization: perhaps if we randomly encode "Correct" or "Incorrect" as 0 or 1, then we can prevent the above problems. But even this is problematic, because the adversary can, as its first set of actions, create faults that set all wires that should contain random values to 0, thereby eliminating the randomization. (In our approach, we are able to combine randomization ideas with other redundant encodings in order to fault-tolerantly detect such behavior.)

Limitations. To motivate the type of fault model we consider, it is instructive to address some inherent limitations on the type of adversarial behavior one could hope to resist, and still obtain a general transformation result that holds for all circuits. These limitations follow from the impossibility of program obfuscation [3]: The most immediate limitation is observed in [16], that it is impossible to protect against an attacker which can simultaneously *read* the values of *all* wires in the circuit. However, a number of natural *fault* models are also equivalent to program obfuscation. For instance, allowing the adversary to cause arbitrary immediate changes to the entire circuit trivially allows it to replace the entire circuit with one that outputs the contents of all memory, thus making the problem equivalent to program obfuscation. Similarly, if the adversary is allowed to insert or replace wires it can mount the same type of attack by undetectably adding wires from all memory cells to the output.

These limitations mean that we must consider attack models in which the adversary is more restricted. We concentrate on models in which the adversary can cause an *unbounded* number of faults over time, where these faults are localized to individual wires *anywhere* in the circuit.

1.2   The Model

We consider reactive functionalities, i.e., functions with an internal state that may be updated at each invocation. Such a functionality can be realized by a stateful boolean circuit C that, given an external input and its current internal state, produces an external output and a new internal state. (The state corresponds to some secret data stored in the circuit's memory.) We think of the interaction of such a circuit with the environment as being paced by *clock cycles*, where in each cycle the circuit receives an input, produces an output, and updates its internal state.[1] We would like to protect the secrecy of the internal state against an adversary that can induce faults in an unbounded number of wires. That is, in each clock cycle (or epoch) the adversary can adaptively choose t wires and permanently "set" (to 1), "reset" (to 0), or "toggle" the value of each wire. Then, the adversary can feed a new input to the modified circuit and observe the resulting output. By inducing such faults, the adversary's hope is to extract more information about the circuit's internal state than is possible via black-box access to the circuit. For instance, if the circuit implements an encryption or a signature scheme, the adversary may try to learn some nontrivial information about the secret key.

Our goal is to prevent the adversary from gaining any advantage by mounting the above type of attack. We formalize this requirement via a simulation-based

---

[1] An attacker may also try to tamper with the clock. To counter such attacks, we envision the use of volatile memory, such as DRAM, to implement memory cells whose contents fade over time if not refreshed regularly. In our main result, we need to assume that the attacker cannot induce too many faults within a single "epoch" defined by the amount of time it takes for the volatile memory to lose its value. If the adversary is tampering with the clock, then we define a clock cycle to the lesser of the actual clock cycle and one epoch.

definition (in the spirit of similar definitions from [16, 12]). Specifically, we say that a stateful circuit C' is a secure private (or self-destructing) implementation of C if there is a (randomized, efficient) transformation from an initial state s_0 and circuit C to an initial state s_0' and circuit C' such that:

1. $C'[s_0']$ realizes the same functionality as $C[s_0]$.
2. Whatever the adversary can observe by interacting with $C'[s_0']$ and adaptively inducing an unbounded number of wire faults, can be simulated by only making a *black-box* use of $C[s_0]$, without inducing any faults.

1.3 Our Contribution

We present general feasibility results, showing how to efficiently transform an arbitrary (reactive, stateful) circuit C into an equivalent self-destructing circuit C'. Specifically:

1. In the case of *reset only* wire faults (set wires or memory cells to 0), the circuit C' is secure against an unbounded number of adaptive faults. Security is either statistical, if C' is allowed to produce fresh randomness in each cycle, or is computational otherwise.
2. In the case of *arbitrary* wire faults (set wires or memory cells to 1, set to 0, or toggle the value), we can get the same results as above except that we limit the adversary to performing only a bounded number of faults per clock cycle.[2] Since the adversary can choose the faults to be permanent, the overall number of faults is still unbounded.

In all cases, the circuit C' is proven secure under the conservative simulation-based definition outlined above. Our techniques in both constructions can also yield privacy against a bounded number of probing attacks per cycle as per [16].

Our Techniques. A central high-level idea behind our constructions is the following. Given a circuit C, we compile it into a similar circuit C' which can be viewed as a "randomized minefield". As long as C' is not tampered with, it has the same functionality as C. However, any tampering with C' will lead with high probability to "exploding a mine", triggering an automatic self-destruction and rendering C' useless to the adversary.

Implementing the above approach is far from being straightforward. One problem that needs to be dealt with is preventing the adversary from learning some useful partial information by merely getting "lucky" enough to not land on a mine. This problem is aggravated by the fact that the adversary may possess partial information about the values of internal circuit wires, implied by the observed inputs and outputs. Another problem, already discussed above, is that of

---

[2] The complexity of the constructions depends linearly on the parameter t bounding the number of faults. In fact, our constructions resist attacks that can involve an *arbitrary* number of simultaneous faults, provided that no more than t faults are (simultaneously) concentrated in the same area. Thus, the task of mounting such a coordinated attack within a single clock cycle does not become easier as the size of the circuit grows.

protecting the self-destruction mechanism itself from being destroyed. This difficulty is overcome through a novel distributed and randomized self-destruction mechanism.

Combining the techniques in this paper with the results in [16], one can construct self-destructing circuits which simultaneously resist probing attacks in addition to fault attacks. (As in [16], and as discussed above, we need to consider a limited number of probes in each clock cycle.)

1.4   Related Work

As noted above, negative results for program obfuscation [3] rule out the possibility of defeating adversaries who can observe *all* values propagating through the circuit, for all circuits. This observation motivated the study of "private circuits", withstanding a *limited* number of such probing attacks [16]. The results of [16] do not consider active faults of the type considered here, yet are used as an essential building block in our main constructions.

A more general study of security against passive attacks was taken in [22] under an elegant framework of "physically observable cryptography". In contrast to [16], the focus of [22] is on obtaining model-independent *reductions* between physically secure primitives rather than implement them with respect to a specific attack model.

Most relevant to our work is the work of Gennaro et al. [12], who considered the problem of achieving security when an adversary can tamper with hardware. In contrast to the current work, they make the (seemingly necessary) assumption that there are parts of the circuitry that are *totally tamper-proof*. Indeed, as discussed above, the typical use in [12] is to have a signature stored in memory that is verified by the tamper-proof hardware. We stress that in our model, *no part of the circuitry* is free from tampering. In particular, all wires and internal memory cells can be affected. Thus, if an approach like the above is attempted, the adversary can tamper with the signature-checking portion of the circuitry (e.g., permanently fixing the output bit of the signature checker to indicate success). To the best of our knowledge, our work is the first that allows every portion of the hardware to be tampered with, at the level of individual wires between logical gates.

We note that [12] consider a more general type of tampering attack, albeit in a more restricted setting, in which the adversary can apply an arbitrary polynomial-time computable function to the contents of the memory. Defending against this general type of attacks is, in general, impossible in our setting (where no wire is free from tampering). Indeed, if the attacker could simply set the value of a wire to some arbitrary function of the other wires, then the impossibility result based on program obfuscation [3] would still hold.

Finally, it is instructive to contrast the positive results we achieve with a negative result from [12]. In the model of [12] it is shown that an attacker can recover the secret information stored in the memory, say a signature key, by sequentially setting or resetting bits of the memory and observing the effects of these changes on the output. Our model gets around this impossibility by allowing to feed values back into the memory. This form of feedback, which prevails in real-world computing devices, is essential for realizing the strong notion of privacy considered in this work.

1.5   Future Work

In this work we initiate the study of a fascinating question — can a circuit keep a secret even when *all* parts of the circuit are open to tampering? We give the first positive results, for an unbounded number of individual wire faults to any set of wires in the circuit. We believe the theory of private circuits, and private cryptographic implementations more generally, is still in its infancy and there are many more questions to address. Most notably, what other fault models allow for general positive results? As discussed above, negative results on obfuscation [3] give rise to severe restrictions on such fault models.

2   Preliminaries

Physical Model. We consider clocked circuits with memory gates. Specifically, our model is as follows:

- A memory gate has one input wire and one output wire: in each clock cycle, the output value of the memory gate becomes the input value from the previous clock cycle. The memory can be initialized with some data, which gets updated in each clock cycle. We shall denote a circuit C initialized with data D by $C[D]$.
- In addition to the memory gates, the circuit can have AND, OR and NOT gates, as well as input wires and output wires.
- The adversary can set each input wire to 0 or 1, and can read output wires.
- The adversary can also cause *faults* in the circuit. We consider the following kinds of faults: (1) setting a wire to 1 (which we call a "set" attack), (2) setting a wire to 0 (which we call a "reset" attack), or (3) toggling the value on a wire.
- We assume that wires are conducting: that is, with a single fault on a wire the adversary simultaneously causes faults everywhere that wire goes. In our construction in Section 5 we use NOT gates which are reversible (see e.g. [33]), so that faults on the output side of a NOT gate propagate to the input side. For AND and OR gates (as well as NOT gates in the construction in Section 4), faults can be introduced on input and output wires independently of each other.

Circuit Transformations. We shall refer to transformations which take a (circuit, data) pair to another (circuit, data) pair. It will always be the case that these are two separate transformations carried out independently of each other, one for the circuit and one for the data. However, for convenience and brevity we shall use a single transformation to denote these two transformations.

Definition 1. *A transformation between (circuit, data) pairs $T^{(k)}$ is called functionality preserving if for any pair (C, D), if $T^{(k)}(C, D) \mapsto (C_1, D_1)$ then $C[D]$ and $C_1[D_1]$ have the same input-output behavior.*

ISW Transformation and Security Definition. The starting point for our constructions is a transformation $T_{\text{ISW}}^{(k)}$ from [16]. The transformation yields a circuit which uses standard gates and some randomness gates (which output fresh random bits in each clock cycle). $T_{\text{ISW}}^{(k)}$ ensures that *reading* (but not tampering with) "a few wires" of the circuit in each clock cycle does not leak any information about the initial data in the memory (beyond what the output reveals). This is achieved using a (proactive) secret-sharing scheme, which shares each bit among k or more wires. Here we will not need any particulars of that construction, beyond the properties summarized below.

$T_{\text{ISW}}^{(k)}(C, D) \mapsto (C', D')$, is a functionality preserving transformation where each wire in C' is assigned at most two indices from $[k]$. To define the security guarantees of $T_{\text{ISW}}^{(k)}$ we define two adversaries: an "ideal" adversary which has only black-box access to C and a "real" adversary which can probe the internals of C'. For future reference we define these classes of adversaries $\mathcal{A}_{\text{IDEAL}}$ and $\mathcal{A}_{\text{ISW}}^{(k)}$ more formally, below.

- If $\mathcal{A} \in \mathcal{A}_{\text{IDEAL}}$ is given a circuit $C[D]$, then in every clock cycle $\mathcal{A}$ can feed inputs to the circuit and observe the outputs. This kind of access to the circuit is considered legitimate (ideal).
- If $\mathcal{A} \in \mathcal{A}_{\text{ISW}}^{(k)}$ is given a circuit $C'[D']$ with wires indexed from $[k]$, then in each cycle it can feed inputs to the circuit, read the outputs *and* probe wires in the circuit such that no more than $k - 1$ indices are covered by the wires probed in that clock cycle.[3]

Without loss of generality, all adversaries are considered to output a single bit at the end of the interaction with the circuit.

Lemma 1 (Properties of the ISW Transformation). *[16] There exists a functionality preserving transformation $T_{\text{ISW}}^{(k)}(C, D) \mapsto (C', D')$, where C' uses AND gates, XOR gates, NOT gates and "randomness gates," and each wire is assigned at most two indices from $[k]$, such that the following hold:*

1. *Values on any $k - 1$ wires in C' (excluding wires in the input and output phases), such that no two wires share an index, are distributed so that the following condition holds (distribution being as determined by the distribution of the outputs of the randomness gates during that clock cycle): any bit, even conditioned on all other $k - 2$ bits and all other information obtained by any $\mathcal{A}' \in \mathcal{A}_{\text{ISW}}^{(k)}$ in previous clock cycles, has entropy at least c for a fixed $c > 0$.*

---

[3] To be precise about the counting, we should consider the values on the wires that go into the memory at a clock cycle same as the values that come out of the memory at the next clock cycle. Thus probing one of these wires in one clock cycle counts towards probes in both clock cycles.

2. $\forall C$, $\exists \mathcal{S}_{\mathrm{ISW}}$ *(a universal simulator), such that* $\forall D$, $\forall \mathcal{A}' \in \mathcal{A}_{\mathrm{ISW}}^{(k)}$, *we have* $\mathcal{S}' = \mathcal{S}_{\mathrm{ISW}}^{\mathcal{A}'} \in \mathcal{A}_{\mathrm{IDEAL}}$, *and* $\mathcal{S}'$ *after interacting with* $C[D]$ *outputs 1 with almost the same probability as* $\mathcal{A}'$ *outputs 1 after interacting with* $C'\,[D'\,]$ *(the difference in probabilities being negligible in the security parameter* k*).*

We remark that in [16] these properties are not explicitly stated in this form. In particular in the second property above, [16] is interested only in restricting $\mathcal{A}'$ to probing at most $(k-1)/2$ wires. However to employ $T_{\mathrm{ISW}}^{(k)}$ within our transformations we shall use the fact that the construction allows $\mathcal{A}'$ to probe any number of wires as long as they cover at most $k-1$ indices.

3   Security When Circuits Are Completely Tamperable

In the next two sections we present our constructions which do not require any untamperable components (except the topology of the circuit and the atomic gates (AND, NOT, OR)). The adversary is allowed to change the values in any of the wires in the circuit. We give constructions for two scenarios:

1. **Tamper-resistance against "reset" attacks:** In this case the only kind of faults that the adversary can introduce into the circuit are "resets." That is, it can change the value of any wire to zero (but not to one). In each clock cycle, the adversary can set the input values, reset any number of wires of its choice and observe the outputs. We call this class of adversaries $\mathcal{A}_{\mathrm{RESET}}$.
2. **Tamper-resistance against "set, reset and toggle" attacks:** Here the adversary is allowed to set or reset the wires. That is, it can change the value of any wire to one or zero. Also, it can toggle the value in a wire (if the value prior to attack is zero, change it to one, and vice versa). There is an *a priori* bound on the number of new wires it can attack (set, reset or toggle) at each clock cycle. However, it is allowed to introduce *persistent (or permanent) faults* to any wire it attacks (such a fault will not be counted as a new fault in every cycle). Hence, after multiple clock cycles, the adversary can potentially have faults in *all* the wires in the circuit simultaneously. We call this class of adversaries $\mathcal{A}_{\mathrm{TAMPER}}^{(t)}$, where t is the bound on the number of wires the adversary can attack in each clock cycle.

The two constructions use similar techniques. First we introduce our basic construction techniques and proof ideas for the reset-only case, and then explain the extensions used to make the construction work for the general case.

4   Tamper-Resistance Against Reset Attacks

We present our construction as two transformations $T_1^{(k)}$ and $T_2^{(k)}$. The complete transformation consists of applying $T_1^{(k)}$ followed by $T_2^{(k)}$. The first transformation converts any given circuit to a private circuit which uses "encoded randomness gates" (which output fresh random bits in every cycle, but each bit of the output is encoded into a pair of bits as explained later). The second transformation converts the resulting circuit to a standard deterministic circuit (using only AND, NOT and OR gates), while preserving the security property. The formal security statements for $T_1^{(k)}$ and $T_2^{(k)}$ follow.

Lemma 2. *There is a polynomial time (in input size and security parameter k) functionality preserving transformation $T_1^{(k)}(C, D) \mapsto (C_1, D_1)$, where C_1 uses "encoded randomness gates," such that $\forall C$, $\exists S_1$ (a universal simulator), such that $\forall D$, $\forall A_1 \in A_{\text{RESET}}$, we have $S = S_1^{A_1} \in A_{\text{IDEAL}}$ and the following two experiments output 1 one with almost the same probability (the difference in probabilities being negligible in the security parameter k):*

- *Experiment A: S outputs a bit after interacting with $C[D]$.*
- *Experiment B: A_1 outputs a bit after interacting with $C_1[D_1]$.*

Lemma 3. *There is a polynomial time (in input size and security parameter k) functionality preserving transformation $T_2^{(k)}(C_1, D_1) \mapsto (C_2, D_2)$, where C_1 may use encoded randomness gates, such that $\forall C_1$, $\exists S_2$ (a universal simulator), such that $\forall D_1$, $\forall A \in A_{\text{RESET}}$, we have $A_1 = S_2^A \in A_{\text{RESET}}$ and the following two experiments output 1 with almost the same probability (the difference in probabilities being negligible in the security parameter k):*

- *Experiment B: A_1 outputs a bit after interacting with $C_1[D_1]$.*
- *Experiment C: A outputs a bit after interacting with $C_2[D_2]$.*

Theorem 1. *There is a polynomial time (in input size and security parameter k) functionality preserving transformation $T_{\text{RESET}}^{(k)}(C, D) \mapsto (C_2, D_2)$, such that $\forall C$, $\exists S_0$ (a universal simulator), such that $\forall D$, $\forall A \in A_{\text{RESET}}$, we have $S = S_0^A \in A_{\text{IDEAL}}$ and experiment A and experiment C output 1 with almost the same probability (the difference in probabilities being negligible in the security parameter k).*

Proof. This follows from the above two lemmas, by setting $T_{\text{RESET}}^{(k)}(C, D) = T_2^{(k)}(T_1^{(k)}(C, D))$ and $S_0^A = S_1^{S_2^A}$.

4.1   Proof of Lemma 2

As proof of Lemma 2 we first present the transformation $T_1^{(k)}$. We then will demonstrate a universal simulator as required in the Lemma and show the correctness of simulation.

The Transformation $T_1^{(k)}$. The transformation $T_1^{(k)}$ is carried out in two stages. In the first step, we apply the transformation $T_{\text{ISW}}^{(k)}$ from [16] to (C, D) to obtain (C', D').

Next we shall transform (C', D') further so that the following "encoding" gets applied to all the data: the bit 0 is mapped to a pair of bits 01 and the bit 1 is mapped to 10. We shall refer to this encoding as the *Manchester* encoding.

Encoding D' to get D_1 is straight-forward: we simply replace 0 and 1 by 01 and 10 respectively (thereby doubling the size of the memory and doubling the number of wires connecting the memory to the circuit). C' is transformed to get C_1 as follows:

1. The input data is passed through a simple encoding gadget, which converts 0 to 01 and 1 to 10. The encoding simply involves fanning out the signal into two: the first output wire and an input to a NOT gate whose output is the second output wire.
2. The "core" of the circuit C_1 is derived from C' as follows: every wire in C' is replaced by a pair of wires. Then the input wire pairs are connected to the outputs from the encoding gates (described above), and the output wire pairs are fed into the decoding phase (below). The gates to which the wires are connected are modified as follows:
   (a) Each randomness gate is replaced by an encoded randomness gate.
   (b) XOR and AND gates in C' are replaced by the gadgets shown in Figure 1. NOT gates are replaced by a gadget which simply swaps the two wires in the pair.
3. An "error cascade" stage (described below) is added before the output stage (including the output from the circuit to the memory).
4. A simple decoding stage is added just before the final output wires (excluding the wires going into the memory): the decoding is done by simply ignoring the second wire in the encoding of each signal.

Error Cascading. The circuit will be designed to "detect" reset attacks, and if an attack is detected, to erase all the data in the memory. (Such self-destruction is not *required* by Lemma 2, but it is a desirable property that is achieved by our construction.) The basic step in this is to ensure that if a detectable error is produced at some point in the circuit, it is propagated all the way to an "error cascading stage" (such an error propagation will be ensured by the gadgets in Figure 1). Then, the cascading stage will ensure that all the data in the memory and output is erased.

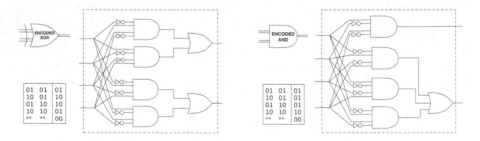

Fig. 1. XOR and AND Gadgets used by $T_1^{(k)}$. Note that the outputs of the gadgets are implemented as OR of ANDs of input wires and their NOTs. It is important that the gadgets do not have NOT gates except at the input side (to maintain the invariant that the encoding 11 does not appear in the circuit).

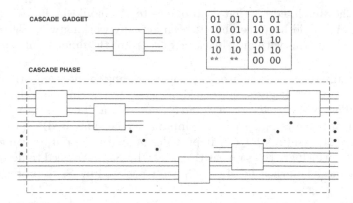

Fig. 2. The error-cascade phase and the truth table of the cascade gadget used by $T_1^{(k)}$

The detectable error referred to above is the invalid encoding 00 in a pair of wires corresponding to a single wire in C'. Recall that the only valid encodings are 01 and 10. We shall denote the encoding 00 by the symbolic value $\perp$. (We will show that the transformed circuit is so designed that the other invalid encoding, namely 11, will never occur in the circuit, even after any number of reset attacks.) As sketched in Figure 2, the error cascading phase is built using small cascade gadgets which take two encoded inputs and convert them both to $\perp$ if either of them is $\perp$. (These gadgets are implemented similar to the gadgets in Figure 1, with NOT gates only at the input side.) It is easy to see that if any of the input pairs has the value $\perp$, then after the cascade phase all output pairs will encode $\perp$.

All the wire pairs going to the output or back to the memory are passed through the cascade phase. In addition, for simplicity, we shall have all the wire pairs coming from the input and from the memory also go into the cascade phase. This will ensure that if a $\perp$ value appears anywhere in the memory or input, the entire memory is erased in the next round (even if the wire pair where $\perp$ originally appears does not influence the output or memory otherwise).

Note that in Figure 2 the inputs to the cascade stage are ordered top to bottom. The wire pairs corresponding to the output signals are fed into the cascade phase first and the other signals are fed into the cascade phase below them. This will ensure that even if a $\perp$ is introduced *within* the cascade phase, but in one of the wires going into the memory, the entire output is erased in the same clock cycle (and the memory contents will get erased in the next clock cycle). We shall use the convention that the wires within the cascade phase, except those corresponding to the output wires, are considered as part of the core of the circuit.

The Simulator for $T_1^{(k)}$. The universal simulator S_1 is constructed as follows: S_1 runs A_1, simulating to it $C_1[D_1]$. This means A_1 can request a particular input value for C_1, and expect to be provided the output from C_1 for that input. Further A_1 can request that a particular reset attack be applied to C_1. The output of the simulated C_1 that $\mathcal{S}_1$ provides to $\mathcal{A}_1$ should be indistinguishable from what $\mathcal{A}_1$ would have seen had it been actually working with $C_1[D_1]$. However, $\mathcal{S}_1$ has to do this with only black-box access to (the functionally equivalent) $C[D]$.

$\mathcal{S}_1$ internally runs the simulator $\mathcal{S}_{\text{ISW}}$ corresponding to the transformation $T_{\text{ISW}}^{(k)}$ which takes C to C'. It gives $\mathcal{S}_{\text{ISW}}$ blackbox access to C, which can then simulate (non-blackbox, probing) access to C'. So $\mathcal{S}_1$ can work as if it had (probing) access to C'. It then requests $\mathcal{S}_{\text{ISW}}$ to set the inputs to (the simulated) C' to the same as the input $\mathcal{A}_1$ requests for C_1. It provides $\mathcal{A}_1$ with the output values of the simulated C_1 which are the same as that given by $\mathcal{S}_{\text{ISW}}$. We need to describe how the reset attacks launched by $\mathcal{A}_1$ on C_1 are translated by $\mathcal{S}_1$ to probing attacks on C'.

Recall that C_1 is composed of an encoding of C', and a few extra phases (input encoding, error cascading and output decoding phases). $\mathcal{S}_1$ maintains a flag called destroyed which is initially set to false. While the flag remains set to false, at each clock cycle, $\mathcal{S}_1$ receives an attack pattern (i.e., a collection of reset attacks) from $\mathcal{A}_1$ and applies it to the simulated circuit as follows.

1. *For attacked wires that belong to the core of the circuit:* $\mathcal{S}_1$ will determine a set of indices $I \subset [k]$, as follows. I is initialized to the empty set. Then each attacked wire is considered as below:

    (a) *If the attacked wire is outside of the gadgets in Figure 1:* Then the wire is one of a wire pair in C_1 which encodes a single a wire in C'. Recall that all the wires in C' are indexed by values in $[k]$. $\mathcal{S}_1$ will add the index of the wire corresponding to the wire that is reset, to I. (If the wire has two index values, both are added to I.)[4]

    (b) *If the attacked wire is inside a gadget in Figure 1:* The gadget corresponds to a gate (AND, XOR or NOT) in C'. For each input wire to this gate in C', $\mathcal{S}_1$ checks if that wire has an *influence* on whether the attack creates a $\perp$ value or not. A wire is said to have such an influence if there are two settings of the values of the input to the gate such that in one a $\perp$ value is created and in the other it is not. $\mathcal{S}_1$ adds the indices of the wires with influence to I.

    Once I is determined by considering all attacks in the core of the circuit, $\mathcal{S}_1$ (acting as the adversary $\mathcal{A}'$) will make probes into the circuit simulated by $\mathcal{S}_{\text{ISW}}$ on all wires with indices in I. From the values obtained from this probe, it can check if the reset attacks produce a $\perp$. If they do, then $\mathcal{S}_1$ will set the flag destroyed.

    Note that $\mathcal{S}_{\text{ISW}}$ allows I to have at most $k - 1$ indices. If I is of size k (i.e., $I = [k]$), then $\mathcal{S}_1$ sets the flag destroyed (without querying $\mathcal{S}_{\text{ISW}}$).

2. *For attacked wires that belong to the encoding phase:* Such a wire is an input to an encoding gate (the outputs from an encoding gate are considered part

---

[4] C' has input encoding and output decoding phases. Recall that the wires in these phases are not indexed by values in $[k]$ and the Manchester encodings of these wires are not considered to belong to the core of the transformed circuit.

of the core of the circuit), or equivalently an input to the simulated C_1. In this case the corresponding input to C' is set to zero.

3. *For attacked wires that belong to the cascade phase:* The cascade phase consists of pairs of wire which correspond to wires going to the output phase in C'. (The wires in the cascade phase that correspond to the wires going to the memory are considered part of the core of the circuit). S_1 obtains the values on these from the simulated C_1 and determines how the attack affects the output from the cascade phase.

4. *For attacked wires is in the decode phase:* A reset attack on the second wire in a pair does not have any influence on the output while a reset attack on the first wire causes the corresponding output to be reset.

Once the flag `destroyed` is set S_1 simply produces the all zero output in every round.

The Proof. We start by observing what reset attacks can do in C_1. An invariant maintained in C_1 is that no wire pair carries the value 11: this is true for the data in the memory and also for the inputs coming out of the encoding stage; a reset attack on a wire which is 00, 10 or 01 cannot generate 11; further each gadget ensures that the invariant is maintained from inputs to outputs even when the internal wires of the gadget are attacked. (This follows from an analysis of the gadgets of the form of "OR of ANDs of signals and their NOTs.") Not having 11 in the wires has the following consequences:

- *Impossibility of changing a signal to a non-$\perp$ value:* Reset attacks can either leave a wire pair unchanged, or convert it to a $\perp$, but not generate a new non-$\perp$ value.
- $\perp$ *Propagation and Self-destruction:* The gadgets shown in Figure 1 are "$\perp$-propagating." That is, if any input wire pair encodes $\perp$ the output will be $\perp$ too. Thus any $\perp$ introduced by an attack in the core of the circuit will reach the cascade stage, which will ensure that even a single $\perp$ will result in the entire memory being erased and the outputs zeroed out.
  Thus, the output of the circuit will either be correct (or contain resets introduced after the cascade stage), or will be all zeroes. If a $\perp$ is introduced it will result in the entire memory being erased as well. If the $\perp$ is introduced after the cascade phase, this will happen in the next round.

Now we turn to the simulation by S_1. S_1 simply uses S_{isw} to get the outputs and also to check if a $\perp$ is created by the resets.

First, suppose that the set of indices I determined by S_1 is of size at most $k - 1$ in each round. In this case we observe the simulation by S_1 is perfect. This is because, in the simulation, C_1 as simulated by S_1 can be considered to encode the values in C' as simulated by S_{isw}. Since in this case for each reset S_{isw} allows all the indices to be queried, S_1 can precisely determine if a $\perp$ value is created or not in the core of C_1. When $\perp$ is not created, the output is simply the correct output (with any modifications caused by reset attacks in or after the cascade phase). When $\perp$ is created in the core, the output will be all zeroes in all subsequent clock cycles. Note that if a $\perp$ is created in the cascade phase in a signal going to the memory (which is considered part of the core), though the output is zeroed out in the same clock cycle, the memory may be zeroed out only in the next clock cycle.

Now we consider the case when $I = [k]$ in some round. $\mathcal{S}_1$ sets the flag destroyed but it is possible that in C_1 corresponding to C' as simulated by $\mathcal{S}_{\mathrm{ISW}}$, $\perp$ is not produced. However the probability of the latter happening is negligible in k. To see this, note that in building I, whenever a reset attack causes new indices to be added to I, there is a constant probability (independent of values of wires of indices already added to I) that a $\perp$ is produced by that attack. Further for each attack at most four indices are added (at most two inputs to a gate, each with at most two indices). Thus having added indices for $\Omega(k)$ attacks, the probability that none of the attacks produce a $\perp$ is exponentially small in k.

Thus in either case the simulation is good.

4.2   Proof of Lemma 3

The transformation $T_2^{(k)}$ removes the "encoded randomness gates" from a circuit. If $(C_2, D_2) = T_2^{(k)}(C_1, D_1)$ we need to show that an adversary $\mathcal{A}$ cannot gain any advantage in Experiment C in Lemma 3 than it will when employed by a simulator $\mathcal{S}_2$ in Experiment B.

The plan is to replace the encoded randomness gates with some sort of a pseudorandom generator (PRG) circuit, with an initial seed built into the memory. However, since the PRG circuit itself is open to attack from the adversary, it needs to be somehow protected. First we introduce a transformation which gives a weak protection. Then we show how multiple PRG units protected by such a transformation can be put together to obtain a PRG implementation which will also be secure against the reset attacks.

Lemma 4. Weak Protection Against Reset Attacks: *There is a polynomial time (in input size and security parameter k) transformation $T_{\mathrm{WEAK}}^{(k)}(C_P, D_P)$ $\mapsto (C_Q, D_Q)$, such that the following properties hold for all C_P and D_P:*

- *$C_Q[D_Q]$ is functionally equivalent to $C_P[D_P]$, except that the output of C_Q is Manchester encoded.*
- *Consider any adversary $\mathcal{A} \in \mathcal{A}_{\mathrm{RESET}}$ interacting with $C_Q[D_Q]$. If it resets even one wire inside C_Q (not an input or output wire), with probability at least q (for some constant $q > 0$), at least one of the output signals of C_Q becomes $\perp$.*

$T_{\mathrm{WEAK}}^{(k)}$ differs from $T_1^{(k)}$ in that the resulting circuit (C_Q, above) does not contain any encoded randomness gates. It is just a conventional deterministic circuit. On the other hand, the guarantee given by the transformation is much weaker: it guarantees introducing a $\perp$ into the output only with some positive probability.

The basic idea behind the construction is to randomize all the signals in the circuit, so that a reset attack has a constant probability of causing a $\perp$. The construction is essentially the same as $T_1^{(2)}$ (i.e., with security parameter 2), but without using randomness gates. Instead we use built-in randomness (i.e., it is stored in the memory). This will be sufficient to guarantee that the first time the circuit is attacked, there is a constant probability of producing a $\perp$. Also for this transformation we do not need the cascade stage and the output decoding stage of $T_1^{(2)}$.

Transformation $T_2^{(k)}$. Now we are ready to describe $T_2^{(k)}$. Suppose the input circuit requires n encoded random bits. Let C_P be a PRG circuit, which at every round, outputs n freshly generated pseudorandom bits, as well as refreshes its seed (kept in the memory). Consider k such circuits $C_P[D_P^i]$, D_P^i being a uniformly and independently drawn seed for the PRG. Let $(C_Q, D_Q^i) = T_{\text{WEAK}}^{(k)}(C_P, D_P^i)$. $T_2^{(k)}$ replaces the collection of all n encoded randomness gates by the following: the outputs of $C_Q[D_Q^i]$ $(i = 1, \ldots, k)$, are XOR-ed together using $k - 1$ encoded XOR gadgets (from Figure 1).

The proof that the above transformation indeed satisfies the properties required in Lemma 3 is included in the full version of this paper. The proof depends on the fact that as long as the adversary has attacked fewer than k of $C_Q[D_Q^i]$ in C_2, a careful simulation can reproduce the effect of this attack in C_1. On the other hand, if the adversary attacks all k of $C_Q[D_Q^i]$, then due to constant probability of each attack resulting in a $\perp$, except with negligible probability at least one $\perp$ value will indeed be generated which will propagate to the cascade stage and the output of the circuit (and hence can be easily simulated).

5   General Attacks on Wires

Next we turn to more general attacks in which the adversary can set the values in the wires to 1 or 0, as well as toggle the values in the wires. We shall impose a bound on the number of wires it can attack at each cycle, but allow the attacks to be persistent. That is, the wires set or reset in any one cycle are stuck at that value until explicitly released by the adversary; similarly toggled wires retain the toggling fault until released. There is no limit on the number of wires the adversary can release at any cycle.

Theorem 2. *There is a polynomial time (in input size and security parameter k) functionality preserving transformation $T_{\text{FULL}}^{(k)}(C, D) \mapsto (C^*, D^*)$, such that $\forall C$, $\exists S_0$ (a universal simulator), such that $\forall D$, $\forall \mathcal{A} \in \mathcal{A}_{\text{TAMPER}}^{(t)}$, we have $S_0^{\mathcal{A}} \in \mathcal{A}_{\text{IDEAL}}$ and the following two experiments output 1 with almost the same probability (the difference in probabilities being negligible in the security parameter k):*

– *Experiment A: $S_0^{\mathcal{A}}$ outputs a bit after interacting with C.*
– *Experiment B: $\mathcal{A}$ outputs a bit after interacting with C^*.*

5.1   Proof Sketch of Theorem 2

The construction of $T_{\text{FULL}}^{(k)}$, the simulation and proof of simulation roughly follow that in the reset-only case. The construction first applies the transformation

$T_{\text{ISW}}^{(k)}$, then changes the circuit to use some sort of encoding for each bit, adds an error cascade stage, and finally replaces all encoded randomness gates by a psuedo-randomness generator (PRG) circuit.

In the construction for reset attacks, it was crucial that the adversary cannot set a wire to 1, thereby being unable to change an encoded wire pair to anything but $\perp$. Here, however, the adversary is allowed to set as well as reset the wires. Nevertheless, using the fact that it can launch only t attacks per cycle, and using a longer encoding (instead of using a pair of wires) to encode each bit, we can ensure that if the adversary attacks any encoding, it will either leave it unchanged or change it to an invalid encoding. Below we sketch the details of this.

Encoding. Each bit is encoded by $2kt$ wires, where k is the security parameter and t is the bound on the number of attacks that the adversary can make per cycle. 0 is encoded as 0^{2kt} and 1 as 1^{2kt}. All other values are invalid ($\perp$); a special value $\perp^*$ is defined as $0^{kt}1^{kt}$.

Transformation. First $T_{\text{ISW}}^{(k)}$ is applied to get a circuit using randomness gates. Then the circuit is modified to use the encoded values. The core of the circuit is derived by replacing each wire by $2kt$ wires, and each of the atomic gates (AND, XOR and NOT) by gates shown in Figure 3. Input encoding and output decoding are straightforward. The error cascading stage is the same as shown in Figure 2, but using the cascade gadget from Figure 3. In implementing these gadgets, each bit of the output is generated by a circuit of the form OR of ANDs

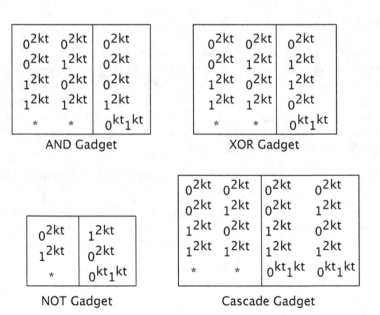

0^{2kt}	0^{2kt}	0^{2kt}
0^{2kt}	1^{2kt}	0^{2kt}
1^{2kt}	0^{2kt}	0^{2kt}
1^{2kt}	1^{2kt}	1^{2kt}
$*$	$*$	$0^{kt}1^{kt}$

AND Gadget

0^{2kt}	0^{2kt}	0^{2kt}
0^{2kt}	1^{2kt}	1^{2kt}
1^{2kt}	0^{2kt}	1^{2kt}
1^{2kt}	1^{2kt}	0^{2kt}
$*$	$*$	$0^{kt}1^{kt}$

XOR Gadget

0^{2kt}	1^{2kt}
1^{2kt}	0^{2kt}
$*$	$0^{kt}1^{kt}$

NOT Gadget

0^{2kt}	0^{2kt}	0^{2kt}	0^{2kt}
0^{2kt}	1^{2kt}	0^{2kt}	1^{2kt}
1^{2kt}	0^{2kt}	1^{2kt}	0^{2kt}
1^{2kt}	1^{2kt}	1^{2kt}	1^{2kt}
$*$	$*$	$0^{kt}1^{kt}$	$0^{kt}1^{kt}$

Cascade Gadget

Fig. 3. Truthtables for the gadgets used by $T_{\text{FULL}}^{(k)}$. The gadgets can be implemented using atomic gates: AND, OR and NOT gates. The AND gates used have $4kt$ input wires and the NOT gates are reversible (see below).

of input wires or their NOTs, or NOT of such a circuit. The AND gates involved have $4kt$ input wires. (These AND gates are the only components in this work which are not of constant size.) Note that to keep the circuit size polynomial it is important to allow the output to be of the form NOT of OR of ANDs, as well as OR of ANDs. In contrast, in the reset-only case it was important that the gadgets did not have NOT gates except at the input side. However, with general faults such a restriction is not helpful, nor used. The NOT gates used are reversible, so that faults on the output side of a NOT gate propagate to the input side. (This has the effect that NOT gates appearing immediately before the input to the AND gates can be considered part of the atomic AND gates.)

Finally, the encoded randomness gates can be replaced by a PRG circuit similar to that in Section 4.2.

Simulation. We sketch the simulator $\mathcal{S}_1$ (analogous to the simulator described in Section 4.1) for the part of the transformation before replacing the randomness gates by the PRG circuit. (The details of the simulation for the latter part can be found in the full version.) As in Section 4.1, $\mathcal{S}_1$ will internally run the simulator $\mathcal{S}_{\text{ISW}}$ corresponding to the transformation $T_{\text{ISW}}^{(k)}$ and translates attacks on C_1 to probing attacks on C'. However now the attacks are not just reset attacks but set, reset or toggle attacks. Further, each wire in C' is represented by a bundle of $2kt$ wires in C_1 (instead of just a pair of wires). $\mathcal{S}_1$ maintains the flag destroyed and calculates the set of indices I as before. The one additional action now is that *at any clock cycle if kt or more wires in any bundle are subject to attack, then the flag* destroyed *is set*. Here, attacking a wire inside any of the gadgets of Figure 3 (i.e., output of any of the AND or OR gates inside a gadget) is considered as an attack on the unique wire in the output of the gadget affected by the attack. Another difference is that, even after the flag destroyed is set, the simulator here continues to output non-zero values, but these values can be determined just from the attacks (in the cascade and output phases).

To analyze this new simulator we follow the same arguments as in Section 4.1, but with the following modifications.

- *Impossibility of changing a signal to a non-⊥ value:* We claim that as long as the flag destroyed is not set, an attack can either leave the values in a bundle corresponding to a signal in C' unchanged or convert it to a ⊥. To see this note that to produce a new non-⊥ signal the adversary must have attacked at least kt wires in a bundle. (These attacks may be direct attacks on the wires or attacks inside a gadget from which the wire emanates.) This is because the minimum distance between the signals that occur in an unattacked circuit (namely $0^{2kt}, 1^{2kt}$ and $\perp^* = 0^{kt}1^{kt}$), and each valid signal is kt. But when the adversary attacks kt or more wires in a single bundle (directly or by attacking a wire inside a gadget), then the simulator sets the flag destroyed.

- *⊥ Propagation and Self-destruction:* If a ⊥ (an encoding which is not 0^{2kt} or 1^{2kt}) is produced in any input signal to (the gadget corresponding to) a gate, it results in a $\perp^*$ being generated by the gate. Since $\perp^*$ is far from a valid encoding, the adversary cannot change it to a valid signal in the same cycle. So the $\perp$ value will reach the cascade stage, which ensures that all information in the circuit is lost. (If the $\perp$ is introduced after the cascade phase, this will happen in the next round.)

Now to prove that the simulation is good, first we observe that the probability that **destroyed** is set due to at least kt wires in a bundle being under attack is negligible. This is because at any cycle, the adversary can set/reset at most t wires, and so it will need at least k cycles to attack kt wires in a bundle. But during these cycles if the signal value in that bundle changes, then a $\perp$ is necessarily produced (resulting in the flag **destroyed** being set). Since each signal value is randomized by $T_{\text{ISW}}^{(k)}$ (Lemma 1), except with probability $2^{-\Omega(k)}$ this will indeed happen. The argument extends to the case when some of the attacks are on wires inside the gadgets as well, by observing that all the internal wires have influence on the output of the gadget, and the randomization ensures that with constant probability the input signals to the gadget will take values causing the attack to influence the output wire of the gadget.Here by internal wires in a gadget we refer to the outputs of the AND gates used in the gadgets; the only other wires inside a gadget are all connected to wires external to the gadget either directly or through a *reversible* NOT gate, and as such are accounted for by attacks on the wires external to the gadget. (This is where we require that the NOT gates be reversible; attacks on either side of a NOT gate propagates to the other side as well.)

Given this, the rest of the analysis of this simulator follows that in Section 5.1.

References

1. R. Anderson, M. Kuhn, "Tamper Resistance—A Cautionary Note," *USENIX E-Commerce Workshop*, USENIX Press, 1996, pp.1–11.
2. R. Anderson, M. Kuhn, "Soft Tempest: Hidden Data Transmission Using Electromagnetic Emanations," *Proc. 2nd Workshop on Information Hiding*, Springer, 1998.
3. B. Barak, O. Goldreich, R. Impagliazzo, S. Rudich, A. Sahai, S. Vadhan, and K. Yang. On the (im)possibility of obfuscating programs. CRYPTO 2001, 2001.
4. M. Ben-Or, S. Goldwasser, and A. Widgerson. Completeness theorems for non-cryptographic fault-tolerant distributed computation. In *Proc. of 20th STOC*, 1988.
5. E. Biham and A. Shamir, "Differential fault analysis of secret key cryptosystems," *CRYPTO '97*.
6. D. Boneh, R.A. Demillo, R.J. Lipton, "On the Importance of Checking Cryptographic Protocols for Faults," *EUROCRYPT'97*, Springer-Verlag, 1997, pp.37–51.
7. S. Chari, C.S. Jutla, J.R. Rao, P. Rohatgi, "Towards Sound Approaches to Counteract Power-Analysis Attacks," *CRYPTO'99*, Springer-Verlag, 1999, pp.398–412.
8. D. Chaum, C. Crepeau, and I. Damgård. Multiparty unconditional secure protocols. In *Proc. of 20th STOC*, 1988.
9. J.-S. Coron, L. Goubin, "On Boolean and Arithmetic Masking against Differential Power Analysis," *CHES'00*, Springer-Verlag, pp.231–237.
10. J. Daemen, V. Rijmen, "Resistance Against Implementation Attacks: A Comparative Study of the AES Proposals," *AES'99*, Mar. 1999.
11. K. Gandolfi, C. Mourtel, F. Olivier, "Electromagnetic Analysis: Concrete Results," *CHES'01*, LNCS 2162, Springer-Verlag, 2001.

12. R. Gennaro, A. Lysyanskaya, T. Malkin, S. Micali, and T. Rabin. Algorithmic Tamper-Proof (ATP) Security: Theoretical Foundations for Security against Hardware Tampering. Proceedings of *Theory of Cryptography Conference,* 2004.
13. O. Goldreich. *Foundations of Cryptography: Basic Applications.* Cambridge University Press, 2004.
14. O. Goldreich, S. Micali, and A. Wigderson. How to play any mental game (extended abstract). In *Proc. of 19th STOC,* 1987.
15. L. Goubin, J. Patarin, "DES and Differential Power Analysis—The Duplication Method," *CHES'99,* Springer-Verlag, 1999, pp.158–172.
16. Y. Ishai, A. Sahai, and D. Wagner, "Private Circuits: Protecting Hardware against Probing Attacks," *Proceedings of Crypto '03,* pages 462-479, 2003.
17. D. Kahn, *The Codebreakers,* The MacMillan Company, 1967.
18. J. Kelsey, B. Schneier, D. Wagner, "Side Channel Cryptanalysis of Product Ciphers," *ESORICS'98,* LNCS 1485, Springer-Verlag, 1998.
19. P. Kocher, "Timing Attacks on Implementations of Diffie-Hellman, RSA, DSS, and Other Systems," *CRYPTO'96,* Springer-Verlag, 1996, pp.104–113.
20. P. Kocher, J. Jaffe, B. Jun, "Differential Power Analysis," *CRYPTO'99,* Springer-Verlag, 1999, pp.388–397.
21. T.S. Messerges, "Securing the AES Finalists Against Power Analysis Attacks," *FSE'00,* Springer-Verlag, 2000.
22. S. Micali and L. Reyzin. Physically Observable Cryptography. In *Proc. of TCC '04,* pages 278-286, 2004.
23. R. Ostrovsky and M. Yung. How to Withstand Mobile Virus Attacks (Extended Abstract). In *Proc. of PODC '91,* pages 51-59, 1991.
24. D. Page, "Theoretical Use of Cache Memory as a Cryptanalytic Side-Channel," Tech. report CSTR-02-003, Computer Science Dept., Univ. of Bristol, June 2002.
25. B. Pfitzmann, M. Schunter and M. Waidner, "Secure Reactive Systems", IBM Technical report RZ 3206 (93252), May 2000.
26. N. Pippenger, "On Networks of Noisy Gates," in *Proc. of FOCS '85,* pages 30-38.
27. J.-J. Quisquater, D. Samyde, "Eddy current for Magnetic Analysis with Active Sensor," *Esmart 2002,* Sept. 2002.
28. J.-J. Quisquater, D. Samyde, "ElectroMagnetic Analysis (EMA): Measures and Counter-Measures for Smart Cards," *Esmart 2001,* LNCS 2140, Springer-Verlag, 2001.
29. J.R. Rao, P. Rohatgi, "EMpowering Side-Channel Attacks," IACR ePrint 2001/037.
30. US Air Force, *Air Force Systems Security Memorandum 7011—Emission Security Countermeasures Review,* May 1, 1998.
31. W. van Eck, "Electromagnetic Radiation from Video Display Units: An Eavesdropping Risk," *Computers & Security,* v.4, 1985, pp.269–286.
32. D. Wright, *Spycatcher,* Viking Penguin Inc., 1987.
33. S.G. Younis and T. F. Knight, Jr. Asymptotically Zero Energy Split-Level Charge Recovery Logic. *Proceedings of 1994 International Workshop on Low Power Design,* Napa, CA, 1994.

Composition Implies Adaptive Security in Minicrypt

Krzysztof Pietrzak*

Département d'informatique, Ecole Normale Supérieure, Paris, France
pietrzak@di.ens.fr

Abstract. To prove that a secure key-agreement protocol exists one must at least show $P \neq NP$. Moreover any proof that the sequential composition of two non-adaptively secure pseudorandom functions is secure against at least two adaptive queries must falsify the decisional Diffie-Hellman assumption, a standard assumption from public-key cryptography. Hence proving any of this two seemingly unrelated statements would require a significant breakthrough. We show that *at least one* of the two statements is true.

To our knowledge this gives the first *positive* cryptographic result (namely that composition implies some weak adaptive security) which holds in Minicrypt, but not in Cryptomania, i.e. under the assumption that one-way functions exist, but public-key cryptography does not.

1 Introduction

A pseudorandom function (PRF) is a function which cannot be distinguished from a uniformly random function by any efficient adversary. One can give different security definitions for PRFs depending on how the attacker can access the function: a *non-adaptive* adversary must choose all his queries to the function at once, whereas a (more powerful) *adaptive* adversary must only decide on the i'th query after receiving the $i-1$'th output. As a generalisation we define k-adaptive adversaries which can choose k blocks of queries to be made, where the k'th block must be chosen at once but only after receiving the outputs to the $k-1$'th block (in particular 1-adaptive means non-adaptive, and ∞-adaptive means adaptive). Consider the following two statements:

$\mathfrak{K}_k$: There exists a secure k-pass key-agreement protocol.
$\mathfrak{C}_k$: The sequential composition of two $(k-1)$-adaptively secure PRFs is k-adaptively secure.

The main result of this paper is that either composition of PRFs always increases the security in the sense that the cascade is k-adaptive secure whenever the components are $k-1$ secure OR that key agreement exists.

---

* Most of this work was done while the author was a PhD student at ETH where he was supported by the Swiss National Science Foundation, project No. 200020-103847/1. Part of this work is supported by the Commission of the European Communities through the IST program under contract IST-2002-507932 ECRYPT.

S. Vaudenay (Ed.): EUROCRYPT 2006, LNCS 4004, pp. 328–338, 2006.

Theorem 1. *For any $k \geq 2$:* $\qquad \mathfrak{C}_k \vee \mathfrak{K}_{2k-1}$

This theorem has a nice interpretation in terms of Impagliazzo's five possible worlds as described in the survey paper "A Personal View of Average-Case Complexity" [8]. Here "possible world" means that with our current knowledge we cannot rule out it as being reality. As each world does exists relative to an oracle, showing equivalence of two worlds would require non-relativizing techniques, and in the ten years that passed since this survey none has been resolved.[1] This five worlds are Algorithmica (where $P = NP$), Heuristica ($NP \neq P$ but NP is tractable on average), Pessiland (NP is hard on average but one-way functions do not exist), Minicrypt (one-way functions exist) and Cryptomania (Public-key cryptography exists, this is probably the real world). In this view, the theorem states that for any $k \geq 2$ the statement $\mathfrak{C}_k$ holds in Minicrypt but not in Cryptomania. As the naming suggests, Cryptomania is cryptographers paradise, but our result somewhat challenges this viewpoint, as cryptographers interested only in symmetric cryptography might well prefer to live in Minicrypt rather than in Cryptomania, as some results (in particular $\mathfrak{C}_k$) only can be found there.

But let us stress that there are known (black-box) constructions of adaptively secure PRFs from non-adaptively secure PRFs [4], but these constructions are inefficient as they need a linear (in the security parameter) number of calls to the underlying primitive on each invocation. Thus we do not show that adaptively secure PRF exists in Minicrypt (as this is known), but rather that here adaptive security can be achieved by probably most straight forward and efficient construction: cascading two functions.

We prove Theorem 1 by constructing a $2k - 1$-pass key-agreement protocol from any pseudorandom functions which provides a counterexample for $\mathfrak{C}_k$, i.e. from any $(k - 1)$-adaptively secure pseudorandom functions $\mathsf{F}(\cdot)$ and $\mathsf{G}(\cdot)$ where there exists an efficient k-adaptive D which can distinguish $\mathsf{G}(\mathsf{F}(\cdot))$ from a random function.

There is a gap between what is generally considered a successful distinguisher (or any other kind of an adversary) and what one expects from a protocol: a system is usually considered broken even if only a *non-uniform* advantage exists, whereas a protocol should be *uniform* and achieve its task with *overwhelming*[2] probability to be considered useful. The key-agreement protocol we construct uses D as a black-box, and only if D is *uniform* and has *noticeable* advantage in distinguishing $\mathsf{G}(\mathsf{F}(\cdot))$ from random, we will get a useful (as described above) key-agreement protocol. But if D in non-uniform, also the key-agreement protocol will be non-uniform. Furthermore if D has only *non-negligible* (but not noticeable) advantage, then our key-agreement protocol will only work (i.e. have overwhelming success probability) for infinitely many values of the security parameter (and not as usually for all).

---

[1] But several new worlds, in particular between Minicrypt and Cryptomania [3], have been added. Recently Harnik and Naor [5] proposed an interesting approach to show Minicrypt=Cryptomania. Wee investigates Pessiland in [17]. A classical result due to Rudich [15] oracle separates $\mathfrak{K}_k$ from $\mathfrak{K}_{k+1}$ for every k.

[2] $\tau(\cdot)$ is overwhelming if $1 - \tau(\cdot)$ is negligible.

1.1   What Is Known?

It is known that under the decisional Diffie–Hellman (DDH) assumption two-pass key-agreement (i.e. public-key encryption) exists [1, 2], and in [13] it is shown that under the same assumption $\neg \mathfrak{C}_2$ holds, i.e. that composition does not imply adaptive security.[3] Thus [13] shows a negative result for private-key systems under a standard assumption from public-key cryptography. By Theorem 1 this is not just an artificial property of the counterexample given in [13], but in fact any falsification of $\mathfrak{C}_2$ implies (and thus must either assume or unconditionally prove) the existence of the central public-key primitive key-agreement.

Interestingly the equivalent of $\mathfrak{C}_2$ in the information theoretic setting is true: the cascade of two functions, each having security ϵ against *non-adaptive* (computationally unbounded) distinguishers making at most q queries, has security 2ϵ against any *adaptive* distinguisher making q queries [11]. Therefore the reason why composition does imply adaptive security in the information-theoretic but probably not in the computational setting is closely related to the fact that public-key cryptography cannot exist in the information theoretic setting [16, 10] but is believed to exist in the real world [1]. We'll muse further on the implications of Theorem 1 in Section 4.

2   Basic Definitions

Throughout we denote by $n \in \mathbb{N}$ a security parameter. An algorithm is efficient if it can be implemented by a probabilistic Turing machine whose expected running time is polynomial in the input length (which for us will always mean polynomial in n). We use a SANS-SERIF font for efficient entities and a $\mathcal{CALLIGRAPHIC}$ font for idealised systems like uniform random functions.

NEGLIGIBLE. A function $\mu : \mathbb{N} \to [0, 1]$ is negligible if for any $c > 0$ there is an n_0 such that $\mu(n) \le 1/n^c$ for all $n \ge n_0$. And contrarily μ is non-negligible if for any $c > 0$ we have $\mu(n) \ge 1/n^c$ for infinitely many n.

NOTICEABLE. A function $\phi : \mathbb{N} \to [0, 1]$ is noticeable if for some $c > 0$ there is an n_0 such that $\phi(n) \ge 1/n^c$ for all $n \ge n_0$.

Note that non-negligible is not the same as noticeable, for example $\mu(n) \stackrel{\text{def}}{=} n \bmod 2$ is non-negligible but not noticeable.

Unless stated otherwise, all characters that appear below are probabilistic efficient Turing machines.

BIT-AGREEMENT. Bit-agreement is a protocol between two efficient parties, let's call them Amélie and Benoît . They get as a common input the security

---

[3] In [13] a $\mathsf{F}(\cdot)$ and $\mathsf{G}(\cdot)$ are constructed which are non-adaptively secure under the DDH assumption, but where *three* (and not two as required for $\neg \mathfrak{C}_2$) adaptive queries are enough to learn the whole key when querying $\mathsf{G}(\mathsf{F}(\cdot))$. But after two adaptive queries one already learns the key of G and thus can distinguish $\mathsf{G}(\mathsf{F}(\cdot))$ from random, and this is all we need to get $\neg \mathfrak{C}_2$. Previous to [13] is was already known that there is no black-box proof for $\mathfrak{C}_2$ as Myers [12] has constructed an oracle relative to which $\neg \mathfrak{C}_2$.

parameter n in unary (denoted 1^n) and can communicate over an authentic channel. Finally Amélie and Benoît output a bit b_A and b_B respectively. The protocol has correlation ϵ if for all n

$$\Pr[b_A = b_B] \geq \frac{1 + \epsilon(n)}{2}$$

and the protocol is δ-secure if for any efficient adversary E which can observe the whole communication C we have for all n

$$\Pr[E(1^n, C) \to b_A] \leq 1 - \frac{\delta(n)}{2}$$

KEY-AGREEMENT. If $\epsilon(\cdot)$ and $\delta(\cdot)$ are overwhelming then such a protocol achieves key-agreement. Any protocol which achieves bit-agreement with a noticeable correlation $\epsilon(\cdot)$ and overwhelming security $\delta(\cdot)$ can be turned into a key-agreement protocol by sequential composition, and using parallel repetition this can even be done without increasing the number of rounds [6, 7].

If $\epsilon(\cdot)$ is only non-negligible (i.e. for any $c > 0 : \epsilon(n) \geq 1/n^c$ for all $n \in S_c \subset \mathbb{Z}$ where $|S_c|$ is infinite), then also the key-agreement protocol will only achieve correctness for security parameters $n \in S_c$ (one can choose any constant c here, the running time of the key-agreement protocol will then basically grows as n^{2c}).

DISTINGUISHER. By a k-adaptive *distinguisher* we denote an efficient oracle algorithm which at the end of the computation outputs a decision bit. He may query the oracle an arbitrary number of times, but the queries must come in k blocks where he must settle for a whole block before reading any outputs on queries from that block.

This definition is not standard, but note that a 1-adaptive distinguisher is just a standard non-adaptive distinguisher and a ∞-adaptive distinguisher is a standard adaptive distinguisher.

As we only consider stateless systems (which always give the same answer on the same query) w.l.o.g we always *can and will assume that a distinguisher never makes the same query twice*. Moreover we *require the distinguishers themselves to be stateless*. This can be done w.l.o.g. if we always provide the previous outputs of the system queried as an input to the distinguisher when he must come up with the next query or the final decision bit (note that we need not to provide the previous inputs to the system as the distinguisher can compute this inputs himself).

PSEUDORANDOM FUNCTION/PERMUTATION. A pseudorandom function (PRF) is a pair of efficient algorithms F and KeyGen_F where for any $n \in \mathbb{N}$ we have $\mathsf{KeyGen}_F : 1^n \to \mathcal{K}_n$ and $F : \mathcal{K}_n \times \{0,1\}^n \to \{0,1\}^n$. Let $F_k(\cdot) \stackrel{\text{def}}{=} F(k, \cdot)$. Let $\mathcal{R}_n : \{0,1\}^n \to \{0,1\}^n$ be a uniform random function, then F is ℓ-adaptive secure if for any efficient ℓ-adaptive distinguisher D

$$|\Pr[D^{F_k(\cdot)}(1^n) \to 1 | k \leftarrow \mathsf{KeyGen}_F(1^n)] - \Pr[D^{\mathcal{R}_n(\cdot)}(1^n) \to 1]| = \tau(n).$$

for some negligible τ. Pseudorandom permutations (PRP) are defined similarly, but here one additionally requires that for any k, $F_k(\cdot)$ is a permutation.

SEQUENTIAL COMPOSITION. For two functions F and G we denote by G∘F their
sequential composition.

$$G \circ F(x) \overset{\text{def}}{=} G(F(x)).$$

For a set S we denote by $x \overset{\$}{\leftarrow} S$ that x is assigned a value from S uniformly at
random.

3   The Reduction

In this section we prove the statement $\neg\mathfrak{C}_k \Rightarrow \mathfrak{K}_{2k-1}$ of Theorem 1. Actually, we
only show that $\neg\mathfrak{C}_k$ implies a $(2k-1)$-pass *bitagreement* protocol with noticeable
correlation and overwhelming security, but as said in the previous section, this
is equivalent to $\mathfrak{K}_{2k-1}$.

For the clarity of exposition we prove only the special case $k = 2$ and we
assume that $\neg\mathfrak{C}_2$ holds in a strong sense, namely that the cascade considered
can be distinguished by an adversary which makes only two adaptive queries,
this is a special case of a general 2-adaptive distinguisher which can make two
blocks of arbitrary many queries (where he must settle for whole blocks at once).
At the end of this section we will show how the reduction must be extended to
cover the general case (and thus to prove Theorem 1).

Let F, KeyGen_F and G, KeyGen_G be two pseudorandom functions, each secure
against non-adaptive distinguishers, but which can be distinguished with two
adaptive queries. This means that there exists an efficient D and a non-negligible
ϕ such that

$$\Pr[b_2 = 1] - \Pr[b_1 = 1] \geq \phi(n) \tag{1}$$

where b_1 and b_2 are bits whose distribution is defined by Games 1 and 2 below
where D either queries the sequential composition (Game 1) or a random function
(Game 2) with two adaptive queries.

Game 1	Game 2	Game 3
$k_1 \leftarrow \mathsf{KeyGen}_F(1^n)$		
$k_2 \leftarrow \mathsf{KeyGen}_G(1^n)$		$k \leftarrow \mathsf{KeyGen}_G(1^n)$
$x_1 \leftarrow D(1^n)$	$x_1 \leftarrow D(1^n)$	$z_1 \overset{\$}{\leftarrow} \{0,1\}^n$
$y_1 \leftarrow G_{k_2} \circ F_{k_1}(x_1)$	$y_1 \leftarrow \mathcal{R}_n(x_1)$	$y_1 \leftarrow G_k(z_1)$
$x_2 \leftarrow D(y_1)$	$x_2 \leftarrow D(y_1)$	$z_2 \overset{\$}{\leftarrow} \{0,1\}^n$
$y_2 \leftarrow G_{k_2} \circ F_{k_1}(x_2)$	$y_2 \leftarrow \mathcal{R}_n(x_2)$	$y_2 \leftarrow G_k(z_2)$
$b_1 \leftarrow D(y_1, y_2)$	$b_2 \leftarrow D(y_1, y_2)$	$b_3 \leftarrow D(y_1, y_2)$

In Game 2 the y_1, y_2 are just uniform random values whereas in Game 3 the
y_1, y_2 are computed by G on random inputs. From the non-adaptive security of
G it also follows that for some negligible δ_{23}

$$|\Pr[b_2 = 1] - \Pr[b_3 = 1]| \leq \delta_{23}(n). \tag{2}$$

PROTOCOL BitAgreement(n)

Amélie   Benoît

$$b_A \overset{\$}{\leftarrow} \{0,1\}$$

$k_A \leftarrow \mathsf{KeyGen_F}(1^n)$ $k_B \leftarrow \mathsf{KeyGen_G}(1^n)$

$$x_1 \leftarrow \mathsf{D}(1^n)$$

if $b_A = 0$ then $z_1 \leftarrow \mathsf{F}_{k_A}(x_1)$

otherwise $z_1 \overset{\$}{\leftarrow} \{0,1\}^n$ $z_1 \rightarrow y_1 \leftarrow \mathsf{G}_{k_B}(z_1)$

$$\leftarrow y_1$$

$$x_2 \leftarrow \mathsf{D}(y_1)$$

if $b_A = 0$ then $z_2 \leftarrow \mathsf{F}_{k_A}(x_2)$

otherwise $z_2 \overset{\$}{\leftarrow} \{0,1\}^n$ $z_2 \rightarrow y_2 \leftarrow \mathsf{G}_{k_B}(z_2)$

$$b_B \leftarrow \mathsf{D}(y_1, y_2)$$

Fig. 1. 3-pass BitAgreement protocol from a 2-adaptive D

With such an F, G and D we can construct a bit-agreement protocol with non-negligible correlation and overwhelming security (and thus get key-agreement) as shown in Figure 1. If D is randomised we need Amélie and Benoît to use the same random coins for D in BitAgreement. Here Amélie can simply choose the random coins initially and send them to Benoît .

Claim 1. BitAgreement(n) *has correlation* $\phi - \delta_{23}$.

Proof. Note that if $b_A = 0$ ($b_A = 1$) then the distribution of b_B is the same as the distribution of b_1 (b_3) in game 1 (game 3), now as (1) and (2) imply

$$\Pr[b_3 = 1] - \Pr[b_1 = 1] \geq \phi(n) - \delta_{23}(n)$$

we get

$$\Pr[b_A = b_B] = \Pr[b_A = 0]\Pr[b_B = 0|b_A = 0] + \Pr[b_A = 1]\Pr[b_B = 1|b_A = 1]$$
$$= \frac{1 - \Pr[b_1 = 1]}{2} + \frac{\Pr[b_3 = 1]}{2}$$
$$\geq \frac{1 + \phi(n) - \delta_{23}(n)}{2} \qquad\qquad \square$$

Claim 2. BitAgreement(n) *is δ-secure for an overwhelming δ.*

Proof. We must show that there is an overwhelming δ such that for all efficient D

$$\Pr[\mathsf{D}(z_1, y_1, z_2) \rightarrow b_A] \leq 1 - \frac{\delta(n)}{2}$$

We consider six more games which all define a distribution for the values (z_1, y_1, z_2). The distribution of (z_1, y_1, z_2) in game 4 and 9 is the same as in BitAgreement(n) conditioned on $b_A = 0$ and $b_A = 1$ respectively.

Game 4	Game 5	Game 6
$k_1 \leftarrow \mathsf{KeyGen}_\mathsf{F}(1^n)$	$k_1 \leftarrow \mathsf{KeyGen}_\mathsf{F}(1^n)$	$k_1 \leftarrow \mathsf{KeyGen}_\mathsf{F}(1^n)$
$k_2 \leftarrow \mathsf{KeyGen}_\mathsf{G}(1^n)$		
$x_1 \leftarrow \mathsf{D}(1^n)$	$x_1 \leftarrow \mathsf{D}(1^n)$	$x_1 \leftarrow \mathsf{D}(1^n)$
$z_1 \leftarrow \mathsf{F}_{k_1}(x_1)$	$z_1 \leftarrow \mathsf{F}_{k_1}(x_1)$	$y_1 \xleftarrow{\$} \{0,1\}^n$
$y_1 \leftarrow \mathsf{G}_{k_2}(z_1)$	$y_1 \xleftarrow{\$} \{0,1\}^n$	$x_2 \leftarrow \mathsf{D}(y_1)$
$x_2 \leftarrow \mathsf{D}(y_1)$	$x_2 \leftarrow \mathsf{D}(y_1)$	$z_1 \leftarrow \mathsf{F}_{k_1}(x_1)$
$z_2 \leftarrow \mathsf{F}_{k_1}(x_2)$	$z_2 \leftarrow \mathsf{F}_{k_1}(x_2)$	$z_2 \leftarrow \mathsf{F}_{k_1}(x_2)$

Game 7	Game 8	Game 9
		$k_2 \leftarrow \mathsf{KeyGen}_\mathsf{G}(1^n)$
$x_1 \leftarrow \mathsf{D}(1^n)$	$x_1 \leftarrow \mathsf{D}(1^n)$	$x_1 \leftarrow \mathsf{D}(1^n)$
$y_1 \xleftarrow{\$} \{0,1\}^n$	$z_1 \leftarrow \mathcal{R}_n(x_1)$	$z_1 \leftarrow \mathcal{R}_n(x_1)$
$x_2 \leftarrow \mathsf{D}(y_1)$	$y_1 \xleftarrow{\$} \{0,1\}^n$	$y_1 \leftarrow \mathsf{G}_{k_2}(z_1)$
$z_1 \leftarrow \mathcal{R}_n(x_1)$	$x_2 \leftarrow \mathsf{D}(y_1)$	$x_2 \leftarrow \mathsf{D}(y_1)$
$z_2 \leftarrow \mathcal{R}_n(x_2)$	$z_2 \leftarrow \mathcal{R}_n(x_2)$	$z_2 \leftarrow \mathcal{R}_n(x_2)$

With $\mathrm{Pr}_{Gi}[E]$ we denote the probability of the event E in game i, and δ_{ij} is defined by

$$|\mathrm{Pr}_{Gi}[\mathsf{D}(z_1, y_1, z_2) \to 1] - \mathrm{Pr}_{Gj}[\mathsf{D}(z_1, y_1, z_2) \to 1]| = \delta_{ij}(n)$$

Game 4 differs from Game 5 only by the computation of y_1 which is computed by G and random respectively. As G is non-adaptively secure (and a single query is always non-adaptive) δ_{45} is negligible. For the same reason δ_{89} is negligible. Game 6 differs from Game 7 only by the computation of z_1 and z_2 which in Game 6 are non-adaptively computed by F and in Game 7 by $\mathcal{R}$, so from F's non-adaptive security it follows that δ_{67} is also negligible. Finally δ_{56} and δ_{78} are 0 as Game 5 is equivalent to Game 6 (only the order of the commands is changed to emphasis that in Game 5 the F is in fact queried non-adaptively) and Game 7 is equivalent to Game 8.

Using the triangle inequality we see that $\delta_{49} \leq \sum_{i=4}^{8} \delta_{i\,i+1}$ is negligible, and thus $\delta \stackrel{\mathrm{def}}{=} 1 - \delta_{49}$ is overwhelming. We can now conclude the proof of the claim as

$$\Pr[\mathsf{D}(z_1, y_1, z_2) \to b_A]$$
$$= \Pr[b_A = 0]\Pr[\mathsf{D}(z_1, y_1, z_2) \to 0 | b_A = 0] +$$
$$\quad \Pr[b_A = 1]\Pr[\mathsf{D}(z_1, y_1, z_2) \to 1 | b_A = 1]$$
$$= (1 - \Pr[\mathsf{D}(z_1, y_1, z_2) \to 1 | b_A = 0] + \Pr[\mathsf{D}(z_1, y_1, z_2) \to 1 | b_A = 1])/2$$
$$= (1 - \mathrm{Pr}_{G4}[\mathsf{D}(z_1, y_1, z_2) \to 1] + \mathrm{Pr}_{G9}[\mathsf{D}(z_1, y_1, z_2) \to 1])/2$$
$$\leq (1 + \delta_{49})/2$$
$$= 1 - \delta/2$$

This concludes the proof of $\neg \mathfrak{C}_k \Rightarrow \mathfrak{K}_{2k-1}$ for the case $k = 2$ with the additional assumption that the cascade can be broken by a distinguisher D which makes two adaptive queries (and not a general 2-adaptive distinguisher).   □

We first explain how to adapt the reduction so that if works for any 2-adaptive distinguisher and not just for two adaptive queries. Then we show how to adapt it so that it works for any $k \geq 2$ which will then conclude the proof of Theorem 1.

PROTOCOL BITAGREEMENT(n)

Amélie   Benoît

$$b_A \overset{\$}{\leftarrow} \{0,1\}$$
$$k_A \leftarrow \mathsf{KeyGen_F}(1^n) \qquad k_B \leftarrow \mathsf{KeyGen_G}(1^n)$$

for $i = 1$ to $k - 1$ do

$$X_i \leftarrow \mathsf{D}''(Y_1, \ldots, Y_{i-1})$$
$$\text{if } b_A = 0\text{then } Z_i \leftarrow \mathsf{F}_{k_A}(X_i)$$
$$\text{otherwise } Z_i \overset{\$}{\leftarrow} \{0,1\}^n \; Z_i \rightarrow Y_i \leftarrow \mathsf{G}_{k_B}(Z_i)$$
$$\leftarrow Y_i$$

od;

$$X_k \leftarrow \mathsf{D}''(Y_1, \ldots, Y_{k-1})$$
$$\text{if } b_A = 0 \text{ then } Z_k \leftarrow \mathsf{F}_{k_A}(X_k)$$
$$\text{otherwise } Z_k \overset{\$}{\leftarrow} \{0,1\}^n \; Z_k \rightarrow Y_k \leftarrow \mathsf{G}_{k_B}(Z_k)$$
$$b_B \leftarrow \mathsf{D}''(Y_1, \ldots, Y_k)$$

Fig. 2. $(2k - 1)$-pass BITAGREEMENT protocol from a k-adaptive D''

REDUCTION FROM 2-ADAPTIVE D'. Let D' be any 2-adaptive distinguisher which can distinguish $\mathsf{F}_{k_1} \circ \mathsf{G}_{k_2}$ from random. From such a D' we can construct a 3-pass bitagreement protocol almost like from the D which made only two queries. If $q = q(n)$ denotes (an upper bound on) the size of the blocks requested by D', then just replace all occurrences of $x_1, x_2, y_1, y_2, z_1, z_2$ by appropriate q-tuples $X_1, X_2, Y_1, Y_2, Z_1, Z_2$ in the bitagreement protocol. For example replace $x_1 \leftarrow \mathsf{D}(1^n)$ with $X_1 = (x_1^1, x_1^2, \ldots, x_1^q)$ where $X_1 \leftarrow \mathsf{D}'(1^n)$, similarly replace $y_1 \leftarrow \mathsf{F}_{k_1} \circ \mathsf{G}_{k2}(x_1)$ by $Y_1 \leftarrow \mathsf{F}_{k_1} \circ \mathsf{G}_{k2}(X_1)$ and so on.

REDUCTION FROM k-ADAPTIVE D''. For any $k \geq 2$, let D'' be any k-adaptive distinguisher for $\mathsf{F}_{k_1} \circ \mathsf{G}_{k2}$ from random. To construct a bitagreement from such a distinguisher we can proceed similarly to the $k = 2$ case, only the number of rounds must be increased as now D'' must be fed with k and not just 2 input blocks.

The construction of $(2k-1)$-pass bitagreement from a k-adaptive D'' is shown in Figure 2. It is straight forward (and we omit it) to adapt the Claims 1 and 2 and their proofs for this protocol.

4 Discussion

Does Theorem 1 $\mathfrak{C}_k \vee \mathfrak{K}_{2k-1}$ have any practical meaning? After all, DDH is believed to be true in the real world, so $\mathfrak{K}_2$ is true [1] and $\mathfrak{C}_2$ is wrong [13]. Even if someday $(2k-1)$-pass key-agreement turns out to be impossible, having $\mathfrak{C}_k$ instead is a cold comfort.

But one can see $\mathfrak{C}_k \vee \mathfrak{K}_{2k-1}$ as a positive result, even when assuming that DDH is true: Composition of k-adaptively secure pseudorandom functions implies $(k+1)$-adaptive security[4], *unless the pseudorandom functions themselves have some public-key functionality* in the sense that they can be turned into a key-agreement protocol by a black-box (BB for short) reduction. Of course that was more an intuitive argument than a result that can be actually applied. In the next section we prove a first positive composition result for PRFs whose security can be BB-reduced to the security of a one-way function.

4.1   Black-Box Breaks

Combining Theorem 1 with the Impagliazzo-Rudich result [9] that key-agreement cannot be BB-reduced to one-way functions we can prove a first positive result in the direction that composition sometimes does imply adaptive security (or rather, that the adaptive security cannot be broken in a generic way) even in the computational setting. Before we can state the theorem we first need some definitions.

$F^{(\cdot)}$ is an oracle PRF whose k-adaptive security can be BB-reduced to the one-wayness of the oracle if the following is true: There exists an efficient $B^{(\cdot)}$ such that for any (not necessarily efficient) k-adaptive adversary $\mathcal{A}^{(\cdot)}$ and any f (for simplicity we assume f is $\{0,1\}^* \to \{0,1\}^*$ and length preserving) for which

$$\left| \Pr[k \leftarrow \mathsf{KeyGen}_F^f(n); \mathcal{A}^{F_k^f} \to 1] - \Pr[\mathcal{A}^{\mathcal{R}_n} \to 1] \right|$$

is k-negligible (note that this means that $\mathcal{A}$ breaks the k-adaptive pseudorandomness of F^f), $B^{\mathcal{A},f}$ breaks the one-wayness of f, this means that then also

$$\Pr[x \xleftarrow{\$} \{0,1\}^n; B^{\mathcal{A},f}(f(x)) \in f^{-1}(x)]$$

is non-negligible. This definition of BB-reduction is standard and called a fully-BB reduction in the taxonomy from [14]. The definition of a BB-break given below is not standard.

We say that the k-adaptive security of $F^{(\cdot)}$ can be *BB-broken* if there exists an efficient k-adaptive $C^{(\cdot)}$ where

$$\left| \Pr[k \leftarrow \mathsf{KeyGen}_F^f(n); C^{F_k^f, f} \to 1] - \Pr[C^{\mathcal{R}_n, f} \to 1] \right|$$

is noticeable for all f; So C can distinguish F^f from $\mathcal{R}$ for every f, i.e. C breaks the the security of the construction $F^{(\cdot)}$ and not some particular instantiation.

---

[4] And in particular composition of non-adaptively secure pseudorandom functions implies 2-adaptive security.

Note that if the k-adaptive security of $F^{(\cdot)}$ can be BB-broken, then it obviously cannot be BB-reduced to the one-wayness of the oracle, but the converse is not true in general.

Theorem 2. *If the k-adaptive security of the PRFs $F^{(\cdot)}$ and $G^{(\cdot)}$ can be BB-reduced to the one-wayness of the oracle, then the $(k+1)$-adaptive security of $G^{(\cdot)} \circ F^{(\cdot)}$ cannot be BB-broken.*

Proof. The proof is by contradiction: assume there is $(k+1)$-adaptive distinguisher $C^{(\cdot)}$ which can distinguish $G^f \circ F^f$ from a random function with noticeable advantage for any f. With such a $C^{(\cdot)}, F^f, G^f$ we can construct a key-agreement protocol.[5] The security of this protocol can be BB-reduced to the k-adaptive security of F^f and G^f whose security can again be BB-reduced to the one-wayness of f. So we have a BB-reduction from key-agreement to one-way functions which is not possible [9].   □

Note that the theorem does not claim that the $k+1$-adaptive security of $F^f \circ G^f$ can be BB-reduced to the one-wayness of f, but something weaker. Namely that there is no single efficient $C^{(\cdot)}$ which breaks the $(k+1)$-adaptive security for all f.

4.2   Outlook

Are there other interesting statements that one can we prove to be true only under the assumption that public-key cryptography does not exist? It seems unlikely that our composition result is an isolated example.

As shown in Theorem 2 given such a statement one might well be able to prove a weaker version of it without making the (unlikely) assumption that public-key crypto does not exist. But what does "BB-broken" as used in Theorem 2 actually mean? Can one strengthen this theorem and replace "BB-broken" with "BB-reduced to the one-wayness of the oracle" or show that this is not possible.

Can we strengthen Theorem 1? For example can we show that key-agreement (via a BB-reduction) exists when the composition of two $(k-2)$-adaptive PRFs secure PRFs is k-adaptive secure[6]? We think this is not true,[7] but we believe that Theorem 2 holds with an infinite gap, i.e. where k-adaptive is replaced by non-adaptive and $(k+1)$-adaptive by adaptive. To show this one would have to show that there exists some statement $\mathfrak{L}$ such that $\mathfrak{L}$ is implied by the statement "the composition of two non-adaptive PRFs is not adaptively secure" and where $\mathfrak{L}$ cannot be BB-reduced to one-way functions.

---

[5] As shown in Section 3 for the special case $k = 1$ and where each of the $k+1$ blocks contained only one message.

[6] This is statement $\mathfrak{C}_k$ with an increased gap, i.e. $k - 2$ instead of $(k-1)$.

[7] Because there seems to be an oracle relative to which no key-agreement exists and cascading $(k-2)$-adaptive PRFs does not give k-adaptive security. But we didn't check all details.

Acknowledgments

I'd like to thank Ueli Maurer for insightful discussions on this topic and Thomas Holenstein for several clarifying conversations on key- and bit-agreement.

References

1. Whitfield Diffie and Martin E. Hellman. New directions in cryptography. *IEEE Transactions on Information Theory*, IT-22(6):644–654, 1976.
2. Taher El-Gamal. A public key cryptosystem and a signature scheme based on discrete logarithms. *IEEE Transactions on Information Theory*, 31(4):469–472, 1985.
3. Yael Gertner, Sampath Kannan, Tal Malkin, Omer Reingold, and Mahesh Viswanathan. The Relationship between Public Key Encryption and Oblivious Transfer. In *FOCS*, pages 325–335, 2000.
4. Oded Goldreich, Shafi Goldwasser, and Silvio Micali. How to construct random functions. *J. ACM*, 33(4):792–807, 1986.
5. Danny Harnik and Moni Naor. On the Compressibility of NP instances and Cryptographic Applications, 2005. Manuscript.
6. Thomas Holenstein, 2005. Personal Communication.
7. Thomas Holenstein. *Immunization of key-agreement schemes, PhD.thesis*. PhD thesis, ETH Zürich, 2006. to appear.
8. Russell Impagliazzo. A personal view of average-case complexity. In *Structure in Complexity Theory Conference*, pages 134–147, 1995.
9. Russell Impagliazzo and Steven Rudich. Limits on the Provable Consequences of One-way Permutations. In *Proc, 21th ACM Symposium on the Theory of Computing (STOC)*, pages 44–61, 1989.
10. Ueli M. Maurer. Secret key agreement by public discussion from common information. *IEEE Transactions on Information Theory 39(3)*, pages 733-742, 1993
11. Ueli Maurer, Krzysztof Pietrzak, and Renato Renner. Indistinguishability Amplification, 2006. Manuscript.
12. Steven Myers. Black-box composition does not imply adaptive security. In *Advances in Cryptology — EUROCRYPT 04*, volume 3027 of *Lecture Notes in Computer Science*, pages 189–206, 2004.
13. Krzysztof Pietrzak. Composition does not imply adaptive security. In *Advances in Cryptology — CRYPTO '05*, volume 3621 of *Lecture Notes in Computer Science*, pages 55–65, 2005.
14. Omer Reingold, Luca Trevisan, and Salil P. Vadhan. Notions of reducibility between cryptographic primitives. In *TCC*, pages 1–20, 2004.
15. Steven Rudich. The use of interaction in public cryptosystems (extended abstract). In *CRYPTO*, pages 242–251, 1991.
16. Claude E. Shannon. A mathematical theory of communication. *Bell Systems Technical Journal*, 27:373–423 and 27:623–656, 1948.
17. Hoeteck Wee. Finding pessiland. In *TCC*, pages 429–442, 2006.

Perfect Non-interactive Zero Knowledge for NP

Jens Groth*, Rafail Ostrovsky**, and Amit Sahai***

UCLA, Computer Science Department,
4732 Boelter Hall,
Los Angeles, CA 90095, USA
{jg, rafail, sahai}@cs.ucla.edu

Abstract. Non-interactive zero-knowledge (NIZK) proof systems are fundamental cryptographic primitives used in many constructions, including CCA2-secure cryptosystems, digital signatures, and various cryptographic protocols. What makes them especially attractive, is that they work equally well in a concurrent setting, which is notoriously hard for interactive zero-knowledge protocols. However, while for interactive zero-knowledge we know how to construct statistical zero-knowledge argument systems for all NP languages, for non-interactive zero-knowledge, this problem remained open since the inception of NIZK in the late 1980's. Here we resolve two problems regarding NIZK:

- We construct the first perfect NIZK argument system for any NP language.
- We construct the first UC-secure NIZK argument for any NP language in the presence of a dynamic/adaptive adversary.

While it is already known how to construct efficient prover computational NIZK *proofs* for any NP language, the known techniques yield large common reference strings and large proofs. Another contribution of this paper is NIZK proofs with much shorter common reference string and proofs than previous constructions.

Keywords: Non-interactive zero-knowledge, universal composability, non-malleability.

1 Introduction

In this paper, we resolve a central open problem concerning Non-Interactive Zero-Knowledge (NIZK) protocols: how to construct *statistical* NIZK arguments for any NP language. While for *interactive* zero knowledge (ZK), it has long been known how to construct statistical zero-knowledge argument systems for all NP languages [5], for NIZK this question has remained open for nearly two decades.

---

* Supported by NSF Cybertrust ITR grant No. 0456717.
** Supported in part by a gift from Teradata, Intel equipment grant, NSF Cybertrust grant No. 0430254, OKAWA research award, B. John Garrick Foundation and Xerox Innovation group Award.
*** Supported by NSF Cybertrust ITR grant No. 0456717, an equipment grant from Intel, and an Alfred P. Sloan Foundation Research Fellowship.

S. Vaudenay (Ed.): EUROCRYPT 2006, LNCS 4004, pp. 339–358, 2006.

IN CONTEXT WITH PREVIOUS WORK – STATISTICAL ZERO KNOWLEDGE: Blum, Feldman, and Micali [3] introduced the notion of NIZK in the common random string model and showed how to construct *computational* NIZK proof systems for proving a single statement about any NP language. The first computational NIZK proof system for multiple theorems was constructed by Blum, De Santis, Micali, and Persiano [2]. Both [3] and [2] based their NIZK systems on certain number-theoretic assumptions (specifically, the hardness of deciding quadratic residues modulo a composite number). Feige, Lapidot, and Shamir [18] showed how to construct computational NIZK proofs based on any trapdoor permutation.

The above work, and the plethora of research on NIZK that followed, mainly considered NIZK where the zero-knowledge property was only true *computationally*; that is, a computationally bounded party cannot extract any information beyond the correctness of the theorem being proven. In the case of *interactive* zero knowledge, it has long been known that all NP statements can in fact be proven using *statistical* (in fact, perfect) zero knowledge arguments [6,5]; that is, even a computationally unbounded party would not learn anything beyond the correctness of the theorem being proven, though we must assume that the prover, *only during the execution of the protocol*, is computationally bounded to ensure soundness[1].

Achieving statistical NIZK has been an elusive goal. The original work of [3] showed how a computationally unbounded prover can prove to a polynomially bounded verifier that a number is a quadratic-residue, where the zero-knowledge property is perfect. Statistical ZK (including statistical NIZK[2]) for any nontrivial language for both proofs and arguments were shown to imply the existence of a one-way function by Ostrovsky [29]. Statistical NIZK proof systems were further explored by De Santis, Di Crescenzo, Persiano, and Yung [14] and Goldreich, Sahai, and Vadhan [23], who gave complete problems for the complexity class associated with statistical NIZK proofs. However, these works came far short of working for all NP languages, and in fact NP-complete languages cannot have (even interactive) statistical zero-knowledge proof systems unless the polynomial hierarchy collapses [19, 1][3]. Unless our computational complexity beliefs are wrong, this leaves open only the possibility of argument systems.

Do there exist *statistical* NIZK arguments for all NP languages? Despite nearly two decades of research on NIZK, the answer to this question was not known. In this paper, we answer this question in the affirmative, based on a number-theoretic complexity assumption introduced in [4].

---

[1] Such systems where the soundness holds computationally have come to be known as *argument systems*, as opposed to *proof systems* where the soundness condition must hold unconditionally.

[2] We note that the result of [29] is for *honest-verifier* SZK, and does not require the simulator to produce Verifier's random tape, and therefore it includes NIZK, even for the common reference string which is not uniform. See also [31] for an alternative proof.

[3] See also [22] appendix regarding subtleties of this proof, and [33] for an alternative proof.

OUR RESULTS. Our main results, which we describe in more detail below, are:

- Significantly more efficient NIZK *proofs* for circuit satisfiability.
- Perfect NIZK *arguments* for any NP language.
- UC-secure perfect NIZK arguments for any NP language, secure against adaptive/dynamic adversaries.

In our second result, we prove that our perfect NIZK argument has non-adaptive soundness, i.e., it is infeasible to forge a proof for a false statement that is chosen independently of the common reference string. In our third result, however, we construct a UC-secure perfect NIZK argument (i.e., in which the adversary sees the common reference string first, and can adaptively choose theorems and try to forge arguments afterwards)[4].

NIZK PROOFS. As a building block we start by constructing a simple and efficient computational NIZK proof of knowledge for circuit satisfiability, based on the subgroup decision problem introduced in [4]. To the best of our knowledge, our techniques are completely different from all previous constructions of NIZK proofs, which use the hidden bits model. In this NIZK proof system, the size of the common reference string is $\mathcal{O}(k)$, where k is the security parameter; thus it is independent of the size of the NP statements. The NIZK proofs have size $\mathcal{O}(k|C|)$, where $|C|$ is the size of the circuit. This is a significant result in its own right; the most efficient NIZK proof systems for an NP-complete problem with efficient provers previously known [27] required a reference string of size at least $\mathcal{O}(k^3)$ and the NIZK proofs of size at least $\mathcal{O}(|C|k^2)$. For comparison with the most efficient previous work, please see Table 1.

Table 1. Comparison of CRS size and NIZK proof size for efficient-prover NIZK proof systems for circuit satisfiability

Reference	CRS size	Proof Size	Assumption
Damgård [9]	$O(|C|k^2 + k^3)$	$O(|C|k^2 + k^3)$	Quadratic Residuosity
Kilian-Petrank [27]	$O(|C|k^2)$	$O(|C|k^2)$	Trapdoor Permutations
Kilian-Petrank [27]	$O(k^3)$	$O(|C|k^3)$	Trapdoor Permutations
De Santis et al. [12, 13]	$O(k + |C|^\varepsilon)$	$\text{poly}(|C|k)$	NIZK & One-Way Functions
This paper	$O(k)$	$O(|C|k)$	Subgroup Decision [4]

PERFECT NIZK ARGUMENTS. The NIZK proofs we construct are built using encryptions of the bits in the circuit. However, by a slight modification to only

---

[4] In general, there is an interesting and subtle technical point regarding the UC framework that should be pointed out here: while the UC framework does rule out proofs of false theorems in the ideal model, it does not explicitly rule out the possibility of proofs of false theorems in the real world. Instead, since the ideal and real world executions are indistinguishable, the UC framework rules out the possibility that the adversary (or the environnement) can gain any *advantage* from proofs of false theorems that it manages to generate in the real world.

the reference string, we effectively transform the cryptosystem into a perfectly hiding commitment scheme. With this transformation, we obtain a perfect NIZK argument for circuit satisfiability.

UC-SECURE PERFECT NIZK ARGUMENT. We generalize our techniques to construct perfect NIZK arguments that satisfy Canetti's UC definition of security. Canetti introduced the universal composability (UC) framework [7] as a general method to argue security of protocols in an arbitrary environment. It is a strong security definition; in particular it implies non-malleability [17], and security when arbitrary protocols are executed concurrently.

We define NIZK arguments in the UC framework and construct a NIZK argument that satisfies the UC security definition. From the theory behind the UC framework, this means that we can plug in our NIZK argument in arbitrary settings and maintain soundness and privacy. At the same time, we can prove that our UC NIZK argument enjoys a perfect zero-knowledge property.

We stress that our result of prefect NIZK in the UC framework holds even in the setting of dynamic/adaptive adversaries without erasures: where the adversary can corrupt parties adaptively, and upon corruption of a party, it learns the entire history of the internal state of this party. Prior to our result, no NIZK protocol was known to be UC-secure against dynamic/adaptive adversaries. In [8], it was observed that De Santis et al. [11] achieve UC-security, but only for the setting with *static* adversaries (in the non-erasure model).

2   Non-interactive Zero-Knowledge

Let R be an efficiently computable binary relation. For pairs $(x, w) \in R$ we call x the statement and w the witness. Let L be the language consisting of statements in R.

An argument system for a relation R consists of a key generation algorithm K, a prover P and a verifier V. The key generation algorithm produces a common reference string σ. The prover takes as input (σ, x, w) and checks whether $(x, w) \in R$. In that case, it produces a proof or argument π, otherwise it outputs failure. The verifier takes as input (σ, x, π) and outputs 1 if the proof is acceptable and 0 if rejecting the proof. We call (K, P, V) an argument system for R if it has the completeness and soundness properties described below.

(PERFECT) COMPLETENESS. For all $(x, w) \in R$, we have

$$\Pr\left[\sigma \leftarrow K(1^k); \pi \leftarrow P(\sigma, x, w) : V(\sigma, x, \pi) = 1 \text{ if } (x, w) \in R\right] = 1.$$

(NON-ADAPTIVE COMPUTATIONAL) SOUNDNESS. For all non-uniform polynomial time adversaries $\mathcal{A}$ and $x \notin L$, we have

$$\Pr\left[\sigma \leftarrow K(1^k); \pi \leftarrow \mathcal{A}(x, \sigma) : V(\sigma, x, \pi) = 0\right] \approx 1.$$

We call (K, P, V) a proof system for R if soundness also holds for computationally unbounded adversaries.

(Perfect) Knowledge Extraction. We call (K, P, V) a proof of knowledge for R if there exists a knowledge extractor $E = (E_1, E_2)$ with the properties described below.

For all non-uniform polynomial time adversaries $\mathcal{A}$ we have

$$\Pr\left[\sigma \leftarrow K(1^k) : \mathcal{A}(\sigma) = 1\right] = \Pr\left[(\sigma, \tau) \leftarrow E_1(1^k) : \mathcal{A}(\sigma) = 1\right]$$

For all non-uniform polynomial time adversaries $\mathcal{A}$ we have

$$\Pr\left[(\sigma, \tau) \leftarrow E_1(1^k); (x, \pi) \leftarrow \mathcal{A}(\sigma); w \leftarrow E_2(\sigma, \tau, x, \pi) : \right.$$
$$\left. V(\sigma, x, \pi) = 0 \text{ or } (x, w) \in R\right] = 1.$$

Since perfect knowledge extraction implies the existence of a witness for the statement being proven, it implies perfect adaptive soundness.

(Adaptive Multi-theorem) Zero-Knowledge. We call (K, P, V) a NIZK argument or NIZK proof for R if there exists a simulator $S = (S_1, S_2)$ with the following zero-knowledge property. For all non-uniform polynomial time adversaries $\mathcal{A}$ we have

$$\Pr\left[\sigma \leftarrow K(1^k) : \mathcal{A}^{P(\sigma, \cdot, \cdot)}(\sigma) = 1\right] \approx \Pr\left[(\sigma, \tau) \leftarrow S_1(1^k) : \mathcal{A}^{S'(\sigma, \tau, \cdot, \cdot)}(\sigma) = 1\right],$$

where $S'(\sigma, \tau, x, w) = S_2(\sigma, \tau, x)$ for $(x, w) \in R$ and outputs `failure` if $(x, w) \notin R$.

Honest prover state reconstruction. In modeling adaptive UC security without erasures, an honest prover may be corrupted at some time. To handle such cases, we want to extend the zero-knowledge property such that not only can we simulate an honest party making a proof, we also want to be able to simulate how it constructed the proof. In other words, once the party is corrupted, the adversary will learn the witness and the randomness used; we want to create convincing randomness so that it looks like the simulated proof was constructed by an honest prover using this randomness.

We say a NIZK argument or proof for R has honest prover state reconstruction if there exists a simulator $S = (S_1, S_2, S_3)$ so for all $\mathcal{A}$ we have

$$\Pr\left[\sigma \leftarrow K(1^k) : \mathcal{A}^{PR(\sigma, \cdot, \cdot)}(\sigma) = 1\right] \approx \Pr\left[(\sigma, \tau) \leftarrow S_1(1^k) : \mathcal{A}^{SR(\sigma, \tau, \cdot, \cdot)}(\sigma) = 1\right],$$

where $PR(\sigma, x, w)$ runs $r \leftarrow \{0, 1\}^{\ell_P(k)}$; $\pi \leftarrow P(\sigma, x, w; r)$ and returns π, r, and where SR runs $\rho \leftarrow \{0, 1\}^{\ell_S(k)}$; $\pi \leftarrow S_2(\sigma, \tau, x; \rho)$; $r \leftarrow S_3(\sigma, \tau, x, w, \rho)$ and returns π, r, both of the oracles outputting `failure` if $(x, w) \notin R$.

Perfect completeness, soundness, knowledge extraction and zero-knowledge. We speak of perfect completeness, perfect soundness, perfect knowledge extraction, perfect zero-knowledge and perfect honest prover state reconstruction if for sufficiently large security parameters we have equalities in the respective definitions.

3   The Boneh-Goh-Nissim Cryptosystem

Boneh, Goh and Nissim [4] suggest a cryptosystem with interesting homomorphic properties. The BGN-cryptosystem is the main building block in the paper.

BILINEAR GROUPS. We use two cyclic groups $\mathbb{G}, \mathbb{G}_1$ of order n, where $n = pq$ and p, q are primes. We make use of a bilinear map $e : \mathbb{G} \times \mathbb{G} \to \mathbb{G}_1$. I.e., for all $u, v \in \mathbb{G}$ and $a, b \in \mathbb{Z}$ we have $e(u^a, v^b) = e(u, v)^{ab}$. We require that $e(g, g)$ is a generator of $\mathbb{G}_1$ if g is a generator of $\mathbb{G}$. We also require that group operations, group membership, sampling of a random generator for $\mathbb{G}$ and the bilinear map be efficiently computable.

[4] suggest the following example. Pick large primes p, q and let $n = pq$. Find the smallest ℓ so $P = \ell n - 1$ is prime and equal to 2 modulo 3. Consider the points on the elliptic curve $y^2 = x^3 + 1$ over $\mathbb{F}_P$. This curve has $P + 1 = \ell n$ points, so it has a subgroup $\mathbb{G}$ of order n. We let $\mathbb{G}_1$ be the order n subgroup of $\mathbb{F}_{P^2}^*$ and $e : \mathbb{G} \times \mathbb{G} \to \mathbb{G}_1$ be the modified Weil-pairing.

THE SUBGROUP DECISION PROBLEM. Let $\mathcal{G}$ be an algorithm that takes a security parameter as input and outputs $(p, q, \mathbb{G}, \mathbb{G}_1, e)$ such that p, q are primes, $n = pq$ and $\mathbb{G}, \mathbb{G}_1$ are descriptions of groups of order n and $e : \mathbb{G} \times \mathbb{G} \to \mathbb{G}_1$ is a bilinear map.

Let $\mathbb{G}_q$ be the subgroup of $\mathbb{G}$ of order q. The subgroup decision problem is to distinguish elements of $\mathbb{G}$ from elements of $\mathbb{G}_q$. Let $\mathbb{G}_{gen}$ be the generators of $\mathbb{G}$.

Definition 1. *The subgroup decision assumption holds for generator $\mathcal{G}$ if there exists a negligible function $\nu_{SD} : \mathbb{N} \to [0; 1]$ so for any non-uniform polynomial time adversary $\mathcal{A}$ we have*

$$\Pr\Big[(p, q, \mathbb{G}, \mathbb{G}_1, e) \leftarrow \mathcal{G}(1^k); n = pq; g, h \leftarrow \mathbb{G}_{gen} : \mathcal{A}(n, \mathbb{G}, \mathbb{G}_1, e, g, h) = 1\Big]$$

$$- \Pr\Big[(p, q, \mathbb{G}, \mathbb{G}_1, e) \leftarrow \mathcal{G}(1^k); n = pq; g \leftarrow \mathbb{G}_{gen}, h \leftarrow \mathbb{G}_q \setminus \{1\} :$$

$$\mathcal{A}(n, \mathbb{G}, \mathbb{G}_1, e, g, h) = 1\Big] \quad < \quad \nu_{SD}(k).$$

We remark that we have changed the wording of the subgroup decision problem slightly in comparison with [4], but the definitions are equivalent.

THE BGN-CRYPTOSYSTEM. We generate a public key by running $(p, q, \mathbb{G}, \mathbb{G}_1, e) \leftarrow \mathcal{G}(1^k)$, setting $n = pq$, selecting g as a random generator of $\mathbb{G}$ and h as a random generator of $\mathbb{G}_q$. The public key is $(n, \mathbb{G}, \mathbb{G}_1, e, g, h)$ while the decryption key is p, q.

To encrypt a message m of length $\mathcal{O}(\log k)$ using randomness $r \leftarrow \mathbb{Z}_n^*$ we compute the ciphertext $c = g^m h^r$. To decrypt we compute $c^q = g^{mq} h^{mq} = (g^q)^m$ and exhaustively search for m.

By the subgroup decision assumption, we could indistinguishably select h to be a random generator of $\mathbb{G}$ as well. In this case, we do not have a cryptosystem but rather a perfectly hiding trapdoor commitment scheme.

4 Non-interactive Zero-Knowledge Proof

4.1 NIZK Proof That c Encrypts 0 or 1

We will construct a NIZK proof of knowledge for circuit satisfiability in Section 4.2. As a building block in this NIZK proof, we will encrypt the truth-values of the wires in the circuit. We need to convince the verifier that these ciphertexts have been correctly formed. We therefore start by constructing a NIZK proof that a BGN-ciphertext has either 0 or 1 as plaintext.

We observe that a ciphertext c contains 0 or 1, if and only if $c \in \mathbb{G}_q$ or $cg^{-1} \in \mathbb{G}_q$. Our strategy is therefore to show that $e(c, cg^{-1})$ has order q. If we know m, w so $c = g^m h^w$ then $m = 0$ implies $e(c, cg^{-1}) = e(h^w, g^{-1}h^w) = e(h, (g^{-1}h^w)^w)$ and $m = 1$ means $e(c, cg^{-1}) = e(gh^w, h^w) = e(h, (gh^w)^w)$. In both cases we get $e(c, cg^{-1}) = e(h, (g^{2m-1}h^w)^w)$. Since h has order q, revealing the two components will immediately convince the verifier that $e(c, cg^{-1})$ has order q, however may not be zero-knowledge.

Instead, we make a NIZK proof for $e(c, cg^{-1})$ having order q as follows. We choose a random exponent r and compute $e(c, cg^{-1}) = e(h^r, (g^{2m-1}h^w)^{wr^{-1}})$. We reveal these two components, and must convince the verifier that the first element $\pi_1 = h^r$ has order q. For this purpose, we show him the element g^r. Since $e(\pi_1, g) = e(h^r, g) = e(h, g^r)$ the verifier can now tell that π_1 has order q.

To argue zero-knowledge we change the public key. Instead of having h of order q, we use h of order n and select g so we know the discrete logarithm. Now all ciphertexts are perfectly hiding commitments so we can create all of them as encryptions of 0. We can simulate the revelation of g^r because we know the discrete logarithm.

Common reference string:
    1. $(p, q, \mathbb{G}, \mathbb{G}_1, e) \leftarrow \mathcal{G}(1^k)$
    2. $n = pq$
    3. g random generator of $\mathbb{G}$
    4. h random generator of $\mathbb{G}_q$
    5. Return $\sigma = (n, \mathbb{G}, \mathbb{G}_1, e, g, h)$.
Statement: The statement is an element $c \in \mathbb{G}$. The claim is that there exists a pair $(m, w) \in \mathbb{Z}^2$ so $m \in \{0, 1\}$ and $c = g^m h^w$.
Proof: Input $(\sigma, c, (m, w))$.
    1. Check $m \in \{0, 1\}$ and $c = g^m h^w$. Return **failure** if check fails.
    2. $r \leftarrow \mathbb{Z}_n^*$
    3. $\pi_1 = h^r, \pi_2 = (g^{2m-1}h^w)^{wr^{-1}}, \pi_3 = g^r$
    4. Return $\pi = (\pi_1, \pi_2, \pi_3)$
Verification: Input $(\sigma, c, \pi = (\pi_1, \pi_2, \pi_3))$.
    1. Check $c \in \mathbb{G}$ and $\pi \in \mathbb{G}^3$
    2. Check $e(c, cg^{-1}) = e(\pi_1, \pi_2)$ and $e(\pi_1, g) = e(h, \pi_3)$
    3. Return 1 if both checks pass, else return 0

Fig. 1. NIZK proof of plaintext being zero or one

Theorem 1. *The protocol in Figure 1 is a NIZK proof that $c \in \mathbb{G}$ has plaintext $m \in \{0, 1\}$. The NIZK proof has perfect completeness, perfect soundness and computational zero-knowledge and honest prover state reconstruction.*

Proof. PERFECT COMPLETENESS. We know that $c = g^m h^w$, where $m \in \{0, 1\}$. This gives us $e(c, cg^{-1}) = e(h, (g^{2m-1}h^w)^w) = e(h^r, (g^{2m-1}h^w)^{wr^{-1}}) = e(\pi_1, \pi_2)$. Furthermore, $e(\pi_1, g) = e(h^r, g) = e(h, g^r) = e(h, \pi_3)$.

PERFECT SOUNDNESS. We have $e(\pi_1^q, g) = e(\pi_1, g)^q = e(h, \pi_3)^q = e(h^q, \pi_3) = e(1, \pi_3) = 1$. Therefore, π_1 must have order 1 or q. This means $e(c, cg^{-1})^q = e(\pi_1, \pi_2)^q = e(\pi_1^q, \pi_2) = 1$, implying that c or cg^{-1} has order 1 or q.

COMPUTATIONAL ZERO-KNOWLEDGE AND HONEST PROVER STATE RECONSTRUCTION. First, we describe the simulator $S = (S_1, S_2, S_3)$. S_1 runs the algorithm for generating the common reference string with the following modification. It selects h to be a random generator for $\mathbb{G}$ and sets $g = h^\gamma$, where $\gamma \leftarrow \mathbb{Z}_n^*$. During the generation of the common reference string the simulator also learns p, q. S_1 outputs $(\sigma, \tau) = ((n, \mathbb{G}, \mathbb{G}_1, e, g, h), (p, q, \gamma))$.

S_2 on input (σ, τ, c) simulates a proof as follows. Either c, cg^{-1}, or both are generators for $\mathbb{G}$. The simulator picks $r \leftarrow \mathbb{Z}_n^*$. If c is a generator it sets $\pi_1 = c^r, \pi_2 = (cg^{-1})^{r^{-1}}$ and $\pi_3 = \pi_1^\gamma$. If c is not a generator for the group, then the simulator sets $\pi_1 = (cg^{-1})^r, \pi_2 = c^{r^{-1}}, \pi_3 = \pi_1^\gamma$.

S_3 is given the witness (m, w) so $c = g^m h^w$ and $m \in \{0, 1\}$ and wishes to reconstruct how the prover could have come up with the proof π. Since it knows γ it can write $c = h^{\gamma m + w}$. Consider first the case where c is a generator for $\mathbb{G}$, then we have $\gcd(n, \gamma m + w) = 1$. So we can write the proof as $\pi_1 = h^{r(\gamma m + w)}, \pi_2 = (g^{2m-1}h^w)^{w(r(\gamma m + w))^{-1}}, \pi_3 = g^{r(\gamma m + w)}$. We return $r(\gamma m + w) \bmod n$ as the prover's simulated randomness that would cause it to produce π. In case c is not a generator, we know that cg^{-1} is a generator and we write the proof as $\pi_1 = h^{r(\gamma(m-1)+w)}, \pi_2 = (g^{2m-1}h^w)^{w(r(\gamma(m-1)+w))^{-1}}, \pi_3 = g^{r(\gamma(m-1)+w)}$ and return $r(\gamma(m-1) + w) \bmod n$ as the prover's simulated randomness.

To argue computational zero-knowledge we consider a hybrid experiment, where we use S_1 to generate the common reference string σ, but implement the simulation oracle using the real prover P. We first show that for all non-uniform polynomial time adversaries $\mathcal{A}$ we have

$$\left| \Pr\left[\sigma \leftarrow K(1^k) : \mathcal{A}^{PR(\sigma, \cdot, \cdot)}(\sigma) = 1\right] - \Pr\left[(\sigma, \tau) \leftarrow S_1(1^k) : \mathcal{A}^{PR(\sigma, \cdot, \cdot)}(\sigma) = 1\right] \right| < \nu_{SD}(k),$$

where $PR(\sigma, (\sigma, c), (m, w))$ runs $r \leftarrow \mathbb{Z}_n^*; \pi \leftarrow P(\sigma, (\sigma, c), (m, w); r)$ and returns π, r, and outputs `failure` if $m \notin \{0, 1\}$ or $c \neq g^m h^w$.

The only difference between the two experiments is the choice of h. In one case, h is a random generator of $\mathbb{G}$ in the other case it is a generator of $\mathbb{G}_q$. We do not use the knowledge of p, q or the discrete logarithm of g with respect to h in either experiment. Consider now a subgroup decision problem challenge $(n, \mathbb{G}, \mathbb{G}_1, e, g, h)$. The challenges correspond exactly to common reference strings produced by respectively K and S_1. The advantage of $\mathcal{A}$ is therefore bounded by $\nu_{SD}(k)$.

Next, we go from the hybrid experiment to the simulation. For all $\mathcal{A}$ we have

$$\Pr\left[(\sigma,\tau) \leftarrow S_1(1^k) : \mathcal{A}^{PR(\sigma,\cdot,\cdot)}(\sigma) = 1\right]$$
$$= \Pr\left[(\sigma,\tau) \leftarrow S_1(1^k) : \mathcal{A}^{SR(\sigma,\tau,\cdot,\cdot)}(\sigma) = 1\right],$$

where SR runs $\rho \leftarrow \mathbb{Z}_n^*; \pi \leftarrow S_2(\sigma,\tau,(\sigma,c);\rho); r \leftarrow S_3(\sigma,\tau,(\sigma,c),(m,w),\rho)$ and returns π, r, or failure if $m \notin \{0,1\}$ or $c \neq g^m h^w$.

A simulated proof $\pi = (\pi_1, \pi_2, \pi_3)$ uniquely defines the randomness $r \in \mathbb{Z}_n^*$ so $\pi_1 = h^r$, and it is indeed this randomness S_3 outputs. We therefore just need to argue that simulated proofs have the same distribution as real proofs in the hybrid experiment. In case c is a generator for $\mathbb{G}$, S_2 selects $r \leftarrow \mathbb{Z}_n^*$ at random and set $\pi_1 = c^r$, which gives us a random generator of $\mathbb{G}$. In a real prover's proof π_1 is also a random generator of $\mathbb{G}$ when h has order n. Since π_1 uniquely defines π_2 and π_3, we see that the two distributions are identical. If c is not a generator for $\mathbb{G}$, then cg^{-1} and since a simulated $\pi_1 = (cg^{-1})^r$ for $r \leftarrow \mathbb{Z}_n^*$ is a random generator of $\mathbb{G}$, we can use a similar argument to show that also in this case we get a perfect simulation. $\square$

4.2   NIZK Proof of Knowledge for Circuit Satisfiability

Suppose we have a circuit C and want to prove that there exists w so $C(w) = 1$. Since any circuit can be linearly reduced to a circuit built only from NAND-gates, we will without loss of generality focus on this simpler case.

To prove satisfiability of C we encrypt the bit value of each wire, when the circuit is evaluated on the input bits in w. Using the NIZK proof in Figure 1 it is straightforward to prove that all ciphertexts contain a plaintext in $\{0,1\}$. We form the output ciphertext with randomness 0 so it is straightforward for the verifier to check that the output of the circuit is 1.

The only thing left is to prove that all the encrypted output wires do indeed evaluate the NAND-gates correctly. We make the following observation, leaving the proof to the reader.

Lemma 1. *Let $b_0, b_1, b_2 \in \{0,1\}$.*

$$b_0 + b_1 + 2b_2 - 2 \in \{0,1\} \text{ if and only if } b_2 = \neg(b_0 \wedge b_1).$$

Given ciphertexts c_0, c_1, c_2 containing plaintexts b_0, b_1, b_2 we can use the homomorphic property to form the ciphertext $c_0 c_1 c_2^2 g^{-2}$. A NIZK proof that $c_0 c_1 c_2^2 g^{-2}$ contains a plaintext in $\{0,1\}$ implies $b_2 = \neg(b_0 \wedge b_1)$, as required. We make such a NIZK proof for each NAND-gate in the circuit.

Theorem 2. *The protocol in Figure 2 is a NIZK proof of knowledge of circuit satisfiability. It has perfect completeness, perfect soundness, perfect knowledge extraction and computational zero-knowledge and honest prover state reconstruction.*

Common reference string:
1. $(p, q, \mathbb{G}, \mathbb{G}_1, e) \leftarrow \mathcal{G}(1^k)$
2. $n = pq$
3. g random generator of $\mathbb{G}$
4. h random generator of $\mathbb{G}_q$
5. Return $\sigma = (n, \mathbb{G}, \mathbb{G}_1, e, g, h)$.

Statement: The statement is a circuit C built from NAND-gates. The claim is that there exist input bits w so $C(w) = 1$.

Proof: The prover has a witness w consisting of input bits so $C(w) = 1$.
1. Extend w to contain the bits of all wires in the circuit.
2. Encrypt each bit w_i as $c_i = g^{w_i} h^{r_i}$, with $r_i \leftarrow \mathbb{Z}_n^*$.
3. For all c_i make a NIZK proof of existence of w_i, r_i so $w_i = \{0, 1\}$ and $c_i = g^{w_i} h^{r_i}$.
4. For the output of the circuit we let the ciphertext be $c_{\text{output}} = g$, i.e., an easily verifiable encryption of 1.
5. For all NAND-gates, we do the following. We have input ciphertexts c_{i_0}, c_{i_1} and output ciphertexts c_{i_2}. We wish to prove the existence of $w_{i_0}, w_{i_1}, w_{i_2} \in \{0, 1\}$ and $r_{i_0}, r_{i_1}, r_{i_2}$ so $w_2 = \neg(w_0 \wedge w_1)$ and $c_{i_j} = g^{w_{i_j}} h^{r_{i_j}}$. To do so we make a NIZK proof that there exist m, r with $m \in \{0, 1\}$ so $c_{i_0} c_{i_1} c_{i_2}^2 g^{-2} = g^m h^r$.
6. Return π consisting of all the ciphertexts and NIZK proofs.

Verification: The verifier given a circuit C and a proof π.
1. Check that all wires have a corresponding ciphertext and that the output wire's ciphertext is g.
2. Check that all ciphertexts have a NIZK proof of the plaintext being 0 or 1.
3. Check that all NAND-gates have a valid NIZK proof of compliance.
4. Return 1 if all checks pass, else return 0.

Fig. 2. NIZK proof for circuit satisfiability

Proof. PERFECT COMPLETENESS. Knowing a satisfying assignment w for C, we can compute truth-values for all wires that are consistent with the NAND-gates and make the circuit have 1 as output. Perfect completeness follows from the perfect completeness of the NIZK proofs of plaintexts being either 0 or 1.

PERFECT SOUNDNESS. Since we prove for each wire that the encrypted plaintext is either 0 or 1, we have made a perfectly binding commitment to a bit for each wire. By Lemma 1, the NIZK proofs for the gates imply that all encrypted wire-bits respect the NAND-gates. Finally, we know that the output ciphertext is g, so the output bit is 1.

PERFECT KNOWLEDGE EXTRACTION. The extractor sets up the common reference string by running the key generator for the NIZK proof. In the process it learns p, q. This allows it to decrypt the ciphertexts containing the input-bits. Since the NIZK proof has perfect soundness, these input bits must correspond to a witness w so $C(w) = 1$.

COMPUTATIONAL ZERO-KNOWLEDGE AND HONEST PROVER STATE RECONSTRUCTION. Let S_1 be the simulator of the NIZK proof for a ciphertext having 0 or 1 as plaintext. We use the same algorithm to create the common reference string for simulation of circuit satisfiability NIZK proofs. In other words, both g, h are random generators of $\mathbb{G}$ and the simulator knows $\gamma \in \mathbb{Z}_n^*$ so $g = h^\gamma$.

S_2 starts by choosing the ciphertexts for the wires: The output wire gets the ciphertext g. For all other wires, it selects a ciphertext $c_i = h^{r_i}$ with $r_i \leftarrow \mathbb{Z}_n^*$. Later, when S_3 learns a witness w, it can compute the corresponding messages $m_i \in \{0, 1\}$ for all these ciphertexts, and open them as $c_i = g^{m_i} h^{r_i - m_i \gamma^{-1}}$.

For all these ciphertexts S_2 simulates a NIZK proof that they contain 0 or 1 as the plaintext. Also for all NAND-gates with input wires i_0, i_1 and output wire i_2 it simulates a NIZK proof that $c_{i_0} c_{i_1} c_{i_2}^2 g^{-2}$ contains a plaintext that is 0 or 1. Later, upon learning the witness w, S_3 knows the plaintexts $w_{i_j} \in \{0, 1\}$ and randomizers $r_{i_j} - w_{i_j} \gamma^{-1}$ that constitute a satisfactory encryption of the wires of a satisfied circuit. For each NIZK proof of a plaintext being 0 or 1, S_3 can run the honest prover state reconstructor to get convincing randomness that would make the prover produce this proof.

To prove that this is a good simulation, we first consider a hybrid experiment where we use the simulator to create the common reference string, but use the real prover to create the NIZK proofs. As in the proof of Theorem 1, we can argue that for all non-uniform polynomial time adversaries $\mathcal{A}$ we have

$$\left| \Pr \left[\sigma \leftarrow K(1^k) : \mathcal{A}^{PR(\sigma, \cdot, \cdot)}(\sigma) = 1 \right] \right.$$
$$\left. - \Pr \left[(\sigma, \tau) \leftarrow S_1(1^k) : \mathcal{A}^{PR(\sigma, \cdot, \cdot)}(\sigma) = 1 \right] \right| < \nu_{SD}(k),$$

where $PR(\sigma, C, w)$ runs $\pi \leftarrow P(\sigma, C, w; r)$ and returns π, r.

Next, we modify the way we create proofs. Instead of running the real prover, we create the encryptions of the wires c_i as the real prover, but simulate the NIZK proofs of 0 or 1 being the plaintext and simulate the NIZK proofs for the NAND-gates as well. From the proof of Theorem 1 we get that this modification does not increase $\mathcal{A}$'s probability of outputting 1. We have

$$\Pr \left[(\sigma, \tau) \leftarrow S_1(1^k) : \mathcal{A}^{PR(\sigma, \cdot, \cdot)}(\sigma) = 1 \right]$$
$$= \Pr \left[(\sigma, \tau) \leftarrow S_1(1^k) : \mathcal{A}^{PSR(\sigma, \tau, \cdot, \cdot)}(\sigma) = 1 \right],$$

where $PSR(\sigma, \tau, C, w)$ creates ciphertexts c_i correctly but simulates NIZK proofs for 0- or 1-plaintexts and the randomness involved, and outputs `failure` if $C(w) \neq 1$.

Finally, we go to the full simulation. For all $\mathcal{A}$ we have

$$\Pr \left[(\sigma, \tau) \leftarrow S_1(1^k) : \mathcal{A}^{PSR(\sigma, \tau, \cdot, \cdot)}(\sigma) = 1 \right]$$
$$= \Pr \left[(\sigma, \tau) \leftarrow S_1(1^k) : \mathcal{A}^{SR(\sigma, \tau, \cdot, \cdot)}(\sigma) = 1 \right],$$

where SR runs $\pi \leftarrow S_2(\sigma, \tau, C; \rho); r \leftarrow S_3(\sigma, \tau, C, w, \rho)$ and returns π, r, and outputs `failure` if $C(w) \neq 1$. The only difference here is in the way we create the ciphertexts, but since they are perfectly hiding, we cannot distinguish the two experiments. □

5   Non-interactive Statistical Zero-Knowledge Argument

In this section, we construct a NIZK argument of circuit satisfiability with perfect zero-knowledge. The main idea is a simple modification of the NIZK proof for circuit satisfiability in Figure 2. Instead of choosing h of order q, we let h be a random generator of $\mathbb{G}$. This way $g^m h^r$ is no longer an encryption of m, but a perfectly hiding commitment to m. It corresponds to using S_1 restricted to the first half of its outputs as key generator. Completeness is obvious and the proof of Theorem 2 reveals that the argument is perfect zero-knowledge.

Soundness is also simple enough. Suppose we have circuit $C \notin L$ generated independently of the common reference string. We can argue that no adversary can distinguish an h of order n from an h of order q, and therefore by Theorem 2 has negligible probability of making an acceptable NIZK argument.

Let S_σ be the simulator S_1 from the proof of Theorem 2 restricted to its first output. We have the following theorem

Theorem 3. (S_σ, P, V) *is a NIZK argument for circuit satisfiability.*

Proof. As in the proof of Theorem 2, we can show that the protocol has perfect completeness. Perfect zero-knowledge and honest prover state reconstruction follows from the proof of Theorem 2. This leaves us with the question of soundness.

SOUNDNESS. We first demonstrate that the NIZK argument is sound, i.e., for any fixed false statement, all adversaries have negligible probability of generating a valid proof of this statement.

Consider any unsatisfiable circuit C and a polynomial time adversary $\mathcal{A}$ that with probability So-Adv$_{\mathcal{A}}(1^k)$ breaks the soundness property. In other words, $\mathcal{A}$ is given a common reference string and proceeds to output a valid argument π. We will construct an adversary $\mathcal{B}$ that decides the subgroup decision problem with probability SD-Adv$_{\mathcal{B}}(1^k)$ =So-Adv$_{\mathcal{A}}(1^k)$.

$\mathcal{B}$ gets a challenge $(n, \mathbb{G}, \mathbb{G}_1, e, g, h)$ and has to decide whether h has order n or not. This corresponds to a common reference string generated by either K or S_σ. So we can give it to $\mathcal{A}$ and output 1 if and only if $\mathcal{A}$ forms a valid argument for C being true.

In case h has order n, the common reference string produced by $\mathcal{B}$ is distributed exactly as in a real argument. The adversary therefore has probability So-Adv$_{\mathcal{A}}(1^k)$ of generating an acceptable argument.

On the other hand, in case h has order q the common reference string produced by $\mathcal{B}$ is distributed as the reference string in the previously described NIZK proof. Since the NIZK proof has perfect soundness, the probability of $\mathcal{A}$ producing a valid argument is 0. □

6   Universally Composable Non-interactive Zero-Knowledge

6.1   Modeling Non-interactive Zero-Knowledge Arguments

The universal composability (UC) framework (see [7] for a detailed description) is a strong security model capturing security of a protocol under concurrent execution of arbitrary protocols. We model all other things not directly related to the protocol through a polynomial time environment. The environment can at its own choosing give inputs to the parties running the protocol, and according to the protocol specification, the parties can give outputs to the environment. In addition, there is a non-uniform polynomial time adversary $\mathcal{A}$ that attacks the protocol. $\mathcal{A}$ can communicate freely with the environment. It can also corrupt parties, in which case it learns the entire history of that party and gains complete control over the actions of this party.

To model security we use a simulation paradigm. We specify the functionality $\mathcal{F}$ that the protocol should realize. The functionality $\mathcal{F}$ can be seen as a trusted party that handles the entire protocol execution and tells the parties what they would output if they executed the protocol correctly. In the ideal process, the parties simply pass on inputs from environment to $\mathcal{F}$ and whenever receiving a message from $\mathcal{F}$ they output it to the environment. In the ideal process, we have an ideal process adversary $\mathcal{S}$. $\mathcal{S}$ does not learn the content of messages sent from $\mathcal{F}$ to the parties, but is in control of when, if ever, a message from $\mathcal{F}$ is delivered to the designated party. $\mathcal{S}$ can corrupt parties, at the time of corruption it will learn all inputs the party has received and all outputs it has sent to the environment. As the real world adversary, $\mathcal{S}$ can freely communicate with the environment.

We now compare these two models and say that it is secure if no environment can distinguish between the two worlds. This means, the protocol is secure, if for any $\mathcal{A}$ running in the real world, there exists an $\mathcal{S}$ running in the ideal process with $\mathcal{F}$ so no environment can distinguish between the two worlds.

The standard zero-knowledge functionality $\mathcal{F}_{ZK}$ as defined in [7] goes as follows: On input $(\mathbf{prove}, P, V, sid, ssid, x, w)$ from P the functionality $\mathcal{F}_{ZK}$ checks that $(x, w) \in R$ and in that case sends $(\mathbf{proof}, P, V, sid, ssid, x)$ to V. It is thus part of the model that the prover will send the proof to a particular receiver and that this receiver will learn who the prover is. This is a very reasonable model when we talk about interactive zero-knowledge proofs of knowledge. We remark that with only small modifications in the UC NIZK argument that we are about to suggest we could securely realize this functionality.

Parameterized with relation R and running with parties $P_1, \ldots, P_n$ and adversary $\mathcal{S}$.

Proof: On input $(\mathbf{prove}, sid, ssid, x, w)$ from party P ignore if $(x, w) \notin R$. Send $(\mathbf{prove}, x)$ to $\mathcal{S}$ and wait for answer $(\mathbf{proof}, \pi)$. Upon receiving the answer store (x, π) and send $(\mathbf{proof}, sid, ssid, \pi)$ to P.

Verification: On input $(\mathbf{verify}, sid, ssid, x, \pi)$ from V check whether (x, π) is stored. If not send $(\mathbf{verify}, x, \pi)$ to $\mathcal{S}$ and wait for an answer $(\mathbf{witness}, w)$. Upon receiving the answer, check whether $(x, w) \in R$ and in that case, store (x, π). If (x, π) has been stored return $(\mathbf{verification}, sid, ssid, 1)$ to V, else return $(\mathbf{verification}, sid, ssid, 0)$.

Fig. 3. NIZK argument functionality $\mathcal{F}_{\text{NIZK}}$

However, when we talk about NIZK arguments we do not always know who is going to receive the NIZK argument. We simply create a string π, which is the NIZK argument. We may create this string in advance and later decide to whom to send it. Furthermore, anybody who intercepts the string π can verify the truth of the statement and can use the string to convince others about the truth of the statement. The NIZK argument is not deniable; quite on the contrary, it is transferable [30]. For this reason, and because the protocol and the security proof becomes a little simpler, we suggest a different functionality $\mathcal{F}_{\text{NIZK}}$ to capture the essence of NIZK arguments.

6.2   Tools

We will need a few cryptographic tools to securely realize $\mathcal{F}_{\text{NIZK}}$.

PERFECTLY HIDING COMMITMENT SCHEME WITH EXTRACTION. A perfectly hiding commitment scheme with extraction (first used in [16] in the setting of perfectly hiding non-malleable commitment) has the following property. We can run a key generation algorithm $hk \leftarrow K_{\text{hiding}}(1^k)$ to get a hiding key hk, or we can alternatively run a key generation algorithm $(hk, xk) \leftarrow K_{\text{extract}}(1^k)$ in which case we get both a hiding key hk and an extraction key xk. $(K_{\text{hiding}}, \text{com})$ constitute a perfectly hiding commitment scheme. On the other hand, $(K_{\text{extract}}, \text{com}, \text{dec})$ constitute a public key cryptosystem with errorless decryption, i.e.,

$$\Pr\left[(hk, xk) \leftarrow K_{\text{extract}}(1^k) : \forall(m, r) : \text{dec}_{xk}(\text{com}_{hk}(m; r)) = m\right] \approx 1.$$

We demand that no non-uniform polynomial time adversary $\mathcal{A}$ can distinguish between the two key generation algorithms. This implies that the cryptosystem is semantically secure against chosen plaintext attack since the perfectly hiding commitment does not reveal what the message is.

We have already seen one example of a perfectly hiding commitment scheme with extraction. We can set up the BGN-cryptosystem with a public key, where h has full order n. In this case, the cryptosystem is a perfectly hiding commitment scheme. We can also set it up with h having order q, in this case, the cryptosystem has errorless decryption. The subgroup decisional assumption implies that no non-uniform polynomial time adversary can distinguish commitment keys from cryptosystem keys.

PSEUDORANDOM CRYPTOSYSTEM. A cryptosystem $(K_{\text{pseudo}}, E, D)$ has pseudorandom ciphertexts of length $\ell_E(k)$ if for all non-uniform polynomial time adversaries $\mathcal{A}$ we have

$$\Pr\left[(pk, dk) \leftarrow K_{\text{pseudo}}(1^k) : \mathcal{A}^{E_{pk}(\cdot)}(pk) = 1\right]$$
$$\approx \Pr\left[(pk, dk) \leftarrow K_{\text{pseudo}}(1^k) : \mathcal{A}^{R_{pk}(\cdot)}(pk) = 1\right],$$

where $R_{pk}(m)$ runs $c \leftarrow \{0, 1\}^{\ell_E(k)}$ and returns c. We require that the cryptosystem have errorless decryption as defined earlier.

Trapdoor permutations imply pseudorandom cryptosystems, we can use the Goldreich-Levin hard-core bit [21] of a trapdoor permutation to make a one-time pad. In the concrete case of the BGN cryptosystem, we observe that it implies hardness of factorization and it is possible to transform Rabin-encryption into a pseudorandom cryptosystem. When working over elliptic curves, there are also more direct constructions of pseudorandom cryptosystems based on the subgroup decision assumption.

TAG-BASED SIMULATION-SOUND TRAPDOOR COMMITMENT A tag-based commitment scheme has four algorithms. The key generation algorithm $K_{\text{tag-com}}$ produces a commitment key ck as well as a trapdoor key tk. There is a commitment algorithm that takes as input the commitment key ck, a message m and any tag tag and outputs a commitment $c = \text{commit}_{ck}(m, tag; r)$. To open a commitment c with tag tag we reveal m and the randomness r. Anybody can now verify whether indeed $c = \text{commit}_{ck}(m, tag; r)$. As usual, the commitment scheme must be both hiding and binding.

In addition, to these two algorithms there are also a couple of trapdoor algorithms Tcom, Topen that allow us to create an equivocal commitment and later open this commitment to any value we prefer. We create an equivocal commitment and an equivocation key as $(c, ek) \leftarrow \text{Tcom}_{ck,tk}(tag)$. Later we can open it to any message m as $r \leftarrow \text{Topen}_{ck,ek}(c, m, tag)$, such that $c = \text{commit}_{ck}(m, tag; r)$. We require that equivocal commitments and openings are indistinguishable from real openings. For all non-uniform polynomial time adversaries $\mathcal{A}$ we have

$$\Pr\left[(ck, tk) \leftarrow K_{\text{tag-com}}(1^k) : \mathcal{A}^{\mathcal{R}(\cdot,\cdot)}(ck) = 1\right]$$
$$\approx \Pr\left[(ck, tk) \leftarrow K_{\text{tag-com}}(1^k) : \mathcal{A}^{\mathcal{O}(\cdot,\cdot)}(ck) = 1\right],$$

where $\mathcal{R}(m, tag)$ returns a randomly selected randomizer and $\mathcal{O}(m, tag)$ computes $(c, ek) \leftarrow \text{Tcom}_{ck,tk}(m, tag); r \leftarrow \text{Topen}_{ck,ek}(c, m, tag)$ and returns r and $\mathcal{A}$ does not submit the same tag twice to the oracle.

Tag-based simulation-sound trapdoor commitments were first implicitly constructed in [15], and explicitly in [16, 28]. The tag-based simulation soundness property is based on the notion of simulation soundness introduced by Sahai [32] for NIZK proofs. Aside from [15, 16, 28], other constructions of tag-based simulation sound commitments or schemes that can easily be transformed into tag-based simulation-sound commitments have appeared in [11, 8, 20, 10, 24, 25]. The tag-based simulation-soundness property means that a commitment using tag remains binding even if we have made equivocations for commitments using different tags. For all non-uniform polynomial time adversaries $\mathcal{A}$ we have

$$\Pr\left[(ck, tk) \leftarrow K(1^k); (c, tag, m_0, r_0, m_1, r_1) \leftarrow \mathcal{A}^{\mathcal{O}(\cdot)}(ck) : tag \notin Q \text{ and}\right.$$
$$\left. c = \text{commit}_{ck}(m_0, tag; r_0) = \text{commit}_{ck}(m_1, tag; r_1) \text{ and } m_0 \neq m_1\right] \approx 0,$$

where $\mathcal{O}(commit, tag)$ computes $(c, ek) \leftarrow \text{Tcom}_{ck,tk}(tag)$, returns c and stores (c, tag, ek), and $\mathcal{O}(open, c, m, tag)$ returns $r \leftarrow \text{Topen}_{ck,ek}(c, m, tag)$ if (c, tag, ek) has been stored, and where Q is the list of tags for which equivocal commitments have been made by $\mathcal{O}$.

STRONG ONE-TIME SIGNATURES. We remind the reader that strong one-time signatures allow a non-uniform polynomial time adversary to ask an oracle for a signature on one arbitrary message. Then it must be infeasible to forge a signature on any different message and infeasible to come up with a different signature on the same message. Strong one-time signatures can be constructed from one-way functions.

6.3   UC NIZK

The standard technique to prove that a protocol securely realizes a functionality in the UC framework is to show that the ideal model adversary S can simulate everything that happens on top of the ideal functionality. In our case, there are two tricky parts. First, S may learn that a statement C has been proved and has to simulate a UC NIZK argument π without knowing the witness. Furthermore, if this honest prover is corrupted later then we learn the witness but must now simulate the randomness of the prover that would lead it to produce π. The second problem is that whenever S sees an acceptable UC NIZK argument π for a statement C, then an honest verifier V will accept. We must therefore, input a witness w to $\mathcal{F}_{\text{NIZK}}$ so it can instruct V to accept.

The main idea in overcoming these hurdles is to commit to the witness w and make a NIZK proof that indeed we have committed to a witness w so $C(w) = 1$. If the NIZK proof has the honest prover state reconstruction property, then we can simulate NIZK proofs and the prover's random coins.

This leaves us with the commitment scheme. On one hand, when we simulate UC NIZK arguments we want to make equivocal commitments that can be opened to anything since we do not know the witness yet. On the other hand, when we see a UC NIZK argument that we did not construct ourselves we want to be able to extract the witness, since we have to give it to $\mathcal{F}_{\text{NIZK}}$.

We will construct such a commitment scheme from the tools specified in the previous section. We use a tag-based simulation-sound trapdoor commitment scheme to commit to each bit of w. If w has length ℓ this gives us commitments $c_1, \ldots, c_\ell$. For honest provers we can use the trapdoor key tk to create equivocal commitments that can be opened to any bit we like. This enables us to simulate the commitments of the honest provers, and when we learn w upon corruption, we can simulate the randomness they could have used to commit to the witness w.

We still have an extraction problem, it is not clear that we can extract a witness from tag-based commitments created by a malicious adversary. To solve this problem we choose to encrypt the openings of the commitments. Now we can extract witnesses, but we have reintroduced the problem of equivocation. In a simulated commitment we may know two different openings of a commitment c_i to respectively 0 and 1, however, if we encrypt the opening then we are stuck with one possible opening. This is where the pseudorandomness property of the cryptosystem comes in handy. We can simply make two encryptions, one of an opening to 0 and one of an opening to 1. Since the ciphertexts are pseudorandom, we can open the ciphertext containing the opening we want and claim that the other ciphertext was chosen as a random string. To recap, the idea so far to commit to a bit b is to make a commitment c_i to this bit, and create a ciphertext $c_{i,b}$ containing an opening of c_i to b, while choosing $c_{i,1-b}$ as a random string.

The commitment scheme is equivocable, however, again we must be careful that we can extract a message from an adversarial commitment. The problem is that since we equivocate commitments for honest provers it may be the case that the adversary can produce equivocable commitments. This means, the adversary can produce some simulation sound commitment c_i and encryptions $c_{i,0}, c_{i,1}$ of openings to respectively 0 and 1. To resolve this issue we will select the tags for the commitments in a way so the adversary is forced to use a tag that has not been used to make an equivocable commitment. When an honest prover is making a commitment, we will select keys for a strong one-time signature scheme $(vk, sk) \leftarrow K_{\mathrm{sign}}(1^k)$. We will use $tag = (vk, C)$ when making the commitment c_i. The verification key vk will be published together with the commitment, and we will sign the commitment (as well as something else) using this key. Since the adversary cannot forge signatures, it must use a different tag, and therefore the commitment is binding and only one of the ciphertexts can contain an opening of c_i. This allows us to establish simulation soundness.

If the adversary corrupts a party that has used vk earlier, then it may indeed sign messages using vk and can therefore use vk in the tag for commitments. However, since we also include the statement C in the tag for the commitment using vk, the adversary can only create an equivocable commitment in a UC NIZK argument for the same statement C. We will observe that in this particular case we do not need to extract the witness w, because we can get it during the corruption of the prover.

Finally, in order to make the UC NIZK argument perfect zero-knowledge we wrap all the commitments c_i and the ciphertexts $c_{i,b}$ inside a perfectly hiding commitment c. In the simulation, however, we generate the key for this commitment scheme in a way such that it is instead a cryptosystem and we can extract the plaintext. This last step is only added to make the UC NIZK argument perfect zero-knowledge, it can be omitted if perfect zero-knowledge is not needed.

The resulting protocol can be seen in Figure 4. We use the notation from Section 6.2.

We prove the following theorems in the full paper [26].

Theorem 4. *The protocol in Figure 6 securely realizes* $\mathcal{F}_{\mathrm{NIZK}}$ *in the* $\mathcal{F}_{CRS}$-*hybrid model.*

Theorem 5. *The UC NIZK argument in Figure 4 is perfect zero-knowledge.*

Corollary 1. *Bilinear groups as described in Section 3 for which the decisional subgroup assumption holds imply the existence of a non-interactive perfect zero-knowledge protocol that securely realizes* $\mathcal{F}_{\mathrm{NIZK}}$.

CRS generation:
1. $hk \leftarrow K_{\text{hiding}}(1^k)$
2. $(ck, tk) \leftarrow K_{\text{tag-com}}(1^k)$
3. $(pk, dk) \leftarrow K_{\text{pseudo}}(1^k)$
4. $(\sigma, \tau) \leftarrow S_1(1^k)$
5. Return $\Sigma = (hk, ck, pk, \sigma)$

Statement: A circuit C and a claim that there exists input wires w so $C(w) = 1$.

Proof: On input (Σ, C, w).
1. Check $C(w) = 1$ and return `failure` if not
2. $(vk, sk) \leftarrow K_{\text{sign}}(1^k)$
3. For $i = 1$ to ℓ select r_i at random and let $c_i = \text{commit}_{ck}(w_i, (vk, C); r_i)$
4. For $i = 1$ to ℓ select R_{w_i} at random and set $c_{i,w_i} = E_{pk}(r_i; R_{w_i})$ and choose $c_{i,1-w_i}$ as a random string.
5. Choose r at random and let $c = \text{com}_{hk}(c_1, c_{1,0}, c_{1,1}, \ldots, c_\ell, c_{\ell,0}, c_{\ell,1}; r)$
6. Create a NIZK argument π for the statement that there exists w such that $C(w) = 1$ and there exists randomness so c has been produced as described in steps 3,4 and 5.
7. $s \leftarrow \text{sign}_{sk}(C, vk, c, \pi)$
8. Return $\Pi = (vk, c, \pi, s)$

Verification: On input (Σ, C, Π)
1. Parse $\Pi = (vk, c, \pi, s)$
2. Verify that s is a signature on (C, vk, c, π) under vk.
3. Verify the NIZK argument π
4. Return 1 if all checks work out, else return 0

Fig. 4. UC NIZK argument

Common reference string: On input (**start**,sid) run $\Sigma \leftarrow K(1^k)$.
Send (**crs**,sid,Σ) to all parties and halt.

Fig. 5. Protocol for UC NIZK common reference string generation

Proof: Party P waits until receiving (**crs**,sid, Σ) from $\mathcal{F}_{CRS}$.
On input (**prove**,sid, $ssid$, C, w) run $\Pi \leftarrow P(\Sigma, C, w)$. Output (**proof**,$sid$, $ssid$, Π).

Verification: Party V waits until receiving (**crs**,sid, Σ) from $\mathcal{F}_{CRS}$.
On input (**verify**,sid, $ssid$, C, Π) run $b \leftarrow V(\Sigma, C, \Pi)$. Output (**verification**,sid, $ssid$, b).

Fig. 6. Protocol for UC NIZK argument

References

1. William Aiello and Johan Håstad. Perfect zero-knowledge languages can be recognized in two rounds. In *Proceedings of FOCS '87*, pages 439–448, 1987.
2. Manuel Blum, Alfredo De Santis, Silvio Micali, and Giuseppe Persiano. Noninteractive zero-knowledge. *SIAM Jornal of Computation*, 20(6):1084–1118, 1991.
3. Manuel Blum, Paul Feldman, and Silvio Micali. Non-interactive zero-knowledge and its applications. In *proceedings of STOC '88*, pages 103–112, 1988.

4. Dan Boneh, Eu-Jin Goh, and Kobbi Nissim. Evaluating 2-dnf formulas on ciphertexts. In *proceedings of TCC '05, LNCS series, volume 3378*, pages 325–341, 2005.
5. Gilles Brassard, David Chaum, and Claude Crèpeau. Minimum disclosure proofs of knowledge. *JCSS*, 37(2):156–189, 1988.
6. Gilles Brassard and Claude Crèpeau. Non-transitive transfer of confidence: A perfect zero-knowledge interactive protocol for sat and beyond. In *Proceedings of FOCS '86*, pages 188–195, 1986.
7. Ran Canetti. Universally composable security: A new paradigm for cryptographic protocols. In *proceedings of FOCS '01*, pages 136–145, 2001. Full paper available at http://eprint.iacr.org/2000/067.
8. Ran Canetti, Yehuda Lindell, Rafail Ostrovsky, and Amit Sahai. Universally composable two-party and multi-party secure computation. In *proceedings of STOC '02*, pages 494–503, 2002. Full paper available at http://eprint.iacr.org/2002/140.
9. Ivan Damgård. Non-interactive circuit based proofs and non-interactive perfect zero-knowledge with proprocessing. In *proceedings of EUROCRYPT '92, LNCS series, volume 658*, pages 341–355, 1992.
10. Ivan Damgård and Jens Groth. Non-interactive and reusable non-malleable commitment schemes. In *proceedings of STOC '03*, pages 426–437, 2003.
11. Alfredo De Santis, Giovanni Di Crescenzo, Rafail Ostrovsky, Giuseppe Persiano, and Amit Sahai. Robust non-interactive zero knowledge. In *proceedings of CRYPTO '01, LNCS series, volume 2139*, pages 566–598, 2002.
12. Alfredo De Santis, Giovanni Di Crescenzo, and Giuseppe Persiano. Non-interactive zero-knowledge: A low-randomness characterization of np. In *proceedings of ICALP '99, LNCS series, volume 1644*, pages 271–280, 1999.
13. Alfredo De Santis, Giovanni Di Crescenzo, and Giuseppe Persiano. Randomness-optimal characterization of two np proof systems. In *proceedings of RANDOM '02, LNCS series, volume 2483*, pages 179–193, 2002.
14. Alfredo De Santis, Giovanni Di Crescenzo, Giuseppe Persiano, and Moti Yung. Image density is complete for non-interactive-szk. In *proceedings of ICALP '98, LNCS series, volume 1443*, pages 784–795, 1998.
15. Giovanni Di Crescenzo, Yvail Ishai, and Rafail Ostrovsky. Non-interactive and non-malleable commitment. In *proceedings of STOC '98*, pages 141–150, 1998.
16. Giovanni Di Crescenzo, Jonathan Katz, Rafail Ostrovsky, and Adam Smith. Efficient and non-interactive non-malleable commitment. In *proceedings of EUROCRYPT '01*, pages 40–59, 2001.
17. Danny Dolev, Cynthia Dwork, and Moni Naor. Non-malleable cryptography. *SIAM J. of Computing*, 30(2):391–437, 2000. Earlier version at STOC '91.
18. Uriel Feige, Dror Lapidot, and Adi Shamir. Multiple non-interactive zero knowledge proofs under general assumptions. *SIAM J. Comput.*, 29(1):1–28, 1999. Earlier version entitled Multiple Non-Interactive Zero Knowledge Proofs Based on a Single Random String appeared at FOCS '90.
19. Lance Fortnow. The complexity of perfect zero-knowledge. In *Proceedings of STOC '87*, pages 204–209, 1987.
20. Juan A. Garay, Philip D. MacKenzie, and Ke Yang. Strengthening zero-knowledge protocols using signatures. In *proceedings of EUROCRYPT '03, LNCS series, volume 2656*, pages 177–194, 2003. Full paper available at http://eprint.iacr.org/2003/037.
21. Oded Goldreich and Leonid A. Levin. A hard-core predicate for all one-way functions. In *proceedings of STOC '89*, pages 25–32, 1989.

22. Oded Goldreich, Rafail Ostrovsky, and Erez Petrank. Computational complexity and knowledge complexity. *SIAM J. Comput.*, 27:1116–1141, 1998.
23. Oded Goldreich, Amit Sahai, and Salil P. Vadhan. Can statistical zero knowledge be made non-interactive? or on the relationship of szk and niszk. In *CRYPTO '99, LNCS series, volume 1666*, pages 467–484, 1999.
24. Jens Groth. Honest verifier zero-knowledge arguments applied. Dissertation Series DS-04-3, BRICS, 2004. PhD thesis. xii+119 pp.
25. Jens Groth. Cryptography in subgroups of $\mathbb{Z}_n^*$. In *proceedings of TCC '05, LNCS series, volume 3378*, pages 50–65, 2005.
26. Jens Groth, Rafail Ostrovsky, and Amit Sahai. Perfect non-interactive zero-knowledge for np. ECCC Report TR05-097, `http://eccc.uni-trier.de/eccc-reports/2005/TR05-097/index.html`, 2005.
27. Joe Kilian and Erez Petrank. An efficient noninteractive zero-knowledge proof system for np with general assumptions. *Journal of Cryptology*, 11(1):1–27, 1998.
28. Philip D. MacKenzie and Ke Yang. On simulation-sound trapdoor commitments. In *proceedings of EUROCRYPT '04, LNCS series, volume 3027*, pages 382–400, 2004. Full paper available at `http://eprint.iacr.org/2003/252`.
29. Rafail Ostrovsky. One-way functions, hard on average problems, and statistical zero-knowledge proofs. In *Proceedings of Structure in Complexity Theory Conference*, pages 133–138, 1991.
30. Rafael Pass. On deniability in the common reference string and random oracle model. In *proceedings of CRYPTO '03, LNCS series, volume 2729*, pages 316–337, 2003.
31. Rafael Pass and Abhi Shelat. Characterizing non-interactive zero-knowledge in the public and secret parameter models. In *proceedings of CRYPTO '05, LNCS series*, 2005.
32. Amit Sahai. Non-malleable non-interactive zero-knowledge and adaptive chosen-ciphertext security. In *proceedings of FOCS '01*, pages 543–553, 2001.
33. Amit Sahai and Salil P. Vadhan. A complete problem for statistical zero knowledge. *J. ACM*, 50(2):196–249, 2003.

Language Modeling and
Encryption on Packet Switched Networks*

Kevin S. McCurley

Google

Abstract. The holy grail of a mathematical model of secure encryption
is to devise a model that is both faithful in its description of the real
world, and yet admits a construction for an encryption system that fulfills
a meaningful definition of security against a realistic adversary. While
enormous progress has been made during the last 60 years toward this
goal, existing models of security still overlook features that are closely
related to the fundamental nature of communication. As a result there
is substantial doubt in this author's mind as to whether there is any
reasonable definition of "secure encryption" on the Internet.

1   Introduction

In any area of science there is a fundamental tension between the desire to describe the real world with a model that is accurate in detail, vs. the desire to
use models that facilitate precise mathematical reasoning. In the case of cryptology, if a model fails to describe the real-world application, it leaves room for
attacks in the real world that were not anticipated by the model itself. This
has been highlighted in recent years by the discovery of multiple "side-channel
attacks" that are very effective against real-world systems, but usually fall outside the scope of existing security models. Examples include algorithmic timing
analysis [12], differential power analysis [13], protocol fault analysis [4], and differential fault analysis[3]. There is at least anecdotal evidence that many other
side channel attacks exist (e.g., RF and acoustic attacks) for popular models of
security. Unfortunately, a security model is successful only to the extent that
it accurately describes the process and the capabilities of the adversaries; any
omission, oversight, or ambiguity in this may properly be regarded as a weakness
of the model itself.

Micali and Reyzin [14] have recently sought to address some of the deficiencies
in current models of encryption by devising a corresponding model for *physically
observable cryptography*. They proposed an extension to the complexity-theoretic
model to embrace the notion that cryptographic algorithms are typically executed in a physical environment of a computer. In so doing, they sought to
address the deficiencies that have arisen from the aforementioned side-channel
attacks.

In this work I will address a different deficiency of current models, namely
the failure to model conveyance of semantic meaning through the physical act

---

* Updates to this paper may appear at `http://mccurley.org/papers/traffic/`.

S. Vaudenay (Ed.): EUROCRYPT 2006, LNCS 4004, pp. 359–372, 2006.

of communication. Just as the physical act of computation has side effects that are usable to a cryptanalyst, so too does the physical act of communication produce features that can be used to the advantage of the cryptanalyst. I will develop several examples of the phenomenon to demonstrate my point, but one obvious example is the encoding of communication into packets for transmission on a packet-switched network. It has been observed by multiple others (see Section 6) that the packetization of communication often leaks information about the content. While it may not leak the exact contents of the packets themselves, it leaks knowledge about the communication, and provides a tempting target for cryptanalysis.

The goal of this work is perhaps more modest than that of Micali and Reyzin, since I do not put forward any reasonable model under which a secure cryptosystem could be constructed. Instead, I will advance the view that *the structure of the Internet as we know it may actually preclude the existence of any reasonable model for completely secure encryption.* Given the degree to which society has come to depend upon the Internet, this is a startling possibility. Moreover, I will give examples to suggest that the phenomena is more general than just packet switched networks, and arises from many forms of communication via language.

While optimistic cryptologists should and will continue their quest for perfectly secure systems, there is no a priori reason why such a thing has to exist. Indeed, the entire framework of complexity-based security arguments would be radically changed if it turns out that P=NP, though it may also be argued that a polynomial separation between the capabilities of the legitimate user and the adversaries is sufficient for practical considerations. Moreover, the entire approach of complexity-theoretic security was an attempt to get around the limitation imposed by Shannon's result on perfect secrecy, and has proved to be remarkably effective in practice. It remains to be seen whether there is a similar approach that will mitigate the effects induced by the process of communication.

The point of view taken in this paper is partly historical and partly philosophical. An outline of the paper is as follows. In the next section we shall consider the definition of communication, after which we will present some examples of communication and how the process can leak knowledge. Following that we will propose a framework from which a partial security model can be constructed, without dwelling on the details. In fact, due to space and time constraints, I make no attempt describe a complete security model, but focus instead on the nature of the problem and why it may be impossible to construct such a model. I lay no claims on theorems regarding the possible existence or nonexistence of provably secure encryption. My hope is that this work will at least point the way toward better understanding of the underlying process of communication that we seek to model in the science of cryptology.

2   Mathematical Models of Encryption

During the last 60 years of mathematical research in cryptology, remarkable progress has been made in advancing cryptology to a science from what was once a black art. Most of the fundamental work has centered on the analysis of three models, namely information-theoretic security [19], complexity-based security [8], and quantum-theoretic security [2]. The goal of these is to construct a mathematical model of security, characterize the capabilities of an adversary, and (hopefully), provide a system that achieves some level of security under reasonable assumptions.

In both the information-theoretic model and the complexity-theoretic models of security, a secure cryptosystem is typically defined as a family of functions $E_k : M \to C$ that maps plaintexts $m \in M$ to ciphertexts $c \in C$. The family of encryption functions is indexed by the key $k \in K$ for some set of keys K. In Shannon's original formulation [19], an encryption system is said to have perfect secrecy if the adversary gains no more information about the plaintext from observing the ciphertext, i.e., $P_k(p|c) = P_k(p)$ for all keys $k \in K$. The major result that Shannon proved about this is that perfect secrecy requires that the key have as much entropy as the plaintext. This is often cited as a negative result, as it implies that substitution of secrecy of one piece of information (the plaintext) for another (the key) does not effectively result in any savings for the amount of secret information. This fact has motivated a lot of the research that has followed.

The example of the one-time pad is generally held up as the prototypical example of an encryption system that satisfies the perfect secrecy requirement, but in fact this holds only for messages that have constant length. The reason for this is obvious; the plaintext and the ciphertext are in fact the same length, so knowledge of the length of the ciphertext immediately reveals the length of the plaintext. While most theoreticians sweep this problem aside by simply assuming that all messages are the same size, I believe that this problem is in fact related to an important weakness in existing models.

In practice, the limitation of the one-time pad to message spaces in which all messages have the same length is at least as troublesome as the requirement for a large source of secure key bits. Moreover, the limitation is inherent to the definition of perfect secrecy, as is evidenced by the seminal observation of Chor and Kushilevitz [5] that it is impossible to construct an encryption system over a countably infinite message space that has information-theoretically perfect secrecy. Note that Shannon's original formulation incorporates an underlying probability distribution on plaintexts. This was reformulated in [5] by stating that for every pair of plaintexts p_1, p_2, $P(c|p_1) = P(c|p_2)$, or in other words, the probability of observing a given ciphertext is independent of the plaintext that generates it. Under this definition, they proved that there is no encryption system over a countably infinite message space that can achieve perfect secrecy. The implicit suggestion is that this is due to leakage of the length of the plaintext, and that this is unavoidable.

A primary driving force in the development of the complexity-theoretic models of security was to address the fact that just because information about the plaintext would be present in the ciphertext need not compromise the plaintext, provided the adversary was constrained in their ability to compute the implicit information. A major breakthrough in this line of research was the construction by Goldwasser and Micali[10] of a *semantically secure* encryption system. A semantically secure encryption system is based on the notion of indistinguishability of ciphertexts; given two plaintexts it should be infeasible to distinguish which of them gave rise to a given ciphertext. A fundamental part of their construction was the realization that randomness is necessarily a part of any secure encryption system.

Unfortunately, it was proved by Oded Goldreich[9] that *a semantically secure encryption scheme must also leak information about the length of the plaintext.* A related problem lies at the heart of indistinguishability, namely that it does not address the issue of whether the eavesdropper can determine whether communication takes place, but only which message was sent. The mere fact that an eavesdropper observes the communication of bits from one party to another is in itself information, and knowledge about the number of bits is simply further leakage. The problem of leaking the size of the plaintext is often swept aside in mathematical treatments with the casual remark to simply pad or packetize all messages to be the same length. This approach was refuted in [9], but the problem has been largely ignored since then. It should be noted that Shannon [19] also chose not to address the problem of hiding the existence of communication, though he explicitly mentioned the distinction.

3   The Nature of Communication

In addressing the original problem of providing a reasonable definition for secure encryption of communication, it is prudent to consider what constitutes communication in the first place. In its purest form, communication is an amorphous concept, since the term is used to describe a variety of physical behaviors and other features in addition to the encoding of symbols. Moreover, it's not even clear what is being transferred in the act of communication. The problem of defining communication cuts very close to the often-cited DIKW hierarchy of data, information, knowledge, and wisdom. The definitions of such terms are hotly debated, lying on the boundary between philosophy, mathematics, and computer science. For a philosopher, knowledge is a topic in epistemology, and consists of thoughts that are true, believed, and justified. For a mathematician, knowledge is a concept in modal or temporal logic. For a computer scientist, knowledge represents the inference from and application of data and information, whereas information contains only answers to "who what where" questions. For followers of artificial intelligence, knowledge represents a degree of uncertainty. All of these points of view are probably relevant to the study of cryptology.

Consider the sentence "Why are you doing that?". At one level it can be thought of as a string of symbols (data). At a higher layer it consists of a sequence of words representing concepts (information). At an even higher layer, it has meaning as a question, though only within a context. The mere presence of the symbol '?' indicates that it is a question, but the mapping of interpretations from one layer to the next is seldom this transparent. The use of the term "that" indicates that the sentence only makes sense in a broader context, with reference either through physical proximity or through reference to an earlier information state.

Communication is often associated with action. If this is a sentence uttered from one person to another, then we probably should expect a response from the other party to shortly follow. If a response does follow, then we might expect it to be a response to the question. If the question is sent over a radio broadcast, then no such response is likely, since the channel does not support it. If the speaker of this sentence is waving their arms wildly then it probably has a different meaning than if the person is simply arching an eyebrow. All of these nuances can be considered elements of a model of communication, and all are potentially relevant to a cryptanalyst.

Unfortunately, models of secure encryption typically assume that the cryptanalyst is restricted to only the encoded symbols, or at best, to the concepts represented by the grouping of symbols. Cryptologic research has typically taken information theory as the the starting point for characterizing communication, starting from the seminal work of Claude Shannon. In this characterization, messages are emitted as blocks of symbols by the sender according to some known probability distribution, and that the problem is simply to conceal *which* of the possible messages was emitted. In practice, communication is much more complicated than this. Shannon's original model of communication was first published as a paper [17], in which he said:

> The fundamental problem of communication is that of reproducing at one point either exactly or approximately a message selected at another point. Frequently the messages have meaning; that is they refer to or are correlated according to some system with certain physical or conceptual entities. These semantic aspects of communication are irrelevant to the engineering problem.

Shannon's paper was republished the following year as part of a book, with introductory material by his coauthor, Warren Weaver [18]. The introductory material by Weaver alludes to the limitations of Shannon's definition for communication, and states that:

> In fact, two messages, one of which is heavily loaded with meaning and the other of which is pure nonsense, can be exactly equivalent, from the present viewpoint, as regards information. It is this, undoubtedly, that Shannon means when he says that "the semantic aspects of communication are irrelevant to the engineering aspects."

Weaver casts Shannon's theory as one layer of a more complex set of phenomena. Many of the advancements of the information age can be traced to the fundamental contributions of Shannon; this separation of meaning from encoding and transport is what allowed engineers to concentrate on a useful paradigm for technology, while ignoring the incredibly complex nuances of the underlying conveyance of concepts. Weaver identified three layers of problems in communication:

Level A. How accurately can the symbols of communication be transmitted? (the technical problem)

Level B. How precisely do the transmitted symbols convey the desired meaning? (the semantic problem)

Level C. How effectively does the received meaning affect conduct in the desired way? (the effectiveness problem)

Shannon and Weaver's separation of the communication problem into layers is analogous to the invention of written language, where complex human communication processes were reduced to a sequence of symbols on a page. The result was enormously powerful in influencing the nature of human communication because it eliminated the need for humans to be in physical proximity in order to communicate, but at the same time, something was lost in communication by the conversion to symbols. For example, a sentence that is spoken while waving arms wildly in the air has different semantic meaning than the same sentence that is uttered with arms crossed. Interpersonal communication often applies a secondary semantic interpretation or decoding of the communication that augments and corrects errors and omissions made from the spoken symbols. It should be noted that this nuance of definition for communication is also not limited to communication between humans. For example, it is easy to imagine how a computer will exhibit different characteristics of communication when it is in distress than when it is in a normal mode of operation.

4   The Nature of Cryptanalysis

The purpose of communication is to convey something, and in some cases that is merely to convey data. In this case, communication is thought of as stateless. In other cases, the goal is to convey something more, namely information that can be acted upon. In still other cases, the purpose is to create common knowledge out of knowledge. Our difficulty in defining encrypted communication is probably closely related to this confusion.

By breaking a communication system into layers, Shannon and Weaver were able to separate the problem of conveying ideas from that of conveying a symbolic representation of language. In Shannon's work, the semantic meaning of communication is separated from the problem of conveying it, since this was embodied in an encoding layer that takes place before and after the physical act of communication that was Shannon's focus. Unfortunately, from the point of view of a cryptanalyst, the semantic meaning of the underlying communication may be precisely what they are interested in, and the actual symbols used to convey the ideas may be of only peripheral interest. Cryptanalysis typically has a purpose, and the act of cryptanalysis is the gathering of actionable knowledge for this purpose. Thus while a cryptanalyst may be interested in recovering the credit card of a targeted person, they may also be interested in knowing what the person is buying, or of discovering their social preferences, or of simply knowing that a credit card was used. There is no direct way to quantify the range of semantic concepts that the cryptanalyst may be interested in, and in fact the information content of semantic information that may be derived from context of the communication can be arbitrarily large relative to the amount of information contained in the communication itself.

To illustrate the importance of semantic meaning in the process of cryptanalysis, consider the following questions that a cryptanalyst might ask about communication on the Internet:

- What language is being spoken in a telephone call?
- Does Internet traffic contain VoIP or Skype traffic?
- Does Internet traffic use UDP or TCP?
- Is the same email being sent to multiple recipients?
- What is the nature of the relationships between the two parties in communication? Is one in command?
- What is the likelihood that a buy order will be issued in the next few seconds by a stock trader?

These are completely natural questions for a cryptanalyst to ask, and I claim that in each case there are plausible scenarios where the questions can be answered accurately with high probability, using observations of the physical act of communication.

It is tempting to define cryptanalysis as an attempt to create shared knowledge out of information. Unfortunately, it is completely unclear what falls within the domain of knowledge that is relevant to a given communication, since that requires us to characterize the goals of the eavesdropper relative to the two communicating parties who are his adversaries. It is almost certainly the case that any reasonable definition along these lines will need to take into account the state of knowledge of the eavesdropper before and after the communication, and the way in which it changed (either temporally, logically, or probabilistically). What is clear is that the current approach based only on information seems inadequate for accurately describing many situations.

5   The Use of Fragmentation in Communication

The original motivation for this work was to model the situation of two computers communicating privately over the Internet, and to understand the inherent limitations of using IPSEC to encrypt communication on the Internet. One of the fundamental features of Internet communication is that it is a packet switched network, in which the communication medium is shared between all parties connected to the network, and that communication is fragmented to enable congestion control and buffer management in intermediate routers. This feature of fragmentation of communication also arises in spoken and written language. Such "natural" language is typically composed of a sequence of distinct language elements (paragraphs, sentences, and words) that are themselves encoded into sequences of individual symbols or sounds.

To see why this the process of fragmentation is relevant to cryptanalysis, consider the following illustrative example. Suppose we are given the following fragment of encrypted text in which individual characters are encrypted but word breaks are exposed:

```
    # ####### ### ## ##### #### ###### ### ######## ######## #### #### ####
    #### #### ## # #######
                                                        ###### ########
```

As a cryptanalyst we might begin by noticing that two of the words are only a single letter. If the original text is in English, we might expect these words to be either the letter "I" or the letter "A". We might next notice that the text is arranged visually in a layout that is commonly used for quotations. The last line that is flush right might therefore be guessed to be a name, which greatly restricts the vocabulary. Knowing that this person is likely to be a famous person, we might be able to recover the most popular quotations of such people and apply a process of elimination. Even if the quotation was not in our list, we could apply basic knowledge of common sentence constructions to form a set of most likely candidates. If we hypothesis that the first letter is indeed 'A', then we might further hypothesize that the next word is either an adjective or a noun. By knowing something about the context of the communication, we may form a hypothesis about the candidates for each word, and in the end arrive at a probability distribution on potential plaintexts that has a relatively low entropy from among all possible messages that fit the observed pattern.

There are several observations to be made from this simple example. First, our knowledge of word breaks provides a huge advantage for inferring the actual content of the message. Second, our knowledge of the underlying language and conventions for its usage assists us in identifying a few basic structures. Each of these factors interact with each other and increase our level of confidence in predicting the content of the message.

As another simple example, I took three versions of Tolstoy's novel "Anna Karenina" written in English, French, and German. These should be more or less semantically the same message, with the only difference being that they are expressed in different languages. In order to test them to see if they could be distinguished from each other, I simply calculated the distribution of values of the lengths of the words. The result is shown in Figure 1. The data clearly shows a distinction between French and the other two languages.

Of course one should probably object to the relevance of this experiment since it's hard to imagine a communication system that exposes word boundaries in language. On the other hand, nearly all existing text instant messaging protocols operate on the basis of buffering entire lines, which are often aligned with sentence boundaries. In this case the packets that would be sent would likely reveal the lengths of the sentences. The amount of information being leaked is less in this case, but for all we know it might still be possible to reliably distinguish between the topics of sports vs. travel, or whether the parties are male, or to make a good guess on the age of the sender. Moreover, there are numerous other examples where packetization of communication can reveal knowledge about the application.

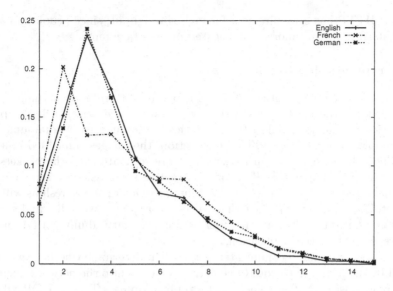

Fig. 1. Distribution of word lengths between three translations of the novel "Anna Karenina". Note that English and German have distinctly longer words on average than French, suggesting that it may be possible to distinguish French from the other two languages if word lengths are exposed through the communication process.

6   Characteristics of Internet Communication

Two of the major characteristics that are present in Internet communications are *layering* and *packetization*. The principal of layering is ubiquitous in engineering of complex tasks such as networking, since it isolates the many different requirements from each other, and provides a layer of abstraction for one layer to address another. By separating the routing, transmission, ordering, buffering, physical device drivers, and error correction into different layers, it simplifies the maintenance of software systems, improves their reliability, and facilitates extensions to new technologies such as wireless.

The principle of packetization is probably the biggest single contributor to the success of the Internet, because the Internet is a shared network that provides transport for all parties who connect to it. By regulating and merging the flow of packets from different sources, the Internet provides congestion control and a degree of fairness in use of the shared network. It also facilitates buffering, error correction, and retransmission. Without packetization there would be no sharing, and by providing a shared network for multiple applications, the Internet greatly increased the efficiency of communication. Much of the value was realized due to the fact that different applications with different quality of service requirements can use the same underlying network infrastructure. Extreme examples of service requirements arise from voice over IP (VoIP) and HTTP. The primary quality of service requirement for VoIP is a high probability of delivery and low latency, since any interruption or delay of voice results in a poor user experience. By contrast, HTTP has a requirement for high throughput, since people would like to download more and more sophisticated pieces of content.

6.1   Cryptographic Implications

The analogy of human natural language to Internet traffic is actually a very strong one, for the simple reason that word breaks are very much like packet boundaries, in that they reflect the semantics of the underlying communication. For example, in a persistent HTTP connection, the images embedded in a page are likely to be transmitted in packets that are separate from the packets containing the HTML page itself. In an interactive ssh session, a screen refresh event will often generate a packet containing as much of the refresh as will fit in a packet. This is due to the fact that within an application, individual `send()`s of information to the network are often broken into natural units of information that are defined by the application.

There are a number of complicated factors that determine whether an individual call by the program to `send()` generates a packet into the network, including the state of existing buffers, whether the application uses TCP or UDP, whether the operating system has properly implemented the TCP PUSH option, and whether the application chose to disable Nagle's algorithm on a TCP connection. In many cases it is still relatively easy to determine from the size and number of packets the number and size of `send()` calls in the application [11]. As a result, the "information breaks" of an application that correspond to word breaks or sentence breaks in natural language are often aligned to packet boundaries, and are therefore visible to an eavesdropper.

One feature of layering is that it provides the ability to address cryptographic requirements at the layer where it is convenient to do so. Examples include SSH at the application layer, SSL at the transport layer, IPSEC at the network layer, and WEP at the media layer. On the other hand, layering tends to introduce cryptographic weaknesses as well. For example, Bellovin [1] has observed that IP and TCP headers contain hints about the nature of the underlying traffic, and this largely results from layering, since quality of service is generally only implemented at the IP layer.

In addition to observations from headers, there are other signals present in packets from their timings, size, and patterns of traffic. For example, observations of SSH interactive login sessions were used to infer keystroke timings in [20], allowing them to mount an effective attack on passwords in SSH. A number of other examples were given by Bellovin [1].

The techniques of classifying traffic by characteristics that are not shielded by encryption have been developed by numerous authors, including Sun et. al. [21], Moore and Zuev [15], Zhang and Paxson [25], and Danezis [7, 6]. and Wright et. al. [24]. In spite of the increasing number of published attacks using characteristics of packet-switched networks, there has been very little discussion of this in the theoretical cryptography literature. Moreover, as was pointed out in section 2, current models of encryption simply avoid the problem of message size.

One application on the Internet that is rapidly gaining in popularity is voice over IP (VoIP). This protocol is extremely sensitive to latency and lost packets, so there are a number of optimizations and quality of service provisions for this service. Unfortunately some of these conflict with the security requirements of personal voice communication [23]. For example, VoIP voice packets are small (10-50 byte payload) and are therefore pretty easily recognized by their length. They are also likely to contain quality of service specifications in their headers. One interesting issue arises from a feature of VoIP that is designed to limit the bandwidth requirement for VoIP. In most phone conversations, only one end of the conversation will be talking at any given time. Hence it is only really important to send data in one direction most of the time, and in order to optimize bandwidth usage, VoIP supports something called silence suppression, where no packets are transmitted from the side that is silent. This feature has significant security implications, since this is precisely the kind of language break that was described in section 5!

7   Mathematical Models of Packetized Communication

Following up from the previous discussion, we can now derive some axioms that any model of packetized communication should follow in order to provide a meaningful model for cryptanalysis.

Axiom 1. A model of communication must include all sources and recipients of transmitted data. Consider for example a two-way communication between two people. A conversation may consist of questions, as well as responses to actions performed on the receipt of previous information. If we neglect to include these in our model, then we neglect a major source of information that is available to the eavesdropper.

Axiom 2. Communication is packetized. One way of looking at this is that communication has two states, namely when information is being transmitted and when it is not. Another way of saying this is that the sender is always sending; either real information or the null symbol, and the transmission of the null symbol is always detectable to the eavesdropper.

Axiom 3. Communication has state associated with it in both sender and receiver. This state changes as a result of receiving information.

Axiom 4. Communication has a temporal dimension, implying both an ordering and a distribution.

Axiom 5. Communication may be coupled to *observable* actions or states of the senders and recipients. In some cases traffic analysis may not be available to determine the source or destination of communication.

A natural model for a bidirectional channel is that of a pair of coupled Markov processes X_i, Y_i where X_i and Y_i are each dependent on $X_j, Y_j, j < i$. Here X_i, Y_i are ternary random variables taking on the values 0,1,null. The question of channel analysis is then to estimate the loss of information about X_i, Y_i when you are told when X_i, Y_i take on null values. More elaborate models would incorporate characteristics that may be observed about the aggregate of values, such as the notification that a packet payload is beginning, or that a packet was fragmented, etc.

7.1 Keeping the Channel Full

The leakage of the length of the message may be regarded as a generalization of the fact that if an adversary observes communication taking place between two parties, then they gain some information that they were not previously in possession of. This fundamental problem lies within a class of attacks commonly referred to as "traffic analysis". In practice this problem has been known for a very long time, and countermeasures are routinely used in modern link encryptors, by making sure that they always send information between sender and receiver, inserting dummy information if necessary [22]. By doing so, they seek to obscure the difference between actual communication and non-communication.

Unfortunately, the approach taken by link encryptors to "keep the channel full" is infeasible on the Internet, due to the requirement that the communication infrastructure serves the needs of multiple parties. In order for the Internet to operate efficiently and in an economically practical way, all parties must abstain from communicating except when they need to. One might ask how much additional bandwidth would be required in order for everyone to "keep the channel full". It has been observed empirically that the topology of the Internet connectivity graph has evolved as a sparse graph (e.g., see[16]), in which the degree distribution follows a power-law distribution. Thus in order to connect an Internet of n nodes, it appears that we require only $O(n)$ edges to provide a robust and scalable infrastructure for communication between potentially any pair of nodes. By contrast, if we adopt the link encryptor approach of masking the existence of communications by always communicating, we could potentially require $\binom{n}{2}$ edges in order for all n parties to be able to speak to each other. This is perhaps a pessimistic number, since the real number is the number of edges that a graph would require in order for there to be a collection of edge-disjoint paths between any bipartite matching of nodes in the graph. Of course even if the paths existed, we would still be left with the problem of finding them for routing purposes; this problem is unfortunately NP-complete. In other words, in order for the Internet to provide edge-disjoint paths that could be kept full between arbitrary matchings of nodes, a substantial increase in investment would be required.

8   Conclusions

The accumulated evidence of cryptanalysis through observation of communication points out that existing models of cryptographic security are lacking for at least two reasons. First, they fail to take into account the physical process of communication, in which the process of packetization is extremely important. It's almost certainly true that without packetization, the Internet could not have had the impact that it has. Yet at the same time, packetization has been seen to introduce numerous cryptographic weaknesses into communication, and there is currently no practical mathematical model to analyze the degree of weakness or within which we could prove anything about mitigating effects.

There is substantial doubt in the author's mind as to whether there is a reasonable balance that can be found between the quality of service demands of Internet applications and the goals of theoretical cryptography to provide an almost perfectly secure encryption methodology. It may turn out that it is inevitable that a cryptanalyst can attain some new knowledge from observing communication in some applications (notably those requiring low latency). If this is the case, then future research will be needed to define and quantify exactly how much knowledge will be leaked.

The second main point about theoretical models of cryptographic security is that they seem to overlook the distinction between knowledge and information. Shannon's achievement was to separate them so that communication engineering could proceed without the need to worry about conveyance of knowledge. Unfortunately, many cryptanalytic attacks take place at the knowledge layer of the DIKM hierarchy, and existing models fail to take this into account.

In many ways, this paper may be regarded as being even more pessimistic than that of Shannon, since I have argued that the nature of Internet communication channels makes it inevitable that cryptanalysts will be able to gain knowledge from passive eavesdropping. I would be happy if I could be proved wrong.

References

[1] Steven M. Bellovin. Probable plaintext cryptanalysis of the IP security protocols. In *Proc. of the Symp. on Network and Distributed System Security*, pages 155–160, 1997.

[2] Charles Bennett, F. Bessctte, Gilles Brassard, L. Salvail, and J. Smolin. Experimental quantum cryptography. *Journal of Cryptology*, 5:3–28, 1992.

[3] Eli Biham and Adi Shamir. Differential fault analysis of secret key cryptosystems. In *Advances in Cryptology, Proc. Crypto 1997*, Lecture Notes in Computer Science, pages 513–525. Springer-Verlag, 1997.

[4] Dan Boneh, Richard A. Demillo, and Richard J. Lipton. On the importance of checking cryptographic protocols for faults. In *Advances in Cryptology, Proc. Eurocrypt 1997*, volume 1233 of *Lecture Notes in Computer Science*, pages 37–51. Springer-Verlag, 1997.

[5] Benny Chor and Eyal Kushilevitz. Secret sharing over infinite domains. In *Proceedings of Crypto '89*, Lecture Notes in Computer Science, pages 299–306, Heidelberg, 1989. Springer-Verlag.

[6] George Danezis. Traffic analysis of the HTTP protocol over TLS. http://homes.esat.kuleuven.be/~gdanezis/TLSanon.pdf.

[7] George Danezis. Introducing traffic analysis: Attacks, defences and public policy issues, 2005. http://homes.esat.kuleuven.be/~gdanezis/TAIntro.pdf.

[8] Whitfield Diffie and Martin Hellman. New directions in cryptography. *IEEE Transactions on Information Theory*, 22:644–654, 1976.

[9] Oded Goldreich. A uniform-complexity treatment of encryption and zero-knowledge. *Journal of Cryptology*, 6:21–53, 1993.

[10] Shafi Goldwasser and Silvio Micali. Probabilistic encryption. *Journal of Computer and System Sciences*, 28:270–299, 1984.

[11] Amit Klein. Detecting and preventing HTTP response splitting and HTTP request smuggling attacks at the TCP level. http://www.securityfocus.com/archive/1/408135.

[12] Paul Kocher. Cryptanalysis of Diffie-Hellman, RSA, DSS, and other cryptosystems using timing attacks. In *Advances in Cryptology, Proc. Crypto '95*, LNCS, pages 171–183. Springer-Verlag, 1995.

[13] Paul Kocher, Joshua Jaffe, and Benjamin Jun. Differential power analysis. In *Advances in Cryptology, Proc. Crypto '99*, LNCS, pages 388–397, Heidelberg, 1999. Springer-Verlag.

[14] Silvio Micali and Leonid Reyzin. Physically observable cryptography. In *Theory of Cryptography Conference*, volume 2951 of *LNCS*, pages 278–296. Springer, 2004.

[15] Andrew W. Moore and Denis Zuev. Internet traffic classification using Bayesian analysis techniques. In *SIGMETRICS '05*, pages 50–60, 2005.

[16] M. E. J. Newman. The structure and function of complex networks. *SIAM Review*, 45:167–256, 2003.

[17] C. E. Shannon. A mathematical theory of communication. *Bell System Technical Journal*, 27:379–423,623–656, 1948.

[18] C. E. Shannon and Warren Weaver. *The Mathematical Theory of Communication*. University of Illinois Press, 1949.

[19] Claude E. Shannon. Communication theory of secrecy systems. *Bell Systems Technical Journal*, pages 656–715, 1949.

[20] Dawn Xiaodong Song, David Wagner, and Xuqing Tian. Timing analysis of keystrokes and timing attacks on ssh. In *Proc. USENIX Security Symposium*, pages 337–352, Washington, D.C., 2001.

[21] Qixiang Sun, Daniel R. Simon, Yi-Min Wang, Wilf Russell, Venkata N. Padmanabhan, and Lili Qiu. Statistical identificatoin of encrypted web browsing traffic. In *Proc. IEEE Security and Privacy Symp.*, pages 19–30, 2002.

[22] V. L. Voydoc and Stephen Kent. Security mechanisms in high-level network protocols. *ACM Computing Surveys*, pages 135–171, 1983.

[23] Thomas J. Walsh and Richard Kuhn. Challenges in security voice over IP. *IEEE Security and Privacy*, pages 44–49, May/June 2005.

[24] Charles Wright, Fabian Monrose, and Gerald M. Masson. HMM profiles for network traffic classification. In *ACM Conference on Computer and Communication Security*, pages 9–15, 2004.

[25] Y. Zhang and V. Paxson. Detecting stepping stones. In *Proc. 9th USENIX Security Symposium*, pages 171–184, 2000.

A Provable-Security Treatment
of the Key-Wrap Problem

Phillip Rogaway[1] and Thomas Shrimpton[2]

[1] Dept. of Computer Science, University of California, Davis, California 95616, USA
[2] Dept. of Computer Science, Portland State University,
Portland, Oregon 97201, USA

Abstract. We give a provable-security treatment for the *key-wrap problem*, providing definitions, constructions, and proofs. We suggest that key-wrap's goal is security in the sense of *deterministic authenticated-encryption* (DAE), a notion that we put forward. We also provide an alternative notion, a *pseudorandom injection* (PRI), which we prove to be equivalent. We provide a DAE construction, SIV, analyze its concrete security, develop a blockcipher-based instantiation of it, and suggest that the method makes a desirable alternative to the schemes of the X9.102 draft standard. The construction incorporates a method to turn a PRF that operates on a string into an equally efficient PRF that operates on a vector of strings, a problem of independent interest. Finally, we consider IV-based authenticated-encryption (AE) schemes that are maximally forgiving of repeated IVs, a goal we formalize as *misuse-resistant AE*. We show that a DAE scheme with a vector-valued header, such as SIV, directly realizes this goal.

1 Introduction

The American Standards Committee Working Group X9F1 has proposed four *key-wrap* schemes in a draft standard known as ANS X9.102, and NIST has promulgated a request for comments on the proposal [13]. The S/MIME working group of the IEEE had earlier adopted a key-wrap scheme [17], and their discussions on this topic go back to at least 1997 [36]. NIST is considering specifying a key-wrap mechanism in their own series of recommendations [M. Dworkin, personal communications]. But despite all this, the key-wrap goal would seem to be essentially unknown to the cryptographic community. No published paper analyzes any key-wrap scheme, and there is no formal definition for key wrap in the literature, let alone any proven-secure scheme. Consequently, the goal of this paper is to put the key-wrap problem on a proper, provable-security footing. In the process, we will learn quite a bit that's new about authenticated-encryption (AE).

Before proceeding it may be useful to give a very informal description of the key-wrap goal, echoing the wording in [13, p. 1]. A key-wrap scheme is a kind of shared-key encryption scheme. It aims to provide "privacy and integrity protection for specialized data such as cryptographic keys, ... without the use of nonces" (meaning counters or random bits). So key-wrap's raison d'être is to remove AE's reliance on a nonce or random bits. At least in the context of transporting cryptographic keys, a deterministic scheme should be just as good as a probabilistic one, anyway. Another goal of key wrap is to provide "integrity protection ... for cleartext associated data, ... which will typically contain control information about the wrapped key" [13, p. 1].

S. Vaudenay (Ed.): EUROCRYPT 2006, LNCS 4004, pp. 373–390, 2006.

CONTRIBUTIONS. We begin by offering a formal definition for what a key-wrap scheme should do, defining a goal we call *deterministic authenticated-encryption* (DAE). A thesis underlying our work is that the goal of a key-wrap scheme *is* DAE. In a DAE scheme, encryption deterministically turns a key, a header, and a message into a ciphertext. The header (which may be absent, a string, or even a vector of strings) is authenticated but not encrypted. To define security, the adversary is presented either a real encryption oracle and a real decryption oracle (both are deterministic), or else a bogus encryption oracle that just returns random bits and a bogus decryption oracle that always returns an indication of invalidity. For a good DAE scheme, the adversary should be unable to distinguish these possibilities. See Section 2.

Next we provide a DAE construction, SIV. (The acronym stands for *Synthetic IV*, where *IV* stands for *Initialization Vector*.) The construction combines a conventional IV-based encryption scheme (eg, CTR mode [27]) and a special kind of pseudorandom function (PRF)—one that takes a vector of strings as input. We prove that SIV is a good DAE, assuming its components are secure. See Section 3.

In practice one would want to realize SIV from a blockcipher, and so we show how to turn a PRF f that operates on a single string into a PRF f^* that takes a vector of strings. Under our S2V construction, the cost of computing the PRF $f^* = S2V[f]$ on a vector $X = (X_1, \ldots, X_n)$ is at most the total cost to compute f on each component X_i, and it can be considerably less, as the contribution from a component X_i can be precomputed if it is to be held constant. See Section 4.

For a concrete alternative to the X9.102 schemes, we suggest to instantiate SIV using modes CTR and CMAC* = S2V[CMAC], where CTR is counter mode [27] and CMAC is an arbitrary-input-length variant of the CBC MAC [28]. The specified mechanism removes unnecessary usage restrictions, improves efficiency, and provides provable security. See Section 5.

Applications of DAEs go beyond the wrapping of keys. Many IV-based encryption schemes, such as CBC, require an adversarially unpredictable IV. Experience has shown that implementers and protocol designers often supply an incorrect IV, such as a constant or counter. In a *misuse-resistant* AE scheme the aim is to do as well as possible with whatever IV is provided. We formalize this goal and show that a DAE scheme that takes a vector-valued header provides an immediate solution: just regard the IV as one component of the header. Adopting this viewpoint, SIV can be regarded as an IV-based AE scheme, one as efficient with respect to blockcipher calls as conventional two-pass AE schemes like CCM [29] but more resilient to IV misuse. See Section 6.

Finally, we investigate the basic properties of DAEs. First, we give an alternative characterization of DAEs. A *pseudorandom injection* (PRI) is like a blockcipher except that the ciphertext may be longer than the plaintext (also, the message space may be richer than $\{0,1\}^n$ for some fixed n, and a header may be provided). We prove PRIs equivalent to DAEs, up to a term that is negligible when the PRI is adequately length-increasing. Next, we explain that the "all-in-one" definition we adopt for DAEs is equivalent to a more conventional, two-requirement (privacy-plus-authenticity) definition. Finally, we sketch a result validating the intuition that DAE-encrypting a message that includes a random key provides semantic security. See Section 7.

WHY THIS GOAL? There are two main reasons to prefer DAE over conventional (probabilistic or stateful) AE. First, DAE saves one from having to introduce random bits or state in contexts where these measures are infeasible or unnecessary. Relatedly, DAE saves on bandwidth, since no nonce or random value need be sent.

That said, in many contexts where one would think to use key wrap, one *can* use a conventional AE scheme, instead. This does not make studying the key-wrap problem pointless. First, it clarifies the relationship between key wrap and conventional AE. Second, DAE leads to misuse-resistant AE, and methods that achieve this aim make practical alternatives to conventional (not misuse-resistant) two-pass AE methods. Finally, practitioners have already "voted" for key-wrap by way of protocol-design and standardization efforts, and it is simply not productive to say "use a conventional AE scheme" after this option has been rejected.

FURTHER RELATED WORK. AE goals were formalized over a series of papers [6, 8, 20, 31, 33]. The idea of binding the encryption process to unencrypted strings is folklore, with recent work in this direction including [23, 31, 35]. Russell and Wong [34] introduce a completely different approach for dealing with the encryption of low-entropy messages, and Dodis and Smith [12] extend this entropy-based approach. Phan and Pointcheval [30] study relationships among security notions for conventional (length-preserving and headerless) ciphers. The SIV construction resembles the AE scheme EAX [9]. A less ambitious relaxation on IV requirements than that formalized as misuse-resistant encryption is given in [32]. A full version of this paper is available from the authors' web pages.

2 DAE Security

NOTATION. For a distribution S let $S \xleftarrow{\$} S$ mean that S is selected randomly from S (if S is a finite set the assumed distribution is uniform). All strings are binary strings. When X and Y are strings we write $X\|Y$ for their concatenation. When $X \in \{0,1\}^*$ is a string $|X|$ is its length and, if $1 \leq i \leq j \leq |X|$, then $X[i..j]$ is the substring running from its ith to j^{th} characters, or the empty string ε otherwise. By a vector we mean a sequence of zero or more strings, and we write $\{0,1\}^{**}$ for the space of all vectors. We write a vector as $X = (X_1, \ldots, X_n)$ where $n = |X|$ is its number of components. If $X = (X_1, \ldots, X_n)$ and $Y = (Y_1, \ldots, Y_m)$ are vectors then X, Y is the vector $(X_1, \ldots, X_n, Y_1, \ldots, Y_m)$. In pseudocode, Boolean variables are silently initialized to false, sets are initialized to the empty set, and partial functions are initialized to everywhere undefined (set to undef). An *adversary* is an algorithm with access to one or more oracles, which we write as superscripts. By $A^{\mathcal{O}} \Rightarrow 1$ we mean the event that adversary A, running with its oracle $\mathcal{O}$, outputs 1. When an adversary has an oracle with an expressed domain D we understand that the oracle returns the distinguished value $\perp$, read as *invalid*, if the adversary asks a query outside of D.

SYNTAX. A scheme for *deterministic authenticated-encryption*, or DAE, is a tuple $\Pi = (\mathcal{K}, \mathcal{E}, \mathcal{D})$. The *key space* $\mathcal{K}$ is a set of strings or infinite strings endowed with a distribution. For a practical scheme there must be a probabilistic algorithm that samples from $\mathcal{K}$, and we identify this algorithm with the distribution it induces. The *encryption algorithm* $\mathcal{E}$ and *decryption algorithm* $\mathcal{D}$ are deterministic algorithms that take an input in $\mathcal{K} \times \{0, 1\}^{**} \times \{0, 1\}^*$ and return either a string or the distinguished value $\perp$. We write $\mathcal{E}_K^H(X)$ or $\mathcal{E}_K(H, X)$ for $\mathcal{E}(K, H, X)$ and $\mathcal{D}_K^H(Y)$ or $\mathcal{D}_K(H, Y)$ for $\mathcal{D}(K, H, Y)$. We assume there are sets $\mathcal{H} \subseteq \{0, 1\}^{**}$, the *header space*, and $\mathcal{X} \subseteq \{0, 1\}^*$, the *message space*, such that $\mathcal{E}_K^H(X) \in \{0, 1\}^*$ iff $H \in \mathcal{H}$ and $X \in \mathcal{X}$. We assume that $X \in \mathcal{X} \Rightarrow \{0, 1\}^{|X|} \subseteq \mathcal{X}$. The *ciphertext space* is $\mathcal{Y} = \{\mathcal{E}_K^H(X) : K \in \mathcal{K}, H \in \mathcal{H}, X \in \mathcal{X}\}$. We require $\mathcal{D}_K^H(Y) = X$ if $\mathcal{E}_K^H(X) = Y$, and $\mathcal{D}_K^H(Y) = \perp$ if there is no such X. It will be our convention that $\mathcal{E}_K^H(\perp) = \mathcal{D}_K^H(\perp) = \perp$ for all $K \in \mathcal{K}$ and $H \in \mathcal{H}$. For any $K \in \mathcal{K}$, $H \in \mathcal{H}$, and $X \in \mathcal{X}$, we assume that $|\mathcal{E}_K^H(X)| = |X| + e(H, X)$ for a function $e : \{0, 1\}^{**} \times \{0, 1\}^* \rightarrow \mathbb{N}$ where $e(H, X)$ depends only on the number of components of H, the length of each of these components, and the length of X. The function e is called the *expansion function* of the DAE scheme. Often we are concerned with the minimum expansion that might arise, and so define the number $s = \min_{H \in \mathcal{H}, X \in \mathcal{X}} \{e(H, X)\}$ as the *stretch* of the scheme.

Among what is formalized above: (1) encryption and decryption are given by algorithms, not just functions; (2) trying to encrypt something outside of the header space or message space returns $\perp$; (3) trying to decrypt something that isn't the encryption of anything returns $\perp$; (4) if you can encrypt a string of some length you can encrypt all strings of that length; and (5) the length of a ciphertext exceeds the length of the plaintext by an amount that depends on, at most, the length of the plaintext and the length of the components of the header.

A DAE is *length-preserving* if $e(H, X) = 0$ for all $H \in \mathcal{H}$, $X \in \mathcal{X}$. An *enciphering scheme* is a length-preserving DAE. A *tweakable blockcipher* is an enciphering scheme where the plaintext space is $\mathcal{X} = \{0, 1\}^n$ for some $n \geq 1$. A *blockcipher* is a tweakable blockcipher where the header space $\mathcal{H} = \{\varepsilon\}$ is a singleton set; as such, we omit mention of it and write $E : \mathcal{K} \times \{0, 1\}^n \rightarrow \{0, 1\}^n$.

SECURITY. We now give our formalization for DAE security.

Definition 1. *Let* $\Pi = (\mathcal{K}, \mathcal{E}, \mathcal{D})$ *be a DAE scheme with header space* $\mathcal{H}$, *message space* $\mathcal{X}$, *and expansion function* e. *The **DAE-advantage** of adversary A in breaking Π is defined as*

$$\mathbf{Adv}_{\Pi}^{\mathrm{dae}}(A) = \Pr\left[K \xleftarrow{\$} \mathcal{K} : A^{\mathcal{E}_K(\cdot, \cdot), \mathcal{D}_K(\cdot, \cdot)} \Rightarrow 1\right] - \Pr\left[A^{\$(\cdot, \cdot), \perp(\cdot, \cdot)} \Rightarrow 1\right] . \quad \blacksquare$$

On query $H \in \mathcal{H}$, $X \in \mathcal{X}$, the adversary's *random-bits* oracle $\$(\cdot, \cdot)$ returns a random string of length $|X| + e(H, X)$. As always, oracle queries outside the specified domain return $\perp$. The $\perp(\cdot, \cdot)$ oracle returns $\perp$ on every input. We assume that the adversary does not ask (H, Y) of its right (ie, second) oracle if some previous left (ie, first) oracle query (H, X) returned Y; does not ask (H, X) of its left oracle if some previous right-oracle query (H, Y) returned X; does not ask left queries outside of $\mathcal{H} \times \mathcal{X}$; and does not repeat a query. The last two assumptions are without loss of generality, as an adversary that violated any of these constraints could be replaced by a more efficient and equally effective adversary (in the $\mathbf{Adv}_{\Pi}^{\mathrm{dae}}$-sense) that did not. The first two assumptions are to prevent trivial wins.

DISCUSSION. The DAE-notion of security directly captures the amalgamation of privacy and authenticity. Assume that $\mathbf{Adv}_{\Pi}^{\mathrm{dae}}(A)$ is insignificantly small for any reasonable adversary. Then, for privacy, we know that any sequence of distinct $\mathcal{E}_K$-queries results in a distribution on outputs resembling a distribution on outputs that depends only on the length of each query (in fact, the outputs look like random strings of the appropriate lengths). For authenticity we have that, despite the ability to perform a chosen-plaintext attack (as provided by the $\mathcal{E}_K$ oracle), we are unable to come up with a new query Y for which $\mathcal{D}_K^H(Y) \neq \perp$.

It is possible to disentangle the privacy and authenticity notions in the DAE definition, defining separate notions for deterministic privacy and deterministic authenticity. While the traditional approach for defining AE has been to split the goal into two separate properties, the unified definition seems to us nicer and more succinct.

We point out that the DAE notion does not formalize the idea that the party that produces a valid ciphertext (a value that decrypts to something other than $\perp$) necessarily *knows* the underlying key K. One could formalize this, but it would not coincide with DAE. Sometimes the key-wrap goal has been described in these terms. We suspect that when security-designers speak of having to know the key in order to produce a valid ciphertext what they typically mean is not a proof of knowledge, but just the inability for a party to produce a valid ciphertext in the absence of the key. It is the latter notion that is well captured by our DAE definition.

3   Building a DAE Scheme: The SIV Construction

CONVENTIONAL IV-BASED ENCRYPTION SCHEMES. Encryption modes like CBC and CTR are what we call *conventional* IV-based encryption schemes. Such a scheme $\Pi = (\mathcal{K}, \mathcal{E}, \mathcal{D})$ is syntactically similar to a DAE but in this context the header space $\mathcal{H}$ is a set of strings and is renamed the *IV space*, $\mathcal{IV}$. We expect only privacy in a conventional IV-based encryption scheme, and demand a random IV. This makes the security notion rather weak, but sufficient for our purposes. The following definition captures the desired notion.

Fix a conventional IV-based encryption scheme $\Pi = (\mathcal{K}, \mathcal{E}, \mathcal{D})$ with IV-space $\mathcal{IV} = \{0,1\}^n$. For simplicity, assume Π is length-preserving. Let $\mathcal{E}^\$$ be the probabilistic algorithm defined from $\mathcal{E}$ that, on input $K \in \mathcal{K}$ and $M \in \{0,1\}^*$, chooses an $IV \xleftarrow{\$} \{0,1\}^n$, computes $C \leftarrow \mathcal{E}_K^{IV}(M)$ and returns $IV \parallel C$. Then we define the advantage of adversary A in violating the privacy of Π by

$$\mathbf{Adv}_\Pi^{\mathrm{priv}\$}(A) = \Pr\left[K \xleftarrow{\$} \mathcal{K}:\ A^{\mathcal{E}_K^\$(\cdot)} \Rightarrow 1\right] - \Pr\left[A^{\$(\cdot)} \Rightarrow 1\right]$$

where the $\$(\cdot)$ oracle, on input M, returns a random string of length $n + |M|$. We assume that the adversary never asks a query M outside of the message space $\mathcal{X}$ of Π.

ARBITRARY-INPUT PSEUDORANDOM FUNCTIONS. Fix nonempty sets $\mathcal{K}$ and $\mathcal{X}$, the first being finite or otherwise endowed with a distribution and the second being finite or countably infinite. A *pseudorandom function* (PRF) is a map $F\colon \mathcal{K} \times \mathcal{X} \to \{0,1\}^n$ for some $n \geq 1$. We write $F_K(X)$ for $F(K, X)$. Let $\mathrm{Func}(\mathcal{X}, \mathcal{Y})$ be the set of all functions from $\mathcal{X}$ to $\mathcal{Y}$ and let $\mathrm{Func}(\mathcal{X}, n) = \mathrm{Func}(\mathcal{X}, \{0,1\}^n)$. Regarding a function as the key, we can consider $\mathrm{Func}(\mathcal{X}, n)$ to be a PRF; to each $X \in \mathcal{X}$ associate a random string in

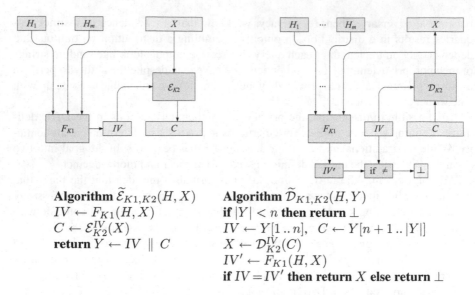

Algorithm $\widetilde{\mathcal{E}}_{K1,K2}(H,X)$

$IV \leftarrow F_{K1}(H,X)$
$C \leftarrow \mathcal{E}_{K2}^{IV}(X)$
return $Y \leftarrow IV \parallel C$

Algorithm $\widetilde{\mathcal{D}}_{K1,K2}(H,Y)$

if $|Y| < n$ **then return** $\perp$
$IV \leftarrow Y[1..n], \quad C \leftarrow Y[n+1..|Y|]$
$X \leftarrow \mathcal{D}_{K2}^{IV}(C)$
$IV' \leftarrow F_{K1}(H,X)$
if $IV = IV'$ **then return** X **else return** $\perp$

Fig. 1. The SIV construction. The left side illustrates and defines encryption, the right side, decryption. The header is $H = (H_1, \ldots, H_m)$, the plaintext is X, the key is $(K1, K2)$, and the ciphertext is $Y = IV \parallel C$. Function $F: \mathcal{K}_1 \times \{0,1\}^{**} \to \{0,1\}^n$ is a PRF and $(\mathcal{K}_2, \mathcal{E}, \mathcal{D})$ is an IV-based encryption scheme, such as CTR mode.

$\{0,1\}^n$. Let A be an adversary. The advantage of A in violating the pseudorandomness of F is

$$\mathbf{Adv}_F^{\mathrm{prf}}(A) = \Pr\left[K \leftarrow \mathcal{K}: A^{F_K(\cdot)} \Rightarrow 1\right] - \Pr\left[\rho \xleftarrow{\$} \mathrm{Func}(\mathcal{X}, n): A^{\rho(\cdot)} \Rightarrow 1\right].$$

It is tacitly assumed that the adversary has a mechanism of naming points in $\mathcal{X}$ by strings; if $\mathcal{X} \subseteq \{0,1\}^*$ then a string names itself, but if $\mathcal{X}$ is not a set of strings then points of $\mathcal{X}$ are encoded as strings in some natural way. Our definition of PRFs is unusual for allowing the input X to be arbitrary (possibly not a string).

THE SIV CONSTRUCTION. Let $F: \mathcal{K}_1 \times \{0,1\}^{**} \to \{0,1\}^n$ be a PRF. Let $\Pi = (\mathcal{K}_2, \mathcal{E}, \mathcal{D})$ be a conventional IV-based encryption scheme with IV-length n and message space $\mathcal{X}$. We write $F_K(H, M)$ instead of $F_K((H, M))$. We construct from (F, Π) a DAE $\widetilde{\Pi} = \mathrm{SIV}[F, \Pi] = (\widetilde{\mathcal{K}}, \widetilde{\mathcal{E}}, \widetilde{\mathcal{D}})$ with header space $\{0,1\}^{**}$ and message space $\mathcal{X}$ where $\widetilde{\mathcal{K}} = \mathcal{K}_1 \times \mathcal{K}_2$ and the encryption and decryption algorithms are as illustrated and defined in Fig. 1. Recall that $Y[n+1..|Y|] = \varepsilon$ if $|Y| < n$.

We will now show that if F is PRF-secure and Π is IND$-secure then $\widetilde{\Pi} = \mathrm{SIV}[F, \Pi]$ is DAE-secure. The intuition behind the proof is this. If any bit of the header H or plaintext X is new then the string IV will look like a random string and so $IV \parallel C$ will be difficult to distinguish from random bits. On decryption, the adversary must create a new (H, Y) where $Y = IV \parallel C$. Let's imagine giving the adversary the corresponding plaintext X for free. Now (H, X) is new because (H, X) determines (H, Y) and the adversary is not allowed to decipher values that it trivially knows the decipherment of. But if (H, X) is new then IV' is adversarially unpredictable and so its chance of being equal to IV is only about 2^{-n}.

In the following result we write $\text{Time}_{\Pi}(\mu)$, where $\Pi = (\mathcal{K}, \mathcal{E}, \mathcal{D})$ is an IV-based encryption scheme and $\mu > 0$ is an integer, for the sum of the worst-case times: to select $K \xleftarrow{\$} \mathcal{K}$, to compute $\mathcal{E}_K^{IV}$ on inputs of total length μ, and to compute $\mathcal{D}_K^{IV}$ on inputs of total length μ. Here, by convention, "time" means actual running time plus program size, all relative to some fixed RAM model of computation.

Theorem 1. *Let* $F\colon \mathcal{K}_1 \times \{0,1\}^{**} \to \{0,1\}^n$ *be a PRF and let* $\Pi = (\mathcal{K}_2, \mathcal{E}, \mathcal{D})$ *be a conventional IV-based encryption scheme with message space* $\mathcal{X}$ *and IV-length* n. *Let* $\widetilde{\Pi} = \text{SIV}[F, \Pi]$. *Let* A *be an adversary (for attacking* $\widetilde{\Pi}$*) that runs in time* t *and asks* q *queries, these of total length* μ. *Then there exists adversaries* B *and* D *such that*

$$\mathbf{Adv}_{\Pi}^{\text{priv\$}}(B) + \mathbf{Adv}_F^{\text{prf}}(D) \geq \mathbf{Adv}_{\widetilde{\Pi}}^{\text{dae}}(A) - q/2^n .$$

What is more, B *and* D *run in time at most* $t' = t + \text{Time}_{\Pi}(\mu) + c\mu$ *for some absolute constant* c *and ask at most* q *queries, these of total length* μ.   ∎

Proof. The proof proceeds in two stages. First we consider the DAE scheme $G = \text{SIV}[\text{Func}(\{0,1\}^{**}, n), \Pi]$ (replacing the function F_{K1} with a random function $\rho \in \text{Func}(\{0,1\}^{**}, n)$). Then we extend this to account for the insecurity of the PRF F.

Denote the forward and reverse algorithms associated to G as $G_{\rho,K2}$ and $G_{\rho,K2}^{-1}$, with $(\rho, K2)$ being the key. Let $\delta = \mathbf{Adv}_G^{\text{dae}}(A)$ and $q = q_L + q_R$ and $\mu = \mu_L + \mu_R$ where q_L and q_R are the number of left and right oracle queries, these totaling μ_L and μ_R bits, respectively. With the obvious simplifications in notation we have

$$\delta = \Pr\left[A^{G_{\rho,K2}(\cdot,\cdot), G_{\rho,K2}^{-1}(\cdot,\cdot)} \Rightarrow 1\right] - \Pr\left[A^{\$(\cdot,\cdot), \perp(\cdot,\cdot)} \Rightarrow 1\right]$$

$$= \left(\Pr\left[A^{G_{\rho,K2}(\cdot,\cdot), G_{\rho,K2}^{-1}(\cdot,\cdot)} \Rightarrow 1\right] - \Pr\left[A^{G_{\rho,K2}(\cdot,\cdot), \perp(\cdot,\cdot)} \Rightarrow 1\right]\right)$$

$$+ \left(\Pr\left[A^{G_{\rho,K2}(\cdot,\cdot), \perp(\cdot,\cdot)} \Rightarrow 1\right] - \Pr\left[A^{\$(\cdot,\cdot), \perp(\cdot,\cdot)} \Rightarrow 1\right]\right) = p_1 + p_2$$

where p_1 and p_2 represent the corresponding parenthesized expressions; it remains to bound these quantities. For p_2 we construct from A an adversary B^g for attacking the priv\$-security of Π. Let B run A. When A asks its left-oracle a query (H, X), let B ask $g(M)$ and return the result to A. When A asks a right-oracle query have B return $\perp$. When A halts with output bit b, let B output b. Notice that if $g = \mathcal{E}_K^{\$}$ then B properly simulates $G_{\rho,K2}(\cdot,\cdot), \perp(\cdot,\cdot)$ oracles for A (here we need the assumption that A never repeats a query). Similarly, if $g = \$$ then B simulates $\$(\cdot,\cdot), \perp(\cdot,\cdot)$ oracles for A. Hence $p_2 \leq \mathbf{Adv}_{\Pi}^{\text{priv\$}}(B)$.

To bound p_1 consider giving the key $K2$ to the adversary and then asking it to carry out its distinguishing task. As this can only make the task easier we may assume

$$p_1 = \Pr\left[A^{G_{\rho,K2}(\cdot,\cdot), G_{\rho,K2}^{-1}(\cdot,\cdot)} \Rightarrow 1\right] - \Pr\left[A^{G_{\rho,K2}(\cdot,\cdot), \perp(\cdot,\cdot)} \Rightarrow 1\right]$$

$$\leq \Pr\left[A(K2)^{G_{\rho,K2}(\cdot,\cdot), G_{\rho,K2}^{-1}(\cdot,\cdot)} \Rightarrow 1\right] - \Pr\left[A(K2)^{G_{\rho,K2}(\cdot,\cdot), \perp(\cdot,\cdot)} \Rightarrow 1\right].$$

We can assume without loss of generality that A halts and outputs 1 as soon as a right-oracle query returns something other than $\perp$. Under this assumption, encryption queries are useless for distinguishing between these two oracle pairs, as prior to the right oracle returning $M \neq \perp$ both pairs behave as $G_{\rho,K2}(\cdot,\cdot), \perp(\cdot,\cdot)$. Hence p_1 is bounded by the probability that A asks a right-oracle query (H, Y) such that $G_{\rho,K2}^{-1}(H, Y) \neq \perp$. Examining the algorithm for $G_{\rho,K2}^{-1}$ we see that this occurs only when $\rho(H, X) = IV$, where $X = \mathcal{D}_{K2}^{IV}(C)$ (with Y having been parsed into IV and C). Since the adversary is given the key $K2$, it can compute $\mathcal{D}_{K2}^{IV}(C)$ for any strings IV, C of its choosing. In particular, when it asks a right-oracle query (H, Y) it knows what is the input to the random function ρ and what is the target output IV. But under our assumption that A never queries its right oracle (H, Y) when some left-oracle query (H, X) returned Y, either the input (H, X) is new, or the target IV is new. Thus, the probability that $\rho(H, X) = IV$ is at most $1/2^n$ for each right-oracle query, and we conclude that $p_1 \leq q_R/2^n$. Since $q_R \leq q$ we have $\delta \leq \mathbf{Adv}_{\Pi}^{\mathrm{priv\$}}(B) + q/2^n$.

For the second part of the proof note that

$$\mathbf{Adv}_{\widetilde{\Pi}}^{\mathrm{dae}}(A) = \delta + \Pr\left[A^{\widetilde{\mathcal{E}}_{K1,K2}(\cdot,\cdot),\widetilde{\mathcal{D}}_{K1,K2}(\cdot,\cdot)} \Rightarrow 1\right] - \Pr\left[A^{G_{\rho,K2}(\cdot,\cdot),G_{\rho,K2}^{-1}} \Rightarrow 1\right]$$

where $\widetilde{\Pi} = (\mathcal{K}1 \times \mathcal{K}2, \widetilde{\mathcal{E}}, \widetilde{\mathcal{D}})$ and we have suppressed the random selections $K1 \xleftarrow{\$} \mathcal{K}_1$ and $K2 \xleftarrow{\$} \mathcal{K}_2$. Let D^g be an adversary for attacking F as a PRF, and let it operate as follows. Adversary D picks $K2 \xleftarrow{\$} \mathcal{K}_2$ and runs A. When A asks a left oracle query (H, X), B answers by setting $IV \leftarrow g(H, X)$, computing $C \leftarrow \mathcal{E}_{K2}^{IV}(X)$ and returning to A the string $IV \| C$. On a right oracle query (H, Y), adversary D parses $IV = Y[1..n]$, $C = Y[n+1..|Y|]$, computes $X \leftarrow \mathcal{D}_{K2}^{IV}(C)$ and tests if $IV = g((H, X))$, returning X to A if so and $\perp$ otherwise. When A halts with output bit b, let D output b. Clearly D correctly simulates $\widetilde{\mathcal{E}}_{K1,K2}(\cdot,\cdot), \widetilde{\mathcal{D}}_{K1,K2}(\cdot,\cdot)$ when its oracle $g = F_{K1}$ for some random key $K1$, and $G_{K1,K2}(\cdot,\cdot), G_{K1,K2}^{-1}(\cdot,\cdot)$ if instead $g = \rho$ for a random $\rho \in \mathrm{Func}(\mathcal{M}, n)$. So, $\mathbf{Adv}_{\widetilde{\mathcal{E}}}^{\mathrm{dae}}(A) \leq \delta + \mathbf{Adv}_F^{\mathrm{prf}}(D)$ and rearranging gives the result.

4   Enriching a PRF to Take Vectors of Strings as Input

THE GOAL. Traditionally, a pseudorandom function (PRF) takes a single string as input: under the control of a key K, a PRF f maps a string $X \in \{0,1\}^*$ into a string $f_K(X)$. But SIV uses a non-traditional PRF—a function F that, under the control of a key K, maps a vector of strings $X = (X_1, \ldots, X_m) \in \{0,1\}^{**}$ into a string $F_K(X)$. Let us call a PRF that takes a string as input an sPRF (string-input PRF) and a PRF that takes a vector of strings as input a vPRF (vector-input PRF). This section is about efficient ways to turn an sPRF f into a vPRF f^*.

At first glance it might seem like there'd be little to say about sPRF-to-vPRF conversion: there's an obvious approach for solving the problem, and it's obviously correct. Namely, encode any vector of strings $X = (X_1, \ldots, X_m)$ into a single string $\langle X \rangle$ and apply the sPRF to that, $f_K^*(X) = f_K(\langle X \rangle)$. By encode we mean any reversible, easily-computed map of a vector of strings into a single one, say $\langle X_1, \ldots, X_m \rangle = X_1 \| N_1 \| \cdots \| X_m \| N_m$ where $N_i = |X_i|_{64}$ is the length of X_i encoded into 64 bits

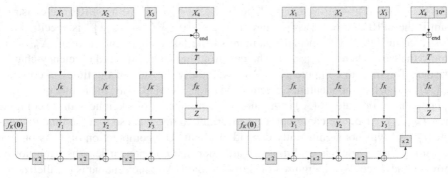

Algorithm $f_K^*(X_1, \ldots, X_m)$ *The S2V Construction,* $f^* = \mathrm{S2V}[f]$

10   **if** $m = 0$ **then return** $f_K(\mathbf{1})$
11    $S \leftarrow f_K(\mathbf{0})$
12   **for** $i \leftarrow 1$ **to** $m - 1$ **do** $S \leftarrow \mathbf{2}S \oplus f_K(X_i)$
13   **if** $|X_m| < n$ **then** $T \leftarrow S \oplus_{\mathrm{end}} X_m$ **else** $T \leftarrow \mathbf{2}S \oplus_{\mathrm{end}} X_m 10^*$
14   **return** $Z \leftarrow f_K(T)$

Fig. 2. The S2V construction makes a PRF $f^*\colon \mathcal{K} \times \{0,1\}^{**} \to \{0,1\}^n$ from a PRF $f\colon \mathcal{K} \times \{0,1\}^* \to \{0,1\}^n$. **Bottom:** Definition of S2V. Strings $X_1, \ldots, X_m \in \{0,1\}^*$ and $m \geq 0$ are arbitrary. **Top:** Illustration of it, computing $Z = f_K^*(X_1, X_2, X_3, X_4)$. The left side shows the case when $|X_4|$ is a nonzero multiple of n bits, the right otherwise.

(assume that $|X_i| < 2^{64}$ for all i). The problem with making a vPRF in such a way is a diminution of efficiency. First, computing $f_K^*(X)$ may take longer than the total time to compute $f_K(X_i)$ for each component X_i since we have added $64m$ bits for length annotation. Second, even if some components of X stay fixed (say X_2 is constant), we must still re-process the entire encoded string each time we compute f_K^* at a new value. Third, the mechanism is not parallelizable; one cannot process X_i until one is done processing X_{i-1}. Fourth, the assumption that $|X_i| < 2^{64}$, while reasonable in practice, is artificial and potentially wasteful, yet use of a stingier encoding will lead to greater complexity. Finally, the given encoding disrupts word alignment: if, for example, the first argument is one byte and all subsequent arguments are multiples of eight bytes, an implementation will now be dealing with non-word-aligned data. Fixing this problem by a smarter encoding will lead to increased complexity. We aim to do sPRF-to-vPRF conversion in a way that fixes the problems above.

NOTATION. Fix a value $n \geq 2$. Let $\mathbf{0} = 0^n$ and $\mathbf{1} = 0^{n-1}1$ and $\mathbf{2} = 0^{n-2}10$. These are regarded as points in finite field $\mathbb{F}_{2^n}$ represented using a primitive polynomial in the customary way. For $S \in \{0,1\}^n$ let $\mathbf{2}S$ mean the n-bit string representing the product of $\mathbf{2}$ and S. This can be computed with a left shift of S followed by a conditional xor. By $\mathbf{2}^i S$ we mean to do this multiplication by $\mathbf{2}$ a total of i times. By $N \oplus_{\mathrm{end}} X$ ("xor-into-the-end") we mean to xor the n-bit string N into the end of the string X, which will have at least n bits; $N \oplus_{\mathrm{end}} X = (0^{x-n}N) \oplus X$ where $x = |X|$. By $X10^*$ we mean $X10^i$ where $i \geq 0$ is the least number such that $|X| + 1 + i$ is divisible by n.

THE S2V CONSTRUCTION. Let $f: \mathcal{K} \times \{0,1\}^* \to \{0,1\}^n$ be an sPRF. We construct from it the vPRF $f^* = \mathrm{S2V}[f]$ where $f^*: \mathcal{K} \times \{0,1\}^{**} \to \{0,1\}^n$ is specified and illustrated in Fig. 2. The special treatment of the last component of input, X_m, is to handle the case where $|X_m| < n$. The construction has the desired efficiency characteristics. The time to compute $f_K^*(X)$ is essentially the sum of the times to compute $f_K(X_i)$ on each component; in particular, when $f = \mathrm{CMAC}$, say, the number of blockcipher calls to compute $f_K^*(X)$ is the sum of the number of blockcipher calls to compute each $f_K(X_i)$. Also, one can preprocess invariant components so that the time to compute $f_K^*(X)$ will not significantly depend on them. The computation of f^* is on-line (assuming that f itself is on-line); in particular, the component lengths need not be known in advance. Word alignment is not disrupted. And the scheme is parallelizable: different arguments can be acted on simultaneously, so f^* will be parallelizable if f is.

In a related effort we have proven the following result. The complexity-theoretic analog of Theorem 2 follows in the usual way.

Theorem 2. *Let* $f = \mathrm{Func}(\{0,1\}^*, n)$ *and* $f^* = \mathrm{S2V}[f]$. *Let A be an adversary that asks at most* $q \geq 3$ *vector-valued queries having p components in all, and each vector having fewer than n components. Then* $\mathbf{Adv}_{f^*}^{\mathrm{prf}}(A) \leq pq/2^n$. ∎

5   The SIV Mode of Operation

SIV MODE. Fix an n-bit blockcipher E and let $\Pi = \mathrm{CTR}$ be counter mode [27] over E, with an incrementing function of $S \mapsto 2S$ (that is, multiply by x in the finite field). Let $F = \mathrm{CMAC}^* = \mathrm{S2V}[\mathrm{CMAC}]$ be the result of applying the S2V construction to the CMAC [28], again with an underlying blockcipher of E. (Recall that CMAC is a NIST-recommended CBC MAC variant. It has a message space $\{0,1\}^*$.) Consider the scheme $\mathrm{SIV}[F, \Pi]$. By combining Theorems 1 and 2 and known results about CMAC and CTR mode [3, 18], the suggested mechanism is a provably secure DAE assuming E is a secure PRP. The proven security falls off, as usual, in $\sigma^2/2^n$ where σ is the total number of blocks asked about. We overload the name SIV and call the mode of operation just described SIV mode. We emphasize that the only thing left unspecified in the definition of SIV mode is the underlying blockcipher, which would typically be AES.

COMMENTS. Comparing SIV-AES and the X9.102 scheme AESKW, say, we note that, with SIV-AES, (1) the message space and header space are now $\{0,1\}^*$ instead of unusual sets; (2) message expansion is now independent of header length and message length; (3) the number of blockcipher calls is reduced by a factor of at least six; (4) vector-valued headers can now be handled, and the contribution of any component can be pre-processed if it is to be held fixed; (5) one now has a provable-security guarantee, falling off in $\sigma^2/2^n$, where σ is the total number of message blocks acted on. On the other hand, there is an effective attack on SIV if one can ask this many message blocks, while we do not know if this is true for AESKW.

In the instantiation of SIV we could have used, in place of CMAC, the composition of a universal hash function that gives n-bit outputs with an n-bit blockcipher. This demonstrates that the DAE goal can be achieved by a single "cryptographic" pass over the plaintext, plus a universal-hash-function computation over the header and plaintext. Similarly, a parallelizable MAC like PMAC [10] could have been used in place of CMAC, illustrating that DAE can be achieved by a parallelizable scheme.

6   Misuse-Resistant AE

This section gives an application of DAEs motivated not by the key-wrap problem but by the goal of constructing symmetric encryption schemes that are resistant to misuse. We are specifically concerned with IV-misuse, meaning that the IV is used in a way other than the way mandated by the scheme; for example, using a counter when the scheme requires a random value, or repeating an IV when the scheme requires it to be a nonce. Experience has shown that IVs are frequently mishandled. An encryption scheme robust against misuse should at least be an AE scheme (as programmers, protocol designers, and even books often assume that encryption provides for authenticity) and so we will treat IV-misuse within the context of authenticated encryption and not privacy-only encryption. The notion is applicable to the latter context, too.

Designing an IV-based AE scheme that is secure when its IV is an arbitrary nonce—not just when it is a random value—is a first move in the direction of making schemes robust against IV-misuse. The current section takes this a step further; we aim for an AE scheme in which if the IV *is* a nonce then one achieves the usual notion for nonce-based AE; and if the IV *does* get repeated then authenticity remains and privacy is compromised only to the extent that some minimal amount of information may be revealed, the information being if this plaintext is equal to a prior one, and even that is revealed only if both the message and its header have been used with this particular IV. Our formalization will capture this intent.

REVISED SYNTAX FOR AN IV-BASED ENCRYPTION SCHEME. Let us update the syntax of a conventional IV-based encryption scheme to accommodate an associated header. In this case an IV-based encryption scheme is a tuple $\Pi = (\mathcal{K}, \mathcal{E}, \mathcal{D})$ where everything is as before except that the encryption algorithm and decryption algorithm take an extra argument: now they are deterministic algorithms that map $\mathcal{K} \times \{0,1\}^{**} \times \{0,1\}^* \times \{0,1\}^*$ to $\{0,1\}^* \cup \{\bot\}$. We write $\mathcal{E}_K(H, IV, X)$ or $\mathcal{E}_K^{H,IV}(X)$ in place of $\mathcal{E}(K, H, IV, X)$ and $\mathcal{D}_K(H, IV, C)$ or $\mathcal{D}_K^{H,IV}(Y)$ in place of $\mathcal{D}(K, H, IV, Y)$. There must be sets $\mathcal{H}$, $\mathcal{IV}$, and $\mathcal{X}$ such that $\mathcal{E}_K^{H,IV}(X) \in \{0,1\}^*$ iff $H \in \mathcal{H}$ and $IV \in \mathcal{IV}$ and $X \in \mathcal{X}$. We call $\mathcal{IV}$ the *IV space* of Π. We require that $\mathcal{D}_K^{H,IV}(Y) = X$ if $\mathcal{E}_K^{H,IV}(X) = Y$ and $\mathcal{D}_K^{H,IV}(Y) = \bot$ if there is no such X.

MISUSE-RESISTANT AE SECURITY. To measure the AE-security of an encryption scheme $\Pi = (\mathcal{K}, \mathcal{E}, \mathcal{D})$ in the face of possible IV-reuse, imagine an adversary that may ask any sequence of encryption queries, even those that repeat IVs, and any sequence of decryption queries, which may likewise repeat IVs. We want the encryption oracle to return bits that look random except when this is impossible—on a repeated triple of (header, IV, message)—and the decryption oracle should return $\bot$ except when the triple is already known to have a valid decryption. For simplicity, assume as before that our IV-based encryption scheme is length-preserving.

Definition 2. *Let* $\Pi = (\mathcal{K}, \mathcal{E}, \mathcal{D})$ *be an IV-based encryption scheme that can handle an associated header and let* A *be an adversary. Then the **MRAE-advantage** of* A *in attacking* Π *is*

$$\mathbf{Adv}_\Pi^{\mathrm{mrae}}(A) = \Pr\left[K \xleftarrow{\$} \mathcal{K}: A^{\mathcal{E}_K(\cdot,\cdot,\cdot),\, \mathcal{D}_K(\cdot,\cdot,\cdot)} \Rightarrow 1\right] - \Pr\left[A^{\$(\cdot,\cdot,\cdot),\, \perp(\cdot,\cdot,\cdot)} \Rightarrow 1\right].$$

The adversary may not repeat a left-query and may not ask a right-query (H, IV, Y) *if some previous left-query* (H, IV, X) *returned* Y. ∎

Of course the $\mathcal{E}_K$ oracle returns $\mathcal{E}_K(H, IV, X)$ on input (H, IV, X) and $\mathcal{D}_K$ returns $\mathcal{D}_K(H, IV, Y)$ on input (H, IV, Y). As before $\$(H, IV, X)$ returns a random string of length $n + |X|$ and $\perp(\cdot, \cdot, \cdot)$ always returns $\perp$.

The MRAE-notion of security trivially implies nonce-based AE-scheme security: the latter is the special case where the adversary is not allowed to repeat an IV to any left query. Note that all proposed AE schemes to date [19,21,26,29,33] do fail should an IV get repeated: existing AE schemes are not MRAE-secure.

BUILDING A MISUSE-RESISTANT AE SCHEME. We can turn a DAE scheme $\Pi = (\mathcal{K}, \mathcal{E}, \mathcal{D})$ with header space $\{0,1\}^{**}$ and message space $\mathcal{X}$ into a misuse-resistant AE scheme $\widetilde{\Pi} = (\mathcal{K}, \widetilde{\mathcal{E}}, \widetilde{\mathcal{D}})$ by regarding the IV as one of the components, say the last component, of the header. In particular, SIV mode can be regarded as an MRAE scheme by asserting that one of the header components, say the last one specified, is an IV.

CORRECTNESS. Correctness of the MRAE scheme described above is nearly immediate. Given an adversary A for breaking the misuse-resistant AE scheme (it distinguishes $\mathcal{E}_K(\cdot,\cdot,\cdot)$, $\mathcal{D}_K(\cdot,\cdot,\cdot)$ from $\$(\cdot,\cdot,\cdot)$, $\perp(\cdot,\cdot,\cdot)$) we get a comparably good adversary B for breaking the DAE, distinguishing $\mathcal{E}_K(\cdot,\cdot)$, $\mathcal{D}_K(\cdot,\cdot)$ from $\$(\cdot,\cdot)$, $\perp(\cdot,\cdot)$: adversary B runs A and maps left queries (H, IV, X) to queries $(\langle H, IV\rangle, X)$, and maps right queries (H, IV, Y) to queries $(\langle H, IV\rangle, Y)$. The syntax and DAE-security notion for a PRI have been designed to "match up" so that there is nothing to do.

COMMENTS. Since all we have done in the construction is to hijack a component of the header as an IV, it seems as though nothing has actually been done. Yet the MRAE goal is conceptually different from the DAE goal, the former employing an IV and gaining for this a stronger notion of security. The header and the IV are conceptually different, the one being user-supplied data that the user wants authenticated, the other being a mechanism-supplied value needed to obtain a strong notion of security.

In retrospect, it is easy to construct an MRAE scheme by a sequence of simple steps. One can achieve this goal in a trivial way from a DAE scheme that takes a vector-valued header. Such a DAE scheme is easily built from a vector-input PRF and an IND$-secure conventional encryption scheme. At least if one is unconcerned with optimizing efficiency, a vector-input PRF is easily made from a string-input PRF. String-input PRFs and IND$-secure conventional encryption schemes can be built from blockciphers by well-known means. So each step along our path is easy or well-known. Still, the direct construction of an MRAE or DAE scheme from a blockcipher is not a simple matter, as evidenced by the long history of buggy or baroque AE schemes. Perhaps simple is how things seem *after* finding the right abstraction boundaries.

7 Properties of DAEs

This section investigates the properties of the DAE notion, looking in three different directions. First we explain the sense in which DAE achieve semantic security, indeed AE, when plaintexts carry a key. Next we give an alternative definition for DAEs, called pseudorandom injections, based on quite different intuition. Finally we show that the "all-in-one" definition of DAE can equivalently be factored into separate privacy and authenticity notions, as is traditionally done in this domain.

DAEs ACHIEVE SEMANTIC SECURITY WHEN PLAINTEXTS CARRY A KEY. A folklore justification for using a key-wrap scheme instead of a conventional AE scheme is that, in the key-wrap setting, one expects the plaintext to carry a random cryptographic key, and so a probabilistic or stateful mechanism should not be needed. We sketch a result that validates this intuition.

A *key-insertion scheme* is a pair of algorithms $\Phi = $ (InsertKey, ExtractKey), the first for inserting a κ-bit random value into a plaintext and the second for extracting it. Algorithm InsertKey, on input of $X \in \{0,1\}^*$, chooses a random $R \xleftarrow{\$} \{0,1\}^\kappa$ and returns $M \xleftarrow{\$} \text{InsertKey}(X)$. An equivalent viewpoint is that InsertKey is deterministic and takes the random string R as input; then we write $M \leftarrow \text{InsertKey}(X, R)$. Algorithm ExtractKey takes $M \in \{0,1\}^*$ and returns $\langle X, R \rangle$ with $|R| = \kappa$. Given a DAE $\Pi = (\mathcal{K}, \mathcal{E}, \mathcal{D})$ define the probabilistic encryption scheme $\widetilde{\Pi} = (\mathcal{K}, \widetilde{\mathcal{E}}, \widetilde{\mathcal{D}})$ by:

Algorithm $\widetilde{\mathcal{E}}_K(H, X)$	**Algorithm** $\widetilde{\mathcal{D}}_K(H, Y)$
$R \xleftarrow{\$} \{0,1\}^\kappa$	$M \leftarrow \mathcal{D}_K(H, Y)$
$M \leftarrow \text{InsertKey}(X, R)$	**if** $M = \perp$ **then return** $\perp$
if $M = \perp$ **return** $\perp$	**return** $\text{ExtractKey}(M)$
return $\mathcal{E}_K(H, M)$	

The encryption scheme is nonstandard insofar as decryption of a ciphertext Y returns not only the underlying plaintext X but also the random bits R that were inserted. Correspondingly, we must adapt the definition of AE to get a variant of $\mathbf{Adv}_{\Pi}^{\text{ae}}(A)$, call it $\mathbf{Adv}_{\Pi}^{\text{kiae}}(A)$, where when the adversary asks for the encryption of X we choose the random R and return it along with the ciphertext. This must look like random bits.

As long as the inserted key is sufficiently long, the algorithm described achieves the security notion we have sketched. We omit further details.

DAEs ARE EQUIVALENT TO PRIs. A secure pseudorandom injection (PRI) resembles a random injective function with the desired amount of length-expansion. We allow a chosen-ciphertext attack in our definition (that is, we focus on a "strong" PRI, analogous to a strong PRP [24]), giving the adversary both the forward and backward direction of the function. We allow the PRI to be tweakable [23], so that the scheme can be used to authenticate an associated header. We allow the domain to be fairly arbitrary—in particular, we consider message spaces that contain strings of various lengths.

Formally, let $\Pi = (\mathcal{K}, \mathcal{E}, \mathcal{D})$ be a DAE with header space $\mathcal{H}$ and message space $\mathcal{X}$. Imagine an adversary A given access to two oracles—one for $\mathcal{E}$ and one for $\mathcal{D}$. We want to say that this pair looks just like a random injection f and its inverse f^{-1}, the random injection f having the same signature as $\mathcal{E}$. For $e: \mathcal{H} \times \mathcal{X} \to \mathbb{N}$ let $\mathsf{Inj}_e^{\mathcal{H}}(\mathcal{X}, \mathcal{Y})$ be the set of all injective functions f from $\mathcal{H} \times \mathcal{X}$ to $\mathcal{Y}$ such that $|f(H, X)| = |X| + e(H, X)$.

Definition 3. *Let $\Pi = (\mathcal{K}, \mathcal{E}, \mathcal{D})$ be a DAE with header space $\mathcal{H}$, message space $\mathcal{X}$, and expansion e. The **PRI-advantage** of adversary A in breaking Π is $\mathbf{Adv}_\Pi^{\mathrm{pri}}(A) =$*

$$\Pr\left[K \xleftarrow{\$} \mathcal{K}: A^{\mathcal{E}_K(\cdot, \cdot),\, \mathcal{D}_K(\cdot, \cdot)} \Rightarrow 1\right] - \Pr\left[f \xleftarrow{\$} \mathsf{Inj}_e^{\mathcal{H}}(\mathcal{X}, \mathcal{Y}): A^{f(\cdot, \cdot),\, f^{-1}(\cdot, \cdot)} \Rightarrow 1\right]. \quad \blacksquare$$

The f^{-1} oracle above, on input (H, Y) returns the point X such that $f(H, X) = Y$; if there is no such point then it returns the distinguished value $\perp$. As before, we may assume without loss of generality that the adversary does not repeat a query, that it does not ask (H, Y) of its right oracle if some previous left oracle query (H, X) returned Y, that it does not ask (H, X) of its left oracle if some previous right-oracle query (H, Y) returned X, and that it does not ask any query (H, X) outside of $\mathcal{H} \times \mathcal{X}$.

Assuming a reasonable amount of stretch, the PRI and DAE notions of security are very close, as the following theorem shows.

Theorem 3. *Let $\Pi = (\mathcal{K}, \mathcal{E}, \mathcal{D})$ be a DAE with header space $\mathcal{H}$, message space $\mathcal{X}$, and stretch s, and let $\tau = \min_{X \in \mathcal{X}}\{|X|\}$ be the length of a shortest plaintext. Let A be an adversary that asks at most q_{L} left-oracle queries, q_{R} right-oracle queries, for a total of $q = q_{\mathrm{L}} + q_{\mathrm{R}}$ queries. Then $\left|\mathbf{Adv}_\Pi^{\mathrm{pri}}(A) - \mathbf{Adv}_\Pi^{\mathrm{dae}}(A)\right| \le q^2/2^{s+\tau+1} + 4q_{\mathrm{R}}/2^s.$* $\quad \blacksquare$

In other words, as the stretch s grows, the DAE and PRI notions converge. The quantitative difference between the measures is small if the stretch is, say, $s = 128$ bits. Among other reasons, it is to achieve this equivalence with PRIs that our definition for them used indistinguishability from random bits rather than, say, indistinguishability from the encryption of random bits.

Proof. Let A be an adversary that has access to two oracles. Let it ask q_{L} queries of its left oracle and q_{R} queries of its right oracle, and let $q = q_{\mathrm{L}} + q_{\mathrm{R}}$. With the obvious notational simplifications we have

$$\left|\mathbf{Adv}_\Pi^{\mathrm{pri}}(A) - \mathbf{Adv}_\Pi^{\mathrm{dae}}(A)\right| = \left|\Pr\left[A^{f(\cdot, \cdot),\, f^{-1}(\cdot, \cdot)} \Rightarrow 1\right] - \Pr\left[A^{\$(\cdot, \cdot),\, \perp(\cdot, \cdot)} \Rightarrow 1\right]\right|$$

$$= \left|\Pr\left[A^{\mathrm{G1}} \Rightarrow 1\right] - \Pr[A^{\mathrm{G0}} \Rightarrow 1]\right|$$

for the games G0 and G1 defined in Fig. 3. Recall that booleans are initialized to `false`, sets are initialized to empty, and partial functions are initialized to everywhere undefined with the symbol `undef`. The set $\mathrm{Image}(f(H, \cdot))$ contains all points $Y \ne$ `undef` such that $f(H, X) = Y$ for some $X \in \mathcal{X}$. Set difference is indicated with a minus sign. Look first at game G0. Much of the code (lines 12–13 and 20–26) is irrelevant to what the adversary sees. Each query $\mathrm{left}(H, X)$ returns a random string of $|X| + e(H, X)$ bits and each query $\mathrm{right}(H, Y)$ returns $\perp$. Thus game G0's (left, right) oracles faithfully simulate a pair of oracles $(\$, \perp)$ and we have that $\Pr[A^{\mathrm{G0}} \Rightarrow 1] = \Pr[A^{\$, \perp} \Rightarrow 1]$.

Game G1 is more subtle. We claim that its (left, right) oracles are simply a lazy evaluation of a pair of oracles (f, f^{-1}) with the desired domain and range. To see this, understand first that the partial function $f(H, \cdot)$ maintains the correspondence $X \mapsto f(H, X)$ for those domain points that we have already assigned values to,

---

On query left(H, X):
10   $c \leftarrow |X| + e(H, X)$
11   $Y \xleftarrow{\$} \{0, 1\}^c$
12   **if** $Y \in \mathrm{Image}(f(H, \cdot)) \cup \mathit{Invalid}^H$ **then**

13      $\mathit{bad} \leftarrow \texttt{true}$, $Y \xleftarrow{\$} \{0, 1\}^c - \mathrm{Image}(f(H, \cdot)) - \mathit{Invalid}^H$
14   **return** $f(H, X) \leftarrow Y$

On query right(H, Y):

20   $c \leftarrow |Y|$
21   $\mathit{EligibleX} \leftarrow \{X \in \{0, 1\}^{\leq c} : |X| + e(H, X) = |Y| \text{ and } f(H, X) = \texttt{undef}\}$
22   $\mathit{EligibleY} \leftarrow \{0, 1\}^c - \mathrm{Image}(f(H, \cdot)) - \mathit{Invalid}^H$
23   $x \xleftarrow{\$} [1 .. |\mathit{EligibleY}|]$
24   **if** $x \in [1..|\mathit{EligibleX}|]$ **then**

25      $\mathit{bad} \leftarrow \texttt{true}$, $X \leftarrow$ the xth string of $\mathit{EligibleX}$, $f(H, X) \leftarrow Y$, **return** X
26   $\mathit{Invalid}^H \leftarrow \mathit{Invalid}^H \cup \{Y\}$
27   **return** $\perp$

Fig. 3. Games used in the proof of Theorem 3. Game G1 is the complete code; game G0 omits the shaded statements.

while the set $\mathit{Invalid}^H$ maintains the set of points Y that have become ineligible to be $f(H, X)$ values, for any X, by virtue of having been asked right(H, Y) and having returned $\perp$, effectively asserting that $f^{-1}(H, Y) = \perp$ and so Y is outside the image of $f(H, \cdot)$. Now, starting at left(H, X) queries, we begin at line 10 by calculating the length c of the ciphertext that we must return. The code at lines 11–14 returns a random string Y of length c subject to the constraint that Y is outside of the image of $f(H, \cdot)$ and not ineligible to be an $f(H, X)$ value by virtue of having asserted that there is no preimage for Y with tweak H. Looking next at right(H, Y) queries, we calculate at line 21 the set $\mathit{EligibleX}$ of values X that could possibly map to Y using tweak H, and we calculate at line 22 the set of strings Y that could, at this moment be paired with strings in $\mathit{EligibleX}$. By our conventions on the adversary making no "pointless" queries, the string Y will necessarily be among the strings in $\mathit{EligibleY}$. Since we aim to randomly and injectively pair points in $\mathit{EligibleX}$ with points in $\mathit{EligibleY}$, the chance that a given point Y in $\mathit{EligibleY}$ has a preimage in $\mathit{EligibleX}$ is just $|\mathit{EligibleX}|/|\mathit{EligibleY}|$. Lines 23 and 24 effectively flip a coin with this bias, deciding if the string $Y \in \mathit{EligibleY}$ should or should not be given a (random) preimage in $\mathit{EligibleX}$. If it is not given a preimage, we record this decision by augmenting $\mathit{Invalid}^H$ at line 26. If it is given a preimage, it is given a random one by lines 23–25, the choice is recorded, and the random preimage is returned. We have thus provided a perfect simulation of an (f, f^{-1}) oracle, and so $\Pr[A^{\mathrm{G1}} \Rightarrow 1] = \Pr[A^{f, f^{-1}} \Rightarrow 1]$.

To bound $|\Pr[A^{\mathrm{G1}} \Rightarrow 1] - \Pr[A^{\mathrm{G0}} \Rightarrow 1]|$ we can now invoke the fundamental lemma of game-playing [7], since games G1 and G0 have been defined to be identical apart from the sequel of statements $\mathit{bad} \leftarrow \texttt{true}$. The lemma assures us that $|\Pr[A^{\mathrm{G1}} \Rightarrow 1] - \Pr[A^{\mathrm{G0}} \Rightarrow 1]| \leq \Pr[A^{\mathrm{G0}} \text{ sets } \mathit{bad}]$.

Let BAD be the event that A^{G0} causes *bad* to get set to true. We must bound the probability of BAD. Remember that the shaded statements have been expunged from the game. Prior to BAD occurring, each left-query adds a single point to a set $\mathrm{Image}(f(H,\cdot))$ but has no impact on any set $\mathit{Invalid}^H$, while each right-query adds a single point to a set $\mathit{Invalid}^H$ but has no impact on any set $\mathrm{Image}(f(H,\cdot))$. If the ith query is left-query then the set $\mathrm{Image}(f(H,\cdot)) \cup \mathit{Invalid}^H$ will have at most $i-1$ points and the chance that *bad* will get set at line 13 will be at most $(i-1)/2^{s+\tau}$ and so, overall, the probability that *bad* gets set at line 13 is at most $\sum_{i=1}^{q}(i-1)/2^{s+\tau} \le q^2/2^{s+\tau+1}$. If the ith query is a right-query then *bad* will be set with probability $|\mathit{EligibleX}|/|\mathit{EligibleY}|$ for the current sets $\mathit{EligibleX}$ and $\mathit{EligibleY}$. How big can $|\mathit{EligibleX}|$ be? Asked a query Y of length c, even if *every* string of length at most $c-s$ (the maximal possible length) is in $\mathit{EligibleX}$, still we will have that $|\mathit{EligibleX}| < 2^{c+1-s}$. Conversely, how small can $|\mathit{EligibleY}|$ be? On the ith query we know that $|\mathit{EligibleY}| > 2^c - i$. So on the ith query we have that $|\mathit{EligibleX}|/|\mathit{EligibleY}| < 2^{c+1-s}/(2^c - i) \le 2^{2-s}$ assuming $i \le 2^{c-1}$ or, more strongly, assuming $q \le 2^{s+\tau-1}$. Summing over all q_{R} right-queries we have that the probability that *bad* gets set at line 25 is at most $4q_{\mathrm{R}}/2^s$. Since the result becomes vacuous when $q > 2^{s+\tau-1}$, we may now drop that technical condition and conclude the theorem.

EQUIVALENCE OF ALL-IN-ONE AND TWO-REQUIREMENT DEFINITIONS. To define DAE-security one could specify separate notions for deterministic privacy, detPriv, and deterministic authenticity, detAuth, and demand both. This "dual-requirement" approach is the one that has been taken in all prior work on AE. In our setting one could let $\mathbf{Adv}_{\Pi}^{\mathrm{detPriv}}(A) = \Pr[A^{\mathcal{E}_K(\cdot,\cdot)} \Rightarrow 1] - \Pr[A^{\$(\cdot,\cdot)} \Rightarrow 1]$ and $\mathbf{Adv}_{\Pi}^{\mathrm{detAuth}}(A) = \Pr[A^{\mathcal{E}_K(\cdot,\cdot),\,\mathcal{D}_K(\cdot,\cdot)}$ forges] where in the first definition A does not repeat a query, and in the second it never asks a right-query (H, Y) having already asked a left-query (H, X) that returned Y. Saying that A forges means that it asks a right-query (H, Y) and gets a response other than $\perp$, and A did not earlier ask a left-query (H, X) that returned Y.

It is straightforward to prove that our all-in-one notion of DAE-security and the two-requirement definition just sketched are equivalent. We omit further details.

The idea above can be extended to other variants of AE: the encryption scheme may be probabilistic, nonce-based, or deterministic; the privacy requirement can be indistinguishability from random bits or conventional indistinguishability; and message headers may be present or absent, strings or vectors. For any of these variants one can give a two-requirement definition or an all-in-one definition. In all cases we have investigated, the results come out as above: the all-in-one definition and the two-requirement definition are equivalent.

All-in-one definitions for AE resemble the definition for chosen-ciphertext-attack (CCA2) security [3, 4]; the definition of AE strengthens CCA2 in a simple and natural way. Perhaps it is only historical accident that our community has come to think of AE as privacy+authenticity and not as "CCA3 security."

Acknowledgments

Many thanks to the X9F1 working group, whose draft standard motivated this paper, and Morris Dworkin, who made this work known to us [13]. Thanks to Jesse Walker for an enormous number of valuable comments; to Susan Langford for noticing a significant error in an earlier draft; to Steve Bellovin for voicing his concerns about IV-misuse at a meeting back in 2000 (his comments ultimately motivated Section 6); Mihir Bellare for his typically perceptive comments; and the Eurocrypt 2006 PC for their comments. Phil Rogaway was supported by NSF 0208842 and a gift from Intel Corp. Much of this paper was written while Rogaway was a visitor to the School of Information Technology at Mae Fah Luang University, Thailand. Many thanks to MFLU and, in particular, to Dr. Thongchai Yooyativong and Dr. Tatsanee Mallanoo, for their generous hospitality.

References

1. J. An and M. Bellare. Does encryption with redundancy provide authenticity? *Advances in Cryptology – Eurocrypt '01*, LNCS vol. 2045, Springer, pp. 512–528, 2001.

2. M. Bellare, A. Boldyreva, L. Knudsen, and C. Namprempre. On-Line ciphers and the Hash-CBC constructions. *Advances in Cryptology – Crypto '01*, LNCS vol. 2139, Springer, pp. 292–309, 2001.

3. M. Bellare, A. Desai, E. Jokipii, and P. Rogaway. A concrete security treatment of symmetric encryption: analysis of the DES modes of operation. *Proc. of the 38th Symposium on Foundations of Computer Science*, IEEE Press, pp. 394–403, 1997.

4. M. Bellare, A. Desai, D. Pointcheval, and P. Rogaway. Relations among notions of security for public-key encryption schemes. *Advances in Cryptology – Crypto'98*, LNCS vol.1462, Springer, pp. 26–45, 1998.

5. M. Bellare, J. Kilian, and P. Rogaway. The security of the cipher block chaining message authentication code. *J. of Computer and System Science (JCSS)*, vol. 61, no. 3, pp. 362–399, Dec 2000.

6. M. Bellare and C. Namprempre. Authenticated encryption: relations among notions and analysis of the generic composition paradigm. *Advances in Cryptology – Asiacrypt '00*, LNCS vol. 1976, Springer, pp. 531–545, 2000.

7. M. Bellare and P. Rogaway. Code-based game-playing proofs and the security of triple encryption. Cryptology ePrint report 2004/331, 2004.

8. M. Bellare and P. Rogaway. Encode-then-encipher encryption: how to exploit nonces or redundancy in plaintexts for efficient encryption. *Advances in Cryptology – Asiacrypt '00*, LNCS vol. 1976, Springer, pp. 317–330, 2000.

9. M. Bellare, P. Rogaway, and D. Wagner. The EAX mode of operation. *Fast Software Encryption (FSE 2004)*, LNCS vol. 3017, Springer, pp. 389–407, 2004.

10. J. Black and P. Rogaway. A block-cipher mode of operation for parallelizable message authentication. *Advances in Cryptology – Eurocrypt '02*, LNCS vol. 2332, Springer, pp. 384-397, 2001.

11. J. Black and P. Rogaway. CBC MACs for arbitrary-length messages: the three-key constructions. *Advances in Cryptology – Crypto '00*, LNCS vol. 1880, Springer, pp. 197–215, 2000.

12. Y. Dodis and A. Smith. Entropic security and the encryption of high entropy messages. *Theory of Cryptography (TCC 2005)*, LNCS vol. 3378, Springer, pp. 556-577, 2005.

13. M. Dworkin. Request for review of key wrap algorithms. Cryptology ePrint report 2004/340, 2004. Contents are excerpts from a draft standard of the Accredited Standards Committee, X9, entitled *ANS X9.102 — Wrapping of Keys and Associated Data*.

14. O. Goldreich, S. Goldwasser, and S. Micali, How to construct random functions. *Journal of the ACM*, vol. 33, no. 4, pp. 210–217, 1986.

15. S. Goldwasser and S. Micali. Probabilistic encryption. *J. Comput. Syst. Sci.*, vol. 28, no. 2, pp. 270–299, 1984.

16. S. Halevi and P. Rogaway. A tweakable enciphering mode. *Advances in Cryptology – Crypto '03*, LNCS vol. 2727, Springer, pp. 482–499, 2003.

17. R. Housley. Triple-DES and RC2 key wrapping. IETF RFC 3217, Dec. 2001. Earlier version in RFC 2630, June 1999.

18. T. Iwata and K. Kurosawa. OMAC: One-key CBC MAC. *Fast Software Encryption (FSE 2003)*, LNCS vol. 2887, Springer, pp. 129–153, 2003.

19. C. Jutla. Encryption modes with almost free message integrity. *Advances in Cryptology – Eurocrypt '01*, LNCS vol. 2045, Springer, pp. 529–544, 2001.

20. J. Katz and M. Yung. Unforgeable encryption and adaptively secure modes of operation. *Fast Software Encryption (FSE 2000)*, LNCS vol. 1978, Springer, pp. 284–299, 2000.

21. T. Kohno, J. Viega, and D. Whiting. CWC: A high-performance conventional authenticated encryption mode. *Fast Software Encryption (FSE 2004)*, LNCS vol. 3017, Springer, pp. 427–445, 2004.

22. H. Krawczyk. The order of encryption and authentication for protecting communications (or: how secure is SSL?) *Advances in Cryptology – Crypto '01*, LNCS vol. 2139, Springer, pp. 310–331, 2001.

23. M. Liskov, R. Rivest, and D. Wagner. Tweakable block ciphers. *Advances in Cryptology – Crypto '02*, LNCS vol. 2442, Springer, pp. 31–46, 2002.

24. M. Luby and C. Rackoff. How to construct pseudorandom permutations from pseudorandom functions. *SIAM J. Comput.*, vol. 17, no. 2, pp. 373–386, 1988.

25. S. Matyas. Key handling with control vectors. *IBM Systems Journal*, vol. 30, no. 2, pp. 151–174, 1991.

26. D. McGrew and J. Viega. The Galois/Counter mode of operation (GCM). Manuscript, May 2005. Available from the NIST website.

27. National Institute of Standards and Technology, M. Dworkin, author. Recommendation for block cipher modes of operation, methods and techniques. NIST Special Publication 800-38A, 2001.

28. National Institute of Standards and Technology, M. Dworkin, author. Recommendation for block cipher modes of operation: the CMAC mode for authentication. NIST Special Publication 800-38B, May 2005.

29. National Institute of Standards and Technology, M. Dworkin, author. Recommendation for block cipher modes of operation: the CCM mode for authentication and confidentiality. NIST Special Publication 800-38C, May 2004.

30. D. Phan and D. Pointcheval. About the security of ciphers (semantic security and pseudorandom permutations). *Selected Areas in Cryptography (SAC 2004)*, LNCS vol 3357, Springer, pp. 182-197, 2004.

31. P. Rogaway. Authenticated-encryption with associated-data. *Proceedings of the 9th Annual Conference on Computer and Communications Security (CCS-9)*, ACM, pp. 98–107, 2002.

32. P. Rogaway. Nonce-based symmetric encryption. *Fast Software Encryption (FSE 2004)*, LNCS vol. 3017, Springer, pp. 348–359, 2004.

33. P. Rogaway, M. Bellare, and J. Black. OCB: A block-cipher mode of operation for efficient authenticated encryption. *ACM Transactions on Information and System Security* (TISSEC), vol. 6, no. 3, pp. 365–403, Aug. 2003.

34. A. Russell and H. Wong. How to fool an unbounded adversary with a short key. *Advances in Cryptology – Eurocrypt '02*, LNCS vol. 2332, Springer, pp. 133–148, 2002.

35. R. Schroeppel. The hasty pudding cipher. AES candidate submitted to NIST, 1998.

36. S/MIME Working Group, IETF. Mailing list archives, 1997. http://www.imc.org/ietf-smime/index.html

Luby-Rackoff Ciphers from Weak Round Functions?

Ueli Maurer[1], Yvonne Anne Oswald[1],
Krzysztof Pietrzak[2,*], and Johan Sjödin[1,**]

[1] Department of Computer Science, ETH Zurich, CH-8092 Zurich, Switzerland
{maurer, sjoedin}@inf.ethz.ch, yoswald@student.ethz.ch
[2] Département d'informatique, Ecole Normale Supérieure, Paris, France
pietrzak@di.ens.fr

Abstract. The Feistel-network is a popular structure underlying many block-ciphers where the cipher is constructed from many simpler rounds, each defined by some function which is derived from the secret key.

Luby and Rackoff showed that the three-round Feistel-network – each round instantiated with a pseudorandom function secure against adaptive chosen plaintext attacks (CPA) – is a CPA secure pseudorandom permutation, thus giving some confidence in the soundness of using a Feistel-network to design block-ciphers.

But the round functions used in actual block-ciphers are – for efficiency reasons – far from being pseudorandom. We investigate the security of the Feistel-network against CPA distinguishers when the only security guarantee we have for the round functions is that they are secure against non-adaptive chosen plaintext attacks (nCPA). We show that in the information-theoretic setting, four rounds with nCPA secure round functions are sufficient (and necessary) to get a CPA secure permutation. Unfortunately, this result does not translate into the more interesting pseudorandom setting. In fact, under the so-called Inverse Decisional Diffie-Hellman assumption the Feistel-network with four rounds, each instantiated with a nCPA secure pseudorandom function, is in general not a CPA secure pseudorandom permutation.

1 Introduction

FEISTEL-NETWORK. The Feistel-network is a popular design approach for block-ciphers where the cipher over $\{0,1\}^{2n}$ is constructed by cascading simpler permutations, each constructed from a round function $\{0,1\}^n \to \{0,1\}^n$. The secret key of the cipher is only used to choose the particular round functions.

---

* Most of this work was done while the author was a PhD student at ETH where he was supported by the Swiss National Science Foundation, project No. 200020-103847/1. Currently the author is partially supported by the Commission of the European Communities through the IST program under contract IST-2002-507932 ECRYPT.

** This work was partially supported by the Zurich Information Security Center. It represents the views of the authors.

S. Vaudenay (Ed.): EUROCRYPT 2006, LNCS 4004, pp. 391–408, 2006.

LUBY-RACKOFF CIPHERS. In their celebrated paper [LR86] Luby and Rackoff prove that the three-round Feistel-network is an adaptive chosen plaintext (CPA) secure block-cipher – i.e. a pseudorandom permutation (PRP) – if each round is instantiated with an independent CPA secure pseudorandom function (PRF), and with one extra round even adaptive chosen ciphertext (CCA) security is achieved.

Besides reducing PRPs to PRFs, this result also gives some confidence in the soundness of using a Feistel-network to design block-ciphers. But unlike in the Luby-Rackoff ciphers, in most block-ciphers based on Feistel-networks the round functions are not independent (in order to keep the secret key short) and also far from being pseudorandom (for efficiency reasons). Instead, the number of rounds is much larger than four. (which was sufficient for the Luby-Rackoff constructions).

In order to achieve more efficient constructions of PRPs from PRFs, many researcher have investigated the security of weakened versions of the Luby-Rackoff ciphers. Several variations of the ciphers were proven to be pseudo-random where for example the round functions were not required to be independent [Pie90], some round functions were replaced by weaker primitives than PRFs [Luc96, NR02] or the distinguisher was given direct oracle access to some of the round functions [RR00]. These results further fortify the confidence in using Feistel-networks to design block ciphers.

All these relaxed constructions need at least some of the round functions to be CPA secure PRFs in order to get a CPA secure PRP. In this paper, we investigate for the first time – to the best of our knowledge – the CPA security of the permutation one gets by a Feistel-network where none of the round functions is guaranteed to be CPA secure. In particular, we investigate the security of the Feistel-network where each round is instantiated with a *non-adaptive* chosen plaintext (nCPA) secure round function. Although nCPA security is still a strong requirement, this was the weakest natural class of attacks we could imagine which does not make the Feistel-network trivially insecure against CPA attackers. For example round functions which are only secure against known-plaintext attacks (KPA), i.e. look random on random inputs, are too weak.[1]

PSEUDO- AND QUASIRANDOMNESS. Informally, a pseudorandom function PRF is a family of functions which can be efficiently computed, and where a random member from the family cannot be distinguished from a uniform random function (URF) by any efficient adversary. Pseudorandom permutations (PRP) are defined analogously. As usual in cryptography, an adversary is efficient if he is in P/poly, i.e. in non-uniform polynomial time (but almost all our results also hold when considering uniform adversaries; the only exception is addressed in Footnote 13). A *quasirandom* function (QRF) (similarly for a quasirandom permutation (QRP)) is defined similar to a *pseudorandom* one but where one does not require the distinguisher or the function to be efficient, only the number of

---

[1] Just consider a function f which satisfies $f(0 \ldots 0) = 0 \ldots 0$ but otherwise looks random. This f is KPA secure as a random query is unlikely to be the all zero string. But a Feistel-network build from such functions will output $0 \ldots 0$ on input $0 \ldots 0$ and thus is easily seen not to be CPA (or even nCPA) secure.

queries the distinguisher is allowed to make is bounded. Quasirandomness can be seen as an extension of the concept of statistically close distributions to systems which can be queried interactively.

In order to prove that some system – which is built from *pseudorandom* components – is pseudorandom itself, it is often enough to prove it to be *quasirandom* when the components are replaced by the corresponding ideal systems. In particular, to prove the security of the original three-round Luby-Rackoff cipher it is enough to prove – the purely information-theoretic result – that the network instantiated with URFs is a CPA secure QRP. It then immediately follows that the construction is a CPA secure PRP when the URFs are replaced by CPA secure PRFs, since if it was not a CPA secure PRP, we could use the distinguisher for it to build a distinguisher for the CPA secure PRF (via a standard hybrid argument). Similarly one can easily show that if the round functions are only nCPA or only KPA secure PRFs, the construction is a PRP, but only against the same class of attacks – i.e. nCPA or KPA.

2 Contributions

Our results and related work are summarized in Fig. 2 on page 395. Due to space limitations, some proofs are provided in the full version of this paper only [MOPS06].

(IN)SECURE RELAXATIONS OF THE THREE-ROUND LUBY-RACKOFF CIPHER. In the pseudo- and quasirandom setting, the three-round Feistel-network is – as mentioned above – ATK $\in$ {CPA, nCPA, KPA} secure when the round functions are ATK secure. Moreover it is known that one can replace the first round with a pairwise independent permutation [Luc96, NR99].[2] We further relax this by showing that the function in the last round only needs to be secure against known plaintext attacks (KPA). This resolves an open question posed by Minematsu and Tsunoo in [MT05]. Furthermore, for ATK = KPA we show that the first round is not necessary – as opposed to when ATK $\in$ {CPA, nCPA} – and that it is sufficient to instantiate the (two) round functions with a single instantiation of a KPA secure function.

But the second round seems to be the crucial one for ATK $\in$ {CPA, nCPA}. We show that for constructing a CPA secure permutation – i.e. PRP or QRP depending on the setting – one cannot in general instantiate the second round with a function which is only nCPA secure by constructing a counter-example, i.e. a nCPA secure function such that the three-round Feistel-network with this function in the second, and any random functions in the first and third round can easily be distinguished from a uniformly random permutation (URP) with only three adaptively chosen queries. Similarly, if one instantiates the second round with a KPA secure function, then the construction will in general not even be nCPA secure.

---

[2] In fact, the permutation must only be such that on any two values, the collision probability on one half of the domain is small. For example one can use one normal Feistel round instantiated with an almost XOR-universal function.

FOUR ROUNDS WITH NON-ADAPTIVE ROUND FUNCTIONS. As a consequence, three rounds with nCPA secure round functions are not enough to get CPA security. On the positive side, we show that one extra nCPA secure round is sufficient (and necessary) in the quasirandom setting. Note that for the translation of a security proof from quasi- to pseudorandom systems – as described at the end of the previous section – it is crucial that we can construct a distinguisher for the components from a distinguisher for the whole system. But here the components have a weaker security guarantee (i.e. nCPA) than what we prove for the whole system (i.e. CPA). So even when we have a CPA distinguisher for the four-round Feistel-network, we cannot construct a nCPA distinguisher for any round function. This is not just a shortcoming of the used approach, but indeed, in the pseudorandom setting the situation is different: we show that here four rounds are not enough to get CPA security. To show this we construct a nCPA secure PRF, such that the four-round Feistel-network with such round functions can easily be distinguished from URP with only three adaptive queries.

This phenomenon – i.e. that some construction implies adaptive security for quasirandom but not for pseudorandom systems – has already been proven [MP04, MPR06, Pie05] for two simple constructions: the sequential composition $f \triangleright g(.) \stackrel{\text{def}}{=} g(f(.))$ and the parallel composition $f \star g(.) \stackrel{\text{def}}{=} f(.) \star g(.)$ (where $\star$ stands for any group operation). The security proofs from [MP04] in the quasirandom setting crucially use the fact that the sequential composition of two permutations is a URPs whenever at least one of the permutations is a URP, similarly the parallel composition of two functions is a URF whenever one of the components is a URF. The Feistel-network does not have this nice property of being ideal whenever one of the components is ideal, and we have to work harder here (using a more general approach from [MPR06]). Our counter-example for the pseudorandom setting – i.e. a four-round Feistel-network with nCPA secure PRFs as round functions that is not a CPA secure PRP – is similar to the counter-examples for sequential and parallel composition shown in [Pie05, Ple05]. In [Ple05], it is shown that the sequential composition of arbitrarily many nCPA secure PRFs will not be a CPA secure PRF in general, whereas for the parallel composition only a counter-example with two components is known [Pie05]. For the Feistel-network we also could only find a counter-example for four rounds. So we cannot rule out the possibility that five or more rounds imply adaptive security. However, if this was the case, then it seems likely that – like for sequential composition [Mye04] – there is no black-box proof for this fact.[3]

UNCONDITIONAL VS. CONDITIONAL COUNTER-EXAMPLES. The counter-example showing that the three-round Feistel-network with a nCPA secure PRF

---

[3] Myers [Mye04] constructs an oracle relative to which there exist PRPs that are nCPA secure, but for which their sequential composition is not a CPA secure PRP. The idea behind this oracle is quite general, and we see no reason (besides being technically challenging) why one should not be able to construct a similar oracle for the Feistel-network, and thus also rule out a black-box proof for showing that the Feistel-network with nCPA secure PRFs as round functions is a CPA secure PRP.

Construction	Quasirandom	Pseudorandom	Reference
$\psi[RRR]$		CPA	[LR86, Mau02]
$\psi[NNN]$		nCPA	§4
$\psi[KKK]$		KPA	§4
$H \triangleright \psi[RR]$		CPA	[Luc96, NR02]
$H \triangleright \psi[RK]$		CPA	§4
$H \triangleright \psi[NK]$		nCPA	§4
$H \triangleright \psi[KK]$		KPA	§4
$\psi[RR]$		KPA (and NOT nCPA)	[MT05] (and §4)
$\psi[K^2]$		KPA	§4
$\psi[RNR]$		NOT CPA	§5
$\psi[RKR]$		NOT nCPA	§5
$\psi[NNNN]$	CPA	NOT CPA (under IDDH)	§6 and §7

Fig. 1. Security of the Feistel-network ψ with various security guarantees on the round functions. Here $\psi[f_1 \cdots f_k](\cdot)$ denotes the k-round Feistel-network with f_i in the i'th round, and $\psi[f^2] \stackrel{\text{def}}{=} \psi[ff]$ – i.e. the same function f in both rounds. Each occurrence of R, N, and K stands for an independent CPA, nCPA, and KPA secure function (i.e. a PRF or a QRF depending on the setting) respectively. The same holds for H which is any "lightweight" permutation from which we only require that the collision probability be small on the left half of the output, an almost pairwise independent permutation or a Feistel round instantiated with an almost XOR-universal function is thus sufficient.

F in the second round is not adaptively secure is unconditional[4] and black-box; with this we mean that we can construct F starting from any (nCPA secure) PRF via a reduction which uses this PRF only as a black-box.[5] As four rounds are enough to get adaptive security for quasirandom systems, there cannot be a black-box counter-example (like for three rounds) for the four (or more) round case. Thus it is not surprising that our counter-example for four rounds is not unconditional. It relies on the so-called Inverse Decisional Diffie-Hellman assumption. The fact that there is no black-box counter-example can be used to show that there is in some sense no "generic" adversary which breaks the adaptive security of the four-round Feistel-network with non-adaptive round functions. What "generic" actually means will not be the topic of this paper, but see Sect. 4 from [Pie06] (in this proceedings) for the corresponding statement for sequential composition.

---

[4] I.e. we make no other assumption besides the trivially necessary one that pseudorandom functions – which are equivalent to one-way functions [HILL99, GGM86] – exist at all.

[5] We build F from a pseudorandom involution (PRI), how to construct a PRI from a PRP (via a black-box reduction) has been shown in [NR02].

3   Basic Definitions and Random Systems

We use capital calligraphic letters like $\mathcal{X}$ to denote sets, capital letters like X to denote random variables and small letters like x denote concrete values. To save on notation we write X^i for $(X_1, X_2, \ldots, X_i)$.

For $x \in \{0,1\}^{2n}$ we denote with $_Lx$ and $_Rx$ the left and right half of x respectively, so $x = {}_Lx\|{}_Rx$. Similarly for any function f with range $\{0,1\}^{2n}$, we denote with $_Lf$ ($_Rf$) the function one gets by ignoring the right (left) half of the output of f. For two functions $f(.)$ and $g(.)$ we denote with $f \triangleright g(.) \stackrel{def}{=} g(f(.))$ the sequential composition of f and g.[6] For a (randomized) function f we denote with $\mathsf{coll}_k(f)$ the collision probability of any fixed k-tuple of distinct inputs, i.e.

$$\mathsf{coll}_k(f) = \max_{x_1, \ldots, x_k} \mathsf{P}(\exists i, j; 1 \leq i < j \leq k : f(x_i) = f(x_j)).$$

If f denotes a uniform random function with range $\{0,1\}^n$, then $\mathsf{coll}_k(f) \leq k^2/2^{n+1}$, this is called the *birthday bound* which we will use quite often.

Definition 1 (Feistel-network). *The (one round) Feistel-network* $\psi[f]$: $\{0,1\}^{2n} \to \{0,1\}^{2n}$ *based on a function* $f : \{0,1\}^n \to \{0,1\}^n$ *is defined as*

$$\psi[f](x) \stackrel{def}{=} (f(_Lx) \oplus {}_Rx)\|{}_Lx.$$

With $\psi[f_1 \cdots f_k] \stackrel{def}{=} \psi[f_1] \triangleright \psi[f_2] \triangleright \cdots \triangleright \psi[f_k]$ *we denote the k-round Feistel-network based on (randomized) round functions* $f_1, \ldots, f_k$, *here the randomness used by any function is always assumed to be independent of the randomness of the other round functions. The k round Feistel-network where the same instantiation of a function* f *is used for all rounds is denoted by* $\psi[f^k] \stackrel{def}{=} \psi[\underbrace{f \cdots f}_{k \text{ times}}]$.

RANDOM SYSTEMS. Many results from this paper are stated and proven in the random systems framework of [Mau02]. A *random system* is a system which takes inputs $X_1, X_2, \ldots$ and generates, for each new input X_i, an output Y_i which depends probabilistically on the inputs and outputs seen so far. We define random systems in terms of the distribution of the outputs Y_i conditioned on $X^i Y^{i-1}$ (i.e. the actual query X_i and all previous input/output pairs $X_1 Y_1, \ldots, X_{i-1} Y_{i-1}$).

Definition 2 (Random systems). *An* $(\mathcal{X}, \mathcal{Y})$-*random system* $\mathbf{F}$ *is a sequence of conditional probability distributions* $\mathsf{P}^{\mathbf{F}}_{Y_i|X^i Y^{i-1}}$ *for* $i \geq 1$. *Here we denote by* $\mathsf{P}^{\mathbf{F}}_{Y_i|X^i Y^{i-1}}(y_i, x^i, y^{i-1})$ *the probability that* $\mathbf{F}$ *will output* y_i *on input* x_i *conditioned on the fact that* $\mathbf{F}$ *did output* y_j *on input* x_j *for* $j = 1, \ldots, i-1$.

As special classes of random systems we will consider *random functions* (which are exactly the stateless random systems) and *random permutations*.

Definition 3 (Random functions and permutations). *A* random function $\mathcal{X} \to \mathcal{Y}$ *(random permutation on* $\mathcal{X}$*) is a random variable which takes as values functions* $\mathcal{X} \to \mathcal{Y}$ *(permutations on* $\mathcal{X}$*).*

---

[6] Note that $f \triangleright g$ is usually denoted with $g \circ f$.

A uniform random function (URF) $\mathbf{R} : \mathcal{X} \to \mathcal{Y}$ *(A uniform random permutation (URP)* $\mathbf{P}$ *on* $\mathcal{X}$*) is a random function with uniform distribution over all functions from* $\mathcal{X}$ *to* $\mathcal{Y}$ *(permutations on* $\mathcal{X}$*). Throughout, the symbols* $\mathbf{R}$ *and* $\mathbf{P}$ *are used for the systems defined above (*$\mathcal{X}, \mathcal{Y}$ *to be understood).*

INDISTINGUISHABILITY OF RANDOM SYSTEMS. The distinguishing advantage of a computationally unbounded distinguisher for two random variables A and B is simply the statistical distance of A and B. It is more intricate to define what we mean by the indistinguishability of random systems as here one must specify how the systems can be accessed. For this we define the concept of a distinguisher.

Definition 4. *A* $(\mathcal{Y}, \mathcal{X})$-*distinguisher is a* $(\mathcal{Y}, \mathcal{X})$-*random system which is one query ahead; i.e. it is defined by* $\mathsf{P}^{\mathbf{D}}_{X_i | Y^{i-1} X^{i-1}}$ *instead of* $\mathsf{P}^{\mathbf{D}}_{X_i | Y^i X^{i-1}}$ *for all* i*. In particular the first output* $\mathsf{P}^{\mathbf{D}}_{X_1}$ *is defined before* $\mathbf{D}$ *is fed with any input.*

We can now consider the random experiment where a $(\mathcal{Y}, \mathcal{X})$-distinguisher queries a $(\mathcal{X}, \mathcal{Y})$-random system

Definition 5. *With* $\mathbf{D} \lozenge \mathbf{F}$ *we denote the random experiment where a distinguisher* $\mathbf{D}$ *interactively queries a compatible random system* $\mathbf{F}$*.*

We divide distinguishers into classes by posing restrictions on how the distinguisher can access his inputs and produce his queries. In particular the following attacks will be of interest to us:

- CPA: Adaptively Chosen Plaintext Attack; here the adversary can choose the i'th query after receiving the $(i-1)$'th output.
- nCPA: Non-Adaptively Chosen Plaintext Attack; here the distinguisher must choose all queries in advance.
- KPA: Known Plaintext Attack; the queries are chosen uniformly at random.

If $\mathbf{F}$ is a permutation, its inverse $\mathbf{F}^{-1}$ is well-defined and we can consider a

- CCA: Chosen Ciphertext Attack.

which is defined like a CPA but where the attacker can additionally make queries from the inverse direction.

Definition 6. *For* $k \geq 1$*, the two random experiments* $\mathbf{D} \lozenge \mathbf{F}$ *and* $\mathbf{D} \lozenge \mathbf{G}$ *define a distribution over* $\mathcal{X}^k \times \mathcal{Y}^k$*. The advantage of* $\mathbf{D}$ *after* k *queries in distinguishing* $\mathbf{F}$ *from* $\mathbf{G}$*, denoted* $\Delta^{\mathbf{D}}_k(\mathbf{F}, \mathbf{G})$*, is the statistical difference between those distributions*[7]

$$\Delta^{\mathbf{D}}_k(\mathbf{F}, \mathbf{G}) \stackrel{def}{=} \frac{1}{2} \sum_{\mathcal{X}^k \times \mathcal{Y}^k} \left| \mathsf{P}^{\mathbf{D} \lozenge \mathbf{F}}_{X^k Y^k} - \mathsf{P}^{\mathbf{D} \lozenge \mathbf{G}}_{X^k Y^k} \right|. \tag{1}$$

---

[7] This definition has a natural interpretation in the random experiment where we first toss a uniform random coin $C \in \{0, 1\}$. Then we let $\mathbf{D}$ (which has no a priori information on C) make k queries to a system $\mathbf{H}$ where $\mathbf{H} \equiv \mathbf{F}$ if $C = 0$ and $\mathbf{H} \equiv \mathbf{G}$ if $C = 1$. Here the expected probability that an optimal guess on C based on the k inputs and outputs of $\mathbf{H}$ will be correct is $1/2 + \Delta^{\mathbf{D}}_k(\mathbf{F}, \mathbf{G})/2$.

The advantage of the best ATK*-distinguisher making* k *queries for* **F** *and* **G** *is*

$$\Delta_k^{\mathsf{ATK}}(\mathbf{F}, \mathbf{G}) \stackrel{def}{=} \max_{\mathsf{ATK}-distinguisher\ \mathbf{D}} \Delta_k^{\mathbf{D}}(\mathbf{F}, \mathbf{G}).$$

PSEUDORANDOMNESS. We denote with $\mathbf{Adv}_{PRP}^{\mathsf{ATK}}(\mathsf{F}, t, k)$ the distinguishing advantage of the best oracle circuit for F from a URP **P** where the circuit must be of size at most t and make at most k ATK-queries to its oracle. So $\mathbf{Adv}$ is defined similarly to Δ but with an additional restriction on the size of the distinguisher. In particular $\mathbf{Adv}_{PRP}^{\mathsf{ATK}}(\mathsf{F}, \infty, k) = \Delta_k^{\mathsf{ATK}}(\mathsf{F}, \mathbf{P})$. $\mathbf{Adv}_{PRF}^{\mathsf{ATK}}$ is defined similarly, but with **P** replaced by **R**.

Informally, a family of keyed functions F indexed by a security parameter $\gamma \in \mathbb{N}$ is an ATK-secure pseudorandom function (PRF) if F (with security parameter γ) can be computed in uniform polynomial (in γ) time, and for any polynomial $p(.)$ the distinguishing advantage $\mathbf{Adv}_{PRF}^{\mathsf{ATK}}(\mathsf{F}, p(\gamma), p(\gamma))$ is negligible in γ (for a key chosen uniformly at random). Pseudorandom permutations (PRP) are defined similarly but using $\mathbf{Adv}_{PRP}^{\mathsf{ATK}}$, and where we additionally require that F (for any security parameter and key) is a permutation.

We usually use sans-serif fonts like F to denote systems which can be efficiently computed (in particular pseudorandom systems), and bold fonts like **F** to denote quasirandom and ideal systems.

4   Relaxations of the Three-Round Luby-Rackoff Cipher

Let us first state some results for the three-round Feistel-network.

Proposition 1. *For any* ATK $\in$ {CPA, nCPA, KPA} *and function* **F**

$$\Delta_k^{\mathsf{ATK}}(\psi_{2n}[\mathbf{FFF}], \mathbf{P}) \le 3 \cdot \Delta_k^{\mathsf{ATK}}(\mathbf{F}, \mathbf{R}) + 2 \cdot \frac{k^2}{2^{n+1}}. \tag{2}$$

The analogous statement also holds in the computational case: for any ATK $\in$ {CPA, nCPA, KPA} *and any efficient function* F

$$\mathbf{Adv}_{PRP}^{\mathsf{ATK}}(\psi_{2n}[\mathsf{FFF}], t, k) \le 3 \cdot \mathbf{Adv}_{PRF}^{\mathsf{ATK}}(\mathsf{F}, t', k) + 2 \cdot \frac{k^2}{2^{n+1}}, \tag{3}$$

where $t' = \mathsf{poly}(t, k)$ *for some polynomial* poly *which accounts for the overhead implied by the reduction we make.*

The classical result of Luby and Rackoff [LR86], states that the Feistel-network with three independent PRF rounds is a CPA secure PRP – i.e (3) for CPA.

Luby and Rackoff proved this result directly. One gets a simpler proof by first showing that the three-round Feistel-network with URFs **R** is a CPA secure QRP as this is a purely information-theoretic statement. In particular it was shown in [Mau02] that[8]

$$\Delta_k^{\mathsf{CPA}}(\psi_{2n}[\mathbf{RRR}], \mathbf{P}) \le 2 \cdot \frac{k^2}{2^{n+1}}, \tag{4}$$

---

[8] This bound has been improved – for various number of rounds – in a series of papers. The latest [Pat04] by Patarin presents the best possible security for up to $k \ll 2^n$ (and not just $k \ll 2^{n/2}$) queries, using five rounds which is also necessary.

from which Proposition 1 directly follows using a standard hybrid argument.[9] Lucks showed [Luc96] (see also [NR02]) that the first round in the three-round Luby-Rackoff cipher can be replaced with a much weaker primitive which only must provide some guarantee on the collision probability on the left half of the output (for any two fixed inputs). In particular, an almost pairwise independent permutation or a Feistel-round with an almost XOR-universal function will do.

Proposition 2. *For any* ATK $\in \{$CPA, nCPA, KPA$\}$, *any functions* **F**, **G**, *and any permutation* **H**

$$\Delta_k^{\text{ATK}}(\text{H} \triangleright \psi_{2n}[\mathbf{FG}], \mathbf{P}) \leq \Delta_k^{\text{ATK}}(\mathbf{F}, \mathbf{R}) + 2 \cdot \Delta_k^{\text{KPA}}(\mathbf{G}, \mathbf{R}) + \text{coll}_k(_L\text{H}) + 2 \cdot \frac{k^2}{2^{n+1}}. \quad (5)$$

The analogous statement also holds in the computational case: for any ATK $\in \{$CPA, nCPA, KPA$\}$, *any efficient functions* F, G, *and any efficient permutation* H

$$\mathbf{Adv}_{PRP}^{\text{ATK}}(\text{H} \triangleright \psi_{2n}[\text{FG}], t, k) \quad (6)$$
$$\leq \mathbf{Adv}_{PRF}^{\text{ATK}}(\text{F}, t', k) + 2 \cdot \mathbf{Adv}_{PRF}^{\text{KPA}}(\text{G}, t', k) + \text{coll}_k(_L\text{H}) + 2 \cdot \frac{k^2}{2^{n+1}},$$

where $t' = t + \text{poly}(n, k)$ *for some polynomial* poly *which accounts for the overhead implied by the reduction we make.*

Let us stress that (6) does *not* directly follow from (5).[10] The proof of Proposition 2 is given in the full version of this paper [MOPS06].

We relax the construction further for ATK = KPA by showing that the first round can be removed completely (as opposed to when ATK $\in \{$CPA, nCPA$\}$)[11]. The round functions can also be replaced by a *single* instantiation of a KPA secure function. Note that the resulting construction is an involution, i.e. has the structural property of being self inverse. This result also generalizes Lemma 2.2 of [MT05] which states that the two round Feistel-network with CPA secure PRFs is a KPA secure PRP.

---

[9] The argument goes as follows for pseudorandom systems: suppose there is an efficient ATK $\in \{$CPA, nCPA, KPA$\}$ distinguisher A for $\psi_{2n}[\text{FFF}]$ and $\mathbf{P}$, then by (4) this A will also distinguish $\psi_{2n}[\text{FFF}]$ from $\psi_{2n}[\text{RRR}]$. Consider the hybrids $H_0 = \psi_{2n}[\text{FFF}], H_1 = \psi_{2n}[\text{RFF}], \ldots, H_3 = \psi_{2n}[\text{RRR}]$. By the triangle inequality there is an $0 \leq i \leq 2$ (say $i = 1$) such that A can distinguish H_i from H_{i+1}. Now, the distinguisher which – with access to an oracle G (implementing either F or **R**) – simulates $A \Diamond \psi_{2n}[\text{RGF}]$ and outputs the output of A is an efficient ATK-distinguisher for F with the same advantage as A's advantage for H_1 and H_2. The corresponding argument also holds in the quasirandom setting.

[10] The reason why a reduction – like the simple argument to show that Proposition 1 follows from (4) – fails here, is that the KPA security guarantee for one of the components is weaker than the CPA security for the whole construction. But fortunately the *proof* of (5) is such that it easily translates to the pseudorandom setting.

[11] $\psi_{2n}[\mathbf{RR}]$ can be distinguish from $\mathbf{P}$ with two non-adaptively chosen queries: query $0^n \| 0^n \mapsto {}_Ly\|_Ry$ and $0^n \| 1^n \mapsto {}_Ly'\|_Ry'$, and output 1 if ${}_Ry \oplus {}_Ry' = 1^n$ and 0 otherwise.

Proposition 3. *For any function* $\mathbf{F}$

$$\Delta_k^{\mathsf{KPA}}(\psi_{2n}[\mathbf{F}^2], \mathbf{P}) \leq \Delta_{2k}^{\mathsf{KPA}}(\mathbf{F}, \mathbf{R}) + 4 \cdot \frac{k^2}{2^{n+1}}. \tag{7}$$

The analogous statement also holds in the computational case: for any function $\mathbf{F}$
(in particular any efficient function $\mathbf{F}$*)*

$$\mathbf{Adv}_{PRP}^{\mathsf{KPA}}(\psi_{2n}[\mathbf{F}^2], t, k) \leq \mathbf{Adv}_{PRF}^{\mathsf{KPA}}(\mathbf{F}, t', 2k) + 4 \cdot \frac{k^2}{2^{n+1}}, \tag{8}$$

where $t' = t + \mathsf{poly}(n, k)$ *for some polynomial* poly *which accounts for the overhead*
implied by the reduction we make.

The proof is in the full version of this paper [MOPS06]. Note that unlike in the
previous propositions, here we do not require the round function $\mathbf{F}$ to be efficient
in the computational case (the reason is that in the proof we do not need the
distinguisher to simulate any round function).

5   The Second Round Is Crucial

In the previous section we have seen that in the classical three-round Luby-
Rackoff cipher the first and third round function need not be CPA secure. In this
section we will see that the security requirements for the second round cannot be
relaxed. We only give proof sketches for the propositions of this section. Detailed
proofs can be found in the full version.

The following proposition states that to achieve CPA security in general it is
not sufficient that the second round function is nCPA secure. There exists a nCPA
secure function, such that the three-round Feistel-network with this function in
the second, and any random functions in the first and third round, is not CPA
secure.

Proposition 4. *There exists a function* $\mathbf{F}$ *such that for any functions* $\mathbf{G}$ *and*
$\mathbf{G}'$ *(in particular for* $\mathbf{G} = \mathbf{R}$ *and* $\mathbf{G}' = \mathbf{R}$*)*

$$\Delta_k^{\mathsf{nCPA}}(\mathbf{F}, \mathbf{R}) \leq 4 \cdot \frac{k^2}{2^{n+1}} \quad \text{and} \quad \Delta_2^{\mathsf{CPA}}(\psi_{2n}[\mathbf{GFG}'], \mathbf{P}) \geq 1 - 2^{-n+1}.$$

The analogous statement also holds in the computational case: (informal) there
is a nCPA secure PRF $\mathbf{F}$ *such that* $\psi_{2n}[\mathbf{GFG}']$ *is not a CPA secure PRP for any*
(not necessarily efficient) functions $\mathbf{G}$ *and* $\mathbf{G}'$*.*

Proof (sketch). Let us first consider the quasirandom statement. Let $\mathbf{I}$ be a
uniform random involution, i.e. $\mathbf{I}(\mathbf{I}(x)) = x$ for all x. Now, $\mathbf{F}$ is simply defined
as $\mathbf{F}(x) = x \oplus \mathbf{I}(x)$, note that this $\mathbf{F}$ satisfies $\mathbf{F}(x) = \mathbf{F}(x \oplus \mathbf{F}(x))$ for all x.

The nCPA security of $\mathbf{F}$ (which is simply the nCPA security of $\mathbf{I}$) can be
bounded as stated in the proposition by standard techniques. Furthermore,
$\psi_{2n}[\mathbf{GFG}']$ can easily be distinguished from $\mathbf{P}$ with two adaptively chosen

queries as follows. After a first query $0^n \| 0^n$, the output $_L Y \| Z$ contains the output Z of the internal function $\mathbf{F}$. Now make a second query $0^n \| Z$. If the (unknown) input to $\mathbf{F}$ in the first query was some value V, then in this query it will be $V \oplus Z$, and as $\mathbf{F}$ satisfies $\mathbf{F}(V) = \mathbf{F}(V \oplus \mathbf{F}(V)) = \mathbf{F}(V \oplus Z)$, the output of $\mathbf{F}$ will again be Z, and the overall output will be $(_L Y \oplus Z) \| Z$. The proposition follows as the output of $\mathbf{P}$ will satisfy such a relation with probability at most 2^{-n+1}.

The corresponding statement for the pseudorandom setting is proven almost identically. The only difference is that we need to use a CPA secure pseudorandom involution instead of the uniform random involution. It is shown in [NR02] how to construct a pseudorandom involution from any CPA secure PRF. $\qquad\square$

The next proposition states that the network will in general not (even) be nCPA secure when the second round function is only secure against KPAs.

Proposition 5. *There exists a function $\mathbf{F}$ such that for any functions $\mathbf{G}$ and $\mathbf{G}'$*

$$\Delta_k^{\mathsf{KPA}}(\mathbf{F}, \mathbf{R}) \leq \frac{k^2}{2^{n+1}}, \quad and \quad \Delta_2^{\mathsf{nCPA}}(\psi_{2n}[\mathbf{GFG}'], \mathbf{P}) \geq 1 - 2^{-n+2}.$$

The analogous statement also holds in the computational case: (informal) there is a KPA secure PRF F such that $\psi_{2n}[\mathbf{GFG}']$ is not a nCPA secure PRP for any (not necessarily efficient) functions $\mathbf{G}$ and $\mathbf{G}'$.

Proof (sketch). Let us first consider the statement in the quasirandom setting. Let $\mathbf{F}$ be a URF which ignores the first input bit, i.e. for all $x \in \{0,1\}^{n-1}$ we have $\mathbf{F}(0\|x) = \mathbf{F}(1\|x)$. The KPA security of $\mathbf{F}$ follows from the fact that $\mathbf{F}$ looks completely random unless we happen to query two queries of the form $0\|x$ and $1\|x$. By the birthday bound the probability that this happens after k queries is at most $\frac{k^2}{2^{n+1}}$. Furthermore, $\psi_{2n}[\mathbf{GFG}']$ can be distinguished from $\mathbf{P}$ with two non-adaptively chosen queries. For instance on input $0^n \| 0^n$ and $0^n \| (1\|0^{n-1})$, the right half of the output will be identical.

The corresponding statement in the pseudorandom setting is proven exactly as above, except that we have to use a PRF F instead of $\mathbf{F}$. $\qquad\square$

6 Four nCPA Secure Rounds, the Quasirandom Case

In this section we will show that the four-round Feistel-network with nCPA secure QRFs is a CPA secure QRP. This is also the best possible as in Sect. 5 we showed that four rounds are also necessary. The theorem is even stronger as the third and fourth round function must only be KPA secure QRFs.

Theorem 1. *For any functions $\mathbf{F}$ and $\mathbf{G}$*

$$\Delta_k^{\mathsf{CPA}}(\psi_{2n}[\mathbf{FFGG}], \mathbf{P}) \leq 4 \cdot \Delta_k^{\mathsf{nCPA}}(\mathbf{F}, \mathbf{R}) + 3 \cdot \Delta_k^{\mathsf{KPA}}(\mathbf{G}, \mathbf{R}) + 9 \cdot \frac{k^2}{2^{n+1}}.$$

To prove this theorem we use Theorem 2 from [MPR06] which, for the special case of the four-round Feistel-network, is given as Proposition 6 below. The proposition bounds the security of a composition against a "strong" attacker sATK (in particular CPA) in terms of the security of the components against "weak" attackers wATK$_i$ (in particular nCPA or KPA).

The proposition uses the concept of conditions defined for random systems which we only define informally here (see [MPR06] for a formal definition): With $\mathbf{F}^{\mathcal{A}}$ we denote the random system $\mathbf{F}$, but which additionally defines an internal binary random variable after each query (called a condition). Let $A_i \in \{0,1\}$ denote the condition after the i'th query. We set $A_0 = 0$ and require the condition to be monotone which means that $A_i = 1 \Rightarrow A_{i+1} = 1$ (i.e. when the condition failed, it will never hold again). Let $\bar{a}_i$ denote the event $A_i = 1$, then

$$\nu^{\mathsf{ATK}}(\mathbf{F}^{\mathcal{A}}, \bar{a}_k) \stackrel{\text{def}}{=} \max_{\mathsf{ATK}-distinguisher\ \mathbf{D}} \mathsf{P}^{\mathbf{D}\Diamond \mathbf{F}^{\mathcal{A}}}_{\bar{a}_k}, \tag{9}$$

denotes the advantage of the best ATK distinguisher to make the condition fail after at most k queries to $\mathbf{F}^{\mathcal{A}}$.

Proposition 6. *If for any $(\{0,1\}^n, \{0,1\}^n)$-random system with a condition $\mathbf{F}^{\mathcal{A}}$*

$$\nu^{\mathsf{sATK}}(\psi_{2n}[\mathbf{F}^{\mathcal{A}}\mathbf{RRR}], \bar{a}_k) \leq \nu^{\mathsf{wATK}_1}(\mathbf{F}^{\mathcal{A}}, \bar{a}_k) + \alpha_1 \tag{10}$$

$$\nu^{\mathsf{sATK}}(\psi_{2n}[\mathbf{RF}^{\mathcal{A}}\mathbf{RR}], \bar{a}_k) \leq \nu^{\mathsf{wATK}_2}(\mathbf{F}^{\mathcal{A}}, \bar{a}_k) + \alpha_2 \tag{11}$$

$$\nu^{\mathsf{sATK}}(\psi_{2n}[\mathbf{RRF}^{\mathcal{A}}\mathbf{R}], \bar{a}_k) \leq \nu^{\mathsf{wATK}_3}(\mathbf{F}^{\mathcal{A}}, \bar{a}_k) + \alpha_3 \tag{12}$$

$$\nu^{\mathsf{sATK}}(\psi_{2n}[\mathbf{RRRF}^{\mathcal{A}}], \bar{a}_k) \leq \nu^{\mathsf{wATK}_4}(\mathbf{F}^{\mathcal{A}}, \bar{a}_k) + \alpha_4 \tag{13}$$

for some attacks wATK$_1$, wATK$_2$, wATK$_3$, wATK$_4$, sATK *and some* $\alpha_1, \alpha_2, \alpha_3,$ $\alpha_4 \geq 0$, *then for any* $\mathbf{F}_1, \mathbf{F}_2, \mathbf{F}_3, \mathbf{F}_4$

$$\Delta_k^{\mathsf{sATK}}(\psi_{2n}[\mathbf{F}_1\mathbf{F}_2\mathbf{F}_3\mathbf{F}_4], \psi_{2n}[\mathbf{RRRR}]) \leq \sum_{i=1}^{4} (\Delta_k^{\mathsf{wATK}_i}(\mathbf{F}_i, \mathbf{R}) + \alpha_i).$$

To apply this proposition we must show that equations (10), (11), (12) and (13) hold for some attack wATK$_i$ and α_i for $i = 1, 2, 3, 4$.

In the full version [MOPS06] we prove the following claim, from which Theorem 1 now follows.

Claim 1. *Equation (10) - (13) are satisfied for any function with a condition $\mathbf{F}^{\mathcal{A}}$, sATK = CPA, and*

$$(\mathsf{wATK}_i, \alpha_i) = \begin{cases} \left(\mathsf{nCPA}, 2 \cdot \frac{k^2}{2^{n+1}}\right) & \text{if} \quad i = 1 \\[2mm] \left(\mathsf{nCPA}, 2 \cdot \frac{k^2}{2^{n+1}} + 2 \cdot \Delta_k^{\mathsf{nCPA}}(\mathbf{F}, \mathbf{R})\right) & \text{if} \quad i = 2 \\[2mm] \left(\mathsf{KPA}, 3 \cdot \frac{k^2}{2^{n+1}} + \Delta_k^{\mathsf{KPA}}(\mathbf{F}, \mathbf{R})\right) & \text{if} \quad i = 3 \\[2mm] \left(\mathsf{KPA}, 2 \cdot \frac{k^2}{2^{n+1}}\right) & \text{if} \quad i = 4. \end{cases}$$

7   Four nCPA Secure Rounds, the Pseudorandom Case

In this section we again investigate the CPA security of the four-round Feistel-network with nCPA secure round functions, but this time for *pseudorandom* systems. We show that here the situation is dramatically different from the quasirandom setting by constructing a nCPA secure PRF where the four-round Feistel-network with this PRF as round function is not CPA secure.

This PRF is defined over some group, and to prove the nCPA security we assume that the so-called *inverse decisional Diffie-Hellman* (IDDH) is hard in this group. Informally, the IDDH assumption requires that for a generator g and random x, y it is hard do distinguish the triple (g, g^x, g^y) from $(g, g^x, g^{x^{-1}})$.

Theorem 2. *(Informal) Under the IDDH assumption there exists a* nCPA *secure PRF* F *such that the four-round Feistel-network where each round is instantiated with* F *(with independent keys) is not a* CPA *secure pseudorandom permutation.*

This theorem follows from Lemma 1 below which states that there exist nCPA secure PRFs F_1, F_2, F_3 such that the left half of the *three* round Feistel-network $_L\psi_{2n}[F_1F_2F_3]$ is not a CPA secure PRF. This implies that also $\psi_{2n}[F_1F_2F_3G]$ is not a CPA secure PRP for any G (and thus proves Theorem 2) as follows. By the so-called PRF/PRP Switching Lemma any CPA secure PRP P is also a CPA secure PRF. Clearly, then also $_LP$ must be a CPA secure PRF. Now, by Lemma 1 $_L\psi_{2n}[F_1F_2F_3] = {}_R\psi_{2n}[F_1F_2F_3G]$ is not a CPA secure PRF, so $\psi_{2n}[F_1F_2F_3G]$ cannot be a CPA secure PRP.[12]

Lemma 1. *Under the IDDH-assumption there exist* nCPA *secure PRFs* F_1, F_2, F_3 *such that* $_L\psi_{2n}[F_1F_2F_3]$ *is not a* CPA *secure PRF: it can be distinguished efficiently from a URF with only three (adaptive) queries with high probability.*

OUTLINE FOR THIS SECTION. In §7.1 we give a more formal definition of the IDDH assumption. Then, in §7.2 we first show the construction from [Ple05] of a nCPA secure PRF whose sequential composition will not be CPA secure. This extremely simple and intuitive construction is the basis for the (more involved) counter-example for the Feistel-network (i.e. Lemma 1) given in §7.3.

7.1   The Non-uniform IDDH Assumption

Below we define the IDDH assumption which is similar (and easily seen to imply) the well known decisional Diffie-Hellman assumption. Throughout, we will work with hardness assumptions in a non-uniform model of computation (i.e. we

---

[12] The lemma talks about three different F_i's (and in the proof we really construct a different F_i for every round), but the theorem is stated for a single F. This does not really make a difference. For example this single F can be defined as behaving like F_i with probability $1/3$ for $i \in \{1, 2, 3\}$. Then with constant probability 3^{-3} the $\psi_{2n}[FFF]$ behaves like $\psi_{2n}[F_1F_2F_3]$.

require hardness against polynomial size circuit families and not just any fixed Turing machine).[13]

Let $\mathcal{G}$ denote an efficiently computable family of groups indexed by a security parameter $n \in \mathbb{N}$. By efficiently computable we mean that one can efficiently (i.e. in time polynomial in n) sample a group (together with a generator) from the family, and efficiently compute the group operations therein. Abusing notation we denote with $(G, g) = \mathcal{G}(n)$ any group/generator pair for security parameter n.

The IDDH assumption is hard in $\mathcal{G}$ if for $(G, g) = \mathcal{G}(n)$ polynomial size circuits have negligible advantage guessing whether for a given triple (g, g^x, g^y) the y is random or computed as $y = x^{-1}$, more formally

Definition 7 (non-uniform IDDH). *For a group G and a generator g of G*

$$\mathbf{Adv}_{IDDH}(G, g, s) \overset{def}{=} \max_{C, |C| \le s} \left| \Pr_x \left[C(g, g^x, g^{x^{-1}}) = true \right] - \Pr_{x,y} [C(g, g^x, g^y) = true] \right|,$$

where the probability is over the random choice of $x, y \in [1, \ldots, |G|]$. We say that IDDH is hard in $\mathcal{G}$ if for any polynomial $p(.)$

$$\mathbf{Adv}_{IDDH}(\mathcal{G}(n), p(n)) = \mathsf{negl}(n).$$

7.2   Counter-Example for Sequential Composition from [Ple05]

In this section we construct a simple PRF F, but where the sequential composition of (arbitrary many) such F (with independent keys) is not CPA secure.

F is based on some prime order cyclic group $(G, g) = \mathcal{G}(n)$ where the IDDH problem is hard and where the elements of the group can be efficiently and densely encoded into $\{0, 1\}^n$ (with dense we mean that all but a negligible fraction of the strings should correspond to an element of the group).[14] For example we can take the subgroup of prime order q of $\mathbb{Z}_p^*$ where p is a safe prime (i.e. $2q + 1$) and q is close to 2^n ([Dam04] describes how to embed such a G into $\{0, 1\}^n$).

Let $[.] : \mathcal{G}(n) \to \{0, 1\}^n$ denote an (efficient) embedding of $\mathcal{G}$ into bitstrings (to save on notation we let $[a, b]$ denote the concatenation of $[a]$ and $[b]$). Let

---

[13] In cryptography security usually means security against non-uniform (and not just uniform) adversaries, and thus also the hardness assumptions used are usually non-uniform, though this is sometimes not explicitly stated as the security proofs work in both settings – i.e. a uniform (non-uniform) assumption implies hardness against uniform (non-uniform) adversaries. But here this is not quite the case, we do not know how to prove a uniform version of Lemma 1. (But one can do so under a somewhat stronger assumption than IDDH. Loosely speaking, this assumption is IDDH but where the attacker can also choose the generators to be used in the challenge.)

[14] For this construction we actually do not need this embedding, we could define F directly over the group. But we will need it (or more precisely, the fact that if X is in the range of F, also $X \oplus R$ for a random bitstring R is in the range with overwhelming probability) when we extend this construction to get the counter-example for the Feistel-network in the next section.

$R : \mathcal{K} \times \{0,1\}^{4n} \to \mathbb{Z}_q^4$ be any nCPA secure PRF. Now consider the following definition of a nCPA secure PRF $F : \{0,1\}^{4n} \to \{0,1\}^{4n}$ with secret key ($k_1 \in \mathcal{K}, x \in \mathbb{Z}_q^*$).

The first thing F does on input $(\alpha, \beta, \gamma, \delta) \in \{0,1\}^{4n}$ is to generate some pseudorandom values using R, i.e.

$$(r_1, r_2, r_3, r_4) \leftarrow R(k_i, \alpha, \beta, \gamma, \delta). \tag{14}$$

Further, if there exists $(a, b, c, d) \in G^4$ s.t. $\alpha = [a], \beta = [b], \gamma = [c], \delta = [d]$ then F outputs (here x^{-1} is the inverse of x in $\mathbb{Z}_q^*$)

$$F([a, b, c, d]) \to ([a^{xr_1}, b^{r_1}, c^{x^{-1}r_2}, d^{r_2}]), \tag{15}$$

with r_1, r_2 generated as in (14). On the remaining inputs (which are a negligible fraction of $\{0,1\}^{4n}$) F outputs just the (pseudo) random values $[g^{r_1}, g^{r_2}, g^{r_3}, g^{r_4}]$.

Now consider the cascade $F' \triangleright F'' \triangleright F'''$ of three independent F's (with corresponding keys (x_1, k_1), (x_2, k_2), and (x_3, k_3)). Make a first query $[g, g, g, g]$

$$F' \triangleright F'' \triangleright F'''([g, g, g, g]) \to [g^{x_1 x_2 x_3 r}, g^r, g^{x_1^{-1} x_2^{-1} x_3^{-1} r'}, g^{r'}].$$

Then the output will have the form $g^{x_1 x_2 x_3 r}, g^r, g^{x_1^{-1} x_2^{-1} x_3^{-1} r'}, g^{r'}$ for some r, r'. Now exchange the right and the left half of this output and use it as the second query

$$F' \triangleright F'' \triangleright F'''([g^{x_1^{-1} x_2^{-1} x_3^{-1} r'}, g^{r'}, g^{x_1 x_2 x_3 r}, g^r]) \to [g^{r''}, g^{r''}, g^{r'''}, g^{r'''}]$$

so the output is of the form $[u, u, v, v]$ for some u, v and thus can be distinguished from random. Therefore $F' \triangleright F'' \triangleright F'''$ is not a CPA secure PRF. This proves that the sequential composition of nCPA secure PRFs does not yield a CPA secure function in general. Note that this distinguishing attack works for any number of rounds, not just three. In the full version of this paper [MOPS06] we prove the following lemma which states that F is an nCPA secure PRF if IDDH is hard in $\mathcal{G}$ and R is a nCPA secure PRF.

Lemma 2. *Let g be any generator of the group over which F is defined, then*

$$\mathbf{Adv}_{PRF}^{nCPA}(F, k, s) \leq 6k \cdot \mathbf{Adv}_{IDDH}(F, g, s') + \mathbf{Adv}_{PRF}^{nCPA}(R, k, s'),$$

where $s' = s + \mathsf{poly}(k, n)$ for some polynomial poly which accounts for the overhead implied by the reduction we make.

7.3 Proof of Lemma 1

The Feistel-network can be seen as a sequential composition of the round functions, but where one additionally XORs the input to the i'th round function to the output of the $(i + 1)$'th round function. So it is not surprising that we can use F_i's similar to the F from the previous section to prove Lemma 1. But the $F_1, F_2,$ and F_3 (from the statement of the lemma) are a bit more complicated as we have to "work around" this additional XORs. Like F, each F_i has a $k_i \in \mathcal{K}$ as part of its secret key. Moreover F_1 has a $x \in \mathbb{Z}_q^*$ and $s, t \in \{0, 1\}^n$, F_2 has a $y \in \mathbb{Z}_q^*$, and F_3 a $z \in \mathbb{Z}_q^*$ as keys. On input $(\alpha, \beta, \gamma, \delta) = [a, b, c, d]$ the F_i's are defined as (with the r_i's generated as in (14))

$$F_1([a, b, c, d]) \rightarrow \begin{cases} [g^{xr_1}, g^{r_1}], s, t & \text{if } [a, b, c, d] = [0, 0, 0, 0]; \\ [0, 0, 0, 0] & \text{elseif } c = d^x; \\ [g^{xr_1}, g^{r_1}, ([\gamma \oplus s]^{-1})^{x^{-1}r_2}, ([\delta \oplus t]^{-1})^{r_2}] & \text{elseif } [a, b] = [0, 0]; \\ [g^{r_1}, g^{r_2}, g^{r_3}, g^{r_4}] & \text{otherwise.} \end{cases}$$

$$F_2([a, b, c, d]) \rightarrow [c^{y^{-1}r_1}, d^{r_1}, a^{yr_2}, b^{r_2}]$$

$$F_3([a, b, c, d]) \rightarrow \begin{cases} [0, 0, 0, 0] & \text{if } b^z = a; \\ [a^{z^{-1}r_1}, b^{r_1}, c^{zr_2}, d^{r_2}] & \text{otherwise.} \end{cases}$$

Proof (of Lemma 1). The lemma follows from Claim 2 and 3 below.   □

Claim 2. *One can distinguish $_L\psi_{2n}[F_1 F_2 F_3]$ from a URF with three adaptively chosen queries with advantage almost 1.*

Proof (sketch). In Fig. 2 we demonstrate an adaptive three query distinguishing attack on $_L\psi_{2n}[F_1 F_2 F_3]$. In the figure, values which are not relevant for the attack are denoted by $*$. All r_i' values are random, but not necessarily equal to a random value generated by a round function (i.e. as in (14)).[15] To see that this is a legal attack note that every query Q_i can be computed from the previous output O_{i-1}. That the values will really have the form as described in the attack can be verified from the definition of the F_i's.[16] Since the third output starts with $[0, 0]$ it can be distinguished from a random output with high probability.   □

Claim 3. *$F_1, F_2,$ and F_3 are nCPA secure PRFs if IDDH is hard in $\mathcal{G}$.*

Proof (sketch). The nCPA security of the F_i's follows from the nCPA security of F from the previous section as stated in Lemma 2: F_2 is exactly F, so there is nothing else to prove here. The function F_3 behaves exactly as F unless it is queried on an input $[a, b, c, d]$ which satisfies $b^z = a$ for a random z. The probability that this happen on any (non-adaptive) query is just $|G|^{-1}$ (and thus exponentially small even after taking the union bound over all polynomially many queries). For the somewhat longer argument for F_1, we refer to the full version [MOPS06].   □

---

[15] For instance, r_1' is the first random value generated by F_1 and r_2' is the product of r_1' and the second random value generated by F_2.

[16] Actually, there is an exponentially small probability that the values will not have that form, namely when the input to some round function "by chance" satisfies a condition that is checked. E.g. when R_1^3 is of the form $[b^z, b, c, d]$, then the "$b^z = a$" case of F_3 applies, which is only supposed to happen in the second and third query.

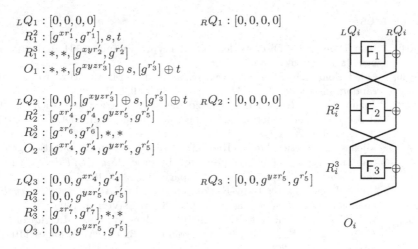

$_LQ_1 : [0,0,0,0]$ $_RQ_1 : [0,0,0,0]$
$R_1^2 : [g^{xr_1'}, g^{r_1'}], s, t$
$R_1^3 : *, *, [g^{xyr_2'}, g^{r_2'}]$
$O_1 : *, *, [g^{xyzr_3'}] \oplus s, [g^{r_3'}] \oplus t$

$_LQ_2 : [0,0], [g^{xyzr_3'}] \oplus s, [g^{r_3'}] \oplus t$ $_RQ_2 : [0,0,0,0]$
$R_2^2 : [g^{xr_4'}, g^{r_4'}, g^{yzr_5'}, g^{r_5'}]$
$R_2^3 : [g^{zr_6'}, g^{r_6'}], *, *$
$O_2 : [g^{xr_4'}, g^{r_4'}, g^{yzr_5'}, g^{r_5'}]$

$_LQ_3 : [0,0, g^{xr_4'}, g^{r_4'}]$ $_RQ_3 : [0,0, g^{yzr_5'}, g^{r_5'}]$
$R_3^2 : [0,0, g^{yzr_5'}, g^{r_5'}]$
$R_3^3 : [g^{zr_7'}, g^{r_7'}], *, *$
$O_3 : [0,0, g^{yzr_5'}, g^{r_5'}]$

Fig. 2. An adaptive three query distinguishing attack for $_L\psi_{2n}[F_1F_2F_3]$

8   Some Remarks on CCA Security

We have shown that the four-round Feistel-network with nCPA secure round functions is CPA secure in the information-theoretic, but in general not in the computational setting. A natural question is to ask how many rounds are necessary/not sufficient to achieve CCA security. In this section we state some observations. The full version of this paper addresses this question in more detail.

In order to get a CCA secure quasirandom permutations (QRP), it is enough – by the following statement (taken from [MPR06]) – to cascade two nCPA secure QRPs (the second in inverse direction)

$$\Delta_k^{CCA}(\mathbf{F} \triangleright \mathbf{G}^{-1}, \mathbf{P}) \leq \Delta_k^{nCPA}(\mathbf{F}, \mathbf{P}) + \Delta_k^{nCPA}(\mathbf{G}, \mathbf{P}).$$

With this and Proposition 1 we directly get that six rounds with nCPA secure QRFs give a CCA secure QRP, i.e.

$$\Delta_k^{CCA}(\psi_{2n}[\mathbf{FFFFFF}], \mathbf{P}) \leq 6 \cdot \Delta_k^{nCPA}(\mathbf{F}, \mathbf{R}) + \frac{k^2}{2^{n-1}}.$$

So six nCPA secure round functions are sufficient to get CCA security, and by Proposition 4 we know that at least four rounds are necessary. Using Proposition 5 we can further relax the requirements for the round functions as

$$\Delta_k^{CCA}(\mathsf{H} \triangleright \psi_{2n}[\mathbf{FGGF}] \triangleright \mathsf{H}^{-1}, \mathbf{P})$$

$$\leq 2 \cdot \Delta_k^{nCPA}(\mathbf{F}, \mathbf{R}) + 4 \cdot \Delta_k^{KPA}(\mathbf{G}, \mathbf{R}) + 2 \cdot \mathsf{coll}_k(_L H) + 2 \cdot \frac{k^2}{2^{n+1}}.$$

As to the (in)security of the Feistel-network with nCPA secure round-functions in the computational setting, we do not know anything beyond what is already implied by CPA security alone, i.e. four rounds are not enough to get CCA security (as it is not enough to get CPA security by Theorem 2).

References

[Dam04]   Ivan Damgård. Discrete log based cryptosystems, 2004. Manuscript, www.daimi.au.dk/ivan/DL.pdf.

[GGM86]   Oded Goldreich, Shafi Goldwasser, and Silvio Micali. How to construct random functions. *J. ACM*, 33(4):792–807, 1986.

[HILL99]   Johan Håstad, Russell Impagliazzo, Leonid A. Levin, and Michael Luby. A pseudorandom generator from any one-way function. *SIAM J. Comput.*, 28(4):1364–1396, 1999.

[LR86]   Michael Luby and Charles Rackoff. Pseudo-random permutation generators and cryptographic composition. In *Proc, 18th ACM Symposium on the Theory of Computing (STOC)*, pages 356–363, 1986.

[Luc96]   Stefan Lucks. Faster Luby-Rackoff ciphers. In *Fast Software Encryption*, volume 3557 of *LNCS*, pages 189–203. Springer-Verlag, 1996.

[Mau02]   Ueli Maurer. Indistinguishability of random systems. In *Advances in Cryptology — EUROCRYPT '02*, volume 2332 of *LNCS*, pages 110–132. Springer-Verlag, 2002.

[MOPS06]   For the full version of this paper see www.crypto.ethz.ch/publications

[MP04]   Ueli Maurer and Krzysztof Pietrzak. Composition of random systems: When two weak make one strong. In *Theory of Cryptograpy — TCC '04*, volume 2951 of *LNCS*, pages 410–427. Springer-Verlag, 2004.

[MPR06]   Ueli Maurer, Krzysztof Pietrzak, and Renato Renner. Indistinguishability amplification, 2006. Manuscript.

[MT05]   Kazuhiko Minematsu and Yukiyasu Tsunoo. Hybrid symmetric encryption using known-plaintext attack-secure components. In *ICISC '05, LNCS*. Springer-Verlag, 2005.

[Mye04]   Steven Myers. Black-box composition does not imply adaptive security. In *Advances in Cryptology — EUROCRYPT '04*, volume 3027 of *LNCS*, pages 189–206. Springer-Verlag, 2004.

[NR99]   Moni Naor and Omer Reingold. On the construction of pseudorandom permutations: Luby-Rackoff revisited. *J. Cryptology*, 12(1):29–66, 1999.

[NR02]   Moni Naor and Omer Reingold. Constructing pseudo-random permutations with a prescribed structure. *J. Cryptology*, 15(2):97–102, 2002.

[Pat04]   Jacques Patarin. Security of random feistel schemes with 5 or more rounds. In *Advances of Cryptology — CRYPTO '04*, volume 3152 of *LNCS*, pages 106–122. Springer-Verlag, 2004.

[Pie90]   Josef Pieprzyk. How to construct pseudorandom permutations from single pseudorandom functions. In *Advances in Cryptology — EUROCRYPT '90*, volume 537 of *LNCS*, pages 140–150. Springer-Verlag, 1990.

[Pie05]   Krzysztof Pietrzak. Composition does not imply adaptive security. In *Advances in Cryptology — CRYPTO '05*, volume 3621 of *LNCS*, pages 55–65. Springer-Verlag, 2005.

[Pie06]   Krzysztof Pietrzak. Composition implies adaptive security in minicrypt. In *Advances in Cryptology — EUROCRYPT '06, LNCS*. Springer-Verlag, 2006.

[Ple05]   Patrick Pletscher. Adaptive security of composition, 2005. Semester Thesis. www.pletscher.org/eth/minor/adapt_sec.pdf

[RR00]   Zulfikar Ramzan and Leonid Reyzin. On the round security of symmetric-key cryptographic primitives. In *Advances in Cryptology — CRYPTO '00*, volume 1880 of *LNCS*, pages 376–393. Springer-Verlag, 2000.

The Security of Triple Encryption and a Framework for Code-Based Game-Playing Proofs

Mihir Bellare[1] and Phillip Rogaway[2]

[1] Dept. of Computer Science & Engineering University of California at San Diego, 9500 Gilman Drive, La Jolla, California 92093 USA

[2] Dept. of Computer Science, University of California, Davis, California 95616, USA

Abstract. We show that, in the ideal-cipher model, triple encryption (the cascade of three independently-keyed blockciphers) is more secure than single or double encryption, thereby resolving a long-standing open problem. Our result demonstrates that for DES parameters (56-bit keys and 64-bit plaintexts) an adversary's maximal advantage against triple encryption is small until it asks about 2^{78} queries. Our proof uses code-based game-playing in an integral way, and is facilitated by a framework for such proofs that we provide.

1 Introduction

TRIPLE ENCRYPTION. Given a blockcipher $E \colon \{0,1\}^k \times \{0,1\}^n \to \{0,1\}^n$ with inverse D consider blockciphers $\mathsf{Cascade}_E^{\mathrm{eee}}(K_0 K_1 K_2, X) = E_{K_2}(E_{K_1}(E_{K_0}(X)))$ and $\mathsf{Cascade}_E^{\mathrm{ede}}(K_0 K_1 K_2, X) = E_{K_2}(D_{K_1}(E_{K_0}(X)))$. Our results are the same for both constructions. Following [14,6,9], we model E as a family of random permutations, one for each key, and we provide the adversary with oracle access to the blockcipher $E(\cdot, \cdot)$ and its inverse $E^{-1}(\cdot, \cdot)$. Given such oracles, the adversary is asked to distinguish between (a) $\mathsf{Cascade}_E^{\mathrm{eee}}(K_0 K_1 K_2, \cdot)$ and its inverse, for a random key $K_0 K_1 K_2$, and (b) a random permutation on n bits and its inverse. We show that the adversary's advantage in making this determination, $\mathbf{Adv}_{k,n}^{\mathrm{eee}}(q)$, remains small until it asks about $q = 2^{k+0.5 \min\{k,n\}}$ queries (the actual expression is more complex). The bound we get is plotted as the rightmost curve of Fig. 1 for DES parameters $k = 56$ and $n = 64$. In this case an adversary must ask more than $2^{78.5}$ queries to get advantage 0.5. Also plotted are the security curves for single and double encryption, where the adversary must ask 2^{55} and $2^{55.5}$ queries to get advantage 0.5. For a blockcipher with $k = n = 64$, the adversary must ask more than 2^{89} queries to get advantage 0.5. As there are matching attacks and security bounds for single and double encryption [4,1] our result proves that, in the ideal-cipher model, triple encryption is much more secure than single or double encryption.

As background for the above, note that the security of the cascade construction, where two or more independently keyed blockciphers are composed with one another, is a long-standing open problem [4,12]. Even and Goldreich refer to it as a "critical question" in cryptography [5, p. 109]. They showed that the

S. Vaudenay (Ed.): EUROCRYPT 2006, LNCS 4004, pp. 409–426, 2006.

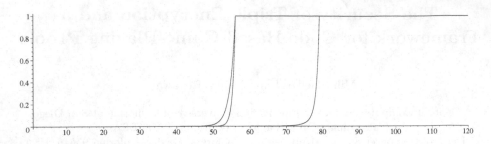

Fig. 1. Upper bound on adversarial advantage (proven security) verses $\log_2 q$ (where q=number of queries) for the cascade construction, assuming key length $k = 56$ and block length $n = 64$. Single encryption is the leftmost curve, double encryption is the middle curve [1], and triple encryption in the rightmost curve, as given by Theorem 1.

cascade of ciphers is at least as strong as the *weakest* cipher in the chain [5], while Maurer and Massey showed that, in a weaker attack model, it is at least as strong as the *first* cipher in the chain [11]. Aiello, Bellare, Di Crescenzo, and Venkatesan [1] prove the security of double-encryption (the two-stage cascade) in the same model as we use in this paper, showing that the maximal adversary advantage in q queries is $q^2/2^{2k}$. The meet-in-the-middle attack [4] implies that this is the best possible. So the adversary's advantage for double-encryption is only small until $q \approx 2^k$, just as for single encryption (although it grows as a slower rate). Thus triple encryption is the shortest potentially "good" cascade. And, indeed, triple DES is the cascade that is widely standardized and used [13].

The best published attack on three-key triple-encryption is due to Lucks [10]. He does not work out an explicit lower bound for $\mathbf{Adv}_{k,n}^{eee}(q)$, but in the case of triple-DES, the adversary's advantage becomes large at around $q = 2^{90}$ queries (using enormous time and memory, too). We prove security to about 2^{78} queries, so there is no contradiction.

As for the cascade of $\ell \geq 4$ blockciphers, the maximal advantage in our attack model is no worse than it is for triple encryption, so our result proves that cascade "works" for all $\ell \geq 3$. It is open if security increases with increasing ℓ.

GAME-PLAYING FRAMEWORK. Our proof for triple-encryption uses game-playing in an integral way, first to recast the advantage we wish to bound via a simpler game, and later to analyze that game via others. Ultimately one is left with a game where conventional probabilistic reasoning can be applied.

What constitutes a game-playing proof is a matter of perspective. To some, a game-playing proof in cryptography is any proof where one conceptualizes the adversary's interaction with its environment as a kind of game, the proof proceeding by constructing a "chain" of such games. Viewed in this way, game-playing proofs have their origin in the earliest hybrid arguments, which began with Goldwasser and Micali [7] and Yao [16]. Bellare and Goldwasser [2] provide an early example of the use of a game-chain to prove security of a construction that uses multiple different cryptographic primitives.

In our treatment, games are code (ie, programs), not abstract environments; as we develop it, game-playing centers around making disciplined transformations to code. This approach begins with Kilian and Rogaway [9], and was used by Rogaway in many subsequent works. The framework of Section 2 develops this approach, and in particular our Fundamental Lemma (Lemma 1) is about the probability that an adversary can distinguish between games (programs) that differ in a certain syntactic way.

Shoup has independently and contemporaneously prepared a manuscript on game playing [15], advocating the use of game chains to make proofs more accessible. Shoup has often used game-playing over the years. His approach is not code-based. Shoup's [15, Lemma 1] functions like our Fundamental Lemma, but the former is cast in terms of conditional probabilities while the latter talks of programs that differ only after the setting of a flag *bad*.

Following the web distribution of this paper, Halevi argues for the creation of an automated tool to help write and verify game-based proofs [8]. We agree. The possibility for such tools has always been one of our motivations, and one of the reasons we focus on code-based games.

Our broader paper [3] contains further illustrations of game-based proofs (the PRP/PRF Switching Lemma, the CBC MAC, and OAEP) and a discussion of general techniques for code-based game-playing.

2   The Game-Playing Framework

Games are programs, written in pseudocode or in some formalized programming language. We describe some elements of the language we use. The semantics of a boolean variable, which we will also call a *flag*, is that once true it stays true. A *random-assignment* statement has the form $s \xleftarrow{\$} S$ where S is a finite set. This is the only source of randomness in programs. A game consists of an *initialization procedure* (Initialize), a *finalization procedure* (Finalize), and named *oracles* (each a procedure). The adversary, which we also regard as code, makes calls to the oracles, passing in values from some finite domain associated to each oracle. The initialization or finalization procedures may be absent, and often are, and there may be any number of oracles, including none. All variables in a game are global, and they are not visible to the adversary.

We can *run* a game G with an adversary A. To begin, variables are given initial values. Integer variables are initialized to 0; boolean variables are initialized to false; string variables are initialized to the empty string ε; set variables are initialized to the empty set $\emptyset$; and array variables hold the value undefined at every point. These conventions often enable omitting explicit initialization code. When used in a boolean expression, undefined values are regarded as false.

The Initialize procedure is the first to execute, possibly producing an output *inp*. This is provided as input to the Adversary procedure A, which now runs. The adversary code can make oracle queries via statements of the form $y \leftarrow P(\cdots)$ for any oracle P that has been defined in the game. The result is to assign to y the value returned by the procedure call. We assume that the game and adversary match syntactically, meaning that all the oracle calls made by the adversary are to oracles specified in the game, and with arguments that match in quantity and type. The semantics of a call is call-by-value; the only way for an oracle to return a value to the adversary is via a return statement. When adversary A halts, possibly with some *adversary output*, we call Finalize, providing it any such output. The Finalize procedure returns a string that we call the *game output*. If we omit specifying Initialize or Finalize, or their return-statements, it means that the procedure returns its input. The game output and adversary output are often the same, because Finalize (or a return-statement for it) is unspecified.

The adversary and game outputs can be regarded as random variables. We write $\Pr[A^G \Rightarrow 1]$ for the probability that the adversary output is 1 when we run game G with adversary A, and $\Pr[G^A \Rightarrow 1]$ for the probability that the game output is 1 when we run game G with adversary A.

ADVANTAGES. If G and H are games and A is an adversary, let $\mathbf{Adv}(A^G, A^H) = \Pr[A^G \Rightarrow 1] - \Pr[A^H \Rightarrow 1]$ and $\mathbf{Adv}(G^A, H^A) = \Pr[G^A \Rightarrow 1] - \Pr[H^A \Rightarrow 1]$. These represent the advantage of the adversary in distinguishing the games, the first measured via adversary output and the second via game output. We refer to the first as the *adversarial advantage* and the second as the *game advantage*. We say that G, H are *adversarially indistinguishable* if for any adversary A it is the case that $\mathbf{Adv}(A^G, A^H) = 0$, and *equivalent* if for any adversary A it is the case that $\mathbf{Adv}(G^A, H^A) = 0$. We will often use the fact that

$$\mathbf{Adv}(A^G, A^I) = \mathbf{Adv}(A^G, A^H) + \mathbf{Adv}(A^H, A^I) \tag{1}$$
$$\mathbf{Adv}(G^A, I^A) = \mathbf{Adv}(G^A, H^A) + \mathbf{Adv}(H^A, I^A) \tag{2}$$

for any games G, H, I and any adversary A. These follow simply from the fact that $(a - b) + (b - c) = a - c$. These will be referred to as the *triangle equalities*.

THE FUNDAMENTAL LEMMA. Let G and H be games and let *bad* be a flag that occurs in both of them. Then we say that G and H are *identical-until-bad* if their code is the same except that there might be places where G has a statement $\underline{bad \leftarrow \text{true}, S}$ while game H has a corresponding statement $\underline{bad \leftarrow \text{true}, T}$ for some T that may be different from S. (One could also say that G and H are are identical-until-*bad* if one has the statement $\underline{\text{if } bad \text{ then } S}$ where the other has the empty statement, for this can be rewritten in the form above.) The identical-until-*bad* predicate is an equivalence relation.

We write $\Pr[A^G$ sets *bad*$]$ or $\Pr[G^A$ sets *bad*$]$ to refer to the probability that the flag *bad* is true at the end of the execution of the adversary A with game G (that is, when Finalize terminates). The fundamental lemma says that the advantage that an adversary can obtain in distinguishing a pair of identical-until-*bad* games is at most the probability that its execution sets *bad* in one of them.

Lemma 1. [Fundamental lemma of game-playing] *Let G and H be identical-until-bad games and let A be an adversary. Then*

$$\mathbf{Adv}(A^G, A^H) \leq \Pr[A^G \text{ sets } bad] \text{ and } \mathbf{Adv}(G^A, H^A) \leq \Pr[G^A \text{ sets } bad].$$

More generally, if G, H, and I are identical-until-bad games then

$$\left|\mathbf{Adv}(A^G, A^H)\right| \leq \Pr[A^I \text{ sets } bad] \text{ and } \left|\mathbf{Adv}(G^A, H^A)\right| \leq \Pr[I^A \text{ sets } bad]. \quad \blacksquare$$

One of the most common manipulations of games along a game chain is to change what happens after *bad* gets set to true. Any modification following the setting of *bad* leaves unchanged the probability of setting *bad*.

Proposition 1. [After *bad* is set, nothing matters] *Let G and H be identical-until-bad, and A an adversary. Then* $\Pr[G^A \text{ sets } bad] = \Pr[H^A \text{ sets } bad]$. $\quad \blacksquare$

3   The Security of Three-Key Triple-Encryption

DEFINITIONS. Let $E\colon \{0,1\}^k \times \{0,1\}^n \to \{0,1\}^n$ be a blockcipher with key length k and block length n. For $K \in \{0,1\}^k$ and $X \in \{0,1\}^n$ let $E_K(X) = E(K, X)$. Let $E^{-1}\colon \{0,1\}^k \times \{0,1\}^n \to \{0,1\}^n$ be the blockcipher that is the inverse of E. We associate to E two blockciphers formed by composition; denoted $\mathsf{Cascade}_E^{\mathrm{eee}}, \mathsf{Cascade}_E^{\mathrm{ede}}\colon \{0,1\}^{3k} \times \{0,1\}^n \to \{0,1\}^n$, these were defined in Section 1. These blockciphers have key length $3k$ and block length n and are sometimes referred to as the *three-key* forms of triple encryption. We will call the two methods EEE and EDE, respectively. There is also a *two-key* variant of triple encryption, obtained by setting $K_0 = K_2$, but we do not investigate it since the method admits comparatively efficient attacks [12].

We will be working in the ideal-blockcipher model, as in works like [6,9,1]. Let $\mathrm{Bloc}(k,n)$ be the set of all blockciphers $E\colon \{0,1\}^k \times \{0,1\}^n \to \{0,1\}^n$. Thus $E \xleftarrow{\$} \mathrm{Bloc}(k,n)$ means that $E_K\colon \{0,1\}^n \to \{0,1\}^n$ is a random permutation on n-bit strings for each $K \in \{0,1\}^k$. We consider an adversary A that can make four types of oracle queries: $\boldsymbol{T}(X)$, $\boldsymbol{T}^{-1}(Y)$, $\boldsymbol{E}(K,X)$, and $\boldsymbol{E}^{-1}(K,Y)$, where $X, Y \in \{0,1\}^n$ and $K \in \{0,1\}^k$. (As for our syntax, $\boldsymbol{T}, \boldsymbol{T}^{-1}, \boldsymbol{E}, \boldsymbol{E}^{-1}$ are formal symbols, not specific functions.) The *advantage of A against EEE* and the *maximal advantage against EEE obtainable using q queries* are defined as

$$\mathbf{Adv}_{k,n}^{\mathrm{eee}}(A) = \mathbf{Adv}(A^{C_0}, A^{R_0}) \quad \text{and} \quad \mathbf{Adv}_{k,n}^{\mathrm{eee}}(q) = \max_A \left\{ \mathbf{Adv}_{k,n}^{\mathrm{eee}}(A) \right\}$$

where the games C_0, R_0 are shown in Fig. 2 and the maximum is over all adversaries A that make at most q oracle queries (that is, a total of q across all four oracles). The advantage of A measures its ability to tell whether $\boldsymbol{T}(\cdot)$ is a random permutation or is $\mathsf{Cascade}_E^{\mathrm{eee}}(K_0 K_1 K_2, \cdot)$ for $K_0 K_1 K_2$ chosen independently at random from $\{0,1\}^{3k}$ and where $\boldsymbol{E}$ realizes a random blockcipher and $\boldsymbol{T}^{-1}, \boldsymbol{E}^{-1}$ realize inverses of $\boldsymbol{T}, \boldsymbol{E}$, respectively.

Define the *query threshold* $\mathbf{QTh}_{1/2}^{\mathrm{eee}}(k,n)$ as the largest integer q for which $\mathbf{Adv}_{k,n}^{\mathrm{eee}}(q) \leq 1/2$. We will regard EEE as being secure up to $\mathbf{QTh}_{1/2}^{\mathrm{eee}}(k,n)$ queries. Let $\mathbf{Adv}_{k,n}^{\mathrm{ede}}(A), \mathbf{Adv}_{k,n}^{\mathrm{ede}}(q)$, and $\mathbf{QTh}_{1/2}^{\mathrm{ede}}(k,n)$ be defined in the analogous way.

RESULTS. We are now ready to state our result about the security of triple encryption.

Theorem 1. [Security of triple-encryption] *Let* $k, n \geq 2$. *Let* $\alpha = \max(2e2^{k-n}, 2n+k)$. *Then*

$$\mathbf{Adv}_{k,n}^{\mathrm{eee}}(q) \leq 4\alpha \frac{q^2}{2^{3k}} + 10.7 \left(\frac{q}{2^{k+n/2}}\right)^{2/3} + \frac{12}{2^k} . \qquad\blacksquare\ (3)$$

We display the result graphically in Fig. 1 for DES parameters $k = 56$ and $n = 64$. Our bound implies that $\mathbf{QTh}_{1/2}^{\mathrm{eee}}(k, n)$ is, very roughly, about $2^{k+\min(k,n)/2}$, meaning that EEE is secure up to this many queries.

For EDE the result is the same, meaning that $\mathbf{Adv}_{k,n}^{\mathrm{ede}}(q)$ is also bounded by the quantity on the right-hand-side of (3). This can be shown by mostly-notational modifications to the proof of Theorem 1.

4   Proof of Theorem 1

OVERVIEW. The first step in our proof reduces the problem of bounding the advantage of an adversary A against EEE to bounding certain quantities that relate to a different, *simplified* adversary B. By a simplified adversary we mean one that makes no $\boldsymbol{T}(\cdot)$, $\boldsymbol{T}^{-1}(\cdot)$ queries, meaning it only has oracles $\boldsymbol{E}(\cdot, \cdot)$ and $\boldsymbol{E}^{-1}(\cdot, \cdot)$. We will consider two games, both involving random, distinct keys K_0, K_1, K_2. In one game (R_3) E_{K_2} is random, while in the other (D_S), it is correlated to E_{K_0}, E_{K_1}. The quantities we will need to bound are the ability of our simplified adversary to either distinguish these games without extending a 2-chain, or to extend a 2-chain in one of the games, where what it means to extend a 2-chain is explained below. We will be able to provide these two bounds via two lemmas. The first considers a simplified game in which an adversary has only three permutation oracles, either all random or one correlated to the rest, and has to distinguish them without extending a 2-chain. The second bounds the probability that the adversary can extend a 2-chain in R_3.

CONVENTIONS. We begin with some conventions. An adversary A against EEE can make oracle queries $\boldsymbol{T}(X)$, $\boldsymbol{T}^{-1}(Y)$, $\boldsymbol{E}(K, X)$, or $\boldsymbol{E}^{-1}(K, Y)$ for any $X, Y \in \{0, 1\}^n$ and $K \in \{0, 1\}^k$. We will assume that any adversary against EEE is deterministic and never makes a *redundant* query. A query is redundant if it has been made before; a query $\boldsymbol{T}^{-1}(Y)$ is redundant if A has previously received Y in answer to a query $\boldsymbol{T}(X)$; a query $\boldsymbol{T}(X)$ is redundant if A has previously received X in answer to a query $\boldsymbol{T}^{-1}(Y)$; a query $\boldsymbol{E}^{-1}(K, Y)$ is redundant if A has previously received Y in answer to a query $\boldsymbol{E}(K, X)$; a query $\boldsymbol{E}(K, X)$ is redundant if A has previously received X in answer to a query $\boldsymbol{E}^{-1}(K, Y)$. Assuming A to be deterministic and not to ask redundant queries is without loss of generality in the sense that for any A that asks q queries there is an A' asking at most q queries that satisfies these assumptions and achieves the same advantage as A. Our general conventions about games imply that A never asks a query with arguments outside of the intended domain, meaning $\{0, 1\}^k$ for keys and $\{0, 1\}^n$ for messages.

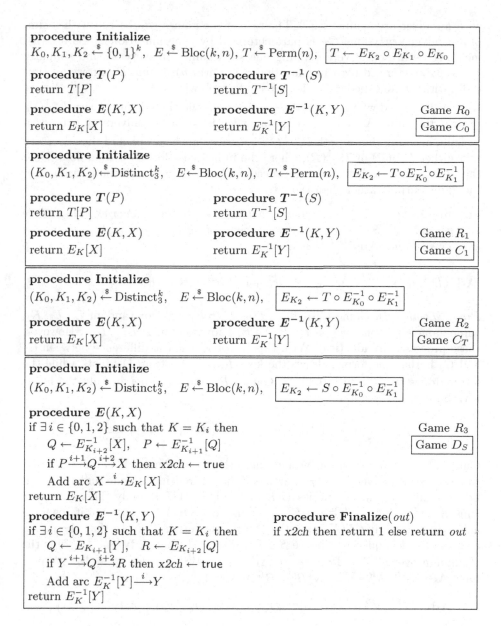

Fig. 2. The C_X or D_X games include the boxed statements while the R_i games do not

4.1 Reduction to Simplified Adversary

Consider the games in Fig. 2. The R-games (where R stands for random) omit the boxed assignment statements while the C-games and D-game include them. Distinct$_3^k$ denotes the set of all triples $(K_0, K_1, K_2) \in (\{0,1\}^k)^3$ such that $K_0 \neq K_1$ and $K_1 \neq K_2$ and $K_0 \neq K_2$. Games R_0, R_1, C_0, C_1 will be run with an adversary against EEE. The rest of the games will be run with a simplified adversary. Game C_T is parameterized by a permutation $T \in \text{Perm}(n)$, meaning we are effectively defining one such game for every T, and similarly D_S is parameterized by a permutation $S \in \text{Perm}(n)$. Game D_S grows an (initially without edges) edge-labeled directed graph with vertex set $\{0,1\}^n$. An arc $X \xrightarrow{i} Y$ is created when a query $\mathbf{E}_{K_i}(X)$ returns the value Y or a query $E_{K_i}^{-1}(Y)$ returns the value X. The boolean flag x2ch is set if the adversary extends a 2-chain, meaning that a path $P \xrightarrow{i+1} Q \xrightarrow{i+2} R$ exists in the graph and the adversary asks either $E_{K_i}(R)$ or $E_{K_i}^{-1}(P)$, where the indicated addition is modulo 3. Note that D_S has an explicit Finalize procedure, indicating we will be interested in the game output rather than the adversary output.

Lemma 2. *Let A be an adversary that makes at most q queries. Then there is a permutation $S \in \text{Perm}(n)$ and a simplified adversary B making at most q queries such that $\mathbf{Adv}_{k,n}^{\text{eee}}(A)$ is at most*

$$\mathbf{Adv}(D_S^B, R_3^B) + \Pr\left[D_S^B \text{ sets x2ch}\right] + \Pr\left[R_3^B \text{ sets x2ch}\right] + \frac{6}{2^k} \ . \qquad \blacksquare$$

Proof (Lemma 2). Game C_0 defines T as $E_2 \circ E_1 \circ E_0$ for random E_0, E_1, E_2, while game C_1 defines E_2 as $T \circ E_{K_0}^{-1} \circ E_{K_1}^{-1}$ for random T, E_{K_0}, E_{K_1}. However, these processes are identical. With this factored out, the difference between C_1 and C_0 is that the former draws the keys K_0, K_1, K_2 from Distinct_3^k while the latter draws them from $(\{0,1\}^k)^3$. Games R_1 and R_0 differ in only the latter way. So using (1) we have

$$\mathbf{Adv}_{k,n}^{\text{eee}}(A) = \mathbf{Adv}(A^{C_0}, A^{R_0}) \le \mathbf{Adv}(A^{C_1}, A^{R_1}) + \frac{6}{2^k} \ .$$

Game C_T is parameterized by a permutation $T \in \text{Perm}(n)$. For any such T we consider an adversary A_T that has T hardwired in its code and is simplified, meaning can make queries $\mathbf{E}(K, X)$ and $\mathbf{E}^{-1}(K, Y)$ only. This adversary runs A, answering the latter's $\mathbf{E}(K, X)$ and $\mathbf{E}^{-1}(K, Y)$ queries via its own oracles, and answering $\mathbf{T}(X)$ and $\mathbf{T}^{-1}(Y)$ queries using T. Note that A_T makes at most q oracle queries. Choose $S \in \text{Perm}(n)$ such that $\mathbf{Adv}(A_S^{C_S}, A_S^{R_2})$ is the maximum over all $T \in \text{Perm}(n)$ of $\mathbf{Adv}(A_T^{C_T}, A_T^{R_2})$ and let $B = A_S$. We now have $\mathbf{Adv}(A^{C_1}, A^{R_1}) \le \mathbf{Adv}(B^{C_S}, B^{R_2})$. Now by (2) we have

$$\mathbf{Adv}(B^{C_S}, B^{R_2}) \le \mathbf{Adv}(C_S^B, D_S^B) + \mathbf{Adv}(D_S^B, R_3^B) + \mathbf{Adv}(R_3^B, R_2^B) \ .$$

Game C_S (resp. game R_2) can be easily transformed into an equivalent game such that this game and game D_S (resp. R_3) are identical-until-x2ch, so by the Fundamental Lemma we have $\mathbf{Adv}(C_S^B, D_S^B) \le \Pr[D_S^B \text{ sets x2ch}]$ and $\mathbf{Adv}(R_3^B, R_2^B) \le \Pr[R_3^B \text{ sets x2ch}]$. Putting all this together completes the lemma's proof. $\quad\blacksquare$

Letting $p = \Pr\left[R_3^B \text{ sets x2ch}\right]$, we now need to bound

$$\mathbf{Adv}(D_S^B, R_3^B) + (\Pr\left[D_S^B \text{ sets x2ch}\right] - p) + 2p \ . \tag{4}$$

We will be able to bound the first two terms by bounding the advantages of a pair B_1, B_2 of adversaries, related to B, in distinguishing between a pair of games that involve only three permutation oracles, the first two random, and the third either random or correlated to the first two. We will bound p separately via a combinatorial argument. We now state the lemmas we need, conclude the proof of Theorem 1 using them in Section 4.4, and then return to provide the proofs of the two lemmas.

4.2   Pseudorandomness of Three Correlated Permutations

We posit a new problem. Consider games G and H defined in Fig. 3. Game G grows an edge-labeled graph, which we shall describe shortly. An adversary may make queries $\Pi(i, X)$ or $\Pi^{-1}(i, Y)$ where $i \in \{0, 1, 2\}$ and $X, Y \in \{0, 1\}^n$. The oracles realize three permutations and their inverses, the function realized by $\Pi^{-1}(i, \cdot)$ being the inverse of that realized by $\Pi(i, \cdot)$. In both games permutations π_0, π_1 underlying $\Pi(0, \cdot)$ and $\Pi(1, \cdot)$ are uniform and independent. In game G the permutation π_2 underlying $\Pi(2, \cdot)$ is also uniform and independent of π_0 and π_1, but in game H it is equal to $\pi_1^{-1} \circ \pi_0^{-1}$.

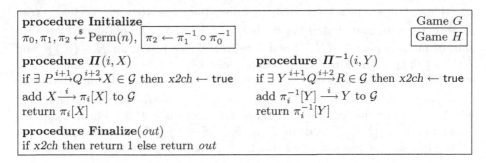

Fig. 3. Game H includes the boxed statement, game G does not

Notice that it is easy for an adversary to distinguish between games G and H by making queries that form a "chain" of length three: for any $P \in \{0, 1\}^n$, let the adversary ask and be given $Q \leftarrow \pi_0(P)$, then $R \leftarrow \pi_1(Q)$, then $P' \leftarrow \pi_2(R)$, and then have the adversary output 1 if $P = P'$ (a "triangle" has been found) or 0 if $P \neq P'$ (the "three-chain" is not in fact a triangle). What we will establish is that, apart from such behavior—extending a known "2-chain"—the adversary is not able to gain much advantage. To capture this, as the adversary A makes its queries and gets replies, the games grow an (initially without edges) edge-labeled directed graph $\mathcal{G}$ with vertex set. An arc $X \xrightarrow{i} Y$ is created when a query $\Pi(i, X)$ returns the value Y or a query $\Pi^{-1}(i, Y)$ returns the value X. The boolean flag x2ch is set in the games if the adversary extends a 2-chain, meaning that a path $P \xrightarrow{i+1} Q \xrightarrow{i+2} R$ exists in the graph and the adversary asks either $\Pi(i, R)$ or $\Pi^{-1}(i, P)$, where the indicated addition is modulo 3. We will be interested in the game outputs rather than the adversary outputs. Again using a game-based proof, we prove the following in Section 4.5:

procedure $E(K, X)$ procedure $E^{-1}(K, Y)$ Game L

return $E_K[X] \stackrel{\$}{\leftarrow} \overline{\mathrm{image}}(E_K)$ $E_K^{-1}[Y] \stackrel{\$}{\leftarrow} \overline{\mathrm{domain}}(E_K)$

procedure Finalize

$K_0, K_1, K_2 \stackrel{\$}{\leftarrow} \{0,1\}^k$

if $(\exists P) \; [E_{K_2}[E_{K_1}[E_{K_0}[P]]]]$ then $bad \leftarrow$ true

Fig. 4. Game L captures improbability of making three chains

Lemma 3. *If* $\Pr\left[B^G \text{ makes} \geq h \text{ oracle queries}\right] \leq \delta$ *then* $\mathbf{Adv}(H^B, G^B) \leq 2.5\, h^2/2^n + \delta.$ ∎

We remark that the lemma makes no (explicit) assumption about the probability that B^H makes h or more oracle queries.

4.3 The Improbability of Forming a 3-Chain

Consider an adversary B that can make $\boldsymbol{E}(K, X)$ or $\boldsymbol{E}^{-1}(K, Y)$ queries. Game L of Fig. 3 implements the oracles as a random blockcipher and its inverse, respectively, but samples these lazily, defining points as they are needed. Write $X \xrightarrow{K} Y$ to mean that that B has made query $\boldsymbol{E}(K, X)$ and obtained Y as a result, or made query $\boldsymbol{E}^{-1}(K, Y)$ and obtained X as a result, for $K \in \{0,1\}^k$ and $X, Y \in \{0,1\}^n$. The Finalize procedure picks keys K_0, K_1, K_2 at random, and sets bad if the adversary's queries have formed a three chain, meaning that there exist points $P, Q, R, S \in \{0,1\}^n$ such that $P \xrightarrow{K_0} Q \xrightarrow{K_1} R \xrightarrow{K_2} S$: the conditional which is the last line of Finalize means that there is a P for which $E_{K_0}[P]$ is defined and $E_{K_1}[E_{K_0}[P]]$ is defined and $E_{K_2}[E_{K_1}[E_{K_0}[P]]]$ is defined. Our next lemma bounds the probability of this happening. The proof is in Section 4.6.

Lemma 4. *Let* $k, n \geq 1$. *Let* B *be an adversary that asks at most* q *queries. Let* $\alpha = \max(2e\, 2^{k-n}, 2n + k)$. *Then* $\Pr[B^L \text{ sets } bad] < 2\alpha\, q^2/2^{3k}.$ ∎

4.4 Putting Together the Pieces to Conclude Theorem 1

Let B be a simplified adversary and $S \in \mathrm{Perm}(n)$ a permutation. We associate to B, S a pair of adversaries $B_{S,1}$ and $B_{S,2}$ that make $\boldsymbol{\Pi}(i, X)$ or $\boldsymbol{\Pi}^{-1}(i, Y)$ queries, where $i \in \{0, 1, 2\}$ and $X, Y \in \{0,1\}^n$, as follows. For $b \in \{1, 2\}$, adversary $B_{S,b}$ picks (K_0, K_1, K_2) at random from $\mathrm{Distinct}_3^k$ and picks E at random from $\mathrm{Bloc}(k, n)$. It then runs B, replying to its oracle queries as follows. If B makes a query $\boldsymbol{E}(K, X)$, adversary $B_{S,b}$ returns $E_K(X)$ if $K \notin \{K_0, K_1, K_2\}$; returns $\boldsymbol{\Pi}(i, X)$ if $K = K_i$ for $i \in \{0, 1\}$; and returns $S \circ \boldsymbol{\Pi}(2, X)$ if $K = K_2$. Similarly, if B makes a query $\boldsymbol{E}^{-1}(K, Y)$, adversary $B_{S,b}$ returns $E_K^{-1}(Y)$ if $K \notin \{K_0, K_1, K_2\}$; returns $\boldsymbol{\Pi}^{-1}(i, Y)$ if $K = K_i$ for $i \in \{0, 1\}$; and returns $\boldsymbol{\Pi}^{-1}(2, Y) \circ S^{-1}$ if $K = K_2$. Adversaries $B_{S,1}, B_{S,2}$ differ only in their output, the first always returning 0 and the second returning the output out of B.

Lemma 5. *Let B be a simplified adversary that makes at most q oracle queries, and let $S \in \mathrm{Perm}(n)$. Let $B_{S,1}, B_{S,2}$ be defined as above. Let $K = 2^k$. Then for $b \in \{1, 2\}$ and any $c > 0$, $\Pr[B_{S,b}^G$ makes $\geq 3cq/K$ oracle queries$] \leq 1/c$.* ∎

Proof (Lemma 5). The oracles B sees when it is run by $B_{S,b}$ are exactly a random block cipher and its inverse. (A random permutation composed with a fixed one is still random so the composition by S does not change anything.) Now let X be the random variable that is the number of queries by B that involve keys K_0, K_1, or K_2 in the experiment where we first run B with oracles E, E^{-1} for $E \xleftarrow{\$} \mathrm{Bloc}(k, n)$ and then pick $(K_0, K_1, K_2) \xleftarrow{\$} \mathrm{Distinct}_3^k$. Then the probability that $B_{S,b}^G$ makes $\geq 3cq/K$ oracle queries is exactly the probability that $X \geq 3cq/K$. Now assume wlog that B always makes exactly q distinct oracle queries rather than at most q. Then

$$\mathbf{E}[X] = q \cdot \left[1 - \left(1 - \frac{1}{K}\right)\left(1 - \frac{1}{K-1}\right)\left(1 - \frac{1}{K-2}\right) \right]$$

$$= q \cdot \left[1 - \frac{K-1}{K}\frac{K-2}{K-1}\frac{K-3}{K-2} \right] = q \cdot \left[1 - \frac{K-3}{K} \right] = \frac{3q}{K} .$$

We can conclude via Markov's inequality. ∎

Proof (Theorem 1). Let A be an adversary against EEE that makes at most q oracle queries. Let B be the simplified adversary, and S the permutation, given by Lemma 2, and let $p = \Pr\left[R_3^B \text{ sets x2ch}\right]$. Let $B_{S,1}, B_{S,2}$ be the adversaries associated to B as described above. Note that

$$\Pr[D_S^B \text{ sets x2ch}] = \Pr[H^{B_{S,1}} \Rightarrow 1] \quad \text{and} \quad \Pr[R_3^B \text{ sets x2ch}] = \Pr[G^{B_{S,1}} \Rightarrow 1]$$
$$\Pr[D_S^B \Rightarrow 1] = \Pr[H^{B_{S,2}} \Rightarrow 1] \quad \text{and} \quad \Pr[R_3^B \Rightarrow 1] = \Pr[G^{B_{S,2}} \Rightarrow 1] . \tag{5}$$

Combining (4) and (5) we have:

$$\mathbf{Adv}_{k,n}^{\mathrm{eee}}(A) \leq 2p + \mathbf{Adv}(H^{B_{S,1}}, G^{B_{S,1}}) + \mathbf{Adv}(H^{B_{S,2}}, G^{B_{S,2}}) + \frac{6}{2^k} . \tag{6}$$

Let $\alpha = \max(2e2^{k-n}, 2n + k)$ and let c be any positive real number. Since the probability that R_3^B extends a 2-chain is at most the probability that L^B forms a 3-chain we have $p \leq 3 \cdot 2^{-k} + \Pr[B^L \text{ sets } bad]$. (The extra term is because L picks the keys K_0, K_1, K_2 independently at random while R_3 picks them from $\mathrm{Distinct}_3^k$.) Applying Lemma 4 we get $p \leq 3 \cdot 2^{-k} + 2\alpha q^2 \cdot 2^{-3k}$. Applying Lemma 3 in conjunction with Lemma 5 we have

$$\mathbf{Adv}(H^{B_{S,b}}, G^{B_{S,b}}) \leq \frac{2.5}{2^n}\left(\frac{3cq}{2^k}\right)^2 + \frac{1}{c}$$

for both $b = 1$ and $b = 2$. Putting everything together we have

$$\mathbf{Adv}_{k,n}^{\mathrm{eee}}(A) \leq 2\left(\frac{3}{2^k} + 2\alpha\frac{q^2}{2^{3k}}\right) + \frac{5}{2^n}\left(\frac{3cq}{2^k}\right)^2 + \frac{2}{c} + \frac{6}{2^k} .$$

Now, since the above is true for any $c > 0$, we pick a particular one that minimizes the function $f(c) = 45\, c^2 q^2\, 2^{-n-2k} + 2c^{-1}$. The derivative is $f'(c) = 90\, cq^2\, 2^{-n-2k} - 2c^{-2}$, and the only real root of the equation $f'(c) = 0$ is $c = (2^{n+2k}/45q^2)^{1/3}$, for which we have $f(c) = 3(45q^2/2^{n+2k})^{1/3}$. Plugging this into the above yields (3) and concludes the proof of Theorem 1.   ∎

4.5   Proof of Lemma 3

We prove Lemma 3 as a corollary of:

Lemma 6. *If A asks at most q queries then* $\left|\mathbf{Adv}(G^A, H^A)\right| \leq 2.5\, q^2/2^n$.   ∎

Proof (Lemma 3). We construct an adversary A that has the same oracles as B. Adversary A runs B, answering B's oracle queries via its own oracles. It also keeps track of the number of oracle queries that B makes. If this number hits h, it stops and outputs 1; else it outputs whatever B outputs. Then we note that $\Pr[H^B \Rightarrow 1] \leq \Pr[H^A \Rightarrow 1]$ and $\Pr[G^A \Rightarrow 1] \leq \Pr[G^B \Rightarrow 1] + \delta$. Thus we have

$$\begin{aligned}
\mathbf{Adv}(H^B, G^B) &= \Pr[H^B \Rightarrow 1] - \Pr[G^B \Rightarrow 1] \\
&\leq \Pr[H^A \Rightarrow 1] - \left(\Pr[G^A \Rightarrow 1] - \delta\right) \;=\; \mathbf{Adv}(H^A, G^A) + \delta\,.
\end{aligned}$$

As A makes $\leq h$ queries, conclude by applying Lemma 6 to A with $q = h$.   ∎

Proof (Lemma 6). We assume that the adversary A never repeats a query, never asks a query $\boldsymbol{\Pi}^{-1}(i, Y)$ having asked some $\boldsymbol{\Pi}(i, X)$ that returned Y, and never asks a query $\boldsymbol{\Pi}(i, X)$ having asked some $\boldsymbol{\Pi}^{-1}(i, Y)$ that returned X. Call an adversary *valid* if it never extends a two-chain.

We begin by noting that to bound A's advantage in distinguishing games G and H we may assume that A is valid. Why? Because for any adversary A^* making at most q_0 queries there exists a valid A that makes at most q_0 queries and the advantage of A is at least that of A^*. Adversary A runs A^*, answering A^*'s oracle queries via its own oracles, but at any point that A^* would extend a two chain, adversary A simply halts and outputs 1. So now assuming A's validity, our task is to show that $\left|\mathbf{Adv}(A^{G_1}, A^{H_1})\right| \leq 2.5\, q^2/2^n$ where the games G_1, H_1 are shown in Fig. 5. We show that games G_1 and H_1 are close by showing that both are close to game G_3 (defined in the same figure). First, we claim that $\left|\mathbf{Adv}(A^{G_1}, A^{G_3})\right| \leq 0.5\, q^2/N$ where, here and in the rest of this proof, $N = 2^n$. Rewrite game G_1 to game $G_{1.5}$ (not shown) by lazily growing π_0, π_1, π_2, setting the flag *bad* whenever there is a collision; that is, game $G_{1.5}$ is identical to game G_2 except, after setting *bad* at line 211, set $Y \xleftarrow{\$} \overline{\text{image}}(\pi_i)$, and after setting *bad* at line 221, set $X \xleftarrow{\$} \overline{\text{domain}}(\pi_i)$. Then modify game $G_{1.5}$ to *not* re-sample after setting *bad*, obtaining game G_2. Now $\left|\mathbf{Adv}(A^{G_1}, A^{G_3})\right| = \left|\mathbf{Adv}(A^{G_{1.5}}, A^{G_3})\right| = \left|\mathbf{Adv}(A^{G_{1.5}}, A^{G_2})\right| \leq \Pr[A^{G_2} \text{ sets } bad]$. Then note that on the ith query the probability that *bad* will be set in game G_2 is at most $(i-1)/N$ since the size of domain(π_j) and image(π_j) will be at most $i-1$ for each $j \in \{0, 1, 2\}$. So

procedure **Initialize**		Game G_1
100 $\pi_0, \pi_1, \pi_2 \xleftarrow{\$} \mathrm{Perm}(n)$, $\boxed{\pi_2 \leftarrow \pi_1^{-1} \circ \pi_0^{-1}}$		$\boxed{\text{Game } H_1}$
procedure $\Pi(i, X)$	procedure $\Pi^{-1}(i, Y)$	
110 return $\pi_i[X]$	120 return $\pi_i^{-1}[Y]$	

procedure $\Pi(i, X)$	procedure $\Pi^{-1}(i, Y)$	Game G_2
210 $Y \xleftarrow{\$} \{0,1\}^n$	220 $X \xleftarrow{\$} \{0,1\}^n$	
211 if $Y \in \mathrm{image}(\pi_i)$ then $bad \leftarrow$ true	221 if $X \in \mathrm{domain}(\pi_i)$ then $bad \leftarrow$ true	
213 $\pi[X] \leftarrow Y$	223 $\pi[X] \leftarrow Y$	
214 return Y	224 return X	

procedure $\Pi(i, X)$	procedure $\Pi^{-1}(i, Y)$	Game G_3
310 return $Y \xleftarrow{\$} \{0,1\}^n$	320 return $X \xleftarrow{\$} \{0,1\}^n$	

procedure $\Pi(i, X)$	Game G_4
410 if $\exists\,(i, X, Y) \in \mathcal{C}$ then return Y	
411 $X_i \leftarrow X$	
412 $X_{i+1} \xleftarrow{\$} \{0,1\}^n$, if $X_{i+1} \in S_{i+1}$ then $bad \leftarrow$ true, $X_{i+1} \xleftarrow{\$} \{0,1\}^n \setminus S_{i+1}$	
413 $X_{i+2} \xleftarrow{\$} \{0,1\}^n$, if $X_{i+2} \in S_{i+2}$ then $bad \leftarrow$ true, $X_{i+2} \xleftarrow{\$} \{0,1\}^n \setminus S_{i+2}$	
414 $S_i \leftarrow S_i \cup \{X_i\}$, $S_{i+1} \leftarrow S_{i+1} \cup \{X_{i+1}\}$, $S_{i+2} \leftarrow S_{i+2} \cup \{X_{i+2}\}$	
415 $\mathcal{C} \leftarrow \mathcal{C} \cup \{(i, X_i, X_{i+1}), (i+1, X_{i+1}, X_{i+2}), (i+2, X_{i+2}, X_i)\}$	
416 return X_{i+1}	
procedure $\Pi^{-1}(i, Y)$	
420 if $\exists\,(i, X, Y) \in \mathcal{C}$ then return X	
421 $X_{i+1} \leftarrow Y$	
422 $X_i \xleftarrow{\$} \{0,1\}^n$, if $X_i \in S_{i+1}$ then $bad \leftarrow$ true, $X_i \xleftarrow{\$} \{0,1\}^n \setminus S_{i+1}$	
423 $X_{i+2} \xleftarrow{\$} \{0,1\}^n$, if $X_{i+2} \in S_{i+2}$ then $bad \leftarrow$ true, $X_{i+2} \xleftarrow{\$} \{0,1\}^n \setminus S_{i+2}$	
424 $S_i \leftarrow S_i \cup \{X_i\}$, $S_{i+1} \leftarrow S_{i+1} \cup \{X_{i+1}\}$, $S_{i+2} \leftarrow S_{i+2} \cup \{X_{i+2}\}$	
425 $\mathcal{C} \leftarrow \mathcal{C} \cup \{(i, X_i, X_{i+1}), (i+1, X_{i+1}, X_{i+2}), (i+2, X_{i+2}, X_i)\}$	
426 return X_i	

Fig. 5. Games for bounding the probability of distinguishing (π_0, π_1, π_2) and $(\pi_0, \pi_1, \pi_1^{-1} \circ \pi_0^{-1})$ by an adversary that never extends a two-chain

over all q queries, the probability that bad ever gets set in game G_2 is at most $0.5q(q-1)/N \leq 0.5q^2/N$. To establish Lemma 6 we now claim that

$$\left| \mathbf{Adv}(A^{H_1}, A^{G_3}) \right| \leq 2\,q^2/N \,. \tag{7}$$

First rewrite game H_1 as game G_4 (again in Fig. 5). Addition ($+1$ and $+2$) is again understood to be modulo 3. Game G_4 uses a form of lazy sampling, but it is not maximally lazy; on each query, not only is its answer chosen, but answers for some related queries are chosen and stored. In particular, the game maintains a set $\mathcal{C}$ of *commitments*. Initially there are no commitments, but every time a query $\Pi(i, X)$ or $\Pi^{-1}(i, Y)$ is asked, one of two things happens: if a commitment has already been made specifying how to answer this query, we answer according to that commitment; else we not only answer the query asked, but commit ourselves to all of the queries in a "triangle" containing the queried point. In greater detail, $(i, X, Y) \in \mathcal{C}$ (for $i \in \{0, 1, 2\}$ and $X, Y \in \{0, 1\}^n$) means that it has already been decided that $\pi_i(X) = Y$, so a forward query $\Pi(i, X)$ will need to be answered by Y and a backward query $\Pi^{-1}(i, Y)$ will need to be answered by X. In effect, we grow permutations π_0, π_1, and π_2 but store their values in $\mathcal{C}$ and their domains in S_0, S_1, and S_2.

We claim that games H_1 and G_4 are adversarially indistinguishable even by an adversary that is *not* valid and asks all $6N$ possible queries. From this we know that $\Pr[A^{G_4} \Rightarrow 1] = \Pr[A^{H_1} \Rightarrow 1]$. To show this equivalence we claim that whether the queries are answered by game G_4 or by game H_1 the adversary gets the same view: any of $(N!)^2$ possible outcomes, each with probability $1/(N!)^2$, the answers correspond to a pair of permutations π_0, π_1 along with $\pi_2 = \pi_1^{-1} \circ \pi_0^{-1}$. This is obviously the case when playing game H_1; we must show it is so for game G_4. Note that sets S_0, S_1, S_2, and $\mathcal{C}$ begin with no points in them, then they grow to 1, 1, 1, and 3 points; then to 2, 2, 2, and 6 points; and so forth, until they have N, N, N, and $3N$ points. Not every query changes the sizes of these sets; it either leaves the sets unaltered or changes them as indicated. The first query that augments $\mathcal{C}$ extends the partial functions (π_0, π_1, π_2) in any of N^2 different ways, each with the same probability; the second query that augments $\mathcal{C}$ extends (π_0, π_1, π_2) in any of $(N - 1)^2$ different ways, each with the same probability; and so forth, until we have extended (π_0, π_1, π_2) in any of $(N!)^2$ different ways, each with the same probability. This establishes the claim.

Now let us go back to assuming that the adversary is valid. We make a change to game G_4 to arrive at game G_5, shown in Fig. 6. In the transition, we drop the first commitment from each group of three, since our assumptions about the adversary's behavior mean that these queries cannot be asked. We also drop the sequels to *bad* getting set at lines 412, 413, 422, and 423. More interestingly, in game G_5 we maintain a set of "poisoned" queries $\mathcal{P}$. As with game G_4, when the adversary asks $\Pi(i, X_i)$ we return a random X_{i+1}, and when the adversary asks $\Pi^{-1}(i, X_{i+1})$ we return a random X_i, and in either case we choose a random X_{i+2} and "complete the triangle" using this point. We don't expect the adversary to ask about X_{i+2}, and, what is more, his asking will cause problems. So we record the unlikely but problematic queries involving X_{i_2} in $\mathcal{P}$. If the adversary makes a poisoned query then we set *bad*. The changes we have made can only increase the probability that *bad* gets set: $\Pr[A^{G_4} \text{ sets } bad] \leq \Pr[A^{G_5} \text{ sets } bad]$.

We claim that game G_5 is adversarially indistinguishable from game G_3. Remember that our adversary is valid: it does not ask queries whose answers are trivially known and it does not ask to extend any 2-chain. Suppose first that the adversary asks a query whose answer has not been memoized in a commitment. Then for a forward query, we choose a uniform value X_{i+1} at line 514 and return it at line 519. Likewise for a backward query, we choose a uniform value X_i at line 524 and return it at line 529. So consider instead a query for which a commitment has been memoized. The code executes at lines 511–512 or lines 521–522. If the memoized query was poisoned—added to set $\mathcal{P}$ by an earlier execution of lines 518 or 528—then we return a random string (at line 511 or 521). If the

```
procedure Π(i, X)                                                    Game G₅
510   if ∃ (i, X, Y) ∈ C then
511       if (+1, i, X) ∈ P then bad ← true,  Y ←$ {0,1}ⁿ
512       return Y
513   Xᵢ ← X
514   X_{i+1} ←$ {0,1}ⁿ,  if X_{i+1} ∈ S_{i+1} then bad ← true
515   X_{i+2} ←$ {0,1}ⁿ,  if X_{i+2} ∈ S_{i+2} then bad ← true
516   Sᵢ ← Sᵢ ∪ {Xᵢ},  S_{i+1} ← S_{i+1} ∪ {X_{i+1}},  S_{i+2} ← S_{i+2} ∪ {X_{i+2}}
517   C ← C ∪ {(i+1, X_{i+1}, X_{i+2}), (i+2, X_{i+2}, Xᵢ)}
518   P ← P ∪ {(1, i+2, X_{i+2}), (-1, i+1, X_{i+2})}
519   return X_{i+1}

procedure Π⁻¹(i, Y)
520   if ∃ (i, X, Y) ∈ C then
521       if ∃ (-1, i, Y) ∈ P then bad ← true,  X ←$ {0,1}ⁿ
522       return X
523   X_{i+1} ← Y
524   Xᵢ ←$ {0,1}ⁿ,   if Xᵢ ∈ S_{i+1} then bad ← true
525   X_{i+2} ←$ {0,1}ⁿ, if X_{i+2} ∈ S_{i+2} then bad ← true
526   Sᵢ ← Sᵢ ∪ {Xᵢ},  S_{i+1} ← S_{i+1} ∪ {X_{i+1}},  S_{i+2} ← S_{i+2} ∪ {X_{i+2}}
527   C ← C ∪ {(i+1, X_{i+1}, X_{i+2}), (i+2, X_{i+2}, Xᵢ)}
528   P ← P ∪ {(1, i+2, X_{i+2}), (-1, i+1, X_{i+2})}
529   return Xᵢ
```

Fig. 6. Game G_5

memoized query was not poisoned, then we are extending a 1-chain, providing a value X_{i+2} that was selected uniformly from $\{0,1\}^n$ at an earlier execution of line 515 or 525, with this value not yet having influenced the run. Thus we return a uniform random value, independent of all oracle responses so far, and $\Pr[A^{G_5} \Rightarrow 1] = \Pr[A^{G_3} \Rightarrow 1]$.

Finally, we must bound the probability that bad gets set in game G_5. The probability that bad ever gets set at any of lines 514, 515, 524, or 525 is at most $2(1 + 2 + \cdots + (q-1))/N \leq q^2/N$. The probability that it gets set at lines 511 or 521 is at most $2(1 + 2 + \cdots + (q-1))/N$ because no information about the poisoned query is surfaced to the adversary. Overall we have that $\Pr[A^{G_5} \text{ sets } bad] \leq 2q^2/N$. Putting everything together we have (7) and the proof of the lemma is complete. ∎

4.6   Proof of Lemma 4

To prove this lemma we can assume without loss of generality that B is deterministic. For any particular blockcipher $E \in \text{Bloc}(k, n)$ we consider the game in which B is executed with oracles E, E^{-1}, which it queries, adaptively, until it halts. Note that there is no randomness involved in this game, since E is fixed and B is deterministic. Recall that $X \xrightarrow{K} Y$ means that B has either made query $E(K, X)$ and obtained Y as a result, or it has made query $E^{-1}(K, Y)$ and obtained X as a result, for $K \in \{0,1\}^k$ and $X, Y \in \{0,1\}^n$. Now we let

$$\mathsf{Ch}_3^{E,B} = \left| \{ (K_0, K_1, K_2, P) \; : \; \exists\, Q, R, S \, [\, P \xrightarrow{K_0} Q \xrightarrow{K_1} R \xrightarrow{K_2} S\,]\, \} \right| \; .$$

This is the number of *3-chains* created by B's queries. Here $K_0, K_1, K_2 \in \{0,1\}^k$ are keys, and $P, Q, R, S \in \{0,1\}^n$. As the notation indicates, $\mathsf{Ch}_3^{E,B}$ is a number that depends on E and B. Regarding it as a random variable over the choice of E we have the following lemma, from which Lemma 4 will follow.

Lemma 7. *Let* $\alpha = \max(2e2^{k-n}, 2n + k)$. *Then* $\mathbf{E}[\mathsf{Ch}_3^{E,B}] < 2\alpha \cdot q^2$, *the expectation over* $E \xleftarrow{\$} \mathrm{Bloc}(k, n)$. ∎

Proof (Lemma 4). Consider the following game L^E parameterized by a blockcipher $E \in \mathrm{Bloc}(k, n)$: adversary B is executed with oracles E, E^{-1} until it halts, then K_0, K_1, K_2 are chosen at random from $\{0,1\}^k$, and flag *bad* is set if there exist P, Q, R, S such that $P \xrightarrow{K_0} Q \xrightarrow{K_1} R \xrightarrow{K_2} S$. Let $p^{E,B} = \Pr[L_B^E \text{ sets } bad]$, the probability being over the random choices of K_0, K_1, K_2. Then for any $E \in \mathrm{Bloc}(k, n)$ we have

$$
\begin{aligned}
p^{E,B} &= \Pr\left[\, \exists\, P, Q, R, S \; : \; P \xrightarrow{K_0} Q \xrightarrow{K_1} R \xrightarrow{K_2} S \,\right] \\[4pt]
&= \frac{|\{\, (K_0, K_1, K_2) \; : \; \exists\, P, Q, R, S \; : \; P \xrightarrow{K_0} Q \xrightarrow{K_1} R \xrightarrow{K_2} S \,\}|}{2^{3k}} \\[4pt]
&\leq \frac{\sum_P |\{\, (K_0, K_1, K_2) \; : \; \exists\, Q, R, S \; : \; P \xrightarrow{K_0} Q \xrightarrow{K_1} R \xrightarrow{K_2} S \,\}|}{2^{3k}} = \frac{\mathsf{Ch}_3^{E,B}}{2^{3k}} \; .
\end{aligned}
$$

By Lemma 7 we have $\Pr[B^L \text{ sets } bad] = \mathbf{E}[p^{E,B}] \leq \mathbf{E}[\mathsf{Ch}_3^{E,B}] \cdot 2^{-3k} < 2\alpha\, q^2\, 2^{-3k}$ where $\alpha = \max(2e2^{k-n}, 2n + k)$ and the expectation is over $E \xleftarrow{\$} \mathrm{Bloc}(k, n)$. ∎

Towards the proof of Lemma 7, for $E \in \mathrm{Bloc}(k, n)$ and $Q, R \in \{0,1\}^n$ we let

$$\mathsf{Keys}^E(Q, R) = |\{\, K \; : \; E(K, Q) = R\,\}| \quad \text{and} \quad \mathsf{Keys}^E = \max_{Q,R}\{\mathsf{Keys}^E(Q, R)\} \; .$$

The first is the number of keys for which Q maps to R under E, and the second is the maximum value of $\mathsf{Keys}^E(Q, R)$ over all $Q, R \in \{0,1\}^n$. No adversary is involved in this definition; Keys^E is simply a number associated to a given blockcipher. Viewing it as a random variable over the choice of blockcipher we have the following.

Lemma 8. *Suppose* $\beta \geq 2e2^{k-n}$. *Then* $\Pr\left[\mathsf{Keys}^E \geq \beta\right] < 2^{2n+1-\beta}$, *where the probability is over* $E \xleftarrow{\$} \mathrm{Bloc}(k, n)$. ∎

Proof (Lemma 8). We claim that for any $Q, R \in \{0,1\}^n$

$$\Pr\left[\mathsf{Keys}^E(Q, R) \geq \beta\right] < 2^{1-\beta} \; . \tag{8}$$

The lemma follows via the union bound. We prove (8) using an occupancy-problem approach. Let $b = \lceil \beta \rceil$. Then

$$\Pr\left[\mathsf{Keys}^E(Q,R) \geq \beta\right] = \sum_{i=b}^{2^k} \binom{2^k}{i} \left(\frac{1}{2^n}\right)^i \left(1 - \frac{1}{2^n}\right)^{2^k - i}$$

$$\leq \sum_{i=b}^{2^k} \left(\frac{2^k e}{i}\right)^i \left(\frac{1}{2^n}\right)^i \leq \sum_{i=b}^{2^k} \left(\frac{2^k e}{2^n b}\right)^i .$$

Let $x = (e/b)2^{k-n}$. The assumption $\beta \geq 2e2^{k-n}$ gives $x \leq 1/2$. So the above is

$$= \sum_{i=b}^{2^k} x^i < x^b \cdot \sum_{i=0}^{\infty} x^i = \frac{x^b}{1-x} \leq \frac{2^{-b}}{1 - 1/2} = 2^{1-b} \leq 2^{1-\beta}$$

as desired. ∎

Proof (Lemma 7). For any $Q, R \in \{0,1\}^n$ we let

$$\mathsf{Ch}_2^{E,B}(R) = |\{\, (K_0, K_1, P) : \exists Q\, [P \xrightarrow{K_0} Q \xrightarrow{K_1} R]\,\}|$$

$$\mathsf{Ch}_1^{E,B}(Q) = |\{\, (K_0, P) : P \xrightarrow{K_0} Q\,\}|$$

$$\mathsf{Ch}_0^{E,B}(R) = |\{\, K_2 : \exists S\, [R \xrightarrow{K_2} S]\,\}| .$$

Then for any $E \in \mathrm{Bloc}(k,n)$ we have

$$\mathsf{Ch}_3^{E,B} = \sum_R \mathsf{Ch}_2^{E,B}(R) \cdot \mathsf{Ch}_0^{E,B}(R)$$

$$\leq \sum_R \left(\sum_Q \mathsf{Ch}_1^{E,B}(Q) \cdot \mathsf{Keys}^E(Q,R)\right) \cdot \mathsf{Ch}_0^{E,B}(R)$$

$$\leq \sum_R \left(\sum_Q \mathsf{Ch}_1^{E,B}(Q) \cdot \mathsf{Keys}^E\right) \cdot \mathsf{Ch}_0^{E,B}(R)$$

$$= \mathsf{Keys}^E \cdot \left(\sum_Q \mathsf{Ch}_1^{E,B}(Q)\right) \cdot \left(\sum_R \mathsf{Ch}_0^{E,B}(R)\right)$$

$$\leq \mathsf{Keys}^E \cdot q \cdot q = q^2 \cdot \mathsf{Keys}^E .$$

Using the above and Lemma 8, we have the following, where the probability and expectation are both over $E \xleftarrow{\$} \mathrm{Bloc}(k,n)$:

$$\mathbf{E}[\mathsf{Ch}_3^{E,B}] < \mathbf{E}\left[\mathsf{Ch}_3^{E,B} \mid \mathsf{Keys}^E < \alpha\right] + \mathbf{E}\left[\mathsf{Ch}_3^{E,B} \mid \mathsf{Keys}^E \geq \alpha\right] \cdot 2^{2n+1-\alpha}$$

$$\leq q^2 \cdot \alpha + q^2 \cdot 2^k \cdot 2^{2n+1-\alpha} .$$

The last inequality above used the fact that Keys^E is always at most 2^k. Since $\alpha = \max(2e2^{k-n}, 2n+k) > 2$ we get $\mathbf{E}[\mathsf{Ch}_3^{E,B}] < q^2\alpha + q^2 \cdot 2 < 2\alpha \cdot q^2$ as desired. ∎

Acknowledgments

We thank the Eurocrypt 2006 PC for their comments. Mihir Bellare was supported by NSF grants CCR-0208842 and CNS-0524765. Phil Rogaway was supported by NSF 0208842 and a gift from Intel Corp. Much of the work on this paper was carried out while Phil was hosted by Chiang Mai University, Thailand.

References

1. W. Aiello, M. Bellare, G. Di Crescenzo, and R. Venkatesan. Security amplification by composition: the case of doubly-iterated, ideal ciphers. *Advances in Cryptology — CRYPTO '98*, Lecture Notes in Computer Science, vol 1462, Springer, pp. 390–407, 1998.
2. M. Bellare and S. Goldwasser. New paradigms for digital signatures and message authentication based on non-interactive zero knowledge proofs. *Advances in Cryptology — CRYPTO '89*, Lecture Notes in Computer Science, vol. 435, Springer, pp. 194–211, 1990.
3. M. Bellare and P. Rogaway. Code-based game-playing proofs and the security of triple encryption. Cryptology ePrint archive report 2004/331, 2006.
4. W. Diffie and M. Hellman. Exhaustive cryptanalysis of the data encryption standard. *Computer*, vol. 10, pp. 74–84, 1977.
5. S. Even and O. Goldreich. On the power of cascade ciphers. *ACM Transactions on Computer Systems*, vol. 3, no. 2, pp. 108–116, 1985.
6. S. Even and Y. Mansour. A construction of a cipher from a single pseudorandom permutation. *Advances in Cryptology — ASIACRYPT '91*, Lecture Notes in Computer Science, vol.739, Springer, pp. 210–224, 1993.
7. S. Goldwasser and S. Micali. Probabilistic encryption. *J. Comput. Syst. Sci.*, vol. 28, no. 2, pp. 270–299, 1984. Earlier version in *STOC '82*.
8. S. Halevi. A plausible approach to computer-aided cryptographic proofs. Cryptology ePrint archive report 2005/181, 2005.
9. J. Kilian and P. Rogaway. How to protect DES against exhaustive key search (an analysis of DESX). *J. of Cryptology*, vol. 14, no. 1, pp. 17–35, 2001. Earlier version in *Crypto '96*.
10. S. Lucks. Attacking triple encryption. *Fast Software Encryption (FSE '98)*, Lecture Notes in Computer Science, vol. 1372, Springer, pp. 239–253, 1998.
11. U. Maurer and J. Massey. Cascade ciphers: the importance of being first. *J. of Cryptology*, vol. 6, no. 1, pp. 55–61, 1993.
12. R. Merkle and M. Hellman. On the security of multiple encryption. *Communications of the ACM*, vol. 24, pp. 465–467, 1981.
13. National Institute of Standards and Technology. FIPS PUB 46-3, Data Encryption Standard (DES), 1999. Also ANSI X9.52, Triple Data Encryption Algorithm modes of operation, 1998, and other standards.
14. C. Shannon. Communication theory of secrecy systems. *Bell Systems Technical Journal*, vol. 28, no. 4, pp. 656–715, 1949.
15. V. Shoup. Sequences of games: a tool for taming complexity in security proofs. Cryptology ePrint archive report 2004/332, 2006.
16. A. Yao. Theory and applications of trapdoor functions. *IEEE Symposium on the Foundations of Computer Science (FOCS 1982)*, IEEE Press, pp. 80–91, 1982.

Compact Group Signatures Without Random Oracles

Xavier Boyen[1] and Brent Waters[2]

[1] Voltage Inc.
xb@boyen.org
[2] SRI International
bwaters@csl.sri.com

Abstract. We present the first efficient group signature scheme that is provably secure without random oracles. We achieve this result by combining provably secure hierarchical signatures in bilinear groups with a novel adaptation of the recent Non-Interactive Zero Knowledge proofs of Groth, Ostrovsky, and Sahai. The size of signatures in our scheme is logarithmic in the number of signers; we prove it secure under the Computational Diffie-Hellman and the Subgroup Decision assumptions in the model of Bellare, Micciancio, and Warinshi, as relaxed by Boneh, Boyen, and Shacham.

1 Introduction

Group signatures allow any member of a group to sign an arbitrary number of messages on behalf of the group, moreover the identity of the signer will be hidden from all members of the system. Preserving the anonymity of the signer can be important in many applications where the signer does not want to be directly identified with the message that he signed. However, there exist situations where it can be deemed desirable to revoke a signer's anonymity. For example, if a signature certified a malicious program, one would want to identify the party that made the malicious statement. Therefore, in group signatures there exists a special party known as the group manager which has the ability to trace the signer of any given signature.

Almost all group signatures schemes are only provably secure in the random oracle model, where we can only make a heuristic argument about security. Additionally, efficient constructions are based on strong assumptions ranging from the Strong Diffie-Hellman [BBS04, BS04] and Strong RSA [ACJT00, AST02, CL02] assumptions to the LRSW [CL04, LRSW99] assumption, which itself has the challenger act as an oracle. The first construction proved secure in the standard model is due to Bellare et. al. [BMW03]. They give a method of constructing group signatures from any signature scheme by using Non-Interactive Zero Knowledge (NIZK) techniques. However, since they use generic NIZK techniques their scheme is too inefficient to be useful in practice.

We approach the problem of group signatures with the goal of creating an efficient group signature scheme that is provably secure without random oracles

S. Vaudenay (Ed.): EUROCRYPT 2006, LNCS 4004, pp. 427–444, 2006.
© International Association for Cryptologic Research 2006

under reasonable assumptions. In particular we at least wish to avoid "oracle-like" assumptions that are difficult to falsify [Nao03], since the value of removing random oracles from the proofs of security while using these types of assumptions is dubious.

In order to solve this problem we combine two recent ideas from pairing-based cryptography. First, we derive our underlying signature scheme from the Waters [Wat05] signature scheme that was proven secure under the computational Diffie-Hellman assumption in bilinear groups. We create a two-level signature scheme where the first level is the signer identity and the second level is the message to be signed. For example, user ID is given a signature on the first level message "ID" as his private key. Group member ID can sign message M by creating the two-level hierarchical signature on "ID.M". Clearly, the signature σ on "ID.M" from the Waters signature scheme will give away the identity of the signer. To protect his anonymity, a signer, in our scheme, will encrypt the signature components of σ using the Boneh-Goh-Nissim [BGN05] encryption system. Additionally, the signer will attach a NIZK proof that the encrypted signature is a signature on "$X.M$" for $1 \leq X \leq 2^k$, where 2^k is the number of signers in the system. Adapting the recent techniques of Groth, Ostrovsky, and Sahai we are able to get efficient NIZKs for our scheme scheme with $O(k)$ complexity in the signature size, signing time, and verification time, i.e., logarithmic in the number of users. We achieve this efficiency by taking advantage of special properties of the NIZK scheme of Groth, Ostrovsky, and Sahai and avoid the general method of circuit construction. The security of these techniques is proven based on the relatively new subgroup decision problem, which was introduced by Boneh, Goh, and Nissim [BGN05]. However, recent work [GOS06a] has shown that the techniques of Groth, Ostrovsky, and Sahai can be generalized to work only with the decision linear assumption, introduced by Boneh, Boyen, and Shacham [BBS04].

1.1   Related Work

Group signatures were first introduced by Chaum and Van Heyst [CvH91] as a way to provide anonymity for signers within a group. The anonymity, however, could be revoked by a special third party if necessary. Since then, there have been several works on this subject [ACJT00, AST02, CL02, CG04, Cam97, Son01, BBS04, KY03, KY05, BSZ05, BMW03].

Until recently, the most efficient group signature constructions [ACJT00, AST02, CL02] were proved secure under the Strong-RSA assumption introduced by Baric and Pfitzman [BP97]. Boneh, Boyen, and Shacham [BBS04] showed how to construct "short" group signatures using bilinear maps under an assumption they introduced called the Strong Diffie-Hellman assumption. Concurrently, Camenish and Lysyanskaya [CL04] gave another group signature scheme that used bilinear maps. Their scheme was proven secure under the interactive LRSW [LRSW99] assumption. All of the above schemes, however, were only proved secure in the random oracle model.

Bellare, Micciancio, and Warinschi [BMW03] gave the first construction that was provably secure in the standard model. Additionally, they provided formal definitions of the security properties of group signatures, which to that point were only informally understood. Since their methods use general NIZK proof techniques, the resulting schemes are inherently too inefficient to be used in practice.

Recently, Ateniese et. al. [ACHdM05] proposed an efficient group signature scheme in the standard model that has the strong exculpability property and is anonymous under CCA attacks. However, they proved their scheme under new strong assumptions.

2   Background

We review a number of useful notions from the recent literature on pairing-based cryptography, which we shall need in later sections. First, we briefly review the properties that constitute a group signature scheme and define its security.

We take this opportunity to clarify once and for all that, in this paper, the word "group" by default assumes its algebraic meaning, except in contexts such as "group signature" and "group manager" where it designates a collection of users. There should be no ambiguity from context. We give a detailed description of the background of group signatures in Appendix A.

2.1   Bilinear Groups of Composite Order

We review some general notions about bilinear maps and groups, with an emphasis on groups of *composite order* which will be used in most of our constructions. We follow [BGN05] in which composite order bilinear groups were first introduced in cryptography.

Consider two finite cyclic groups G and G_T of same order n, in which the respective group operation is efficiently computable and denoted multiplicatively. Assume the existence of an efficiently computable function $e : G \times G \to G_T$, with the following properties:

- (Bilinearity) $\forall u, v \in G$, $\forall a, b \in \mathbb{Z}$, $e(u^a, v^b) = e(u, v)^{ab}$, where the product in the exponent is defined modulo n;
- (Non-degeneracy) $\exists g \in G$ such that $e(g, g)$ has order n in G_T. In other words, $e(g, g)$ is a generator of G_T, whereas g generates G.

If such a function can be computed efficiently, it is called a (symmetric) bilinear map or pairing, and the group G is called a bilinear group. We remark that the vast majority of cryptosystems based on pairings assume for simplicity that bilinear groups have prime order. In our case, it is important that the pairing be defined over a group G containing $|G| = n$ elements, where $n = pq$ has a (hidden) factorization in two large primes, $p \neq q$.

We denote by G_p and G_q the subgroups of G of respective orders p and q.

Complexity Assumptions. We shall make use of two complexity assumptions: the first, computational in the prime order subgroup G_p, the second, decisional in the full group G.

The first one is the familiar Computational Diffie-Hellman assumption in bilinear groups, which states that there is no probabilistic polynomial time (PPT) adversary that, given a triple $(g, g^a, g^b) \in G_p^3$ for random exponents $a, b \in \mathbb{Z}_p$, computes $g^{ab} \in G_p$ with non-negligible probability (i.e., with polynomial probability in the bit-size of the algorithm's input). We shall require the CDH assumption in G_p to remain true when the factorization of n is known.

The second assumption we need is the subgroup decision assumption, introduced in [BGN05]; it is based on the hardness of factoring, and is recalled next.

2.2 Subgroup Decision Assumption

Informally, the subgroup decision assumption posits that for a bilinear group G of composite order $n = pq$, the uniform distribution on G is computationally indistinguishable from the uniform distribution on a subgroup of G (say, G_q, the subgroup of order q). The formal definition is based on the subgroup decision problem, which is as follows [BGN05].

The Subgroup Decision Problem. Consider an "instance generator" algorithm $\mathcal{GG}$ that, on input a security parameter 1^λ, outputs a tuple (p, q, G, G_T, e), in which p and q are independent uniform random λ-bit primes, G and G_T are cyclic groups of order $n = pq$ with efficiently computable group operations (over their respective elements, which must have a polynomial size representation in λ), and $e : G \times G \to G_T$ is a bilinear map. Let $G_q \subset G$ denote the subgroup of G of order q. The subgroup decision problem is as follows:

> On input a tuple $(n = pq, G, G_T, e)$ derived from a random execution of $\mathcal{GG}(1^\lambda)$, and an element w selected at random either from G or from G_q, decide whether $w \in G_q$.

The advantage of an algorithm $\mathcal{A}$ solving the subgroup decision problem is defined as $\mathcal{A}$'s excess probability, beyond $\frac{1}{2}$, of outputting the correct solution. The probability is defined over the random choice of instance and the random bits used by $\mathcal{A}$.

We use composite order groups in order to leverage the recent Non-Interactive Zero Knowledge proof techniques of Groth, Sahai, and Ostrovsky [GOS06b].

2.3 Hierarchical Signatures

In an Λ-level hierarchical signature, a message is a tuple of Λ message components. The crucial property is that a signature on a message, $M_1. \cdots .M_i$, can act as a restricted private key that enables the signing of any extension, $M_1. \cdots .M_i. \cdots .M_j$, of which the original message is a prefix. In a Λ-level signature scheme, the messages must obey the requirement that $1 \le i \le j \le \Lambda$.

We note that this is essentially equivalent to the notion of $(\Lambda - 1)$-hierarchical identity-based signature, or HIBS [GS02], in which the first $\Lambda - 1$ levels are viewed as the components of a hierarchical identity, and the last level (which in HIBS parlance is no longer deemed part of the hierarchy) is for the message proper. Our basic group signature uses a two-level hierarchy, though in Section 5 we shall discuss how additional levels can be used to achieve delegation in group signatures.

3   Group Signature Scheme

In this section, we present our group signature scheme, which is based solely on the CDH and the Subgroup Decision assumptions. It is built upon a two-level hierarchical signature scheme, which we describe first.

3.1   Simple Two-Level Hierarchical Signatures

Waters [Wat05] recently offered an efficient identity-based encryption system provably secure under "full" adaptive attacks. The system generalizes easily to a hierarchical IBE of logarithmically bounded depth $\Lambda \leq O(\log \lambda)$. Here, λ is the security parameter and Λ the maximum depth of the HIBE. It is then a triviality to observe that any Λ-level HIBE scheme also gives an Λ-level hierarchical signature functionality. We describe below the 2-level hierarchical signature scheme (or 1-level IBS) that results from these transformations.

We assume that identities are strings of k bits, and messages strings of m bits. To fix ideas, for group signatures one would have, $k \ll m \approx \lambda$. The description that follows assumes that g is a generator of G_p, so that all elements in G and G_T are in fact in the respective subgroups of prime order p.

Setup(1^λ): To setup the system, first, a secret $\alpha \in \mathbb{Z}_p$ is chosen at random, from which the value $A = e(g, g)^\alpha$ is calculated. Next, two random integers $y' \in \mathbb{Z}_p$ and $z' \in \mathbb{Z}_p$ and two random vectors $\boldsymbol{y} = (y_1, \dots, y_k) \in \mathbb{Z}_p^k$ and $\boldsymbol{z} = (z_1, \dots, z_m) \in \mathbb{Z}_p^m$ are selected. The public parameters of the system and the master secret key are then given by,

$$\mathsf{PP} = \Big(g, u' = g^{y'}, u_1 = g^{y_1}, \dots, u_k = g^{y_k},$$

$$v' = g^{z'}, v_1 = g^{z_1}, \dots, v_m = g^{z_m}, A = e(g, g)^\alpha\Big) \in G^{k+m+3} \times G_T,$$

$$\mathsf{MK} = g^\alpha \in G.$$

The public parameters, PP, also implicitly include k, m, and a description of (p, G, G_T, e).

Extract(PP, MK, ID): To create a private key for a user whose binary identity string is $\mathsf{ID} = (\kappa_1 \dots \kappa_k) \in \{0, 1\}^k$, first select a random $r \in \mathbb{Z}_p$, and return,

$$K_{\mathsf{ID}} = \Big(g^\alpha \cdot \big(u' \prod_{i=1}^k u_i^{\kappa_i}\big)^r, \ g^{-r} \Big) \in G^2.$$

Sign(PP, K_{ID}, M): To sign a message represented as a bit string $M = (\mu_1 \dots \mu_m) \in \{0, 1\}^m$, using a private key $K_{\mathsf{ID}} = (K_1, K_2) \in G^2$, select a random $s \in \mathbb{Z}_p$, and output,

$$S = \left(\ K_1 \cdot (v' \prod_{j=1}^{m} v_j^{\mu_j})^s, \ K_2, \ g^{-s} \ \right)$$

$$= \left(\ g^\alpha \cdot (u' \prod_{i=1}^{k} u_i^{\kappa_i})^r (v' \prod_{j=1}^{m} v_j^{\mu_j})^s, \ g^{-r}, \ g^{-s} \ \right) \in G^3.$$

Verify(PP, ID, M, σ): To verify a signature $S = (S_1, S_2, S_3) \in G^3$ against an identity $\mathsf{ID} = (\kappa_1 \ldots \kappa_k) \in \{0,1\}^k$ and a message $M = (\mu_1 \ldots \mu_m) \in \{0,1\}^m$, verify that,

$$e(S_1, g) \cdot e(S_2, u' \prod_{i=1}^{k} u_i^{\kappa_i}) \cdot e(S_3, v' \prod_{j=1}^{m} v_j^{\mu_j}) \overset{?}{=} A.$$

If the equality holds, output `valid`; otherwise, output `invalid`.

Security from CDH. The scheme's existential unforgeability against adaptive chosen message attacks follows from the Waters's signature scheme. We provide a reduction to CDH in the full version of our paper [BW05].

3.2   Logarithmic-Size Group Signature Scheme

We are now in a position to describe our actual group signature scheme. It is composed of the following algorithms.

Setup(1^λ): The input is a security parameter in unary, 1^λ. Suppose we wish to support up to 2^k signers in the group, and sign messages in $\{0,1\}^m$, where k and m are polynomially related functions of λ.
The setup algorithm first chooses $n = pq$ where p and q are random primes of bit size $\Theta(\lambda)$. Let G be a bilinear group of order n and denote by G_p and G_q its subgroups of respective order p and q. Next, the algorithm chooses generators $g \in G$, and $h \in G_q$. It chooses a random exponent $\alpha \in \mathbb{Z}_n$. Finally, it chooses generators $u', u_1, \ldots, u_k \in G$ and $v', v_1, \ldots, v_m \in G$.
The bilinear group, (n, G, G_T, e), is published together with the public parameters,

$$\mathsf{PP} = \left(\ g, \ h, \ u', \ u_1, \ \ldots, \ u_k, \ v', \ v_1, \ \ldots, \ v_m, \ A = e(g,g)^\alpha \ \right)$$

$$\in G \times G_q \times G^{k+m+2} \times G_T.$$

The master key for user enrollment, MK, and the group manager's tracing key, TK, are,

$$\mathsf{MK} = g^\alpha \in G, \qquad\qquad \mathsf{TK} = q \in \mathbb{Z}.$$

Enroll(PP, MK, ID): Suppose we wish to create a group signature key for user ID where $0 \le \mathsf{ID} < 2^k$. We denote by κ_i the i-th bit of ID. The algorithm chooses a random $s \in \mathbb{Z}_n$ and creates the key for user ID as,

$$K_{\mathsf{ID}} = (K_1, K_2, K_3) = \left(\ g^\alpha \cdot \left(u' \prod_{i=1}^{k} u_i^{\kappa_i} \right)^s, \ g^{-s}, \ h^s \ \right) \in G^3.$$

The key for user ID is essentially a private key for identity ID in the Waters IBS scheme, except that we are working in a bilinear group G of composite order, and are adjoining the additional element $h^s \in G_q$.

Sign(PP, ID, K_{ID}, M): To sign a message $M = (\mu_1 \ldots \mu_m) \in \{0,1\}^m$, user ID first chooses random exponents $t_1, \ldots, t_k \in \mathbb{Z}_n$, and, for all $i = 1, \ldots, k$, it creates,

$$c_i = u_i^{\kappa_i} \cdot h^{t_i}, \qquad\qquad \pi_i = (u_i^{2\kappa_i - 1} \cdot h^{t_i})^{t_i}.$$

The signer also defines $t = \sum_{i=1}^{k} t_i$ and $c = u' \prod_{i=1}^{k} c_i = (u' \prod_{i=1}^{k} u_i^{\kappa_i}) \cdot h^t$. The set of values, c_i and π_i, are proof that c is well formed. It also lets $V = v' \prod_{i=1}^{m} v_i^{\mu_i}$. Then, it picks two random exponents $\tilde{s}_1, s_2 \in \mathbb{Z}_n$, and creates,

$$\sigma_1 = K_1 \cdot K_3^t \cdot c^{\tilde{s}_1} \cdot V^{s_2}, \qquad \sigma_2 = K_2 \cdot g^{-\tilde{s}_1}, \qquad \sigma_3 = g^{-s_2}.$$

If we let $s_1 = \tilde{s}_1 + s$, with s as in the *Enroll* procedure, then we have,

$$\sigma_1 = g^\alpha \cdot \left(u' \prod_{i=1}^{k} u_i^{\kappa_i} \right)^{s_1} \cdot \left(v' \prod_{i=1}^{m} v_i^{\mu_i} \right)^{s_2} \cdot h^{s_1 t} = g^\alpha \cdot c^{s_1} \cdot V^{s_2}, \sigma_2 = g^{-s_1}, \sigma_3 = g^{-s_2}.$$

The final signature is output as:

$$\sigma = \left(\sigma_1, \sigma_2, \sigma_3, \ c_1, \ldots, c_k, \ \pi_1, \ldots, \pi_k \right) \in G^{2k+3}.$$

Verify(PP, ID, M, σ): The verification proceeds in two phases. In the first phase the verifier will reconstruct c and check to make sure that it is well formed. To do this, it computes,

$$c = u' \prod_{i=1}^{k} c_i, \quad \text{and checks that,} \quad \forall i = 1, \ldots, k \ : \ e(c_i, u_i^{-1} c_i) \overset{?}{=} e(h, \pi_i).$$

This proof shows that all $c_i = u_i^{\kappa_i} h^{t_i}$ for $\kappa_i \in \{0,1\}$, and thus that c is well formed. Next, the verifier focuses on the actual signature. To do so, it derives $V = v' \prod_{i=1}^{m} v_i^{\mu_i}$ from the message, and checks that,

$$e(\sigma_1, g) \cdot e(\sigma_2, c) \cdot e(\sigma_3, V) \overset{?}{=} A.$$

This proof shows that $(\sigma_1, \sigma_2, \sigma_3)$ is a valid two-level hierarchical signature, after the blinding factors $h^{s_1 t}$ and h^t cancel each other out in the product after they are respectively paired with g and g^{-s_1}.

If all tests are successful, the verifier outputs `valid`; otherwise, it outputs `invalid`.

Trace(PP, TK, σ): Suppose the tracing algorithm wishes to trace a signature σ, assumed to pass the verification test for some message M that is not needed here. Let κ_i denote the i-th bit of the signer's identity ID that is to be determined. To recover the bits of ID, for each $i = 1, \ldots, k$, the tracer sets,

$$\kappa_i = \begin{cases} 0 & \text{if } (c_i)^q = g^0, \\ 1 & \text{otherwise.} \end{cases}$$

The reconstituted signer identity is output as $\text{ID} = (\kappa_1 \ldots \kappa_k) \in \{0,1\}^k$.

4   Proofs of Security

We now prove the main security properties of our group signature scheme.

4.1   Full Anonymity (Under CPA Attack)

We prove the security of our scheme in the anonymity game against chosen plaintext attacks. We refer to [BMW03] for the game description, which should also be clear from the proof.

Intuitively, our proof follows from two simple arguments. First, we show that an adversary cannot tell whether h is a random generator of G_q or G by reduction from the subgroup decision problem. Next, we show that if h is chosen from G then the identity of a signer is perfectly hidden.

Theorem 1. *Suppose no t-time adversary can solve the subgroup decision problem with advantage at least ϵ_{sd}. Then for every t'-time adversary $\mathcal{A}$ where $t' \approx t$ we have that $\mathrm{Adv}_A < 2\epsilon_{sd}$.*

We first introduce a hybrid game H_1 in which the public parameters are the same as in the original game except that h is chosen randomly from G instead of G_q. We denote the adversary's advantage in this game as Adv_{A,H_1}.

Lemma 1. *For all t'-time adversaries as above, $\mathrm{Adv}_A - \mathrm{Adv}_{A,H_1} < 2\,\epsilon_{sd}$.*

Proof. Consider an algorithm $\mathcal{B}$ that plays the subgroup decision problem. Upon receiving a subgroup decision challenge (n, G, G_T, e, w) the algorithm $\mathcal{B}$ first creates public parameters for our scheme by setting $h = w$ and choosing all other parameters as the scheme does. It then sends the parameters to $\mathcal{A}$ and plays the anonymity game with it. If w is randomly chosen from G_q then the adversary is playing the normal anonymity game, otherwise, if w is chosen randomly from G then it plays the hybrid game H_1. The algorithm $\mathcal{B}$ will be able to answer all chosen plaintext queries—namely, issue private signing keys for, and sign any message by, any user—, since it knows the master key.

At some point the adversary will choose a message M and two identities ID_1 and ID_2 it wishes to be challenged on (under the usual constraints that it had not previously made a signing key query on ID_x or a signature query on $\mathsf{ID}_x.M$). The simulator $\mathcal{B}$ will create a challenge signature on M, and $\mathcal{A}$ will guess the identity of the signer. If $\mathcal{A}$ answers correctly, then $\mathcal{B}$ outputs $b = 1$, guessing $w \in G_q$; otherwise it outputs $b = 0$, guessing $w \in G$.

Denote by Adv_B the advantage of the simulator $\mathcal{B}$ in the subgroup decision game. As we know that $\Pr[w \in G] = \Pr[w \in G_q] = \frac{1}{2}$, we deduce that,

$$\mathrm{Adv}_A - \mathrm{Adv}_{A,H_1} = \Pr[b = 1 | w \in G_q] - \Pr[b = 1 | w \in G]$$
$$= 2\Pr[b = 1, w \in G_q] - 2\Pr[b = 1, w \in G] = 2\,\mathrm{Adv}_B < 2\,\epsilon_{sd},$$

since by our hardness assumption Adv_B must be lesser than ϵ_{sd}, given that $\mathcal{B}$ runs in time $t \approx t'$.

Lemma 2. *For any algorithm $\mathcal{A}$, we have that $\mathrm{Adv}_{A,H_1} = 0$.*

Proof. We must argue that when h is chosen uniformly at random from G, instead of G_q in the real scheme, then the challenge signature is statistically independent of the signer identity, ID, in the adversary's view (which might comprise answers to earlier signature queries on ID). Consider the challenge signature,

$$\sigma = (\, \sigma_1, \sigma_2, \sigma_3, \ c_1, \dots, c_k, \ \pi_1, \dots, \pi_k \,),$$

and let us determine what such an adversary might deduce from σ.

First, observe that σ_2 and σ_3 by themselves do not depend on (any of the bits κ_i comprising) the signer identity ID. However, since the adversary is computationally unbounded, we must assume that they reveal s_1 and s_2.

Next, consider $c_i = u_i^{\kappa_i} h^{t_i}$ and the corresponding $\pi_i = (u_i^{2\kappa_i - 1} h^{t_i})^{t_i} = (u_i^{\kappa_i} u_i^{\kappa_i - 1} h^{t_i})^{t_i}$ for each i. There are two competing hypotheses that may be formulated by the adversary: $\kappa_i = 0 \vee \kappa_i = 1$. For either hypothesis, there is a solution for the ephemeral exponent t_i that explains the observed value of c_i. In other words, in the adversary's view,

$$\forall i \in \{1, \dots, k\} \ : \ \exists \tau_0, \tau_1 \in \mathbb{Z}_n \quad \text{s.t.} \quad (\kappa_i, t_i) = (0, \tau_0) \vee (\kappa_i, t_i) = (1, \tau_1)$$
$$\text{and} \quad c_i = h^{\tau_0} = u_i h^{\tau_1}.$$

Using the last equality we find that the observed value of π_i is compatible with both hypotheses:

$$\left(\pi_i\big|_{\kappa_i = 0}\right) = (u_i^{-1} h^{\tau_0})^{\tau_0} = (u_i^{-1} u_i h^{\tau_1})^{\tau_0} = h^{\tau_0 \tau_1} = (h^{\tau_0})^{\tau_1} = (u_i h^{\tau_1})^{\tau_1} = \left(\pi_i\big|_{\kappa_i = 1}\right).$$

This all means that, the knowledge of c_i and π_i does not disambiguate the relevant bit $\kappa_i \in \{0, 1\}$. Taken together, all the c_i and π_i do not reveal anything about ID.

Last, we consider $\sigma_1 = g^\alpha \cdot c^{s_1} \cdot V^{s_2}$. But this value is just redundant in the eyes of the adversary, since he already knows all the values that determine it, including $\alpha = \log_{e(g,g)} A$.

Therefore, ID is statistically independent of the entire signature σ, which proves the lemma.

4.2 Full Traceability

We show how to reduce the full traceability of our scheme to the two-level signature scheme described in Section 3.1. We create a simulator that will interact with a challenger for the security game of the Waters signature scheme. If the adversary asks for the secret key of user ID, the simulator will simply ask for a first-level signature on ID and give this to the adversary. If the adversary asks for a signature of message M by user ID the simulator will ask the challenger for a second-level signature on ID.M and then blind the signature itself.

The adversary will finally output a signature σ^* on some message M^*. In order for the adversary to be successful the signature will need to verify. By the perfect binding properties of the underlying NIZK techniques, a signature can be traced to some user ID^* and we can recover from it a Waters two-level signature on $\mathsf{ID}^*.M^*$; the simulator will submit this as its forgery in its own attack against the underlying signature scheme. The adversary will only be considered successful if he had not asked for the private key of user ID^* and had not queried for a signature on M^* from user ID^*. However, these are precisely the conditions that the simulator needs to abide by to be successful in its own game.

One tricky point in our reduction is that the simulator will play the signature game in the subgroup G_p, however the parameters for the group signature scheme are to be given in the group G, and so will be the forgery produced by the adversary. In addition, we note that the adversary is effectively given the factorization $n = pq$, as required by the full traceability security definition which demands that the tracing key $\mathsf{TK} = q$ be disclosed for this attack. Our formal reduction follows.

Theorem 2. *If there exists a (t, ϵ) adversary for the full tracing game then there exists a $(\tilde{t}, \epsilon)$ UF-CMA adversary against the two-level signature scheme, where $t \approx \tilde{t}$.*

Proof. Suppose there exists an algorithm $\mathcal{A}$ that is successful in the tracing game of our group signature scheme with advantage ϵ. Then we can create a simulator $\mathcal{B}$ that existentially forges signatures in an adaptive chosen message attack against the two-level signature scheme, with advantage ϵ.

The simulator will be given the factorization $n = pq$ of the group order $|G| = n$. As usual, denote by G_p and G_q the subgroups of G of respective order p and q, and by analogy let G_{Tp} and G_{Tq} be the subgroups of G_T of order p and q. The simulator begins by receiving from its challenger the public parameters of the signature game, all in subgroups of order p,

$$\tilde{\mathsf{PP}} = \Big(\ \tilde{g} \, , \tilde{u}' = \tilde{g}^{y'}, \ \tilde{u}_1 = \tilde{g}^{y_1}, \ldots, \ \tilde{u}_k = \tilde{g}^{y_k},$$

$$\tilde{v}' = \tilde{g}^{z'}, \ \tilde{v}_1 = \tilde{g}^{z_1}, \ldots, \ \tilde{v}_m = \tilde{g}^{z_m}, \ \tilde{A} = e(\tilde{g}, \tilde{g})^{\alpha} \ \Big) \in G_p^{k+m+3} \times G_{Tp}.$$

The simulator then picks random generators $(f, h, \gamma', \gamma_1, \ldots, \gamma_k, \nu', \nu_1, \ldots, \nu_m) \in G_q^{k+m+4}$ and a random exponent $\beta \in \mathbb{Z}_q$. The simulator publishes the group signature public parameters as,

$$\mathsf{PP} = \Big(\ g = \tilde{g} f \, , \ h \, , u' = \tilde{u}' \, \gamma', u_1 = \tilde{u}_1 \, \gamma_1, \ldots, u_k = \tilde{u}_k \, \gamma_k,$$

$$v' = \tilde{v}' \, \nu', v_1 = \tilde{v}_1 \, \nu_1, \ldots, v_m = \tilde{v}_k \, \nu_m, \ A = \tilde{A} \cdot e(f, f)^{\beta} \ \Big).$$

The distribution of the public key is the same is in the real scheme. The simulator also gives the tracing key $\mathsf{TK} = q$ to the adversary.

Suppose the adversary asks for the private key of user ID. To answer the query, the simulator first asks the challenger for a first-level signature on message ID, and receives back $\tilde{K}_{\mathsf{ID}} = (\tilde{K}_1, \tilde{K}_2) \in G_p^2$. As before, κ_i denotes the i-th bit of ID. The simulator then chooses a random $r \in \mathbb{Z}_q$ and creates the requested key as,

$$K_{\mathsf{ID}} = \left(\ K_1 = \tilde{K}_1 \cdot f^\beta \cdot (\gamma' \prod_{i=1}^{k} \gamma_i^{\kappa_i})^r \ , \quad K_2 = \tilde{K}_2 \cdot f^{-r} \ , \quad K_3 = h^{-r} \ \right).$$

This is a well formed private key in our scheme.

Suppose the simulator is asked for a signature on message $M = (\mu_1 \ldots \mu_m) \in \{0,1\}^m$ by user $\mathsf{ID} = (\kappa_1 \ldots \kappa_k) \in \{0,1\}^k$. The simulator starts as in the real scheme, by choosing random $t_1, \ldots, t_k \in \mathbb{Z}_n$, defining $t = \sum_{i=1}^{k} t_i$, and creating the values $c_i = u_i^{\kappa_i} \cdot h^{t_i}$ and $\pi_i = (u_i^{2\kappa_i - 1} \cdot h^{t_i})^{t_i}$ for all $i = 1, \ldots, k$. Next, the simulator requests a two-level signature on $\mathsf{ID}.M$ and receives in return $S = (S_1, S_2, S_3) \in G_p^3$. It then chooses random $r_1, r_2 \in \mathbb{Z}_q$ and creates the remaining components,

$$\sigma_1 = S_1 \cdot f^\beta \cdot (\gamma' \prod_{i=1}^{k} \gamma_i^{\kappa_i})^{r_1} \cdot (\nu' \prod_{i=1}^{m} \nu_i^{\mu_i})^{r_2} \cdot h^{r_1 t}, \qquad \sigma_2 = S_2 \cdot f^{-r_1}, \qquad \sigma_3 = S_3 \cdot f^{-r_2}.$$

The simulator gives the full signature $\sigma = (\sigma_1, \sigma_2, \sigma_3, c_1, \ldots, c_k, \pi_1, \ldots, \pi_k)$ to the adversary. Again, this is a well-formed signature in our scheme.

Finally, the adversary gives the simulator a forgery $\sigma^* = (\sigma_1, \sigma_2, \sigma_3, c_1, \ldots, c_k, \pi_1, \ldots, \pi_k)$ on message $M^* = (\mu_1 \ldots \mu_m)$. The simulator first checks that the signature verifies, otherwise the adversary is not successful and the simulator can abort. Next, it sets out to trace the identity, ID^*, of the forgery. Let κ_i denote the i-th bit of the string ID^* that is to be determined. For each $i = 1, \ldots, k$, the tracer sets $\kappa_i = 0$ if $(c_i)^q = g^0$, and $\kappa_i = 1$ otherwise. It then reconstitutes $\mathsf{ID}^* = (\kappa_1 \ldots \kappa_k)$. If either the key for ID^* or a signature on M^* by ID^* was previously requested by the adversary, the simulator can safely abort since the adversary was not successful. Otherwise the adversary was successful and the simulator must produce its own forgery.

To see how, recall that for all i we have that $e(c_i, u_i^{-1} c_i) = e(h, \pi_i)$, which has order q in G_T. Therefore, either $c_i \in G_q$ or $c_i u_i^{-1} \in G_q$. It follows that $c_i = u_i^{\kappa_i} f^{r_i'}$ for the previously determined $\kappa_i \in \{0,1\}$ for some unknown r_i', and therefore, that $c = u' \prod_{i=1}^{k} c_i = (\tilde{u}' \prod_{i=1}^{k} \tilde{u}_i^{\kappa_i}) f^{r'}$ for some r'. Let then $\delta \in \mathbb{Z}_n$ be an integer which is 0 (mod q) and 1 (mod p). The verification equation entails,

$$e(\sigma_1^\delta, \tilde{g}) \cdot e(\sigma_2^\delta, \tilde{u}' \prod_{i=1}^{k} \tilde{u}_i^{\kappa_i}) \cdot e(\sigma_3^\delta, \tilde{v}' \prod_{j=1}^{m} \tilde{v}_j^{\mu_j})$$

$$= A^\delta = e(\tilde{g}, \tilde{g})^{\alpha \delta} \, e(f, f)^{\beta \delta} = e(\tilde{g}, \tilde{g})^\alpha = \tilde{A}.$$

This, however, leaves $S^* = (\sigma_1^\delta, \sigma_2^\delta, \sigma_3^\delta) \in G_p^3$ as the sought forgery on $\mathsf{ID}^*.M^*$ in the underlying hierarchical signature scheme, which the simulator gives to the challenger. Therefore, our simulator will be successful whenever the adversary is.

5   Extensions

Our framework of creating group signature schemes from hierarchical signature schemes allows us to extend our basic scheme in some interesting ways. We outline a few of these applications in the present section.

5.1 Fast Verification

Perhaps the main drawback of our scheme in terms of practicality, is that, taken at face value, signature verification requires $2k + 3$ pairing computations. However, in all known realizations of the pairing, it turns out that when computing multiple pairings in a product, the cost incurred by adding each extra pairing is significantly lesser than the cost of the first pairing. The reason is because the sequence of doublings in Miller's algorithm [Mil04] can be amortized over all the pairings in a given product, in a very similar way to the multi-exponentiation algorithm. To push this idea further, it is possible to batch the k remaining equations into a single "multi-pairing", using randomization, at the cost of k extra exponentiations in G: to check that $\forall i = 1, \ldots, k : e(c_i, u_i^{-1} c_i) = e(h, \pi_i)$, the verifier would pick $r_1, \ldots, r_k \in \mathbb{Z}_n$, and test,

$$\prod_{i=1}^{k} \left(e(c_i^{r_i}, u_i^{-1} c_i) \cdot e(h^{-r_i}, \pi_i) \right) \stackrel{?}{=} 1.$$

Probabilistic signature verification can thus be performed with a total of 2 multi-pairings and k exponentiations for the c^{r_i}. Notice that since h is constant across all signers for the life of the system, the h^{-r_i} can be computed comparatively very quickly using a few amortized pre-computations.

5.2 Long Messages

Once we have a signature scheme that can sign messages in $\{0, 1\}^m$ for large enough $m = \Theta(\lambda)$, it is easy to sign arbitrary messages with the help of a Universal One-Way Hash Function (UOWHF) family $\mathcal{H}$, a description of which is added to the public key. To sign $M \in \{0, 1\}^*$, first pick a random index h into the family, which determines a function $H_h \in \mathcal{H}$. Next, let $M' = h \| H_h(M)$, and compute $\sigma' = Sign(\mathsf{PP}, \mathsf{ID}, K_{\mathsf{ID}}, M')$, a signature on M' in the initial scheme. The signature on M is then given by $\sigma = (h, \sigma')$.

Since $|h|$ and $|H_h(M)|$ both grow linearly in the security parameter, it suffices to let $m = \Theta(\lambda)$ with a constant factor large enough to accommodate the two. A standard argument shows that the new scheme is existentially unforgeable under adaptive chosen message attacks whenever the old one was. Also, it is easy to see that this transformation does not affect anonymity or tracing, since it operates only on the message, and does so in a "public" way.

5.3 Delegation

Using a hierarchical signature scheme we can allow for a group signature scheme where a signer can delegate its authority down in a hierarchical manner. Suppose we have an $(\Lambda + 1)$-level hierarchical signature scheme where at each level identities can be at most d bits long (except the last level, which must support messages of sufficient size m, as discussed above). Then we can extend the techniques from our basic scheme to create a new group signature scheme that allows for hierarchical identities of up to Λ levels, where someone with an identity at level l can delegate down to a new user at level $l + 1$.

To do this we simply extend our scheme to hide identities at all levels. However, this will come with an $O(\Lambda d + m)$ cost in signing time, verification time, and signature size.

5.4   Revocation

Keys can be revoked in any group signature scheme in a very generic manner, in which the group master sends a revocation message linear in the number of remaining signers. Upon enrollment, each user is assigned an additional, unique, long-lived decryption key. Then, to revoke a user, the group master would re-key the group signature sub-system, and form a public revocation message that contains the new public key as well as the signing key of each remaining user encrypted under that user's long-term key. (Alternatively, the group master could broadcast a constant size revocation message containing only the new PK, and privately communicate a new key to each signer.)

Using an extension of our methods we can have a constant size revocation message along with an $O(r)$ overhead for our group signature scheme, where r is the number of revoked users. Essentially, the idea is for the signers to attach an additional proof for each revoked user that they are not that user.

These two techniques can be used in conjunction. Most revocation messages can be kept short using the second technique. However, when the number of revoked users becomes too large, the group master issues out a long revocation message to re-key the system.

5.5   Partial Revelation of Identities

A user might also wish to selectively reveal parts of his identity. For example, suppose there are two classes of users in a system where one class of users consists of administrators whose extra privilege is important in some, but not all applications. We could then organize the identities in such a way that a user's identity consisted of his class bit followed by a unique bitstring.

For some types signatures it might be important for a user to reveal his privilege, while keeping his identity secret within this class. In our signature scheme a signer can do this by simply not encrypting the class bit of his identity, while hiding all of the other bits.

Using selective revelation will be preferable to the alternative of creating a new group signature scheme for each possible group of users. This type of technique can be generalized to more complicated types of selective revelation by using the NIZK techniques in more complicated ways, although the signature overhead will likely become larger with more complicated proofs.

5.6   Using Prime Order Groups

In our current scheme we work in composite order groups and the anonymity of our scheme rests on the hardness of the Subgroup-Decision problem. An interesting extension of our work would be to apply our techniques to work in prime order groups. This would give us a wider range of underlying elliptic curve implementations to choose from and allow us to explore alternative complexity assumptions. Recent work [GOS06a] has shown that the NIZK techniques of Groth, Sahai, and Ostrovsky can be realized in prime order groups under the Decision-Linear Assumption [BBS04]. We can plug these new NIZK techniques into our group signature framework and realize our scheme in prime order groups.

6   Conclusion

In this paper we presented the first efficient group signature scheme that is provably secure without random oracles, based on bilinear maps. We built our group signature scheme from the Waters two-level hierarchical signatures scheme, where the first level is the identity of the signer and the second level is the signed message. Additionally, we applied the recent NIZK proof techniques of Groth, Ostrovsky, and Sahai in a novel manner to hide the identity of the signer.

We proved the security of our scheme using the subgroup decision and the computational Diffie-Hellman assumptions. Its signing time, verification time, and signature size are all logarithmic in the number of signers.

Our method of using a hierarchical signature scheme allowed us to create clean, modular proofs of security. Additionally, it had the added benefit of allowing for a hierarchical identity structure. We expect our new framework of creating group signatures to enable many other extensions in the future.

Acknowledgments

We thank Dawn Song for suggesting the concept of signature delegation and Dan Boneh for useful comments and suggestions.

References

[ACHdM05] Giuseppe Ateniese, Jan Camenisch, Susan Hohenberger, and Breno de Medeiros.   Practical group signatures without random oracles.   Cryptology ePrint Archive, Report 2005/385, 2005. http://eprint.iacr.org/.

[ACJT00]   Giuseppe Ateniese, Jan Camenisch, Marc Joye, and Gene Tsudik. A practical and provably secure coalition-resistant group signature scheme. In *Proceedings of Crypto 2000*, volume 1880 of *Lecture Notes in Computer Science*, pages 255–70. Springer-Verlag, 2000.

[AST02]   Giuseppe Ateniese, Dawn Song, and Gene Tsudik. Quasi-efficient revocation of group signatures. In *Proceedings of Financial Cryptography 2002*, 2002.

[AT99]   G. Ateniese and G. Tsudik. Some open issues and directions in group signatures. In *Proceedings of Financial Cryptography 1999*, volume 1648 of *Lecture Notes in Computer Science*, pages 196–211. Springer-Verlag, 1999.

[BBS04]   Dan Boneh, Xavier Boyen, and Hovav Shacham. Short group signa-
           tures. In *Advances in Cryptology—CRYPTO 2004*, volume 3152 of *Lec-
           ture Notes in Computer Science*, pages 41–55. Springer-Verlag, 2004.

[BGN05]   Dan Boneh, Eu-Jin Goh, and Kobbi Nissim. Evaluating 2-DNF formulas
           on ciphertexts. In *Proceedings of TCC 2005*, Lecture Notes in Computer
           Science. Springer-Verlag, 2005.

[BMW03]   Mihir Bellare, Daniele Micciancio, and Bogdan Warinschi. Foundations
           of group signatures: Formal definitions, simplified requirements, and a
           construction based on general assumptions. In *Advances in Cryptology—
           EUROCRYPT 2003*, volume 2656 of *Lecture Notes in Computer Science*,
           pages 614–29. Springer-Verlag, 2003.

[BP97]   Niko Baric and Birgit Pfitzman. Collision-free accumulators and fail-
           stop signature schemes without trees. In *Advances in Cryptology—
           EUROCRYPT 1997*, Lecture Notes in Computer Science, pages 480–94.
           Springer-Verlag, 1997.

[BS04]   Dan Boneh and Hovav Shacham. Group signatures with verifier-local
           revocation. In *Proceedings of ACM CCS 2004*, pages 168–77. ACM Press,
           2004.

[BSZ05]   Mihir Bellare, Haixia Shi, and Chong Zhang. Foundations of group signa-
           tures: The case of dynamic groups. In *Proceedings of CT-RSA 2005*, Lec-
           ture Notes in Computer Science, pages 136–153. Springer-Verlag, 2005.

[BW05]   Xavier Boyen and Brent Waters. Compact group signatures without
           random oracles. Cryptology ePrint Archive, Report 2005/381, 2005.
           `http://eprint.iacr.org/`.

[Cam97]   Jan Camenisch. Efficient and generalized group signatures. In *Advances
           in Cryptology—EUROCRYPT 1997*, Lecture Notes in Computer Science,
           pages 465–479. Springer-Verlag, 1997.

[CG04]   Jan Camenisch and Jens Groth. Group signatures: Better efficiency and
           new theoretical aspects. In *Proceedings of SCN 2004*, pages 120–133,
           2004.

[CL02]   Jan Camenisch and Anna Lysyanskaya. Dynamic accumulators and ap-
           plication to efficient revocation of anonymous credentials. In *Advances in
           Cryptology—CRYPTO 2002*, volume 2442 of *Lecture Notes in Computer
           Science*, pages 61–76. Springer-Verlag, 2002.

[CL04]   Jan Camenisch and Anna Lysyanskaya. Signature schemes and anony-
           mous credentials from bilinear maps. In *Advances in Cryptology—
           CRYPTO 2004*, volume 3152 of *Lecture Notes in Computer Science*.
           Springer-Verlag, 2004.

[CvH91]   David Chaum and Eugène van Heyst. Group signatures. In *Advances in
           Cryptology—EUROCRYPT 1991*, volume 547 of *Lecture Notes in Com-
           puter Science*, pages 257–65. Springer-Verlag, 1991.

[GOS06a]   Jens Groth, Rafail Ostrovsky, and Amit Sahai. Non-interactive zaps and
           new techniques for NIZK. Manuscript, 2006.

[GOS06b]   Jens Groth, Rafail Ostrovsky, and Amit Sahai. Perfect non-interactive
           zero knowledge for NP. In *Advances in Cryptology—EUROCRYPT 2006*,
           Lecture Notes in Computer Science. Springer-Verlag, 2006. To appear.

[GS02]   Craig Gentry and Alice Silverberg. Hierarchical ID-based cryptography.
           In *Advances in Cryptology—ASIACRYPT 2002*, Lecture Notes in Com-
           puter Science. Springer-Verlag, 2002.

[KY03]   Aggelos Kiayias and Moti Yung. Extracting group signatures from traitor tracing schemes. In *Advances in Cryptology—EUROCRYPT 2003*, Lecture Notes in Computer Science, pages 630–648. Springer-Verlag, 2003.

[KY04]   Aggelos Kiayias and Moti Yung. Group signatures: Provable security, efficient constructions and anonymity from trapdoor-holders. Cryptology ePrint Archive, Report 2004/076, 2004. http://eprint.iacr.org/.

[KY05]   Aggelos Kiayias and Moti Yung. Group signatures with efficient concurrent join. In *Advances in Cryptology—EUROCRYPT 2005*, Lecture Notes in Computer Science, pages 198–214. Springer-Verlag, 2005.

[LRSW99]   Anna Lysyanskaya, Ron Rivest, Amit Sahai, and Stefan Wolf. Pseudonym systems. In *Proceedings of SAC 1999*, volume 1758 of *Lecture Notes in Computer Science*, pages 184–99. Springer-Verlag, 1999.

[Mil04]   Victor Miller. The Weil pairing, and its efficient calculation. *Journal of Cryptology*, 17(4), 2004.

[Nao03]   Moni Naor. On cryptographic assumptions and challenges. In *Advances in Cryptology—CRYPTO 2003*, Lecture Notes in Computer Science, pages 96–109. Springer-Verlag, 2003.

[Son01]   Dawn Xiaodong Song. Practical forward secure group signature schemes. In *ACM Conference on Computer and Communications Security—CCS 2001*, pages 225–234, 2001.

[Wat05]   Brent Waters. Efficient identity-based encryption without random oracles. In *Advances in Cryptology—EUROCRYPT 2005*, volume 3494 of *Lecture Notes in Computer Science*. Springer-Verlag, 2005.

A   Group Signatures

A group signature scheme consists of a pentuple of PPT algorithms:

- A group setup algorithm, *Setup*, that takes as input a security parameter 1^λ (in unary) and the number of signers in the group, for simplicity taken as a power of two, 2^k, and outputs a public key PK for verifying signatures, a master key MK for enrolling group members, and a tracing key TK for identifying signers.
- An enrollment algorithm, *Enroll*, that takes the master key MK and an identity ID, and outputs a unique identifier s_{ID} and a private signing key K_{ID} which is to be given to the user.
- A signing algorithm, *Sign*, that takes a group member's private signing key K_{ID} and a message M, and outputs a signature σ.
- A (usually deterministic) verification algorithm, *Verify*, that takes a message M, a signature σ, and a group verification key PK, and outputs either valid or invalid.
- A (usually deterministic) tracing algorithm, *Trace*, that takes a valid signature σ and a tracing key TK, and outputs an identifier s_{ID} or the failure symbol $\perp$.

There are four types of entities one must consider:

- The group master, which sets up the group and issues private keys to the users. Often, the group master is an ephemeral entity, and the master key

MK is destroyed once the group is set up. Alternatively, techniques from distributed cryptography can be used to realize the group master functionality without any real party becoming in possession of the master key.

- The group manager, which is given the ability to identify signers using the tracing key TK, but not to enroll users or create new signing keys for existing users.
- Regular member users, or signers, which are each given a distinct private signing key K_{ID}.
- Outsiders, or verifiers, who can only verify signatures using the public key PK.

We require the following correctness and security properties.

Consistency. The consistency requirements are such that, whenever, (for a group of 2^k users),

$$(\text{PK}, \text{MK}, \text{TK}) \leftarrow Setup(1^\lambda, 2^k), \quad (s_{\text{ID}}, K_{\text{ID}}) \leftarrow Enroll(\text{MK}, \text{ID}), \quad \sigma \leftarrow Sign(K_{\text{ID}}, M),$$

we have, (except with negligible probability over the random bits used in *Verify* and *Trace*),

$$Verify(M, \sigma, \text{PK}) = \texttt{valid}, \quad \text{and} \quad Trace(\sigma, \text{TK}) = s_{\text{ID}}.$$

The unique identifier s_{ID} can be used to assist in determining the user ID from the transcript of the *Enroll* algorithm; s_{ID} may but need not be disclosed to the user; it may be the same as ID.

Security. Bellare, Micciancio, and Warinschi [BMW03] characterize the fundamental properties of group signatures in terms of two crucial security properties from which a number of other properties follow. The two important properties are:

Full Anonymity which requires that no PPT adversary be able to decide (with non-negligible probability in excess of one half) whether a challenge signature σ on a message M emanates from user ID_1 or ID_2, where ID_1, ID_2, and M are chosen by the adversary. In the original definition of [BMW03], the adversary is given access to a tracing oracle, which it may query before and after being given the challenge σ, much in the fashion of IND-CCA2 security for encryption.

Boneh, Boyen, and Shacham [BBS04] relax this definition by withholding access to the tracing oracle, thus mirroring the notion of IND-CPA security for encryption. We follow [BBS04] and speak of *CCA2-full anonymity* and *CPA-full anonymity* respectively.

Full Traceability which requires that no coalition of users be able to generate, in polynomial time, a signature that passes the *Verify* algorithm but fails to trace to a member of the coalition under the *Trace* algorithm. According to this notion, the adversary is allowed to ask for the private keys of any user of its choice, adaptively, and is also given the secret key TK meant for tracing—but of course not the enrollment master key MK.

It is noted in [BMW03] that this property implies that of *exculpability* [AT99], which is the requirement that no party, not even the group manager, should be able to frame a honest group member as the signer of a signature he did not make. However, the model of [BMW03] does not consider the possibility of a (long-lived) group master, which could act as a potential framer. To address this problem and achieve the notion of *strong exculpability*, introduced in [ACJT00] and formalized in [KY04, BSZ05], one would need an interactive enrollment protocol, call *Join*, at the end of which only the user himself knows his full private key. We do not further consider exculpability issues in this paper.

We refer the reader mainly to [BMW03] for more precise definitions of these and related notions.

Practical Identity-Based Encryption
Without Random Oracles

Craig Gentry*

Stanford University
cgentry@cs.stanford.edu

Abstract. We present an Identity Based Encryption (IBE) system that is fully secure in the standard model and has several advantages over previous such systems – namely, computational efficiency, shorter public parameters, and a "tight" security reduction, albeit to a stronger assumption that depends on the number of private key generation queries made by the adversary. Our assumption is a variant of Boneh et al.'s decisional Bilinear Diffie-Hellman Exponent assumption, which has been used to construct efficient hierarchical IBE and broadcast encryption systems. The construction is remarkably simple. It also provides recipient anonymity automatically, providing a second (and more efficient) solution to the problem of achieving anonymous IBE without random oracles. Finally, our proof of CCA2 security, which has more in common with the security proof for the Cramer-Shoup encryption scheme than with security proofs for other IBE systems, may be of independent interest.

Keywords: Identity Based Encryption.

1   Introduction

An Identity Based Encryption (IBE) system [25, 8] is a public key encryption system in which a user's public key may be an arbitrary string, such as an email address or other identifier. The user's private key is generated by a trusted authority, called a Private Key Generator (PKG), which applies its master key to the user's identity after the user authenticates itself. Shamir [25] proposed the notion of IBE in 1984 as a way to simplify public key and certificate management. Rather than obtaining the disparate public keys of its intended recipients separately, a message sender who knows the identities of its recipients needs only to obtain the public parameters of the PKG; public key certificates are eliminated altogether.

Boneh and Franklin [8, 9] described the first secure and truly practical IBE system. Their system uses bilinear maps (or "pairings"), and they proved its security in the random oracle model. Canetti et al. [15] presented an IBE system whose security could proven without random oracles, but in a weaker "selective-ID" model, in which the adversary must declare at the beginning of its attack

---

* Supported by the Herbert Kunzel Stanford Graduate Fellowship.

which identity it will target. Boneh and Boyen [4] provided more practical IBE systems in the selective-ID model. Shortly thereafter, Boneh and Boyen [5] presented a fully secure scheme – i.e., one in which the adversary may choose the target identity adaptively – without random oracles. Waters [27] simplified the scheme described in [5], substantially improving its efficiency.

PREVIOUS IBE SYSTEMS. Moni Naor observed that every IBE system secure against an adaptive-ID attack (as defined by Boneh and Franklin in [8]) implies a signature scheme secure against existential forgery under a chosen-message attack. The generic transformation is as follows: the PKG's parameters correspond to the public key of the signature scheme; private key generation queries to the PKG correspond to signature queries. If an adversary of the signature scheme can forge a signature on an unqueried message, it can generate a private key for an unqueried identity, thus breaking the IBE system. So, to design a secure IBE system, one begins (in some sense) by designing a secure signature scheme.

A common strategy for proving the security of a signature scheme in the random oracle model – e.g., for RSA with full-domain-hash – is as follows. The simulator responds to hash queries in such a way that it can generate a signature on most messages, but not all. The simulator aborts if the adversary requests a signature on a message that it cannot sign, or if the adversary's forgery is on a message that the simulator knows how to sign already. One can also use this strategy to design a secure private key generation procedure for an IBE system. Boneh and Franklin [8] did precisely that; the private key generation procedure in their system is essentially equivalent to the BLS signature scheme [12], which uses the proof strategy just described. (Though, inconveniently for our narrative, Boneh and Franklin's IBE system slightly pre-dates its associated signature scheme.)

When Boneh and Boyen [5] and later Waters [27] devised IBE systems fully secure without random oracles, their main innovation was in the private key generation procedures. Each of these procedures corresponds to a signature scheme that is fully secure (i.e., against a chosen-message attack) without random oracles. Interestingly, though, the (implicit) proof strategy for these standard-model signature schemes is still basically the same as above – i.e., the simulator constructs its public key in such a way that it can generate a signature on most messages, but not all. Since, intuitively speaking, the simulator follows the same strategy except for using its control of the public key (or public parameters, for an IBE system) to compensate for not controlling a random oracle, it should not be surprising that the public parameters for these IBE systems are quite large.

Another side effect of the above proof strategy is that the reduction is loose. If δ is the probability that the simulator can generate a private key for a random identity, then the probability that the simulator does not abort is at most $\delta^q(1 - \delta)$, where q is the number of private key generation queries made by the adversary. Setting $\delta \approx 1 - 1/q$ maximizes this probability at $\mathcal{O}(1/q)$. Thus, the reduction loses a multiplicative factor of q. A lossy reduction is not merely a theoretical problem; if we take the lossiness seriously, we should augment the security parameter to compensate, making the system less efficient.

Almost all of the IBE systems since Boneh-Franklin follow the "common strategy" for proving security; consequently, they suffer from long parameters (when security is proven in the standard model) and lossy reductions (in the standard model or the random oracle model). However, we note a couple of exceptions. The IBE systems described in [4] have short parameters and achieve a tight reduction, but this is because they are proven secure only against selective-ID attacks. As noted in [4], one can generally transform a selective-ID scheme into a fully secure scheme by having the simulator guess which identity the adversary will ultimately select, but this transformation loosens the reduction by huge multiplicative factor – namely, by the total number of identities – that is super-polynomial and (much) larger than q. This transformation is also a very unsatisfying approach from a theoretical point of view. A second exception is the IBE system by Katz and Wang [23], which achieves a tight reduction in the random oracle model. In their system, the encryption of M under identity ID effectively consists of two ciphertexts under each of the derived identities $H(\text{ID}, 0)$ and $H(\text{ID}, 1)$ (for hash function H modeled as random oracle). Through its control of the random oracle, the simulator ensures that, for each ID, it knows the private key for exactly one of $H(\text{ID}, 0)$ and $H(\text{ID}, 1)$. It can thus answer any key generation query. The successful adversary partially decrypts the challenge ciphertext with the "wrong" private key with probability $1/2$, giving the simulator useful information. Though this system relies heavily on the random oracle model, it illustrates how a tight reduction for an IBE system can be achieved when the simulator can generate a private key for every identity. A recent paper [2] discusses the Katz-Wang system in detail.

Currently, there is no IBE system that is fully secure without random oracles, yet has short public parameters, or has a tight security reduction. Given this state of affairs, several papers [4, 5, 27] have encouraged work on the open problem of tight security; Waters posed [27] the open problem regarding compact public parameters.

OUR CONTRIBUTIONS. We present an IBE system that is fully secure without random oracles and has several advantages over previous such systems, including:

– Short public parameters (5 group elements for CCA2 security)
– A tight reduction, albeit based on a stronger assumption (see below)
– Recipient-anonymity

Our constructions are simple and efficient. For example, in the construction described in Section 4.1, which we prove secure against adaptive-ID and adaptive chosen-ciphertext attacks, a ciphertext consists of four group elements. Encryption and decryption require only a small constant number of group operations, while user private keys and the PKG's public parameters are compact. Compare, for example, the public parameters in our IBE system (five group elements and a hash function) to those in [27] ($n + 4$ group elements, where an identity is a bitstring of length n).

An IBE system is recipient-anonymous, roughly speaking, if it hard for an eavesdropper to distinguish which identity was used to generate a given ciphertext. Boneh et al. [7] discuss how anonymous IBE is useful in the context of searchable public key encryption; Abdalla et al. [1] propose the open problem of finding an anonymous IBE system secure without random oracles. Boyen and Waters recently presented the first such anonymous IBE system at the rump session of Crypto 2005 (see [14]). Our IBE system represents a second, but more efficient, solution to this problem; it gives recipient-anonymity basically "for free." The security proof for our scheme is also much simpler. However, we note that the Boyen-Waters approach offers *hierarchical* anonymous IBE.

Regarding the open problem of constructing an IBE system with a tight security reduction, our contribution is less clear. Our decision q-ABDHE assumption, discussed in Section 2.3, is related to the q-BDHE assumption, which has been used to construct efficient hierarchical IBE and broadcast encryption systems [6, 10], but it is stronger than the decision BDH assumption used in [5, 27]. We obtain a tight reduction based on q-ABDHE in the sense that the simulator's time complexity and success probability are identical to that of the adversary in breaking the system, except for *additive* factors depending on q. However, since our assumption is stronger, we cannot claim that a tighter reduction is necessarily an improvement. Moreover, it is not obvious what it means to have an asymptotically tight reduction based on the q-ABDHE assumption, since this assumption varies as q varies. However, we can analyze the concrete security of our system for specific values of q, as we do in Section 3.3. One conclusion of this analysis is that if we assume decision q-ABDHE is no easier than decision BDH (which may or may not be true), then our tighter reduction (for specific reasonable values of q) allows us to choose a smaller security parameter, adding to the efficiency advantages of our scheme. But perhaps this is not a very satisfying "solution" to the open problem; certainly, it would be preferable to obtain a tight reduction under a more natural assumption, such as decision BDH.

A final contribution of this paper is our proof technique, which differs substantially from the "common strategy" described above. Interestingly, our proof strategy draws inspiration from the Cramer-Shoup signature scheme [18] (and strong-RSA based signature schemes, generally) for our private key generation procedure, as well as from the Cramer-Shoup encryption scheme [17] for our approach to proving security against chosen-ciphertext attacks.

Strong-RSA based signatures typically achieve a tight reduction and have short public keys. Intuitively, this is related to the fact that, in the reduction, the simulator can produce a signature for any message. Similarly, unlike in previous IBE systems fully secure in the standard model, the simulator in our reduction can generate a private key for any identity. One can view our private key generation procedure as a strongly existentially unforgeable signature scheme that is "tightly" secure in the standard model under the q-strong DH assumption: that it is hard to compute a pair $(c, g^{1/(\alpha-c)})$ given $\{g^{\alpha^i} : i \in [0,q]\}$, where q corresponds to the anticipated number of queries. The savvy reader may notice that this signature scheme has direct analogue based on strong RSA. In the procedure, the PKG (signer) publishes groups $\mathbb{G}$ and $\mathbb{G}_T$, and bilinear map $e : \mathbb{G} \times \mathbb{G} \to \mathbb{G}_T$, along with generators $g, g_1, h \in \mathbb{G}$, where $g_1 = g^\alpha$. A private key for identity $\mathsf{ID} \in \mathbb{Z}_p$ is a pair $(r_\mathsf{ID}, h_\mathsf{ID})$, where $r_\mathsf{ID} \in \mathbb{Z}_p$ and $h_\mathsf{ID} = (hg^{-r_\mathsf{ID}})^{1/(\alpha-\mathsf{ID})}$; if the private key for ID is requested more than once, the PKG uses the same value of r_{ID} each time. In the reduction, the simulator is given $g_i = g^{(\alpha^i)}$ for all $i \in [0, q]$, where q is (roughly) the anticipated number private key generation queries. Given $\{g_i\}$, the simulator computes h by generating a random q-degree polynomial $f(x) \in \mathbb{Z}_p[x]$, and setting $h = g^{f(\alpha)}$. To generate a private key for ID, it sets $r_{\mathsf{ID}} = f(\mathsf{ID})$ and $h_{\mathsf{ID}} = (hg^{-r_{\mathsf{ID}}})^{1/(\alpha - \mathsf{ID})} = g^{(f(\alpha) - f(\mathsf{ID}))/(\alpha - \mathsf{ID})}$; the simulator can compute the latter value from $\{g_i\}$, since $(f(x) - f(\mathsf{ID}))/(x - \mathsf{ID})$ is a $(q-1)$-degree polynomial in x. The values of r_{ID_i} in the simulation for $i \in [1, q]$ appear uniformly random, since $f(x)$ is a random polynomial of degree q. If the adversary can generate a private key $(r'_{\mathsf{ID}}, h'_{\mathsf{ID}})$ for ID for which $r'_{\mathsf{ID}} \neq r_{\mathsf{ID}}$, the simulator can efficiently compute $g^{1/(\alpha - \mathsf{ID})}$.

The fact that the simulator in our system can generate exactly one private key for any identity dovetails nicely with the proof strategy used in the Cramer-Shoup encryption scheme, where the simulator actually knows exactly one valid decryption key: its scalars (x_1, y_1, z_1), along with the dependent values (x_2, y_2, z_2). Roughly speaking, in their proof, Cramer and Shoup show that these scalars remain unconditionally hidden from the adversary (with overwhelming probability), and thus the adversary cannot (except with negligible probability) construct an invalid ciphertext that passes the simulator's validity test, or guess with advantage how the simulator would decrypt its own challenge ciphertext when that challenge ciphertext is incorrectly distributed. How do we adapt their technique to our (multi-user) IBE system? We augment the public parameters to include group elements $h_1, h_2, h_3 \in \mathbb{G}$ (rather than just h), where $h_i = g^{f_i(\alpha)}$ and $f_i(x) \in \mathbb{Z}_p[x]$ is a random and independent q-degree polynomial. The three scalars $r_{\mathsf{ID},i} = f_i(\mathsf{ID})$, which a user receives as part of its private key, play a role analogous to the scalars $z_1, x_1,$ and y_1, respectively, in Cramer-Shoup; the values $r_{\mathsf{ID},2}$ and $r_{\mathsf{ID},3}$ are used in a projective-hash ciphertext validity test. The three scalars remain hidden from the adversary with overwhelming probability, even if the adversary obtains the scalars $r_{\mathsf{ID}',i} = f_i(\mathsf{ID}')$ for less than $q - 1$ identities $\mathsf{ID}' \neq \mathsf{ID}$, since $f_i(x)$ is random and has degree q. Interestingly, previous IBE systems fully secure without random oracles use an entirely different approach to proving chosen-ciphertext security. They employ results by Canetti et al. [16] (later improved by Boneh and Katz [11] and further by Boyen, Mei and Waters [13]) that a chosen-ciphertext-secure IBE system follows from a chosen-plaintext-secure 2-level hierarchical IBE system.

2   Preliminaries

Below, we review the definition of security for an IBE system. We also review the definition of a bilinear map and discuss the complexity assumption on which the security of our system is based.

2.1   Security Model for Identity-Based Encryption

An IBE system consists of four algorithms [25, 8]: *Setup, KeyGen, Encrypt,* and *Decrypt. Setup* establishes the PKG's parameters *params* and a master key *master-key*. *KeyGen* applies the *master-key* to an identity to generate the private key for that identity. *Encrypt* takes a message, an identity and *params* as input, and outputs a ciphertext. *Decrypt* decrypts a ciphertext for an identity using a private key for that identity.

Boneh and Franklin [8, 9] define chosen ciphertext security for IBE systems under a chosen identity attack via the following game.

Setup: The challenger runs *Setup*, and forwards *params* to the adversary.

Phase 1: Proceeding adaptively, the adversary issues queries $q_1, \ldots, q_m$ where q_i is one of the following:
  – Key generation query $\langle ID_i \rangle$: the challenger runs *KeyGen* on ID_i and forwards the resulting private key to the adversary.
  – Decryption query $\langle ID_i, C_i \rangle$. The challenger runs *KeyGen* on ID_i, decrypts C_i with the resulting private key, and sends the result to the adversary.

Challenge: The adversary submits two plaintexts $M_0, M_1 \in \mathcal{M}$ and an identity ID. ID must not have appeared in any key generation query in Phase 1. The challenger selects a random bit $b \in \{0, 1\}$, sets $C = Encrypt(params, \text{ID}, M_b)$, and sends C to the adversary as its challenge ciphertext.

Phase 2: This is identical to Phase 1, except that the adversary may not request a private key for ID or the decryption of (ID, C).

Guess: The adversary submits a guess $b' \in \{0, 1\}$. The adversary wins if $b = b'$.

We call an adversary $\mathcal{A}$ in the above game a IND-ID-CCA adversary.

Definition 1. *An IBE system is* $(t, q_{\text{ID}}, q_C, \epsilon)$ *IND-ID-CCA* *secure if all t-time* IND-ID-CCA *adversaries making at most* q_{ID} *private key queries and at most* q_C *chosen ciphertext queries have advantage at most* ϵ *in winning the above game.*

IND-ID-CPA security is defined similarly, but with the restriction that the adversary cannot make decryption queries.

Definition 2. *An IBE system is* $(t, q_{\text{ID}}, \epsilon)$ *IND-ID-CPA* *secure if it is* $(t, q_{\text{ID}}, 0, \epsilon)$ IND-ID-CCA *secure.*

Recipient-Anonymity. Informally, we say that an IBE system is anonymous if an adversary cannot distinguish the public key ID under which a ciphertext was generated. More formally, we can incorporate anonymity into our game above through the following simple modification. In the Challenge phase, the adversary outputs two identities ID_0 and ID_1 not queried in Phase 1 and two messages M_0 and M_1. The challenger picks two random bits $b, c \in \{0, 1\}$, uses ID_b to encrypt M_c, and sends the resulting ciphertext C to the adversary. Phase 2 is like Phase 1, except that the adversary cannot request a private key for ID_0 or ID_1, or the decryption of C under either identity. Finally, in the Guess phase, the adversary guesses two bits b', c' and wins if $b = b'$ and $c = c'$; we define the adversary's advantage in this game to be $|\Pr[b = b' \wedge c = c'] - \frac{1}{4}|$.

Definition 3. *We say that an IBE system $\mathcal{E}$ is* $(t, q_{\text{ID}}, q_C, \epsilon)$ *ANON-IND-ID-CCA* *secure if all t-time* ANON-IND-ID-CCA *adversaries making at most* q_{ID} *private key queries and at most* q_C *chosen ciphertext queries have advantage at most* ϵ *in the modified game. We define* ANON-IND-ID-CPA *security similarly.*

2.2   Bilinear Maps

We review bilinear maps, using the following standard notation [8, 4, 27]:

1. $\mathbb{G}$ and $\mathbb{G}_T$ are two (multiplicative) cyclic groups of prime order p;
2. g is a generator of $\mathbb{G}$.
3. $e : \mathbb{G} \times \mathbb{G} \to \mathbb{G}_T$ is a bilinear map.

Let $\mathbb{G}$ and $\mathbb{G}_T$ be two groups as above. A bilinear map is a map $e : \mathbb{G} \times \mathbb{G} \to \mathbb{G}_T$ with the following properties:

1. Bilinear: for all $u, v \in \mathbb{G}$ and $a, b \in \mathbb{Z}$, we have $e(u^a, v^b) = e(u, v)^{ab}$.
2. Non-degenerate: $e(g, g) \neq 1$.

We say that $\mathbb{G}$ is a bilinear group if the group action in $\mathbb{G}$ can be computed efficiently and there exists a group $\mathbb{G}_T$ and an efficiently computable bilinear map $e : \mathbb{G} \times \mathbb{G} \to \mathbb{G}_T$ as above. Note that $e(,)$ is symmetric since $e(g^a, g^b) = e(g, g)^{ab} = e(g^b, g^a)$.

2.3   Complexity Assumptions

The security of our system is based on a complexity assumption that we call the decisional augmented bilinear Diffie-Hellman exponent assumption (decisional ABDHE). First, we recall the q-BDHE problem [6, 10], which is as follows: Given a vector of $2q + 1$ elements

$$\left(g', g, g^\alpha, g^{(\alpha^2)}, \ldots, g^{(\alpha^q)}, g^{(\alpha^{q+2})}, \ldots, g^{(\alpha^{2q})}\right) \in \mathbb{G}^{2q+1}$$

as input, output $e(g, g')^{(\alpha^{q+1})} \in \mathbb{G}_T$. Since the input vector is missing the term $g^{(\alpha^{q+1})}$, the bilinear map does not seem to help compute $e(g, g')^{(\alpha^{q+1})}$.

We define the q-ABDHE problem almost identically: Given a vector of $2q + 2$ elements

$$\left(g', g'^{(\alpha^{q+2})}, g, g^\alpha, g^{(\alpha^2)}, \ldots, g^{(\alpha^q)}, g^{(\alpha^{q+2})}, \ldots, g^{(\alpha^{2q})}\right) \in \mathbb{G}^{2q+2}$$

as input, output $e(g, g')^{(\alpha^{q+1})} \in \mathbb{G}_T$. Introducing the additional term $g'^{(\alpha^{q+2})}$ still does not appear to ease the computation of $e(g, g')^{(\alpha^{q+1})}$, since the input vector is missing the term $g^{(\alpha^{-1})}$.

The q-ABDHE problem is actually more than we need for our IBE system. Instead, we can use a truncated version of the q-ABDHE problem, in which the terms $(g^{(\alpha^{q+2})}, \ldots, g^{(\alpha^{2q})})$ are omitted from the input vector. Clearly, the truncated q-ABDHE problem is hard if the q-ABDHE problem is hard. An algorithm $\mathcal{A}$ has advantage ϵ in solving truncated q-ABDHE if

$$\Pr\left[\mathcal{A}\left(g', g'_{q+2}, g, g_1, \ldots, g_q\right) = e(g_{q+1}, g')\right] \geq \epsilon$$

where we use g_i and g'_i to denote $g^{(\alpha^i)}$ and $g'^{(\alpha^i)}$, and where the probability is over the random choice of generators g, g' in $\mathbb{G}$, the random choice of α in $\mathbb{Z}_p$, and the random bits used by $\mathcal{A}$.

The decisional version of truncated q-ABDHE is defined as one would expect. An algorithm $\mathcal{B}$ that outputs $b \in \{0, 1\}$ has advantage ϵ in solving truncated decision q-ABDHE if

$$\left| \Pr\left[\mathcal{B}\big(g', g'_{q+2}, g, g_1, \ldots, g_q, e(g_{q+1}, g')\big) = 0 \right] \right.$$
$$\left. - \Pr\left[\mathcal{B}\big(g', g'_{q+2}, g, g_1, \ldots, g_q, Z\big) = 0 \right] \right| \geq \epsilon$$

where the probability is over the random choice of generators g, g' in $\mathbb{G}$, the random choice of α in $\mathbb{Z}_p$, the random choice of $Z \in \mathbb{G}_T$, and the random bits consumed by $\mathcal{B}$. We refer to the distribution on the left as $\mathcal{P}_{ABDHE}$ and the distribution on the right as $\mathcal{R}_{ABDHE}$.

Definition 4. *We say that the truncated (decision) (t, ϵ, q)-ABDHE assumption holds in $\mathbb{G}$ if no t-time algorithm has advantage at least ϵ in solving the truncated (decision) q-ABDHE problem in $\mathbb{G}$.*

As an aside, we note that the truncated q-ABDHE problem is also closely related to the q-bilinear Diffie-Hellman inversion (q-BDHI) problem, which has been used to construct an IBE system secure without random oracles under a selective-ID attack [4] and a verifiable random function [20]. Specifically, let us define the q-augmented BDHI (q-ABDHI) problem as follows: given a vector of $q + 2$ elements

$$\left(g^{(\alpha^{-q-2})}, g, g^{\alpha}, g^{(\alpha^2)}, \ldots, g^{(\alpha^q)} \right) \in \mathbb{G}^{q+1}$$

as input, output $e(g, g)^{1/\alpha} \in \mathbb{G}_T$. The q-ABDHI problem is identical to the q-BDHI problem, except that the former adds the term $g^{(\alpha^{-q-2})}$ to the input vector, which does not seem to help compute $e(g, g)^{1/\alpha}$. One can reduce (decision) q-ABDHI to truncated (decision) q-ABDHE simply by setting $(g', g'^{\alpha^{q+2}}) = ((g^{(\alpha^{-q-2})})^x, g^x)$ for random $x \in \mathbb{Z}_p^*$, and deriving $e(g, g)^{1/\alpha}$ as $e(g_{q+1}, g')^{1/x}$.

3   Construction I: Chosen-Plaintext Security

We now present an efficient IBE system that is ANON-IND-ID-CPA secure without random oracles under the truncated decision $(q_{\mathsf{ID}} + 1)$-ABDHE assumption. Though this construction is substantially similar to the construction presented in Section 4.1, which is ANON-IND-ID-CCA secure, we present this construction separately because there are applications (such as searchable public key encryption [7, 1]) that only require chosen-plaintext security, and because we believe the reader may benefit from seeing this construction's (relatively) simple proof of security without being distracted by the additional machinery needed to prove chosen-ciphertext security.

3.1 Construction

Let $\mathbb{G}$ and $\mathbb{G}_T$ be groups of order p, and let $e : \mathbb{G} \times \mathbb{G} \to \mathbb{G}_T$ be the bilinear map. The IBE system works as follows.

Setup: The PKG picks random generators $g, h \in \mathbb{G}$ and random $\alpha \in \mathbb{Z}_p$. It sets $g_1 = g^\alpha \in \mathbb{G}$. The public *params* and private *master-key* are given by

$$params = (g, g_1, h) \qquad master\text{-}key = \alpha \ .$$

KeyGen: To generate a private key for identity $\mathsf{ID} \in \mathbb{Z}_p$, the PKG generates random $r_\mathsf{ID} \in \mathbb{Z}_p$, and outputs the private key

$$d_\mathsf{ID} = (r_\mathsf{ID}, \ h_\mathsf{ID}), \quad \text{where} \quad h_\mathsf{ID} = (hg^{-r_\mathsf{ID}})^{1/(\alpha - \mathsf{ID})} \ .$$

If $\mathsf{ID} = \alpha$, the PKG aborts. We require that the PKG always use the same random value r_ID for ID. This can be accomplished, for example, using a PRF or an internal log to ensure consistency.

Encrypt: To encrypt $m \in \mathbb{G}_T$ using identity $\mathsf{ID} \in \mathbb{Z}_p$, the sender generates random $s \in \mathbb{Z}_p$ and sends the ciphertext

$$C = (g_1^s g^{-s \cdot \mathsf{ID}}, \ e(g, g)^s, \ m \cdot e(g, h)^{-s}) \ .$$

Notice that encryption does not require any pairing computations once $e(g, g)$ and $e(g, h)$ have been pre-computed. Alternatively, $e(g, g)$ and $e(g, h)$ can be included in the system parameters, in which case h can be dropped.

Decrypt: To decrypt ciphertext $C = (u, v, w)$ with ID, the recipient outputs

$$m \ = \ w \cdot e(u, h_\mathsf{ID}) v^{r_\mathsf{ID}} \ .$$

Correctness: Assuming the ciphertext is well-formed for ID:

$$e(u, h_\mathsf{ID}) v^{r_\mathsf{ID}} \ = \ e(g^{s(\alpha - \mathsf{ID})}, h^{1/(\alpha - \mathsf{ID})} g^{-r_\mathsf{ID}/(\alpha - \mathsf{ID})}) e(g, g)^{s r_\mathsf{ID}} \ = \ e(g, h)^s \ ,$$

as required.

Intuitively, the recipient can decrypt because it possess a $(\alpha - \mathsf{ID})$-th root of h (after h is perturbed by g^{-r_ID}). When this is paired with u, a $(\alpha - \mathsf{ID})$-th power of g^s, the recipient obtains the mask $e(g, h)^s$ after removing the perturbation.

3.2 Security

We now prove that the above IBE system is ANON-IND-ID-CPA secure under the truncated decision $(q_\mathsf{ID} + 1)$-ABDHE assumption.

Theorem 1. *Let $q = q_\mathsf{ID} + 1$. Assume the truncated decision (t, ϵ, q)-ABDHE assumption holds for $(\mathbb{G}, \mathbb{G}_T, e)$. Then, the above IBE system is $(t', \epsilon', q_\mathsf{ID})$ ANON-IND-ID-CPA secure for $t' = t - \mathcal{O}(t_{exp} \cdot q^2)$ and $\epsilon' = \epsilon + 2/p$, where t_{exp} is the time required to exponentiate in $\mathbb{G}$.*

Proof. Let $\mathcal{A}$ be an adversary that $(t', \epsilon', q_{\mathsf{ID}})$-breaks the ANON-IND-ID-CPA security of the IBE system described above. We construct an algorithm, $\mathcal{B}$, that solves the truncated decision q-ABDHE problem, as follows. $\mathcal{B}$ takes as input a random truncated decision q-ABDHE challenge $(g', g'_{q+2}, g, g_1, \ldots, g_q, Z)$, where Z is either $e(g_{q+1}, g')$ or a random element of $\mathbb{G}_T$ (recall that $g_i = g^{(\alpha^i)}$). Algorithm $\mathcal{B}$ proceeds as follows.

Setup: $\mathcal{B}$ generates a random polynomial $f(x) \in \mathbb{Z}_p[x]$ of degree q. It sets $h = g^{f(\alpha)}$, computing h from $(g, g_1, \ldots, g_q)$. It sends the public key (g, g_1, h) to $\mathcal{A}$. Since g, α, and $f(x)$ are chosen uniformly at random, h is uniformly random and this public key has a distribution identical to that in the actual construction.

Phase 1: $\mathcal{A}$ makes key generation queries. $\mathcal{B}$ responds to a query on $\mathsf{ID} \in \mathbb{Z}_p$ as follows. If $\mathsf{ID} = \alpha$, $\mathcal{B}$ uses α to solve truncated decision q-ABDHE immediately. Else, let $F_{\mathsf{ID}}(x)$ denote the $(q-1)$-degree polynomial $(f(x) - f(\mathsf{ID}))/(x - \mathsf{ID})$. $\mathcal{B}$ sets the private key $(r_{\mathsf{ID}}, h_{\mathsf{ID}})$ to be $(f(\mathsf{ID}), g^{F_{\mathsf{ID}}(\alpha)})$. This is a valid private key for ID, since $g^{F_{\mathsf{ID}}(\alpha)} = g^{(f(\alpha) - f(\mathsf{ID}))/(\alpha - \mathsf{ID})} = (hg^{-f(\mathsf{ID})})^{1/(\alpha - \mathsf{ID})}$, as required. We will describe why this private key appears to $\mathcal{A}$ to be correctly distributed below.

Challenge: $\mathcal{A}$ outputs identities $\mathsf{ID}_0, \mathsf{ID}_1$ and messages M_0, M_1. Again, if $\alpha \in \{\mathsf{ID}_0, \mathsf{ID}_1\}$, $\mathcal{B}$ uses α to solve truncated decision q-ABDHE immediately. Else, $\mathcal{B}$ generates bits $b, c \in \{0, 1\}$, and computes a private key $(r_{\mathsf{ID}_b}, h_{\mathsf{ID}_b})$ for ID_b as in Phase 1. Let $f_2(x) = x^{q+2}$ and let $F_{2,\mathsf{ID}_b}(x) = (f_2(x) - f_2(\mathsf{ID}_b))/(x - \mathsf{ID}_b)$, which is a polynomial of degree $q + 1$. $\mathcal{B}$ sets

$$u = g'^{f_2(\alpha) - f_2(\mathsf{ID}_b)}, \quad v = Z \cdot e(g', \prod_{i=0}^{q} g^{F_{2,\mathsf{ID}_b, i} \alpha^i}) \quad w = M_c/e(u, h_{\mathsf{ID}_b}) v^{r_{\mathsf{ID}_b}} ,$$

where $F_{2,\mathsf{ID}_b,i}$ is the coefficient of x^i in $F_{2,\mathsf{ID}_b}(x)$. It sends (u, v, w) to $\mathcal{A}$ as the challenge ciphertext.

   Let $s = (\log_g g') F_{2,\mathsf{ID}_b}(\alpha)$. If $Z = e(g_{q+1}, g')$, then $u = g^{s(\alpha - \mathsf{ID}_b)}$, $v = e(g, g)^s$, and $M_c/w = e(u, h_{\mathsf{ID}_b}) v^{r_{\mathsf{ID}_b}} = e(g, h)^s$; thus (u, v, w) is a valid ciphertext for (ID_b, M_c) under randomness s. Since $\log_g g'$ is uniformly random, s is uniformly random, and so (u, v, w) is a valid, appropriately-distributed challenge to $\mathcal{A}$.

Phase 2: $\mathcal{A}$ makes key generation queries, and $\mathcal{B}$ responds as in Phase 1.

Guess: Finally, the adversary outputs guesses $b', c' \in \{0, 1\}$. If $b = b'$ and $c = c'$, $\mathcal{B}$ outputs 0 (indicating that $Z = e(g_{q+1}, g')$); otherwise, it outputs 1.

Perfect Simulation: When $Z = e(g_{q+1}, g')$, the public key and challenge ciphertext issued by $\mathcal{B}$ comes from a distribution identical to that in the actual construction; however, we still must show that the private keys issued by $\mathcal{B}$ are appropriately distributed. Let $\mathcal{I}$ be a set consisting of α, ID_b, and the identities queried by $\mathcal{A}$; observe that $|\mathcal{I}| \leq q + 1$. To show that the keys issued by $\mathcal{B}$ are appropriately distributed, it suffices to show that, from $\mathcal{A}$'s view, the values $\{f(a) : a \in \mathcal{I}\}$ are uniformly random and independent. But this follows from the fact that $f(x)$ is a uniformly random polynomial of degree q.

Probability Analysis: If $Z = e(g_{q+1}, g')$, then the simulation is perfect, and $\mathcal{A}$ will guess the bits (b, c) correctly with probability $1/4 + \epsilon'$. Else, Z is uniformly random, and thus (u, v) is a uniformly random and independent element of $\mathbb{G} \times \mathbb{G}_T$. In this case, the inequalities $v \neq e(u, g)^{1/(\alpha - \mathsf{ID}_0)}$ and $v \neq e(u, g)^{1/(\alpha - \mathsf{ID}_1)}$ both hold with probability $1 - 2/p$. When these inequalities hold, the value of

$$e(u, h_{\mathsf{ID}_b})v^{r_{\mathsf{ID}_b}} = e(u, (hg^{-r_{\mathsf{ID}_b}})^{1/(\alpha - \mathsf{ID}_b)})v^{r_{\mathsf{ID}_b}} = e(u, h)^{\alpha - \mathsf{ID}_b}(v/e(u, g)^{1/(\alpha - \mathsf{ID}_b)})^{r_{\mathsf{ID}_b}}$$

is uniformly random and independent from $\mathcal{A}$'s view (except for the value w), since r_{ID_b} is uniformly random and independent from $\mathcal{A}$'s view (except for the value w). Thus, w is uniformly random and independent, and (u, v, w) can impart no information regarding the bits (b, c).

Assuming that no queried identity equals α (which would only increase $\mathcal{B}$'s success probability), we see that $|\Pr[\mathcal{B}(g', g'_{q+2}, g, g_1, \ldots, g_q, Z) = 0] - 1/4| \leq 2/p$ when $(g', g'_{q+2}, g, g_1, \ldots, g_q, Z)$ is sampled from $\mathcal{R}_{ABDHE}$. However, we have that $|\Pr[\mathcal{B}(g', g'_{q+2}, g, g_1, \ldots, g_q, Z) = 0] - 1/4| \geq \epsilon'$ when $(g', g'_{q+2}, g, g_1, \ldots, g_q, Z)$ is sampled from $\mathcal{P}_{ABDHE}$. Thus, for uniformly random g, g', α and Z, we have that

$$\left| \Pr\left[\mathcal{B}\big(g', g'_{q+2}, g, g_1, \ldots, g_q, e(g_{q+1}, g')\big) = 0\right] \right.$$

$$\left. - \Pr\left[\mathcal{B}\big(g', g'_{q+2}, g, g_1, \ldots, g_q, Z\big) = 0\right] \right| \geq \epsilon' - 2/p.$$

Time-Complexity: In the simulation, $\mathcal{B}$'s overhead is dominated by computing $g^{F_{\mathsf{ID}}(\alpha)}$ in response to $\mathcal{A}$'s key generation query on ID, where $F_{\mathsf{ID}}(x)$ is a polynomial of degree $q - 1$. Each such computation requires $\mathcal{O}(q)$ exponentiations in $\mathbb{G}$. Since $\mathcal{A}$ makes at most $q - 1$ such queries, $t = t' + \mathcal{O}(t_{exp} \cdot q^2)$.

This concludes the proof of Theorem 1.   □

3.3   Remarks on the Tightness of the Reduction

In the reduction, $\mathcal{B}$'s success probability and time complexity are the same as $\mathcal{A}$'s, except for *additive* factors depending on q. So, one could say that our IBE system has a tight security reduction in the standard model, addressing an open problem posed in [4, 5, 27]. However, it would be misleading to claim that a tight reduction from decision q-ABDHE is necessarily better than the loose reduction from decision BDH (for the IBE systems described by Boneh and Boyen [4] and Waters [27]), for a couple of reasons. First, decision q-ABDHE is a stronger assumption than decision BDH. Second, it is not even obvious what "a tight reduction from decision q-ABDHE" means, since the assumption is not fixed when q varies; it becomes stronger as the number of queries increases. Given these considerations, let's examine the significance (if any) of the "tight reduction" in closer detail.

Not much is known about the relative hardness of the decision q-ABDHE and decision BDH problems; they could be equally hard, or the former could be significantly easier. Decision q-ABDHE is a new problem, less natural and less well-studied than decision BDH, though it seems closely connected to the decision q-BDHE and decision q-BDHI problems that were used in [6, 10, 4, 20]. Interestingly, Boneh et al. [6] give some evidence that the decision q-ABDHE problem is easier to solve in the generic group model. In particular, Boneh et al. [6] show (roughly) that a generic attacker's advantage in deciding whether an element of $\mathbb{G}_T$ equals $g_1^{f(\alpha)}$ – when given oracle access to the group operation and the values $g \in G$, $g_1 \in \mathbb{G}_T$ and $g^{f_i(\alpha)} \in \mathbb{G}$ for polynomials $f_1, \ldots, f_s$ – is at most $(t + 2s + 2)^2 d/2p$, where p is the group order, t is the number of oracle queries, and $d = \max\{\deg(f), \deg(f_1), \ldots, \deg(f_s)\}$. Since $d = q$ for the decision q-ABDHE problem, Boneh et al.'s result suggests that a generic attacker's advantage in decision q-ABDHE may be about q times greater than in decision BDH (for fixed t and p, and assuming $q \ll t$). This factor of q seems to offset the factor of q that we eliminated by making our reduction tight. On the other hand, this generic-group result doesn't tell us much about relative hardness of the decision q-ABDHE and decision BDH problems in the real world, since the fastest algorithms for solving them are likely non-generic (and sub-exponential). Ultimately, it is unclear whether or not our tighter reduction under a stronger assumption improves security.

However, for the sake of argument, let's try to assess the impact of our tighter reduction under the assumption that the decision q-ABDHE and decision BDH problems are equally hard. Since it is not very useful simply to characterize our reduction as "tight" asymptotically, let's make such a statement more precise by fixing reasonable values of q and assessing the security and efficiency implications concretely. Suppose that we want to choose our security parameter such that, to succeed with probability at least ϵ', the time complexity of $\mathcal{A}$'s attack must be 2^{100}. Suppose also that it is infeasible for $\mathcal{A}$ make more than 2^{30} key generation queries, and that $t_{exp} = 2^{30}$. In this case, we should choose our security parameter such that $t = 2^{100} + \mathcal{O}(t_{exp} \cdot q^2)$. Since 2^{90} is much smaller than 2^{100}, it essentially suffices to choose the security parameter such that $t \approx 2^{100}$.

On the other hand, consider an IBE system whose reduction loses a *multiplicative* factor of q in time-complexity (without much loss in the success probability). In this setting, to ensure that $\mathcal{A}$'s time complexity is 2^{100}, we must choose our security parameter such that $t \approx 2^{130}$. The security parameter in this setting thus must be at least 30% greater (even more if sub-exponential attacks are possible against the system). Assuming, as a rough approximation, that exponentiation takes time proportional to the cube of the security parameter, the increase in the security parameter size more than doubles the time needed to exponentiate, which significantly impacts the computational efficiency of the system. So, our "tight reduction" significantly enhances the efficiency advantages of our system over previous IBE systems that have been proven fully secure in the standard model (under decision BDH), at least when we assume that decision q-ABDHE and decision BDH are equally hard.

Since the relative hardness of decision q-ABDHE and decision BDH is unknown, however, we stress that it remains an excellent open problem to construct an IBE system that has a tight reduction in the standard model under a more natural assumption, such as decision BDH.

4 Construction II: Chosen-Ciphertext Security

We now present an efficient IBE system that is ANON-IND-ID-CCA secure without random oracles under the truncated decision $(q_{\mathsf{ID}} + 2)$-ABDHE assumption.

4.1 Construction

Let $\mathbb{G}$ and $\mathbb{G}_T$ be groups of order p, and let $e : \mathbb{G} \times \mathbb{G} \to \mathbb{G}_T$ be the bilinear map. The IBE system works as follows.

Setup: The PKG picks a random generators $g, h_1, h_2, h_3 \in \mathbb{G}$ and a random $\alpha \in \mathbb{Z}_p$. It sets $g_1 = g^\alpha \in \mathbb{G}$. It chooses a hash function H from a family of universal one-way hash functions. The public *params* and private *master-key* are given by

$$params = (g, g_1, h_1, h_2, h_3, H) \qquad master\text{-}key = \alpha \ .$$

KeyGen: To generate a private key for identity $\mathsf{ID} \in \mathbb{Z}_p$, the PKG generates random $r_{\mathsf{ID},i} \in \mathbb{Z}_p$ for $i \in \{1, 2, 3\}$, and outputs the private key

$$d_{\mathsf{ID}} = \{(r_{\mathsf{ID},i}, h_{\mathsf{ID},i}) : i \in \{1, 2, 3\}\}, \quad \text{where} \quad h_{\mathsf{ID},i} = (h_i g^{-r_{\mathsf{ID},i}})^{1/(\alpha - \mathsf{ID})} \ .$$

If $\mathsf{ID} = \alpha$, the PKG aborts. As before, we require that the PKG always use the same random values $\{r_{\mathsf{ID},i}\}$ for ID.

Encrypt: To encrypt $m \in \mathbb{G}_T$ using identity $\mathsf{ID} \in \mathbb{Z}_p$, the sender generates random $s \in \mathbb{Z}_p$ and sends the ciphertext

$$C = (g_1^s g^{-s \cdot \mathsf{ID}}, \ e(g, g)^s, \ m \cdot e(g, h_1)^{-s}, \ e(g, h_2)^s e(g, h_3)^{s\beta}) \ .$$

Above, for $C = (u, v, w, y)$, we set $\beta = H(u, v, w)$. As before, encryption does not require any pairing computations once $e(g, g)$, and $\{e(g, h_i)\}$ have been pre-computed or alternatively included in *params*.

Decrypt: To decrypt ciphertext $C = (u, v, w, y)$ with ID, the recipient sets $\beta = H(u, v, w)$ and tests whether

$$y \ = \ e(u, h_{\mathsf{ID},2} h_{\mathsf{ID},3}{}^\beta) v^{r_{\mathsf{ID},2} + r_{\mathsf{ID},3}\beta} \ .$$

If the check fails, the recipient outputs $\perp$. Otherwise, it outputs

$$m \ = \ w \cdot e(u, h_{\mathsf{ID},1}) v^{r_{\mathsf{ID},1}} \ .$$

Correctness: Assuming the ciphertext is well-formed for ID:

$$e(u, h_{\mathsf{ID},2} h_{\mathsf{ID},3}{}^\beta) v^{r_{\mathsf{ID},2} + r_{\mathsf{ID},3}\beta}$$
$$= e(g^{s(\alpha - \mathsf{ID})}, (h_2 h_3{}^\beta)^{1/(\alpha - \mathsf{ID})} g^{-(r_{\mathsf{ID},2} + r_{\mathsf{ID},3}\beta)/(\alpha - \mathsf{ID})}) e(g, g)^{s(r_{\mathsf{ID},2} + r_{\mathsf{ID},3}\beta)}$$
$$= e(g^{s(\alpha - \mathsf{ID})}, (h_2 h_3{}^\beta)^{1/(\alpha - \mathsf{ID})}) \ = \ e(g, h_2)^s e(g, h_3)^{s\beta} \ .$$

Thus, the check passes. Moreover, as in the ANON-IND-ID-CPA scheme,

$$e(u, h_{\mathsf{ID},1}) v^{r_{\mathsf{ID},1}} \ = \ e(g^{s(\alpha - \mathsf{ID})}, h_1^{1/(\alpha - \mathsf{ID})} g^{-r_{\mathsf{ID},1}/(\alpha - \mathsf{ID})}) e(g, g)^{s r_{\mathsf{ID},1}} \ = \ e(g, h_1)^s \ ,$$

as required.

4.2 Security

We now prove that the above construction is ANON-IND-ID-CCA secure under the truncated decision $(q_{\mathsf{ID}} + 2)$-ABDHE assumption. We will refer the reader to the proof of Theorem 1 for some portions of the present proof that would otherwise be duplicative.

Theorem 2. *Let* $q = q_{\mathsf{ID}} + 2$. *Assume the truncated decision* (t, ϵ, q)-*ABDHE assumption holds for* $(\mathbb{G}, \mathbb{G}_T, e)$. *Then, the above IBE system is* $(t', \epsilon', q_{\mathsf{ID}}, q_C)$ *ANON-IND-ID-CCA secure for* $t' = t - \mathcal{O}(t_{exp} \cdot q^2)$ *and* $\epsilon' = \epsilon + 4q_C/p$, *where* t_{exp} *is the time required to exponentiate in* $\mathbb{G}$.

Proof. Let $\mathcal{A}$ be an adversary that $(t', \epsilon', q_{\mathsf{ID}}, q_C)$-breaks the ANON-IND-ID-CCA security of the IBE system described above. We construct an algorithm, $\mathcal{B}$, that solves the truncated decision q-ABDHE problem, as follows. $\mathcal{B}$ takes as input a random truncated decision q-ABDHE challenge $(g', g'_{q+2}, g, g_1, \ldots, g_q, Z)$, where Z is either $e(g_{q+1}, g')$ or a random element of $\mathbb{G}_T$. Algorithm $\mathcal{B}$ proceeds as follows.

Setup: $\mathcal{B}$ generates random polynomials $f_i(x) \in \mathbb{Z}_p[x]$ of degree q for $i \in \{1, 2, 3\}$. It sets $h_i = g^{f_i(\alpha)}$. It sends the public key (g, g_1, h_1, h_2, h_3) to $\mathcal{A}$. Since g, α, and $f_i(x)$ for $i \in \{1, 2, 3\}$ are chosen uniformly at random, h_1, h_2, and h_3 are uniformly random and the public key has a distribution identical to that in the actual construction.

Phase 1: $\mathcal{A}$ makes key generation queries. $\mathcal{B}$ responds to a query on $\mathsf{ID} \in \mathbb{Z}_p$ as follows. If $\mathsf{ID} = \alpha$, $\mathcal{B}$ uses α to solve truncated decision q-ABDHE immediately. Else, to generate a pair $(r_{\mathsf{ID},1}, h_{\mathsf{ID},1})$ such that $h_{\mathsf{ID},1} = (h_1 g^{-r_{\mathsf{ID},1}})^{1/(\alpha+\mathsf{ID})}$, $\mathcal{B}$ sets $r_{\mathsf{ID},1} = f_1(\mathsf{ID})$ and computes $h_{\mathsf{ID},1}$ as before (in the proof of Theorem 1). It computes the remainder of the private key similarly. As before, the private key generated for ID in this fashion is valid.

$\mathcal{A}$ also makes decryption queries. To respond to a decryption query on (ID, C), $\mathcal{B}$ generates a private key for ID as above. It then decrypts C by performing the usual *Decrypt* algorithm with this private key.

Challenge: As before, $\mathcal{A}$ outputs identities $\mathsf{ID}_0, \mathsf{ID}_1$ and messages M_0, M_1. If $\alpha \in \{\mathsf{ID}_0, \mathsf{ID}_1\}$, $\mathcal{B}$ uses α to solve truncated decision q-ABDHE immediately. Else, as before, $\mathcal{B}$ generates bits $b, c \in \{0, 1\}$. After computing a private key $\{(r_{\mathsf{ID},i}, h_{\mathsf{ID},i}) : i \in \{1, 2, 3\}\}$ for ID_b, it also computes (u, v, w) as before, using the $(r_{\mathsf{ID}_b,1}, h_{\mathsf{ID}_b,1})$ portion of the key to compute w. After setting $\beta = H(u, v, w)$, $\mathcal{B}$ sets $y = e(u, h_{\mathsf{ID},2} h_{\mathsf{ID},3}{}^{\beta}) v^{r_{\mathsf{ID},2} + r_{\mathsf{ID},3}\beta}$. If $Z = e(g_{q+1}, g')$, then (u, v, w, y) is a valid, appropriately-distributed challenge to $\mathcal{A}$ for essentially the same reason as before.

Phase 2: $\mathcal{A}$ makes key generation and decryption queries, and $\mathcal{B}$ responds as in Phase 1.

Guess: As before.

Now, since the time-complexity analysis is as in the proof of Theorem 1, Theorem 2 follows from the following lemmata.

Lemma 1. *When $\mathcal{B}$'s input is sampled according to $\mathcal{P}_{ABDHE}$, the joint distribution of $\mathcal{A}$'s view and the bits (b, c) is indistinguishable from that in the actual construction, except with probability $2q_C/p$.*

Lemma 2. *When $\mathcal{B}$'s input is sampled according to $\mathcal{R}_{ABDHE}$, the distribution of the bits (b, c) is independent from the adversary's view, except with probability $2q_C/p$.*

Our approach to proving these claims closely follows the proof of security for the Cramer-Shoup encryption scheme [17], in that both proofs rely heavily on the notion of *linear independence*. More specifically, when one expresses the adversary's knowledge (from the public key, queries, etc.) as equations in the simulator's private key variables, one may ask whether a target equation that the adversary is trying to solve is linearly independent to the equations in its knowledge base; if so, then in certain circumstances, the adversary can be said to have an unconditionally negligible probability of finding a solution to the target equation. This will become clearer below.

Proof of Lemma 1: When $\mathcal{B}$'s input is sampled according to $\mathcal{P}_{ABDHE}$, $\mathcal{B}$'s simulation appears perfect to $\mathcal{A}$ if $\mathcal{A}$ makes only key generation queries, as in the proof of Theorem 1. $\mathcal{B}$'s simulation still appears perfect if $\mathcal{A}$ makes decryption queries only on identities for which it queries the private key, since $\mathcal{B}$'s responses give $\mathcal{A}$ no additional information. Furthermore, querying well-formed ciphertexts to the decryption oracle does not help $\mathcal{A}$ distinguish between the simulation and the actual construction, since, by the correctness of *Decrypt*, well-formed ciphertexts will be accepted in either case. Finally, querying a non-well-formed ciphertext (u', v', w', y') for ID for which $v' = e(u', g)^{1/(\alpha - \text{ID})}$ does not help $\mathcal{A}$ distinguish, since this ciphertext will fail the *Decrypt* check under *every* valid private key for ID. Thus, the lemma follows from the following claim:

Claim: The decryption oracle, in the simulation and in the actual construction, rejects all invalid ciphertexts under identities not queried by $\mathcal{A}$, except with probability q_C/p.

We say a ciphertext (u', v', w', y') for ID is "invalid" if $v' \neq e(u', g)^{1/(\alpha - \text{ID})}$.

Let (u', v', w', y') be an invalid ciphertext queried by $\mathcal{A}$ for ID, an identity not queried by $\mathcal{A}$. Let $\{(r_{\text{ID},i}, h_{\text{ID},i}) : i \in \{1, 2, 3\}\}$ be $\mathcal{B}$'s private key for ID. Let $a_{u'} = \log_g u'$, $a_{v'} = \log_{e(g,g)} v'$, and $a_{y'} = \log_{e(g,g)} y'$. For (u', v', w', y') to be accepted, we must have $y' = e(u', h_{\text{ID},2} h_{\text{ID},3}{}^{\beta'}) v'^{r_{\text{ID},2} + r_{\text{ID},3}\beta'} -$ i.e.,

$$a_{y'} = a_{u'}(\log_g h_{\text{ID},2} + \beta' \log_g h_{\text{ID},3}) + a_{v'}(r_{\text{ID},2} + \beta' r_{\text{ID},3}), \tag{1}$$

for $\beta' = H(u', v', w')$. To compute the probability that $\mathcal{A}$ can generate such a y', we must consider the distribution of $\{(r_{\text{ID},i}, h_{\text{ID},i}) : i \in \{2, 3\}\}$ from $\mathcal{A}$'s view.

First, $\mathcal{A}$ knows that

$$\log_g h_1 = (\alpha - \text{ID}) \log_g h_{\text{ID},1} + r_{\text{ID},1} \tag{2}$$

$$\log_g h_2 = (\alpha - \text{ID}) \log_g h_{\text{ID},2} + r_{\text{ID},2} \tag{3}$$

$$\log_g h_3 = (\alpha - \text{ID}) \log_g h_{\text{ID},3} + r_{\text{ID},3} \tag{4}$$

by the construction of the private key. In light of Equations 3 and 4, $\mathcal{A}$'s task may be re-phrased as finding a y' such that

$$a_{y'} = (a_{u'}/(\alpha - \mathsf{ID}))(\log_g h_2 + \beta' \log_g h_3) + (a_{v'} - a_{u'}/(\alpha - \mathsf{ID}))(r_{\mathsf{ID},2} + \beta' r_{\mathsf{ID},3}) . \quad (5)$$

Note that $a_{v'} - a_{u'}/(\alpha - \mathsf{ID}) \neq 0$, since the ciphertext is invalid. Let $z' = a_{v'} - a_{u'}/(\alpha - \mathsf{ID})$.

In the actual construction, the values of $r_{\mathsf{ID},i}$ for $i \in \{2,3\}$ are chosen independently for different identities; however, this is not true in the simulation. Since $f_i(\mathsf{ID}) = r_{\mathsf{ID},i}$, $\mathcal{A}$ could conceivably gain information regarding $(r_{\mathsf{ID},2}, r_{\mathsf{ID},3})$ from its information regarding $(f_2(x), f_3(x))$, which includes the evaluations of $(f_2(x), f_3(x))$ at α (from the public key components (h_2, h_3)) and at $q - 2$ identities (from its key generation queries). We may represent the knowledge gained from these evaluations as a matrix product:

$$[f_{2,0}, f_{2,1}, \ldots, f_{2,q}, f_{3,0}, f_{3,1}, \ldots, f_{3,q}] \begin{bmatrix} 1 & 1 & \cdots & 1 & 0 & 0 & \cdots & 0 \\ x_1 & x_2 & \cdots & x_{q-1} & 0 & 0 & \cdots & 0 \\ \vdots & \vdots & \vdots & \vdots & \vdots & \vdots & \vdots & \vdots \\ x_1^q & x_2^q & \cdots & x_{q-1}^q & 0 & 0 & \cdots & 0 \\ 0 & 0 & \cdots & 0 & 1 & 1 & \cdots & 1 \\ 0 & 0 & \cdots & 0 & x_1 & x_2 & \cdots & x_{q-1} \\ \vdots & \vdots & \vdots & \vdots & \vdots & \vdots & \vdots & \vdots \\ 0 & 0 & \cdots & 0 & x_1^q & x_2^q & \cdots & x_{q-1}^q \end{bmatrix} ,$$

where $f_{i,j}$ is the coefficient of x^j in $f_i(x)$, $x_k \in \mathbb{Z}_p$ is the k-th identity queried by $\mathcal{A}$ to the key generation oracle, and $x_{q-1} = \alpha$. Let $\boldsymbol{f}$ denote the vector on the left and let V denote the matrix on the right. Note that V contains two $(q+1) \times (q-1)$ Vandermonde matrices; its columns are linearly independent. From $\mathcal{A}$'s view, since V has four more rows than columns, the solution space for $\boldsymbol{f}$ is four-dimensional.

Let $\boldsymbol{\gamma}_{\mathsf{ID}}$ denote the vector $(1, \mathsf{ID}, \ldots, \mathsf{ID}^q)$. When we re-phrase Equation 5 in terms of the simulator's private key vector $\boldsymbol{f}$, we obtain:

$$a_{y'} = \text{``public'' terms} + z'(\boldsymbol{f} \cdot \boldsymbol{\gamma}_{\mathsf{ID}} \| \beta' \boldsymbol{\gamma}_{\mathsf{ID}}) , \quad (6)$$

where "$\cdot$" denotes the dot product and $\boldsymbol{\gamma}_{\mathsf{ID}} \| \beta' \boldsymbol{\gamma}_{\mathsf{ID}}$ denotes the $2(q+1)$-dimensional vector formed by concatenating the coefficients of $\boldsymbol{\gamma}_{\mathsf{ID}}$ and $\beta' \boldsymbol{\gamma}_{\mathsf{ID}}$. If $\boldsymbol{\gamma}_{\mathsf{ID}} \| \beta' \boldsymbol{\gamma}_{\mathsf{ID}}$ were in the linear span of V, then potentially $\mathcal{A}$ could use knowledge gained from its key generation queries to compute a solution y' to Equation 6. However, one can easily see that $\boldsymbol{\gamma}_{\mathsf{ID}} \| \beta' \boldsymbol{\gamma}_{\mathsf{ID}}$ is linearly independent. Thus, as in the security proof of Cramer-Shoup, it follows that the decryption oracle will reject (u', v', w', y') for ID with probability $1 - 1/p$ if it is the first invalid ciphertext queried by $\mathcal{A}$, since there is only a $1/p$ chance that $\boldsymbol{f}$ is contained in the 3-dimensional solution space (with p^3 points) defined by Equation 6 and the columns of V, given that $\boldsymbol{f}$ is in the 4-dimensional solution space (with p^4 points) defined by the columns of V.

Each time the decryption oracle rejects an invalid ciphertext in the simulation, the solution space for f is "punctured" in a 3-dimensional space that $\mathcal{A}$ then concludes does not contain f; consequently, the probability that $\mathcal{A}$'s i-th invalid ciphertext is accepted is at most $1/(p - i + 1)$. The probability that q_C invalid ciphertexts (on identities not queried to the key generation oracle) are all rejected is at least $1 - q_C/p$. This bound also holds for the actual construction (where $\mathcal{A}$'s attack is less effective). This concludes the proof of Lemma 1.

Proof of Lemma 2: The lemma follows from the following two claims.

Claim 1: If the decryption oracle rejects all invalid ciphertexts, then $\mathcal{A}$ has advantage at most q_C/p in guessing the bits (b, c).

Claim 2: The decryption oracle rejects all invalid ciphertexts, except with probability q_C/p.

Let $a_u = \log_g u$, $a_v = \log_{e(g,g)} v$ and $a_y = \log_{e(g,g)} y$ for challenge ciphertext (u, v, w, y) on (ID_b, M_c). Since (u, v, w, y) is generated by sampling from $\mathcal{R}_{ABDHE}$ in this case, (a_u, a_v) is a uniformly random element of $\mathbb{Z}_p \times \mathbb{Z}_p$ in $\mathcal{A}$'s view. From the challenge ciphertext and Equations 2-4, $\mathcal{A}$ obtains the equations

$$\log(M_c/w) = (a_u/(\alpha - \mathsf{ID}_b)) \log h_1 + (a_v - a_u/(\alpha - \mathsf{ID}_b)) r_{\mathsf{ID}_b,1} \tag{7}$$

$$a_y = (a_u/(\alpha - \mathsf{ID}_b))(\log_g h_2 + \beta \log_g h_3) + (a_v - a_u/(\alpha - \mathsf{ID}_b))(r_{\mathsf{ID}_b,2} + \beta r_{\mathsf{ID}_b,3}) \tag{8}$$

where $\beta = H(u, v, w)$.

Regarding Claim 1, if no invalid ciphertexts are accepted, then $\mathcal{B}$'s responses to decryption queries leak no information about $r_{\mathsf{ID}_b,1}$. Furthermore, $\mathcal{A}$'s key generation queries do not constrain $r_{\mathsf{ID}_b,1} = f_1(\mathsf{ID}_b)$, since f_1 is of degree q. Thus the distribution of M_c/w – conditioning on (b, c) and everything in $\mathcal{A}$'s view other than w – is uniform. As in Cramer-Shoup, M_c/w serves as a perfect one-time pad; w is uniformly random and independent, and c is independent of $\mathcal{A}$'s view.

The only part of the ciphertext that can reveal information about b is y, since $\mathcal{A}$ views (u, v, w) as a uniformly random and independent element of $\mathbb{G} \times \mathbb{G}_T \times \mathbb{G}_T$. The $2q - 2$ equations corresponding to the columns of V intersect Equation 8 in at least a three-dimensional space in $\mathbb{Z}_p^{2(q+1)}$. $\mathcal{A}$ views f as being contained in one of two three-dimensional spaces, since b has two possible values. By an argument similar to above, each of $\mathcal{A}$'s invalid ciphertext queries punctures each of these three-dimensional spaces in a plane, removing each of the two planes from consideration as containing f. Since no invalid ciphertext is accepted, each three-dimensional space is left with at least $p^3 - q_C p^2$ (out of p^3) candidates. Thus, $\mathcal{A}$ cannot distinguish b, except with advantage at most q_C/p.

Regarding Claim 2, suppose that $\mathcal{A}$ submits an invalid ciphertext (u', v', w', y') for unqueried identity ID, where $(u', v', w', y', \mathsf{ID}) \neq (u, v, w, y, \mathsf{ID}_b)$. Let $\beta' = H(u', v', w')$. There are three cases to consider:

1. $(u', v', w') = (u, v, w)$: In this case, the hashes are also equal. If $\mathsf{ID} = \mathsf{ID}_b$ but $y' \neq y$, the ciphertext will certainly be rejected. If $\mathsf{ID} \neq \mathsf{ID}_b$, $\mathcal{A}$ must

generate a y' that satisfies Equation 6. However, we claim that the vector $\gamma_{\text{ID}} \| \beta\gamma_{\text{ID}}$ (corresponding to Equation 6) is linearly independent in $\mathbb{Z}_p^{2(q+1)}$ to $\gamma_{\text{ID}_b} \| \beta\gamma_{\text{ID}_b}$ (corresponding to the challenge ciphertext) and the columns of V, implying (via arguments analogous to those above) that $\mathcal{A}$ cannot generate such a y' except with probability $1/(p-i+1)$, where (u', v', w', y') is the i-th invalid ciphertext. Let $V_1, \ldots, V_{2q-2}$ be the columns of V. Suppose that there exist integers $(a_1, \ldots, a_{2q})$, not all zero, such that $a_1 V_1 + \cdots + a_{2q-2} V_{2q-2} + a_{2q-1}(\gamma_{\text{ID}} \| \beta\gamma_{\text{ID}}) + a_{2q}(\gamma_{\text{ID}_b} \| \beta\gamma_{\text{ID}_b})$ is the zero vector in $\mathbb{Z}_p^{2(q+1)}$. Then, either $(a_1, \ldots, a_{q-1}, a_{2q-1}, a_{2q})$ or $(a_q, \ldots, a_{2q-2}, a_{2q-1}, a_{2q})$ is not all zeros; wlog, assume the former. The first $q+1$ coordinates of the vectors $(V_1, \ldots, V_{q-1}, \gamma_{\text{ID}}, \gamma_{\text{ID}_b})$ form a Vandermonde matrix (with nonzero determinant), but the first $q+1$ coordinates of $a_1 V_1 + \cdots + a_{q-1} V_{q-1} + a_{2q-1}(\gamma_{\text{ID}} \| \beta\gamma_{\text{ID}}) + a_{2q}(\gamma_{\text{ID}_b} \| \beta\gamma_{\text{ID}_b})$ is the zero vector in $\mathbb{Z}_p^{q+1}$ – a contradiction.

2. $(u', v', w') \neq (u, v, w)$ and $\beta' = \beta$: This violates the universal one-wayness of the hash function H, by an argument analogous to that in Cramer-Shoup.

3. $(u', v', w') \neq (u, v, w)$ and $\beta' \neq \beta$: In this case, $\mathcal{A}$ must generate, for some ID, a y' that satisfies Equation 6. For essentially the same reason as discussed in Item 1, $\mathcal{A}$ can do this with only negligible probability when $\text{ID} \neq \text{ID}_b$. If $\text{ID} = \text{ID}_b$, then $\gamma_{\text{ID}} \| \beta'\gamma_{\text{ID}}$ and $\gamma_{\text{ID}_b} \| \beta\gamma_{\text{ID}_b}$ generate $\gamma_{\text{ID}_b} \| 0^{q+1}$ and $0^{q+1} \| \gamma_{\text{ID}_b}$ since $\beta \neq \beta'$. These vectors are clearly linearly independent to each other and the columns of V, and thus the standard analysis applies.

This completes the proof of Lemma 2.

5 Conclusions and Open Problems

We presented a fully secure IBE system that is quite practical, has very compact public parameters, and has a tight security reduction (though based on a stronger assumption that depends on the anticipated number of private key generation queries). The scheme is recipient-anonymous, and its proof extends Cramer-Shoup-type techniques to IBE systems.

Since a tight reduction based on decision q-ABDHE is not necessarily better than a loose reduction based on decision BDH (or some other natural assumption), it remains an outstanding open problem to construct a fully secure IBE system (without random oracles) that has a tight reduction based on a more natural assumption. Another interesting problem is to construct a hierarchical IBE system that has a reduction based on a reasonable assumption, either in the standard model or the random oracle model, that is polynomial in q and the number of levels.

Acknowledgments

We thank Dan Boneh, Brent Waters and the anonymous reviewers of Eurocrypt 2006 for insightful comments and helpful suggestions.

References

[1] M. Abdalla, M. Bellare, D. Catalano, E. Kiltz, T. Kohno, T. Lange, J. Malone-Lee, G. Neven, P. Paillier, and H. Shi. Searchable Encryption Revisited: Consistency Properties, Relation to Anonymous IBE, and Extensions. In *Advances in Cryptology – Crypto 2005*, volume 3621 of *LNCS*, pages 205–222. Springer-Verlag, 2005.

[2] N. Attrapadung, B. Chevallier-Mames, J. Furukawa, T. Gomi, G. Hanaoka, H. Imai, and R. Zhang. Efficient Identity Based Encryption with Tight Security Reduction. Cryptology ePrint Archive 2005/320.

[3] M. Bellare and P. Rogaway. Random Oracles are Practical: A Paradigm for Designing Efficient Protocols. In *Proc. of ACM CCS*, pages 62–73, 1993.

[4] D. Boneh and X. Boyen. Efficient Selective-ID Identity Based Encryption without Random Oracles. In *Advances in Cryptology – Eurocrypt 2004*, volume 3027 of *LNCS*, pages 223–238. Springer-Verlag, 2004.

[5] D. Boneh and X. Boyen. Secure Identity Based Encryption without Random Oracles. In *Advances in Cryptology – Crypto 2004*, volume 3152 of *LNCS*, pages 443–459. Springer-Verlag, 2004.

[6] D. Boneh, X. Boyen, and E.-J. Goh. Hierarchical Identity Based Encryption with Constant Size Ciphertext. In *Advances in Cryptology – Eurocrypt 2005*, volume 3494 of *LNCS*, pages 440–456. Springer-Verlag, 2005.

[7] D. Boneh, G. Di Crescenzo, R. Ostrovsky, and G. Persiano. Public Key Encryption with Keyword Search. In *Advances in Cryptology – Eurocrypt 2004*, volume 3027 of *LNCS*, pages 506–522. Springer-Verlag, 2004.

[8] D. Boneh and M. Franklin. Identity Based Encryption from the Weil pairing. In *Advances in Cryptology – Crypto 2001*, volume 2139 of *LNCS*, pages 213–229. Springer-Verlag, 2001.

[9] D. Boneh and M. Franklin. Identity Based Encryption from the Weil pairing. *SIAM Journal of Computing*, 32(3):586–615, 2003.

[10] D. Boneh, C. Gentry, and B. Waters. Collusion-Resistant Broadcast Encryption with Short Ciphertexts and Private Keys. In *Advances in Cryptology – Crypto 2005*, volume 3621 of *LNCS*, pages 258–275. Springer-Verlag, 2005.

[11] D. Boneh and J. Katz. Improved Efficiency for CCA-Secure Cryptosystems Built Using Identity Based Encryption. In *Proc. of CT-RSA*, volume 3376 of *LNCS*, pages 87–103. Springer-Verlag, 2005.

[12] D. Boneh, B. Lynn, and H. Shacham, *Short signatures from the Weil pairing*, Advances in Cryptology — Asiacrypt 2001, Lecture Notes in Computer Science 2248 (2001), Springer, 514–532.

[13] X. Boyen, Q. Mei and B. Waters, *Direct Chosen Ciphertext Security from Identity Based Techniques*, In *Proc. of ACM CCS*, pages 320–329, 2005.

[14] X. Boyen and B. Waters. Anonymous Hierarchical Identity-Based Encryption (without Random Oracles). Cryptology ePrint Archive 2006/085.

[15] R. Canetti, S. Halevi, and J. Katz. A Forward-Secure Public-Key Encryption Scheme. In *Advances in Cryptology – Eurocrypt 2003*, volume 2656 of *LNCS*, pages 255–271. Springer-Verlag, 2003.

[16] R. Canetti, S. Halevi, and J. Katz. Chosen-Ciphertext Security from Identity-Based Encryption. In *Advances in Cryptology – Eurocrypt 2004*, volume 3027 of *LNCS*, pages 207–222. Springer-Verlag, 2004.

[17] R. Cramer and V. Shoup. A Practical Public Key Cryptosystem Provably Secure Against Adaptive Chosen Ciphertext Attacks. In *Advances in Cryptology – Crypto 1998*, volume 1462 of *LNCS*, pages 13–25. Springer-Verlag, 1998.

[18] R. Cramer and V. Shoup. Signature Schemes Based on the Strong RSA Assumption. In *Proc. of ACM CCS*, pages 46–51, 1999.

[19] Y. Dodis. Efficient Construction of (Distributed) Verifiable Random Functions. In *Proc. of Public Key Cryptography*, volume 2567 of *LNCS*, pages 1–17. Springer-Verlag, 2002.

[20] Y. Dodis and A. Yampolskiy. A Verifiable Random Function with Short Proofs and Keys. In *Proc. of Public Key Cryptography*, volume 3386 of *LNCS*, pages 416–431. Springer-Verlag, 2005.

[21] C. Gentry and A. Silverberg. Hierarchical ID-Based Cryptography. In *Advances in Cryptology – Asiacrypt 2002*, volume 2501 of *LNCS*, pages 548–566. Springer-Verlag, 2002.

[22] J. Horwitz and B. Lynn. Toward Hierarchical Identity-Based Encryption. In *Advances in Cryptology – Eurocrypt 2002*, volume 2332 of *LNCS*, pages 466–481. Springer-Verlag, 2002.

[23] J. Katz and N. Wang. Efficiency Improvements for Signature Schemes with Tight Security Reductions. In *Proc. of ACM CCS*, pages 155–164, 2003.

[24] K. Kurosawa and Y. Desmedt. A New Paradigm of Hybrid Encryption Scheme. In *Advances in Cryptology – Crypto 2004*, volume 3152 of *LNCS*, pages 426–442. Springer-Verlag, 2004.

[25] A. Shamir. Identity-Based Cryptosystems and Signature Schemes. In *Advances in Cryptology – Crypto 1984*, volume 196 of *LNCS*, pages 47–53. Springer-Verlag, 1984.

[26] V. Shoup. Lower Bounds for Discrete Logarithms and Related Problems. In *Advances in Cryptology – Eurocrypt 1997*, volume 1233 of *LNCS*, pages 256–266. Springer-Verlag, 1997.

[27] B. Waters. Efficient Identity-Based Encryption without Random Oracles. In *Advances in Cryptology – Eurocrypt 2005*, volume 3494 of *LNCS*, pages 114–127. Springer-Verlag, 2005.

Sequential Aggregate Signatures and Multisignatures Without Random Oracles

Steve Lu[1,*], Rafail Ostrovsky[2,**], Amit Sahai[3,***],
Hovav Shacham[4], and Brent Waters[5,†]

[1] UCLA
stevelu@math.ucla.edu
[2] UCLA
rafail@cs.ucla.edu
[3] UCLA
sahai@cs.ucla.edu
[4] Weizmann Institute of Science
hovav.shacham@weizmann.ac.il
[5] SRI International
bwaters@csl.sri.com

Abstract. We present the first aggregate signature, the first multisignature, and the first verifiably encrypted signature provably secure without random oracles. Our constructions derive from a novel application of a recent signature scheme due to Waters. Signatures in our aggregate signature scheme are sequentially constructed, but knowledge of the order in which messages were signed is not necessary for verification. The aggregate signatures obtained are shorter than Lysyanskaya et al. sequential aggregates and can be verified more efficiently than Boneh et al. aggregates. We also consider applications to secure routing and proxy signatures.

1 Introduction

In this paper we present an aggregate signature scheme, a multisignature scheme, and a verifiably encrypted scheme. Unlike previous such schemes, our constructions are provably secure without random oracles. A series of papers beginning with the uninstantiability result of Canetti, Goldreich, and Halevi [10] has cast some doubt on the soundness of the random oracle methodology, making random-oracle–free schemes more attractive. Moreover, our proposed schemes are quite practical, and in some cases outperform the most efficient random-oracle–based schemes.

---
* Supported in part by NSF grant DMS-0502315
** Supported in part by a gift from Teradata, Intel equipment grant, NSF Cybertrust grant No. 0430254, OKAWA research award, B. John Garrick Foundation and Xerox Innovation group Award.
*** Supported in part by grants from the NSF ITR and Cybertrust programs, a generous equipment grant from Intel, and an Alfred P. Sloan Foundation Fellowship.
† Supported by DHS and DOI contract No. NBCHF040146. Views expressed in this paper do not necessarily reflect those of DHS and DOI.

S. Vaudenay (Ed.): EUROCRYPT 2006, LNCS 4004, pp. 465–485, 2006.

An aggregate signature scheme allows a collection of signatures to be able to be compressed into one short signature. Aggregate signatures are useful for applications such as secure route attestation and certificate chains where the space requirements for a sequence of signatures can impact practical application performance.

Boneh et al. [8] presented the first aggregate signature scheme, which was based on the BLS signature [9] in groups with efficiently computable bilinear maps. Subsequently, Lysyanskaya et al. [20] presented a sequential RSA-based scheme that, while more limited, could be instantiated using more general assumptions. In a sequential aggregate signature scheme the aggregate signature must be constructed sequentially, with each signer modifying the aggregate signature in turn. However, most known applications are sequentially constructed anyway. One drawback of both schemes is that they are provably secure only in the random oracle model and thus there is only a heuristic argument for their security.

We present the first aggregate signature scheme that is provably secure without random oracles. Our signatures are sequentially constructed, however, unlike the scheme of Lysyanskaya et al., a verifier need not know the order in which the aggregate signature was created. Additionally, our signatures are shorter than those of Lysyanskaya et al. and can be verified more efficiently than those of Boneh et al.

In addition, we present the first multisignature scheme that is provably secure without random oracles. In a multisignature scheme, a single short object – the multisignature – can take the place of n signatures by n signers, all on the *same* message. (Aggregate signatures can be thought of as a multisignature without this restriction.) Boldyreva [6] gave the first multisignature scheme in which multisignature generation does not require signer interaction, based on BLS signatures.

Finally, we present the first verifiably encrypted signature scheme that is provably secure without random oracles. A verifiably encrypted signature is an object that anyone can confirm contains the encryption of a signature on some message, but from which only the party under whose key it was encrypted can recover the signature. Such a primitive is useful in contract signing. Boneh et al. [8] gave the first verifiably encrypted signature scheme, based on BLS signatures.

All our constructions derive from novel adaptations of the signature scheme of Waters [28], which follows from his Identity-Based Encryption scheme.

2   Preliminaries

In this section we first present some background on groups with efficiently computable bilinear maps. Next, we recall the definition of existentially unforgeable signatures. Then we present the Waters [28] signature algorithm.

2.1   Groups with Efficiently Computable Bilinear Maps

We briefly review the necessary facts about bilinear maps and bilinear map groups. (For more detail, see, e.g., [13, 27].) Consider the following setting:

- $\mathbb{G}$ and $\mathbb{G}_T$ are multiplicative cyclic groups of order p;
- the group action on $\mathbb{G}$ and $\mathbb{G}_T$ can be computed efficiently;
- g is a generator of $\mathbb{G}$;
- $e : \mathbb{G} \times \mathbb{G} \to \mathbb{G}_T$ is an efficiently computable map with the following properties:
  - Bilinear: for all $u, v \in \mathbb{G}$ and $a, b \in \mathbb{Z}$, $e(u^a, v^b) = e(u, v)^{ab}$;
  - Non-degenerate: $e(g, g) \neq 1$.

We say that $\mathbb{G}$ is a bilinear group if it satisfies these requirements.

The security of our scheme relies on the hardness of the Computational Diffie-Hellman (CDH) problem in bilinear groups. We state the problem and our assumption as follows. Define the success probability of an algorithm $\mathcal{A}$ in solving the Computational Diffie-Hellman problem on $\mathbb{G}$ as

$$\mathbf{Adv}_{\mathcal{A}}^{\mathrm{cdh}} \stackrel{\mathrm{def}}{=} \Pr\left[\mathcal{A}(g, g^a, h) = h^a : g, h \stackrel{\mathrm{R}}{\leftarrow} \mathbb{G}, a \stackrel{\mathrm{R}}{\leftarrow} \mathbb{Z}_p\right] \ .$$

The probability is over the uniform random choice of g and h from $\mathbb{G}$, of a from $\mathbb{Z}_p$, and the coin tosses of $\mathcal{A}$. We say that an algorithm $\mathcal{A}$ (t, ϵ)-breaks Computational Diffie-Hellman on $\mathbb{G}$ if $\mathcal{A}$ runs in time at most t, and $\mathbf{Adv}_{\mathcal{A}}^{\mathrm{cdh}}$ is at least ϵ. The (t, ϵ)-Computational Diffie-Hellman assumption on $\mathbb{G}$ is that no adversary (t, ϵ)-breaks Computational Diffie-Hellman on $\mathbb{G}$.

Asymmetric Pairings and Short Representations. It is a simple (though tedious) matter to rewrite our schemes to employ an asymmetric pairing $e : \mathbb{G}_1 \times \mathbb{G}_2 \to \mathbb{G}_T$. Signatures will then include elements of $\mathbb{G}_1$, while public keys will include elements of G_2 and $\mathbb{G}_T$. This setting allows us to take advantage of curves due to Barreto and Naehrig [3]. With these curves, elements of $\mathbb{G}_1$ have a 160-bit representation at the 1024-bit security level.[1] In this case, security follows from the Computational co-Diffie-Hellman problem [9].

2.2   The Waters Signature Scheme

We describe the Waters signature scheme [28]. In our description the messages will be signatures on bitstrings of the form $\{0, 1\}^k$ for some fixed k. However, in practice one could apply a collision-resistant hash function $H_k : \{0, 1\}^* \to \{0, 1\}^k$ to sign messages of arbitrary length.

The scheme requires, besides the random generator $g \in \mathbb{G}$, $k + 1$ additional random generators $u', u_1, \ldots, u_k \in \mathbb{G}$. In the basic scheme, these can be generated at random as part of system setup and shared by all users. In some of the variants below, each user has generators $(u', u_1, \ldots, u_k)$ of her own, which must be included in her public key. We will draw attention to this in introducing the individual schemes.

The Waters signature scheme is a three-tuple of algorithms $\mathcal{W} = (\mathsf{Kg}, \mathsf{Sig}, \mathsf{Vf})$. These behave as follows.

---

[1] By "1024-bit security," we mean parameters such that the conjectured complexity of computing discrete logarithms is roughly comparable to the complexity of factoring 1024-bit numbers. For a more refined analysis see Koblitz and Menezes [19].

W.Kg. Pick random $\alpha \overset{R}{\leftarrow} \mathbb{Z}_p$ and set $A \leftarrow e(g,g)^\alpha$. The public key pk is $A \in \mathbb{G}_T$. The private key sk is α.

W.Sig(sk, M). Parse the user's private key sk as $\alpha \in \mathbb{Z}_p$ and the message M as a bitstring $(m_1, \ldots, m_k) \in \{0,1\}^k$. Pick a random $r \overset{R}{\leftarrow} \mathbb{Z}_p$ and compute

$$S_1 \leftarrow g^\alpha \cdot \left(u' \prod_{i=1}^k u_i^{m_i}\right)^r \qquad \text{and} \qquad S_2 \leftarrow g^r \;. \tag{1}$$

The signature is $\sigma = (S_1, S_2) \in \mathbb{G}^2$.

W.Vf(pk, M, σ). Parse the user's public key pk as $A \in \mathbb{G}_T$, the message M as a bitstring $(m_1, \ldots, m_k) \in \{0,1\}^k$, and the signature σ as $(S_1, S_2) \in \mathbb{G}^2$. Verify that

$$e(S_1, g) \cdot e\left(S_2, u' \prod_{i=1}^k u_i^{m_i}\right)^{-1} \overset{?}{=} A \tag{2}$$

holds; if so, output `valid`; if not, output `invalid`.

This signature is existentially unforgeable under a chosen-message attack – the standard notion of signature security, due to Goldwasser, Micali, and Rivest [14] – if CDH is hard. We give a roundabout proof of this as Corollary 1.

3   Sequential Aggregate Signatures

In a sequential aggregate signature, as in an ordinary aggregate signature, a single short object – called the aggregate – takes the place of n signatures by n signers on n messages. Thus aggregate signatures are a generalization of multisignatures. Sequential aggregates differ from ordinary aggregates in that the aggregation operation is performed by each signer in turn, rather than by an unrelated party after the fact.

Aggregate signatures have many applications, as noted by Boneh et al. [8] and Lysyanskaya et al. [20]. Below, we consider two: Secure BGP route attestation and proxy signatures.

In BGP, routers generate and forward route attestations to other routers to advertise the routes which should be used to reach their networks. Secure BGP solves the problem of attestation forgery by having each router add its signature to a valid attestation before forwarding it to its neighbors. Because of the size of route attestations is limited, aggregate signatures are useful in reducing the overhead of multiple signatures along a path. Nicol, Smith, and Zhao [24] gave a detailed analysis of the application of aggregate signatures to the Secure BGP routing protocol [18]. Our sequential aggregate signature scheme is well suited for improving SBGP. Since all of the incoming route attestations need to be verified anyway, the fact that our signing algorithm requires a verification adds no overhead. Additionally, our signature scheme can have signatures that are smaller than those of Lysyanskaya et al. and verification will be faster than that of the Boneh et al. scheme.

A proxy signature scheme allows a user, called the *designator*, to delegate signing authority to another user, the *proxy signer*. This signature primitive, introduced by Mambo, Usada, and Okamoto [21], has been discussed and used in several practical applications. Boldyreva, Palacio, and Warinschi [7] show how to construct a secure proxy signature scheme from any aggregate (or sequential aggregate) signature scheme. Instantiating the Boldyreva-Palacio-Warinschi construction with our scheme, we obtain a practical proxy signature secure without random oracles.

3.1 Definitions

A sequential aggregate signature scheme includes three algorithms. The first, Kg, is used to generate public-private keypairs. The second, ASig, takes not only a private key and a message to sign, as does an ordinary signing algorithm, but also an aggregate-so-far by a set of l signers on l corresponding messages; it folds the new signature into the aggregate, yielding a new aggregate signature by $l + 1$ signers on $l + 1$ messages. The third algorithm, AVf, takes a purported aggregate signature, along with l public keys and l corresponding messages, and decides whether the the aggregate is valid.

The Sequential Aggregate Certified-Key Model. Because our aggregate signature behaves like a sequential aggregate signature from the signers' viewpoint, but like standard aggregate signature from the verifiers' viewpoint, we describe a security model for it that is a hybrid of the sequential aggregate chosen key model of Lysyanskaya et al. [20] and the aggregate chosen key model of Boneh et al. [8]. In both models, the adversary is given a single challenge key, along with an appropriate signing oracle for that key. His goal is to generate a sequential aggregate that frames the challenge user. The adversary is allowed to choose all the keys in that forged aggregate but the challenge key.

We prove our scheme in a more restricted model that requires that the adversary certify that the public keys it includes in signing oracle queries and in its forgery were properly generated. This we handle by having the adversary hand over the private keys before using the public keys. We could also extract the keys by rewinding or, if this is impossible, using the NIZKs proposed by Groth, Ostrovsky, and Sahai [15].

Formally, the advantage of a forger $\mathcal{A}$ in our model is the probability that the challenger outputs 1 in the following game:

Setup. Initialize the list of certified public keys $C \leftarrow \emptyset$. Choose $(pk, sk) \stackrel{\mathrm{R}}{\leftarrow}$ Kg. Run algorithm $\mathcal{A}$ with pk as input.

Certification Queries. Algorithm $\mathcal{A}$ provides a keypair (pk', sk') in order to certify pk'. Add pk' to C if sk' is its matching private key.

Signing Queries. Algorithm $\mathcal{A}$ requests a sequential aggregate signature, under the challenge key pk, on a message M. In addition, it supplies an aggregate-so-far σ' on messages $\boldsymbol{M}$ under keys $\boldsymbol{pk}$. Check that the signature σ' verifies; that each key in $\boldsymbol{pk}$ is in C; that pk does not appear in $\boldsymbol{pk}$;

and that $|pk| < n$. If any of these fails to hold, answer `invalid`. Otherwise respond with $\sigma = \mathsf{ASig}(sk, M, \sigma', \boldsymbol{M}, \boldsymbol{pk})$.

Output. Eventually, $\mathcal{A}$ halts, outputting a forgery σ^* on messages $\boldsymbol{M}$ under keys $\boldsymbol{pk}$. This forgery must verify as valid under AVf; each key in $\boldsymbol{pk}$ (except the challenge key) must be in C; and $|\boldsymbol{pk}| \leq n$ must hold. In addition, the forgery must be nontrivial: the challenge key pk^* must appear in $\boldsymbol{pk}$, wlog at index 1 (since signature verification in our scheme has no inherent order), and the corresponding message $\boldsymbol{M}[1]$ must not have been queried by $\mathcal{A}$ of its sequential aggregate signing oracle. Output 1 if all these conditions hold, 0 otherwise.

We say that an aggregate signature scheme is $(t, q_C, q_s, n, \epsilon)$ secure if no t-time adversary making q_C certification queries and q_s signing queries can win the above game with advantage more than ϵ, where n is an upper bound on the length of the sequential aggregates involved.

3.2   Our Scheme

We start by giving some intuition for our scheme. Each signer in our scheme will have a unique public key from the Waters signature scheme

$$u', \boldsymbol{u} = (u_1, \ldots, u_k), A \leftarrow e(g, g)^\alpha.$$

While in the original signature scheme the private key consists only of g^α, in our aggregate signature scheme it is important that the private key holder will additionally choose and remember the discrete logs of $u', \boldsymbol{u} = (u_1, \ldots, u_k)$. In the Waters signature scheme, signatures are made of two group elements S_1 and S_2. At a high level, we can view S_2 as some randomness for the signature and S_1 as the signature on a message relative to that randomness.

An aggregate signature in our scheme also consists of group elements S_1', S_2'. The second element S_2' again consists of some "shared" randomness for the signature. When a signer wishes to add his signature on a message to an aggregate (S_1', S_2'), he simply figures out what his S_1 component would be in the underlying signature scheme given S_2' as the randomness. In order to perform this computation the signer must know the discrete log values of all of his public generators. He then then then multiplies this value into S_1' and finally re-randomizes the signature.

We now formally describe the sequential aggregate obtained from the Waters signature.

Our sequential aggregate scheme is a three-tuple of algorithms $\mathcal{WSA} = (\mathsf{Kg}, \mathsf{ASig}, \mathsf{AVf})$. These behave as follows.

WSA.Kg. Pick random $\alpha, y' \xleftarrow{\mathrm{R}} \mathbb{Z}_p$ and a random vector $\boldsymbol{y} = (y_1, \ldots, y_k) \xleftarrow{\mathrm{R}} \mathbb{Z}_p^k$. Compute

$$u' \leftarrow g^{y'} \quad \text{and} \quad \boldsymbol{u} = (u_1, \ldots, u_k) \leftarrow (g^{y_1}, \ldots, g^{y_k}) \quad \text{and} \quad A \leftarrow e(g, g)^\alpha .$$

The user's private key is $sk = (\alpha, y', \boldsymbol{y}) \in \mathbb{Z}_p^{k+2}$. The public key is $pk = (A, u', \boldsymbol{u}) \in \mathbb{G}_T \times \mathbb{G}^{k+1}$; it must be certified to ensure knowledge of the corresponding private key.

WSA.ASig$(sk, M, \sigma', \boldsymbol{M}, \boldsymbol{pk})$. The input is a private key sk, to be parsed as $(\alpha, y', y_1, \ldots, y_k) \in \mathbb{Z}_p^{k+2}$; a message M to sign, parsed as $(m_1, \ldots, m_k) \in \{0,1\}^k$; and an aggregate-so-far σ' on messages $\boldsymbol{M}$ under public keys $\boldsymbol{pk}$. Verify that σ' is valid by calling $\mathsf{AVf}(\sigma', \boldsymbol{M}, \boldsymbol{pk})$; if not, output `fail` and halt. Check that the public key corresponding to sk does not already appear in $\boldsymbol{pk}$; if it does, output `fail` and halt. (We revisit the issue of having one signer sign multiple messages below.)

Otherwise, parse σ' as $(S_1', S_2') \in \mathbb{G}^2$. Set $l \leftarrow |\boldsymbol{pk}|$. Now, for each i, $1 \leq i \leq l$, parse $\boldsymbol{M}[i]$ as $(m_{i,1}, \ldots, m_{i,k}) \in \{0,1\}^k$, and parse $\boldsymbol{pk}[i]$ as $(A_i, u_i', u_{i,1}, \ldots, u_{i,k}) \in \mathbb{G}_T \times \mathbb{G}^{k+1}$. Compute

$$w_1 \leftarrow S_1' \cdot g^\alpha \cdot (S_2')^{(y' + \sum_{j=1}^k y_j m_j)} \quad \text{and} \quad w_2 \leftarrow S_2' \ . \tag{3}$$

The values (w_1, w_2) form a valid signature on $\boldsymbol{M} \| M$ under keys $\boldsymbol{pk} \| pk$, but this signature needs to be re-randomized: otherwise whoever created σ' could learn the user's private key g^α. Choose a random $\tilde{r} \in \mathbb{Z}_p$, and compute

$$S_1 \leftarrow w_1 \cdot \Big(u' \prod_{j=1}^k u_j^{m_j}\Big)^{\tilde{r}} \cdot \prod_{i=1}^l \Big(u_i' \prod_{j=1}^k u_{i,j}^{m_{i,j}}\Big)^{\tilde{r}} \quad \text{and} \quad S_2 \leftarrow w_2 g^{\tilde{r}} \ . \tag{4}$$

It is easy to see that $\sigma = (S_1, S_2)$ is also a valid sequential aggregate signature on $\boldsymbol{M} \| M$ under keys $\boldsymbol{pk} \| pk$, with randomness $r + \tilde{r}$, where $w_2 = g^r$; output it and halt.

WSA.AVf$(\sigma, \boldsymbol{M}, \boldsymbol{pk})$. The input is a purported sequential aggregate σ on messages $\boldsymbol{M}$ under public keys $\boldsymbol{pk}$. Parse σ as $(S_1, S_2) \in \mathbb{G}$. If any key appears twice in $\boldsymbol{pk}$, if any key in $\boldsymbol{pk}$ has not been certified, or if $|\boldsymbol{pk}| \neq |\boldsymbol{M}|$, output `invalid` and halt.

Otherwise, set $l \leftarrow |\boldsymbol{pk}|$. If $l = 0$, output `valid` if $S_1 = S_2 = 1$, `invalid` otherwise.

Now, for each i, $1 \leq i \leq l$, parse $\boldsymbol{M}[i]$ as $(m_{i,1}, \ldots, m_{i,k}) \in \{0,1\}^k$, and parse $\boldsymbol{pk}[i]$ as $(A_i, u_i', u_{i,1}, \ldots, u_{i,k}) \in \mathbb{G}_T \times \mathbb{G}^{k+1}$. Finally, verify that

$$e(S_1, g) \cdot e\Big(S_2, \prod_{i=1}^l \Big(u_i' \prod_{j=1}^k u_{i,j}^{m_{i,j}}\Big)\Big)^{-1} \stackrel{?}{=} \prod_{i=1}^l A_i \tag{5}$$

holds; if so, output `valid`; if not, output `invalid`.

Signature Form. Consider a sequential aggregate signature on l messages $\boldsymbol{M}$ under l public keys $\boldsymbol{pk}$. For each i let $\boldsymbol{M}[i]$ be $(m_{i,1}, \ldots, m_{i,k})$ and let $\boldsymbol{pk}[i]$ be $(A_i, u_i', u_{i,1}, \ldots, u_{i,k})$ with corresponding private key $(\alpha_i, y_i', y_{i,1}, \ldots, y_{i,k})$. A well-formed sequential aggregate signature $\sigma = (S_1, S_2)$ in this case has the form

$$S_1 = \prod_{i=1}^l g^{\alpha_i} \cdot \prod_{i=1}^l \Big(u_i' \prod_{j=1}^k u_{i,j}^{m_{i,j}}\Big)^r \quad \text{and} \quad S_2 = g^r \ .$$

Additionally, we consider $\sigma = (1, 1)$ to be a valid signature on an empty set of signers. Notice that (S_1, S_2) is the product of Waters signatures all sharing the same randomness r.

Even though in our description we did not allow a signer to sign twice in an aggregate signature, a simple trick allows for this. Suppose a signer wishes to add his signature on message M to a sequential aggregate signature that already contains his signature on another message M'. He need simply first remove his signature on M' from the aggregate, essentially by dividing it out of S_1, and multiply in a signature on $M' : M$, which is a message that attests to both M' and M.

Performance. Verification in our signatures is fast, taking approximately $k/2$ multiplications per signer in the aggregate, and only two pairings regardless of how many signers are included. In contrast, the aggregate signatures of Boneh et al. [8] take $l + 1$ pairings to verify when the aggregate includes l signers.

3.3   Proof of Security

Theorem 1. *The WSA sequential aggregate signature scheme is $(t, q_C, q_S, n, \epsilon)$-unforgeable if the W signature scheme is (t', q', ϵ')-unforgeable on $\mathbb{G}$, where*

$$t' = t + O(q_C + nq_S + n) \quad\quad and \quad\quad q' = q_S \quad\quad and \quad\quad \epsilon' = \epsilon .$$

Proof. Suppose that there exists an adversary $\mathcal{A}$ that succeeds with advantage ϵ. We build a simulator $\mathcal{B}$ to play the forgeability game against the W signature scheme. Given the challenge W-signature public key $pk^* = (A, u', u_1, \ldots, u_k)$, simulator $\mathcal{B}$ interacts with $\mathcal{A}$ as follows.

Setup. Algorithm $\mathcal{B}$ runs $\mathcal{A}$ supplying it with the challenge key pk^*.

Certification Queries. Algorithm $\mathcal{A}$ wishes to certify some public key $pk = (A, u', u_1, \ldots, u_k)$, providing also its corresponding private key $sk = (\alpha, y', y_1, \ldots, y_k)$. Algorithm $\mathcal{B}$ checks that the private key is indeed the correct one and if so registers (pk, sk) in its list of certified keypairs.

Aggregate Signature Queries. Algorithm $\mathcal{A}$ requests a sequential aggregate signature, under the challenge key, on a message M. In addition, it supplies an aggregate-so-far σ' on messages $\boldsymbol{M}$ under keys $\boldsymbol{pk}$. The simulator first checks that the signature σ' verifies; that each key in $\boldsymbol{pk}$ has been certified; that the challenge key does not appear in $\boldsymbol{pk}$; and that $|\boldsymbol{pk}| < n$. If any of these conditions does not hold, $\mathcal{B}$ returns `fail`.

Otherwise, $\mathcal{B}$ queries its own signing oracle for key pk^*, obtaining a signature σ on message M, which we view as a sequential aggregate on messages (M) under keys (pk^*). The simulator now constructs the rest of the required aggregate by adding to σ, for each signer $\boldsymbol{pk}[i]$, the appropriate signature on message $\boldsymbol{M}[i]$ using algorithm ASig. It can do this because it knows – by means of the certification procedure – the private key corresponding to each public key in $\boldsymbol{pk}$. The result is an aggregate signature σ' on messages $\boldsymbol{M}\|M$ under keys $\boldsymbol{pk}\|pk^*$. This reconstruction method works because signatures are re-randomized after each aggregate signing operation and because our signatures have no inherent verification order.

Output. Eventually, $\mathcal{A}$ halts, outputting a forgery, $\sigma^* = (S_1^*, S_2^*)$ on messages $\boldsymbol{M}$ under keys $\boldsymbol{pk}$. This forgery must verify as valid under AVf; each key in $\boldsymbol{pk}$ (except the challenge key) must have been certified; and $|\boldsymbol{pk}| \leq n$ must hold. In addition, the forgery must be nontrivial: the challenge key pk^* must appear in $\boldsymbol{pk}$, wlog at index 1 (since signature verification in our scheme has no inherent order), and the corresponding message $\boldsymbol{M}[1]$ must not have been queried by $\mathcal{A}$ of its sequential aggregate signing oracle. If the adversary was not successful we can quit and disregard the attempt.

Now, for each i, $1 \leq i \leq l = |\boldsymbol{pk}| = |\boldsymbol{M}|$, parse $\boldsymbol{pk}[i]$ as $(A_i, u_i', u_{i,1}, \ldots, u_{i,k})$ and $\boldsymbol{M}[i]$ as $(m_{i,1}, \ldots, m_{i,k}) \in \{0,1\}^k$. Note that we have $pk^* = (A_1, u_1', u_{1,1}, \ldots, u_{1,k})$. Furthermore, for each i, $2 \leq i \leq l$, let $(\alpha_i, y_i', y_{i,1}, \ldots, y_{i,k})$ be the private key corresponding to $\boldsymbol{pk}[i]$. Algorithm $\mathcal{B}$ computes

$$S_1 \leftarrow S_1^* \cdot \prod_{i=2}^{l} \left(g^{\alpha_i} \cdot (S_2^*)^{\left(y_i' + \sum_{j=1}^{k} y_{i,j} m_{i,j}\right)}\right)^{-1} \quad \text{and} \quad S_2 \leftarrow S_2^* \ .$$

We now have

$$e(S_1, g) \cdot e\left(S_2, u_1' \prod_{j=1}^{k} u_{1,j}^{m_{1,j}}\right)^{-1}$$

$$= e(S_1^*, g) \cdot e\left(S_2^*, u_1' \prod_{j=1}^{k} u_{1,j}^{m_{1,j}}\right)^{-1}$$

$$\times \prod_{i=2}^{l} e(g^{\alpha_i}, g)^{-1} \cdot \prod_{i=2}^{l} e\left((S_2^*)^{\left(y_i' + \sum_{j=1}^{k} y_{i,j} m_{i,j}\right)}, g\right)^{-1}$$

$$= e(S_1^*, g) \cdot e\left(S_2^*, u_1' \prod_{j=1}^{k} u_{1,j}^{m_{1,j}}\right)^{-1}$$

$$\times \prod_{i=2}^{l} A_i^{-1} \cdot \prod_{i=2}^{l} e\left(S_2^*, u_i' \prod_{j=1}^{k} u_{i,j}^{m_{i,j}}\right)^{-1}$$

$$= e(S_1^*, g) \cdot \prod_{i=1}^{l} e\left(S_2^*, u_i' \prod_{j=1}^{k} u_{i,j}^{m_{i,j}}\right)^{-1} \cdot \prod_{i=2}^{l} A_i^{-1}$$

$$= \prod_{i=1}^{l} A_i \cdot \prod_{i=2}^{l} A_i^{-1} = A_1 = A \ .$$

So (S_1, S_2) is a valid W signature on $M^* = \boldsymbol{M}[1] = (m_{1,1}, \ldots, m_{1,k})$ under key $\boldsymbol{pk}[1] = pk^*$. The last line follows from the sequential aggregate verification equation. Moreover, since $\mathcal{A}$ did not make an aggregate signing query at M^*, $\mathcal{B}$ did not make a signing query at M^*, so $\sigma = (S_1, S_2)$ is a nontrivial W signature forgery. Algorithm $\mathcal{B}$ returns it and halts.

Algorithm $\mathcal{B}$ is successful whenever $\mathcal{A}$ is. Algorithm $\mathcal{B}$ makes as many signing queries as $\mathcal{A}$ makes sequential aggregate signing queries. Algorithm $\mathcal{B}$'s running time is that of $\mathcal{A}$, plus the overhead in handling $\mathcal{A}$'s queries, and computing the final result. Each certification query can be handled in $O(1)$ time; each aggregate signing query can be handled in $O(n)$ time; and the final result can also be computed from $\mathcal{A}$'s forgery in $O(n)$ time.

4   Multisignatures

In a multisignature scheme, a single multisignature – the same size as one ordinary signature – stands for l signatures on a message M. Multisignatures were introduced by Itakura and Nakamura [17], and have been the subject of much research [26, 25, 6]. The first multisignatures in which signatures could be combined into a multisignature without interaction was proposed by Boldyreva [6], based on BLS signatures [9]. Below, we present another non-interactive multisignature scheme, based on the Waters signature, which is provably secure without random oracles.

Security Model. Micali, Ohta, and Reyzin [22] gave the first formal treatment of multisignatures. We prove security in a variant of the Micali-Ohta-Reyzin model due to Boldyreva [6]. In this model, the adversary is given a single challenge public key pk, and a signing oracle for that key. His goal is to output a forged multisignature σ^* on a message M^* under keys $pk_1, \ldots, pk_l$. Of these keys, pk_1 must be the challenge key pk. For the forgery to be nontrivial, the adversary must not have queried the signing oracle at M^*. The adversary is allowed to choose the remaining keys, but must prove knowledge of the private keys corresponding to them. For simplicity, Boldyreva handles this by having the adversary hand over the private keys; in a more complicated proof of knowledge, the keys could be extracted by rewinding, with the same result.

4.1   Our Scheme

We describe the multisignature obtained from the Waters signature. In this scheme, all users share the same random generators $u', u_1, \ldots, u_k$, which are included in the system parameters. Our scheme is a five-tuple of algorithms $\mathcal{WM} = (\mathsf{Kg}, \mathsf{Sig}, \mathsf{Vf}, \mathsf{Comb}, \mathsf{MVf})$, which behave as follows.

WM.Kg, WM.Sig, WM.Vf. Same as W.Kg, W.Sig, and W.Vf, respectively.
WM.Comb($\{pk_i, \sigma_i\}_{i=1}^l, M$). For each user in the multisignature the algorithm takes as input a public key pk_i and a signature σ_i. All these signatures are on a single message M. For each i, parse user i's public key pk_i as $A_i \in \mathbb{G}_T$ and her signature σ_i as $(S_1^{(i)}, S_2^{(i)}) \in \mathbb{G}^2$; parse the message M as a bitstring $(m_1, \ldots, m_k) \in \{0,1\}^k$. Verify each signature using Vf; if any is invalid, output fail and halt. Otherwise, compute

$$S_1 \leftarrow \prod_{i=1}^{l} S_1^{(i)} \qquad \text{and} \qquad S_2 \leftarrow \prod_{i=1}^{l} S_2^{(i)} . \tag{6}$$

The multisignature is $\sigma = (S_1, S_2)$; output it and halt.

WM.MVf$(\{pk_i\}_{i=1}^l, M, \sigma)$. For each user in the multisignature, the algorithm takes a public key pk_i. The algorithm also takes a purported multisignature σ on a message M. Parse user i's public key pk_i as $A_i \in \mathbb{G}_T$, the message M as a bitstring $(m_1, \ldots, m_k) \in \{0,1\}^k$, and the multisignature σ as $(S_1, S_2) \in \mathbb{G}^2$. Verify that

$$e(S_1, g) \cdot e(S_2, u' \prod_{i=1}^k u_i^{m_i})^{-1} \overset{?}{=} \prod_{i=1}^l A^{(i)} \tag{7}$$

holds; if so, output `valid`; if not, output `invalid`.

It is clear that if all signatures verify individually, the multisignature formed by their product also verifies according to (7). Note that we have

$$(S_1, S_2) = \left(g^{\sum_{i=1}^l \alpha^{(i)}} \cdot \left(u' \prod_{j=1}^k u_j^{m_j}\right)^{\sum_{i=1}^l r^{(i)}}, \, g^{\sum_{i=1}^l r^{(i)}} \right),$$

where $r^{(i)}$ is the randomness used by User i to generate her signature.

Proof of Security. The WM scheme is unforgeable if W signatures are unforgeable. The proof is given in Appendix A.

5   Verifiably Encrypted Signatures

A verifiably encrypted signature on some message attests to two facts:

- that the signer has produced an ordinary signature on that message; and
- that the ordinary signature can be recovered by the third party under whose key the signature is encrypted.

Such a primitive is useful for contract signing, in a protocol called optimistic fair exchange [1, 2]. Suppose both Alice and Bob wish to sign some contract. Neither is willing to produce a signature without being sure that the other will. But Alice can send Bob a verifiably encrypted signature on the contract. Bob can now send Alice his signature, knowing that if Alice does not respond with hers he can take Alice's verifiably encrypted signature and the transcript of his interaction with Alice to the third party – called the adjudicator – who will reveal Alice's signature.

Boneh et al. [8] introduced verifiably encrypted signatures, gave a security model for them, and constructed a scheme satisfying the definitions, based on the BLS short signature [9].

We describe the verifiably encrypted signature scheme obtained from the Waters signature scheme. Unlike the scheme of Boneh et al., ours is secure without random oracles.

Security Model. Boneh et al. specify two properties (besides correctness) that a verifiably encrypted signature scheme must satisfy: unforgeability and opacity. Both are defined in games. In each, the adversary is given a signer's public key pk

and an adjudicator's public key apk. He is allowed to make verifiably encrypted signing queries of the form $\mathsf{ESig}(sk, apk, \cdot)$ and adjudication queries of the form $\mathsf{Adj}(ask, pk, \cdot, \cdot)$. In the unforgeability game, his goal is to output (M^*, η^*) such that he didn't query his signing oracle at M^*; in the opacity game his goal is to output (M^*, σ^*) such that he didn't query his *adjudication* oracle at M^*. An adversary can thus win the opacity game either by creating a forgery for the underlying signature scheme directly or by recovering the ordinary signature from an encrypted signature without the adjudicator's help.

5.1   Our Scheme

Our scheme is a seven-tuple of algorithms $\mathcal{WVES} = (\mathsf{Kg}, \mathsf{Sig}, \mathsf{Vf}, \mathsf{AKg}, \mathsf{ESig}, \mathsf{EVf}, \mathsf{Adj})$ that behave as follows.

WVES.Kg, WVES.Sig, WVES.Vf. These are the same as W.Kg, W.Sig, and W.Vf, respectively.

WVES.AKg. Pick $\beta \xleftarrow{\mathrm{R}} \mathbb{Z}_p$, and set $v \leftarrow g^\beta$. The adjudicator's public key is $apk = v$; the adjudicator's private key is $ask = \beta$.

WVES.ESig(sk, apk, M). Parse the user's private key sk as $\alpha \in \mathbb{Z}_p$ and the adjudicator's public key apk as $v \in \mathbb{G}$. To sign the message $M = (m_1, \ldots, m_k)$, compute a signature $(S_1, S_2) \xleftarrow{\mathrm{R}} \mathsf{Sig}(sk, M)$. Pick a random $s \xleftarrow{\mathrm{R}} \mathbb{Z}_p$, and compute

$$K_1 \leftarrow S_1 \cdot v^s \qquad \text{and} \qquad K_2 \leftarrow S_2 \qquad \text{and} \qquad K_3 \leftarrow g^s \ .$$

The verifiably encrypted signature η is the tuple (K_1, K_2, K_3).

WVES.EVf(pk, apk, M, η). Parse the user's public key pk as $A \in \mathbb{G}_T$, the adjudicator's public key apk as $v \in \mathbb{G}$, and the verifiably encrypted signature η as $(K_1, K_2, K_3) \in \mathbb{G}^3$. Accept if the following equation holds:

$$e(K_1, g) \cdot e(K_2, u' \prod_{i=1}^{k} u_i^{m_i})^{-1} \cdot e(K_3, v)^{-1} \overset{?}{=} A \ , \tag{8}$$

where $M = (m_1, \ldots, m_k)$.

WVES.Adj(ask, pk, M, η). Parse the adjudicator's private key ask as $\beta \in \mathbb{Z}_p$. Parse the user's public key pk as $A \in \mathbb{G}_T$, and check that it has been certified. Parse the message M as $(m_1, \ldots, m_k) \in \{0, 1\}^k$. Verify (using EVf) that the verifiably encrypted signature η is valid, and parse it as $(K_1, K_2, K_3) \in \mathbb{G}^3$. Compute

$$S_1 \leftarrow K_1 \cdot K_3^{-\beta} \qquad \text{and} \qquad S_2 \leftarrow K_2 \ ;$$

re-randomize (S_1, S_2) by choosing $s \xleftarrow{\mathrm{R}} \mathbb{Z}_p$ and computing

$$S_1' \leftarrow S_1 \cdot \left(u' \prod_{i=1}^{k} u_i^{m_i} \right)^s \qquad \text{and} \qquad S_2' \leftarrow S_2 \cdot g^s \ ;$$

and output the signature (S_1', S_2').

It is easy to see that this scheme is valid, since if all parties are honest we have, for a verifiably encrypted signature (K_1, K_2, K_3),

$$e(K_1, g) \cdot e(K_2, u' \prod_{i=1}^{k} u_i^{m_i})^{-1} \cdot e(K_3, v)^{-1}$$

$$= \left(e(S_1, g) \cdot e(v^s, g)\right) \cdot e(S_2, u' \prod_{i=1}^{k} u_i^{m_i})^{-1} \cdot e(g^s, v)^{-1}$$

$$= e(S_1, g) \cdot e(S_2, u' \prod_{i=1}^{k} u_i^{m_i})^{-1} = A \ ,$$

as required; and if (K_1, K_2, K_3) is a valid verifiably encrypted signature then

$$e(S_1, g) \cdot e(S_2, u' \prod_{i=1}^{k} u_i^{m_i})^{-1} = \left(e(K_1, g) \cdot e(K_3^{-\beta}, g)\right) \cdot e(K_2, u' \prod_{i=1}^{k} u_i^{m_i})^{-1}$$

$$= e(K_1, g) \cdot e(K_2, u' \prod_{i=1}^{k} u_i^{m_i})^{-1} \cdot e(K_3, v)^{-1} = A \ ,$$

so the adjudicated signature is indeed a valid one.

Proofs of Security. The WVES scheme is unforgeable if W signatures are unforgeable, and opaque if CDH is hard on $\mathbb{G}$. The proofs are given in Appendix B.

5.2   VES from General Assumptions

Recent work has shown that group signatures [4] and ring signatures [5] can be built from general assumptions using Non-Interactive Zero Knowledge (NIZK) proofs. We note that verifiably encrypted signatures can also be realized from general assumptions. Roughly, the signer signs a message, encrypts the signature to the adjudicator and then attaches a NIZK proof that this was performed correctly.

6   Comparison to Previous Work

In this section, we compare the schemes we have presented to previous schemes in the literature. For the comparison, we instantiate pairing-based schemes using Barreto-Naehrig curves [3] with 160-bit point representation. Note that BLS-based constructions must compute, for signing and verification, a hash function onto $\mathbb{G}$. This is an expensive operation [9, Sect. 3.2].

Sequential Aggregate Signatures. We compare our sequential aggregate signature scheme to the aggregate scheme of Boneh et al. [8] (BGLS) and to the sequential aggregate signature scheme of Lysyanskaya et al. [20] (LMRS).

Table 1. Comparison of aggregate signature schemes. Signatures are by l signers; k is the output length of a collision resistant hash function; "R.O." denotes if the security proof uses random oracles.

Scheme	R.O.	Sequential	Key Model	Size	Verification	Signing
BGLS	YES	NO	Chosen	160 bits	$l + 1$ pairings	1 exp.
LMRS-1	YES	YES	Chosen	1024 bits	$2l$ exp.	verify + 1 exp.
LMRS-2	YES	YES	Registered	1024 bits	$4l$ mult.	verify + 1 exp.
Ours	NO	YES	Registered	320 bits	2 pairings, $lk/2$ mult.	verify + 1 exp.

We instantiate the LMRS scheme using the RSA-based permutation family with common domain devised by Hayashi, Okamoto, and Tanaka [16]. With this permutation family LMRS signatures do not grow by 1 bit with each signature, as is the case with the RSA-based instantiation given by Lysyanskaya et al. [20]; but evaluating the permutation requires two applications of the underlying RSA function. Lysyanskaya et al. give two variants of their scheme. One places constraints on the format of the RSA keys, thereby avoiding key certification; we call this variant LMRS-1. The other uses ordinary RSA keys and can have public exponent $e = 3$ for fast verification, but requires key certification, like our scheme; we call this variant LMRS-2.

We present the comparisons in Table 1. The size column gives signature length at the 1024-bit security level. The Verification and Signing columns give the computational costs of those operations; l is the number of signatures in an aggregate, and k is the output length of a collision-resistant hash function.

One drawback of our scheme is that a user's public key will be quite large. If we use a 160-bit collision resistant hash function, then keys will be approximately 160 group elements and take around 10KB to store. While it is desirable to achieve smaller public keys, this will be acceptable in many settings such as SBGP where achieving the signature size is a much more important consideration than the public key size. Additionally, Naccache [23] and Chatterjee and Sarkar [11] independently proposed ways to achieve shorter public keys in the Waters signature scheme. Using these methods we can also achieve considerably shorter public keys.

Multisignatures. We compare our multisignature scheme to the Boldyreva multisignature [6]. We present the comparisons in Table 2. The size column gives signature length at the 1024-bit security level. The Verification and Signing columns give the computational costs of those operations; l is the number of signatures in a multisignature, and k is the output length of a collision-resistant hash function.

Verifiably Encrypted Signatures. We compare our verifiably encrypted signature scheme to that of Boneh et al. [8] (BGLS). We present the comparisons in Table 3. The size column gives signature length at the 1024-bit security level. The Verification and Generation columns give the computational costs of those operations; k is the output length of a collision-resistant hash function.

Table 2. Comparison of multisignature schemes. Multisignatures are by l signers; k is the output length of a collision resistant hash function; "R.O." denotes if the security proof uses random oracles.

Scheme	R.O.	Key Model	Size	Verification	Signing
Boldyreva	YES	Registered	160 bits	2 pairings	1 exp.
Ours	NO	Registered	320 bits	2 pairings, $k/2$ mult.	1 exp.

Table 3. Comparison of verifiably encrypted signature schemes. We let k be the output length of a collision resistant hash function. "R.O." specifies whether the security proof uses random oracles.

Scheme	R.O.	Key Model	Size	Verification	Generation
BGLS	YES	Registered	320 bits	3 pairings	3 exp.
Ours	NO	Registered	480 bits	3 pairings, $k/2$ mult.	4 exp.

7 Conclusions and Open Problems

In this paper we gave the first aggregate signature scheme which is provably secure without random oracles; the first multisignature scheme which is provably secure without random oracles; and the first verifiably encrypted signature scheme which is provably secure without random oracles. All our constructions derive from the recent signature scheme due to Waters [28]. All our constructions are quite practical.

Signatures in our aggregate signature scheme are sequentially constructed, but knowledge of the order in which messages are signed is not necessary for verification. Additionally, our scheme gives shorter signatures than in the LMRS sequential aggregate signature scheme [20] and has a more efficient verification algorithm than the BGLS aggregate signature scheme [8]. That this gives some interesting tradeoffs for practical applications such as secure routing and proxy signatures.

Some interesting problems remain open for random-oracle–free aggregate signatures:

1. To find a scheme which supports full aggregation, in which aggregate signature do not need to be sequentially constructed. While many applications only require sequential aggregation, having a more general capability is desirable.
2. To find a sequential aggregate signature scheme provably secure in the chosen-key model.
3. To find a sequential aggregate signature scheme with shorter user keys. The size of public keys in our system reflects the size of keys in the underlying Waters signature scheme. Naccache [23] and Chatterjee and Sarkar [11] have proposed ways to shorten the public keys of the Waters IBE/signature scheme by trading off parameter size with tightness in the security reduction. It would be better to have a solution in which the public key is just a few group elements.

The last two are particularly important for certificate chain compression, proposed by Boneh et al. [8] as an application for aggregate signatures. If keys need to be registered with an authority then a chaining application is impractical, and having large public keys negates any benefit from reducing the signature size in a certificate chain, since the keys must be included in the certificates.

References

[1] N. Asokan, V. Shoup, and M. Waidner. Optimistic fair exchange of digital signatures. *IEEE J. Selected Areas in Comm.*, 18(4):593–610, Apr. 2000.

[2] F. Bao, R. Deng, and W. Mao. Efficient and practical fair exchange protocols with offline TTP. In P. Karger and L. Gong, editors, *Proceedings of IEEE Security & Privacy*, pages 77–85, May 1998.

[3] P. Barreto and M. Naehrig. Pairing-friendly elliptic curves of prime order. In B. Preneel and S. Tavares, editors, *Proceedings of SAC 2005*, volume 3897 of *LNCS*, pages 319–31. Springer-Verlag, 2006.

[4] M. Bellare, D. Micciancio, and B. Warinschi. Foundations of group signatures: Formal definitions, simplified requirements, and a construction based on general assumptions. In E. Biham, editor, *Proceedings of Eurocrypt 2003*, volume 2656 of *LNCS*, pages 614–29. Springer-Verlag, May 2003.

[5] A. Bender, J. Katz, and R. Morselli. Ring signatures: Stronger definitions, and constructions without random oracles. In S. Halevi and T. Rabin, editors, *Proceedings of TCC 2006*, volume 3876 of *LNCS*, pages 60–79. Springer-Verlag, Mar. 2006.

[6] A. Boldyreva. Threshold signature, multisignature and blind signature schemes based on the gap-Diffie-Hellman-group signature scheme. In Y. Desmedt, editor, *Proceedings of PKC 2003*, volume 2567 of *LNCS*, pages 31–46. Springer-Verlag, Jan. 2003.

[7] A. Boldyreva, A. Palacio, and B. Warinschi. Secure proxy signature schemes for delegation of signing rights. Cryptology ePrint Archive, Report 2003/096, 2003. http://eprint.iacr.org/.

[8] D. Boneh, C. Gentry, B. Lynn, and H. Shacham. Aggregate and verifiably encrypted signatures from bilinear maps. In E. Biham, editor, *Proceedings of Eurocrypt 2003*, volume 2656 of *LNCS*, pages 416–32. Springer-Verlag, May 2003.

[9] D. Boneh, B. Lynn, and H. Shacham. Short signatures from the Weil pairing. *J. Cryptology*, 17(4):297–319, Sept. 2004. Extended abstract in *Proceedings of Asiacrypt 2001*.

[10] R. Canetti, O. Goldreich, and S. Halevi. The random oracle methodology, revisited. *J. ACM*, 51(4):557–94, July 2004.

[11] S. Chatterjee and P. Sarkar. Trading time for space: Towards an efficient IBE scheme with short(er) public parameters in the standard model. In D. Won and S. Kim, editors, *Proceedings of ICISC 2005*, LNCS. Springer-Verlag, Dec. 2005. To appear.

[12] J.-S. Coron and D. Naccache. Boneh et al.'s k-element aggregate extraction assumption is equivalent to the Diffie-Hellman assumption. In C. S. Laih, editor, *Proceedings of Asiacrypt 2003*, volume 2894 of *LNCS*, pages 392–7. Springer-Verlag, Dec. 2003.

[13] S. Galbraith. Pairings. In I. F. Blake, G. Seroussi, and N. Smart, editors, *Advances in Elliptic Curve Cryptography*, volume 317 of *London Mathematical Society Lecture Notes*, chapter IX, pages 183–213. Cambridge University Press, 2005.

[14] S. Goldwasser, S. Micali, and R. Rivest. A digital signature scheme secure against adaptive chosen-message attacks. *SIAM J. Computing*, 17(2):281–308, 1988.

[15] J. Groth, R. Ostrovsky, and A. Sahai. Perfect non-interactive zero knowledge for NP. In S. Vaudenay, editor, *Proceedings of Eurocrypt 2006*, LNCS. Springer-Verlag, May 2006. This volume.

[16] R. Hayashi, T. Okamoto, and K. Tanaka. An RSA family of trap-door permutations with a common domain and its applications. In F. Bao, R. H. Deng, and J. Zhou, editors, *Proceedings of PKC 2004*, volume 2947 of *LNCS*, pages 291–304. Springer-Verlag, Mar. 2004.

[17] K. Itakura and K. Nakamura. A public-key cryptosystem suitable for digital multisignatures. *NEC J. Res. & Dev.*, 71:1–8, Oct. 1983.

[18] S. Kent, C. Lynn, and K. Seo. Secure border gateway protocol (Secure-BGP). *IEEE J. Selected Areas in Comm.*, 18(4):582–92, April 2000.

[19] N. Koblitz and A. Menezes. Pairing-based cryptography at high security levels. In N. Smart, editor, *Proceedings of Cryptography and Coding 2005*, volume 3796 of *LNCS*, pages 13–36. Springer-Verlag, Dec. 2005.

[20] A. Lysyanskaya, S. Micali, L. Reyzin, and H. Shacham. Sequential aggregate signatures from trapdoor permutations. In C. Cachin and J. Camenisch, editors, *Proceedings of Eurocrypt 2004*, volume 3027 of *LNCS*, pages 74–90. Springer-Verlag, May 2004.

[21] M. Mambo, K. Usuda, and E. Okamoto. Proxy signatures for delegating signing operation. In L. Gong and J. Stearn, editors, *Proceedings of CCS 1996*, pages 48–57. ACM Press, Mar. 1996.

[22] S. Micali, K. Ohta, and L. Reyzin. Accountable-subgroup multisignatures (extended abstract). In P. Samarati, editor, *Proceedings of CCS 2001*, pages 245–54. ACM Press, Nov. 2001.

[23] D. Naccache. Secure and practical identity-based encryption. Cryptology ePrint Archive, Report 2005/369, 2005. http://eprint.iacr.org/.

[24] D. Nicol, S. Smith, and M. Zhao. Evaluation of efficient security for BGP route announcements using parallel simulation. *Simulation Modelling Practice and Theory*, 12:187–216, 2004.

[25] K. Ohta and T. Okamoto. Multisignature schemes secure against active insider attacks. *IEICE Trans. Fundamentals*, E82-A(1):21–31, 1999.

[26] T. Okamoto. A digital multisignature scheme using bijective public-key cryptosystems. *ACM Trans. Computer Systems*, 6(4):432–41, November 1988.

[27] K. Paterson. Cryptography from pairings. In I. F. Blake, G. Seroussi, and N. Smart, editors, *Advances in Elliptic Curve Cryptography*, volume 317 of *London Mathematical Society Lecture Notes*, chapter X, pages 215–51. Cambridge University Press, 2005.

[28] B. Waters. Efficient identity-based encryption without random oracles. In R. Cramer, editor, *Proceedings of Eurocrypt 2005*, volume 3494 of *LNCS*, pages 114–27. Springer-Verlag, May 2005.

A   WM Proof of Security

Theorem 2. *The WM multisignature scheme is (t, q, ϵ)-unforgeable if the W signature scheme is (t', q', ϵ')-unforgeable, where*

$$t' = t + O(q) \qquad and \qquad q' = q \qquad and \qquad \epsilon' = \epsilon \ .$$

Proof. Suppose $\mathcal{A}$ is an adversary that can forge multisignatures, and (t, q, ϵ)-breaks the WM scheme. We show how to construct an algorithm $\mathcal{B}$ that (t', q, ϵ)-breaks the W scheme. Algorithm $\mathcal{B}$ is given a W public key $A = e(g, g)^\alpha$. It interacts with $\mathcal{A}$ as follows.

Setup. Simulator $\mathcal{B}$ invokes $\mathcal{A}$, providing to it the public key A.

Signature queries. Algorithm $\mathcal{A}$ requests a signature on some message M under the challenge key A. Algorithm $\mathcal{B}$ requests a signature on M in turn from its own signing oracle, and returns the result to the adversary.

Output. Finally, $\mathcal{A}$ halts, having output a signature (S_1^*, S_2^*) on some message M^*, along with public keys $A^{(1)}, \ldots, A^{(l)}$ for some l, where $A^{(1)}$ equals A, the challenge key. It must not previously have requested a signature on M^*. In addition, it outputs the private keys $\alpha^{(2)}, \ldots, \alpha^{(l)}$ for all keys except the challenge key. Algorithm $\mathcal{B}$ sets $S \leftarrow S_1^* / \prod_{i=2}^{l} g^{\alpha^{(i)}}$. Then we have

$$e(S, g) \cdot e(S_2, u' \prod_{i=1}^{k} u_i^{m_i})^{-1} = e(S_1, g) \cdot e(S_2, u' \prod_{i=1}^{k} u_i^{m_i})^{-1} \cdot \prod_{i=2}^{l} e(g, g)^{-\alpha^{(i)}}$$

$$= \prod_{i=1}^{l} A^{(i)} \cdot \prod_{i=2}^{l} A^{-(i)} = A^{(1)} = A ,$$

so (S, S_2) is a valid W signature on M^* under the challenge key A. Since $\mathcal{A}$ did not make a signing query to the challenger at M^*, neither did $\mathcal{B}$ make a signing query to its own signing oracle at M^*, and the forgery is thus nontrivial. Algorithm $\mathcal{B}$ outputs (S, S_2) and halts.

Thus $\mathcal{B}$ succeeds whenever $\mathcal{A}$ does. Algorithm $\mathcal{B}$ makes exactly as many signing queries as $\mathcal{A}$ does. Its running time is the same as $\mathcal{A}$'s, plus the time required for setup and output – both $O(1)$ – and to handle $\mathcal{A}$'s signing queries – $O(1)$ for each of at most q queries.

B WVES Proofs of Security

B.1 Unforgeability

Theorem 3. *The WVES verifiably encrypted signature scheme is (t, q_S, q_A, ϵ)-unforgeable if the W signature scheme is (t', q', ϵ')-unforgeable, where*

$$t' = t + O(q_S + q_A) \qquad and \qquad q' = q_S \qquad and \qquad \epsilon' = \epsilon .$$

Proof. We show how to turn a verifiably-encrypted signature forger $\mathcal{A}$ into a forger $\mathcal{B}$ for the underlying Waters signature scheme.

Algorithm $\mathcal{B}$ is given a Waters signature public key $A = e(g, g)^\alpha$. It picks $\beta \xleftarrow{\text{R}} \mathbb{Z}_p$, sets $v \leftarrow g^\beta$, and provides the adversary $\mathcal{A}$ with A and v.

When $\mathcal{A}$ requests a verifiably encrypted signature on some message M, the challenger $\mathcal{B}$ requests a signature on M from its own signing oracle, obtaining a signature (S_1, S_2). It picks $s \xleftarrow{\text{R}} \mathbb{Z}_p$ and computes

$$K_1 \leftarrow S_1 \cdot v^s \qquad \text{and} \qquad K_2 \leftarrow S_2 \qquad \text{and} \qquad K_3 \leftarrow g^s \ .$$

The tuple (K_1, K_2, K_3) is a valid verifiably encrypted signature on M. Algorithm $\mathcal{B}$ provides $\mathcal{A}$ with it. (Here $\mathcal{B}$ is simply evaluating ESig, except that it uses its signing oracle instead of evaluating Sig directly.)

When algorithm $\mathcal{A}$ requests adjudication of a verifiably encrypted signature (K_1, K_2, K_3) on some message M under the challenge key A, $\mathcal{B}$ responds with $\mathsf{Adj}(\beta, A, M, (K_1, K_2, K_3))$. Note that $\mathcal{B}$ knows the adjudicator's private key β.

Finally, $\mathcal{A}$ outputs a forged verifiably-encrypted signature (K_1^*, K_2^*, K_3^*) on some message $M^* = (m_1^*, \ldots, m_k^*)$. Algorithm $\mathcal{A}$ must never have made a verifiably encrypted signing query at M^*.

The challenger $\mathcal{B}$ computes

$$S_1^* \leftarrow K_1^* \cdot (K_3^*)^{-\beta} \qquad \text{and} \qquad S_2^* \leftarrow K_2^*.$$

Then we have

$$e(S_1^*, g) \cdot e\big(S_2^*, u' \prod_{i=1}^{k} u_i^{m_i^*}\big)^{-1}$$

$$= \Big[e(K_1^*, g) \cdot e\big(K_2^*, u' \prod_{i=1}^{k} u_i^{m_i^*}\big)^{-1} \Big] \cdot e\big((K_3^*)^{-\beta}, g\big)$$

$$= e(K_1^*, g) \cdot e\big(K_2^*, u' \prod_{i=1}^{k} u_i^{m_i^*}\big)^{-1} \cdot e(K_3^*, v)^{-1} = A \ ,$$

and (S_1^*, S_2^*) is therefore a valid Waters signature on M^*. The last equality follows from equation (8). Because $\mathcal{A}$ did not make a verifiably encrypted signing query at M^*, neither did $\mathcal{B}$ make a signing query at M^*, and the forgery is thus nontrivial. The challenger $\mathcal{B}$ outputs (S_1^*, S_2^*) and halts.

Algorithm $\mathcal{B}$ thus succeeds whenever $\mathcal{A}$ does. Its running time overhead is $O(1)$ for each of $\mathcal{A}$'s verifiably encrypted signing and adjudication queries, and for computing the final output.

B.2 Opacity

For convenience, we prove opacity by reduction from the aggregate extraction assumption: given $(g^\alpha, g^\beta, g^\gamma, g^\delta, g^{\alpha\gamma+\beta\delta})$, computing $g^{\alpha\gamma}$ is hard. Coron and Naccache [12] showed that this assumption, introduced by Boneh et al. [8], is equivalent to CDH.

Theorem 4 (Coron–Naccache [12]). *The aggregate extraction and Computational Diffie-Hellman problems are Karp reducible to each other with $O(1)$ computation.*[2]

Theorem 5. *The WVES verifiably encrypted signature scheme is (t, q_S, q_A, ϵ)-opaque if aggregate extraction is (t', ϵ')-hard on $\mathbb{G}$, where*

$$t' = t + O(q_s + q_A) \quad and \quad q' = q_s \quad and \quad \epsilon' = 4kq_A\epsilon \ .$$

Proof. Given an algorithm $\mathcal{A}$ that breaks the opacity of the scheme, we show how to construct an algorithm $\mathcal{B}$ that breaks the aggregate extraction assumption.

The challenger $\mathcal{B}$ is given values g^α, g^β, g^γ, and g^δ, along with $g^{\alpha\gamma+\beta\delta}$; its goal is to produce $g^{\alpha\gamma}$. It sets $v \leftarrow g^\beta$, $g_1 \leftarrow g^\alpha$, and $g_2 \leftarrow g^\gamma$. It computes $A \leftarrow e(g_1, g_2) = e(g, g)^{\alpha\gamma}$.

Let $\lambda = 2q_A$. Algorithm $\mathcal{B}$ picks $\kappa \stackrel{R}{\leftarrow} \{0, \ldots, k\}$, $x', x_1, \ldots, x_k \stackrel{R}{\leftarrow} \mathbb{Z}_\lambda = \{0, \ldots, \lambda - 1\}$ and $y', y_1, \ldots, y_k \stackrel{R}{\leftarrow} \mathbb{Z}_p$ and sets

$$u' \leftarrow g_2^{x'-\kappa\lambda} g^{y'} \quad and \quad u_i \leftarrow g_2^{x_i} g^{y_i} \quad for\ i = 1, \ldots, k \ .$$

It then interacts with $\mathcal{A}$ as follows.

Setup. Algorithm $\mathcal{B}$ gives to $\mathcal{A}$ the system parameters $(g, u', u_1, \ldots u_k)$, the signer's public key A, and the adjudicator's public key v. Note that the private signing key is $\alpha\gamma$.

Verifiably Encrypted Signing Queries. $\mathcal{A}$ requests a verifiably-encrypted signature on $M = (m_1, \ldots, m_k) \in \{0, 1\}^k$ under challenge key A and adjudicator key v. Define $F = -\kappa\lambda + x' + \sum_{i=1}^{k} x_i m_i$ and $J = y' + \sum_{i=1}^{k} y_i m_i$. If $F \neq 0 \bmod p$ algorithm $\mathcal{B}$ proceeds as follows. It picks $r \stackrel{R}{\leftarrow} \mathbb{Z}_p$ and sets

$$S_1 \leftarrow g_1^{-J/F} \Big(u' \prod_{i=1}^{k} u_i^{m_i} \Big)^r \quad and \quad S_2 \leftarrow g_1^{-1/F} g^r \ .$$

This is a valid W signature with randomness $\tilde{r} = r - \alpha/F$: observing that $u' \prod_{i=1}^{k} u_i^{m_i} = g_2^F g^J$, we see that

$$S_1 = g_1^{-J/F} \Big(u' \prod_{i=1}^{k} u_i^{m_i} \Big)^r = g_2^\alpha (g_2^F g^J)^{-\alpha/F} (g_2^F g^J)^r = g^{\alpha\gamma} \Big(u' \prod_{i=1}^{k} u_i^{m_i} \Big)^{\tilde{r}} \ ,$$

where for the second equality we have multiplied and divided by g_2^α. Algorithm $\mathcal{B}$ then encrypts (S_1, S_2) by choosing $s \stackrel{R}{\leftarrow} \mathbb{Z}_p$ and setting

$$K_1 \leftarrow S_1 \cdot v^s \quad and \quad K_2 \leftarrow S_2 \quad and \quad K_3 \leftarrow g^s \ .$$

If $F = 0$, however, $\mathcal{B}$ picks $r, s \stackrel{R}{\leftarrow} \mathbb{Z}_p$ and sets

---

[2] Strictly speaking, the amount of work is poly-logarithmic in the security parameter since the group element representations grow. The number of algebraic operations is constant.

$$K_1 \leftarrow (g^{\alpha\gamma+\gamma\delta}) \cdot (g^\gamma)^s \cdot \left(u' \prod_{i=1}^{k} u_i^{m_i}\right)^r \quad \text{and} \quad K_2 \leftarrow g^r \quad \text{and} \quad K_3 \leftarrow (g^\delta) \cdot g^s \ .$$

This is a W signature with randomness r, encrypted with randomness $\delta + s$.
In either case, $\mathcal{B}$ returns to $\mathcal{A}$ the verifiably encrypted signature (K_1, K_2, K_3).
Adjudication Queries. Suppose $\mathcal{A}$ requests adjudication on (K_1, K_2, K_3) for message $M = (m_1, \ldots, m_k)$. Algorithm $\mathcal{B}$ first verifies that (K_1, K_2, K_3) is valid and rejects it otherwise. Define $F = -\kappa\lambda + x' + \sum_{i=1}^{k} x_i m_i$ and $J = y' + \sum_{i=1}^{k} y_i m_i$ as before. If $F = 0 \bmod p$, $\mathcal{B}$ declares failure and halts. Otherwise, it picks $r \xleftarrow{\text{R}} \mathbb{Z}_p$ and computes

$$S_1 \leftarrow g_1^{-J/F}\left(u' \prod_{i=1}^{k} u_i^{m_i}\right)^r \quad \text{and} \quad S_2 \leftarrow g_1^{-1/F} g^r$$

as above, returning (S_1, S_2) to $\mathcal{A}$.
(Note that $\mathcal{A}$ must previously have made a verifiably encrypted signing query at M, since otherwise we could use it to break the unforgeability of WVES.)
Output. Finally, algorithm $\mathcal{A}$ outputs a signature (S_1^*, S_2^*) on a message $M^* = (m_1^*, \ldots, m_k^*)$; it must not have queried its adjudication oracle at M^*. Define $F^* = -\kappa\lambda + x' + \sum_{i=1}^{k} x_i m_i^*$ and $J^* = y' + \sum_{i=1}^{k} y_i m_i^*$. If $F^* \neq 0 \bmod p$, $\mathcal{B}$ declares failure and exits. Otherwise, we have $u' \prod_{i=1}^{k} u_i^{m_i^*} = g^{J^*}$, so that

$$e(g_1, g_2) = A = e(S_1^*, g) \cdot e\left(S_2^*, u' \prod_{i=1}^{k} u_i^{m_i^*}\right)^{-1}$$

$$= e(S_1^*, g) \cdot e(S_2^*, g^{J^*})^{-1} = e(S_1^*(S_2^*)^{-J^*}, g) \ ,$$

and $S_1^*(S_2^*)^{-J^*}$ equals $g^{\alpha\gamma}$, which is the solution to the aggregate extraction challenge; $\mathcal{B}$ outputs it and halts.

The probability that $\mathcal{B}$ doesn't abort in any adjudication query is at least $1 - 1/\lambda$; since there are at most $q_A = \lambda/2$ such queries, $\mathcal{B}$ manages to answer all queries without aborting with probability at least $1/2$. Having done so, $\mathcal{B}$ then receives a forgery such that $F^* = 0 \bmod p$ with probability at least $1/(\kappa\lambda) \geq 1/(2kq_A)$. Thus $\mathcal{B}$ succeeds with probability at least $\epsilon/(4kq_A)$. (For more detailed probability analysis, see Waters' original proof [28].) Algorithm $\mathcal{B}$'s run-time overhead is $O(1)$ to answer each of $\mathcal{A}$'s queries and to compute the final output.

Security of the Waters Signature. The reduction above did not require that $\mathcal{A}$ had requested a verifiably encrypted signature at M^*. It is easy to convert an algorithm $\mathcal{A}'$ that forges the underlying W signature to a WVES opacity breaker of this sort: simulate a W signing oracle by a call to the verifiably encrypted signing oracle followed by a call to the adjudication oracle. Combining this insight with Theorems 5 and 4 immediately gives the following corollary:

Corollary 1 (Waters [28]). *The Waters signature scheme is (t, q, ϵ)-unforgeable if Computational Diffie-Hellman is $(t + O(q), 4kq\epsilon)$-hard on $\mathbb{G}$. Here q is the number of signing queries.*

Our Data, Ourselves: Privacy Via Distributed Noise Generation

Cynthia Dwork[1], Krishnaram Kenthapadi[2,4,5], Frank McSherry[1],
Ilya Mironov[1], and Moni Naor[3,4,6]

[1] Microsoft Research, Silicon Valley Campus
{dwork, mcsherry, mironov}@microsoft.com
[2] Stanford University
kngk@cs.stanford.edu
[3] Weizmann Institute of Science
moni.naor@weizmann.ac.il

Abstract. In this work we provide efficient distributed protocols for generating shares of random noise, secure against malicious participants. The purpose of the noise generation is to create a distributed implementation of the privacy-preserving statistical databases described in recent papers [14, 4, 13]. In these databases, privacy is obtained by perturbing the true answer to a database query by the addition of a small amount of Gaussian or exponentially distributed random noise. The computational power of even a simple form of these databases, when the query is just of the form $\sum_i f(d_i)$, that is, the sum over all rows i in the database of a function f applied to the data in row i, has been demonstrated in [4]. A distributed implementation eliminates the need for a trusted database administrator.

The results for noise generation are of independent interest. The generation of Gaussian noise introduces a technique for distributing shares of many unbiased coins with fewer executions of verifiable secret sharing than would be needed using previous approaches (reduced by a factor of n). The generation of exponentially distributed noise uses two shallow circuits: one for generating many arbitrarily but identically biased coins at an amortized cost of two unbiased random bits apiece, independent of the bias, and the other to combine bits of appropriate biases to obtain an exponential distribution.

1 Introduction

A number of recent papers in the cryptography and database communities have addressed the problem of *statistical disclosure control* – revealing

---

[4] Part of the work was done in Microsoft Research, Silicon Valley Campus.

[5] Supported in part by NSF Grant ITR-0331640. This work was also supported in part by TRUST (The Team for Research in Ubiquitous Secure Technology), which receives support from the National Science Foundation (NSF award number CCF-0424422) and the following organizations: Cisco, ESCHER, HP, IBM, Intel, Microsoft, ORNL, Qualcomm, Pirelli, Sun and Symantec.

[6] Incumbent of the Judith Kleeman Professorial Chair. Research supported in part by a grant from the Israel Science Foundation.

accurate statistics about a population while preserving the privacy of individuals [1, 2, 15, 11, 14, 5, 6, 4, 13]. Roughly speaking, there are two computational models; in a non-interactive solution the data are somehow sanitized and a "safe" version of the database is released (this may include histograms, summaries, and so on), while in an interactive solution the user queries the database through a privacy mechanism, which may alter the query or the response in order to ensure privacy. With this nomenclature in mind the positive results in the literature fall into three broad categories: non-interactive with trusted server, non-interactive with untrusted server – specifically, via *randomized response*, in which a data holder alters her data with some probability before sending it to the server – and interactive with trusted server. The current paper provides a *distributed* interactive solution, replacing the trusted server with the assumption that strictly fewer than one third of the participants are faulty (we handle Byzantine faults). Under many circumstances the results obtained are of provably better quality (accuracy and conciseness, i.e., number of samples needed for correct statistics to be computed) than is possible for randomized response or other non-interactive solutions [13]. Our principal technical contribution is in the cooperative generation of shares of noise sampled from in one case the Binomial distribution (as an approximation for the Gaussian) and in the second case the Poisson distribution (as an approximation for the exponential).

Consider a database that is a collection of rows; for example, a row might be a hospital record for an individual. A query is a function f mapping rows to the interval $[0, 1]$. The *true answer* to the query is the value obtained by applying f to each row and summing the results. By responding with an appropriately perturbed version of the true answer, privacy can be guaranteed. The computational power of this provably private "noisy sums" primitive is demonstrated in Blum et al. [4], where it was shown how to carry out accurate and privacy-preserving variants of many standard data mining algorithms, such as k-means clustering, principal component analysis, singular value decomposition, the perceptron algorithm, and anything learnable in the statistical queries (STAT) learning model[1].

Although the powerful techniques of secure function evaluation [25, 17] may be used to emulate any privacy mechanism, generic computations can be expensive. The current work is inspired by the combination of the simplicity of securely computing sums and the power of the noisy sums. We provide *efficient* methods allowing the parties holding their own data to act *autonomously* and without a central trusted center, while simultaneously preventing malicious parties from interfering with the utility of the data.

The approach to decentralization is really very simple. For ease of exposition we describe the protocol assuming that every data holder participates in every query and that the functions f are predicates. We discuss relaxations of these assumptions in Section 5.

---

[1] This was extended in [13] to handle functions f that operate on the database as a whole, rather than on individual rows of the database.

Structure of ODO (Our Data, Ourselves) Protocol

1. **Share Summands:** On query f, the holder of d_i, the data in row i of the database, computes $f(d_i)$ and shares out this value using a *non-malleable verifiable secret sharing scheme* (see Section 2), $i = 1, \ldots, n$. The bits are represented as $0/1$ values in $GF(q)$, for a large prime q. We denote this set $\{0, 1\}_{GF(q)}$ to make the choice of field clear.
2. **Verify Values:** Cooperatively verify that the shared values are *legitimate* (that is, in $\{0, 1\}_{GF(q)}$, when f is a predicate).
3. **Generate Noise Shares:** Cooperatively generate shares of appropriately distributed random noise.
4. **Sum All Shares:** Each participant adds together all the shares that it holds, obtaining a share of the noisy sum $\sum_i f(d_i) + \textbf{noise}$. All arithmetic is in $GF(q)$.
5. **Reconstruct:** Cooperatively reconstruct the noisy sum using the reconstruction technique of the verifiable secret sharing scheme.

Our main technical work is in Step 3. We consider two types of noise, *Gaussian* and *scaled symmetric exponential*. In the latter distribution the probability of being at distance $|x|$ from the mean is proportional to $\exp(-|x|/R)$, the scale R determining how "flat" the distribution will be. In our case the mean will always be 0. Naturally, we must approximate these distributions using finite-precision arithmetic. The Gaussian and exponential distributions will be approximated, respectively, by the Binomial and Poisson distributions.

The remainder of this paper is organized as follows. In Section 2 we review those elements from the literature necessary for our work, including definitions of randomness extractors and of privacy. In Sections 3 and 4 we discuss implementations of Step 3 for Gaussian and Exponential noise, respectively. Finally, various generalizations of our results are mentioned in Section 5.

2   Cryptographic and Other Tools

Model of Computation. We assume the standard synchronous model of computation in which n processors communicate by sending messages via point-to-point channels and up to $t \leq \lfloor \frac{n-1}{3} \rfloor$ may fail in an arbitrary, Byzantine, adaptive fashion. If the channels are secure, then the adversary may be computationally unbounded. However, if the secure channels are obtained by encryption then we assume the adversary is restricted to probabilistic polynomial time computations.

We will refer to several well-known primitive building blocks for constructing distributed protocols: Byzantine Agreement [20], Distributed Coin Flipping [22], Verifiable Secret Sharing (VSS) [8], Non-Malleable VSS, and Secure Function Evaluation (SFE) [18].

A VSS scheme allows any processor distribute shares of a secret, which can be verified for consistency. If the shares verify, the honest processors can always reconstruct the secret regardless of the adversary's behavior. Moreover, the faulty processors by themselves cannot learn any information about the secret. A non-malleable VSS scheme ensures that the values shared by a non-faulty processor are completely independent of the values shared by the other processors; even exact copying is prevented.

Throughout the paper we will use the following terminology. Values that have been shared and verified, but not yet reconstructed, are said to be *in shares*. Values that are publicly known are said to be *public*.

A *randomness extractor* [21] is a method of converting a non-uniform input distribution into a near-uniform distribution on a smaller set. In general, an extractor is a randomized algorithm, which additionally requires a perfect source of randomness, called the seed. Provided that the input distribution has sufficiently high min-entropy, a good extractor takes a short seed and outputs a distribution that is statistically close to the uniform. Formally,

Definition 1. *Letting the min-entropy of a distribution $\mathcal{D}$ on X be denoted $H_\infty(\mathcal{D}) = -\log\max_{x \in X} \mathcal{D}(x)$, a function $F \colon X \times Y \mapsto \{0,1\}^n$ is a (δ, ϵ, n)-extractor, if for any distribution $\mathcal{D}$ on X such that $H_\infty(\mathcal{D}) > \delta$,*

$$|\{F(x,y) \colon x \in_{\mathcal{D}} X, y \in_U Y\} - U_n| < \epsilon,$$

where $|\cdot|$ is the statistical distance between two distributions, U_n is the uniform distribution on $\{0,1\}^n$, and $x \in_{\mathcal{D}} X$ stands for choosing $x \in X$ according to $\mathcal{D}$.

Optimal extractors can extract $n = \delta - 2\log(1/\epsilon) + O(1)$ nearly-random bits with the seed length $O(\log|X|)$ (see [23] for many constructions matching the bound).

While in general the presence of a truly random seed cannot be avoided, there exist *deterministic* extractors (i.e. without Y) for sources with a special structure [7,9,24,19,16] where the randomness is concentrated on k bits and the rest are fixed. Namely,

Definition 2. *A distribution $\mathcal{D}$ over $\{0,1\}^N$ is an (N, k) oblivious bit-fixing source if there exists $S = \{i_1, \ldots, i_k\} \subset [N]$, such that $X_{i_1}, \ldots, X_{i_k}$ are uniformly distributed in $\{0,1\}^k$, and the bits outside S are constant.*

For any (N, k) bit-fixing source and any constant $0 < \gamma < 1/2$ Gabizon et al. [16] give an explicit deterministic (k, ϵ)-extractor that extracts $m = k - N^{1/2+\gamma}$ bits of entropy with $\epsilon = 2^{-\Omega(n^\gamma)}$ provided that $k \gg \sqrt{N}$. In our case $N = 2n$ (n is the number of participants), and strictly more than $2/3$ of the input bits will be good. Thus, $k > 2N/3$, and so we extract more than $N/2 = n$ high quality bits by taking $\gamma < 1/2$.

A *privacy mechanism* is an interface between a user and data. It can be interactive or non-interactive.

Assume the database consists of a number n of rows, $d_1, \ldots, d_n$. In its simplest form, a query is a predicate $f \colon Rows \to \{0,1\}$. In this case, the true answer is simply $\sum_i f(d_i)$. Slightly more generally, f may map $[n] \times Rows \to [0,1]$, and the true answer is $\sum_i f(i, d_i)$. Note that we are completely agnostic about the domain $Rows$; rows can be Boolean, integers, reals, tuples thereof, or even strings or pictures.

A mechanism gives ϵ-*indistinguishability* [13] if for any two data sets that differ on only one row, the respective output random variables (query responses) τ and τ' satisfy for all sets S of responses:

$$\Pr[\tau \in S] \le \exp(\epsilon) \times \Pr[\tau' \in S] \ . \tag{1}$$

This definition ensures that seeing τ instead of τ' can only increase the probability of any event by at most a small factor. As a consequence, there is little incentive for any one participant to conceal or misrepresent her value, as so doing could not substantially change the probability of any event.

Similarly, we say a mechanism gives δ-*approximate* ϵ-*indistinguishability* if for outputs τ and τ' based, respectively, on data sets differing in at most one row,

$$\Pr[\tau \in S] \le \exp(\epsilon) \times \Pr[\tau' \in S] + \delta \ .$$

The presence of a non-zero δ permits us to relax the strict relative shift in the case of events that are not especially likely. We note that it is inappropriate to add non-zero δ to the statement of ϵ-indistinguishability in [13], where the sets S are constrained to be singleton sets.

Historically, the first strong positive results for output perturbation added noise drawn from a Gaussian distribution, with density function $\Pr[x] \propto \exp(-x^2/2R)$. A slightly different definition of privacy was used in [14, 4]. In order to recast those results in terms of indistinguishability, we show in Section 2.1 that the addition of Gaussian noise gives δ-approximate ϵ-indistinguishability for the noisy sums primitive when $\epsilon > [\log(1/\delta)/R]^{1/2}$. In a similar vein, Binomial noise, where n tosses of an unbiased ± 1 coin are tallied and divided by 2, also gives δ-approximate ϵ-indistinguishability so long as the number of tosses n is at least $64 \log(2/\delta)/\epsilon^2$.

Adding, instead, exponential noise results in a mechanism that can ensure ϵ-indistinguishability (that is, $\delta = 0$) [4,13]. If the noise is distributed as $\Pr[x] \propto \exp(-|x|/R)$, then the mechanism gives $1/R$-indistinguishability (cf. $\epsilon > [\log(1/\delta)/R]^{1/2}$ for Gaussian noise). Note that although the Gaussian noise is more tightly concentrated around zero, giving somewhat better accuracy for any given choice of ϵ, the exponential noise allows $\delta = 0$, giving a more robust solution.

2.1   Math for Gaussians and Binomials

We extend the results in [13] by determining the values of ϵ and δ for the Gaussian and Binomial distributions for which the noisy sums primitive yields δ-approximate ϵ-indistinguishability. Consider an output τ on a database D and query f. Let $\tau = \sum_i f(i, d_i) + \mathbf{noise}$, so replacing D with D' differing only in one row changes the summation by at most 1. Bounding the ratio of probabilities that τ occurs with inputs D and D' amounts to bounding the ratio of probabilities that $\mathbf{noise} = x$ and $\mathbf{noise} = x + 1$, for the different possible ranges of values for x. Thus, we first determine the largest value of x such that a relative bound of $\exp(\epsilon)$ holds, and then integrate the probability mass outside of this interval.

Recall the Gaussian density function: $p(x) \propto \exp(-x^2/2R)$. The ratio of densities at two adjacent integral points is

$$\frac{\exp(-x^2/2R)}{\exp(-(x+1)^2)/2R} = \exp(x/R + 1/2R).$$

This value remains at most $\exp(\epsilon)$ until $x = \epsilon R - 1/2$. Provided that $R \geq 2\log(2/\delta)/\epsilon^2$ and that $\epsilon \leq 1$, the integrated probability beyond this point will be at most

$$\Pr[x > \epsilon R - 1/2] \leq \frac{\exp(-(\epsilon R)^2/2R)}{(\epsilon R)\sqrt{\pi}} \leq \delta.$$

As a consequence, we get δ-approximate ϵ-indistinguishability when R is at least $2\log(2/\delta)/\epsilon^2$.

For the Binomial noise with bias $1/2$, whose density at $n/2 + x$ is

$$\Pr[n/2 + x] = \binom{n}{n/2+x}1/2^n,$$

we see that the relative probabilities are

$$\frac{\Pr[n/2+x]}{\Pr[n/2+x+1]} = \frac{n/2+x+1}{n/2-x}.$$

So long as x is no more than $\epsilon n/8$, this should be no more than $(1+\epsilon) < \exp(\epsilon)$. Of course, a Chernoff bound tells us that for such x the probability that a sample exceeds it is

$$\Pr[y > n/2 + \epsilon n/8] = \Pr[y > (1 + \epsilon/4)n/2]$$
$$\leq \exp(-(\epsilon^2 n/64)).$$

We get δ-approximate ϵ-indistinguishability so long as n is chosen to be at least $64\log(2/\delta)/\epsilon^2$. This exceeds the estimate of the Gaussian due to approximation error, and general slop in the analysis, though it is clear that the form of the bound is the same.

2.2 Adaptive Query Sequences

One concern might be that after multiple queries, the values of ϵ and δ degrade in an inelegant manner. We now argue that this is not the case.

Theorem 1. *A mechanism that permits T adaptive interactions with a δ-approximate ϵ-indistinguishable mechanism ensures δT-approximate ϵT-indistinguishability.*

Proof. We start by examining the probability that the transcript, written as an ordered T-tuple, lands in a set S.

$$\Pr[x \in S] = \prod_{i \leq T} \Pr[x_i \in S_i | x_1, \ldots, x_{i-1}].$$

As the noise is independent at each step, the conditioning on $x_1, \ldots, x_{i-1}$ only affects the predicate that is asked. As a consequence, we can substitute

$$\prod_{i \leq T} \Pr[x_i \in S_i | x_1, \ldots, x_{i-1}] \leq \prod_{i \leq T} \left(\exp(\epsilon) \times \Pr[x_i' \in S_i | x_1, \ldots, x_{i-1}] + \delta \right).$$

If we look at the additive contribution of each of the δ terms, of which there are T, we notice that they are only ever multiplied by probabilities, which are at most one. Therefore, each contributes at most an additive δ.

$$\prod_{i \leq T} \Pr[x_i \in S_i | x_1, \ldots, x_{i-1}] \leq \prod_{i \leq T} \left(\exp(\epsilon) \times \Pr[x_i' \in S_i | x_1, \ldots, x_{i-1}] \right) + \delta T$$

$$= \exp(\epsilon T) \times \prod_{i \leq T} \left(\Pr[x_i' \in S_i | x_1, \ldots, x_{i-1}] \right) + \delta T$$

$$= \exp(\epsilon T) \times \Pr[x' \in S] + \delta T .$$

The proof is complete.   □

3   Generating Gaussian Noise

Were we not concerned with malicious failures, a simple approach would be to have each participant i perturb $f(d_i)$ by sampling from a Gaussian with mean zero and variance $\frac{3}{2}\mathbf{var}/n$, where $\mathbf{var}$ is a lower bound on the variance needed for preserving privacy (see Section 2). The perturbed values would be shared out and the shares summed, yielding $\sum_i f(d_i) + \mathbf{noise}$ in shares. Since, as usual in the Byzantine literature, we assume that at least $2/3$ of the participants will survive, the total variance for the noise would be sufficient (but not excessive). However, a Byzantine processor might add an outrageous amount of noise to its share, completely destroying the integrity of the results. We now sketch the main ideas in our solution for the Byzantine case.

Recall that the goal is for the participants to obtain the noise in shares. As mentioned earlier, we will approximate the Gaussian with the Binomial distribution, so if the participants hold shares of sufficiently many unbiased coins they can sum these to obtain a share of (approximately) correctly generated noise. Coin flipping in shares (and otherwise) is well studied, and can be achieved by having each participant non-malleably verifiably share out a value in $GF(2)$, and then locally summing (in $GF(2)$) the shares from all n secret sharings.

This suggests a conceptually straightforward solution: Generate many coins in shares, convert the shares from $GF(2)$ to shares of values in a large field $GF(q)$ (or to shares of integers), and then sum the shares. In addition to the conversion costs, the coins themselves are expensive to generate, since they require $\Omega(n)$ executions of verifiable secret sharing per coin, which translates into $\Omega(nc)$ secret sharings for c coins[2]. To our knowledge, the most efficient scheme for generating

---

[2] When a single player shares out many values (not the case for us), the techniques of Bellare, Garay, and Rabin [3] can be used to reduce the cost of verifying the shared out values. The techniques in [3] complement ours; see Section 5.

random bits is due to Damgård et al. [10], which requires n sharings and two multiplications per coin.

We next outline a related but less expensive solution which at no intermediate or final point uses the full power of coin-flipping. The solution is cost effective when c is sufficiently large, i.e., $c \in \Omega(n)$. As a result, we will require only $\Omega(c)$ sharings of values in $\mathrm{GF}(2)$ when $c \in \Omega(n)$. Let n denote both the number of players and the desired number of coins[3].

1. Each player i shares a random bit by sharing out a value $b_i \in \{0,1\}_{\mathrm{GF}(q)}$, using a non-malleable verifiable secret sharing scheme, where q is sufficiently large, and engages in a simple protocol to prove that the shared value is indeed in the specified set. (The verification is accomplished by distributively checking that $x^2 = x$ for each value x that was shared, in parallel. This is a single secure function evaluation of a product, addition of two shares, and a reconstruction, for each of the n bits b_i.) This gives a sequence of low-quality bits in shares, as some of the shared values may have been chosen adversarially. (Of course, the faulty processors know the values of the bits they themselves have produced.)

2. Now, suppose for a moment that we have a public source of unbiased bits, $c_1, c_2, \ldots, c_n$. By XORing together the corresponding b's and c's, we can transform the low quality bits b_i (in shares) into high-quality bits $b_i \oplus c_i$, in shares. (Again, the faulty processors know the values of the (now randomized) bits they themselves have produced.) The XORing is simple: if $c_i = 0$ then the shares of b_i remain unchanged. If $c_i = 1$ then each share of b_i is replaced by one minus the original share.

3. Replace each share s by $2s - 1$, all arithmetic in $\mathrm{GF}(q)$. This maps shares of 0 to shares of -1, and shares of 1 to (different) shares of 1.

4. Finally, each participant sums her shares to get a share of the Binomial noise.

We now turn to the generation of the c_i. Each participant randomly chooses and non-malleably verifiably shares out two bits, for a total of $2n$ low-quality bits in shares. This is done in $\mathrm{GF}(2)$, so there is no need to check legitimacy. Let the low-quality source be $b_1', b_2', \ldots, b_{2n}'$. The b_i' are then reconstructed, so that they become public. The sequence $b_1' b_2' \ldots b_{2n}'$ is a bit-fixing source: some of the bits are biased, but they are independent of the other bits (generated by the good participants) due to the non-malleability of the secret sharing. The main advantage of such a source is that it is possible to apply a *deterministic* extractor on those bits and have the output be very close to uniform. Since the bits $b_1' \ldots b_{2n}'$ are public, this extraction operation can be done by each party individually with no additional communication. In particular we may use, say, the currently best known deterministic extractor of [16], which produces a number $m > n$ of nearly unbiased bits. The outputs of the extractor are our public coins $c_1 \ldots c_m$.

---

[3] If the desired number of coins is $o(n)$, we can generate $\Theta(n)$ coins and keep the unused ones in reserve for future executions of the protocol. If $m \gg n$ coins are needed, each processor can run the protocol m/n times.

The principal costs are the multiplications for verifying membership in $\{0,1\}_{GF(q)}$ and the executions of verifiable secret sharing. Note that all the verifications of membership are performed simultaneously, so the messages from the different executions can be bundled together. The same is true for the verifications in the VSS. The total cost of the scheme is $\Theta(n)$ multiplications and additions in shares, which can be all done in a constant number of rounds.

4   Generating Exponential Noise

Recall that in the exponential distribution the probability of obtaining a value at distance $|x|$ from the mean is proportional to $\exp(-|x|/R)$, where R is a scaling factor. For the present discussion we take $R = 1/(\ln 2)$, so that $\exp(-|x|/R) = 2^{-|x|}$. We approximate the exponential distribution with the Poisson distribution. An intuitively simple approach is to generate a large number of unbiased[4] random bits in shares, and then find (in shares) the position ℓ of the first 1. The value returned by this noise generation procedure is $\pm\ell$ (we flip one additional bit to get the sign). If there is no 1, then the algorithm fails, so the number of bits must be sufficiently large that this occurs with negligible probability. All the computation must be done in shares, and we can't "quit" once a 1 has been found (this would be disclosive). This "unary" approach works well when $R = 1/(\ln 2)$ and the coins are unbiased. For much larger values of R, the case in high-privacy settings, the coins need to be heavily biased toward 0, flattening the curve. This would mean more expected flips before seeing a 1, potentially requiring an excessive number of random bits.

Instead, we take advantage of the special structure of the exponential distribution, and see that we can generate the *binary* representation of an exponential variable using a number of coins that is independent of the bias. Let us return to the question of the location ℓ of the first 1 in a sequence of randomly generated bits. We can describe ℓ one bit at a time by answering the following series of questions:

1. What is the parity of ℓ? That is, $\ell = 2^i$ for some $i \geq 0$? (We begin counting the positions at 0, so that ℓ will be the number of 0's preceding the first 1.)
2. Is ℓ in the left half or the right half of a block of 4 positions, i.e., is it the case that $2^2 i \leq \ell < 2^2 i + 2$ for some $i \geq 0$?
3. Is ℓ in the left half or the right half of a block 8 positions, i.e., is it the case that $2^3 i \leq \ell < 2^3 i + 2^2$ for some $i \geq 0$?
4. And so on.

We generate the distribution of ℓ "in binary" by generating the answers to the above questions. (For some fixed d we simply assume that $\ell < 2^d$, so only a finite number of questions need be answered.)

To answer the questions, we need to be able to generate biased coins. The probability that ℓ is even (recall that we begin counting positions with 0) is

---

[4] For values of $R \neq 1/(\ln 2)$ we would need to use biased bits.

$(1/2)\sum_{i=0}^{\infty}(2^{-2i})$. Similarly, the probability that ℓ is odd is $(1/2)\sum_{i=0}^{\infty}(2^{-(2i+1)})$. Thus,

$$\Pr[\ell \text{ odd}] = (1/2)\Pr[\ell \text{ even}].$$

Since the two probabilities sum to 1, the probability that ℓ is even is $2/3$. Similar analyses yield the necessary biases for the remaining questions.

The heart of the technical argument is thus to compute coins of arbitrary bias in shares in a manner that consumes on average a constant number of unbiased, completely unknown, random bits held in shares. We will construct and analyze a shallow circuit for this. In addition, we will present two incomparable probabilistic constructions. In any distributed implementation these schemes would need to be implemented by general secure function evaluation techniques. The circuits, which only use Boolean and finite field arithmetic, allow efficient SFE implementation.

4.1 Poisson Noise: The Details

In this section we describe several circuits for generating Poisson noise. The circuits will take as input random bits (the exact number depends on the circuit in question). In the distributed setting, the input would be the result of a protocol that generates (many) unbiased bits in shares. The circuit computation would be carried out in a distributed fashion using secure function evaluation, and would result in *many* samples, in shares, of **noise** generated according to the Poisson distribution. This fits into the high-level ODO protocol in the natural way: shares of the noise are added to the shares of $\sum_i f(i, d_i)$ and the resulting noisy sum is reconstructed.

For the remainder of this section, we let n denote the number of coins to be generated. It is unrelated to the number of participants in the protocol.

Recall the discussion in the Introduction of the exponential distribution, where $\Pr[x] \propto \exp(-|x|/R)$. Recall that one interpretation is to flip a (possibly biased) coin until the first 1 is seen, and then to output the number ℓ of 0's seen before the 1 occurs. Recall also that instead of generating ℓ in unary, we will generate it in binary.

We argue that the bits in the binary representation of the random variable ℓ are independent, and moreover we can determine their biases analytically. To see the independence, consider the distribution of the ith bit of ℓ:

$$\ell_i = \begin{cases} 0 \text{ w.p. } \Pr[0 \times 2^i \leq \ell < 1 \times 2^i] + \Pr[2 \times 2^i \leq \ell < 3 \times 2^i] + \ldots \\ 1 \text{ w.p. } \Pr[1 \times 2^i \leq \ell < 2 \times 2^i] + \Pr[3 \times 2^i \leq \ell < 4 \times 2^i] + \ldots \end{cases}$$

Notice that corresponding terms in the two summations, eg $\Pr[0 \times 2^i \leq \ell < 1 \times 2^i]$ and $\Pr[1 \times 2^i \leq \ell < 2 \times 2^i]$, are directly comparable; the first is exactly $\exp(2^i/R)$ times the second. This holds for every corresponding pair in the sums, and as such the two sums share the same ratio. As the two sum must total to one, we have additionally that

$$1 - \Pr[\ell_i] = \exp(2^i/R) \times \Pr[\ell_i].$$

Solving, we find that

$$\Pr[\ell_i] = 1/(1 + \exp(2^i/R)) \,.$$

Recall as well that the observed ratio applied equally well to each pair of intervals, indicating that the bias is independent of the more significant bits. The problem of producing an exponentially distributed ℓ is therefore simply a matter of flipping a biased coin for each bit of ℓ. The circuit we will construct will generate many ℓ's according to the desired distribution, at an expected low amortized cost (number of input bits) per bit position in the binary expansion of ℓ. The circuit is a collection of circuits, each for one bit position, with the associated bias hard-wired in. It suffices therefore to describe the circuitry for one of these smaller circuits (Section 4.3). We let p denote the hard-wired bias.

A well-known technique for flipping a single coin of arbitrary bias p is to write p in binary, examine random bits until one differs from the corresponding bit in p, and then emit the complement of the random bit. To achieve a high fidelity to the original bias p, a large number d of random bits must be available. However, independent of p, the expected number of random bits consumed is at most 2. This fact will be central to our constructions.

In the sequel we distinguish between unbiased *bits*, which are inputs to the algorithm, and the generated, biased, *coins*, which are the outputs of the algorithm.

4.2   Implementation Details: Finite Resources

With finite randomness we will not be able to perfectly emulate the bias of the coins. Moreover, the expectation of higher order bits in the binary representation of ℓ diminishes at a doubly exponential rate (because the probability that $\ell \geq 2^i$ is exponentially small in 2^i), quickly giving probabilities that simply can not be achieved with any fixed amount of randomness.

To address these concerns, we will focus on the statistical difference between our produced distribution and the intended one. The method described above for obtaining coins with arbitrary bias, truncated after d bits have been consumed, can emulate any biased coin within statistical difference at most 2^{-d}. Accordingly, we set all bits of sufficiently high order to zero, which will simplify our circuit. The remaining output bits – let us imagine there are k of them – will result in a distribution whose statistical difference is at most $k2^{-d}$ from the target distribution. We note that by trimming the distribution to values at most 2^d in magnitude, we are introducing an additional error, but one whose statistical difference is quite small. There is an $\exp(-2^d/R)$ probability mass outside the $[-2^d, 2^d]$ interval that is removed and redistributed inside the interval. This results in an additional $2\exp(-2^d/R)$ statistical difference that should be incorporated into δ. For clarity, we absorb this term into the value k.

Using our set of coins with statistical difference at most $k2^{-d}$ from the target distribution, we arrive at a result akin to (1), though with an important difference. For response variables τ and τ' as before (based on databases differing it at most one row),

$$\forall S \subseteq U : \ \Pr[\tau \in S] \leq \Pr[\tau' \in S] \times \exp(1/R) + k2^{-d} \,.$$

As before, the probability of any event increases by at most a factor of $\exp(1/R)$, but now with an additional additive $k2^{-d}$ term. This term is controlled by the parameter d, and can easily be made sufficiently small to allay most concerns.

We might like to remove the additive $k2^{-d}$ term, which changes the nature of the privacy guarantee. While this seems complicated at first, notice that it is *possible* to decrease the relative probability associated with each output coin arbitrarily, by adding more bits (that is, increasing d). What additional bits can not fix is our assignment of zero probability to noise values outside the permitted range (i.e., involving bits that we do not have circuitry for).

One pleasant resolution to this problem, due to Adam Smith, is to constrain the output range of the sum of noise plus signal. If the answer plus noise is constrained to be a k-bit number, and conditioned on it lying in that range the distribution looks exponential, the same privacy guarantees apply. Guaranteeing that the output will have only k bits can be done by computing the sum of noise and signal using $k + 1$ bits, and then if there is overflow, outputting the noise-free answer. This increases the probability that **noise** $= 0$ by a relatively trivial amount, and ensures that the output space is exactly that of k-bit numbers.

4.3   A Circuit for Flipping Many Biased Coins

We are now ready to construct a circuit for flipping a large number of independent coins with common bias. By producing many $(\Omega(n))$ coins at once, we could hope to leverage the law of large numbers and consume, with near certainty, a number of input bits that is little more than $2n$ and depends very weakly on d. For example, we could produce the coins sequentially, consuming what randomness we need and passing unused random bits on to the next coin. The circuit we now describe emulates this process, but does so in a substantially more parallel manner.

The circuit we construct takes 2^i unbiased input bits and produces 2^i output coins, as well as a number indicating how many of the coins are actually the result of the appropriate biased flips. That is, it is unlikely that we will be able to produce fully 2^i coins, and we should indicate how many of the coins are in fact valid. The construction is hierarchical, in that the circuit that takes 2^i inputs will be based on two level $i - 1$ circuits, attached to the first and second halves of its inputs.

To facilitate the hierarchical construction, we augment the outputs of each circuit with the number of bits at the end of the 2^i that were consumed by the coin production process, but did not diverge from the binary representation of p. Any process that wishes to pick up where this circuit has left off should start under the assumption that the first coin is in fact this many bits into its production. For example, if this number is r then the process should begin by comparing the next random bit to the $(r+1)$st bit in the expansion of p. Bearing this in mind, we "bundle" d copies of this circuit together, each with a different assumption about the initial progress of the production of their first coin.

For each value $1 \leq j \leq d$ we need to produce a vector of 2^i coins c_j, a number of coins n_j, and d_j, a measure of progress towards the last coin. We imagine that we have access to two circuits of one level lower, responsible for the left and right half of our 2^i input bits, and whose corresponding outputs are superscripted by L and R. Intuitively, for each value of j we ask the left circuit for d_j^L, which we use to select from the right circuit. Using index j for the left circuit and d_j^L for the right circuit, we combine the output coins using a shift of n_j^L to align them, and add the output counts n_j^L and $n_{d_j^L}^R$. We simply pass $d_{d_j^L}^R$ out as the appropriate value for d_j.

$$c_j = c_j^L \mid (c_{d_j^L}^R >> n_j^L)$$
$$n_j = n_j^L + n_{d_j^L}^R$$
$$d_j = d_{d_j^L}^R$$

The operation of subscripting is carried out using a multiplexer, and shifts, bitwise ors, and addition are similarly easily carried out in logarithmic depth.

The depth of each block is bounded by $\Theta(\log(nd))$, with the size bounded by $\Theta(2^i d(\log(n) + d)$, as each of d outputs must multiplex d possible inputs (taking $\Theta(d)$ circuitry) and then operate on them (limited by $\Theta(\log(n)2^i)$ for the barrel shifter). All told, the entire circuit has depth $\Theta(\log(nd)^2)$, with size $\Theta(nd(\log(n) + d)\log(n))$.

4.4   Probabilistic Constructions with Better Bounds

We describe two probabilistic constructions of circuits that take as input unbiased bits and produce as output coins of arbitrary, *not necessarily identical*, bias. Our first solution is optimal in terms of depth ($\Theta(\log d)$) but expensive in the gate count. Our second solution dramatically decreases the number of gates, paying a modest price in depth ($O(\log(n+d))$) and a logarithmic increase in the number of input bits.

A module common to both constructions is the comparator – a circuit that takes two bit strings $b_1, \ldots, b_d$ and $p^{(1)} \ldots p^{(d)}$ and outputs 0 if and only if the first string precedes the second string in the lexicographic order. Equivalently, the comparator outputs $\bar{b}_i$, where i is the index of the earliest occurrence 1 in the sequence $b_1 \oplus p^{(1)}, \ldots, b_d \oplus p^{(d)}$, or 1 if the two strings are equal. Based on this observation, a circuit of depth $\Theta(\log d)$ and size $\Theta(d)$ can be designed easily. Notice that the result of comparison is independent of the values of the strings beyond the point of divergence.

Brute Force Approach. Assume that we have nd independent unbiased bits $b_i^{(j)}$, for $1 \le i \le n$ and $1 \le j \le d$. To flip n independent coins, each with its own bias p_i, whose binary representation is $0.p_i^{(1)} \ldots p_i^{(d)}$, we run n comparators in parallel on inputs $(b_1^{(1)}, \ldots, b_1^{(d)}, p_1^{(1)}, \ldots, p_1^{(d)}), \ldots, (b_n^{(1)}, \ldots, b_n^{(d)}, p_n^{(1)}, \ldots, p_n^{(d)})$.

Our goal is to get by with many fewer than nd unbiased input bits of the brute force approach, since each of these requires an unbiased bit in shares. Intuitively, we may hope to get away with this because, as mentioned previously, the average number of bits consumed per output coin is 2, independent of the bias of the coin. Let c_i for $1 \leq i \leq n$ be the smallest index where $b_i^{(c_i)} \neq p_i^{(c_i)}$, and $d+1$ if the two strings are equal. The number c_i corresponds to the number of bits "consumed" during computation of the ith coin. Let $C = \sum_{i=1}^{n} c_i$. On expectation $E[C] = 2n$, and except with a negligible probability $C < 4n$.

Rather than having the set $\{b_i^{(j)}\}_{i,j}$ be given as input (too many bits), we will compute the set $\{b_i^{(j)}\}_{i,j}$ from a much smaller set of input bits. The construction will ensure that the *consumed* bits are independent except with negligible probability. Let the number of input bits be D, to be chosen later.

We will construct the circuit probabilistically. Specifically, we begin by choosing nd binary vectors $\{r_i^{(j)}\}_{i,j}$, $1 \leq i \leq n$ and $1 \leq j \leq d$, uniformly from $\{0,1\}^D$ to be hard-wired into the circuit. Let $b \in_R \{0,1\}^D$ be the uniformly chosen random input to the circuit.

The circuit computes the inner products of each of the hard-wired vectors $r_i^{(j)}$ with the input b. Let $b_i^{(j)} = \langle r_i^{(j)}, b \rangle$ denote the resulting bits. These are the $\{b_i^{(j)}\}_{i,j}$ we will plug into the brute force approach described above. Note that although much randomness was used in defining the circuit, the input to the circuit requires only D random bits.

Although the nd vectors are not linearly independent, very few of them – $O(n)$ – are actually used in the computation of our coins, since with overwhelming probability only this many of the $b_i^{(j)}$ are actually consumed. A straightforward counting argument therefore shows that the set of vectors actually used in generating consumed bits will be linearly independent, and so the coins will be mutually independent.

We claim that if $D > 4C$, then the consumed bits are going to be independent with high probability. Conditional on the sequence $c_1, \ldots, c_n$, the vectors $r_i^{(j)}$ for $1 \leq i \leq n$ and $1 \leq j \leq c_i$ are independent with probability at least $1 - C2^{C-D} < 1 - 2^{-2C}$, where the probability space is the choice of the r's. For fixed C the number of possible $c_1, \ldots, c_n$ is at most $\binom{C}{n} < 2^C$. Hence the probability that for some $C < 4n$ and some $c_1, \ldots, c_n$, such that $c_1 + \cdots + c_n = C$ the vectors $r_i^{(j)}$ are linearly independent is at least than $1 - 4n2^{-C}$. Finally, we observe that if the vectors are linearly independent, the bits $b_i^{(j)}$ are independent as random variables. The depth of this circuit is $\Theta(\log D)$, which is the time it takes to compute the inner product of two D-bit vectors. Its gate count is $\Theta(ndD)$, which is clearly suboptimal.

Using low weight independent vectors. Our second solution dramatically decreases the number of gates by reducing the weight (the number of non-zero elements) of the vectors r from the expected value $D/2$ to $s^2 \lceil \log(n+1) \rceil$, where s is a small constant. To this end we adopt the construction from [12] that converts an expander-like graph into a set of linearly independent vectors.

The construction below requires a field with at least nd non-zero elements. Let $\nu = \lceil \log(nd + 1) \rceil$. We use $GF(2^\nu)$, representing its elements as ν-bit strings.

Consider a bipartite graph G of constant degree s connecting sets $L = \{u_1, \ldots, u_n\}$, where the u's are distinct field elements, and $R = \{1, \ldots, \Delta\}$.

The degree s can be as small as 3. Define matrix M of size $n \times s\Delta$ as follows: if $(u_i, \tau) \in G$, the elements $M[i][s(\tau - 1), s(\tau - 1) + 1, \ldots, s\tau - 1] = u_i, u_i^2, \ldots, u_i^s$, and $(0, \ldots, 0)$ (s zeros) otherwise. Thus, each row of the matrix has exactly s^2 non-zero elements.

For any set $S \subseteq L$, let $\Gamma(S) \subseteq R$ be the set of neighbors of S in G. The following claim is easily obtained from the proof of Lemma 5.1 in [12]. It says that if for a set of vertices $T \in L$ all of T's subsets are sufficiently expanding, then the rows of M corresponding to vertices in T are linearly independent.

Theorem 2. *Let $T \subseteq L$ be any set for which $\forall S \subseteq T$, $|\Gamma(S)| > (1 - \frac{1}{s+1})|S|$. Then the set of vectors $\{M[u] : u \in T\}$ is linearly independent.*

Consider a random bipartite graph with nd/ν elements in one class and $2C$ elements in the other. Associate the elements from the first class with bits $b_i^{(j)}$'s, grouped in ν-tuples. Define the bits as the results of the inner product of the corresponding rows of the matrix M from above with the input vector of length $2s^2C$ that consists of random elements from $GF(2^\nu)$. Observe that the random graph G satisfies the condition of Theorem 2 for all sets of size less than C with high probability if $C > (nd/\nu)^{1/(s-1)}$.

The depth of the resulting circuit is $\Theta(\log(n + d))$, the gate count is $\Theta(nds^2 \log(n + d))$, and the size of the input is $2n \log(n + d)$.

5   Generalizations

In this section we briefly discuss several generalizations of the basic scheme.

5.1   Alternatives to Full Participation

The main idea is to use a set of facilitators, possibly a very small set, but one for which we are sufficiently confident that fewer than one third of the members are faulty. Let $\mathcal{F}$ denote the set of facilitators. To respond to a query f, participant i shares $f(i, d_i)$ among the facilitators, and takes no further part in the computation.

To generate the noise, each member of $\mathcal{F}$ essentially takes on the work of $n/|\mathcal{F}|$ participants. When $|\mathcal{F}|$ is small, the batch verification technique of [3] may be employed to verify the secrets shared out by each of the players (that is, one batch verification per member of $\mathcal{F}$), although this technique requires that the faulty players form a smaller fraction of the total than we have been assuming up to this point.

5.2   When f Is Not a Predicate

Suppose we are evaluating f to k bits of precision, that is, k bits beyond the binary point. Let q be sufficiently large, say, at least $q > n2^k$. We will work in $GF(q)$. Participant i will share out $2^k f(i, d_i)$, *one bit at a time*. Each of these is checked for membership in $\{0, 1\}_{GF(q)}$. Then the shares of the most significant bit are multiplied by 2^{k-1}, shares of the next most significant are multiplied by

2^{k-2} and so on, and the shares of the binary representation of $f(i, d_i)$ are then summed. The noise generation procedure is amplified as well. Details omitted for lack of space.

5.3 Beyond Sums

We have avoided the case in which f is an arbitrary function mapping the entire database to a (tuple of) value(s), although the theory for this case has been developed in [13]. This is because without information about the structure of f we can only rely on general techniques for secure function evaluation of f, which may be prohibitively expensive.

One case in which we can do better is in the generation of *privacy-preserving histograms*. A histogram is specified by a partition of the domain Rows; the true response to the histogram query is the exact number of elements in the database residing in each of the cells of the histogram. Histograms are *low sensitivity* queries, in that changing a single row of the database changes the counts of at most two cells in the histogram, and each of these two counts changes by at most 1. Thus, as discussed in [13], ϵ-indistinguishable histograms may be obtained by adding exponential noise with $R = 1/2\epsilon$ to each cell of the histogram. A separate execution of ODO for each cell solves the problem. The executions can be run concurrently. All participants in the histogram query must participate in each of the concurrent executions.

5.4 Individualized Privacy Policies

Suppose Citizen C has decided she is comfortable with a *lifetime* privacy loss of, say $\epsilon = 1$. Privacy erosion is cumulative: any time C participates in the ODO protocol she incurs a privacy loss determined by R, the parameter used in noise generation. C has two options: if R is fixed, she can limit the number of queries in which she participates, *provided the decision whether or not to participate is independent of her data*. If R is not fixed in advance, but is chosen by consensus (in the social sense), she can propose large values of R, or to use large values of R for certain types of queries. Similarly, queries could be submitted with a stated value of R, and dataholders could choose to participate only if this value of R is acceptable to them for this type of query. However, the techniques will all fail if the set of participants is more than one-third faulty; so the assumption must be that this bound will always be satisfied. This implicitly restricts the adversary.

6   Summary of Results

This work ties together two areas of research: the study of privacy-preserving statistical databases and that of cryptographic protocols. It was inspired by the combination of the computational power of the noisy sums primitive in the first area and the simplicity of secure evaluation of sums in the second area. The effect is to remove the assumption of a trusted collector of data, allowing individuals control over the handling of their own information.

In the course of this work we have developed distributed algorithms for generation of Binomial and Poisson noise in shares. The former makes novel use of extractors for bit-fixing sources in order to reduce the number of secret sharings needed in generating massive numbers of coins. The latter examined for the first time distributed coin-flipping of coins with arbitrary bias.

References

1. D. Agrawal and C. Aggarwal. On the design and quantification of privacy preserving data mining algorithms. In *Proceedings of the 20th ACM SIGMOD-SIGACT-SIGART Symposium on Principles of Database Systems*, pages 247–255, 2001.
2. R. Agrawal and R. Srikant. Privacy-preserving data mining. In *Proceedings of the ACM SIGMOD International Conference on Management of Data*, pages 439–450, May 2000.
3. M. Bellare, J. A. Garay, and T. Rabin. Distributed pseudo-random bit generators— a new way to speed-up shared coin tossing. In *Proceedings of the 15th ACM Symposium on Principles of Distributed Computing*, pages 191–200, 1996.
4. A. Blum, C. Dwork, F. McSherry, and K. Nissim. Practical privacy: The SuLQ framework. In *Proceedings of the 24th ACM SIGMOD-SIGACT-SIGART Symposium on Principles of Database Systems*, pages 128–138, June 2005.
5. S. Chawla, C. Dwork, F. McSherry, A. Smith, and H. Wee. Toward privacy in public databases. In *Proceedings of the 2nd Theory of Cryptography Conference*, pages 363–385, 2005.
6. S. Chawla, C. Dwork, F. McSherry, and K. Talwar. On the utility of privacy-preserving histograms. In *Proceedings of the 21st Conference on Uncertainty in Artificial Intelligence*, 2005.
7. B. Chor, O. Goldreich, J. Håstad, J. Friedman, S. Rudich, and R. Smolensky. The bit extraction problem of t-resilient functions. In *Proceedings of the 26th IEEE Symposium on Foundations of Computer Science*, pages 429–442, 1985.
8. B. Chor, S. Goldwasser, S. Micali, and B. Awerbuch. Verifiable secret sharing and achieving simultaneity in the presence of faults. In *Proceedings of the 26th Annual IEEE Symposium on Foundations of Computer Science*, pages 383–395, 1985.
9. A. Cohen and A. Wigderson. Dispersers, deterministic amplification, and weak random sources. In *Proceedings of the 30th Annual IEEE Symposium on Foundations of Computer Science*, pages 14–19, 1989.
10. I. Damgård, M. Fitzi, E. Kiltz, J.B. Nielsen, and T. Toft. Unconditionally secure constant-rounds multi-party computation for equality, comparison, bits and exponentiation. In *Proceedings of the 3rd Theory of Cryptography Conference*, pages 285–304, 2006.
11. I. Dinur and K. Nissim. Revealing information while preserving privacy. In *Proceedings of the 22nd ACM SIGMOD-SIGACT-SIGART Symposium on Principles of Database Systems*, pages 202–210, 2003.
12. C. Dwork, J. Lotspiech, and M. Naor. Digital signets for protection of digital information. In *Proceedings of the 28th annual ACM symposium on Theory of computing*, pages 489–498, 1996.

13. C. Dwork, F. McSherry, K. Nissim, and A. Smith. Calibrating noise to sensitivity in private data analysis. In *Proceedings of the 3rd Theory of Cryptography Conference*, pages 265–284, 2006.

14. C. Dwork and K. Nissim. Privacy-preserving datamining on vertically partitioned databases. In *Advances in Cryptology: Proceedings of Crypto*, pages 528–544, 2004.

15. A. Evfimievski, J. Gehrke, and R. Srikant. Limiting privacy breaches in privacy preserving data mining. In *Proceedings of the 22nd ACM SIGMOD-SIGACT-SIGART Symposium on Principles of Database Systems*, pages 211–222, June 2003.

16. Ariel Gabizon, Ran Raz, and Ronen Shaltiel. Deterministic extractors for bit-fixing sources by obtaining an independent seed. In *Proceedings of the 45th IEEE Symposium on Foundations of Computer Science*, pages 394–403, 2004.

17. O. Goldreich, S. Micali, and A. Wigderson. How to play any mental game or A completeness theorem for protocols with honest majority. In *Proceedings of the 19th Annual ACM Symposium on Theory of Computing*, pages 218–229, 1987.

18. Oded Goldreich. *Foundations of Cryptography - Basic Applications*, volume 2. Cambridge University Press, 2004.

19. J. Kamp and D. Zuckerman. Deterministic extractors for bit-fixing sources and exposure-resilient cryptography. In *Proceedings of the 44th Annual IEEE Symposium on Foundations of Computer Science*, pages 92–101, 2003.

20. L. Lamport, R. Shostak, and M. Pease. The Byzantine generals problem. *ACM Transactions on Programming Languages and Systems*, 4(3):382–401, 1982.

21. N. Nisan and D. Zuckerman. Randomness is linear in space. *J. Comput. Syst. Sci.*, 52(1):43–52, 1996.

22. Michael O. Rabin. Randomized Byzantine generals. In *Proceedings of the 24th IEEE Symposium on Foundations of Computer Science*, pages 403–409, 1983.

23. Ronen Shaltiel. Recent developments in explicit constructions of extractors. *Bulletin of the EATCS*, 77:67–95, 2002.

24. L. Trevisan and S. Vadhan. Extracting randomness from samplable distributions. In *Proceedings of the 41st Annual IEEE Symposium on Foundations of Computer Science*, pages 32–42, 2000.

25. A. Yao. Protocols for secure computations (extended abstract). In *Proceedings of the 23rd IEEE Symposium on Foundations of Computer Science*, pages 160–164, 1982.

On the (Im-)Possibility of Extending Coin Toss

Dennis Hofheinz[1], Jörn Müller-Quade[2], and Dominique Unruh[2]

[1] CWI, Cryptology and Information Security Group, Prof. Dr. R. Cramer
`Dennis.Hofheinz@cwi.nl`
[2] IAKS, Arbeitsgruppe Systemsicherheit, Prof. Dr. Th. Beth, Universität Karlsruhe
`{muellerq, unruh}@ira.uka.de`

Abstract. We consider the cryptographic two-party protocol task of extending a given coin toss. The goal is to generate n common random coins from a single use of an ideal functionality which gives $m < n$ common random coins to the parties. In the framework of Universal Composability we show the impossibility of securely extending a coin toss for statistical and perfect security. On the other hand, for computational security the existence of a protocol for coin toss extension depends on the number m of random coins which can be obtained "for free".

For the case of stand-alone security, i.e., a simulation based security definition without an environment, we present a novel protocol for unconditionally secure coin toss extension. The new protocol works for superlogarithmic m, which is optimal as we show the impossibility of statistically secure coin toss extension for smaller m.

Combining our results with already known results, we obtain a (nearly) complete characterization under which circumstances coin toss extension is possible.

Keywords: coin toss, universal composability, reactive simulatability, cryptographic protocols.

1 Introduction

Manuel Blum showed in [5] how to flip a coin over the telephone line. His protocol guaranteed that even if one party does not follow the protocol, the other party still gets a uniformly distributed coin toss result. This general concept of generating common randomness in a way such that no dishonest party can dictate the result proved very useful in cryptography, e.g., in the construction of protocols for general secure multi-party computation.

Here we are interested in the task of *extending* a given coin toss. That is, suppose that two parties already have the possibility of making a single m-bit coin-toss. Is it possible for them to get $n > m$ bits of common randomness? The answer we come up with is basically: "it depends."

The first thing the extensibility of a given coin toss depends on is the required security type. One type of security requirement (which we call "stand-alone simulatabiliy" here) can simply be that the protocol imitates an ideal coin toss

S. Vaudenay (Ed.): EUROCRYPT 2006, LNCS 4004, pp. 504–521, 2006.

functionality in the sense of [13], where a simulator has to invent a realistic protocol run after learning the outcome of the ideal coin-toss. A stronger type of requirement is to demand universal composability, which basically means that the protocol imitates an ideal coin toss functionality even in arbitrary protocol environments. Security in the latter sense can conveniently be captured in a simulatability framework like the Universal Composability framework [6, 8] or the Reactive Simulatability model [16, 3].

Orthogonal to this, one can vary the level of fulfilment of each of these requirements. For example, one can demand stand-alone simulatability of the protocol with respect to polynomial-time adversaries in the sense that real protocol and ideal functionality are only computationally indistinguishable. This specific requirement is already fulfilled by the protocol of Blum. Alternatively, one can demand, e.g., universal composability of the protocol with respect to unbounded adversaries. This would then yield statistical or even perfect security. We show that whether such a protocol exists depends on the asymptotic behaviour of m.

Our results are summarized in the table below. A "yes" or "no" indicates whether a protocol for coin toss extension exists in that setting. "Depends" means that the answer depends on the size of the seed (the m-bit coin toss available by assumption), and **boldface** indicates novel results.

Security type ↓ / level →	Computational	Statistical	Perfect
stand-alone simulatability	yes	**depends**[1]	no
universal composability	**depends**[2]	no	no

Known results in the perfect and statistical case. A folklore theorem states, that (perfectly non-trivial) statistically secure coin-toss is impossible from scratch (even in very lenient security models). By Kitaev, this result was extended even to protocols using quantum communication (cf. [1]). [4] first investigated the problem of extending a coin-toss. They presented a statistically secure protocol for extending a given coin-toss (pre-shared using a VSS), if less than $\frac{1}{6}$ of the parties are corrupted. Note that their main attention was on the efficiency of the protocol, since in that scenario arbitrary multi-party computations and therefore in particular coin-toss from scratch are known to be possible. The result does not apply to the two-party case.

Our results in the perfect and statistical case. Our results in the perfect case are most easily explained. For the perfect case, we show impossibility of *any* coin toss extension, no matter how (in-)efficient. We show this for stand-alone simulatability (Coro. 1) and for universal composability. Now for the statistical case. When demanding only stand-alone simulatability, the situation depends on the number of the already available common coins. Namely, we give an efficient protocol to extend m common coins to any polynomial number (in the security

---

[1] Coin toss extension is possible if and only if the seed has superlogarithmic length.

[2] Coin toss extension is impossible if the seed does not have superlogarithmic length. The possibility result depends on the complexity assumption we use, cf. Section 3.1.

parameter), if m is superlogarithmic (Th. 5). Otherwise, we show that there can even be no protocol that derives $m + 1$ common random coins (Coro. 1). In the universal composability setting, the situation is more clear: we show that there simply is no protocol that derives from m common coins $m + 1$ coins, no matter how large m is (Th. 6). (However, here we restrict to protocols that run in a polynomial number of rounds.)

Known results in the computational case. The possibility of coin tossing (in a non-simulation based model) was first shown by [5] and this protocol can be proven secure in a stand-alone security model. For the UC framework coin-toss was proven to be impossible in [9], unless a helping functionality like a CRS is given. In [12], the task of coin-toss is considered in a scenario slightly different from ours: in [12], protocol participants may not abort protocol execution without generating output. In that setting, [12] show that coin-toss is generally not possible even against computationally limited adversaries. However, to the best of our knowledge, an *extension* of a given coin toss has not been considered so far in the computational setting.

Our results in the computational case. We answer the question concerning the minimal size necessary for a coin-toss to be extensible: If an m-bit coin-toss functionality is given, and m is not superlogarithmic, then it is already impossible for the parties to derive $m + 1$ common random coins (in a universally composable way) from it (Th. 2). However, we also show that under strengthened computational assumptions, there are protocols that extend m to any polynomial number (in the security parameter) of common random coins, if m is superlogarithmic (Th. 1). In that sense, we give the remaining parts for a complete characterization of the computational case.

Notation
- A function f is *negligible*, if for any $c > 0$, $f(k) \leq k^{-c}$ for sufficiently large k (i.e., $f \in k^{-\omega(1)}$).
- f is *polynomially bounded*, if for some $c > 0$, $f(k) \leq k^c$ for sufficiently large k (i.e., $f \in k^{O(1)}$).
- f is *polynomially-large*, if there is a $c > 0$ s.t. $f(k)^c \geq k$ for sufficiently large k (i.e., $f \in k^{\Omega(1)}$).
- f is *superpolynomial*, if for any $c > 0$, $f(k) > k^c$ for sufficiently large k (i.e., $f \in k^{\omega(1)}$).
- f is *superlogarithmic*, if $f / \log k \to \infty$ (i.e., $f \in \omega(\log k)$). It is easy to see that f is superlogarithmic if and only if 2^{-f} is negligible.
- f is *superpolylogarithmic*, if for any $c > 0$, $f(k) > (\log k)^c$ for sufficiently large k (i.e., $f \in (\log k)^{\omega(1)}$).
- f is *exponentially-small*, if there exists a $c > 1$, s.t. $f(k) \leq c^{-k}$ for sufficiently large k (i.e., $f \in \Omega(1)^{-k} = 2^{-\Omega(k)}$).
- f is *subexponential*, if for any $c > 1$, $f(k) < c^k$ for sufficiently large k (i.e., $f \in o(1)^k = 2^{o(k)}$).

2   Security Definitions

In this section we roughly sketch the security definitions used throughout this paper. We distinguish between two notions: stand-alone simulatability as defined in [13],[3] and Universal Composability (UC) as defined in [6].

Stand-Alone Simulatability. In [13] a definition for the security of two-party secure function evaluations is given (called *security in the malicious model*). We will give a sketch, for more details we refer to [13].

A protocol consists of two parties that alternatingly send messages to each other. The parties may also invoke an ideal functionality, which is given as an oracle (in our cases, they invoke a smaller coin-toss to realise a larger one).

We say the protocol π stand-alone simulatably realises a probabilistic function f, if for any efficient adversary A that may replace none or a single party, there is an efficient simulator S s.t. for all inputs the following random variables are computationally indistinguishable:

- *The real protocol execution.* This consists of the view of the corrupted parties upon inputs x_1 and x_2 for the parties and the auxiliary input z for the adversary, together with the outputs I of the parties.
- *The ideal protocol execution.* Here the simulator first learn the auxiliary input z and possibly the input for the corrupted party (the simulator must corrupt the same party as the adversary). Then he can choose the input of the corrupted party for the probabilistic function f, the other inputs are chosen honestly (i.e., the first input is x_1 if the first party is uncorrupted, and the second input x_2 if the second party is).
  Then the simulator learns the output I of f (we assume the output to be equal for all parties). It may now generate a fake view v of the corrupted parties. The ideal protocol execution then consists of v and I.

Of course, in our case the probabilistic function f (the coin-toss) has no input, so the above definition gets simpler.

What we have sketched above is what we call *computational* stand-alone simulatability. We further define *statistical* stand-alone simulatability and *perfect* stand-alone simulatability. In these cases we do not consider efficient adversaries and simulators, but unlimited ones. In the case of statistical stand-alone simulatability we require the real and ideal protocol execution to be statistically indistinguishable (and not only computationally), and in the perfect case we even require these distributions to be identical.

Universal Composability. In contrast to stand-alone simulatability, Universal Composability [6] is a much stricter security notion. The main difference is the existence of an environment, that may interact with protocol and adversary (or with ideal functionality and simulator) and try to distinguish between real

---

[3] In fact, [13] does not use the name stand-alone simulatability but simply speaks about *security in the malicous model*. We adopt the name stand-alone simulatability for this paper to be able to better distinguish the different notions.

and ideal protocol. This additional strictness brings the advantage of a versatile composition theorem (the Universal Composition Theorem [6]).

We only sketch the model here and refer to [6] for details.

A protocol consists of several machines that may (a) get input from the environment, (b) give output to the environment (both also during the execution of the protocol), and (c) send messages to each other.

The *real protocol execution* consists of a protocol π, an adversary $\mathcal{A}$ and an environment $\mathcal{Z}$. Here the environment may freely communicate with the adversary, and the latter has full control over the network, i.e., it may deliver, delay or drop messages sent between parties. We assume the authenticated model in this paper, so the adversary learns the content of the messages but may not modify it. When $\mathcal{Z}$ terminates, it gives a single bit of output. The adversary may choose to corrupt parties at any point in time.[4]

The *ideal protocol execution* is defined analogously, but instead of a protocol π there is an *ideal functionality* $\mathcal{F}$ and instead of the adversary there is a simulator $\mathcal{S}$. The simulator can only learn and influence protocol data, if (a) the functionality explicitly allows this, or (b) it corrupts a party (note that the simulator may only corrupt the same parties as the adversary). In the latter case, the simulator can choose inputs into the functionality in the name of that party and gets the outputs appartaining to that party. In the case of uncorrupted parties, the environment is in control of the corresponding in- and output of the ideal functionality.

We say a protocol π universally composably (UC)-implements an ideal functionality $\mathcal{F}$ (or short π is universally composable if $\mathcal{F}$ is clear from the context), if for any efficient adversary $\mathcal{A}$, there is an efficient simulator $\mathcal{S}$, s.t. for all efficient environments $\mathcal{Z}$ and all auxiliary inputs z for $\mathcal{Z}$, the distributions of the output-bit of $\mathcal{Z}$ in the real and the ideal protocol execution are indistinguishable.

What has been sketched above we call *computational* UC. We further define *statistical* and *perfect* UC. In these notions, we allow adversary, simulator and environment to be unlimited machines. Further, in the case of perfect UC, we require the distributions of the output-bit of $\mathcal{Z}$ to be *identical* in real and ideal protocol execution.

The Ideal Functionality for Coin Toss. To describe the task of implementing a universally composable coin-toss, we have to define the ideal functionality of n-bit coin-toss.

In the following, let n denote a positive integer-valued function.

Below is an informal description of our ideal functionality for a n-bit coin toss. First, the functionality waits for initialization inputs from both parties P_1 and P_2. As soon as both parties have this way signalled their willingness to start, the functionality selects n coins in form of an n-bit string κ uniformly and sends this κ to the adversary. (Note that a coin toss does not guarantee secrecy of any kind.)

---

[4] It is then called an *adaptive* adversary. If the adversary can only corrupt parties before the start of the protocol, we speak of *static corruption*. All results in this paper hold for both variants of the security definition.

If the functionality now sent κ directly and without delay to the parties, this behaviour would not be implementable by any protocol (this would basically mean that the protocol output is immediately available, even without interaction). So the functionality lets the adversary decide when to deliver κ to each party. Note however, that the adversary may not in any way influence the κ that is delivered.

A more detailed description follows:

Ideal functionality CT_n (n-bit Coin Toss)

1. Wait until there have been "init" inputs from P_1 and P_2. Ignore messages from the adversary, but immediately inform the adversary about the init.
2. Select $\kappa \in \{0,1\}^n$ uniformly and send κ to the adversary. From now on:
   - on the first (and only the first) "deliver to 1" message from the adversary, send κ to P_1,
   - on the first (and only the first) "deliver to 2" message from the adversary, send κ to P_2.

Using CT_n, we can also formally express what we mean by *extending* a coin toss. Namely:

Definition 1. *Let $n = n(k)$ and $m = m(k)$ be positive, polynomially bounded and computable functions such that $m(k) < n(k)$ for all k. Then a protocol is a* universally composable $(m \rightarrow n)$-coin toss extension protocol *if it securely and non-trivially implements* CT_n *by having access only to* CT_m. *This security can be computational, statistical or perfect.*

By a "non-trivial" implementation we mean a protocol that, with overwhelming probability, guarantees outputs if no party is corrupted and all messages are delivered. (Alternatively, one may also consider protocols that provide output with overwhelming probability.) This requirement is useful since without it, a trivial protocol that does not generate any output formally implements every functionality. (Cf. [10] and [2, Section 5.1] for more discussion and formal definitions of "non-triviality.")

On Unlimited Simulators. Following [3], we have modelled statistical and perfect stand-alone and UC security using unlimited simulators. Another approach is to require the simulators to be polynomial in the running-time of the adversary. All our results apply also to that case: For the impossibility results, this is straightforward, since the security notion gets stricter when the simulators become more restricted. The only possibility result for statistical/perfect security is given in Theorem 5. There, the simulator we construct is in fact polynomial in the runtime of the adversary.

In the following sections, we investigate the existence of such coin toss extension protocols, depending on the desired security level (i.e., computational / statistical / perfect security) and the parameters n and m.

3   The Computational Case

3.1   Universal Composability

In the following, we need the assumption of enhanced trapdoor permutations with dense public descriptions (called ETD henceforth). Roughly, these are trapdoor permutations with the additional properties that (i) one can choose the public key in an oblivious fashion, i.e., even given the coin tosses we used it is infeasible to invert the function, and (ii) the public keys are computationally indistinguishable from random strings. We also need the notion of exponentially-hard ETD, which are secure even against subexponential-time adversaries. For detailed definitions, cf. the full version [14].

Lemma 1. *There is a constant $d \in \mathbb{N}$ s.t. the following holds:*

Assume that ETD exist, s.t. the size of the circuits describing the ETD is bounded by $s(k)$ for security parameter k.[5]

Then there is a protocol π using a uniform common reference string (CRS) of length $s(k)^d$, s.t. π securely UC-realises a bit commitment that can be used polynomially many times.

A protocol for realising bit commitment using a CRS has been given in [10]. To show this lemma, we only need to review their construction to see, that a CRS of length s^d is indeed sufficient. For details, see the full version [14].

Lemma 2. *Let $s(k)$ be a polynomially bounded function, that is computable in time polynomial in k.*

Assume one of the following holds:
- *ETD exist and s is a polynomially-large function.*
- *Exponentially-hard ETD exist and s is a superlogarithmic function.*

Then there also exist a constant $e \in \mathbb{N}$ independent of s and ETD, s.t. the size of the circuits describing the ETD is bounded by $s(k)^e$ for security parameter k.

This is shown by scaling the security parameter of the original ETD. The proof is given in the full version [14].

Theorem 1. *Let $n = n(k)$ and $m = m(k)$ be polynomially bounded and efficiently computable functions. Assume one of the following conditions holds:*
- *m is polynomially-large and ETD exist, or*
- *m is superpolylogarithmic and exponentially-hard ETD exist.*

Then there is a polynomial-time computationally universally composable protocol π for $(m \rightarrow n)$-coin toss extension.

---

[5] By the size of the circuits we means the total size of the circuits describing both the key generation and the domain sampling algorithm. Note that then trivially also the size of the resulting keys and the amount of randomness used by the domain sampling algorithm are bounded by $s(k)$.

Proof. Let d be as in Lemma 1. Let further e be as in Lemma 2. If m is polynomially-large or superpolylogarithmic, then $s := m^{1/(de)}$ is polynomially-large or superlogarithmic, resp. So, by Lemma 2 there are ETD, s.t. the size of the circuits describing the ETD is bounded by $s^e = m^{1/e}$. Then, by Lemma 1 there is a UC-secure protocol for implementing n bit commitments using an $(m^{1/d})^d = m$-bit CRS.

It is straightforward to see that using n UC-bit-commitments one can UC-securely implement an n-bit coin-toss using the protocol from [5]. Furthermore, an m-bit CRS can be trivially implemented using an m-bit coin-toss. Using the Composition Theorem we can put the above constructions together and get a protocol that UC-realises an n-bit coin-toss using an m-bit coin-toss.   □

Note that given stronger, but possibly unrealistic assumptions, the lower bound for m in Theorem 1 can be decreased. If we assume that for any superlogarithmic m, there are ETD s.t. the size of their circuits is bounded by $m^{1/d}$ (where d is the constant from Lemma 1), we get coin-toss extension even for superlogarithmic m (using the same proof as for Theorem 1, except that instead of Lemma 2 we use the stronger assumption).

However, we cannot expect an even better lower bound for m, as the following theorem shows:

Theorem 2. *Let $n = n(k)$ and $m = m(k)$ be functions with $n(k) > m(k) \geq 0$ for all k, and assume that m is not superlogarithmic (i.e., 2^{-m} is non-negligible). Then there is no non-trivial polynomial-time computationally universally composable protocol for $(m \to n)$-coin toss extension.*

Proof (sketch). Assume for contradiction that protocol π, with parties P_1 and P_2 using CT_m, implements CT_n (with m, n as in the theorem statement). Let $\mathcal{A}_1$ be an adversary on π that, taking the role of a corrupted party P_1, simply reroutes all communication of P_1 (with either P_2 or CT_m) to the protocol environment $\mathcal{Z}_1$ and thus lets $\mathcal{Z}_1$ take part as P_1 in the real protocol.

Imagine a protocol environment $\mathcal{Z}_1$, running with π and $\mathcal{A}_1$ as above, that keeps and internal simulation $\overline{P_1}$ of P_1 and lets this simulation take part in the protocol (through $\mathcal{A}_1$). After a protocol run, $\mathcal{Z}_1$ inspects the output $\overline{\kappa_1}$ of $\overline{P_1}$ and compares it to the output κ_2 of the uncorrupted P_2.

In a real protocol run with π, $\mathcal{A}_1$, and $\mathcal{Z}_1$, we will have $\overline{\kappa_1} = \kappa_2$ with overwhelming probability since π non-trivially implements CT_n, and CT_n guarantees common outputs. So a simulator $\mathcal{S}_1$, running in the ideal model with CT_n and $\mathcal{Z}_1$, must be able to achieve that the ideal output κ_2 (that is ideally chosen by CT_n and cannot be influenced by $\mathcal{S}_1$) is identical to what the simulation $\overline{P_1}$ of P_1 inside $\mathcal{Z}_1$ outputs. In that sense, $\mathcal{S}_1$ must be able to "convince" $\overline{P_1}$ to also output κ_2. To this end, $\mathcal{S}_1$ may—and must—fake a complete real protocol communication as $\mathcal{A}_1$ would deliver it to $\mathcal{Z}_1$ (and thus, to $\overline{P_1}$).

However, then we can construct another protocol environment $\mathcal{Z}_2$ that expects to take the role of party P_2 in a real protocol run (just like $\mathcal{Z}_1$ expected to take the role of P_1). To this end, an adversary $\mathcal{A}_2$ on π with corrupted P_2 is employed that forwards all communication of P_2 with either P_1 or CT_n to $\mathcal{Z}_2$. Internally,

$\mathcal{Z}_2$ now simulates $\mathcal{S}_1$ (and not P_2!) from above and an instance $\overline{\mathsf{CT}_n}$ of the trusted host $\mathsf{CT_n}$. Recall that $\mathcal{S}_1$, given a target string κ by $\mathsf{CT_n}$, mimics an uncorrupted P_2 along with an instance of CT_m. In that situation, $\mathcal{S}_1$ can convince an honest P_1 with overwhelming probability to eventually output κ.

Chances are 2^{-m} that the CT_m-instance made up by $\mathcal{S}_1$ outputs the same seed as the real CT_m in a run of $\mathcal{Z}_2$ with π and $\mathcal{A}_2$. So with probability at least $2^{-m} - \mu$ for negligible μ, in such a run, $\mathcal{Z}_2$ observes a P_1-output κ that is identical to the output of the internally simulated $\overline{\mathsf{CT}_n}$. But then, by assumption about the security of π, there is also a simulator $\mathcal{S}_2$ for $\mathcal{A}_2$ and $\mathcal{Z}_2$ that provides $\mathcal{Z}_2$ with an indistinguishable view. In particular, in an ideal run with $\mathcal{S}_2$ and CT_n, $\mathcal{Z}_2$ observes equal outputs from CT_n and $\overline{\mathsf{CT}_n}$ with probability at least $2^{-m} - \mu'$ for negligible μ'. This is a contradiction, as both outputs are uniformly and independently chosen n-bit strings, and $n \geq m + 1$.   □

4   Statistical and Perfect Cases

4.1   Stand-Alone Simulatability

We start off with a negative result:

Theorem 3. *Let $m < n$ be functions in the security parameter k. If m is not superlogarithmic, there is no two-party n-bit coin-toss protocol π (not even an inefficient one) that uses an m-bit coin-toss and has the following properties:*

- *Non-triviality. If no party is corrupted, the probability that the parties give different, invalid or no output is negligible (by invalid output we mean output not in $\{0, 1\}^n$).*
- *Security. For any (possibly unbounded) adversary corrupting one of the parties there is a negligible function μ, s.t. for every security parameter k and every $c \in \{0, 1\}^n$, the probability for protocol output c is at most $2^{-n} + \mu(k)$.*

If we require perfect non-triviality (the probability for different or no outputs is 0) and perfect security (the probability for a given output c is at most 2^{-n}), such a protocol π does not exist, even if m is superlogarithmic.

Proof (sketch). It is sufficient to consider the case $n = m + 1$.

Without loss of generality, we can assume that the available m-bit coin toss is only used at the end of the protocol. Similarly, we can assume that in the honest case, the parties never output distinct values. A detailed proof for these statements can be found in the full proof.

To show the theorem, we first consider "complete transcripts" of the protocol. By a complete transcript we mean all messages sent during the run of a protocol, excluding the value of the m-bit coin-toss. We distinguish three sets of complete transcripts: the set $\mathfrak{A}$ of transcripts having non-zero probability for the protocol output 0^n, the set $\mathfrak{B}$ of transcripts having zero probability of output 0^n and zero probability that the protocol gives no output, and the set $\mathfrak{C}$ of transcripts having non-zero probability of giving no output. Note that, since for a complete transcript, the protocol output only depends on the m-bit coin-toss, any of the above non-zero probabilities is at least 2^{-m}.

For any partial transcript p (i.e., a situation *during* the run of the protocol), we define three values α, β, γ. The value α denotes the probability with which a corrupted Alice can enforce a transcript in $\mathfrak{A}$ starting from p, the value β denotes the probability with which a corrupted Bob can enforce a transcript in $\mathfrak{B}$, and the value γ denotes the probability that the complete protocol transcript will lie in $\mathfrak{C}$ if no-one is corrupted. We show inductively that for any partial transcript p, $(1 - \alpha)(1 - \beta) \leq \gamma$. In particular, this holds for the beginning of the protocol. For simplicity, we assume that 2^{-m} is not only non-negligible, but noticeable (in the full proof, the general case is considered). Since a transcript in $\mathfrak{C}$ gives no output with probability at least 2^{-m}, the probability that the protocol generates no output (in the uncorrupted case) is at least $2^{-m}\gamma$. By the non-triviality condition, this probability is negligible, so γ must be negligible, too. So $(1 - \alpha)(1 - \beta)$ is negligible, too. Therefore $\max\{1 - \alpha, 1 - \beta\}$ must be negligible. For now, we assume that $1 - \alpha$ is negligible or $1 - \beta$ is negligible (for the general case, see the full proof).

If $1 - \alpha$ is negligible, the probability for output 0^n is at least $2^{-m}\alpha$. Since α is overwhelming and 2^{-m} noticeable, this is greater than $2^{-n} = \frac{1}{2}2^{-m}$ by a noticeable amount which contradicts the security property.

If $1 - \beta$ is negligible, we consider the maximum probability a corrupted Bob can achieve that the protocol output is not 0^n. By the security property, this probability should be at most $(2^n - 1)2^{-n}$ plus a negligible amount, which is not overwhelming. However, since every transcript in $\mathfrak{B}$ gives such an output with probability 1, the probability of such is β, which is overwhelming, in contradiction of the security property.

The perfect case is proven similarly.   □

The full proof is given in the full version [14].

Corollary 1. *By a non-trivial coin-toss protocol we mean a protocol s.t. (in the uncorrupted case) the probability that the parties give no or different output is negligible. By a perfectly non-trivial coin-toss protocol where this probability is zero.*

Let m be not superlogarithmic and $n > m$. Then there is no non-trivial protocol realising n-bit coin-toss using an m-bit coin-toss in the sense of statistical stand-alone simulatability.

Let m be any function (possibly superlogarithmic) and $n > m$. Then there is no perfectly non-trivial protocol realising n-bit coin-toss using an m-bit coin-toss in the sense of perfect stand-alone simulatability.

Proof. A statistically secure protocol would have the security property from Theorem 3 and thus, if non-trivial, contradict Theorem 3. Analogously for perfect security.   □

However, not all is lost:

Now we will prove that there exists a protocol for coin toss extension from m to n bit which is statistically stand-alone simulatably secure. The basic idea is to have the parties P_1 and P_2 contribute random strings to generate one string with sufficiently large min-entropy (the min-entropy of a random variable X is defined as $\min_x - \log \Pr[X = x]$). The randomness from this string is then extracted using a randomness extractor. Interestingly the amount of perfect randomness (i.e., the size of the m-bit coin-toss) one needs to invest is smaller than the amount extracted. This makes coin toss extension possible.

To obtain the coin toss extension we need a result about randomness extractors able to extract one bit of randomness while leaving the seed reusable like a catalyst.

Lemma 3. *For every m there exists a function $h_m : \{0,1\}^m \times \{0,1\}^{m-1} \to \{0,1\}, (s,x) \mapsto r$ such that for a uniformly distributed s and for an x with a min-entropy of at least t the statistical distance of $s\|h_m(s,x)$ and the uniform distribution on $\{0,1\}^{m+1}$ is at most $2^{-t/2}/\sqrt{2}$.*

Proof. Let $h_m(s,x) := \langle s_1 \ldots s_{m-1}, x \rangle \oplus s_m$. Here $\langle \cdot, \cdot \rangle$ denotes the inner product and $\oplus$ the addition over $GF(2)$. It is easy to verify that $h_m(s,\cdot)$ constitutes a family of universal hash functions [11], where s is the index selecting from that family. Therefore the Leftover Hash Lemma [15, 17] guarantees that the statistical distance between $s\|h_m(s,x)$ and the uniform distribution on $\{0,1\}^{m+1}$ is bounded by $\frac{1}{2}\sqrt{2 \cdot 2^{-t}} = 2^{-t/2}/\sqrt{2}$. □

With this function h_m a simple protocol is possible which extends $m(k)$ coin tosses to $m(k) + 1$ if the function $m(k)$ is superlogarithmic.

Theorem 4. *Let $m(k)$ be a superlogarithmic function, then there exists a constant round statistically stand-alone simulatable protocol that realises an $(m+1)$-bit coin-toss using an m-bit coin-toss.*

Proof. Let h_m be as in Lemma 3. Then the following protocol realises a coin toss extension by one bit. Assume $m := m(k)$ where k is the security parameter.

1. P_1 uniformly chooses $a \in \{0,1\}^{\lfloor \frac{m-1}{2} \rfloor}$ and sends a to P_2
2. P_2 uniformly chooses $b \in \{0,1\}^{\lceil \frac{m-1}{2} \rceil}$ and sends b to P_1
3. If one party fails to send a string of appropriate length or aborts then this string is assumed by the other party to be an all-zero string of the appropriate length
4. P_1 and P_2 invoke the m-bit coin toss functionality and obtain a uniformly distributed $s \in \{0,1\}^m$. If one party P_i fails to invoke the coin toss functionality or aborts, then the other party chooses s at random
5. Both P_1 and P_2 compute $s\|h_m(s,a\|b)$ and output this string.

Similar to construction 7.4.7 in [13] the protocol is constructed in a way that the adversary is not able to abort the protocol (not even by not terminating). Hence we can safely assume that the adversary will send some message of the correct length and will invoke the coin toss functionality. We assume the adversary to corrupt P_2, corruption of P_1 is handled analogously. Further we assume the random tape of $\mathcal{A}$ to be fixed in the following. Due to these assumptions there exists a function $f_{\mathcal{A}} : \{0,1\}^{\lfloor m/2 \rfloor} \to \{0,1\}^{\lceil m/2 \rceil}$ for each real adversary $\mathcal{A}$ such that the message b sent in step 2 of the protocol equals $f_{\mathcal{A}}(a)$. There is no loss in generality if we assume the view of the parties to consists of just a, b, s and the protocol output to be $s \| h_m(s, a \| b)$.

Now for a specific adversary $\mathcal{A}$ with fixed random tape the output distribution of the real protocol (i.e., view and output) is completely described by the following experiment: choose $a \stackrel{R}{\in} \{0, 1\}^{\lfloor m/2 \rfloor}$, let $b \leftarrow f_{\mathcal{A}}(a)$, choose $s \stackrel{R}{\in} \{0, 1\}^{m(k)}$, let $r \leftarrow s \| h_m(s, a \| b)$ and return $((a, b, s), r)$.

We now describe the simulator. To distinguish the the random variables in the ideal model from their real counterparts, we decorate them with a $\sim$, e.g., $\tilde{a}, \tilde{b}, \tilde{s}$. The simulator in the ideal model obtains a string $\tilde{r} \stackrel{R}{\in} \{0, 1\}^{m+1}$ from the ideal n-bit coin-toss functionality and sets $\tilde{s} = r_1 \ldots r_m$. Then the simulator chooses $\tilde{a} \stackrel{R}{\in} \{0, 1\}^{\lfloor \frac{m-1}{2} \rfloor}$ and computes $\tilde{b} = f_{\mathcal{A}}(\tilde{a})$ by giving $\tilde{a}$ to a simulated copy of the real adversary. If $h_m(\tilde{s}, \tilde{a} \| \tilde{b}) = \tilde{r}_{m+1}$ then the simulator gives $\tilde{s}$ to the simulated real adversary expecting the coin toss. Then the simulator outputs the view $(\tilde{a}, \tilde{b}, \tilde{s})$. If however, $h_m(\tilde{s}, \tilde{a} \| \tilde{b}) \neq \tilde{r}_{m+1}$ then the simulator *rewinds* the adversary, i.e., the simulator chooses a fresh $\tilde{a} \stackrel{R}{\in} \{0, 1\}^{\lfloor \frac{m-1}{2} \rfloor}$ and again computes $\tilde{b} = f_{\mathcal{A}}(a)$. If now $h_m(\tilde{s}, \tilde{a} \| \tilde{b}) = \tilde{r}_{m+1}$ the simulator outputs $(\tilde{a}, \tilde{b}, \tilde{s})$. If again $h_m(\tilde{s}, \tilde{a} \| \tilde{b}) \neq \tilde{r}_{m+1}$ then the simulator rewinds the adversary again. If after k invocations of the adversary no triple $(\tilde{a}, \tilde{b}, \tilde{s})$ was output, the simulator aborts and outputs *fail*.

To show that the simulator is correct, we have to show that the following to distributions are statistically indistinguishable: $((a, b, s), r)$ as defined in the real model, and $((\tilde{a}, \tilde{b}, \tilde{s}), \tilde{r})$.

By construction of the simulator, it is obvious that the two distributions are identical under the condition that $r_m = 0$, $\tilde{r}_m = 0$ and that the simulator does not fail. The same holds given $r_m = 1$, $\tilde{r}_m = 1$ and that the simulator does not fail. Therefore it is sufficient to show two things: (i) the statistical distance between r and the uniform distribution on n bits is negligible, and (ii) the probability that that the simulator fails is negligible. Property (i) is shown using the properties of the randomness extractor h_m. Since a is chosen at random, the min-entropy of a is at least $\lfloor \frac{m-1}{2} \rfloor \geq \frac{m}{2} - 1$, so the min-entropy of $a \| b$ is also at least $\frac{m}{2} - 1$. Since s is uniformly distributed, it follows by Lemma 3 that the statistical distance between $r = s \| h_m(s, a \| b)$ is bounded by $2^{-m/4 - 1/2} / \sqrt{2} = (2^{-m})^{1/4} / 2$. Since for superlogarithmic m it is 2^{-m} negligible, this statistical distance is negligible.

Property (ii) is then easily shown: From (i) we see, that after each invocation of the adversary the distribution of $h_m(\tilde{s}, \tilde{a} \| \tilde{b})$ is negligibly far from uniform. So the probability that $h_m(\tilde{s}, \tilde{a} \| \tilde{b}) \neq \tilde{r}_m$ is at most negligibly higher than $\frac{1}{2}$. Since the $h_m(\tilde{s}, \tilde{a} \| \tilde{b})$ in the different invokations of the adversary are independent, the probability that $h_m(\tilde{s}, \tilde{a} \| \tilde{b}) \neq \tilde{r}_m$ after each activation is negligibly far from 2^{-k}. So the simulator fails only with negligible probability.

It follows that the real and the ideal protocol execution are indistinguishable, and the protocol stand-alone simulatably implements an $(m+1)$-bit coin-toss. $\square$

The idea of the one bit extension protocol can be extended by using an extractor which extracts a larger amount of randomness (while not necessarily treating the seed like a catalyst). This yields constant round coin toss extension protocols.

However, the simulator needed for such a protocol does not seem to be efficient, even if the real adversary is. To get a protocol that also fulfils both the property of computational stand-alone simulatabiliy and of statistical stand-alone simulatabiliy, we need a simulator that is efficient if the adversary is.

Below we give such a coin toss extension protocol for superlogarithmic $m(k)$ which is statistically secure *and* computationaly secure, i.e., the simulator for polynomial adversaries is polynomially bounded, too. The basic idea here is to extract one bit at a time in polynomially many rounds.

Theorem 5. *Let $m(k)$ be superlogarithmic, and $p(k)$ be a positive polynomially-bounded function, then there exists a statistically and computationally stand-alone simulatable protocol that realises an $(m + p)$-bit coin-toss using an m-bit coin-toss.*

Proof. Let h_m be as in Lemma 3. Then the following protocol realises a coin toss extension by $p(k)$ bits.

1. for $i = 1$ to $p(k)$ do
   (a) P_1 uniformly chooses $a_i \in \{0,1\}^{\lfloor \frac{m-1}{2} \rfloor}$ and sends a_i to P_2
   (b) P_2 uniformly chooses $b_i \in \{0,1\}^{\lceil \frac{m-1}{2} \rceil}$ and sends b_i to P_1
   (c) If one party fails to send a string of appropriate length or aborts then this string is assumed by the other party to be an all-zero string of the appropriate length
2. P_1 and P_2 invoke the m-bit coin toss functionality and obtain a uniformly distributed $s \in \{0,1\}^m$. If one party P_i fails to invoke the coin toss functionality or aborts, then the other party chooses s at random ·
3. P_1 and P_2 compute $s\|h_m(s, a_1\|b_1)\| \ldots \|h_m(s, a_{p(k)}\|b_{p(k)})$ and output this string.

We only roughly sketch the differences to the proof of Theorem 4. For each protocol round the simulator follows the strategy described in the proof of Theorem 4 (i.e., the simulator rewinds the adversary by *one* round, if the coin-toss produced is not the correct one.) Then using standard hybrid techniques it can be shown that this simulator indeed gives an indistinguishable ideal protocol run. Here it is only noteworthy that we use the fact that $s\|h_m(s, a_1\|b_1)\| \ldots \|h_m(s, a_{p(k)}\|b_{p(k)})$ is statistically indistinguishable from the uniform distribution on $m + p$ bits. However, this follows directly from Lemma 3 and the fact that each $a_i\|b_i$ has min-entropy at least $\lfloor \frac{m-1}{2} \rfloor$ even given the values of all $a_\mu\|b_\mu$ for $\mu < i$. □

4.2   Universal Composability (Statistical/Perfect Case)

In the case of statistical security, adversary and protocol environment are allowed to be computationally unbounded. In that case, we show that there is no simulatably secure coin toss extension protocol that runs in a polynomial number of rounds. This is forced by requiring the parties to halt after a polynomial number of activations. However, note that we do not impose any restrictions on the amount of computational work these parties perform in one of those activations.

The proof of this statement is done by contradiction. Furthermore, the proof is split up into an auxiliary lemma and the actual proof. In the auxiliary lemma, we show that without loss of generality, a protocol for statistically universally composable coin toss extension has a certain outer form. Then we show that any such protocol (of this particular outer form) is insecure.

For the following statements, we always assume that $m = m(k), n = n(k)$ are arbitrary functions, only satisfying $0 \leq m(k) < n(k)$ for all k. We also restrict to protocols that proceed in a polynomial number of rounds. That is, by a "protocol" we mean in the following one in which each party halts after at most $p(k)$ activations, where $p(k)$ is a polynomial which depends only on the protocol. (As stated above, the parties are still unbounded in each activation.) We start with a helping lemma whose proof is available in the full version [14].

Lemma 4. *If there is a statistically universally composable protocol for* $(m \rightarrow n)$-*coin toss extension, then there is also one in which each party*

- *has only one connection to the other party and one connection to* CT_m,
- *in each activation sends either an "init" message to* CT_m *or some message to the other party,*
- *sends in each protocol run at most one message to* CT_m, *and this is always an "init" message,*
- *the internal state of each of the two parties consists only of the view that this party has experienced so far, and*
- *after* P_i *sends "init" to* CT_m, *it does not further communicate with* P_{3-i} *(for* $i = 1, 2$ *and in case of no corruptions).*

We proceed with

Lemma 5. *There is no statistically universally composable protocol for* $(m \rightarrow n)$-*coin toss extension which meets the requirements from Lemma 4.*

Proof. Assume for contradiction that π, using CT_m, is a statistically universally composable implementation of CT_n, and also satisfies the requirements from Lemma 4.

Assume a fixed environment $\mathcal{Z}_0$ that gives both parties "init" input and then waits for both parties to output a coin toss result. Consider an adversary $\mathcal{A}_0$ that delivers all messages between the parties immediately. The resulting setting D_0 is depicted in Figure 1.

Denote the protocol communication in a run of D_0, i.e., the ordered list of messages sent between P_1 and P_2, by *com*. Denote by κ_1 and κ_2 the final outputs of the parties. For $M \subseteq \{0, 1\}^n$ and a possible protocol communication prefix $\bar{c}$, let $\mathsf{E}(M, \bar{c})$ be the probability that the protocol outputs are identical and in M, provided that the protocol communication starts with $\bar{c}$, i.e.,

$$\mathsf{E}(M, \bar{c}) := \Pr[\kappa_1 = \kappa_2 \in M \mid \bar{c} \leq com],$$

where $x \leq y$ means that x is a prefix of y.

Note that the parties have, apart from their communication *com*, only the seed $\omega \in \{0, 1\}^m$ provided by CT_m for computing their final output κ. So we may assume that there is a deterministic function f for which $\kappa_1 = \kappa_2 = f(com, \omega)$ with overwhelming probability.

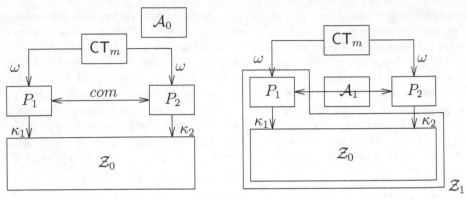

Fig. 1. *Left:* The initial setting D_0 for the statistical case. (Some connections which are not important for our proof have been omitted.) *Right:* Setting D_1 with a corrupted P_1. Setting D_2 (with P_2 corrupted instead of P_1) is defined analogously.

For a fixed protocol communication $com = c$, consider the set

$$M_c := \{0,1\}^n \setminus \{ f(c,s) \mid s \in \{0,1\}^m \}$$

of "improbable outputs" after communication c. Then obviously $|M_c| \geq 2^n - 2^m \geq 2^{n-1}$. By definition of the ideal output (i.e., the output of CT_n in the ideal model), this implies that for sufficiently large security parameters k, the probability that $\kappa_1 = \kappa_2 \in M_c$ is at least $2/5$. (Here, any number strictly between 0 and $1/2$ would have done as well.) Otherwise, an environment could distinguish real and ideal model by testing for $\kappa_1 = \kappa_2 \in M_c$. Since $\mathsf{E}(M_c, \varepsilon)$ is exactly that probability, we have $\mathsf{E}(M_c, \varepsilon) \geq 2/5$ for sufficiently large k. Also, $\mathsf{E}(M_c, c)$ is negligible by definition, so M_c satisfies

$$\mathsf{E}(M_c, \varepsilon) - \mathsf{E}(M_c, c) \geq \frac{1}{3} \qquad (1)$$

for sufficiently large k.

Since the protocol consists by assumption only of polynomially many rounds, c is a list of size at most $p(k)$ for a fixed polynomial p. This means that there is a prefix $\bar{c}$ of c and a single message m (either sent from P_1 to P_2 or vice versa) such that $\bar{c}m \leq c$ and

$$\mathsf{E}(M_c, \bar{c}) - \mathsf{E}(M_c, \bar{c}m) \geq \frac{1}{3p(k)} \qquad (2)$$

for sufficiently large k. Intuitively, this means that at a certain point during the protocol run, a single message m had a significant impact on the probability that the protocol output is in M_c.

Note that such an m must be either sent by P_1 or P_2. So there is a $j \in \{1,2\}$, such that for infinitely many k, party P_j sends such an m with probability at least $1/2$. We describe a modification D_j of setting D_0. In setting D_j, party P_j is corrupted and simulated (honestly) inside $\mathcal{Z}_j$. Furthermore, adversary $\mathcal{A}_j$ simply relays all communication between this simulation inside $\mathcal{Z}_j$ and the uncorrupted party P_{3-j}. For supplying inputs to the simulation of P_j and to the uncorrupted P_{3-j}, a simulation of $\mathcal{Z}_0$ is employed inside $\mathcal{Z}_j$. The situation (for $j = 1$) is depicted in Figure 1.

Since D_j is basically only a re-grouping of D_0, the random variables com, ω, and κ_i are distributed exactly as in D_0, so we simply identify them. In particular, in D_j, for infinitely many k, there is with probability at least $1/2$ a prefix $\bar{c}$ and a message m *sent by P_j* of com that satisfy (2).

Now we slightly change the environment $\mathcal{Z}_j$ into an environment $\mathcal{Z}'_j$. Each time the simulated P_j sends a message m to P_{3-j}, $\mathcal{Z}'_j$ checks for *all* subsets M of $\{0,1\}^n$ whether

$$\exists M \subseteq \{0,1\}^n : \quad \mathsf{E}(M,\bar{c}) - \mathsf{E}(M,\bar{c}m) \geq \frac{1}{3p(k)}, \tag{3}$$

where $\bar{c}$ denotes the communication between P_j and P_{3-j} so far.

If (3) holds at some point for the first time, then $\mathcal{Z}'_j$ tosses a coin b uniformly at random, and proceeds as follows: if $b = 0$, then $\mathcal{Z}'_j$ keeps going just as $\mathcal{Z}_j$ would have. In particular, $\mathcal{Z}'_j$ then lets P_j send m to P_{3-j}. However, if $b = 1$, then $\mathcal{Z}'_j$ rewinds the simulation of P_j to the point *before* that activation, and activates P_j again with fresh randomness, thereby letting P_j send a possibly different message m'. In the further proof, $\bar{c}$, m, and M refer to these values for which (3) holds.

In any case, after having tossed the coin b once, $\mathcal{Z}'_j$ remembers the set M from (3), and does not check (3) again. After the protocol finishes, $\mathcal{Z}_j$ outputs either $(\perp, \perp)$ (if (3) was never fulfilled), or (b, β) for the evaluation β of the predicate $[\kappa_1 = \kappa_2 \in M]$ (i.e., $\beta = 1$ iff the protocol gives output, the protocol outputs match and lie in M).

Now by our choice of j, $\Pr[b \neq \perp] \geq 1/2$ for infinitely many k.

Also, Lemma 4 guarantees that the internal state of the parties at the time of tossing b consists only of $\bar{c}$. So, when $\mathcal{Z}'_j$ has chosen $b = 1$, and rewound the simulated P_j, the probability that at the end of the protocol $\kappa_1 = \kappa_2 \in M$ is the same as the probability of that event in the setting D_j under the condition that the communication com begins with $\bar{c}$. This probability again is exactly $\mathsf{E}(M,\bar{c})$ by definition.

Similarly, when $\mathcal{Z}'_j$ has chosen $b = 0$, the probability that at the end of the protocol $\kappa_1 = \kappa_2 \in M$ is the same as the probability of that event in the setting D_j under the condition that the communication com begins with $\bar{c}m$, i.e. $\mathsf{E}(M,\bar{c}m)$.

Therefore just before $\mathcal{Z}'_j$ chooses b (i.e., when $\bar{c}$ and M are already determined), the probability that at the end we will have $\beta = 1 \wedge b = 1$ is $\frac{1}{2}\mathsf{E}(M,\bar{c})$ and the probability of $\beta = 1 \wedge b = 0$ is $\frac{1}{2}\mathsf{E}(M,\bar{c}m)$. Therefore the difference between these probabilities is at least $\frac{1}{2}\big(\mathsf{E}(M,\bar{c}) - \mathsf{E}(M,\bar{c}m)\big) \geq \frac{1}{3p(k)}$.

Since this bound on the difference of the probabilities always holds when $b \neq \perp$, by averaging we get

$$\Pr[\beta = 1 \wedge b = 1 \mid b \neq \perp] - \Pr[\beta = 1 \wedge b = 0 \mid b \neq \perp] \geq \frac{1}{3p(k)}$$

and using the fact that $\Pr[b \neq \perp] \geq \frac{1}{2}$ for infinitely many k we then have that

$$\Pr[\beta = 1 \wedge b = 1] - \Pr[\beta = 1 \wedge b = 0] \geq \frac{1}{6p(k)} \tag{4}$$

for infinitely many k when $\mathcal{Z}'_j$ runs with the real protocol as described above.

We show that no simulator $\mathcal{S}_j$ can achieve property (4) in the ideal model, where $\mathcal{Z}'_j$ runs with CT_n and $\mathcal{S}_j$. To distinguish random variables during a run of $\mathcal{Z}'_j$ in the ideal model from those in the real model, we add a tilde to a random variable in a run of $\mathcal{Z}'_j$ in the ideal model, e.g., $\tilde{b}$, $\tilde{\beta}$.

For any $\mathcal{S}_j$ achieving indistinguishability of real and ideal model, this can happen only with negligible probability, so we can assume without losing generality that $\mathcal{S}_j$ always delivers outputs.

By construction of $\tilde{b}$ and κ, the variable $\tilde{b}$ and the tuple $(\tilde{M}, \kappa)$ are independent given $\tilde{b} \neq \perp$. Hence, since $\tilde{\beta}$ is a function of $\tilde{M}$ and κ,

$$\Pr\left[(\tilde{b}, \tilde{\beta}) = (0, 1)\right] = \Pr\left[(\tilde{b}, \tilde{\beta}) = (1, 1)\right]. \tag{5}$$

So comparing (4) and and (5), $\mathcal{Z}'_j$'s output distribution differs non-negligibly in real and ideal model. So no simulator $\mathcal{S}_j$ can simulate attacks carried out by $\mathcal{Z}'_j$ and $\mathcal{A}_j$, which gives the desired contradication.   □

Combining the above Lemmas 4 and 5 we therefore get:

Theorem 6. *There is no non-trivial statistically universally composable protocol for $(m \to n)$-coin toss extension that proceeds in a polynomial of rounds.*

The case of perfect security is shown analogously.

Acknowledgements. This work was partially supported by the projects PRO-SECCO (IST-2001-39227) and SECOQC of the European Commission. Part of this work was done while the first author was with the IAKS, Universität Karlsruhe. Further, we thank the anonymous referees for valuable comments.

References

1. Andris Ambanis, Harry Buhrman, Yevgeniy Dodis, and Heinz Röhrig. Multiparty quantum coin flipping. In *19th Annual IEEE Conference on Computational Complexity, Proceedings of CoCo 2002*, pages 250–259. IEEE Computer Society, 2004.
2. Michael Backes, Dennis Hofheinz, Jörn Müller-Quade, and Dominique Unruh. On fairness in simulatability-based cryptographic systems. In *3rd ACM Workshop on Formal Methods in Security Engineering, Proceedings of FMSE 2005*, pages 13–22. ACM Press, 2005.

3. Michael Backes, Birgit Pfitzmann, and Michael Waidner. Secure asynchronous reactive systems. IACR ePrint Archive, March 2004.
4. Mihir Bellare, Juan A. Garay, and Tal Rabin. Distributed pseudo-random bit generators – a new way to speed-up shared coin tossing. In *Fifteenth Annual ACM Symposium on Principles of Distributed Computing, Proceedings of PODC 1996*, pages 191–200. ACM Press, 1996.
5. Manuel Blum. Coin flipping by telephone. In Allen Gersho, editor, *Advances in Cryptology, A report on CRYPTO 81*, number 82-04 in ECE Report, pages 11–15. University of California, Electrical and Computer Engineering, 1982.
6. Ran Canetti. Universally composable security: A new paradigm for cryptographic protocols. In *42th Annual Symposium on Foundations of Computer Science, Proceedings of FOCS 2001*, pages 136–145. IEEE Computer Society, 2001.
7. Ran Canetti. Universally composable security: A new paradigm for cryptographic protocols. In *42th Annual Symposium on Foundations of Computer Science, Proceedings of FOCS 2001*, pages 136–145. IEEE Computer Society, 2001.
8. Ran Canetti. Universally composable security: A new paradigm for cryptographic protocols. IACR ePrint Archive, January 2005. Full and revised version of [7].
9. Ran Canetti and Marc Fischlin. Universally composable commitments. In Joe Kilian, editor, *Advances in Cryptology, Proceedings of CRYPTO 2001*, number 2139 in Lecture Notes in Computer Science, pages 19–40. Springer-Verlag, 2001.
10. Ran Canetti, Yehuda Lindell, Rafail Ostrovsky, and Amit Sahai. Universally composable two-party and multi-party secure computation. In *34th Annual ACM Symposium on Theory of Computing, Proceedings of STOC 2002*, pages 494–503. ACM Press, 2002. Extended abstract.
11. J. Lawrence Carter and Mark N. Wegman. Universal classes of hash functions. *Journal of Computer and System Sciences*, 18(2):143–154, April 1979.
12. Richard Cleve. Limits on the security of coin flips when half the processors are faulty. In *Eighteenth Annual ACM Symposium on Theory of Computing, Proceedings of STOC 1986*, pages 364–369. ACM Press, 1986.
13. Oded Goldreich. *Foundations of Cryptography – Volume 2 (Basic Applications)*. Cambridge University Press, May 2004.
14. Dennis Hofheinz, Jörn Müller-Quade, and Dominique Unruh. On the (im-)possibility of extending coin toss. IACR ePrint Archive, 2006. Full version of this paper.
15. Russell Impagliazzo, Leonid A. Levin, and Michael Luby. Pseudo-random generation from one-way functions. In *Twenty-First Annual ACM Symposium on Theory of Computing, Proceedings of STOC 1989*, pages 12–24. ACM Press, 1989. Extended abstract.
16. Birgit Pfitzmann and Michael Waidner. A model for asynchronous reactive systems and its application to secure message transmission. In *IEEE Symposium on Security and Privacy, Proceedings of SSP 2001*, pages 184–200. IEEE Computer Society, 2001.
17. Douglas R. Stinson. Universal hash families and the leftover hash lemma, and applications to cryptography and computing. *Journal of Combinatorial Mathematics and Combinatorial Computing*, 42:3–31, 2002.

Efficient Binary Conversion for Paillier Encrypted Values

Berry Schoenmakers[1] and Pim Tuyls[2]

[1] Dept. of Mathematics and Computing Science, TU Eindhoven,
P.O. Box 513, 5600 MB Eindhoven, The Netherlands
berry@win.tue.nl
[2] Philips Research Labs
Prof. Holstlaan 4, 5656 AA Eindhoven, The Netherlands
pim.tuyls@philips.com

Abstract. We consider the framework of secure n-party computation based on threshold homomorphic cryptosystems as put forth by Cramer, Damgård, and Nielsen at Eurocrypt 2001. When used with Paillier's cryptosystem, this framework allows for efficient secure evaluation of any arithmetic circuit defined over $\mathbb{Z}_N$, where N is the RSA modulus of the underlying Paillier cryptosystem.

In this paper, we extend the scope of the framework by considering the problem of converting a given Paillier encryption of a value $x \in \mathbb{Z}_N$ into Paillier encryptions of the bits of x. We present solutions for the general case in which x can be any integer in $\{0, 1, \ldots, N - 1\}$, and for the restricted case in which $x < N/(n2^\kappa)$ for a security parameter κ. In the latter case, we show how to extract the ℓ least significant bits of x (in encrypted form) in time proportional to ℓ, typically saving a factor of $\log_2 N/\ell$ compared to the general case.

Thus, intermediate computations that rely in an essential way on the binary representations of their input values can be handled without enforcing that the *entire* computation is done bitwise. Typical examples involve the relational operators such as $<$ and $=$. As a specific scenario we will consider the setting for (approximate) matching of biometric templates, given as bit strings.

1 Introduction

We consider secure n-party computation in the framework based on threshold homomorphic cryptosystems, as put forth by Cramer, Damgård, and Nielsen [CDN01]. To evaluate a given n-ary function f securely, one expresses f as an arithmetic circuit C composed of elementary gates, such as addition gates and multiplication gates. Given ciphertexts $[\![x_1]\!], \ldots, [\![x_n]\!]$, the gates are evaluated one by one, ultimately resulting in a ciphertext $[\![f(x_1, \ldots, x_n)]\!]$. Here, $[\![x]\!]$ denotes a probabilistic encryption of x in the underlying threshold homomorphic cryptosystem, which will be the Paillier cryptosystem [Pai99] throughout most of this paper. The homomorphic property ensures that evaluation of addition gates

S. Vaudenay (Ed.): EUROCRYPT 2006, LNCS 4004, pp. 522–537, 2006.

is essentially for free, as addition may simply be done by multiplying two cipher-texts. For multiplication gates, however, one needs a joint protocol involving at least one threshold decryption in the underlying cryptosystem.

An important feature of this type of protocols for secure computation is that the communication complexity, which is the dominating complexity measure, is $O(nk|C|)$ bits, where k is a security parameter, and $|C|$ is the number of (multiplication) gates of circuit C. Moreover, if Paillier is used as the underlying cryptosystem, the result of [CDN01] is particularly efficient in handling arithmetic with *large* numbers. Each arithmetic gate handles integer values modulo N, where N is the RSA modulus used for Paillier, at a cost *independent* of the size of these integers.

Arithmetic circuits for functions that also involve some bit-oriented steps tend to be inefficient, as can be seen from the following simple but somewhat contrived example. Consider the function $ws(x_1, \ldots, x_n)$ which outputs the Hamming weight of the binary representation of the sum $\sum_{i=1}^{n} x_i$ of the inputs x_i, which are assumed to be from a bounded range, $0 \leq x_i < 2^{64}$ say. Lacking an efficient secure conversion of an integer value into its binary presentation, the natural thing to do would be to require that the inputs x_i are given bitwise (hence as 64 encrypted bits each), and perform the entire computation bitwise. This way, however, no advantage is taken from the potential of arithmetic circuits.

In this paper, we address this shortcoming by means of efficient protocols for securely converting an integer into its binary representation. Such a protocol can be viewed as a new type of gate, next to the addition/subtraction gates and multiplication/division gates that are already known from [CDN01]. We call it the BITREP gate. So, on input $[\![x]\!]$, $0 \leq x < 2^m$, a BITREP gate outputs $[\![x_0]\!], \ldots, [\![x_{m-1}]\!]$, of course, without leaking any information on x.

A moment's thought will reveal that an efficient BITREP gate is only feasible if a cryptosystem such as Paillier is used as the underlying cryptosystem. For instance, the approach of [ST04], which is based on the homomorphic ElGamal cryptosystem, cannot lead to an efficient BITREP gate: given $[\![x]\!]$, $0 \leq x < 2^m$, extracting x cannot be done in time polynomial in m, as it involves solving a discrete log problem even when one knows the private key for $[\![\cdot]\!]$ (if $[\![x]\!] = (g^r, h^r g^x)$, one needs to recover x from g^x). For a Paillier ciphertext of the form $g^x r^N \bmod N^2$, there is no such contradiction as x can be recovered efficiently if one knows the private key.

As a related problem, we will also provide an efficient protocol for computing the least significant bits of an encrypted value, called an LSB gate. Clearly, a cascade of LSB gates can also be used to implement a BITREP gate, but our solutions for these gates will be of independent interest. The complexity of the LSB gate is independent of m, for input $[\![x]\!]$, $0 \leq x < 2^m$, under a mild restriction on m in terms of N, such as $m + 100 < \log_2 N$ (in general, $m + \kappa + \log_2 n < \log_2 N$, for a security parameter κ).

As a further motivation for studying these problems we like to point out the following application. Consider a constrained device which must send a certain number of bits (ranging from a few hundred to a few thousand), which will be processed further as inputs to a secure computation. The obvious approach of sending a separate Paillier encryption for each of these many bits, however, may be completely infeasible. Given the BITREP gate, it suffices if the device packs the bits into one or more large integers, and sends these out using a single Paillier encryption per integer.

A particularly interesting case of this scenario can be seen in the context of biometric authentication. A tamper-resistant measuring device will obtain a biometric sample, e.g., an IrisCode® consisting of 512 bytes, which it sends out using a couple of Paillier encryptions. At the server side, the BITREP gate may then be applied to obtain the encrypted bits, which are subsequently used to evaluate securely whether the sample matches a stored (encrypted) biometric template, which was previously obtained during enrollment. In such a system, biometric details are never exposed in the clear, except in the measuring device during capture. The relevance of such an approach has been argued, e.g. in [DRS04, TG04, KAMR04].

Our Contributions

We present new n-party protocols for several variants of the problem of securely computing the binary representation of an integer value. In each case we make essential use of the Paillier cryptosystem (or variations thereof). The most general solution, which we call the BITREP gate, handles input values that can be *any* integer in $\{0, \ldots, N-1\}$, where N is an RSA modulus of $k = \log_2 N$ bits. The broadcast complexity of the BITREP gate ranges between $O(nk^2)$ and $O(n^2 k^2 \log k)$ depending on which subprotocols are used (e.g., for the generation of jointly random bits).

Since k is large, ranging from 1024 to 2048 say, the cost of the BITREP gate is high even if the actual inputs are known to lie in a much smaller range than $\{0, \ldots, N-1\}$. Therefore, we also consider the case that the input values are from a limited range $\{0, \ldots, 2^m - 1\}$, $m < k - \kappa - \log_2 n$, where κ is a security parameter to be set such that $2^{-\kappa}$ is negligible. In this case, we show how to reduce the broadcast complexity by a factor of k/m compared to the general case. The bound on m is not a severe restriction as typically $\kappa + \log_2 n$ is much smaller than k.

Finally, again for input values in $\{0, \ldots, 2^m - 1\}$, $m < k - \kappa - \log_2 n$, we show how to extract the least significant bit in time *independent* of m (LSB gate). More generally, we show how to extract the ℓ least significant bits in time proportional to ℓ (LSBs gate). The broadcast complexity is reduced accordingly, by a factor k/ℓ compared to the general solution, independent of m.

Apart from these new protocols (which rely on a special application of $O(1)$ interval proofs), we also show how to integrate the security proofs for these new gates in the framework of [CDN01]. First, we observe that the security proof of [CDN01] actually achieves a *tight* reduction, namely that a distinguisher of the simulated vs. real protocol is transformed into a distinguisher for the underlying cryptosystem *without* loss in success probability. Secondly, the proof of [CDN01] is *modular* in the sense that a statistically indistinguishable simulation is given for each of the basic gates separately: e.g., the multiplication gate is simulated given inputs $[\![x]\!]$, $[\![y]\!]$ and corresponding output $[\![xy]\!]$.

To retain the tightness and modularity, however, we need to state the security for a gate such as a BITREP gate in a particular way. It turns out to be impossible to simulate our BITREP gate given only input $[\![x]\!]$ and corresponding outputs $[\![x_0]\!], \ldots, [\![x_{m-1}]\!]$. The reason is that the protocol for the BITREP gate produces additional encryptions such as $[\![x_0 x_1]\!]$ and encryptions of other monomials as intermediate results, which cannot be computed by the simulator! This is an interesting phenomenon and we show how to extend the framework of [CDN01] to handle it. The LSB gate can easily be simulated, though, given just the input $[\![x]\!]$ and corresponding output $[\![x_0]\!]$.

We also highlight some applications of the new gates. A similar tradeoff as mentioned above for biometric authentication is possible in the context of electronic voting protocols, where one would like to minimize the effort required of the voter in casting an encrypted ballot (e.g., when the voter's client software needs to run on a simple mobile phone). This problem has already been considered in [DJ02], where incidentally Paillier encryption is used as well. Using a BITREP gate one gets an interesting alternative to [DJ02]. To vote for a single candidate x, $0 \le x < 2^m$ say, one simply lets the voter release an encryption $[\![x]\!]$. To tally these votes we use the radix M representation of [CFSY96], where M is an integer larger than the number of voters. A vote x will thus contribute M^x to the final tally, represented as an integer in radix M. To compute this exponentiation securely, one first computes $[\![x_0]\!], \ldots, [\![x_{m-1}]\!]$ from $[\![x]\!]$, using the BITREP gate. Noting that $M^{x_0} = 1 + x_0(M-1)$ and so on, it follows that $[\![M^x]\!]$ can be computed securely, using $O(m)$ multiplication gates. The so-obtained encryptions $[\![M^x]\!]$ (for all voters) are then multiplied together and one gets, upon decryption, the election result in radix M representation.

Related Work

Independent of our work, the problem of securely computing the bits of an integer value has been studied recently by Damgård et al. [DFK$^+$06]. An obvious difference is they consider the unconditional setting, whereas we consider the cryptographic model. This is reflected by the use of sharings (for an underlying linear secret sharing scheme) in their case vs. the use of encryptions (for an underlying threshold homomorphic cryptosystem) in our case.

The shared values as well as the shares and the arithmetic circuit in [DFK$^+$06] are all defined over $\mathbb{F}_p$, for a prime p, where the security of the protocol does not depend on the size of p. Prime p can thus be chosen freely to fit an application. For example, to make things practical, one may use moderately large primes p of 64 bits say (e.g., to ensure that inputs of a reasonable size can be handled and, also, to easily exclude some failure events that happen with $O(1/p)$ probability). In our case, however, the arithmetic circuit will be defined over $\mathbb{Z}_N$, where N is the RSA modulus of a Paillier cryptosystem. Necessarily, the number N is therefore very large, say 1024 upto 2048 bits.

The protocol of [DFK$^+$06] for handling input values in $\{0, \ldots, p-1\}$ and our protocol for handling input values in $\{0, \ldots, N-1\}$ (the BITREP gate) follow the same pattern. The emphasis in [DFK+06] is on constant round complexity (which is achieved by solving various subproblems in constant rounds). In this paper, we focus on techniques for handling inputs from a restricted range efficiently, to limit the consequences of the fact that all arithmetic is done modulo N, where N is necessarily large: informally, to extract the ℓ least significant bits, the complexity of our protocol is *independent* of the size of the gap $N - 2^\ell$.

The problem of securely computing the bits of an integer value has also been considered by Algesheimer et al. [ACS02], in the unconditional setting, but restricted to the passive case. Their approach follows a more complicated pattern, involving an addition circuit for securely adding n numbers bitwise (rather than just 2 numbers as we and [DFK+06] do), which does not readily seem to extend to the problems and solutions considered in this paper and in [DFK+06].

2   Preliminaries

In this section, we introduce some notation and present the basic tools that we need throughout the paper.

We assume the framework for secure computation based on threshold homomorphic cryptosystems of [CDN01], used with the Paillier cryptosystem. This framework allows one to securely evaluate arithmetic circuits, composed of several types of basic gates, as listed below. For simplicity, we assume that the parties $P_1, \ldots, P_n$ evaluating the circuit coincide with the parties running the underlying $(t + 1, n)$-threshold Paillier cryptosystem. Here, t denotes the maximum number of statically, but actively corrupted parties tolerated by the circuit evaluation protocol. If a party fails to complete a step during any of the (sub)protocols (e.g., if a proof fails), then that party is simply discarded and the (sub)protocol is rerun by the remaining parties; we will not describe this explicitly for the protocols in this paper.

The bounds on t are as in previous papers. For $0 \leq t < n/2$, the case of a dishonest minority, robustness is achieved directly using the protocol of [CDN01]. For $n/2 \leq t < n$, the case of a dishonest majority, the protocol of [CDN01] can be extended in a modular way to achieve various degrees of fairness: the new property of "resource-fairness", as introduced and studied in [GMPY06]), can be achieved under additional intractability assumptions such as the strong-RSA assumption; also, as described in [ST04], a strictly weaker form of fairness can be achieved without requiring additional assumptions, using a simple gradual release approach.

2.1   Paillier Cryptosystem

The Paillier cryptosystem is a probabilistic, additively homomorphic encryption scheme, known to be semantically secure under the *Decisional Composite Residuosity Assumption* [Pai99]. Several variations and generalizations of the basic Paillier cryptosystem have been proposed since, see, e.g., [FPS00, DJ01, CS02, DJ03]. Below, we summarize the threshold variant of a generalized Paillier cryptosystem, as introduced in [DJ01, Section 4.1], but any other variant providing threshold decryption can be used as well for our purposes.

The public key consists of an RSA modulus $N = pq$ of bit length k (security parameter), where p, q are safe primes, $p = 2p' + 1$ and $q = 2q' + 1$. The set of plaintexts is given by the additive group $\mathbb{Z}_{N^s}$, and an encryption of a message $x \in \mathbb{Z}_{N^s}$ takes the form $[\![x]\!] = (N + 1)^x r^{N^s} \bmod N^{s+1}$ for random $r \in \mathbb{Z}^*_{N^{s+1}}$ (case $s = 1$ corresponds to the basic Paillier cryptosystem).

The private key is given by the unique value $d \in \mathbb{Z}_{\tau N^s}$ satisfying $d = 0 \bmod \tau$ and $d = 1 \bmod N^s$, where $\tau = p'q'$. In the threshold case, polynomial shares d_i for $i = 1, \ldots, n$ of d are generated and distributed parties $P_1, \ldots, P_n$. Decryption of a ciphertext $[\![x]\!]$ is done by means of a (t, n)-threshold decryption protocol requiring the cooperation of t or more parties, $1 \leq t \leq n$. We refer to [DJ01] for further details, noting that it is also proved that, given a ciphertext and the corresponding plaintext, the decryption protocol can be simulated statistically indistinguishable for an active, static adversary corrupting at most t parties.

Throughout this paper, the results are described for the case $s = 1$, as in the original Paillier cryptosystem, such that the plaintext space is simply the additive group $\mathbb{Z}_N$ (but the results can readily be extended to the case $s > 1$). In this case, the cryptosystem is additively homomorphic over $\mathbb{Z}_N$: given $[\![x]\!]$ and $[\![y]\!]$ with $x, y \in \mathbb{Z}_N$ we have $[\![x + y]\!] = [\![x]\!][\![y]\!]$, where multiplication of ciphertexts is done modulo N^2. Note that this implies that, for any $c \in \mathbb{Z}_N$, $[\![cx]\!] = [\![x]\!]^c \bmod N^2$.

2.2 Efficient Proofs

In order to withstand active attacks, we use several standard types of zero-knowledge proofs (Σ-protocols). We will assume the *random oracle model*, so that all the proofs are non-interactive and can be simulated easily. We use standard proofs for proving plaintext knowledge, equality of plaintexts, and that a plaintext is a bit for given Paillier encryptions.

In particular, we will use efficient *range proofs* by which a party that generated a Paillier encryption $[\![x]\!]$ can prove that x belongs to given interval using $O(1)$ modular exponentiations only. Although the hidden constant is quite large, the $O(1)$ methods certainly pay off for intervals of length 2^{100} and up, as we need (hence, the size of the non-interactive versions of these proofs is $O(k)$ for security parameter k, in the random oracle model). Efficient proofs for showing that a committed integer value belongs to a given interval were introduced in [Bou00], and later refined in [Lip03]. The underlying integer commitment scheme of [FO97, DF02] relies on the *strong RSA assumption*. By proving equality of such a committed value and an encrypted value (see, e.g., [DJ02]), one can thus prove that an encrypted value belongs to a given interval.

2.3 Basic Gates and Circuits

In this section, we briefly describe the secure subprotocols (gates and circuits) we need for our protocols. The distinction between gates and circuits may be somewhat arbitrary.

We describe the standard gates for multiplication and inversion modulo N, followed by two gates (or circuits) for jointly generating random values. Finally, we describe circuits for the bitwise operations used in our protocols. These are the basic arithmetic operators for addition and subtraction of integers, and the basic relational operators for comparison of integers.

The exact details of these gates and circuits are immaterial to the validity of our constructions, but we do give an indication of the broadcast and round complexities.

Multiplication and Inversion Gates. The general multiplication gate developed in [CDN01] allows n parties to compute an encryption $[\![xy]\!]$ given encryptions $[\![x]\!]$ and $[\![y]\!]$, where $x, y \in \mathbb{Z}_N$, in a constant number of rounds.

In case one of the inputs, say input x, is private to one of the parties, say party P_i, the following simplified multiplication protocol can be used, requiring no interaction at all. Party P_i computes $[\![xy]\!]$ directly from $[\![y]\!]$ using the homomorphic property $[\![xy]\!] = [\![y]\!]^x$. Party P_i also generates a Σ-proof showing that it computed $[\![xy]\!]$ correctly w.r.t. $[\![x]\!]$ and $[\![y]\!]$. We will refer to this gate as a *private-multiplier gate*.

An inversion gate allows n parties to compute an encryption $[\![x^{-1}]\!]$ given encryption $[\![x]\!]$, with $x \in \mathbb{Z}_N^*$, in a constant number of rounds [CDN01].

The broadcast complexity is $O(nk)$ bits for a multiplication gate and an inversion gate, and $O(k)$ bits for a private-multiplier gate.

Random Invertible Element Gates. All parties choose a random element $r_i \in_R \mathbb{Z}_N^*$ and broadcast an encryption $[\![r_i]\!]$ together with a Σ-proof for knowledge of the plaintext r_i. Finally, $[\![r]\!] = [\![\sum_{i=1}^{n} r_i]\!]$ is publicly computed. Note that this gate fails with negligible probability.

The broadcast complexity is $O(nk)$ bits.

Random-Bit Gates. We list three protocols for jointly generating encrypted random bits. Each protocol starts the same. For $i = 1, \ldots, n$, party P_i generates a uniformly random bit $b_i \in \{0, 1\}$ and broadcasts $[\![b_i]\!]$ together with a Σ-proof to show knowledge of b_i and that b_i is indeed a bit. To combine bits $[\![b_i]\!]$ into a joint random bit $[\![b]\!]$, with $b = \oplus_{i=1}^{n} b_i$, we mention three options:

- use of an unbounded fan-in multiplication gate to compute $[\![b]\!]$ in constant rounds, see [CDN01];
- use of $O(n)$ multiplication gates to compute $[\![b]\!]$ in $O(\log n)$ rounds;
- use of $O(n)$ private-multiplier gates to compute $[\![b]\!]$ in $O(n)$ rounds.

The broadcast complexity is $O(n^2 k)$ bits for the first two options (but the hidden constant is significantly higher for the constant rounds protocol), and $O(nk)$ bits for the last option. Note that the joint random bits can be computed in preprocessing, if so desired, creating an opportunity to use more rounds for a lower broadcast complexity (and for a lower computational complexity too).

Addition and Subtraction Circuits. Given encrypted bit representations $[\![x_0]\!], \ldots, [\![x_{m-1}]\!]$ and $[\![y_0]\!], \ldots, [\![y_{m-1}]\!]$ of two numbers x, y, an addition circuit essentially computes the bits of $x+y$, given by $[\![z_0]\!], \ldots, [\![z_{m-1}]\!], [\![c_{m-1}]\!]$ as follows:

$$z_i = x_i + y_i + c_{i-1} - 2c_i$$
$$c_{-1} = 0, \quad c_i = x_i y_i + x_i c_{i-1} + y_i c_{i-1} - 2 x_i y_i c_{i-1}.$$

A similar circuit can be used for subtraction.

Depending on the scenario such a circuit can be refined in various ways. We mention two extreme options. One can optimize for broadcast complexity and use $O(m)$ rounds to compute an addition with broadcast complexity $O(mnk)$ bits. At the other extreme, one can optimize for round complexity and use $O(1)$ rounds as shown in [DFK+06]; their $O(1)$-depth circuit for bitwise addition can also be used in the threshold homomorphic setting, resulting in a broadcast complexity of $O(m \log mnk)$ bits.

Equality and Comparison Circuits Let $[C]$ denote the Iverson bracket defined by $[C] = 1$ if $C \Leftrightarrow$ true and $[C] = 0$ otherwise. Given encrypted bit representations $[\![x_0]\!], \ldots, [\![x_{m-1}]\!]$ and $[\![y_0]\!], \ldots, [\![y_{m-1}]\!]$ of two numbers x, y, equality and comparison circuits compute $[\![[x = y]]\!]$ or $[\![[x < y]]\!]$, respectively. A $O(\log m)$-depth circuit for equality is straightforward, leading to round complexity of $O(\log m)$ and broadcast complexity of $O(mnk)$ bits. Interestingly, $O(1)$ rounds and $O(mnk)$ bits are also achievable as shown in [DFK+06]; again their methods carry over to the threshold homomorphic setting.

3 LSB Gate

We present a protocol for securely computing the least significant bit, which on input $[\![x]\!]$, outputs an encryption $[\![x_0]\!]$. To obtain a particularly efficient solution, we assume that x is a bounded value, that is, $0 \le x < 2^m$ where the value of m is restricted as a function of N, the number of parties n, and a security parameter κ. The parameter κ is chosen such that $2^{-\kappa}$ is negligible. The restriction on m is that $m + \kappa + \log_2 n < \log_2 N$. In practice this is not a severe restriction. For example, if N is a 1024-bit modulus, $\kappa = 100$, and $n = 16$, then m is bounded above by 920.

3.1 Protocol

The idea is to jointly generate a random value $[\![r]\!]$ and to decrypt $[\![x + r]\!]$ such that (i) $y = x + r$ is statistically indistinguishable from random and (ii) $[\![x_0]\!]$ can be recovered from y_0 and $[\![r_0]\!]$. To this end, the random value r will be generated in the form $r = r_0 + 2r_*$, where r_0 is a bit and r_* is an integer value from a sufficiently large range.

LSB Gate

1. The parties jointly generate a random bit $[\![r_0]\!]$, using a random-bit gate. In parallel, each party P_i chooses $r_{*,i} \in_R \{0, \ldots, 2^{m+\kappa-1} - 1\}$ and broadcasts $[\![r_{*,i}]\!]$ accompanied with a range proof that the encryption is correctly formed. The encryption $[\![r_*]\!]$ with $r_* = \sum_{i=1}^{n} r_{*,i}$ is publicly computed.

2. The encryption $[\![x]\!][\![r_0]\!][\![r_*]\!]^2$ is formed and jointly decrypted to reveal the value $y = x + r$, where $r = r_0 + 2r_*$.
3. The output is $[\![r_0 \oplus y_0]\!]$, which can be computed publicly from $[\![r_0]\!]$ and y_0, as $r_0 \oplus y_0 = r_0 + y_0 - 2r_0y_0$.

The broadcast complexity incurred by the range proofs in the first step of the protocol is limited to $O(nk)$ bits, assuming an efficient range proof is used, as mentioned in Section 2.2. The broadcast complexity of the entire protocol depends on the broadcast complexity of the random-bit gate used for generating $[\![r_0]\!]$, and varies between $O(n^2k)$ bits (and $O(1)$ round complexity) and $O(nk)$ bits (and $O(n)$ round complexity).

We note that once x_0 is computed, the next bit of x can be computed by applying the protocol to $[\![x_*]\!]$, with $x_* = (x - x_0)/2$. Indeed, the homomorphic property implies $[\![x_*]\!] = ([\![x]\!]/[\![x_0]\!])^{1/2}$, where $1/2 = (N+1)/2$ is the inverse of 2 modulo N. This way all of the bits of x can be recovered.

3.2   Security

The value output by the protocol is correct because $y_0 = x_0 \oplus r_0$, as the addition $y = x + r$ is computed over the integers (the limited size of x and r ensures that $0 \le x + r < N$).

Next, we show that the protocol can be simulated, assuming that the random-bit gate can be simulated when given an output $[\![\tilde{r}_0]\!]$. Of course, the bit $\tilde{r}_0$ should be distributed uniformly.

Theorem 1. *On input $[\![x]\!]$ and $[\![x_0]\!]$, where $0 \le x < 2^m$, the LSB gate can be simulated in a statistically indistinguishable manner.*

Proof. Let $x_* = (x - x_0)/2$. Then $[\![x_*]\!] = ([\![x]\!]/[\![x_0]\!])^{1/2}$.

To argue that no additional information on x is leaked we present the following simulation of the protocol. The simulation takes as input encryptions $[\![x]\!]$ and $[\![x_0]\!]$. Given this information, the simulator is able to generate a complete transcript for the protocol, for which the distribution is exactly the same as in real executions of the protocol. Note that we have to ensure that the simulator "knows" the plaintext for the joint decryption step in the middle of the protocol, as the simulator for the threshold decryption protocol needs both a ciphertext and the corresponding plaintext as input.

Assume w.l.o.g. that parties $P_1, \ldots, P_t$ are corrupted. The simulator chooses $\tilde{y}_0 \in_R \{0, 1\}$. We use the simulator for the random-bit gate to obtain a simulation for output $[\![\tilde{r}_0]\!]$ with $\tilde{r}_0 = x_0 \oplus \tilde{y}_0$. This encryption can be computed from the given encryption $[\![x_0]\!]$, using $\tilde{y}_0$. In parallel, the simulator lets the adversary run the first step of the protocol for parties $P_1, \ldots, P_t$, each time rewinding the proofs of knowledge used to extract the values $r_{*,1}, \ldots, r_{*,t}$. Subsequently, the simulator runs the first step of the protocol for parties $P_{t+1}, \ldots, P_{n-1}$ as in the real protocol, resulting in the values $r_{*,t+1}, \ldots, r_{*,n-1}$. For party P_n, however, the simulator first generates $s \in_R \{0, \ldots, 2^{m+\kappa-1} - 1\}$, and sets $\tilde{r}_{*,n} = s - x_* - (x_0 + \tilde{r}_0 - \tilde{y}_0)/2$. Then $[\![\tilde{r}_{*,n}]\!]$ can be computed using $[\![x_0]\!]$ and $[\![x_*]\!]$. The range proof for $[\![\tilde{r}_{*,n}]\!]$ is generated using the simulator for these proofs. Finally, the simulator sets $\tilde{r}_* = \sum_{i=1}^{n-1} r_{*,i} + \tilde{r}_{*,n}$.

As a result, we have that

$$[\![x+\tilde{r}_0+2\tilde{r}_*]\!] = [\![x+\tilde{r}_0+2\sum_{i=1}^{n-1} r_{*,i}+2s-2x_*-(x_0+\tilde{r}_0-\tilde{y}_0)]\!] = [\![\tilde{y}_0+2(\sum_{i=1}^{n-1} r_{*,i}+s)]\!],$$

for which the simulator knows the decryption, as required.

By construction, the values are all consistent. We need to argue that all the probability distributions for these values are correct too. Clearly, the values for parties $P_1, \ldots, P_{n-1}$ follow the right distribution. Since $\tilde{y}_0$ is a uniformly random bit, so is $\tilde{r}_0$. Finally, the distribution of $r_{*,n} = s - x_* - (x_0 + \tilde{r}_0 - \tilde{y}_0)/2$ is statistically indistinguishable from the uniform distribution on $\{0, \ldots, 2^{m+\kappa-1} - 1\}$, because $s \in_R \{0, \ldots, 2^{m+\kappa-1} - 1\}$ and the term $x_* + (x_0 + \tilde{r}_0 - \tilde{y}_0)/2 = x_* + x_0(1 - \tilde{y}_0)$ is much smaller than it (bounded above by 2^m). More precisely, the statistical distance is bounded by $2^{-\kappa+1}$ (see Appendix A).

4   LSBs Gate

Next, we consider the more general case of extracting the ℓ least significant bits of x rather than just the least significant one. We describe the protocol for the case that *all* of the bits of x are computed (case $\ell = m$); the case $\ell < m$ can be handled by combining the techniques of this section and the previous section.

4.1   Protocol

Let $m + \kappa + \log_2 n < \log_2 N$, as before. On input $[\![x]\!]$, where $0 \leq x < 2^m$, the following protocol computes $[\![x_0]\!], \ldots, [\![x_{m-1}]\!]$ securely. The idea is to jointly generate a random value $[\![r]\!]$ and to decrypt $[\![x + r]\!]$ such that (i) $y = x + r$ is statistically indistinguishable from random and (ii) $[\![x_0]\!], \ldots, [\![x_{m-1}]\!]$ can be recovered from $y_0, \ldots, y_{m-1}$ and $[\![r_0]\!], \ldots, [\![r_{m-1}]\!]$. To this end, the value r will be generated in the form $r = \sum_{j=0}^{m-1} r_j 2^j + 2r_*$, where $r_0, \ldots, r_{m-1}$ are bits and r_* is an integer value from a sufficiently large range.

For technical reasons, we will actually compute $y = x - r$ in (i) and use an addition circuit to perform step (ii).

LSBs Gate

1. The parties jointly generate random bits $[\![r_0]\!], \ldots, [\![r_{m-1}]\!]$, using m random-bit gates. In parallel, each party P_i chooses $r_{*,i} \in_R \{0, \ldots, 2^{m+\kappa-1} - 1\}$ and broadcasts $[\![r_{*,i}]\!]$ accompanied with a range proof that the encryption is correctly formed. The encryption $[\![r_*]\!]$ with $r_* = \sum_{i=1}^n r_{*,i}$ is publicly computed.
2. The encryption $[\![x - r]\!]$ is formed and jointly decrypted to reveal the signed value $y = x - r \in (-n/2, n/2)$, where $r = \sum_{j=0}^{m-1} r_j 2^j + r_* 2^m$. The signed value y is computed such that $y \equiv x - r \pmod{n}$.

3. Let $y_0, \ldots, y_{m-1}$ denote the binary representation of $y \bmod 2^m$. An addition circuit for inputs $y_0, \ldots, y_{m-1}$ (public) and $[\![r_0]\!], \ldots, [\![r_{m-1}]\!]$ is used to produce an output of m encrypted bits (ignoring the final carry bit, hence computing modulo 2^m).

Note that the probability that $y \geq 0$ at step 4.1 is negligible.

The broadcast complexity of the protocol depends on the broadcast complexity of the random-bit gate used for generating $[\![r_0]\!], \ldots, [\![r_{m-1}]\!]$, and varies between $O(mn^2k)$ bits (and $O(1)$ round complexity) and $O(mnk)$ bits (and $O(n)$ round complexity).

4.2   Security

The main proof obligation is to show that the protocol can be simulated. We would like to do so given just a matching pair of inputs and outputs, which in this case consists of the encryptions $[\![x]\!]$ and $[\![x_0]\!], \ldots, [\![x_{m-1}]\!]$. Unfortunately, such a simulation will not succeed for the LSBs protocol because the protocol is releasing *additional* encryptions apart from the encrypted bits of x. The problem lies in the addition circuit, as used in the final step of the protocol.

The problem is that an addition circuit (see Section 2.3) releases encrypted carry bits, next to the encrypted output bits. Even with full knowledge of r, these encrypted carry bits can, in general, not be computed from $[\![x_0]\!], \ldots, [\![x_{m-1}]\!]$, as this would imply that encryptions such as $[\![x_0 x_1]\!]$, or other encrypted monomials, can be computed in polynomial time. What is more, the number of monomials that are possibly needed in a simulation is equal to 2^m, hence exponential m.

We observe, however, that to prove simulatability in the framework of [CDN01] it suffices to perform a simulation for input/output pairs of a special form. This is a consequence of the fact that the security proof of [CDN01] (full version) centers around the construction of the so-called YAD^b distribution, which is defined as a function of an encrypted bit $[\![b]\!]$.

In terms of the YAD^b distribution, the structure of the security proof is as follows, following an ideal/real approach. The YAD^0 distribution is identical to the distribution in the ideal case, whereas the YAD^1 distribution is statistically indistinguishable from the distribution in the real case. Consequently, if an adversary is able to distinguish the ideal/real cases, it follows that the adversary is able to distinguish the YAD^0 distribution from the YAD^1 distribution. But the choice between these two distributions is entirely determined by the value of the encrypted bit b. Hence, a distinguisher for the ideal/real cases implies a distinguisher for the underlying cryptosystem, and it does so in a *tight* way (without loss in success probability for the distinguisher).

The special form for the input/output pairs is given by $[\![\tilde{x}]\!] = [\![(1-b)x + bx']\!]$, where x and x' are given in the clear, but bit b is only given as an encryption $[\![b]\!]$. The values x and x' correspond to the values arising in the YAD^0 case and the YAD^1 case, respectively, and are *both* known to the simulator. The x-values correspond to fake values used to set up a consistent simulated view ($b = 0$) and the x'-values correspond to the values used to set up a consistent real view ($b = 1$).

Thus, the security is stated in a less general way, but still sufficiently general to match the (adapted) framework of [CDN01]. Stating the security of a gate (or, sub-circuit) in this way allows one to capture the security in a modular way, while retaining the tightness of the overall reduction. Clearly, this idea is more widely applicable, beyond the gates we are considering in this paper.

Theorem 2. *Given input values x, x' with $0 \leq x, x' < 2^m$ and an encryption $[\![b]\!]$ with $b \in \{0, 1\}$, the LSBs gate can be simulated statistically indistinguishably for input $[\![\tilde{x}]\!] = [\![(1 - b)x + bx']\!]$.*

Proof. The goal is to generate a simulated transcript for input $[\![\tilde{x}]\!] = [\![(1 - b)x + bx']\!]$, which is to the adversary statistically indistinguishable from a real transcript. The values of x, x' and $[\![b]\!]$ are available to the simulator.

Assume w.l.o.g. that parties $P_1, \ldots, P_t$ are corrupted. Pick $\tilde{r}_0, \ldots, \tilde{r}_{m-1} \in_R \{0, 1\}$ and simulate the generation of these random bits. The simulator extracts/generates the values $r_{*,i}$ for the corrupted/honest parties P_i ($1 \leq i < n$). The simulator chooses $s \in_R \{0, \ldots, 2^{m+\kappa} - 1\}$ and sets $r_{*,n} = s - x - \sum_{j=0}^{m-1} \tilde{r}_j 2^j$ and $r'_{*,n} = s - x' - \sum_{j=0}^{m-1} \tilde{r}_j 2^j$. The distributions will be statistically close to the uniform distribution on $\{0, \ldots, 2^{m+\kappa} - 1\}$.

By construction, the simulator knows how to decrypt $x - r$ and $x' - r'$, where $r = \sum_{j=0}^{m-1} \tilde{r}_j 2^j + r_* 2^m$ and $r' = \sum_{j=0}^{m-1} \tilde{r}_j 2^j + r'_* 2^m$, and where $r_* = \sum_{i=1}^{n-1} r_{*,i} + r_{*,n}$ and $r'_* = \sum_{i=1}^{n-1} r_{*,i} + r'_{*,n}$

To complete the proof we need to assign the correct values to all the wires in the addition circuit, consistent with either $b = 0$ or $b = 1$. In both cases, the first input to the addition circuit is $s_0, \ldots, s_{m-1}$. The second input is either $r_0, \ldots, r_{m-1}$ or $r'_0, \ldots, r'_{m-1}$, and the corresponding output is $x_0, \ldots, x_{m-1}$ or $x'_0, \ldots, x'_{m-1}$. Now, for each wire, two values can be computed from these values. If values v and v' are thus computed for such a wire, we assign the encryption $[\![(1 - b)v + bv']\!]$ to the wire.

This leads to a consistent assignment of encryptions to all of the wires. And, therefore the simulators for the multiplication gates constituting the addition circuit can be used to complete the simulation.

The values generated by the simulator are all consistent. Moreover, the distribution of $-s$ is statistically indistinguishable from the value of y used in the real protocol.

5   BITREP Gate

In this section we consider the case that x is any value in the range $[0, N)$, where N is the value of the Paillier modulus. We first present a protocol for jointly generating a random value $r \in_R [0, N)$, given by its bits $[\![r_0]\!], \ldots, [\![r_{m-1}]\!]$, with m denoting the bit length of N. Subsequently, we present the BITREP gate, which converts $[\![x]\!]$ into $[\![x_0]\!], \ldots, [\![x_{m-1}]\!]$.

The protocol for generating a random value $r \in_R [0, N)$ uses the basic protocol for jointly generating m random bits between parties $P_1, \ldots, P_n$. We then test whether the integer represented by these m bits is in the range $[0, N)$:

1. The parties jointly generate random bits $[\![r_0]\!], \ldots, [\![r_{m-1}]\!]$, using m random-bit gates.
2. A comparison circuit for encrypted inputs $[\![r_0]\!], \ldots, [\![r_{m-1}]\!]$ and public inputs $N_0, \ldots, N_{m-1}$, denoting the bits of N, is used to compute $[\![[r < N]]\!]$, where $r = \sum_{j=0}^{m-1} r_j 2^j$.
3. The value $[\![[r < N]]\!]$ is decrypted to see if r is in range; if not, go back to the first step.

The average number of iterations is bounded above by 2.

The protocol for converting $[\![x]\!]$ into $[\![x_0]\!], \ldots, [\![x_{m-1}]\!]$, where $0 \le x < N$, now runs as follows.

BITREP Gate

1. The parties generate encrypted bits $[\![r_0]\!], \ldots, [\![r_{m-1}]\!]$ of a random number $0 \le r < N$.
2. The parties compute $[\![x]\!] \prod_{j=0}^{m-1} [\![r_j]\!]^{2^j}$ and perform a threshold decryption to obtain $y = x + r \bmod N$, $0 \le y < N$.
3. Using a subtraction circuit with $y_0, \ldots, y_{m-1}$ and $[\![r_0]\!], \ldots, [\![r_{m-1}]\!]$ as inputs, the parties determine the bit representation $[\![z_0]\!], \ldots, [\![z_m]\!]$ of the value $z = x$ or $z = x - N$, where z_m is a sign bit.
4. The parties reduce the value of z modulo N, by adding $N z_m$ to z using an addition circuit with inputs $[\![(N z_m)_0]\!], \ldots, [\![(N z_m)_{m-1}]\!]$ and $[\![z_0]\!], \ldots, [\![z_{m-1}]\!]$.

Note that the equality $y = x + r$ holds in $\mathbb{Z}_n$ but not necessarily in $\mathbb{Z}$. But if $y \ne x + r$ over the integers, then it follows that $y = x + r - N$ must hold over the integers, since $0 \le x < N$ and $0 \le r < N$. In step 3, the case $z = x$ occurs exactly when $y = x + r$ over the integers, and the case $z = x - N$ occurs when $y = x + r - N$.

The security is proved similar as in the previous section.

Theorem 3. *Given input values x, x' with $0 \le x, x' < N$ and an encryption $[\![b]\!]$ with $b \in \{0, 1\}$, the BITREP gate can be simulated statistically indistinguishably for input $[\![\tilde{x}]\!] = [\![(1 - b)x + bx']\!]$.*

6   Applications

The result of [CDN01] shows how to efficiently evaluate arithmetic circuits composed of addition/subtraction gates and multiplication/division gates defined over $\mathbb{Z}_N$. This way large numbers can be handled, practically independent of the size of the numbers. This contrasts favorably with approaches based on Boolean circuits, where arithmetic is done in a bitwise fashion. However, as argued in the introduction, the potential of arithmetic circuits is limited when some (inherently) bitwise operations are required as well. Without binary conversion gates, one is forced to perform the entire computation using a (large) Boolean circuit.

The relational operators such as $<$ and $=$ are typically handled by representing the numbers in binary form. Another important example is exponentiation, where one wishes to compute $[\![x^y]\!]$ securely, given $[\![x]\!]$ and $[\![y]\!]$. Using a method of repeated squaring one may compute $[\![x^y]\!]$ using $O(m)$ multiplication gates, once $[\![y]\!]$ is converted into binary, given by its encrypted bits $[\![y_0]\!], \ldots, [\![y_{m-1}]\!]$, say.

As an interesting application we close this paper with a discussion of private biometric authentication. This topic received quite some attention during the last years, see e.g., [DRS04, TG04, KAMR04]. The goal of private biometric authentication is to identify or authenticate people based on their physical characteristics (fingerprint, iris, ...) without revealing any information on these personal characteristics to the verifier or an attacker. This problem has been investigated in the information theoretical setting by several authors [TG04, DRS04]. They gave general constructions (using "helper data" and "fuzzy extractors" resp.). At the same time, it was shown that perfect privacy cannot be achieved from an information theoretic point of view. It is therefore natural to explore whether a full privacy preserving and efficient biometric authentication scheme can be constructed in the cryptographic setting. Below, we briefly describe how this is achieved.

The heart of the system is formed by a set of servers which correspond to the parties $P_1, \ldots, P_n$ sharing the private key for a threshold Paillier cryptosystem. These servers will match encrypted biometric templates as obtained during enrollment of the users against encrypted biometric templates as measured by sensors as part of an authentication protocol. Since the sensors are typically lightweight devices, the goal is to minimize the computational load for the sensors.

Authentication of a user will succeed if the biometric template $Y = (y_1, \ldots, y_m)$ measured by a sensor is sufficiently close to the biometric template $X = (x_1, \ldots, x_m)$ obtained during enrollment. For instance, assuming that the biometric templates are actually bit strings in $\{0, 1\}^m$, a possible similarity measure is the Hamming distance between the bit strings: if $d_H(X, Y) < T$, where T is a predetermined threshold, then X and Y are said to match.

The BITREP gate can now be used as follows. To minimize the work for the sensor we let the sensor first convert the measured biometric template Y to an integer $y \in \{0, \ldots, 2^m - 1\}$, using an obvious mapping. The sensor is then required to release the encrypted value $[\![y]\!]$ together with the claimed identifier for the person being authenticated. At the server side, the encrypted bits of Y are recovered using a BITREP gate (or, an LSBs gate if appropriate), before running a secure matching protocol on the encrypted bits of X and Y.

Acknowledgements. We would like to thank the anonymous referees for their helpful comments.

References

[ACS02]   J. Algesheimer, J. Camenisch, and V. Shoup. Efficient computation modulo a shared secret with application to the generation of shared safe-prime products. In *Advances in Cryptology—CRYPTO '02*, volume 2442 of *Lecture Notes in Computer Science*, pages 417–432, Berlin, 2002. Springer-Verlag.

[Bou00]   F. Boudot. Efficient proofs that a committed number lies in an interval. In *Advances in Cryptology—EUROCRYPT '00*, volume 1807 of *Lecture Notes in Computer Science*, pages 431–444, Berlin, 2000. Springer-Verlag.

[CDN01]   R. Cramer, I. Damgård, and J.B. Nielsen. Multiparty computation from threshold homomorphic encryption. In *Advances in Cryptology—EUROCRYPT '01*, volume 2045 of *Lecture Notes in Computer Science*, pages 280–300, Berlin, 2001. Springer-Verlag. Full version eprint.iacr.org/2000/055, October 27, 2000.

[CFSY96]   R. Cramer, M. Franklin, B. Schoenmakers, and M. Yung. Multi-authority secret ballot elections with linear work. In *Advances in Cryptology—EUROCRYPT '96*, volume 1070 of *Lecture Notes in Computer Science*, pages 72–83, Berlin, 1996. Springer-Verlag.

[CS02]   R. Cramer and V. Shoup. Universal hash proofs and a paradigm for adaptive chosen ciphertext secure public-key encryption. In *Advances in Cryptology—EUROCRYPT '02*, volume 2332 of *Lecture Notes in Computer Science*, pages 45–64, Berlin, 2002. Springer-Verlag.

[DF02]   I. Damgård and E. Fujisaki. A statistically-hiding integer commitment scheme based on groups with hidden order. In *Advances in Cryptology—ASIACRYPT '02*, volume 2501 of *Lecture Notes in Computer Science*, pages 125–142, Berlin, 2002. Springer-Verlag.

[DFK+06]   I. Damgård, M. Fitzi, E. Kiltz, J.B. Nielsen, and T. Toft. Unconditionally secure constant-rounds multi-party computation for equality, comparison, bits and exponentiation. In *Proc. 3rd Theory of Cryptography Conference, TCC 2006*, volume 3876 of *Lecture Notes in Computer Science*, pages 285–304, Berlin, 2006. Springer-Verlag.

[DJ01]   I. Damgård and M. Jurik. A generalisation, a simplification and some applications of Paillier's probabilistic public-key system. In *Public Key Cryptography—PKC '01*, volume 1992 of *Lecture Notes in Computer Science*, pages 119–136, Berlin, 2001. Springer-Verlag.

[DJ02]   I. Damgård and M. Jurik. Client/server tradeoffs for online elections. In *Public Key Cryptography—PKC '02*, volume 2274 of *Public Key Cryptography—PKC '*, pages 125–140, Berlin, 2002. Springer-Verlag.

[DJ03]   I. Damgård and M. Jurik. A length-flexible threshold cryptosystem with applications. In *ACISP 2003*, volume 2727 of *Lecture Notes in Computer Science*, pages 350–364, Berlin, 2003. Springer-Verlag.

[DRS04]   Y. Dodis, M. Reyzin, and A. Smith. Fuzzy extractors: How to generate strong keys from biometrics and other noisy data. In *Advances in Cryptology—EUROCRYPT '04*, volume 3027 of *Lecture Notes in Computer Science*, pages 523–540, Berlin, 2004. Springer-Verlag.

[FO97]   E. Fujisaki and T. Okamoto. Statistical zeroknowledge protocols to prove modular polynomial relations. In *Advances in Cryptology—CRYPTO '97*, volume 1294 of *Lecture Notes in Computer Science*, pages 16–30, Berlin, 1997. Springer-Verlag.

[FPS00]   P.-A. Fouque, G. Poupard, and J. Stern. Sharing decryption in the context of voting or lotteries. In *Financial Cryptography 2000*, volume 1962 of *Lecture Notes in Computer Science*, pages 90–104, Berlin, 2000. Springer-Verlag.

[GMPY06]   J. Garay, P. MacKenzie, M. Prabhakaran, and K. Yang. Resource fairness and composability of cryptographic protocols. In *Proc. 3rd Theory of Cryptography Conference, TCC 2006*, volume 3876 of *Lecture Notes in Computer Science*, pages 404–428, Berlin, 2006. Springer-Verlag.

[KAMR04] F. Kerschbaum, M.J. Atallah, D. M'Raïhi, and J.R. Rice. Private fingerprint verification without local storage. In *Proceedings of the first International Conference on Biometric Authentication*, volume 3072 of *Lecture Notes in Computer Science*, pages 387–394, Berlin, 2004. Springer-Verlag.

[Lip03] H. Lipmaa. On diophantine complexity and statistical zero-knowledge arguments. In *Advances in Cryptology—ASIACRYPT '03*, volume 2894 of *Lecture Notes in Computer Science*, pages 398–415, Berlin, 2003. Springer-Verlag.

[Pai99] P. Paillier. Public-key cryptosystems based on composite degree residuosity classes. In *Advances in Cryptology—EUROCRYPT '99*, volume 1592 of *Lecture Notes in Computer Science*, pages 223–238, Berlin, 1999. Springer-Verlag.

[ST04] B. Schoenmakers and P. Tuyls. Practical two-party computation based on the conditional gate. In *Advances in Cryptology—ASIACRYPT '04*, volume 3329 of *Lecture Notes in Computer Science*, pages 119–136, Berlin, 2004. Springer-Verlag.

[TG04] P. Tuyls and J. Goseling. Capacity and examples of template protecting biometric authentication systems. In *Proceedings of Biometric Authentication Workshop*, volume 3087 of *Lecture Notes in Computer Science*, pages 158–170, Berlin, 2004. Springer-Verlag.

A Statistical Distance

We use elementary results on statistical distance in our proofs. As an illustration we prove the following one.

Lemma 1. *Let M and K be positive integers, $M \leq K$. Let random variable X take on values in $\{0, \ldots, M-1\}$, and let random variable U be uniform on $\{0, \ldots, K-1\}$. Then $\Delta(U, X+U) \leq (M-1)/K$, and this upper bound is tight.*

Proof. For any w we have that $\Pr[X+U=w] = \sum_{a=0}^{M-1} \Pr[X=a]\Pr[U=w-a]$, hence that $0 \leq \Pr[X+U=w] \leq 1/K$, and, if $M-1 \leq w < K$, that $\Pr[X+U=w] = 1/K$. Thus,

$$\Delta(U, X+U) = \tfrac{1}{2} \sum_{w=0}^{M+K-2} |\Pr[U=w] - \Pr[X+U=w]|$$
$$\leq \tfrac{1}{2} \left(\sum_{w=0}^{M-2} |1/K - 0| + \sum_{w=M-1}^{K-1} |1/K - 1/K| + \sum_{w=K}^{M+K-2} |0 - 1/K| \right)$$
$$= (M-1)/K.$$

Take $X = M - 1$ (constant) to conclude that this bound is tight.

Hence, $\Delta(U, X+U)$ is small if $K \gg M$. For instance, if K is exponential in a security parameter k and M is polynomial in k, then the statistical distance is negligible in k. In particular, if $M = 2^m$ and $K = 2^{m+\kappa}$, then $\Delta < 2^{-\kappa}$.

Information-Theoretic Conditions for Two-Party Secure Function Evaluation

Claude Crépeau[1,*], George Savvides[1,*], Christian Schaffner[2,**],
and Jürg Wullschleger[3,***]

[1] McGill University, Montréal, QC, Canada
{crepeau, gsavvi1}@cs.mcgill.ca
[2] BRICS, University of Århus, Denmark
chris@brics.dk
[3] ETH Zürich, Switzerland
wjuerg@inf.ethz.ch

Abstract. The standard security definition of unconditional secure function evaluation, which is based on the ideal/real model paradigm, has the disadvantage of being overly complicated to work with in practice. On the other hand, simpler ad-hoc definitions tailored to special scenarios have often been flawed. Motivated by this unsatisfactory situation, we give an information-theoretic security definition of secure function evaluation which is very simple yet provably equivalent to the standard, simulation-based definitions.

1 Introduction

1.1 Secure Function Evaluation

Secure function evaluation is a cryptographic task originally introduced by Yao in [30]. In essence, this task enables a set of mutually distrustful parties without access to a trusted intermediary to jointly compute the output of a function f without any party revealing any information about its input or output to the other parties beyond what these parties can infer from their own inputs and outputs. Goldreich, Micali and Wigderson [21] showed how to achieve this for any function f in a computationally secure way. Schemes ensuring unconditional security were subsequently provided by Ben-Or, Goldwasser and Wigderson [3] and independently by Chaum, Crépeau and Damgård [12].

Micali and Rogaway [25] and Beaver [2] proposed formal security definitions for secure function evaluation. Both definitions were inspired by the simulation paradigm used by Goldwasser, Micali and Rackoff [22] to define zero-knowledge proofs of knowledge. In a nutshell, to each real protocol computing f we associate a two-step procedure in an *ideal* model, where each party simply forwards

---

* Supported in part by NSERC, MITACS, and CIAR.
** Supported by the EC-Integrated Project SECOQC, No: FP6-2002-IST-1-506813.
*** Supported by Canada's NSERC, Québec's FQRNT, and Switzerland's SNF.

S. Vaudenay (Ed.): EUROCRYPT 2006, LNCS 4004, pp. 538–554, 2006.

its input to a trusted party which in turn computes f and distributes the relevant outputs to the parties. The real protocol is deemed secure if any adversary attacking the protocol has a counterpart in the ideal model that achieves a similar result simply by processing the input prior to forwarding it to the trusted party, and then by processing the output it receives from it. In other words, a protocol is secure if any attack can be simulated in the much more restrictive ideal model. Such protocols secure in *the ideal/real model paradigm* were later shown to be *sequentially composable* in the sense that the composition of two or more secure protocols is itself a secure protocol. The sequential composability of secure protocols was further explored by Canetti [9, 10] and Goldreich [20].

Canetti [11] also defined *universal composability*, an even stronger security requirement that guarantees that protocols satisfying it can be securely composed *concurrently* in any environment. A similar security definition was provided independently by Backes, Pfitzmann and Waidner [1]. Unfortunately, however appealing the properties of these security definitions may be, they are too strong to allow even basic tasks such as bit commitment to be realized without further assumptions. For this reason, we will limit ourselves to the simpler definition given by Goldreich [20].

1.2 Oblivious Transfer

1-out-of-n string oblivious transfer, denoted $\binom{n}{1}$-OT^k, is a primitive that allows a sender Alice to send one of n binary strings of length k to a receiver Bob. The primitive allows Bob to receive the string of his choice while concealing this choice from (possibly dishonest) Alice. On the other hand, the primitive guarantees that (any dishonest) Bob cannot obtain information about more than one of the strings, including partial joint information on two or more strings.

The first variant of oblivious transfer was introduced by Wiesner [28]. Independently, Rabin re-introduced oblivious transfer in [27] and demonstrated its potential as a cryptographic tool. Its applicability to multi-party computation was shown by Even, Goldreich and Lempel in [19]. It has since been proved that oblivious transfer is in fact sufficient by itself to securely compute any function [23]. More completeness results followed in [14], [15] and [24].

1.3 Contributions

The motivation behind our work was to come up with a general information-theoretic security definition to replace the various ad-hoc definitions proposed in the past for specific cryptographic primitives. To this end, we adopt the standard security definition based on the ideal/real model paradigm of Goldreich [20] for computationally-bounded parties, and adapt it to a model where the parties are allowed to be computationally unbounded and to use independent sources of randomness such as channels. We then distill the relevant security properties of the ideal model into a set of information-theoretic conditions, which we use as a basis for constructing our new formal definition of security. We prove that despite its apparent simplicity, our definition is in fact *equivalent* to the original based on the ideal/real model paradigm. We then examine the important special case of oblivious transfer, and show that in this case, the resulting security requirements can be significantly simplified. We also revisit some of the information-theoretic definitions of security used in the past and point out subtle flaws that some of them contain. As an illustration of the usefulness of our definitions, we give a simple proof for the protocol presented in [29] that optimally inverts $\binom{2}{1}$-OT.

1.4   Shortcomings of Previously Proposed Security Definitions for Oblivious Transfer

We revisit some information-theoretic definitions for oblivious transfer that appear in the literature and list some of their shortcomings. Our examples demonstrate that coming up with the 'right' information-theoretic definition is a delicate task, which is the reason why in this paper we aim for a security definition provably equivalent to the standard definition based on the ideal/real model paradigm.

Random Inputs. In [18], only oblivious transfer with random inputs is considered, thereby restricting the scope of the proposed definitions to only a few special cases.

Problems with the Security for the Receiver. In [5, 26], the definition of security for the receiver requires that the sender's view be independent of the receiver's input. This is often unattainable: in the most general case, where we assume that there is a known dependency between the inputs, no protocol can satisfy the above security condition since the sender's input, which is always part of his own view, will be correlated with the input of the receiver. The definition should instead require that the two variables be independent *given the sender's input*.

Problems with the Security for the Sender. The security for the sender is more difficult to formalize correctly. In addition to problems analogous to the ones presented above for the definition of security for the receiver ([5, 26]), there are several commonly encountered difficulties:

- In [6, 16] a dishonest receiver is only allowed to change his input in a *deterministic* way. Specifically, the random variable C' indicating the receiver's *effective input* (i.e., the bit he eventually obtains) must be a deterministic function of the input C, in contrast to the ideal model where C' can be chosen probabilistically by the dishonest receiver.
- In [7] the random variable C' may depend on the honest sender's input, which is impossible in the ideal model. Furthermore, the view V of the dishonest receiver is required to be independent of the honest sender's input X conditioned on the original input C and the receiver's output $X_{C'}$, *but not on C'*. This definition will hence admit some clearly insecure protocols.

For example, suppose the dishonest receiver picks $C' \in \{0, 1\}$ uniformly at random (independently of C) and the protocol allows him to output $V = (X_{C'}, X_{1-C'} \oplus C')$. While it is true that V is independent of X_0, X_1 given C, X'_C, no such protocol can be simulated in the ideal model since both inputs can be deduced from C' and V.

Abort. In [6, 16, 7], the honest player is allowed to abort the protocol. However, it is possible that the dishonest player gets some information *before* the honest player aborts, or that the fact of aborting itself provides information about the honest player's inputs.

A correct definition is given in [17] in the context of the bounded-storage model. However, this definition is overly complicated and requires a special *setup stage*, which is in general not present in OT protocols.

1.5   Preliminaries

Let X, Y, and Z be three random variables. We will often use expressions of the form

$$I(X; Y \mid Z) = 0 ,$$

where I is the conditional mutual Shannon information. This means that X *and Y are independent, given Z*. The same condition can also be expressed by saying that X, Y *and Z form a Markov-chain*,

$$X \leftrightarrow Z \leftrightarrow Y ,$$

or by

$$P_{Y|ZX} = P_{Y|Z} .$$

By the *chain rule* for mutual information we have

$$I(X; YW \mid Z) = I(X; W \mid Z) + I(X; Y \mid WZ) .$$

The *information processing inequality* says that local computation cannot increase mutual information. In other words, for any probabilistic f we have

$$I(X; Y \mid Z) \geq I(f(X); Y \mid Z) .$$

The *statistical distance* or *variational distance* between the distributions of two random variables X and Y over the same domain $\mathcal{V}$ is defined as

$$\delta(X, Y) = \frac{1}{2} \sum_{v \in \mathcal{V}} \left| \Pr[X = v] - \Pr[Y = v] \right|.$$

We also use the notation $X \equiv_\varepsilon Y$ for $\delta(X, Y) \leq \varepsilon$. If X and Y have the *same* distribution, i.e., $\delta(X, Y) = 0$, we write $X \equiv Y$. The statistical distance can alternatively be expressed as:

$$\delta(X, Y) = \max_S \left(\Pr[X \in S] - \Pr[Y \in S] \right) .$$

From this expression it is easy to see that the optimal algorithm distinguishing the two distributions can succeed with probability exactly $\frac{1}{2} + \delta(X, Y)$. Another important property of the statistical distance is that for any random variables X and Y, there exists a random variable $\tilde{X}$ with the same distribution as Y satisfying $\Pr[\tilde{X} \neq X] = \delta(X, Y)$.

2   Definition of Secure Function Evaluation

In this section we provide a definition of secure function evaluation. We follow Definition 7.2.10 of [20] (see also [10]) but modify the associated model as follows:

i) We allow the adversary to be *computationally unbounded*.
ii) We require that the output distributions of the ideal and the real model be either *perfectly indistinguishable* or *statistically indistinguishable* (as opposed to computationally indistinguishable).
iii) We consider the input alphabet to be *fixed*.
iv) We allow randomized players that use independent sources of randomness, rather than supplying randomness to otherwise deterministic players.
v) We allow both players to have an output.

Note that ii) and iii) are just consequences of i) while iv) is used to simplify notation and v) simplifies the model by making it symmetric and generalizes it to allow functions such as coin flipping by telephone [4] where both players have an output, but which can be implemented without allowing either party to abort the protocol. In Section 6 we also discuss the model of Definition 7.2.6 of [20], i.e., the model where the first party is allowed to abort the protocol after receiving its result but before the second party receives its own.

We use the following notation: $x \in \mathcal{X}$ denotes the input of the first party, $y \in \mathcal{Y}$ the input of the second party and $z \in \{0, 1\}^*$ represents an additional auxiliary input available to both parties but assumed to be ignored by all honest parties. A *g-hybrid protocol* is a pair of (randomized) algorithms $\Pi = (A_1, A_2)$ which can interact by exchanging messages and which additionally have access to the functionality g. More precisely, for a (randomized) function $g : \mathcal{X} \times \mathcal{Y} \to \mathcal{U} \times \mathcal{V}$ the two parties can send x and y to a trusted party and receive u and v, respectively, where $(u, v) = g(x, y)$. Note that a default value is used if a player refuses to send a value. A pair of algorithms $\overline{A} = (\overline{A}_1, \overline{A}_2)$ is called *admissible* for protocol A if either $\overline{A}_1 = A_1$ or $\overline{A}_2 = A_2$, i.e., if at least one of the parties is honest and uses the algorithm defined by the protocol Π.

Definition 1 (Real Model). *Let $\Pi = (A_1, A_2)$ be a g-hybrid protocol and let $\overline{A} = (\overline{A}_1, \overline{A}_2)$ be an admissible pair of algorithms for the protocol Π. The joint execution of Π under $\overline{A}$ on input pair $(x, y) \in \mathcal{X} \times \mathcal{Y}$ and auxiliary input $z \in \{0, 1\}^*$ in the real model, denoted by*

$$\mathrm{REAL}^g_{\Pi, \overline{A}(z)}(x, y) ,$$

is defined as the output pair resulting from the interaction between $\overline{A}_1(x, z)$ and $\overline{A}_2(y, z)$ using the functionality g.

The *ideal model* defines the optimal scenario where the players have access to an ideal functionality f corresponding to the function they wish to compute. A malicious player may therefore only change (1) his input to the functionality and (2) the output he obtains from the functionality.

Definition 2 (Ideal Model). *The trivial f-hybrid protocol $B = (B_1, B_2)$ is defined as the protocol where both parties send their inputs x and y unchanged to the functionality f and output the values u and v received from f unchanged. Let $\overline{B} = (\overline{B}_1, \overline{B}_2)$ be an admissible pair of algorithms for B. The joint execution of f under $\overline{B}$ in the ideal model on input pair $(x, y) \in \mathcal{X} \times \mathcal{Y}$ and auxiliary input $z \in \{0, 1\}^*$, denoted by*

$$\text{IDEAL}_{f, \overline{B}(z)}(x, y) \,,$$

is defined as the output pair resulting from the interaction between $\overline{B}_1(x, z)$ and $\overline{B}_2(y, z)$ using the functionality f.

Any admissible protocol $\overline{B}$ in the ideal model can be expressed in the following way: the first party receives input (x, z) and the second party receives input (y, z). The two parties produce $(x', z_1) = \overline{B}_1^{\text{in}}(x, z)$ and $(y', z_2) = \overline{B}_2^{\text{in}}(y, z)$, from which x' and y' are inputs to a trusted third party, and z_1 and z_2 are some auxiliary output. The trusted party computes $(u', v') = f(x', y')$ and sends u' to the first party and v' to the second party. The two parties are now given the outputs v' and u' and the auxiliary inputs z_1 and z_2, respectively. The first party outputs $u = \overline{B}_1^{\text{out}}(u', z_1)$ while the second party outputs $v = \overline{B}_2^{\text{out}}(v', z_2)$. Note that if the first party is honest, we have $\overline{B}_1^{\text{in}}(x, z) = (x, \perp)$ and $\overline{B}_1^{\text{out}}(u', z_1) = u'$ and similarly for the second party.

Now, to show that a g-hybrid protocol Π securely computes a functionality f, we have to show that anything an adversary can do in the real model can also be done in the ideal model.

Definition 3 (Perfect Security). *A g-hybrid protocol Π securely computes f perfectly if for every pair of algorithms $\overline{A} = (\overline{A}_1, \overline{A}_2)$ that is admissible in the real model for the protocol Π, there exists a pair of algorithms $\overline{B} = (\overline{B}_1, \overline{B}_2)$ that is admissible in the ideal model for protocol B (and where the same players are honest), such that for all $x \in \mathcal{X}$, $y \in \mathcal{Y}$, and $z \in \{0, 1\}^*$, we have*

$$\text{IDEAL}_{f, \overline{B}(z)}(x, y) \equiv \text{REAL}^g_{\Pi, \overline{A}(z)}(x, y) \,.$$

It is sometimes not possible to achieve perfect security. The following definition captures the situation where the simulation has a (small) error ε, defined as the maximal statistical distance between the output distributions in the real and ideal model.

Definition 4 (Statistical Security). *A g-hybrid protocol Π securely computes f with an error of at most ε if for every pair of algorithms $\overline{A} = (\overline{A}_1, \overline{A}_2)$ that is admissible in the real model for the protocol Π, there exists a pair of algorithms $\overline{B} = (\overline{B}_1, \overline{B}_2)$ that is admissible in the ideal model for protocol B*

(and where the same players are honest), such that for all $x \in \mathcal{X}$, $y \in \mathcal{Y}$, and $z \in \{0,1\}^*$, we have

$$\text{IDEAL}_{f,\overline{B}(z)}(x,y) \equiv_\varepsilon \text{REAL}^g_{\Pi,\overline{A}(z)}(x,y) \ .$$

The statistical distance is used because it has nice properties and intuitively measures the error of a computation: a protocol Π which securely computes f with an error of at most ε, computes f *perfectly* with probability at least $1 - \varepsilon$.

A very important property of the above definitions is that they imply *sequential composition*. The following theorem has been proven in [10].

Theorem 1. *If an h-hybrid protocol Γ securely computes g with an error of at most γ and a g-hybrid protocol Π securely computes f with an error of at most π, then the composed protocol Π^Γ, namely the protocol Π where every call to g is replaced by Γ, is an h-hybrid protocol that securely computes f with an error of at most $\pi + t\gamma$, where t is the number of calls of Π to g.*

2.1   Efficient Simulation

So far, we have not been talking about *efficiency*. Indeed, if we live in a world where every participant has unlimited computer power, efficiency is not an issue, and our security definitions work well. In the world of zero-knowledge interactive proof systems [22] we have learned that "perfect zero-knowledge" is a more powerful notion than "zero-information" because the former also imposes computational conditions that require an efficient simulator. In this paper we choose to focus on the latter because in the context of two-party secure function evaluation, even in the simplest case security is not yet properly defined. When considering computationally bounded adversaries, the situation is different: It might be the case that, even though an attack in the ideal model is possible in principle, simulation is infeasable, because it takes much more time than to attack the real protocol. This problem can be solved by requiring that the running time of the ideal adversary is polynomial in the running time of the real adversary. We do not consider efficient simulation any further in this paper.

3   Secure Function Evaluation from an Information-Theoretic Point of View

In this section, we adopt an information-theoretic view of the security definition. We change our notation slightly to make it more suitable to the information-theoretic domain. We let X, Y and Z be random variables denoting the inputs, distributed according to an unknown distribution. Likewise, we let U and V be random variables denoting the outputs of the two parties. Hence, for specific inputs x, y, z we have

$$(U, V) = \text{REAL}^g_{\Pi,\overline{A}(z)}(x,y)$$

and

$$(\overline{U}, \overline{V}) = \text{IDEAL}_{f,\overline{B}(z)}(x,y) \ .$$

Note that the condition of Definition 3, namely that for all $x \in \mathcal{X}$, $y \in \mathcal{Y}$, and $z \in \{0,1\}^*$, we have

$$\mathrm{IDEAL}_{f,\overline{B}(z)}(x,y) \equiv \mathrm{REAL}^g_{\Pi,\overline{A}(z)}(x,y) ,$$

can equivalently be expressed as

$$P_{UV|XYZ} = P_{\overline{UV}|XYZ} .$$

We now state our main theorem. It gives an information-theoretic condition for the security of a real protocol, *without the use of an ideal model*. Intuitively, the security condition for player 1 (and its counterpart for player 2) says the following: Since $I(X;Y' \mid ZY) = 0$, we have $P_{Y'|YZX} = P_{Y'|YZ}$. Therefore, Y' could have been created without knowing X. The condition

$$P_{UV'|XY'YZ}(u,v' \mid x,y',y,z) = \Pr[(u,v') = f(x,y')]$$

ensures that the distributions of U and V' are the same as those of the outputs of f on input X and Y'. Finally, $I(UX;V \mid ZYY'V') = 0$ ensures that V could have been constructed out of Z, Y, Y' and V', without the help of X and U. Therefore, these conditions ensure that the resulting distribution in the real model could also have been obtained in the ideal model.

Theorem 2. *A g-hybrid protocol Π securely computes f perfectly if and only if for every pair of algorithms $\overline{A} = (\overline{A}_1, \overline{A}_2)$ that is admissible in the real model for the protocol Π and for all inputs (X, Y) and auxiliary input Z, $\overline{A}$ produces outputs (U, V), such that the following conditions are satisfied:*

- (Correctness) *If both players are honest, we have*

$$P_{UV|XYZ}(u,v,x,y,z) = \Pr[(u,v) = f(x,y)] .$$

- (Security for Player 1) *If player 1 is honest, then there exist random variables Y' and V' such that we have*

$$I(X;Y' \mid ZY) = 0 ,$$

$$P_{UV'|XY'YZ}(u,v' \mid x,y',y,z) = \Pr[(u,v') = f(x,y')] ,$$

  and

$$I(UX;V \mid ZYY'V') = 0 .$$

- (Security for Player 2) *If player 2 is honest, then there exist random variables X' and U' such that we have*

$$I(Y;X' \mid ZX) = 0 ,$$

$$P_{U'V|X'YXZ}(u',v \mid x',y,x,z) = \Pr[(u',v) = f(x',y)] ,$$

  and

$$I(VY;U \mid ZXX'U') = 0 .$$

Proof. Let us first assume that the protocol Π securely computes f. Then there exists an admissible pair of algorithms $\overline{B} = (\overline{B}_1, \overline{B}_2)$ for the ideal model such that for all $x \in \mathcal{X}$, $y \in \mathcal{Y}$, and $z \in \{0, 1\}^*$, we have

$$\mathrm{IDEAL}_{f, \overline{B}(z)}(x, y) \equiv \mathrm{REAL}^g_{\Pi, \overline{A}(z)}(x, y) ,$$

or equivalently,

$$P_{UV|XYZ} = P_{\overline{U}\overline{V}|XYZ} .$$

If both players are honest we have $\overline{B} = B$. B_1 and B_2 forward their inputs (X, Y) unchanged to the trusted third party, get back $(\overline{U}', \overline{V}') := f(X, Y)$ and output $(\overline{U}, \overline{V}) = (\overline{U}', \overline{V}')$. This establishes the correctness condition.

Without loss of generality, let player 1 be honest and player 2 be malicious. Let us look at the execution of $\overline{B} = (\overline{B}_1, \overline{B}_2)$. The malicious $\overline{B}_2$ can be modeled by the two conditional probability distributions $P_{\overline{Y}'\overline{Z}_2|YZ}$ computing the input to the ideal functionality and some internal data $\overline{Z}_2$, and $P_{\overline{V}|\overline{V}'\overline{Z}_2}$ computing the output. Note that we can write $P_{\overline{Y}'\overline{Z}_2|YZ} = P_{\overline{Y}'|YZ} P_{\overline{Z}_2|YZ\overline{Y}'}$, i.e., we can say that $\overline{Y}'$ is computed from X and Z, and that $\overline{Z}_2$ is computed from Y, Z, and $\overline{Y}'$. Clearly, we have

$$I(X; \overline{Y}' \mid ZY) = 0 .$$

The honest B_1 always sends X to the trusted party, which computes $(\overline{U}', \overline{V}') = f(X, \overline{Y}')$ and sends the results to B_1 and $\overline{B}_2$. Since B_1 always outputs $\overline{U} = \overline{U}'$, we have

$$P_{\overline{U}\overline{V}'|X\overline{Y}'YZ}(u, v' \mid x, y', y, z) = \Pr[(u, v') = f(x, y')] .$$

$\overline{B}_2$'s output $\overline{V}$ only depends on $\overline{V}'$ and $\overline{Z}_2$, which only depends on Y, Z and $\overline{Y}'$. It follows that

$$I(\overline{U}X; \overline{V} \mid ZY\overline{Y}'\overline{V}') = 0 .$$

Since the probability distributions $P_{UV|XYZ}$ and $P_{\overline{U}\overline{V}|XYZ}$ are identical, there must exist random variables satisfying the same properties for the output of protocol Π in the real model. Consequently, there must exist random variables Y' and V', such that

$$I(X; Y' \mid ZY) = 0 ,$$

$$P_{UV'|XY'YZ}(u, v' \mid x, y', y, z) = \Pr[(u, v') = f(x, y')] ,$$

and

$$I(UX; V \mid ZYY'V') = 0 .$$

Now assume that the conditions of Theorem 2 hold. If both players are honest, the correctness condition implies $P_{UV|XYZ} = P_{\overline{U}\overline{V}|XYZ}$. If both players are malicious nothing needs to be shown. Without loss of generality, let player 1 be honest and player 2 be malicious. We will define an admissible protocol $\overline{B} = (B_1, \overline{B}_2)$ in the ideal model that produces the same distribution as the protocol

Π in the real model. Let $\overline{B}_2$ choose his input $\overline{Y}'$ according to $P_{\overline{Y}'|YZ} := P_{Y'|YZ}$, and let him choose his output $\overline{V}$ according to $P_{\overline{V}|YZ\overline{Y}'\overline{V}'} := P_{V|YZY'V'}$. The conditional distribution of the output in the ideal model is given by

$$P_{\overline{UV}|XYZ} = \sum_{y',v'} P_{\overline{Y}'|YZ} P_{\overline{UV}'|X\overline{Y}'} P_{\overline{V}|YZ\overline{Y}'\overline{V}'} \,,$$

where

$$P_{\overline{UV}'|X\overline{Y}'}(u,v' \mid x,y') = \Pr[(u,v') = f(x,y')] \,.$$

From $I(X;Y' \mid ZY) = 0$ and $I(UX;V \mid ZYY'V') = 0$ it follows that $P_{Y'|XYZ} = P_{Y'|YZ}$ and $P_{V|XYZY'UV'} = P_{V|YZY'V'}$. Furthermore, we have $P_{UV'|XY'YZ} = P_{\overline{UV}'|X\overline{Y}'}$. As for the conditional distribution of the output in the real model, we have:

$$\begin{aligned}
P_{UV|XYZ} &= \sum_{y',v'} P_{Y'UV'|XYZ} P_{V|XYZY'UV'} \\
&= \sum_{y',v'} P_{Y'|XYZ} P_{UV'|XYZY'} P_{V|YZY'V'} \\
&= \sum_{y',v'} P_{\overline{Y}'|YZ} P_{\overline{UV}'|X\overline{Y}'} P_{\overline{V}|YZ\overline{Y}'\overline{V}'} \\
&= P_{\overline{UV}|XYZ} \,.
\end{aligned}$$

Therefore, for any admissible $\overline{A}$ in the real model there exists an admissible $\overline{B}$ in the ideal model such that

$$\mathrm{IDEAL}_{f,\overline{B}(z)}(x,y) \equiv \mathrm{REAL}^g_{\Pi,\overline{A}(z)}(x,y) \,,$$

implying that the protocol is perfectly secure. $\square$

Note that the expression

$$P_{UV'|XY'YZ}(u,v' \mid x,y',y,z) = \Pr[(u,v') = f(x,y')]$$

can be replaced by $(U,V') = f(X,Y')$ if f is deterministic. This yields the following corollary for deterministic functionalities.

Corollary 1. *A protocol Π securely computes the deterministic functionality f perfectly, if and only if for every pair of algorithms $\overline{A} = (\overline{A}_1, \overline{A}_2)$ that is admissible in the real model for the protocol Π and for all inputs (X,Y) and auxiliary input Z, $\overline{A}$ produces outputs (U,V), such that the following conditions are satisfied:*

- *(Correctness) If both players are honest, we have $(U,V) = f(X,Y)$.*
- *(Security for Player 1) If player 1 is honest then there exist random variables Y' and V' such that $(U,V') = f(X,Y')$,*

$$I(X;Y' \mid ZY) = 0 \,, \qquad \text{and} \qquad I(UX;V \mid ZYY'V') = 0 \,.$$

- (Security for Player 2) *If player 2 is honest then there exist random variables X' and U' such that $(U', V) = f(X', Y)$,*

$$I(Y; X' \mid ZX) = 0 , \quad and \quad I(VY; U \mid ZXX'U') = 0 .$$

Note that we require the conditions of Theorem 2 and Corollary 1 to hold for all distributions of the inputs (X, Y). In particular, they have to hold for any input distribution $P_{XY|Z=z}$, i.e., given the event that the auxiliary input Z equals z. Since all the requirements are conditioned on Z, it is sufficient to show that the conditions are met for all distributions P_{XY}, ignoring Z in all the expressions.

The information-theoretic security definition of Theorem 2 and Corollary 1 can also be used for protocols which are not perfectly secure. A protocol is *secure with error ε* if for all inputs X, Y, Z, the joint distribution of the outputs has a statistical distance of at most ε from the output of a perfectly secure protocol. In information theory, the distance between distributions is typically expressed using bounds on entropy and mutual information instead of statistical distance. The following inequalities translate such bounds into bounds on statistical distance. Let U be uniformly distributed over the set $\mathcal{X}$.

$$\delta(P_{XYZ}, P_Z P_{X|Z} P_{Y|Z}) \leq \frac{1}{2} \sqrt{2 \ln 2 \, I(X; Y \mid Z)}$$

$$\delta(P_X, P_U) \leq \frac{1}{2} \sqrt{2 \ln 2 (\log |\mathcal{X}| - H(X))}$$

The first inequality can easily be proved from [13], Lemma 16.3.1 while the second inequality was proved in [8], Lemma 3.4.

4   Oblivious Transfer

We now apply our security definition to 1-out-of-n string oblivious transfer, or $\binom{n}{1}$-OT^k for short. The ideal functionality f_{OT} is defined as

$$f_{OT}(X, C) := (\bot, X_C) ,$$

where $\bot$ denotes a constant random variable, $X = (X_0, \ldots, X_{n-1})$, $X_i \in \{0, 1\}^k$ for $i \in \{1, \ldots, n\}$, and $C \in \{1, \ldots n\}$.

Theorem 3. *A protocol Π securely computes $\binom{n}{1}$-OT^k perfectly if and only if for every pair of algorithms $\overline{A} = (\overline{A}_1, \overline{A}_2)$ that is admissible for protocol Π and for all inputs (X, C) and auxiliary input Z, $\overline{A}$ produces outputs (U, V) such that the following conditions are satisfied:*

- (Correctness) *If both players are honest, then $(U, V) = (\bot, X_C)$.*
- (Security for Player 1) *If player 1 is honest, then we have $U = \bot$ and there exists a random variable C', such that*

$$I(X; C' \mid ZC) = 0 , \quad and \quad I(X; V \mid ZCC'X_{C'}) = 0 .$$

– (Security for Player 2) *If player 2 is honest, then we have*

$$I(C; U \mid ZX) = 0 .$$

Proof. We only need to show that the security condition for player 2 is equivalent to the one in Corollary 1:

$$I(C; X' \mid ZX) + I(X'_C C; U \mid ZXX') = 0$$

Since X'_C is a function of C and X',

$$I(X'_C C; U \mid ZXX') = 0 \text{ is equivalent to } I(C; U \mid ZXX') = 0 .$$

From the chain rule it follows that

$$I(C; X' \mid ZX) + I(C; U \mid ZXX') = I(C; X'U \mid ZX) = I(C; U \mid ZX) + I(C; X' \mid ZXU) .$$

Now choose $X' = (X'_0, \ldots, X'_{n-1})$ as follows: for all values i, let X'_i be chosen according to the distribution $P_{V|ZXU,C=i}$ except for X'_C. We set $X'_C = V$. Note that all X'_i, $0 \le i \le n-1$, have distribution $P_{V|ZXU,C=i}$. Thus X' does not depend on C given ZXU, we have $V = X'_C$ and $I(C; X' \mid ZXU) = 0$. So there always exists a suitable X'^1, and the condition simplifies to $I(C; U \mid ZX) = 0$. □

The interpretation of these properties of oblivious transfer is quite intuitive: If player 1 is honest, then she can be confident that anything player 2 can do is basically equivalent to choosing a choice bit C' which is possibly different from C. On the other hand, if player 2 is honest, he can be certain that player 1 does not get to know his input C. Theorem 3 shows that in the case of a dishonest sender in $\binom{n}{1}$-OT^k, *privacy alone implies security*. There *always* exists an input X' that a dishonest sender can use in the ideal model to obtain the same results.

5 An Example

In this section we show how the result from the Section 4 can be used to prove the security of a protocol. Our example will be the protocol from [29], where one instance of $\binom{2}{1}$-OT is implemented using one instance of $\binom{2}{1}$-TO, which is an instance of $\binom{2}{1}$-OT in the opposite direction.

Protocol 1 ([29]). *Let player 1 have input $X = (X_0, X_1) \in \{0,1\} \times \{0,1\}$, and player 2 have input $C \in \{0,1\}$.*

1. *Player 2 chooses $R \in \{0,1\}$ at random.*
2. *The two players execute $\binom{2}{1}$-TO, where player 1 inputs $\overline{C} = X_0 \oplus X_1$, and player 2 inputs $\overline{X}_0 = R$ and $\overline{X}_1 = R \oplus C$.*
3. *Player 1 receives $A = \overline{X}_{\overline{C}}$ and sends $M = X_0 \oplus A$ to the player 2.*
4. *Player 1 outputs $V := R \oplus M$.*

---

[1] Note that these values X' are not necessarily known to a malicious player 1.

Theorem 4. *Protocol 1 perfectly securely reduces $\binom{2}{1}$-OT to one realization of $\binom{2}{1}$-TO.*

Proof. If both parties are honest, the protocol is correct because we have

$$R \oplus M = R \oplus X_0 \oplus (X_0 \oplus X_1)C \oplus R = X_C .$$

Let player 1 be honest, and let $C' := \overline{X}_0 \oplus \overline{X}_1$. Using the data processing inequality,

$$\mathrm{I}(X_0X_1; C' \mid ZC) \leq \mathrm{I}(X_0X_1; \overline{X}_0\overline{X}_1 \mid ZC) \leq \mathrm{I}(X_0X_1; ZC \mid ZC) = 0 .$$

Since $M = X_0 \oplus (X_0 \oplus X_1)(\overline{X}_0 \oplus \overline{X}_1) \oplus \overline{X}_0 = X_{C'} \oplus \overline{X}_0$, the values $\overline{X}_0\overline{X}_1M$, $\overline{X}_0C'M$, and $\overline{X}_0C'X_{C'}$ contain the same information. Thus, using the data processing inequality,

$$\mathrm{I}(X_0X_1; V \mid ZCC'X_{C'}) \leq \mathrm{I}(X_0X_1; CZ\overline{X}_0\overline{X}_1M \mid ZCC'X_{C'})$$
$$= \mathrm{I}(X_0X_1; CZ\overline{X}_0C'X_{C'} \mid ZCC'X_{C'}) = 0 .$$

Now let player 2 be honest. Since $A = R \oplus C\overline{C}$ and R is uniform, we have

$$\mathrm{I}(C; U \mid ZX_0X_1) \leq \mathrm{I}(C; X_0X_1ZA \mid ZX_0X_1) = \mathrm{I}(C; A \mid ZX_0X_1) = 0 .$$

Thus, the protocol is secure.   □

6   Secure Two-Party Computation with Abort

In this section we will briefly discuss the model of Definition 7.2.6 of [20] where the first party is allowed to abort the protocol right after receiving its output but before the second party has received its own. The *ideal model with abort for player 1* is similar to the ideal model from Definition 2, the only difference being that player 1 is given the option of aborting the computation by sending a bit C to the trusted party after having received his output. The trusted party sends to player 2 the corresponding output if $C = 1$, and $\perp$ if $C = 0$. An honest player always sends $C = 1$. The real model and the definition of security are identical to the definition without abort. We call a protocol that satisfies this definition *secure with abort for player 1*.

Theorem 5. *A g-hybrid protocol Π securely computes f perfectly with abort for player 1, if and only if for every pair of algorithms $\overline{A} = (\overline{A}_1, \overline{A}_2)$ that is admissible in the real model for the protocol Π, and for all inputs (X, Y) and auxiliary input Z, $\overline{A}$ produces outputs (U, V), such that the following conditions are satisfied:*

- *(Correctness) If both players are honest, we have*

$$P_{UV|XYZ}(u, v \mid x, y, z) = \Pr[(u, v) = f(x, y)] .$$

- (Security for Player 1) *If player 1 is honest, then there exist random variables* Y' *and* V', *such that we have*

$$\mathrm{I}(X; Y' \mid ZY) = 0 \,,$$

$$P_{UV'|XY'YZ}(u, v' \mid x, y', y, z) = \Pr[(u, v') = f(x, y')] \,,$$

and

$$\mathrm{I}(UX; V \mid ZYY'V') = 0 \,.$$

- (Security for Player 2) *If player 2 is honest, then there exist random variables* X', C *and* U', V', *such that we have*

$$\mathrm{I}(Y; X' \mid ZX) = 0 \,,$$

$$P_{U'V'|X'YXZ}(u', v' \mid x', y, x, z) = \Pr[(u', v') = f(x', y)] \,,$$

$$\mathrm{I}(V'Y; UC \mid ZXX'U') = 0 \,,$$

and $V = V'$ *if* $C = 1$ *and* $V = \perp$ *if* $C = 0$.

Proof. The proof is identical to that of Theorem 2 for the case where player 1 is honest. We therefore only examine the case where player 2 is honest and player 1 is malicious.

Let us assume that the protocol Π securely computes f. Consequently, there exists an admissible pair of algorithms $\overline{B} = (\overline{B}_1, B_2)$ such that for all $x \in \mathcal{X}$, $y \in \mathcal{Y}$, and $z \in \{0, 1\}^*$ we have $P_{UV|XYZ} = P_{\overline{U}\overline{V}|XYZ}$.

The malicious $\overline{B}_1$ can be modeled by the two conditional probability distributions $P_{\overline{X'}\overline{Z}_2|XZ}$ computing the input to the ideal functionality and some internal data $\overline{Z}_2$, and $P_{\overline{UC}|\overline{U'}\overline{Z}_2}$ computing the output $\overline{U}$ and the bit $\overline{C}$. Note that we can write $P_{\overline{X'}\overline{Z}_2|XZ} = P_{\overline{X'}|XZ}P_{\overline{Z}_2|XZ\overline{X'}}$. Clearly, we have

$$\mathrm{I}(Y; \overline{X}' \mid ZX) = 0 \,.$$

The ideal functionality computes $\overline{U}', \overline{V}'$ such that

$$P_{\overline{U'}\overline{V'}|\overline{X'}YXZ}(u', v' \mid x', y, x, z) = \Pr[(u', v') = f(x', y)] \,.$$

B_1 gets back $\overline{U}'$ from the ideal functionality. Based on $X, Z, \overline{X}', \overline{U}'$ he decides to send $\overline{C}$ to the functionality and outputs $\overline{U}$. Hence, we have

$$\mathrm{I}(\overline{V}'Y; \overline{UC} \mid XZ\overline{X}'\overline{U}') = 0 \,.$$

If $C = 1$, the functionality sends $\overline{V} = \overline{V}'$ to B_2, if $C = 0$ it sends $\overline{V} = \perp$. B_2 outputs $\overline{V}$ unchanged. As $P_{UV|XYZ} = P_{\overline{U}\overline{V}|XYZ}$ it must be the case that the same conditions hold in the real model, which implies the security condition for player 2.

Now let the conditions of Theorem 5 hold. We define an admissible protocol $\overline{B} = (\overline{B}_1, B_2)$ in the ideal model that produces the same distribution as the protocol Π in the real model. Let $\overline{B}_1$ choose input $\overline{X}'$ according to $P_{\overline{X}'|XZ} := P_{X'|XZ}$, and $(\overline{U}, \overline{C})$ according to $P_{\overline{UC}|XZ\overline{X}'\overline{U}'} := P_{UC|XZX'U'}$. The conditional distribution of the output in the ideal model is given by

$$P_{\overline{UV}|XYZ} = \sum_{x',c,u',v'} P_{\overline{X}'|XZ} P_{\overline{U}'\overline{V}'|\overline{X}'Y} P_{\overline{UC}|XZ\overline{X}'\overline{U}'} P_{\overline{V}|\overline{V}'\overline{C}} \,,$$

where

$$P_{\overline{U}'\overline{V}'|\overline{X}'Y}(u', v' \mid x', y) = \Pr[(u', v') = f(x', y)] \,.$$

From $I(Y; X' \mid ZX) = 0$ and $I(V'Y; UC \mid XZX'U') = 0$ it follows that $P_{X'|XYZ} = P_{X'|XZ}$ and $P_{UC|XZX'U'V'Y} = P_{UC|XZX'U'}$. Furthermore, we have $P_{U'V'|X'YXZ} = P_{\overline{U}'\overline{V}'|\overline{X}'Y}$ and $P_{V|V'C} = P_{\overline{V}|\overline{V}'\overline{C}}$. We get for the conditional distribution of the output in the real model

$$P_{UV|XYZ} = \sum_{x',c,u',v'} P_{X'|XYZ} P_{U'V'|XYZX'} P_{UCV|XYZX'U'V'}$$

$$= \sum_{x',c,u',v'} P_{X'|XZ} P_{\overline{U}'\overline{V}'|\overline{X}'Y} P_{UC|XYZX'U'V'} P_{V|XYZX'U'V'CU}$$

$$= \sum_{x',c,u',v'} P_{X'|XZ} P_{\overline{U}'\overline{V}'|\overline{X}'Y} P_{UC|XZX'U'} P_{V|V'C}$$

$$= \sum_{x',c,u',v'} P_{\overline{X}'|XZ} P_{\overline{U}'\overline{V}'|\overline{X}'Y} P_{\overline{UC}|XZ\overline{X}'\overline{U}'} P_{\overline{V}|\overline{V}'\overline{C}}$$

$$= P_{\overline{UV}|XYZ} \,.$$

Therefore for any admissible $\overline{A}$ in the real model there exists an admissible $\overline{B}$ in the ideal model such that

$$\text{IDEAL}_{f,\overline{B}(z)}(x, y) \equiv \text{REAL}^g_{\Pi, \overline{A}(z)}(x, y) \,,$$

which means that the protocol is perfectly secure with abort for player 1.   □

7   Conclusion and Open Problems

We have shown that various information-theoretic security definitions for oblivious transfer used in the past contain subtle flaws. We propose a new information-theoretic security definition which is provably equivalent to the security definition based on the ideal/real model paradigm. This not only provides a solid security foundation for most protocols in the literature, which turn out to meet our requirements, but also shows that they are in fact sequentially composable.

An interesting open problem is to generalize our model to various quantum settings, for example to the scenario where two players connected by a quantum channel wish to securely implement a classical functionality.

Acknowledgements

We thank Abdul Ahsan, Serge Fehr and Stefan Wolf for many helpful discussions and the anonymous referees for their comments.

References

1. M. Backes, B. Pfitzmann, and M. Waidner. A universally composable cryptographic library. Cryptology ePrint Archive, Report 2003/015, 2003.
2. D. Beaver. Foundations of secure interactive computing. In *Advances in Cryptology: CRYPTO '91*, pages 377–391, London, UK, 1992. Springer-Verlag.
3. M. Ben-Or, S. Goldwasser, and A. Wigderson. Completeness theorems for non-cryptographic fault-tolerant distributed computation. In *Proceedings of the 20th Annual ACM Symposium on Theory of Computing (STOC '88)*, pages 1–10. Springer-Verlag, 1988.
4. Manuel Blum. Coin flipping by telephone a protocol for solving impossible problems. *SIGACT News*, 15(1):23–27, 1983.
5. C. Blundo, P. D'Arco, A. De Santis, and D. R. Stinson. New results on unconditionally secure distributed oblivious transfer. In *SAC '02: Revised Papers from the 9th Annual International Workshop on Selected Areas in Cryptography*, pages 291–309, London, UK, 2003. Springer-Verlag.
6. G. Brassard, C. Crépeau, and M. Santha. Oblivious transfers and intersecting codes. *IEEETIT: IEEE Transactions on Information Theory*, 42, 1996.
7. G. Brassard, C. Crépeau, and S. Wolf. Oblivious transfers and privacy amplification. *Journal of Cryptology: the journal of the International Association for Cryptologic Research*, 16(4):219–237, 2003.
8. C. Cachin. *Entropy Measures and Unconditional Security in Cryptography*. PhD thesis, No. 12187, ETH Zurich, Switzerland, 1997.
9. R. Canetti. *Studies in Secure Multiparty Computation and Applications*. PhD thesis, Weizmann Institiute of Science, Israel, 1996.
10. R. Canetti. Security and composition of multiparty cryptographic protocols. *Journal of Cryptology*, 13(1):143–202, 2000.
11. R. Canetti. Universally composable security: A new paradigm for cryptographic protocols. Cryptology ePrint Archive, Report 2000/067, 2000.
12. D. Chaum, C. Crépeau, and I. Damgård. Multiparty unconditionally secure protocols (extended abstract). In *Proceedings of the 20th Annual ACM Symposium on Theory of Computing (STOC '88)*, pages 11–19. ACM Press, 1988.
13. T. M. Cover and J. A. Thomas. *Elements of Information Theory*. Wiley-Interscience, New York, USA, 1991.
14. C. Crépeau. Verifiable disclosure of secrets and applications (abstract). In *Advances in Cryptology: EUROCRYPT '89*, Lecture Notes in Computer Science, pages 181 – 191. Springer-Verlag, 1990.
15. C. Crépeau, J. van de Graaf, and A. Tapp. Committed oblivious transfer and private multi-party computation. In *Advances in Cryptology: CRYPTO '95*, Lecture Notes in Computer Science, pages 110–123, 1995.
16. P. D'Arco and D. R. Stinson. Generalized zig-zag functions and oblivious transfer reductions. In *SAC '01: Revised Papers from the 8th Annual International Workshop on Selected Areas in Cryptography*, pages 87–102, London, UK, 2001. Springer-Verlag.

17. Y. Ding, D. Harnik, A. Rosen, and R. Shaltiel. Constant-round oblivious transfer in the bounded storage model. In *In Theory of Cryptography — TCC '04*, volume 2951. Springer-Verlag, 2004.

18. Y. Dodis and S.Micali. Lower bounds for oblivious transfer reductions. In *Advances in Cryptology: EUROCRYPT '97*, 1999.

19. S. Even, O. Goldreich, and A. Lempel. A randomized protocol for signing contracts. *Commun. ACM*, 28(6):637–647, 1985.

20. O. Goldreich. *Foundations of Cryptography*, volume II: Basic Applications. Cambridge University Press, 2004.

21. O. Goldreich, S. Micali, and A. Wigderson. How to play any mental game. In *Proceedings of the 19th Annual ACM Symposium on Theory of Computing (STOC '87)*, pages 218–229. ACM Press, 1987.

22. S. Goldwasser, S. Micali, and C. Rackoff. The knowledge complexity of interactive proof systems. *SIAM J. Comput.*, 18(1):186–208, 1989.

23. J. Kilian. Founding cryptography on oblivious transfer. In *Proceedings of the Twentieth Annual ACM Symposium on Theory of Computing*, pages 20–31, 1988.

24. J. Kilian. More general completeness theorems for secure two-party computation. In *STOC*, pages 316–324, 2000.

25. S. Micali and P. Rogaway. Secure computation (abstract). In *Advances in Cryptology: CRYPTO '91*, pages 392–404, London, UK, 1992. Springer-Verlag.

26. V. Nikov, S. Nikova, B. Preneel, and J. Vandewalle. On unconditionally secure distributed oblivious transfer. In *Progress in Cryptology - INDOCRYPT 2002*, pages 395–408, 2002.

27. M. O. Rabin. How to exchange secrets by oblivious transfer. Technical Report TR-81, Harvard Aiken Computation Laboratory, 1981.

28. S. Wiesner. Conjugate coding. *SIGACT News*, 15(1):78–88, 1983.

29. S. Wolf and J. Wullschleger. Oblivious transfer is symmetric. In *Advances in Cryptology: EUROCRYPT '06*, Lecture Notes in Computer Science. Springer-Verlag, 2006.

30. A. C. Yao. Protocols for secure computations. In *Proceedings of the 23rd Annual IEEE Symposium on Foundations of Computer Science (FOCS '82)*, pages 160–164, 1982.

Unclonable Group Identification

Ivan Damgård, Kasper Dupont, and Michael Østergaard Pedersen

Aarhus University, BRICS

Abstract. We introduce and motivate the concept of unclonable group identification, that provides maximal protection against sharing of identities while still protecting the anonymity of users. We prove that the notion can be realized from any one-way function and suggest a more efficient implementation based on specific assumptions.

1 Introduction

A large body of literature studies the problem of group identification, where one wants to verify that a given user is a member of a certain group, while ensuring that the user's personal identity is not revealed. Particular instances of this include group signatures [5, 3, 22] and identity escrow[16]. In some applications, a dishonest user has an interest in giving away to another person the data that allow him to identify himself as a member of the group - such as password and secret keys. The security problems implied by such a scenario have not been given much attention so far in the literature[1].

In this paper we study this type of problem. As a motivating example, consider the issue of software protection: it is well known that one of the strongest motivating factors in getting people to register as software users is if this enables some functionality that cannot be accessed without registration (and payment). This works particularly well, if the functionality requires access to the vendor's website, since then reverse engineering the software is not sufficient to get unauthorized access to the functionality. In the case of games, for instance, the opportunity to play against others may be available to only registered users, and only through the vendor's website.

Verifying that a user is registered may be done in many different ways. In this paper, we are interested in solutions that work under the following constraints:

- An honest user can connect an unlimited number of times using the same private key material, or at least updates should only be necessary with long time intervals.
- We want to protect users' privacy, i.e., honest users have to identify themselves *only* as registered users and do not have to reveal their personal identities.

---

[1] Some earlier works suggest to discourage this by forcing users to either give away *all* their information, or nothing, but here we are interested in cases where dishonest users in fact have an interest in giving everything away.

S. Vaudenay (Ed.): EUROCRYPT 2006, LNCS 4004, pp. 555–572, 2006.

- We want to do as much as possible to protect against attacks where a user "clones" himself by handing a copy of his personal data (software, secret key(s), etc.) to another person in order to get the benefits of two registrations while only paying for one.

Note that the cloning attack may be easy or very hard to carry out physically, depending on how the user's personal keys are stored, but only in very few cases can it be considered impossible.

Of course, we can only hope to detect cloning if the user and clone actually connect to the vendor's website. A further trivial observation is that if first the user connects, then leaves the site and then the clone connects, we cannot distinguish this from two connections made by an honest user, since he would also use the same private key material in both cases. An event we *can* hope to detect, however, is if both user and clone connect so that they are on the site simultaneously, since this is exactly what cannot occur if the user has been honest. In this case, we not only want to detect the attack, we also want to be able to reveal the identity of the user who cloned himself. Note that, apart from the fact that the above simultaneous scenario is the only one in which we can hope to catch a cloning attack, the scenario is also of practical relevance. For instance, the case of a user who buys one copy of a game and distributes it to all his friends so they can play against each other online, is exactly a case where a number of clones would want to be connected simultaneously.

An *unclonable identification scheme* informally is an identification scheme where honest users can identify themselves anonymously as members of a group, but where clones of users can be detected and have their identities revealed if they identify themselves simultaneously. In this paper, we give a formal definition of this primitive. We show that it can be realized assuming existence of one-way functions (which is clearly a minimal assumption), and we give a more efficient implementation based on specific assumptions. On the technical side, our most efficient solution is based on a new technique for proving in zero-knowledge, given g^x in a group of prime order, that x was chosen pseudorandomly from on a committed secret key.

Of course, before attempting a construction such as we have sketched, one should verify if existing primitives already allow solving the problem. First, one might consider using an anonymous E-cash scheme[17, 6], i.e., some number of electronic coins are issued to each user, and users use them to "pay" for access to the site. This would lead to a functionality that is incomparable to the one we sketched above: Cloning in this case means sharing e-coins with others, and so the cloning attack is exactly double spending and can therefore detected even if the two spendings do not take place simultaneously. But on the other hand, honest users can only use each coin once, and must therefore either possess a very large secure memory, or come back for more coins throughout the life of the system. This reveals information on how often a user connects, and is also not consistent with our goal, namely a solution where you can join a group once and then identify yourself an unlimited number of times using the same key material.

One may also consider using group signatures[5, 3, 22], and have users identify themselves by signing a message chosen by the verifier (using his current system time, for instance). This achieves anonymity but does not protect against cloning. To do this, one would need the property that if the same user signs the same message twice, this would result in signatures that could be detected as coming from the same user. This does not follow from the standard definition of group signatures, and is actually false for known schemes, since these are probabilistic and produce randomly varying signatures even if the message is fixed. A similar comment applies to identity escrow schemes[16].

2   Definition

An unclonable identification scheme involves a *Group Manager GM*, a set of *Verifiers* and some number of *Users*. The idea is that after some initialization, there will be several events, where some set of users prove "at the same time" to a verifier V that they are members of the group managed by GM. Since we want to detect if V is talking to clones of the same user at the same time, every proof should take as input some string α that represents in some sense the current time or phase of the protocol we are in. However, this does not have to be linked to *real time*. What is important is that whenever a set of users want to prove themselves, they should agree with V on a value for α that has not been used before. More precisely, the demands are

- An honest V must be able to ensure that all users he talks to at a given point prove themselves using the same value of α.
- An honest user should be able to ensure that he never executes *Prove* with the same value of α more than once.

One solution that works in the case where V runs a website that users would like to be connected to for some length of time, is as follows: with regular intervals, e.g., each hour each user who is connected must prove himself using the current date and hour as α, as defined by the verifier's system time. This works if there is sufficient agreement on the time between users and V and if users remember at which time they last did a proof. But many other solutions are possible. Therefore, we have chosen to separate the way time is defined from the definition as such by assuming that the entire system proceeds in consecutive *phases*, with a unique number assigned to each phase. In each phase, some subset of users decide to prove themselves to some verifier V, and the number assigned to the current phase will be used as the string α. In the full version of this paper, [10] we propose a way to realize such a scenario without relying on synchronization, or requiring users to keep state.

The system is defined by probabilistic polynomial time algorithms $KeyGen$, $Detect$ and two-party protocols $Join$ and $Prove$. These are used as follows:

- Initially, GM runs $KeyGen$ on input 1^k, to get output public key pk and secret key sk. We assume for simplicity that the set of possible pk's output by $KeyGen(1^k)$ can be recognized in polynomial time.
- When a user U joins the system he runs $Join$ with GM. Common input is pk. Private input to GM is sk. The protocol outputs to GM either "reject"

or a string id. Output to U is "reject" or a membership certificate $cert_U$. We assume $Join$ is executed on a secure channel so that no other entity will have access to the data exchanged.

- To prove he is a member of the group, the user U executes protocol $Prove$ with a verifier V. Common input is the public key pk and the string α assigned to the current phase, U uses $cert_U$ as private input. At the end of the protocol V accepts or rejects. Each user executes $Prove$ at most once in every phase.
- Algorithm $Detect$ gets as input a number of transcripts of executions of $Prove$, done with pk as input in the same phase. It outputs a (possibly empty) list of strings. The intuition is that this algorithm should be able to tell if the result of one or more cloning attacks are among a given set of proofs, and if so, it will output the identities of the involved users.

Definition 1. *The algorithms and protocols in a secure unclonable identification scheme must satisfy the following:*

Completeness. *Assume GM, V and user U are honest. Execution of $KeyGen$, followed by executions of $Join$ and $Prove$ always result in V accepting.*

No Cloning. *Consider an honest GM who executes $(pk, sk) = KeyGen(1^k)$. Consider any probabilistic polynomial time algorithm $\tilde{U}$ who plays the following game on input pk: in any phase, it can issue one or more of the following requests:*

1. *It can ask that a set of honest users execute $Join$ with GM (no data returned to $\tilde{U}$).*
2. *It can ask to execute $Join$ itself with GM.*
3. *It can ask that some number of honest users who already joined the group execute $Prove$ with $\tilde{U}$ acting as verifier, using pk and the current value of α as input.*

Finally, $\tilde{U}$ executes $Prove$ a number of times with an honest verifier V, on input pk and the current value of α.

We now want to capture the idea that in the last step, $\tilde{U}$ can only have proofs accepted by using user identities it got from GM, it must "know" which one of them it is using in each case, and if it uses any of them more then once, the Detect algorithm will catch this.

To this end, we demand that there exists a probabilistic algorithm $Extract$ which gets as input the complete view of $\tilde{U}^2$ and outputs a user identity, for every instance of $Prove$ that V accepted in the last step. The expected time to run $\tilde{U}$ and then $Extract$ must be polynomial.

We require that the following holds except with negligible probability:
All user identities output by $Extract$ are among those that were generated in the conversations between $\tilde{U}$ and GM. Furthermore, the Detect algorithm, when given as input the conversation between $\tilde{U}$ and V, will output exactly those user identities that occur more than once in the output of $Extract$.

---

[2] This means that $Extract$ can rewind $\tilde{U}$ to any state that occurred during the game.

Note that this implies that if $\tilde{U}$ did not execute any Join's, there are no user identities Extract can legally output, so we are then in fact demanding that all $\tilde{U}$'s proofs are rejected except with negligible probability. Thus we do not need a separate soundness condition in the definition demanding that non-members are rejected.

Anonymity. *Consider any probabilistic polynomial time algorithm $\tilde{V}$, who will act as both GM and verifier in an attempt to break the anonymity of honest users. $\tilde{V}$ gets 1^k as input and outputs a valid pk (can be assumed without loss of generality since we assumed that invalid pk's can be easily recognized). It then plays the following game: it interacts with a set of honest users, where in each phase some users execute Join and other users execute Prove with $\tilde{V}$. Of course, no honest user will attempt to do Prove unless he already did Join successfully. At some point $\tilde{V}$ stops and outputs a bit, and we let $p_{real,\tilde{V}}(k)$ be the probability that 1 is output.*

We now want to express the demand that $\tilde{V}$ should only learn what is unavoidable, namely the number of honest users that interact with it in each phase. So we compare the above game to a different one, where $\tilde{V}$ interacts with a simulator M. The simulator gets as input for each phase the number of users who want to execute Join and the number that want to execute Prove in the current phase. These numbers are chosen with the same distribution as in the first game. Let $p_{sim,\tilde{V}}(k)$ be the probability that 1 is output in this case.

We demand that there exists a simulator probabilistic polynomial time simulator M such that for any $\tilde{V}$, $|p_{real,\tilde{V}}(k) - p_{sim,\tilde{V}}(k)|$ is negligible in k.

We note that in this definition, we have for simplicity used the usual two-phase structure of identification schemes to define soundness and non-cloning, where first the adversary talks to the honest users and then tries to fool the honest verifier. Thus we do not allow him to interact with an honest prover and and honest verifier simultaneously. However, this is not a serious restriction, as there are several techniques that allow handling even this concurrent case, such as the so called designated verifier proofs[12, 7]. These techniques can be used with any of the schemes we propose here.

As for the scheduling of the individual protocols in a single phase, we consider two cases: one where in each phase the proofs given to an honest verifier are composed sequentially, and one where the composition may be concurrent, with a scheduling chosen by the adversary. We speak of *sequential* and *concurrent* security, accordingly. On the other hand, we assume that honest users (provers) may interact concurrently with an adversarial verifier.

3   A Theoretical Solution

3.1   Some Tools

We will need a secure string commitment scheme. Such a scheme follows from any one-way function using for instance Naor's construction[18], where there is a public key $Pcom$ which is a random string (of length polynomial in the security parameter k) that can be chosen once and for all by the receiver of commitments. We let $com_{Pcom}(str, r_{str})$ denote a commitment to string str using random coins r_{str}. Such a commitment determines str uniquely except for a negligible fraction of the public keys, and commitments to different strings are polynomially indistinguishable assuming the underlying one-way function is hard to invert.

Based on such a commitment scheme and, for instance, Blum's protocol for Graph Hamiltonicity or the one from [14] for graph 3-colorability, we can build generic proofs of knowledge for any binary relation R that can be checked in polynomial time. The protocol in its basic form is a three move protocol where the second message is a one-bit challenge from the verifier. When we work with security parameter k, we may compose *sequentially* k instances of this protocol, to obtain a zero-knowledge proof of knowledge for R with negligible soundness error. We may also compose *in parallel* k instances of the protocol. This is also a proof of knowledge for R, more precisely, on common input x, the prover proves knowledge of w such that $(x, w) \in R$.

Protocols obtained by this parallel composition are special cases of so-called Σ-*protocols*. By definition, such protocols have three properties: first, conversations are of form (a, e, z), where $a = a(x, w, coins_P)$ is a function of x, w and the prover's random coins, e is a k-bit challenge, and $z = z(x, w, coins_P, e)$ is a function of the prover's private data and the challenge. Based on $x, (a, e, z)$ the verifier decides to accept or reject. Second, the protocol is honest-verifier (computational) zero-knowledge (and is therefore witness indistinguishable). Third, the protocol has the *special soundness property*, i.e., from x and accepting conversations $(a, e, z), (a, e', z')$ with $e \neq e'$, it is easy to compute w such that $(x, w) \in R$.

Using a technique known as the OR-construction[8], one can combine Σ-protocols for two relations R_0, R_1, to obtain a new Σ-protocol, where on input x_0, x_1, the prover proves he knows w such that $(x_0, w) \in R_0$ or $(x_1, w) \in R_1$, without revealing which is the case, i.e., the protocol is witness indistinguishable.

We will need a family of pseudorandom functions[13]. Such a family is indexed by a key s (a random string of length k bits), and can be designed to have any desired (polynomial in k) input and output length, assuming any one-way function. We let $f_s()$ denote such a pseudorandom function. The basic property is that even given oracle access to the function (and not the key), it cannot be efficiently distinguished from a truly random function.

Finally, we will need a secure signature scheme, which can again be built from any one-way function[21]. Such a scheme comes with probabilistic polynomial time algorithms $Gen, Sign, Verify$ for key generation, signing and verifying signatures. $Gen(1^k)$ outputs a key pair $Psign, Ssign$. On input message m and the private key, $Sign$ produces a signature $\sigma = Sign(Ssign, m)$. On input message, signature and public key, $Verify$ produces as output $Verify(Psign, m, \sigma)$ which is *accept* or *reject*.

3.2 The Scheme

We first explain the intuition behind the solution: when joining the group, user U will make a commitment c_U to a random string r_U and will obtain GM's signature σ_U on the commitment. He then proves he is a member of the group by proving that he knows a valid signature σ_U on some message c_U, without revealing either value. Moreover when giving this proof he uses some random coins. These are not chosen at random but pseudorandomly as $f_{r_U}(\alpha)$. That is, he obtains the coins by applying the pseudorandom function to the current α-value, using r_U as key. He also proves that he has done exactly this. Note that this will force a clone of the user to use the same coins if he gives a proof for the same α-value, by security of the commitment and signature schemes. This idea of choosing the randomness for a proof pseudorandomly is somewhat similar to a technique from a completely different context, namely resetable zero-knowledge [15].

The proof given is actually a Σ-protocol, so the transcripts of proofs given by user and clone are of form (a, e, z) and (a', e', z'). But when all inputs and random coins are the same in the two cases, we must have $a = a'$. Furthermore, $e \neq e'$ with overwhelming probability, so if both proofs are accepted, special soundness of the protocol means that one can easily compute the prover's secret, which will immediately identify the user in question.

We now describe the components of our scheme – throughout the descriptions, it is understood that a party who detects an invalid proof or signature will immediately stop and reject:

KeyGen. On input 1^k, it generates keys $(Psign, Ssign)$ for the signature scheme and public key $Pcom$ for the commitment scheme (with security parameter k). Finally, it chooses a random k-bit string R. The public key is $pk = (Psign, Pcom, R)$ while the private key is $sk = Ssign$.

Join. The user U sends $c_U = commit_{Pcom}(r_U, s_U)$ where r_u is a random k-bit string. GM assigns a unique identity id_U to U, and sends to U a signature $\sigma_U = Sign(Ssign, (c_U, id_U))$ on c_U concatenated by id_U. Also, GM proves in zero-knowledge that he knows a signature (valid under $Psign$) on R. This is easy given that GM knows $Ssign$. The output certificate for U is r_U, s_U, σ_U, id_U, while output for GM is id_U.

Prove. Recall that pk and the string α is common input to the protocol. User U first makes commitments C_U, D_U, E_U to c_u, id_U, σ_U, respectively. He will now give a proof of knowledge related to these commitments, the group public key pk and the number α assigned to the current phase. This proof consists of three ingredients. The first is a proof of knowledge, that U knows how to open the commitments C_U, D_U, E_U to strings c_u, id_U, σ_U such that σ_U is GM's signature on (c_U, id_U). While giving this proof, he uses $f_{r_U}(\alpha)$ as random coins. That is, the protocol transcript is (a_1, e_1, z_1), where it should be the case $a_1 = a_1((pk, C_U, D_U, E_U), (c_U, id_U, \sigma_U), f_{r_U}(\alpha))$.

The second ingredient is a proof that U can open C_U to reveal c_U and he knows s_U, r_U such that $c_U = commit(r_U, s_U)$, and the message a_1 from the previous protocol satisfies $a_1 = a_1((pk, C_U, D_U, E_U), (c_U, id_U, \sigma_U), f_{r_U}(\alpha))$. Also this proof is a three move protocol of form (a_2, e_2, z_2), and we are going to do the two proofs in parallel, so that the overall conversation will have form $(a_1, a_2, e_1, e_2, z_1, z_2)$. The final ingredient is a proof of knowledge of GM's signature on the string R that is part of pk. This is combined with the previous ingredients using the OR construction mentioned above, i.e., U is proving that he knows a signature on R, or strings $c_u, id_U, \sigma_U, r_U, s_U$ satisfying the conditions just described [3].

Detect. Looks at all the proofs given in a phase and finds all places where two conversations include tuples of form $(a_1, a_2, e_1, e_2, z_1, z_2)$, respectively $(a_1' a_2', e_1', e_2', z_1', z_2')$ and where $a_1 = a_1'$ and $e_1 \neq e_1'$. For any such case it will use the special soundness property to extract the underlying c_U, id_U, σ_U, and appends id_U to its output list.

Theorem 1. *Assuming one-way functions exist, the above scheme is a secure unclonable identification scheme with sequential security.*

We remark that concurrent security can be obtained under the same assumption in the common reference string model, using a technique similar to the one used in the more efficient protocol we describe later.

The key to the proof of the theorem is

Lemma 1. *The proof of knowledge given by the user during the Prove protocol is witness indistinguishable*

Proof. Recall that the proof given by U is a combination using the OR construction of first a proof of knowledge of a signature on R and second a proof of knowledge of values $c_u, id_U, \sigma_U, r_U, s_U$ satisfying a number of properties. Conversations in the latter protocol are of form $(a_1, a_2, e_1, e_2, z_1, z_2)$. The OR construction leads to a witness-indistinguishable protocol if both protocols used are honest verifier zero-knowledge. This is true for the first protocol, which is just a standard Σ-protocol and so is honest verifier zero-knowledge by construction.

It is therefore enough to show that the second protocol is honest verifier zero-knowledge. Some notation for this: the part (a_1, e_1, z_1) of a conversation will be called *proof 1*. It has the commitments C_U, D_U, E_U and public key pk as public input, while the secret witness is c_U, id_U, σ_U. The rest of the conversation (a_2, e_2, z_2) is called *proof 2*. It has C_U, D_U, E_U, pk, a_1 as public input while the secret witness is $c_U, id_U, \sigma_U, r_U, s_U$.

Both proof 1 and proof 2 are Σ-protocols constructed from generic zero-knowledge techniques as explained above. They therefore have honest verifier simulators M_1, M_2 respectively. However, note that in our context, proof 1 is not done using the normal prover algorithm, we use pseudorandom coins for the prover, and furthermore the key for this pseudorandomness is used as input in proof 2. Hence a proof is required that we can still use M_1, M_2 to simulate. We do this by defining a series of distributions where the first is that of real

---

[3] Of course, the latter is normally the case, the other option is included for proof-technical reasons.

conversations and the last is the one output by the honest verifier simulator we propose. The result will then follow from arguing that each distribution is computationally indistinguishable from the previous one.

The sequence of distributions are produced as follows:

1. Run the honest prover U's algorithm (with known secret witnesses and random challenges).
2. Same as above, but proof 2 is replaced by running the honest verifier simulator $M_2(C_U, D_U, E_U, pk, a_1)$ for proof 2. Note that this requires that r_U is known, to do proof 1 according to the protocol. However, we will still get something indistinguishable from the previous distribution. This is because the output of M_2 is indistinguishable from a real conversation, *even to someone who knows the secret witness for proof 2*. Indeed, M_2 is simulating a protocol constructed from generic techniques based on any commitment scheme as explained earlier. This means that the simulation essentially produces a set of commitments, some of which are opened and some are not. The unopened commitments have contents different from what would be the case in a real conversation, however, this is the only difference. By the hiding property of the commitments, this difference cannot be detected in polynomial time, even knowing what the commitments are supposed to contain.
3. As 2., but the commitment c_U is replaced by a commitment to a random value. This is indistinguishable from 2. by the hiding property of commitments.
4. As 3., but when doing proof 1, instead of using r_U to compute pseudorandom values for the random coins, we use oracle access to the function $f_{r_U}()$. We now do not know r_u explicitly, but we will produce exactly the same distribution as in 3.
5. As in 4., but the oracle access to $f_{r_U}()$ is replaced by oracle access to a random function. This is indistinguishable from 4. by pseudorandomness of the function $f_{r_U}()$.
6. As in 5., but the transcript of proof 1 is now generated by running the honest verifier simulator M_1 for proof 1. This is indistinguishable from 5., since there, we ran proof 1 following the prover's normal algorithm, using real random coins. Summarizing, this last distribution is generated by first running $M_1(C_U, D_U, E_U, pk)$ to get (a_1, e_1, z_1), and running $M_2(C_U, D_U, E_U, pk, a_1)$ to get (a_2, e_2, z_2), and this defines the desired honest verifier simulation.

We can now proceed with the proof of the required properties.

Anonymity: if $\tilde{V}$ behaves such that at least one instance of the *Join* protocol completes successfully with non-negligible probability, then we can extract from the proof of knowledge given by $\tilde{V}$ a signature on R. Note that no attempts to do *Prove* would occur before this point. Given this signature, it is trivial to simulate (without rewinding) all subsequent instances of *Prove* knowing only the number of instances to be done in each phase. This cannot be distinguished from the real game by witness indistinguishability of the underlying proofs of knowledge.

No cloning: we first describe the required *Extract* algorithm. It will, for each proof $\tilde{U}$ had accepted in the last stage of the attack, rewind $\tilde{U}$ to the start of this proof and try to extract the secret witness it is using by the standard rewinding technique of sending random challenges to $\tilde{U}$ until it answers a new challenge correctly. At this point a valid witness can be extracted. Each such witness must include either a signature on R, or a signature σ_U on a pair of form (c_U, id_U). *Extract* outputs id_U in the latter case, and a random string in the former. We put the limitation that the algorithm gives up on a proof and outputs a random string if it rewinds more than 2^k times, where k is the length of challenges.

To estimate the running time of this, note that the probability that $\tilde{U}$ will have a proof accepted, given the state it is in just before the proof, is determined by the number T of challenges it will answer correctly. The probability that we will have to run *Extract* on the proof is $T2^{-k}$, while the number of rewinds we have to do is 0 if $T = 0$, 2^k if $T = 1$ and $2^k/(T-1)$ if $T > 1$. It follows that contribution to the total expected running time from each proof is polynomial. The total expected running time is just the sum of these contributions since we compose sequentially.

To finalize the argument, we need the following

Claim: we may assume that in the output of *Extract*, we will only see triples (c_U, id_U, σ_U) that were obtained earlier by $\tilde{U}$ in some instance of *Join*.

Indeed, if this is false with non-negligible probability, we can break the signature scheme in a chosen message attack: we choose at random to either ask for signatures on all pairs c_U, id_U or a signature on R and use this to simulate the *Join* protocols done by $\tilde{U}$ and all proofs by honest users given to $\tilde{U}$ (without rewinding, we just follow the protocol). Then by witness indistinguishability, $\tilde{U}$'s behaviour will be essentially the same as before, so the knowledge extraction from $\tilde{U}$ will give us a signature on a new message with non-negligible probability.

Consider now any two *Prove* instances where the same triple c_U, id_U, σ_U is extracted. Let $(a_1, a_2, e_1, e_2, z_1, z_2)$, $(a_1', a_2', e_1', e_2', z_1', z_2')$ be the transcripts of the two instances.

Now, soundness of the *Prove* protocol implies that we could extract (except with negligible probability) also two ways of opening c_U from the two instances, that is, two pairs $(r_U, s_U), (r_U', s_U')$ such that

$$c_U = commit_{Pcom}(r_U, s_U) = commit_{Pcom}(r_U', s_U'),$$

and that

$$a_1 = a(pk, (c_U, id_U, \sigma_U), f_{r_U}(\alpha)), a_1' = a(pk, (c_U, id_U, \sigma_U), f_{r_U'}(\alpha)).$$

But we must have $r_U = r_U'$, or the the binding property of the commitment scheme is broken. This immediately implies that $a_1 = a_1'$, and therefore, since $e_1 \neq e_1'$ with overwhelming probability, *Detect* will successfully extract id_U, as required in the definition.

4   A More Efficient Solution

In this section, we present a more efficient unclonable group identification scheme, based on two main ingredients: First a technique recently proposed by Camenisch and Lysyanskaya [3] for digital signatures based on bilinear groups, with protocols for proving knowledge of a signature on a committed value. Second, a new technique for proving that an element in a group is of form g^{ψ} where ψ is a pseudorandom value computed from a committed key. We will borrow some notation from [3] (and several earlier papers): given a public string x, a private witness w and a predicate $pred$,

$$PK\{w : pred(x, w)\}$$

means that we execute a Σ-protocol for the relation $\{(x, w)|\ pred(x, w) = true\}$, that is, a prover convinces a verifier that he knows w such that the predicate on x and w is satisfied. We will also use the following variant:

$$PK(\kappa)\{w : pred(x, w)\}$$

where κ is a bit string. This stands for the following: we execute the underlying Σ-protocol in the normal interactive way, except that the verifier sends as the second message a random string κ, and the challenge the prover has to answer is determined as $H(x, a, \kappa)$, where H is a hash function, modelled as a random oracle and a is the first message in the original protocol. The point of this construction is that it allows simulation of the protocol without rewinding, due to the "programmability" of the random oracle, and (for the same reason) it also allows knowledge extraction by standard rewinding. Since we will need the last point for the proof, we cannot just use the Fiat-Shamir heuristic.

4.1   Proofs of Knowledge with Pseudorandom Exponents

In this subsection, we introduce some tools to be used in our construction. To this end, we consider a group G_p of prime order p. We will assume p is chosen as a safe prime, i.e., $p = 2q + 1$ where q is also prime. G_q will denote the (unique) subgroup of Z_p^* of order q.

We further consider the case where a prover knows exponents $x_1, ..., x_t \in Z_p$ such that $\beta = \alpha_1^{x_1} \cdots \alpha_t^{x_t}$ for publicly known $\beta, \alpha_1, ..., \alpha_t \in G_p$. A standard Σ-protocol for prover P and verifier V can be used to prove knowledge of the x_i's. That is, we want:

$$PK\{(x_1, ..., x_t) : \beta = \alpha_1^{x_1} \cdots \alpha_t^{x_t}\} \tag{1}$$

This Σ-protocol works as follows:

1. P chooses $r_1, ..., r_t \in Z_p$ uniformly at random and sends to V $\tau = \prod_{i=1}^{t} \alpha_i^{r_i}$.
2. V chooses a random challenge $\epsilon \in Z_p$.
3. P responds with $z_i = r_i + \epsilon x_i \bmod p$ for $i = 1..t$. V checks that $\prod_{i=1}^{t} \alpha_i^{z_i} = \tau \beta^{\epsilon}$.

It is well known (and straightforward to show) that this protocol is indeed a Σ-protocol for the underlying relation. It is also well-known that a straightforward variant of the protocol allows us to do the following type of proof:

$$PK\{(x_1, .., x_t, x_1', ..., x_t') : \beta = \alpha_1^{x_1} \cdots \alpha_t^{x_t}, \beta' = {\alpha'}_1^{x_1'} \cdots {\alpha'}_t^{x_t'}, x_1 = x_1'\}$$

Basically, we run the original protocol twice in parallel for the two equations. This would normally involve two independent sets of random numbers $r_1, ..., r_t$ and $r_1', ..., r_t'$. However, to demonstrate that $x_1 = x_1'$ the prover must use $r_1 = r_1'$ and the verifier checks that the responses $z_1, ..., z_t, z_1', ...z_t'$ satisfy $z_1 = z_1'$. We will use this variant later, but for now we stick to the basic version for simplicity. All techniques we describe here can also be applied to the variant in a straightforward way.

We now consider a change to the protocol where P chooses the randomness in the first message according to a pseudorandom function $\Psi_K(i, \alpha, b)$, where K is a key committed to by P, α is a public input, i is a number and b is a bit. We will use a variant of the pseudorandom function of Naor and Reingold, based on the DDH assumption in G_q, so that outputs from Ψ are in G_q. We specify below how the function works and how the key is committed. However, in the previous protocol, the random exponents were chosen in Z_p, whereas the pseudorandom function produces output in the subgroup G_q. To resolve this, we let the exponents be chosen as the difference between two pseudorandom values, which allows us to hit all of Z_p. The modified protocol then works as follows:

1. P sets $r_i = \Psi_K(i, \alpha, 0)$ and $s_i = \Psi_K(i, \alpha, 1)$ and sends to V $\tau = \prod_{i=1}^t \alpha_i^{r_i - s_i}$.
2. V chooses a random challenge $\epsilon \in Z_p$.
3. P responds with $z_i = r_i - s_i + \epsilon x_i \mod p$ for $i = 1..t$. V checks that $\prod_{i=1}^t \alpha_i^{z_i} = \tau \beta^\epsilon$.

To argue that this is a Σ-protocol for the same relation, we need a result by Perron[20]: Let QR_p be the set of quadratic residues mod p. Then for any $a \in Z_p*$, the set $a + QR_p$ contains almost as many quadratic residues as non-residues: the difference is at most 1. Since in our case $G_q = QR_p$, we get from this:

Lemma 2. *The distribution of $u_i - v_i \mod p$ where u_i, v_i are chosen uniformly in G_q, is statistically close to uniform over Z_p.*

Lemma 3. *Under the DDH assumption in G_q, the above protocol is a Σ-protocol for the relation specified in (1).*

Proof. Completeness is trivial, and special soundness follows exactly as for the previous standard protocol. For honest verifier zero-knowledge, we argue as follows: To simulate, we will choose ϵ and z_i at random in their respective domains and then set $\tau = \beta^{-\epsilon} \prod_{i=1}^t \alpha_i^{z_i}$.

Now, assuming K is known only to P, pseudorandomness of Ψ implies that our variant is indistinguishable from a protocol where $\Psi_K(i, \alpha, 0), \Psi_K(i, \alpha, 1)$ are replaced by uniformly random choice u_i, v_i from G_q. This creates a distribution of z_i that is statistically close to the simulated distribution by Lemma 2.

Our goal is now to allow P to prove that he has followed the specified algorithm for choosing the r_i, s_i's pseudorandomly. The first step of this is to have P commit to each individual value under a public key chosen by a third party (which will eventually be the group manager in our case). The public key will be two random elements $\eta, \lambda \in G_p$, and P will make commitments $com_i = \eta^{r_i}\lambda^{\omega_i}$ and $com'_i = \eta^{s_i}\lambda^{\omega'_i}$, for $i = 1..t$ and random ω_i, ω'_i. We can now ask P to prove that he committed to the correct values, that is, execute

$$PK\{(r_i, s_i, \omega_i, \omega'_i, i = 1..t) : \tau = \prod_{i=1}^{t} \alpha^{r_i}(\alpha^{-1})^{s_i},$$

$$com_i = \eta^{r_i}\lambda^{\omega_i}, com'_i = \eta^{s_i}\lambda^{\omega'_i}, i = 1..t\}$$

The Σ-protocol for this is a standard variant of the one we presented above.

The final step is to show that each committed value was chosen according to the pseudorandom function. For this, we need to specify in detail how it works. We assume that input strings to Ψ all have length at most k (where k can in principle be arbitrary). A key to the function is a number $K \in Z_q$. Finally, we will need a hash function H that take a string str of length at most k as input and outputs an element in G_q. We will model this function as a random oracle. The pseudorandom function is now defined as:

$$\Psi_K(str) = H(str)^K \bmod p$$

We note that the function mapping y to $y^K \bmod p$ is a weak pseudorandom function assuming the DDH assumption holds in G_q, i.e., as long as y is randomly chosen and is not controlled by the adversary, the outputs look random. However, in our case, and assuming the random oracle model, the function is only used on values produced by H, and these are guaranteed to be random, even if the adversary chooses the inputs to H. This argument is easily formalized to prove.

Lemma 4. *In the random oracle model, and assuming DDH holds in G_q, $\Psi_K()$ as defined above is a strong pseudorandom function.*

We will assume that the key K is committed to by P in a somewhat non-standard way which, however, fits nicely with the construction we will see in the following. Concretely, we assume that $d = g^{\gamma^K \delta^r}h^u$ is given, for publicly known $g, h \in G_p$ and $\gamma, \delta \in G_q$. With this, we can summarize our goal, namely to give a Σ-protocol implementing

$$PK\{(K, r, u, \omega_i, \omega'_i, i = 1..t) : d = g^{\gamma^K \delta^r}h^u,$$

$$com_i = \eta^{\Psi_K(i,\alpha,0)}\lambda^{\omega_i}, com'_i = \eta^{\Psi_K(i,\alpha,1)}\lambda^{\omega'_i}, i = 1..t\}$$

For this, it will be be enough to show how P can prove that some given commitment com satisfies $com = \eta^{\Psi_K(str)}\lambda^{\omega}$ for public str. Since anyone can compute $\psi = H(str)$, our task reduces to:

$$PK\{(K, r, u, \omega) : d = g^{\gamma^K \delta^r}h^u, com = \eta^{\psi^K}\lambda^{\omega}\} \tag{2}$$

A protocol for this follows here:

1. P chooses $s, w \in Z_q, \nu, \phi \in Z_p$ at random. He sends $v_1 = g^{\gamma^s \delta^w} h^\nu$ and $v_2 = \eta^{\psi^s} \lambda^\phi$ to V.
2. V selects a random bit c.
3. P sends $z_1 = s - cK \bmod q$, $z_2 = w - cr \bmod q$, $z_3 = \nu - cu\gamma^{s-K}\delta^{w-r} \bmod p$, and $z_4 = \phi - c\omega\psi^{s-K} \bmod p$.
   V checks as follows: if $c = 0$, that $g^{\gamma^{z_1}\delta^{z_2}} h^{z_3} = v_1$ and $\eta^{\psi^{z_1}} \lambda^{z_4} = v_2$. If $c = 1$, that $d^{\gamma^{z_1}\delta^{z_2}} h^{z_3} = v_1$ and $com^{\psi^{z_1}} \lambda^{z_4} = v_2$.

Since this protocol only works with a 1-bit challenge, we need to repeat it an appropriate number of times to have a sufficiently small soundness error.

Lemma 5. *The above is a Σ-protocol for the relation specified in (2)*

Proof. Completeness follows by inspection of the protocol. Special soundness: if for given v_1, v_2, the prover can send satisfactory answers z_1, z_2, z_3, z_4 to $c = 0$ and z_1', z_2', z_3', z_4' to $c = 1$, we have by the checks carried out by V that $g^{\gamma^{z_1}\delta^{z_2}} h^{z_3} = v_1$, $\eta^{\psi^{z_1}} \lambda^{z_4} = v_2$. $d^{\gamma^{z_1'}\delta^{z_2'}} h^{z_3'} = v_1$ and $com^{\psi^{z_1'}} \lambda^{z_4'} = v_2$. Combining these equations imply that $com = \eta^{\psi^{z_1 - z_1'}} \lambda^{(z_4 - z_4')\psi^{-z_1'}}$ and $d = \alpha^{\gamma^{z_1 - z_1'}\delta^{z_2 - z_2'}} h^{(z_3 - z_3')\gamma^{-z_1'}\delta^{-z_2'}}$, i.e., a, d are of the required form. Finally, honest verifier ZK is argued by the following simulator: choose z_1, z_2 at random in Z_q, z_3, z_4 at random in Z_p and c as a random bit. If $c = 0$, set $v_1 = g^{\gamma^{z_1}\delta^{z_2}} h^{z_3}$ and $v_2 = \eta^{\psi^{z_1}} \lambda^{z_4}$. If $c = 1$, set $v_1 = d^{\gamma^{z_1}\delta^{z_2}} h^{z_3}$ and $v_2 = com^{\psi^{z_1}} \lambda^{z_4}$. This simulation is seen to be perfect by a standard argument.

4.2   The New Scheme

Our main idea for the scheme is similar to the earlier theoretical one: the user U will commit to a secret key K. When registering with the group manager GM he will obtain a signature on the commitment c_U, using the signature system described in [3] (called scheme A in [3]). He can now prove membership of the group by proving knowledge of a valid signature on c_U (as well as proving knowledge of this value). If he tries to clone his identity we can exploit the special soundness property of the protocol used and extract his identity.

KeyGen. Let GM take a security parameter k and output two groups $G_p = \langle g \rangle$ and $\mathbf{G_p} = \langle \mathbf{g} \rangle$ of prime order $p = \Theta(2^k)$ where $p = 2q + 1$ and q is a prime. Let G_q denote the unique subgroup of Z_p^* of order q. Let γ, δ be random generators of G_q. Let $e : G_p \times G_p \to \mathbf{G_p}$ be an efficiently computable bilinear map, and η, λ be random generators of $\mathbf{G_p}$.
To set up the signature scheme, GM chooses the following values at random: $x \in Z_p$, $y \in Z_p$ and sets $X = g^x$, $Y = g^y$. The secret key for the signature scheme is $S_k = (x, y)$ and the public key is $P_k = (q, G_p, \mathbf{G_p}, g, \mathbf{g}, e, X, Y, \eta, \lambda)$.
Join. The user U chooses at random $r_U \in Z_q$ and a key $K \in Z_q$. U makes a commitment $c_U = \gamma^K \delta^{r_U} \bmod p$ to K and sends it to GM[4].

---

[4] It is not necessary to have U prove that he can open c_U, since later in the *Prove* protocol, he must implicitly show he can open it to have the proof accepted.

GM verifies that U is allowed to join the group and if so, he computes a signature $\sigma = (a, b, c)$ on c_U where a is chosen at random in G_p, $b = a^y$, $c = a^{x+c_U xy}$ and sends it to U. GM considers c_U as the user's id in the following, whereas (K, r_U, a, b, c) serves as the membership certificate.

Prove. Recall that the string α, denoting "the current time", is common input to U and V. U essentially proves that he is a member of a group by proving that he knows a valid message and signature from GM. First U blinds his signature σ by choosing at random $\mu, r' \in Z_p$ and computing $\widetilde{\sigma} = \left(\widetilde{a}, \widetilde{b}, \widehat{c}\right)$ where $\widetilde{a} = a^{r'}$, $\widetilde{b} = b^{r'}$, $\widehat{c} = (c^{r'})^\mu$. Then U makes a commitment $C_U = \eta^{c_U} \lambda^{s_U}$ for random s_U and sends $\widetilde{\sigma}$ and C_U to V. Both compute

$$v_x = e\left(X, \widetilde{a}\right), \quad v_{xy} = e\left(X, \widetilde{b}\right), \quad v_s = e\left(g, \widehat{c}\right)$$

V chooses a k-bit string κ at random, and U proves knowledge of a signature on c_U to V by giving the following proof:

$$PK(\kappa)\{(c_U, \rho, s_U) : C_U = \eta^{c_U} \lambda^{s_U}, v_s^{\rho} = v_x v_{xy}^{c_U}\} \tag{3}$$

Here, as ρ, the honest U uses $\rho = \mu^{-1} \bmod p$. V will accept if this proof is correct and it holds that:

$$e\left(\widetilde{a}, Y\right) = e\left(g, \widetilde{b}\right)$$

Note that it was shown in [3] that the checks carried out by V plus the proof that $v_s^{\rho} = v_x v_{xy}^{c_U}$ together imply that U must know a valid signature on a message. Doing the proof in (3) is a straightforward variant of the general type of proof from lemma 3 as discussed earlier. Consider the underlying Σ-protocol for the part relating to $v_s^{\rho} = v_x v_{xy}^{c_U}$. After specializing it to the concrete scenario here, it will have a first message of form

$$\tau = v_{xy}^{r_1 - s_1} v_s^{r_2 - s_2},$$

Furthermore, we will require that

$$r_1 = \Psi_K(1, \alpha, 0), r_2 = \Psi_K(2, \alpha, 0), s_1 = \Psi_K(1, \alpha, 1), s_2 = \Psi_K(2, \alpha, 1)$$

U must therefore prove that the values of r_1, r_2 and s_1, s_2 were generated pseudorandomly from K. As described earlier, U does this by making commitments

$$com_1 = \eta^{r_1} \lambda^{\omega_1}, com_2 = \eta^{r_2} \lambda^{\omega_2}, com_1' = \eta^{s_1} \lambda^{\omega_1'}, com_2' = \eta^{s_2} \lambda^{\omega_2'},$$

proving that these values are correct with respect to τ and finally using the protocol from Lemma 5 to show that each commitment contains a pseudorandom value of correct form computed from K as committed to by C_U. Note that C_U is in fact a commitment to K of exactly the form needed for that protocol.

All proofs to be given during Prove can be done simultaneously, using the same challenge in all Σ-protocols.

The amount of computation required during Prove is $57 + 68r$ exponentiations and 8 evaluations of the function e, where r is the number of times the protocol from Lemma 5 is repeated. Prove also requires $29k + r(6k + 1)$ bits sent between U and GM, where r is the same as before and k is the security parameter. For example with $r = 16$ and $k = 1024$, Prove requires approximately 1150 exponentiations and needs to communicate $130KB$. It would be interesting to find a protocol that solves the same problem using a constant number of exponentiations and communication.

Detect. Look at all proofs given in a phase and find all places where two conversations include first messages τ, τ' where $\tau = \tau'$. If the two challenge values involved in these two conversations are different, use the special soundness property to extract a witness for the proof in question - this will be a pair of form (c_U, ρ). Output all c_U's found this way.

Theorem 2. *Assuming security of the signature scheme from [3], the DDH assumption in G_q, and Assumption 1, the scheme described above is a secure unclonable identification scheme in the random oracle model, with sequential security. The Join and Prove protocols are constant-round, and have communication complexity $O(k)$ bits, respectively $O(k^2)$ bits.*

The scheme described here is extremely similar in structure to the theoretical solution we gave earlier, so the proof is very similar as well. We only sketch it here. Completeness follows by inspection of the protocols. For no cloning, the required *Extract* algorithm will use standard rewinding to extract witnesses for all proofs given. By a standard argument, this will succeed for all proofs that were accepted by the verifier, with overwhelming probability. Soundness of the proofs means we will extract a set of user id's and corresponding signatures, so security of the signature scheme implies that this forms a subset of the user id's defined in previous *Join* protocols. Now, soundness of the proofs from Lemma 5 and the binding property of the commitment schemes defined by $(\gamma, \delta), (\eta, \lambda)$ imply that the adversary must have used the key involved correctly and consistently, and hence the value of τ will be identical in all instances of subproof (3), where the same key was used. This allows *Detect* to recover the required information. As for anonymity, note that all subproofs except the one from (3) can be replaced by (perfect) simulations without changing the view of the adversary. After this change, the key K is only used to call the pseudorandom function, and no other information on K is present, since the commitment c_U hides K perfectly. We can therefore use Lemma 3 to conclude that also instances of subproofs from (3) can be replaced by simulations without this being detectable by the adversary.

5   On Concurrent Security

For both the theoretical and the more efficient solution, it holds that all the proofs given by honest users can be simulated without rewinding. Hence, the only problem in obtaining concurrent security lies in the *Extract* algorithm that is required for the no cloning property, and which requires rewinding in both solutions.

   To avoid this, we can use, at a small efficiency cost, the technique of Fischlin [11], which shows how to transform any Σ-protocol in the random oracle model into a new one for which there is an on-line extractor, i.e., one can extract the secret witness from a successful prover without rewinding.

Using this transformation on the Σ-protocols underlying our *Prove*-protocol immediately gives a concurrently secure solution.

6 On Membership Revocation and Framing

After discovering the identity of a dishonest user, the group manager needs to act. In some applications it may be sufficient to take some appropriate, say, legal action against the user in question. But it may also be necessary to remove the user out from the group by ensuring that the value c_U can never be used again.

Since the value c_U is unconditionally hidden in the *Prove* protocol, nothing in the above systems prevents a dishonest user from proving membership of the group again at a later point in time. To allow for revocation of memberships, we can extend the protocol with an *dynamic accumulator* as described in [4]. An *accumulator scheme* [1, 2] is an algorithm that allows one to hash large set of values into a short value, called the *accumulator* such that there is a witness that a given input is in the accumulator. A *dynamic accumulator* allows one to efficiently add and remove values from the accumulator. It can be used in the following way.

When the user joins the group and sends c_U, the group manager adds c_U to the accumulator. To prove membership of the group, the user is now required, in addition to the protocol we already have, to prove that the value c_U is in the accumulator. We will omit the details of how this is done, they can be found in [4]. The solutions needs that c_U is committed to, but this is already done in our protocol.

When the identity of a dishonest user is discovered, the group manager removes c_U from the accumulator, which prevents the user or any clones of the user from proving membership of the group, as long as the verifier is aware of the new accumulator value.

An aspect we have not been concerned with in this paper is whether the group manager can *frame* an honest user, that is, create on his own a protocol transcript where the user seems to have cloned himself. For our efficient solution, we believe framing is not possible, since the group manager does not know the user's secret key K. The part of the *Prove* protocol where the user proves knowledge of K can only be simulated without knowing K if one can control the outputs of the random oracle. While we use this to show zero-knowledge in the theoretical analysis, such control is not available to anyone in real life.

References

1. Josh Benaloh, Michael de Mare: *One-Way Accumulators: A Decentralized Alternative To Digital Signatures*, proc. of EUROCRYPT 1993.
2. Josh Benaloh, Michael de Mare: *Collision-free accumulators and fail-stop signature schemes without trees*, proc. of EUROCRYPT 1997.
3. J. Camenisch, A. Lysyanskaya: *Signature Schemes and Anonymous Credentials from Bilinear Maps*, Proc. of Crypto 04, Springer Verlag LNCS 3152.

4. J. Camenisch, A. Lysyanskaya: *Dynamic Accumulators and Application to Efficient Revocation of Anonymous Credentials*, Proc. of Crypto 02, Springer Verlag LNCS 2442.

5. G.Ateniese, J.Camenisch, M.Joye, G.Tsudik: *A practical and provably group signature scheme*, Proc. of Crypto 00, Springer Verlag LNCS 1880.

6. S.Brands: *Untraceable Off-line Cash in Wallets with Observers*, proc. of Crypto 93.

7. Ronald Cramer, Ivan Damgård: *Fast and Secure Immunization Against Adaptive Man-in-the-Middle Impersonation.* EUROCRYPT 1997: 75-87

8. Ronald Cramer, Ivan Damgård, Berry Schoenmakers: *Proofs of Partial Knowledge and Simplified Design of Witness Hiding Protocols*, CRYPTO 1994: 174-187

9. Ivan Damgård, Mads Jurik: *Client/Server Tradeoffs for Online Elections.* Public Key Cryptography 2002: 125-140

10. Damgård, Dupont and Pedersen: *Unclonable Group Identification* the Eprint archive, www.iacr.org.

11. M.Fischlin: *Communication-Efficient Non-Interactive Proofs of Knowledge*, proc. of Crypto 2005, Springer Verlag LNCS 3621.

12. Markus Jakobsson, Kazue Sako, Russell Impagliazzo: *Designated Verifier Proofs and Their Applications.* EUROCRYPT 1996: 143-154.

13. Oded Goldreich, Shafi Goldwasser, Silvio Micali: *How to Construct Random Functions*, FOCS 1984: 464-479.

14. Oded Goldreich, Silvio Micali, Avi Wigderson: *Proofs that Yield Nothing But Their Validity or All Languages in NP Have Zero-Knowledge Proof Systems* J. ACM 38(3): 691-729 (1991).

15. Ran Canetti, Oded Goldreich, Shafi Goldwasser, Silvio Micali: *Resettable zero-knowledge* (extended abstract). STOC 2000: 235-244

16. J.Kilian and E.Petrank: *Identity Escrow*, Proc. of Crypto 98.

17. D.Chaum, A.Fiat, M.Naor: *Untraceable Electronic Cash*, proc. of CRYPTO 88.

18. Moni Naor *Bit Commitment Using Pseudorandomness*, J. Cryptology 4(2): 151-158 (1991).

19. Pascal Paillier: *Public-Key Cryptosystems Based on Composite Degree Residuosity Classes.* EUROCRYPT 1999: 223-238.

20. Perron: *Bemerkungen über die Verteilung der quadratischen Reste*, Math.Z. 56 (1952), pp.122-130.

21. John Rompel: *One-Way Functions are Necessary and Sufficient for Secure Signatures*, STOC 1990: 387-394.

22. Aggelos Kiayias, Moti Yung: *Group Signatures with Efficient Concurrent Join.* EUROCRYPT 2005: 198-214

Fully Collusion Resistant Traitor Tracing with Short Ciphertexts and Private Keys

Dan Boneh[1,*], Amit Sahai[2,**], and Brent Waters[3]

[1] Stanford University
dabo@cs.stanford.edu
[2] U.C.L.A.
sahai@cs.ucla.edu
[3] SRI International
bwaters@csl.sri.com

Abstract. We construct a fully collusion resistant tracing traitors system with sublinear size ciphertexts and constant size private keys. More precisely, let N be the total number of users. Our system generates ciphertexts of size $O(\sqrt{N})$ and private keys of size $O(1)$. We first introduce a simpler primitive we call *private linear broadcast encryption* (PLBE) and show that any PLBE gives a tracing traitors system with the same parameters. We then show how to build a PLBE system with $O(\sqrt{N})$ size ciphertexts. Our system uses bilinear maps in groups of composite order.

1 Introduction

Traitor tracing systems, introduced by Chor, Fiat, and Naor [10], help content distributors identify pirates. Consider a content distributor who broadcasts encrypted content to N legitimate recipients. Recipient i has secret key K_i that it uses to decrypt the broadcast. As a concrete example, imagine an encrypted satellite radio broadcast that should only be played on certified radio receivers. The broadcast is encrypted using a public broadcasting key BK. Any certified player can decrypt using its embedded secret key K_i. Certified players, of course, could enforce digital rights restrictions such as "do not copy" or "play once".

The risk for the distributor is that a pirate will hack a certified player and extract its secret key. The pirate could then build a pirate decoder that will extract the cleartext content and ignore any relevant digital rights restrictions. Even worse, the pirate could make its pirate decoder widely available so that anyone can extract the cleartext content for themselves. DeCSS, for example, is a widely distributed program for decrypting encrypted DVD content.

This is where traitor tracing systems come in — when the pirate decoder is found, the distributor can run a *tracing* algorithm that interacts with the pirate decoder and outputs the index i of at least one of the keys K_i that the pirate

---

* Supported by NSF and the Packard Foundation.
** This research was supported by the NSF Cybertrust and ITR Programs, an Alfred P. Sloan Research Fellowship, and a generous equipment grant from Intel.

S. Vaudenay (Ed.): EUROCRYPT 2006, LNCS 4004, pp. 573–592, 2006.

used to create the pirate decoder. The distributor can then try to take legal action against the owner of this K_i.

We give a precise description of traitor tracing systems in Appendix A. For now we give some intuition that will help explain our results. A traitor tracing system consists of four algorithms *Setup, Encrypt, Decrypt*, and *Trace*. The setup algorithm generates the broadcaster's key BK, a tracing key TK, and N recipient keys $K_1, \ldots, K_N$. The encrypt algorithm encrypts the content using BK and the decrypt algorithm decrypts using one of the K_i. The tracing algorithm is the most interesting — it is an algorithm that takes TK as input and interacts with a pirate decoder, treating it as a black-box oracle. It outputs the index $i \in \{1, \ldots, N\}$ of a key K_i that was used to create the pirate decoder.

In this paper we focus on fully collusion resistant traitor tracing systems. That is, systems that remain secure no matter how many keys are at the disposal of the pirate. Existing traitor tracing systems are not designed to handle arbitrary collusions. When the collusion bound t comes close to N, most existing systems require ciphertext size linear in the number of users, which is no better than the trivial traitor tracing system.

Our results. We construct a practical fully collusion resistant traitor tracing system that has sub-linear size ciphertexts. Our system has the following characteristics:

$$\text{ciphertext-length} = O(\sqrt{N}) \qquad \text{and} \qquad \text{private-key-length} = O(1)$$

Furthermore, decryption time is constant (i.e. depends on the security parameter, but not on N). Other properties of this system include: (1) the broadcaster's key BK is public, but the tracer's key TK must be kept secret, (2) the system is black-box traceable, and (3) is designed for stateless pirate decoders [18]. We give a precise definition of these properties in Appendix A. The system uses bilinear groups of composite order introduced in [5].

We prove security of our tracing algorithm using a tracing technique previously used in [4, 23, 18]. To formalize this technique, we introduce a new primitive called *Private Linear Broadcast Encryption*, or PLBE for short, which is conceptually a simpler primitive than traitor tracing. We show that any secure PLBE gives a (black-box) traitor tracing system. Roughly speaking, a PLBE is a broadcast encryption system [13] that can only broadcast to "linear" sets, that is sets of the form $\{i, i+1, \ldots, N\}$ for some $i = 1, \ldots, N+1$. Thus, a PLBE enables the broadcaster to create ciphertexts that can only be decrypted properly under keys $K_i, K_{i+1}, \ldots, K_N$. A broadcast to everyone, for example, is encrypted using $i = 1$. The main security requirement is that the system should be *private* [1]: a ciphertext should reveal no non-trivial information about the recipient set. That is, a broadcast to users $\{i, \ldots, N\}$ should reveal no non-trivial information about i. We give a precise definition in the next section and show that any secure PLBE gives a secure (black-box) traitor tracing system. In the remainder of the paper we focus on constructing a secure PLBE.

Related work. Traitor tracing systems generally fall into two categories: combinatorial, as in [10, 24, 31, 32, 14, 15, 11, 28, 2, 30, 29, 23], and algebraic, as in [21, 4, 25, 20, 12, 22, 34, 9]. The broadcaster's key BK in combinatorial systems can be either secret or public. Algebraic traitor tracing use public-key techniques and are often more efficient than the public-key instantiations of combinatorial schemes. Some systems, including ours, only provide tracing capabilities. Other systems [25, 23, 17, 16, 12] combine tracing with broadcast encryption to obtain trace-and-revoke features — after tracing, the distributor can revoke the pirate's keys without affecting any other legitimate decoder.

Kiayias and Yung [20] describe a black-box tracing system that achieves constant rate for long messages, where rate is measured as the ratio of ciphertext length to plaintext length. For full collusion resistance, however, the ciphertext size is linear in the number of users N. For comparison, our new system generates ciphertexts of size $O(\sqrt{N})$ and achieves constant rate (rate = 1) for long messages by using hybrid encryption (i.e. encrypting a short message-key using the traitor tracing system and encrypting the long data by using a symmetric cipher with the message-key).

Many traitor tracing systems, including ours, assume that the tracer is a trusted party and require that the tracer's key TK be kept secret. Some exceptions are [26, 27, 35, 19, 9]. Similarly, many traitor tracing systems, including ours, assume that the pirate decoder is stateless. Kiayias and Yung [18] show how to strengthen traitor tracing systems to handle stateful decoders.

Finally, we note that binary fingerprinting codes [8, 33] are closely related to traitor tracing (binary refers to the fact that the code is defined over a binary alphabet). In fact, it is known [6] that any binary fingerprinting code gives rise to a fully collusion-resistant traitor tracing system with *constant* size ciphertexts. The private key size, unfortunately, is quite large. Using [8] the private key size is $\tilde{O}(N^3)$ and using [33] it is $\tilde{O}(N^2)$.

2   Traitor Tracing and Private Linear Broadcast Encryption

In Appendix A we review the precise definition of a traitor tracing system. However, instead of directly building a traitor tracing system we build a simpler primitive called *Private Linear Broadcast Encryption* (PLBE). We first define secure PLBEs below and then briefly explain how a PLBE is used for traitor tracing. The resulting tracing algorithm makes explicit a tracing technique used in [4, 23, 18]. Then in the remainder of the paper we build a secure PLBE.

2.1   Description of Private Linear Broadcast Encryption

A PLBE is comprised of the following four algorithms:

Setup$_{\text{LBE}}(N, \lambda)$. The setup algorithm takes as input N, the number of users in the system, and the security parameter λ. The algorithm runs in polynomial time in λ and outputs a public key PK, a secret key TK, and private keys $K_1, \ldots, K_N$, where K_u is given to user u.

Encrypt$_{\text{LBE}}$(PK, M). Takes as input a public key PK, and a message M and outputs a ciphertext C. This algorithm is used to encrypt a message to all N users.

TrEncrypt$_{\text{LBE}}$(TK, i, M). Takes as input a secret key TK, an integer i satisfying $1 \leq i \leq N + 1$, and a message M. It outputs a ciphertext C. This algorithm encrypts a message to a set $\{i, \ldots, N\}$ and is primarily used for traitor tracing. We will require below that $TrEncrypt_{\text{LBE}}$(TK, 1, M) outputs a distribution on ciphertexts that is indistinguishable from the distribution generated by $Encrypt_{\text{LBE}}$(PK, M).

Decrypt$_{\text{LBE}}$(j, K$_j$, C, PK). Takes as input a private key K$_j$ for user j, a ciphertext C, and the public key PK. The algorithm outputs a message M or $\bot$.

The system must satisfy the following **correctness property**:
for all $i, j \in \{1, \ldots, N + 1\}$, where $j \leq N$, and all messages M:

Let $\left(\text{PK}, \text{TK}, (\text{K}_1, \ldots, \text{K}_N) \right) \overset{\text{R}}{\leftarrow} Setup_{\text{LBE}}(N, \lambda)$

and let $C \overset{\text{R}}{\leftarrow} TrEncrypt_{\text{LBE}}(\text{TK}, i, M)$.

If $\mathbf{j \geq i}$ then $Decrypt_{\text{LBE}}(j, \text{K}_j, C, \text{PK}) = M$.

Security. We define security of a PLBE system using three games. The first game just captures a consistency property which says that $TrEncrypt_{\text{LBE}}$(TK, 1, M) outputs a distribution on ciphertexts that is indistinguishable from the distribution generated by $Encrypt_{\text{LBE}}$(PK, M). The second game is a **message hiding game** and says that a ciphertext created using index $i = N + 1$ is unreadable by anyone. The third game is an **index hiding game** and captures the intuition that a broadcast ciphertext created using index i reveals no non-trivial information about i. We will consider all these games for a fixed number of users N.

Game 1 – Indistinguishability. The first game says that the output of algorithm $TrEncrypt_{\text{LBE}}$(TK, 1, M) is indistinguishable from $Encrypt_{\text{LBE}}$(PK, M). The game proceeds as follows:

- **Setup.** The challenger runs the $Setup_{\text{LBE}}$ algorithm and gives the adversary PK and the set of all private keys $\{\text{K}_1, \ldots, \text{K}_N\}$.
- **Challenge.** The adversary gives the challenger a message M. The challenger flips a coin $\beta \in \{0, 1\}$ and computes

$$c \overset{\text{R}}{\leftarrow} \begin{cases} TrEncrypt_{\text{LBE}}(\text{TK}, 1, M) & \text{if } \beta = 0, \\ Encrypt_{\text{LBE}}(\text{PK}, M) & \text{if } \beta = 1. \end{cases}$$

It gives C to the adversary.
- **Guess.** The adversary returns a guess $\beta' \in \{0, 1\}$ of β.

We define the advantage of adversary $\mathcal{A}$ as $\mathsf{Adv}_{CG} = |\Pr[\beta' = \beta] - 1/2|$.

Game 2 – Message Hiding. The second game says that an adversary cannot break semantic security when encrypting using index $i = N + 1$. The game proceeds as follows:

- **Setup.** The challenger runs the $Setup_{LBE}$ algorithm and gives the adversary PK and all secret keys $\{K_1, \ldots, K_N\}$.
- **Challenge.** The adversary outputs two equal length messages M_0, M_1. The challenger flips a coin $\beta \in \{0, 1\}$ and sets $C \xleftarrow{R} TrEncrypt_{LBE}(TK, N+1, M_\beta)$. The challenger gives C to the adversary.
- **Guess.** The adversary returns a guess $\beta' \in \{0, 1\}$ of β.

We define the advantage of adversary $\mathcal{A}$ as $\mathsf{Adv}_{MH} = |\Pr[\beta' = \beta] - 1/2|$.

Game 3 – Index Hiding. The third game says that an adversary cannot distinguish between an encryption to index i and one to index $i + 1$ without the key K_i. The game takes as input a parameter $i \in \{1, \ldots, N\}$ which is given to both the challenger and the adversary. The game proceeds as follows:

- **Setup.** The challenger runs the $Setup_{LBE}$ algorithm and gives the adversary PK and the set of private keys $\{K_j \text{ s.t. } j \neq i\}$.
- **Challenge.** The adversary outputs a message M. The challenger flips a coin $\beta \in \{0, 1\}$ and computes $C \xleftarrow{R} TrEncrypt_{LBE}(TK, i + \beta, M)$. The challenger returns C to the adversary.
- **Guess.** The adversary returns a guess $\beta' \in \{0, 1\}$ of β.

We define the advantage of adversary $\mathcal{A}$ as $\mathsf{Adv}_{IH}[i] = |\Pr[\beta' = \beta] - 1/2|$.

Now that the three games are established we are ready to define secure PLBE.

Definition 1. *We say that an N-user Private Linear Broadcast System (PLBE) is secure if for all polynomial time adversaries $\mathcal{A}$ we have that Adv_{CG}, and Adv_{MH}, and $\mathsf{Adv}_{IH}[i]$ for $i = 1, \ldots, N$, are negligible functions of λ.*

2.2 Reducing Traitor Tracing to PLBE

We briefly show that a secure PLBE gives a secure traitor tracing system. The complete details and proofs are given in the full version of the paper [7]. Let $\mathcal{E} = (Setup_{LBE}, Encrypt_{LBE}, TrEncrypt_{LBE}, Decrypt_{LBE})$ be a secure PLBE system. The derived traitor tracing system is defined as follows (we use the notation of Appendix A):

- *Setup* simply runs $Setup_{LBE}$ with the same parameters, and outputs PK as the public encryption key, TK as the secret tracing key, and the user keys identically to the PLBE scheme.
- *Encrypt* and *Decrypt* run algorithms $Encrypt_{LBE}$ and $Decrypt_{LBE}$ respectively with the same parameters.
- $Trace^{\mathcal{D}}(TK, \epsilon)$, when called with oracle $\mathcal{D}$, and inputs TK and $\epsilon > 0$, does the following:

1. For $i = 1$ to $N + 1$, do the following:
   (a) The algorithm repeats the following $8(N \ln N)\lambda/\epsilon$ times:
      i. Sample M from the finite message space at random.
      ii. Let $C \xleftarrow{\text{R}} TrEncrypt_{\text{LBE}}(\text{TK}, i, M)$.
      iii. Call oracle $\mathcal{D}$ on input C, and compare the output of $\mathcal{D}$ to M.
   (b) Let $\hat{p}_i$ be the fraction of times that $\mathcal{D}$ decrypted the ciphertexts correctly.
2. Let S be the set of all $i \in \{1, \ldots, N\}$ for which $\hat{p}_i - \hat{p}_{i+1} \geq \epsilon/(4N)$.
3. Output the set S as the set of guilty colluders.

Note that the running time of *Trace* is quadratic in N. It can be made $O(N \log N)$ using binary search instead of a linear scan.

Security. We prove that this traitor tracing scheme is secure. We argue that the system is semantically secure and provides secure tracing. Note that we did not explicitly require that a PLBE be semantically secure against a chosen plaintext attack to an outsider who possess no secret keys. Nevertheless, semantic security does follow straightforwardly from the three games used to define PLBE using a hybrid argument by means of the Index Hiding game.

We now briefly explain why traceability against arbitrary collusion follows from the security of the PLBE scheme. We show that the probability of winning the traceability game defined in Appendix A is negligible.

Let $p_i = \Pr[\mathcal{D}(TrEncrypt_{\text{LBE}}(\text{TK}, i, M)) = M]$. We know that that $p_1 \geq \epsilon$ and p_{N+1} is negligible. The former follows from the fact that $\mathcal{D}$ is a useful decoder. The later follows directly from the PLBE message hiding game. Then there must exist some $j \in \{1, \ldots, N\}$ such that $p_j - p_{j+1} \geq \epsilon/(2N)$. By the Chernoff bound it follows that with overwhelming probability, $\hat{p}_j - \hat{p}_{j+1} \geq \epsilon/(4N)$. Hence, the set S output by $Trace^{\mathcal{D}}(S_{\mathcal{D}}, \text{TK}, \epsilon)$ is non-empty.

Using the notation of Game 2 from Appendix A, it remains to show that whenever $\hat{p}_j - \hat{p}_{j+1} > \epsilon/(4N)$ we have that $j \in T$. For such j we know, by Chernoff, that with overwhelming probability $p_j - p_{j+1} \geq \epsilon/(8N)$. Hence, $\mathcal{D}$ is able to distinguish $TrEncrypt_{\text{LBE}}(\text{TK}, j, M)$ from $TrEncrypt_{\text{LBE}}(\text{TK}, j+1, M)$ for random M. But since the PLBE is secure, the index hiding game implies that these two distributions are indistinguishable, unless one has K_j. It follows that the pirate who built $\mathcal{D}$ must have had K_j and therefore $j \in T$, as required. We give the full proof details in the full version of the paper.

3   Background and Complexity Assumptions

3.1   Bilinear Maps

We review some general notions about bilinear maps and groups, with an emphasis on groups of *composite order* which will be used in our construction. We follow [5] in which composite order bilinear groups were first introduced.

Consider two finite cyclic groups $\mathbb{G}$ and $\mathbb{G}_T$ of same order $n = pq$, where p and q are distinct primes, and in which the respective group operation is efficiently computable and denoted multiplicatively. Assume the existence of an efficiently computable function $e : \mathbb{G} \times \mathbb{G} \to G_T$, with the following properties:

- (Bilinear) $\forall u, v \in G$, $\forall a, b \in \mathbb{Z}$, $e(u^a, v^b) = e(u, v)^{ab}$, where the product in the exponent is defined modulo n;
- (Non-degenerate) $\exists g \in G$ such that $e(g, g)$ has order n in G_T. In other words, $e(g, g)$ is a generator of G_T, whereas g generates G.

We will use the notation $\mathbb{G}_p, \mathbb{G}_q$ to denote the respective subgroups of order p and order q of $\mathbb{G}$.

We now review three assumptions we will use for proving our security. The first two assumptions are in prime order subgroups and the last two are over a composite group $\mathbb{G}$.

3.2 Decision 3-Party Diffie-Hellman Assumption

The decision 3-party Diffie-Hellman problem is stated as follows. Given a group $\mathbb{G}_p$ of prime order p and random elements $g_p, A = g_p^a, B = g_p^b, C = g_p^c$ of $\mathbb{G}$ distinguish between $T = g_p^{abc}$ and $T = g_p^z$, where z is random in $\mathbb{Z}_p$.

We say that an algorithm $\mathcal{A}$ has advantage ϵ in solving the problem if

$$\left| \Pr[\mathcal{A}(g_p, g_p^a, g_p^b, g_p^c, g_p^{abc}) = 1] - \Pr[\mathcal{A}(g_p, g_p^a, g_p^b, g_p^c, g_p^z) = 1] \right| \geq \epsilon$$

The (t, ϵ)-decision 3-party Diffie-Hellman assumption (D3DH) is that no t-time adversary has advantage more than ϵ. Note that the decision 3-party Diffie-Hellman assumption implies the decision Bilinear Diffie-Hellman assumption. It also implies the standard linear assumption defined in [3].

3.3 Subgroup Decision Problem

The Subgroup Decision (SD) problem is stated as follows. Given a group $\mathbb{G}$ of composite order $n = pq$, where p, q are distinct (unknown) primes, and generators $g_p \in \mathbb{G}_p$ and $g \in \mathbb{G}$, distinguish between whether an element T is a random member of the subgroup $\mathbb{G}_p$ or a random element of the full group $\mathbb{G}$. That is distinguish whether T is a random element of $\mathbb{G}_p$ or $\mathbb{G}$.

We say that an algorithm $\mathcal{A}$ has advantage ϵ in solving the Subgroup Decision Problem if

$$\left| \Pr[\mathcal{A}(n, g_p, g, T) = 1 \; : \; T \overset{\text{R}}{\leftarrow} \mathbb{G}_p] - \Pr[\mathcal{A}(n, g_p, g, T) = 1 \; : \; T \overset{\text{R}}{\leftarrow} \mathbb{G}] \right| \geq \epsilon.$$

The (t, ϵ)-subgroup decision assumption is that no t-time adversary has advantage more than ϵ.

3.4 Bilinear Subgroup Decision Problem

The Bilinear Subgroup Decision (BSD) problem is stated as follows. Given a group $\mathbb{G}$ of composite order $n = pq$, where p, q are distinct (unknown) primes, and generators $g_p \in \mathbb{G}_p$ and $g_q \in \mathbb{G}_q$, distinguish a random order p element in the group $\mathbb{G}_T$ from a uniform element in the group $\mathbb{G}_T$. More precisely, we say that an algorithm $\mathcal{A}$ has advantage ϵ in solving the problem if

$$\left| \Pr[\mathcal{A}(n, g, g_p, g_q, \ e(T, g)) = 1 \ : \ T \overset{R}{\leftarrow} \mathbb{G}_p] - \right.$$

$$\left. \Pr[\mathcal{A}(n, g, g_p, g_q, \ e(T, g)) = 1 \ : \ T \overset{R}{\leftarrow} \mathbb{G}] \right| \geq \epsilon.$$

The (t, ϵ)-bilinear subgroup decision assumption is that no t-time adversary has advantage more than ϵ.

4   A $\sqrt{N}$ Size Private Linear Broadcast Encryption System

In this section we show how to construct a Private Linear Broadcast Encryption (PLBE) system with $O(\sqrt{N})$ size ciphertext. We can then apply the results of Section 2 and use this to build a traitor tracing scheme with $O(\sqrt{N})$ size ciphertexts.

Before we describe our construction we give some intuition as to why constructing PLBE systems with sublinear ciphertext size is difficult and describe the framework for which we will construct our PLBE system.

PLBE with Sublinear Ciphertext Size. The primary difficulty in constructing a PLBE system is to provide the Index Hiding property. Using linear size ciphertexts this is easy: each user has a unique portion of the ciphertext assigned to them, which is used to encrypt the message (or session key) to just that user. If an encryptor replaces the ciphertext component of user u with a random encryption, only user u can tell the difference. All other users will be associated with a completely different portions of the ciphertext and changing u's component has no effect on their ability to decrypt.

To construct a PLBE system with sublinear size ciphertexts we must use a fundamentally different approach than the one above. Since the ciphertexts are sublinear in size, we cannot let every user have a component of the ciphertext that is dedicated for them alone. Intuitively, ciphertext components must be "shared" amongst users. Therefore, we cannot use the simple strategy of completely randomizing a portion of the ciphertext to prevent a particular user u from decrypting, since this will inherently effect the ability of other users to decrypt.

Our Framework. We now give a framework for our PLBE system. We assume that the number of users, N in the system equals m^2 for some m. If the number of real users is not a square we can add "dummy" users to pad out to the next square. We arrange the users in an $m \times m$ matrix. Each user is assigned and identified by an unique tuple (x, y) where $1 \leq x, y \leq m$.

Since we will be constructing a Private Linear Broadcast Encryption system, we must have a linear ordering of the users that we can traverse. The first user in the system will be the user at matrix position $(1,1)$ and from there we will order the users by traversing one row at a time. More precisely, the user at matrix position (x,y) will have the index $u = (x-1)m + y$ in our ordering. We can think of this as a "row-major" ordering.

We can now refer to our Private Linear Broadcast Encryption scheme in terms of positions on the matrix. An encryption to position (i,j) means that a user at position (x,y) will be able to decrypt the message if either $x > i$ or both $x = i$ and $y \geq j$. With this notation, the Index Hiding game property states that:

- For $j < m$ it is difficult to distinguish between an encryption of a message to (i,j) from $(i,j+1)$ without the key of user $(x = i, y = j)$.
- For $j = m$ it is difficult to distinguish an encryption of a message to position $(i, j = m)$ to that of one to $(i+1, j = 1)$ without the key of user $(i, j = m)$.

The use of pairwise notation for referring to users and encryptions will be a purely notational convenience for describing our system.

4.1   Our Construction

Our construction makes use of bilinear maps of composite order n, where $n = pq$ and p and q are primes. In describing our scheme we will often use p or q in a subscript to denote if a group element is in the subgroup of order p or order q. *The key algebraic fact that underlies our scheme is that if g_p is any element from the order p subgroup (which we call $\mathbb{G}_p$) and g_q is any element from the order q subgroup (which we call $\mathbb{G}_q$), then we have: $e(g_p, g_q) = 1$.*

When the $TrEncrypt_{\mathrm{LBE}}$ algorithm encrypts to an index (i,j) it creates ciphertext components for every column and every row. The keys of user (x,y) are structured in such a way that in order to decrypt he must pair the ciphertext components from row x, with the ciphertext components from column y. The encryption algorithm works by creating ciphertexts in the following way.

Column Ciphertext Components. (1) Ciphertexts for columns greater than or equal to j are "well formed" in both subgroups. (2) However, for a column that is less than j, the encryption algorithm will create a ciphertext that is well formed in the $\mathbb{G}_q$ subgroup, but random in the $\mathbb{G}_p$ subgroup.

Row Ciphertext Components. (1) Ciphertexts for rows less than i are completely random. Therefore, any user whose row index is less than x will not be able to decrypt. (2) The ciphertext components for row i are well formed in both subgroups. A user with row index i will be able to decrypt if his column index is greater than or equal to j. If it is less than j, the randomized ($\mathbb{G}_p$) part of the column ciphertext will scramble the result of pairing the row and column ciphertexts together. (3) Finally, for rows greater than i the ciphertext components will be well formed elements in the $\mathbb{G}_q$ subgroup only. A user with row index greater than i will be able to decrypt no matter what his column is, because the pairing will "cancel out" the randomized ($\mathbb{G}_p$) part of any column ciphertext component with the row ciphertext component that lives in $\mathbb{G}_q$.

The decryption algorithm for a user (x, y) will attempt to decrypt a ciphertext in the same manner no matter what the target index (i, j) is. The structure of the ciphertext will restrict decryption to only be successful for a user (x, y) if $x > i$ or $x = i$ and $y \geq j$. Additionally, since the attempted decryption procedure is independent of (i, j) a user can only learn whether his decryption was successful or not and the system will be private.

We describe the four algorithms that compose our PLBE system:

$Setup_{\text{LBE}}(N = m^2, 1^\kappa)$. The setup algorithm takes as input the number of users N and a security parameter κ. It first generates an integer $n = pq$ where p, q are random primes (whose size is determined by the security parameter). The algorithm creates a bilinear group $\mathbb{G}$ of composite order n. It next creates random generators $g_p, h_p \in \mathbb{G}_p$ and $g_q, h_q \in \mathbb{G}_q$ and sets $g = g_p g_q, h = h_p h_q \in \mathbb{G}$. Next it chooses random exponents $r_1, \ldots, r_m, c_1, \ldots, c_m, \alpha_1, \ldots, \alpha_m \in \mathbb{Z}_n$ and $\beta \in \mathbb{Z}_q$.

The public key PK includes the description of the group and the following elements:

$$\left[\begin{array}{l} g, h, E = g^\beta, E_1 = g_q^{\beta r_1}, \ldots, E_m = g_q^{\beta r_m}, F_1 = h_q^{\beta r_1}, \ldots, F_m = h_q^{\beta r_m}, \\ G_1 = e(g_q, g_q)^{\beta \alpha_1}, \ldots, G_m = e(g_q, g_q)^{\beta \alpha_m}, H_1 = g^{c_1}, \ldots, H_m = g^{c_m} \end{array} \right]$$

The private key for user (x, y) is generated as $K_{x,y} = g^{\alpha_x} g^{r_x c_y}$. Finally, the authority's secret key K includes factors p, q along with exponents used to generate the public key.

$TrEncrypt_{\text{LBE}}(K, M, (i, j))$. The $TrEncrypt_{\text{LBE}}$ algorithm is a secret key algorithm used by the tracing authority. The algorithm encrypts a message M to the subset of receivers that have row values greater than i or both row value equal to i and column values greater than or equal to j.

The encryption algorithm will take as input the secret key, a message $M \in \mathbb{G}_T$ and an index i, j. The encryption algorithm first chooses random $t \in \mathbb{Z}_n$, $w_1, \ldots, w_m, s_1, \ldots, s_m \in \mathbb{Z}_n$, $z_{p,1}, \ldots, z_{p,j-1} \in \mathbb{Z}_p$, and $(v_{1,1}, v_{1,2}, v_{1,3})$, $\ldots$, $(v_{i-1,1}, v_{i-1,2}, v_{i-1,3}) \in \mathbb{Z}_n^{(3)}$.

For each row x we create four ciphertext components $(R_x, \tilde{R}_x, A_x, B_x)$ as follows:

$$\begin{array}{llll} \text{if } x > i : R_x = g_q^{s_x r_x} & \tilde{R}_x = h_q^{s_x r_x} & A_x = g_q^{s_x t} & B_x = Me(g_q, g)^{\alpha_x s_x t} \\ \text{if } x = i : R_x = g^{s_x r_x} & \tilde{R}_x = h^{s_x r_x} & A_x = g^{s_x t} & B_x = Me(g, g)^{\alpha_x s_x t} \\ \text{if } x < i : R_x = g^{v_{x,1}} & \tilde{R}_x = h^{v_{x,1}} & A_x = g^{v_{x,2}} & B_x = e(g, g)^{v_{x,3}} \end{array}$$

For each column y the algorithm creates values $C_y, \tilde{C}_y$ as:

$$\begin{array}{ll} \text{if } y \geq j : C_y = g^{c_y t} h^{w_y} & \tilde{C}_y = g^{w_y} \\ \text{if } y < j : C_y = g^{c_y t} g_p^{z_{p,y}} h^{w_y} & \tilde{C}_y = g^{w_y} \end{array}$$

Note that the ciphertext contains $5\sqrt{N}$ elements in $\mathbb{G}$ and $\sqrt{N}$ elements of $\mathbb{G}_T$.

In the above description there are three classes of rows. A row $x > i$ will have all its elements in the $\mathbb{G}_q$ subgroup, while the "target" row i will have its components in the full group $\mathbb{G}$. A row $x < i$ will essentially have its group elements randomly chosen. A column $y \geq j$ will be well formed, while a column $y < j$ will be well formed in the $\mathbb{G}_q$ subgroup, but not in the $\mathbb{G}_p$ subgroup.

$Encrypt_{\text{LBE}}$(PK, M). The $Encrypt_{\text{LBE}}$ algorithm is used by an encryptor to encrypt a message such that all the recipients can receive it. This algorithm is used during normal (non-tracing) operation to distribute content to all the receivers. The $Encrypt_{\text{LBE}}$ algorithm should produce ciphertexts that are indistinguishable from $TrEncrypt_{\text{LBE}}$ algorithm to the index $(1, 1)$ for the same message.

The encryption algorithm first chooses random $t \in \mathbb{Z}_n$, $w_1, \ldots, w_m, s_1, \ldots, s_m \in \mathbb{Z}_n$. For each row x the algorithm creates the four ciphertext components $(R_x, \tilde{R}_x, A_x, B_x)$ as follows:

$$R_x = E_x^{s_x} \qquad \tilde{R}_x = F_x^{s_x} \qquad A_x = E^{s_x t} \qquad B_x = MG_x^{s_x t}$$

For each column j the algorithm creates $C_y, \tilde{C}_y$ as:

$$C_y = H_y^t h^{w_y} \qquad \tilde{C}_y = g^{w_y}$$

$Decrypt_{\text{LBE}}$$((x, y), K_{x,y}, C)$ User (x, y) uses key $K_{x,y}$ to decrypt by computing:

$$B_x \cdot \left(e(K_{x,y}, A_x)e(\tilde{R}_x, \tilde{C}_y)/e(R_x, C_y) \right)^{-1}.$$

We observe that if the ciphertext was created from the tracing algorithm $TrEncrypt_{\text{LBE}}$ with parameters (i, j) then the result is M if $x > i$ or $x = i$ and $y \geq j$. Additionally, it is easy to observe that if the ciphertext was created as $Encrypt_{\text{LBE}}$(PK, M) then all parties can decrypt and receive M.

4.2   Discussion

Roughly, the size of the ciphertext is $5\sqrt{N}$ elements in $\mathbb{G}$ and $\sqrt{N}$ elements of $\mathbb{G}_T$. In practice, a message will be encrypted with a symmetric key cipher under a key K and our system will be used to transmit the key K to each user. We note that we can actually save in ciphertext size by converting our encryption system into a Key Encapsulation Mechanism (KEM). To do this we do not include the B_x values in the ciphertext, but instead user (x, y) can extract a key $K_x = e(K_{x,y}, A_x)e(\tilde{R}_x, \tilde{C}_y)/e(R_x, C_y)$. The extraction mechanism will actually derive $\sqrt{N}$ different keys $K_1, \ldots K_m$, so key K_x is used to encrypt K to for all users in row x. In practice this would be more space efficient than including $\sqrt{N}$ group elements of $\mathbb{G}_T$.

The $Encrypt_{\text{LBE}}$ algorithm requires $6\sqrt{N}$ exponentiations. The decryption algorithm is surprisingly efficient and simple, requiring only three pairing computations. Thus, decryption time is independent of the number of users in the system.

We constructed a (limited)[1] broadcast encryption system in which decryptors are oblivious as to which set of users the broadcast is targeted for. A set of colluding users will of course be able to learn some information about the target just by testing which one of them was able to decrypt. However, they should not learn anything more than what can naturally be inferred. The key to keeping the broadcast set private is that the decryption algorithm performs the same steps to attempt decryption no matter what the broadcast set is. In the next section we prove this intuition to be correct by showing that our scheme is secure in the Index Hiding game.

5   Security Proof

In this section we prove our Private Linear Broadcast Encryption system secure. We begin with the Index Hiding game, since the proof is the most interesting.

5.1   Proof of Security for Game 3 (Index Hiding)

For the Index Hiding game we must consider two cases. The first is when an adversary tries to distinguish between an encryption to (i, j) and an encryption to $(i, j + 1)$ for $j < m$ and second for when an adversary tries to distinguish between an encryption (i, m) and one to $(i + 1, 1)$.

In the first case we show that the difficulty of this game can be reduced to the 3-party Diffie-Hellman assumption, while the second case is more complicated since the structure of the row ciphertexts are changed. We handle the second case by constructing a sequence of hybrid experiments. Due to space requirements we give the proof of the lemma for the first case in the appendix and refer the reader to our full version of this paper [7] for the proofs of the other claims and lemmas.

Theorem 1. *Suppose that the (t, ϵ_{D3DH})-decision 3-party Diffie-Hellman, (t, ϵ_{BSD})-Bilinear Subgroup Decision, and (t, ϵ_{SD})-Subgroup Decision assumptions hold. Then no $\tilde{t}$-time adversary $\mathcal{A}$ can succeed in the Index-Hiding game with advantage greater than $(2 + m)\epsilon_{D3DH} + \epsilon_{BSD} + \epsilon_{SD}$, where $\tilde{t} \approx t$.*

We first consider the case where an adversary $\mathcal{A}$ attempts to distinguish between an encryption to $(i + j)$ and $(i, j + 1)$ where $j < m$. This is the case when the distinguishing game does not cross rows. We prove the following lemma in Appendix B.

Lemma 1. *Suppose that the (t, ϵ_{D3DH})-decision 3-party Diffie-Hellman, assumption holds. Then no t-time adversary can distinguish between an encryption to (i, j) and $(i, j + 1)$ in the Index Hiding game for $j < m$ with advantage $> \epsilon_{D3DH}$.*

---

[1] A Private Linear Broadcast Encryption system is restricted in the sets of users it can encrypt to — it can only encrypt to sets $\{i, \dots, N\}$ for any i.

We now turn to the more difficult case of when the adversary $\mathcal{A}$ chooses to distinguish between an encryption to (i, m) and one to $(i + 1, 1)$ for some $1 \leq i < m$. This case becomes more complicated because the form of ciphertext rows will change. In our proofs we will refer to the rows with ciphertexts in the $\mathbb{G}_q$ subgroup as "greater than" rows and the the row with well formed ciphertexts in $\mathbb{G}$ as a "target" row. Additionally, when we say we "encrypt to column j" this means that we create ciphertexts for which C_y is well formed in the $\mathbb{G}_p$ subgroup for all $y \geq j$. We state our lemma and then prove it.

Lemma 2. *Suppose the (t, ϵ_{D3DH})-decision 3-party Diffie-Hellman, the (t, ϵ_{BSD}) -Bilinear Subgroup Decision, and the (t, ϵ_{SD})-Subgroup Decision assumptions hold. Then no $\tilde{t}$-time adversary $\mathcal{A}$ can succeed in the Index-Hiding game with advantage greater than $(2 + m)\epsilon_{D3DH} + \epsilon_{BSD} + \epsilon_{SD}$, where $\tilde{t} \approx t$.*

We first define a sequence of hybrid experiments as follows:

- H_1: Encrypt to column m, row i is target row, i+1 is a "greater than" row.
- H_2: Encrypt to column $m + 1$, row i is target row, i+1 is a "greater than" row.
- H_3: Encrypt to column $m + 1$, row i is less than row, i+1 is a "greater than" row (no target row exists).
- H_4: Encrypt to column 1, row i is less than row, i+1 is "greater than" row (no target row exists).
- H_5: Encrypt to column 1, row i is less than row, i+1 is target row.

We prove our lemma by giving reductions for each consecutive pair of hybrid experiments. The proofs are given in [7].

Claim. Suppose that the (t, ϵ_{D3DH})-decision 3-party Diffie-Hellman assumption holds. Then no t-time adversary can distinguish between experiments H_1 and H_2 with advantage greater than ϵ_{D3DH} .

In both experiments we encrypt with row i as the target row and all C_y for $y < m$ random in the $\mathbb{G}_p$ subgroup. The experiment is whether an adversary can tell if the $\mathbb{G}_p$ component of C_m is well-formed without key $K_{i,m}$. This game is exactly the same as the one we proved above and thus we apply the result of Lemma 1.   □

Claim. Suppose that the (t, ϵ_{D3DH})-decision 3-party Diffie-Hellman and the (t, ϵ_{BSD})-Bilinear Subgroup Decision assumptions hold. Then no t-time adversary can distinguish between experiments H_2 and H_3 with advantage greater than $2\epsilon_{D3DH} + \epsilon_{BSD}$.

Claim. Suppose that the (t, ϵ_{D3DH})-decision 3-party Diffie-Hellman assumption holds. Then no t-time adversary can distinguish between experiments H_3 and H_4 with advantage greater than $m \cdot \epsilon_{D3DH}$.

Claim. Suppose that the (t, ϵ_{SD})-Subgroup Decision assumption holds. Then no t-time adversary can distinguish between experiments H_4 and H_5 with advantage greater than ϵ_{SD} .

Lemma 2 follows by summing the maximum adversarial advantages across the hybrid experiments and Theorem 1 follows by observing that the bound of Lemma 1 is included in Lemma 2.   □

5.2 Proof of Security for Game 1

Theorem 2. *Suppose the (t, ϵ_{SD}) Subgroup Decision assumption holds. Then for all messages M no t-time adversary can distinguish between a ciphertext created as $Encrypt_{LBE}(PK, M)$ and one created as $TrEncrypt_{LBE}(K, M, (1, 1))$ with advantage greater than ϵ_{SD}.*

This theorem follows by simply applying the same techniques as in our proof of Claim 5.1, so we omit the details.   □

5.3 Proof of Security for Game 2 (Message Hiding)

Theorem 3. *All adversaries have advantage 0 in playing the Message Hiding game.*

The message hiding theorem is concerned with the adversaries advantage in winning the game when we encrypt to $(m + 1, 1)$. However, this means that all rows will be completely random and independent of the messge, thus an adversary has 0 advantage. Essentially, the inability of the adversary to learn the message when he does not have any of the right keys is actually captured in our Index Hiding experiments. This final theorem shows that at the end the adversary learns now information about the ciphertext.   □

6 Discussion

Our traitor tracing system has a number of possible interesting extensions for future work. In this section we discuss a few of these.

Public Traceability. In our current system the tracing key, TK, is kept secret and only the authority is able to trace pirate boxes. In practice, it might be useful to have a system where the tracing key is public. For example, in a large content distribution system the capturing and tracing of pirate boxes or software will likely be done by different several agents each of which will need the tracing key. We would like our system to remain secure even if one of these agents and his tracing key is compromised.

In our $\sqrt{N}$ PLBE system the tracing algorithm would be public if a user was able to encrypt a message to an arbitrary set of indices (i, j). Then the user could simply run the tracing algorithm in the same way as the authority. In order to this we would need to give the user the capability to form C_y column ciphertext components that were well formed in its $\mathbb{G}_q$ subgroup, but not in the $\mathbb{G}_p$ subgroup. If we simply include an element of $\mathbb{G}_p$ in the public key our scheme will become insecure as an attacker could use this to determine which row index i a broadcast was intended for. Achieving public traceability would seem to require a more complex technique and possibly the use of a stronger assumption.

Stateful Receivers. Like most other tracing traitor solutions our solution solves the tracing traitors problem in the stateless model, where the tracer is allowed to reset the pirate algorithm after each tracing query. However, there are some applications where we would like to consider a stronger model where a pirate box can retain state between each broadcast. In practice, a hardware pirate box might keep state and shut down if it detects that it is being traced.

Kiayias and Yung [18] showed a method which can handle stateful receivers if it were possible to embed watermarks in the distributed content and for a tracer to be able to observe these watermarks when interacting with a pirate algorithm. During *non-tracing* operation the broadcaster encrypts two copies of digital content, each of which has a different watermark embedded in, to a random (and hidden) index u. The encryption is such that all users with index less than u can decrypt the first ciphertext and all users with index greater than u can decrypt the second ciphertext. The decryption algorithm simply tries to decrypt both ciphertexts and uses whichever one results in a well-formed plaintext. The tracing algorithm will create ciphertexts in an *identical* manner to the regular encryption algorithm. The tracer will simply observe which watermarks are embedded in every probing ciphertext and use this information to identify the traitor. Since, the regular broadcast and tracing algorithms are identical a pirate box is unable to leverage its ability to maintain state.

In our current construction, our PLBE scheme is only secure if the pirate constructing the pirate decoder has not seen encryptions to arbitrary indices. However, if we were able to find a new PBLE algorithm that was secure under chosen-plaintext queries to arbitrary indicies then we could implement the techniques of Kiayias and Yung. We would simply set up two PLBE systems in which the users were given the opposite indices in each system. The user with index u in the first system has index $N + 1 - u$ in the second system.

7   Conclusions and Open Problems

We constructed the first fully collusion resistant traitor tracing system with sublinear size ciphertexts and constant size private keys. In particular, our system has ciphertexts of size $O(\sqrt{N})$ where N is the number of users in the system and the time for decryption is independent of N. We achieve our traitor tracing system by first introducing a simpler primitive we call private linear broadcast encryption (PLBE) that we show can give a traitor tracing system. Then, we built an efficient PLBE system by making novel use of bilinear groups of composite order.

One interesting open problem is to create a version of our traitor system that allows for public traceability. This would allow both for the tracer to be untrusted and could be used to give a solution that is secure against stateful receivers. Additionally, it is an open problem to see if one can get smaller than $\sqrt{N}$ size ciphertexts with small private keys.

References

[1] Adam Barth, Dan Boneh, and Brent Waters. Privacy in encrypted content distribution using private broadcast encryption. In *Financial Cryptography '06*, 2006.

[2] O. Berkman, M. Parnas, and J. Sgall. Efficient dynamic traitor tracing. In *Proceedings of SODA '00*, 2000.

[3] Dan Boneh, Xavier Boyen, and Hovav Shacham. Short group signatures. In Matt Franklin, editor, *Proceedings of Crypto 2004*, LNCS. Springer-Verlag, August 2004.

[4] Dan Boneh and Matthew K. Franklin. An efficient public key traitor tracing scheme. In *CRYPTO '99: Proceedings of the 19th Annual International Cryptology Conference on Advances in Cryptology*, pages 338–353, London, UK, 1999. Springer-Verlag.

[5] Dan Boneh, Eu-Jin Goh, and Kobbi Nissim. Evaluating 2-dnf formulas on ciphertexts. In Joe Kilian, editor, *Proceedings of Theory of Cryptography Conference 2005*, volume 3378 of *LNCS*, pages 325–342. Springer, 2005.

[6] Dan Boneh and Moni Naor. Tracing traitors with constant size ciphertext using binary fingerprinting codes. Unpublished, 2002.

[7] Dan Boneh, Amit Sahai, and Brent Waters. Fully collusion resistant traitor tracing with short ciphertexts and private keys. In *Eurocrypt '06*, 2006. Full version available at http://eprint.iacr.org/2006/045.

[8] Dan Boneh and James Shaw. Collusion secure fingerprinting for digital data. *IEEE Transactions on Information Theory*, 44(5):1897–1905, 1998. Extended abstract in Crypto '95.

[9] Hervé Chabanne, Duong Hieu Phan, and David Pointcheval. Public traceability in traitor tracing schemes. In *EUROCRYPT*, pages 542–558, 2005.

[10] Benny Chor, Amos Fiat, and Moni Naor. Tracing traitors. In *CRYPTO '94: Proceedings of the 14th Annual International Cryptology Conference on Advances in Cryptology*, pages 257–270, London, UK, 1994. Springer-Verlag.

[11] Benny Chor, Amos Fiat, Moni Naor, and Benny Pinkas. Tracing traitors. *IEEE Transactions on Information Theory*, 46(3):893–910, 2000.

[12] Yevgeniy Dodis and Nelly Fazio. Public key trace and revoke scheme secure against adaptive chosen ciphertext attack. In *Public Key Cryptography - PKC 2003*, volume 2567 of *LNCS*, pages 100–115, 2003.

[13] A. Fiat and M. Naor. Broadcast encryption. In *Proceedings of Crypto '93*, volume 773 of *LNCS*, pages 480–491. Springer-Verlag, 1993.

[14] Amos Fiat and T. Tassa. Dynamic traitor tracing. In *Proceedings of Crypto '99*, volume 1666 of *LNCS*, pages 354–371, 1999.

[15] Eli Gafni, Jessica Staddon, and Yiqun Lisa Yin. Efficient methods for integrating traceability and broadcast encryption. In *CRYPTO '99: Proceedings of the 19th Annual International Cryptology Conference on Advances in Cryptology*, pages 372–387, London, UK, 1999. Springer-Verlag.

[16] M. T. Goodrich, J. Z. Sun, , and R. Tamassia. Efficient tree-based revocation in groups of low-state devices. In *Proceedings of Crypto '04*, volume 2204 of *LNCS*, 2004.

[17] D. Halevy and A. Shamir. The lsd broadcast encryption scheme. In *Proceedings of Crypto '02*, volume 2442 of *LNCS*, pages 47–60, 2002.

[18] Aggelos Kiayias and Moti Yung. On crafty pirates and foxy tracers. In *ACM Workshop in Digital Rights Management - DRM 2001*, pages 22–39, London, UK, 2001. Springer-Verlag.

[19] Aggelos Kiayias and Moti Yung. Breaking and repairing asymmetric public-key traitor tracing. In Joan Feigenbaum, editor, *ACM Workshop in Digital Rights Management – DRM 2002*, volume 2696 of *Lecture Notes in Computer Science*, pages pp. 32–50. Springer, 2002.

[20] Aggelos Kiayias and Moti Yung. Traitor tracing with constant transmission rate. In *EUROCRYPT '02: Proceedings of the International Conference on the Theory and Applications of Cryptographic Techniques*, pages 450–465, London, UK, 2002. Springer-Verlag.

[21] K. Kurosawa and Y. Desmedt. Optimum traitor tracing and asymmetric schemes. In *Proceedings of Eurocrypt '98*, pages 145–157, 1998.

[22] Shigeo Mitsunari, Ryuichi Sakai, and Masao Kasahara. A new traitor tracing. *IEICE Trans. Fundamentals*, E85-A(2):481–484, 2002.

[23] Dalit Naor, Moni Naor, and Jeffrey B. Lotspiech. Revocation and tracing schemes for stateless receivers. In *CRYPTO '01: Proceedings of the 21st Annual International Cryptology Conference on Advances in Cryptology*, pages 41–62, London, UK, 2001. Springer-Verlag.

[24] Moni Naor and Benny Pinkas. Threshold traitor tracing. In *CRYPTO '98: Proceedings of the 18th Annual International Cryptology Conference on Advances in Cryptology*, pages 502–517, London, UK, 1998. Springer-Verlag.

[25] Moni Naor and Benny Pinkas. Efficient trace and revoke schemes. In *FC '00: Proceedings of the 4th International Conference on Financial Cryptography*, pages 1–20, London, UK, 2001. Springer-Verlag.

[26] B. Pfitzmann. Trials of traced traitors. In *Proceedings of Information Hiding Workshop*, pages 49–64, 1996.

[27] B. Pfitzmann and M. Waidner. Asymmetric fingerprinting for larger collusions. In *Proceedings of the ACM Conference on Computer and Communication Security*, pages 151–160, 1997.

[28] Reihaneh Safavi-Naini and Yejing Wang. Sequential traitor tracing. In *Proceedings of Crypto '00*, volume 1880 of *LNCS*, pages 316–332, 2000.

[29] Alice Silverberg, Jessica Staddon, and Judy L. Walker. Efficient traitor tracing algorithms using list decoding. In *Proceedings of ASIACRYPT '01*, volume 2248 of *LNCS*, pages 175–192, 2001.

[30] Jessica N. Staddon, Douglas R. Stinson, and Ruizhong Wei. Combinatorial properties of frameproof and traceability codes. Cryptology ePrint 2000/004, 2000.

[31] D. Stinson and R. Wei. Combinatorial properties and constructions of traceability schemes and frameproof codes. *SIAM Journal on Discrete Math*, 11(1):41–53, 1998.

[32] D. Stinson and R. Wei. Key preassigned traceability schemes for broadcast encryption. In *Proceedings of SAC '98*, volume 1556 of *LNCS*, 1998.

[33] Gabor Tardos. Optimal probabilistic fingerprint codes. In *Proceedings of STOC '03*, pages 116–125, 2003.

[34] V. To, R. Safavi-Naini, and F. Zhang. New traitor tracing schemes using bilinear map. In *Proceedings of 2003 DRM Workshop*, 2003.

[35] Yuji Watanabe, Goichiro Hanaoka, and Hideki Imai. Efficient asymmetric public-key traitor tracing without trusted agents. In *Proceedings CT-RSA '01*, volume 2020 of *LNCS*, pages 392–407, 2001.

A   Definition of Tracing Traitors

Initially, we view a pirate decoder $\mathcal{D}$ as a probabilistic circuit that takes as input a ciphertext C and outputs some message M or $\perp$. A Traitor-Tracing system, then, consists of the following four algorithms:

Setup(N, λ). The setup algorithm takes as input N, the number of users in the system, and the security parameter λ. The algorithm runs in polynomial time in λ and outputs a public key BK, a secret tracing key TK, and private keys $K_1, \ldots, K_N$, where K_u is given to user u.

Encrypt(BK, M). Encrypts M using the public broadcasting key BK and outputs ciphertext C.

Decrypt(j, K_j, C, BK). Decrypt C using the private key K_j of user j. The algorithm outputs a message M or $\perp$.

Trace$^{\mathcal{D}}(TK, \epsilon)$. The tracing algorithm is an oracle algorithm that is given as input the tracing key TK and a parameter ϵ, and runs in time polynomial in the security parameter λ and $1/\epsilon$. Only values of ϵ that are polynomially related to λ are considered valid inputs to *Trace*. The tracing algorithm queries the pirate decoder $\mathcal{D}$ as a black-box oracle, as defined above. It outputs a set S which is a subset of $\{1, 2, \ldots, N\}$.

The system must satisfy the following **correctness property**:
for all $j \in \{1, \ldots, N\}$ and all messages M:

$$\text{Let} \quad \left(BK, TK, (K_1, \ldots, K_N)\right) \xleftarrow{R} Setup(N, \lambda) \quad \text{and} \quad C \xleftarrow{R} Encrypt(BK, M).$$

$$\text{Then} \quad Decrypt(j, K_j, C, BK) = M.$$

Security. We define security of the traitor tracing scheme in terms of the following two natural game.

Game 1. The first game is the standard **Semantic Security Game**. It says that the system is semantically secure to an outsider who does not possess any of the private keys. Since this is a standard notion we do not give the game details here. We define the advantage of adversary $\mathcal{A}$ in winning this game as $\mathsf{Adv}_{ss} = |\Pr[\beta' = \beta] - 1/2|$.

Game 2. The second game captures the notion of **Traceability against arbitrary collusion**. For a given N, λ and ϵ (where $\epsilon = 1/f(\lambda)$ for some polynomial f), the game proceeds as follows (both challenger and adversary are given N, λ, and ϵ as input):

1. The adversary $\mathcal{A}$ outputs a set $T = \{u_1, u_2, \ldots, u_t\} \subseteq \{1, \ldots, N\}$ of colluding users.
2. The challenger runs $Setup(N, \lambda)$ and provides BK and $K_{u_1}, \ldots, K_{u_t}$ to $\mathcal{A}$. It keeps TK to itself.
3. The adversary $\mathcal{A}$ outputs a pirate decoder $\mathcal{D}$.

4. The challenger now runs $Trace^{\mathcal{D}}(\text{TK}, \epsilon)$ to obtain a set $S \subseteq \{1, \ldots, N\}$. Note that $Trace$ is only given black-box oracle access to $\mathcal{D}$.

We say that the adversary $\mathcal{A}$ wins the game if the following two conditions hold:
- The decoder $\mathcal{D}$ is useful. That is, for a randomly chosen M in the finite message space, we have that

$$\Pr[\mathcal{D}(Encrypt(\text{BK}, M)) = M] \geq \epsilon$$

- The set S is either empty, or is not a subset of T.

We denote by Adv_{TR} the probability that adversary $\mathcal{A}$ wins this game.

Definition 2. *We say that an N-user Traitor Tracing system is secure if for all polynomial time adversaries $\mathcal{A}$ and any constant $\epsilon > 0$ we have that Adv_{MH} and Adv_{TR} are negligible functions of λ.*

We emphasize that Game 2 places no limit on the size of the coalition under the control of the adversary. Furthermore, the pirate decoder need not be perfect. It only needs to play valid content with probability ϵ. Finally, note that we are modeling a stateless (resettable) pirate decoder — the decoder is just an oracle and maintains no state between activations. Non stateless decoders were studied in [18].

In the full version of the paper we describe a more restrictive access model to the pirate decoder $\mathcal{D}$. PLBE enables tracing even in this more restrictive model.

B   Proof of Lemma 1

For this distinguishing experiment we will show that distinguishing between whether an encryption is to position (i, j) or $(i, j + 1)$ is as hard as the 3-party Diffie-Hellman assumption. Since, the assumption is in a prime order group the simulator can know the factorization of n, the order of the group. For this game simulator will run the core part of the simulation in the $\mathbb{G}_p$ subgroup and choose all values in the $\mathbb{G}_q$ subgroup for itself. Our formal proof follows.

Suppose there exists a t-time adversary $\mathcal{A}$ that breaks the Index Hiding game with advantage ϵ. Then we build a simulator as follows. The simulator receives the 3-party Diffie-Hellman challenge from the simulator as:

$$g_p, A = g_p^a, B = g_p^b, C = g_p^c, T.$$

The challenge will be given in the subgroup of prime order p of a composite order group $n = pq$. The simulator is given the factors p, q.

Next, the simulator runs the Init phase and receives the index (i, j) from $\mathcal{A}$. Since the game will be played in the subgroup $\mathbb{G}_p$, the simulator can choose for itself everything in the $\mathbb{G}_q$ subgroup. It chooses random generators $g_q, h_q \in \mathbb{G}_q$ and random exponents $\beta, r_{q,1}, \ldots, r_{q,m}, c_{q,1}, \ldots, c_{q,m} \in \mathbb{Z}_q$. Additionally, it chooses the exponents $\alpha_1, \ldots, \alpha_m \in \mathbb{Z}_n$. It then sets $h_p = B$ and picks blinding factors $r'_{p,1}, \ldots, r'_{p,m}, c'_{p,1}, \ldots, c'_{p,m} \in \mathbb{Z}_p$.

The simulator is now able to create the public and secret keys as follows. It first publishes $g = g_q g_p$ and $h = h_q B$. It creates the public keys:

$$E = g_q^\beta \qquad E_x = g_q^{\beta r_{q,x}} \qquad F_x = h_q^{\beta r_{q,x}}$$

$$G_x = e(g_q, g_q)^{\beta \alpha_x} \qquad H_y = \begin{cases} g_q^{c_{q,y}} g_p^{c'_{p,y}} & : y \neq j \\ g_q^{c_{q,y}} C^{c'_{p,y}} & : y = j \end{cases}$$

Next, it creates the private keys for all users except (i, j) as:

$$K_{x,y} = \begin{cases} g^{\alpha_x} g_q^{r_{q,x} c_{q,y}} g_p^{r'_{p,x} c'_{p,y}} & : x \neq i, y \neq j \\ g^{\alpha_x} g_q^{r_{q,x} c_{q,y}} B^{r'_{p,x} c'_{p,y}} & : x = i, y \neq j \\ g^{\alpha_x} g_q^{r_{q,x} c_{q,y}} C^{r'_{p,x} c'_{p,y}} & : x \neq i, y = j \end{cases}$$

We note that all the simulator creates public and private with the same distribution as the real scheme.

In the challenge phase the adversary first gives the simulator a message $M \in \mathbb{G}_T$. The simulator then chooses exponents $(v_{1,1}, v_{1,2}, v_{1,3}), \ldots, (v_{i-1,1}, v_{i-1,2}, v_{i-1,3}) \in \mathbb{Z}_n^{(3)}$, and exponents $s_{q,i}, \ldots, s_{q,m} \in \mathbb{Z}_q$ and $t_q \in \mathbb{Z}_q$. Additionally, it chooses random $s'_p \in \mathbb{Z}_p$, $z_{p,1}, \ldots, z_{p,j-1} \in \mathbb{Z}_p$, $w'_1, \ldots, w'_m \in \mathbb{Z}_n$.

It then creates the ciphertext as:

$$\begin{aligned}
&\text{if } x > i : R_x = g_q^{s_{q,x} r_{q,x}} && \tilde{R}_x = h_q^{s_{q,x} r_{q,x}} \\
&\qquad\qquad A_x = g_q^{s_{q,x} t_q} && B_x = M e(g_q, g_q)^{\alpha_x s_{q,x} t_q} \\
&\text{if } x = i : R_x = g_q^{s_{q,x} r_{q,x}} g_p^{s'_p r'_{p,x}} && \tilde{R}_x = h_q^{s_{q,x} r_{q,x}} B s'_p r'_{p,x} \\
&\qquad\qquad A_x = g^{s_{q,x} t_q} A^{s'_p} && B_x = M e(g_q, g_q)^{\alpha_x s_{q,x} t_{q,x}} e(g_p, A)^{\alpha_x s'_p} \\
&\text{if } x < i : R_x = g^{v_{x,1}} && \tilde{R}_x = h^{v_{x,1}} \\
&\qquad\qquad A_x = g^{v_{x,2}} && B_x = e(g, g)^{v_{x,3}} \\
&\text{if } y > j : C_y = g_q^{c_{q,y} t_q} h^{w'_y} && \tilde{C}_y = A^{-c'_{p,y}} g^{w'_y} \\
&\text{if } y = j : C_y = g_q^{c_{q,y} t_q} T h^{w'_y} && \tilde{C}_y = g^{w'_y} \\
&\text{if } y < j : C_y = g_q^{c_{q,y} t_q} g_p^{z_{p,y}} h^{w'_y} && \tilde{C}_y = g^{w'_y}
\end{aligned}$$

If T forms a 3-party Diffie-Hellman tuple then the ciphertext is a well-formed encryption to the indices (i, j), otherwise if T is randomly chosen it is a encryption to $(i, j + 1)$. The simulator will receive a guess γ from $\mathcal{A}$ and it will simply repeat this guess as its answer to the 3-party Diffie-Hellman game. The simulator's advantage in the Index Hiding game will be exactly equal to $\mathcal{A}$'s advantage. $\qquad\square$

Simplified Threshold RSA with Adaptive and Proactive Security

Jesús F. Almansa, Ivan Damgård*, and Jesper Buus Nielsen**

BRICS***, Department of Computer Science
University of Aarhus, Denmark
{jfa, ivan, buus}@brics.dk

Abstract. We present the currently simplest, most efficient, optimally resilient, adaptively secure, and proactive threshold RSA scheme. A main technical contribution is a new rewinding strategy for analysing threshold signature schemes. This new rewinding strategy allows to prove adaptive security of a proactive threshold signature scheme which was previously assumed to be only statically secure. As a separate contribution we prove that our protocol is secure in the UC framework.

Keywords: Proactiveness, Adaptive Security, Threshold RSA, Universal Composability.

1 Introduction

The concept of threshold cryptography was first introduced by Desmedt [Des87]. In threshold cryptography n servers run a service in such a way that even if some t servers are corrupted, the service is still available and secure. In a *threshold signature* the servers implement a service for signing messages under a signature key shared between the servers with some threshold t.

The first RSA based threshold signature was given independently by Boyd [Boy89] and Frankel [Fra89]. In both protocols the signing key d is shared additively among the servers. The first RSA threshold scheme was published by Santis et al. [SDFY94], and although the key sharing is polynomial, it does not tolerate actively cheating servers. This restriction was later removed independently by Frankel et al. [FGY96] and Gennaro et al. [GJK96]. All of these protocols are only proved secure against static adversaries, i.e., the set of corrupt parties is fixed before the protocol starts.

In [OY91], Ostrovsky and Yung introduced the notion of *proactive* security, in which the life span of a protocol is divided into separate time periods and we assume that the adversary can corrupt at most t players in each period. However, the set of corrupted players may change from one period to the next,

* Supported by FICS, Foundations in Cryptography and Security, financed by the Danish Research Council.
** Supported by FICS and ECRYPT, European Network of Excellence for Cryptology.
*** Basic Research in Computer Science (www.brics.dk), funded by the Danish National Research Foundation.

so the protocol must remain secure, even though every player may have been corrupt at some point. This can, for instance, be achieved for a protocol based on secret sharing by having players re-randomize the shares they hold between periods, and erase the old shares. This is called *refreshment*.

In [FGMY97b] Frankel et al. published the first *proactive* threshold RSA signature, as a generalization of an unpublished protocol by Jakobsson et al. [JJKY95]. This protocol, does not scale up well: even for a moderately large number of servers it is either highly inefficient or does not tolerate the optimal threshold $t < n/2$, and it is only *statically* secure. Next, in [FGMY97a] Frankel et al. achieved optimal threshold. Later, Rabin [Rab98] gave a simplified static proactive protocol by combining the best of the linear and polynomial sharing techniques. The protocol from [Rab98] was later optimized and simplified by Jarecki and Saxena [JS05], still obtaining a static proactive protocol.

In [CGJ⁺99] Canetti et al. add mechanisms to the protocol from [Rab98], to obtain an *adaptively* secure protocol, and even though adaptive security is only claimed for the non-proactive version of [Rab98], the protocol in [CGJ⁺99] seems to be the first adaptive proactive threshold RSA signature.

Later Frankel et al. [FMY01] gave an adaptively secure version of the protocol from [FGMY97b]. The protocols from [CGJ⁺99, FMY01] seem to be the only published adaptive proactive threshold RSA signatures. Unfortunately both protocols have some practical drawbacks. The linear-sharing based protocol in [FMY01] inherits the problem from [FGMY97b] that it is inefficient or non-optimally secure unless the number of servers is small. The protocol [CGJ⁺99] modifies the protocol from [Rab98] such that before each signature generation the signing-key shares are refreshed. This adds a considerable performance overhead.

Our contributions. The paper has three main contributions.

1. The first contribution is a novel analysis technique which allows to prove that the protocol from [Rab98] is – with minor modifications – *adaptively* secure, contrary to what was previously believed. Indeed, as mentioned above, the authors of [CGJ⁺99] add an expensive mechanism to the protocol to make it adaptively secure. Our first contribution shows that this mechanism is unnecessary. The technical problem we need to solve to do this is explained below.

2. Our second contribution is a technique for avoiding so-called key-share exposure. In the protocol from [Rab98], if a server fails to contribute correctly to the signature generation, the key share of that server is reconstructed and thus exposed in public. Such key-share exposure can degrade security in practice. For instance, if a server fails to contribute only because it is temporarily down, this error will be made worse if the remaining servers expose its key share. Our second contribution shows that key-share exposure is not necessary to make the protocol from [Rab98] actively and adaptively secure.

3. Our final contribution consists of two definitions for security of threshold signature schemes. One is cast in the universally composable (UC) framework, the other one is a more standard definition similar to the one from [CGJ⁺99]. We show that the two are equivalent. This allows simplified proofs that our protocols – and other protocols as well – are secure in the UC framework.

The technical problem we solve with our first contribution is the following: In [CGJ+99] a useful technique known as the *Single Inconsistent Player* technique was introduced. It facilitates proving adaptive security of threshold signature schemes using simulation arguments as follows: we typically want to show that a successful adversary against the protocol can be used to break security of an underlying single server signature scheme. To this end, we build a simulator which does a chosen message attack on the underlying signature scheme while simulating the adversary's view of the protocol with the aim of having him forge a signature. The simulator initially chooses among the currently honest players a single inconsistent player (SIP) and arranges matters such that it knows valid-looking secret keys of *all players except the SIP*. It can therefore simulate successfully even against adaptive corruption, *as long as the SIP is not corrupted*. If there is a non-negligible chance that the SIP stays honest, there is also a non-negligible chance that the adversary produces a forged signature.

It is natural to try using the SIP technique for doing also *proactive* adaptive security. However, in this scenario we must choose a new SIP every time keys are refreshed, since under a proactive attack no single party can hope to stay honest throughout the protocol. But even this will not work: the probability that a single SIP remains honest in a single phase cannot be made arbitrarily close to 1, whence the probability that we are lucky in every phase is negligible. Thus, already with as few as a super-logarithmic number of proactive phases, a simple straight-line simulation will not work. A potential solution is to use rewinding, i.e., every time the SIP is corrupted, rewind back to the last refreshment, choose a new SIP and try again. This turns out to work, if one is willing to refresh keys *every time a message is about to be signed* (as shown in [CGJ+99]).

If we are not willing to pay the price this costs in efficiency, a further technical problem emerges: if we rewind past a point where a message m was given as input, we risk that when we go forward again, the adversary asks us to generate signatures on different messages than before, in particular different from m. The net result of this is that we may end up with a simulated transcript where the adversary apparently breaks the scheme by producing a valid signature on message m, which he did not ask to have signed. But in reality, the simulator had to ask for the signature on m somewhere in the rewinding process, so we did not break the underlying signature scheme after all. In connection with the proof of Theorem 2 we give an example adversary to show that this really is a problem and we show how to solve it. The basic idea is to guess the point in the simulation where the simulator asks for a signature on the "fatal" message, and simply refrain from asking. One then has to show that this does not bias the distribution of the simulation. More details are given later.

How to read this paper. In Section 2.3 we give a sketch of the UC definition of security, and in Section 3 a formal specification of secure threshold signatures as an ideal functionality. Readers who are more interested in the protocol constructions can skip this without loss of continuity, as all protocols are proved according to the more standard definition from Section 3.1, which is equivalent to UC security. However, to read the protocol descriptions and proofs, the notation introduced in Sections 2.2 and 2.1 is necessary.

2   Proactive UC Security

In this section we describe our computational and adversarial model. Our work utilizes the Universal-Composability framework by Canetti, introduced in [Can01], and last revised in [Can]. Among the upgrades in the last revision that are relevant to us, it is now possible to model *erasures*, by allowing a party to leak only partial internal state upon a corruption. Since the composition theorem remains valid, we can cast proactiveness in the framework and make proofs of security, while still being able to use the composition theorem. Likewise, it is also shown that w.l.o.g., one may assume that a single entity (the environment) models all activity external to the protocol, including adversarial activity. We use this technical simplification.

Finally, it is possible to specialize [Can] to the case of synchronous networks, which we will do here. Thus we do not need to define a new synchronous model and reprove the composition theorem as was done in [DN03] and [Nie04].

We now give a brief description of our instance of the UC framework. For a more detailed description of the proactive UC framework, see [Alm05].

2.1   Computations

All entities are PPT Interactive Turing machines (ITM). An n-party protocol π in the $\mathcal{G}$-hybrid model is a set of n ITMs, whose identities $P_1, \ldots, P_n$ are all different, and an ideal functionality $\mathcal{G}$ to which parties are granted use.

In general, $\mathcal{G} = (\mathcal{G}_1, \ldots, \mathcal{G}_m)$, $1 < m$, may include one or more ideal functionalities, and we will assume $\mathcal{G}_1 = \mathcal{F}_{\mathrm{aut}}$ provides authenticated transmission, to be used for communication between parties. In some cases, we will use instead a functionality $\mathcal{F}_{\mathrm{SMT}}$ modeling secure point-to-point channels. These abstractions allow us to focus on the high-level properties of our protocols, yet they can be implemented using well known techniques. Note, however, that for proactive security, care should be taken with refreshing the key material used for message transmission.

The protocol runs while interacting with an *environment*, an ITM $\mathcal{Z}$ that models external (adversarial) activity, and which provides inputs to honest parties and receives their corresponding outputs. We use $\mathrm{HYB}^{\mathcal{G}}_{\pi, \mathcal{Z}}$ to denote the entire process of running π while interacting with $\mathcal{Z}$ and $\mathcal{G}$.

The execution proceeds in communication rounds, that we denote by $r = 0, 1, \ldots$.

A *proactive protocol* proceeds in *phases*. A phase consists of a number of consecutive rounds, and every round belongs to exactly one phase. There are two kinds of phases, *refreshment* and *operational*, which occur alternately. Finally, a *stage* consists of an *opening* refreshment phase, an operational phase in the middle and a *closing* refreshment phase. Thus, each refreshment is the closing of one stage and the opening of another. We use $u = 0, 1, \ldots$ to denote stages.

The intuition is that during the operational phases, the protocol provides whatever service it was designed for, whereas refreshment phases are used to rerandomize various representations of data so that attacks in different phases will not be able to benefit from each other.

We allow $\mathcal{Z}$ to decide when refreshment starts (equivalently, when a new stage begins), by sending a command to each party. Refreshment ends when all honest parties have output a special symbol indicating end of refreshment.

2.2 Adversaries

As mentioned, $\mathcal{Z}$ also models the *adversary*. As such, $\mathcal{Z}$ may corrupt parties adaptively throughout the protocol, subject to the limitation that no more than t parties can be corrupt *in every stage*. In particular, this means that if a party is corrupt during a refreshment phase, he is considered to be corrupt in both of the two stages to which the phase belongs. After corruption, $\mathcal{Z}$ acts on behalf of the corrupted player. Corruption may be passive, where $\mathcal{Z}$ internally executes the correct protocol on behalf of the corrupted player, or active where $\mathcal{Z}$ decides on its own the actions of the corrupted player.

If player P_i is corrupted during an operational phase, $\mathcal{Z}$ is given the view of P_i starting from his state at the beginning of the current operational phase. This models the assumption that all randomness and data used in the previous refreshment phase is erased, except for the information that the protocol specifies should be used afterwards.

If the corruption is made during a refreshment phase, say, the closing refreshment of stage u, $\mathcal{Z}$ receives the view of P_i starting from his state at the beginning of the operational phase of stage u, and P_i is assumed to be corrupt for stage $u+1$.

If P_i is corrupt when a refreshment begins, $\mathcal{Z}$ may decide to *leave* him, which may allow $\mathcal{Z}$ to corrupt new parties, subject to the bound of t corruptions per stage. In this case, we say P_i is *decorrupted*.

A decorrupted player immediately starts taking part in the protocol as any honest player. In the passive corruption case, he starts from the correct state specified by the protocol at this point. In the active corruption case, he starts from a *default state after round r*. This state is application-dependent in general.

2.3 UC Security

Security is defined by comparing protocol π's execution with an *ideal protocol execution*. There, instead of parties, an ideal functionality $\mathcal{F}$ is used to *specify* the desired input/output behavior of π. It also specifies the information allowed to be leaked from π to the environment.

Security loosely speaking means that whatever $\mathcal{Z}$ could achieve by attacking π, it could also achieve by interacting with $\mathcal{F}$. To make this precise, a special ITM $\mathcal{T}$ is introduced. The goal of $\mathcal{T}$ is to *simulate* the adversary's view of π, based only on the information $\mathcal{F}$ is willing to exchange with the environment.

We declare π secure in the $\mathcal{G}$-hybrid model if no environment can distinguish interactions with π from those with $\mathcal{F}$ and $\mathcal{T}$. More formally:

The environment is assumed to always end by outputting a bit which we think of as its guess at whether it works in the ideal or the hybrid scenario. When $\mathcal{Z}$

interacting with π in the $\mathcal{G}$-hybrid model, on security parameter k, auxiliary input z to $\mathcal{Z}$, and the random coins of all machines are uniformly chosen, this output of $\mathcal{Z}$ is a random variable denoted $\text{HYB}^{\mathcal{G}}_{\pi,\mathcal{Z}}(k,z)$. We denote by $\text{HYB}^{\mathcal{G}}_{\pi,\mathcal{Z}}()$ the ensemble $\{\text{HYB}^{\mathcal{G}}_{\pi,\mathcal{Z}}(k,z)\}_{k\in\mathbb{N},z\in\{0,1\}^*}$.

Similarly, $\text{IDEAL}_{\mathcal{F},\mathcal{T},\mathcal{Z}}(k,z)$ and $\text{IDEAL}_{\mathcal{F},\mathcal{T},\mathcal{Z}}()$ are the random variable and ensemble produced when $\mathcal{Z}$ interacts with $\mathcal{F}$ and $\mathcal{T}$ in the ideal process.

Using $\overset{c}{\approx}$ to denote computational indistinguishability, we then have:

Definition 1 (UC Security). *A protocol π proactively t-realizes a functionality $\mathcal{F}$ in the $\mathcal{G}$-hybrid model, if there exists a simulator $\mathcal{T}$ such that for all environments $\mathcal{Z}$ corrupting at most t parties per stage it holds that* $\text{IDEAL}_{\mathcal{F},\mathcal{T},\mathcal{Z}}() \overset{c}{\approx} \text{HYB}^{\mathcal{G}}_{\pi,\mathcal{Z}}()$.

3   Defining Proactive Threshold Signatures

We define threshold signatures by giving a functionality, $\mathcal{F}_{\text{ThSig}}$, that is a version of Canetti's signature functionality[Can04], adapted for the threshold case.

FUNCTIONALITY $\mathcal{F}_{\text{ThSig}}$

Key Generation, initiate Having received the same message $(KeyGen, sid)$ from all honest parties in a set $S = \{S_1, \ldots, S_n\}$ in the same round, and $sid = (S, sid')$ for some sid', send $(KeyGen, sid)$ to $\mathcal{Z}$.

Key Generation, finalize Upon receiving $(KeyGen, sid, v)$ from $\mathcal{Z}$, if $(KeyGen, sid)$ was sent earlier, record v and send (sid, v) to all $S_i \in S$. All further commands that do not contain the sid established here are ignored.

Signature Generation, initiate Having received $(Sign, sid, m)$ from all honest $S_i \in S$ in the same round, store $(Sign, sid, m)$ and send it to $\mathcal{Z}$. There might be several identical $(Sign, sid, m)$ stored.

Signature Generation, finalize Upon receiving $(Signature, sid, m, \sigma)$ from $\mathcal{Z}$, if $(Sign, sid, m)$ is stored and an entry of the form $(m, \sigma, 0)$ was not recorded, delete an entry $(Sign, sid, m)$, record the entry $(m, \sigma, 1)$ and send $(Signature, sid, m, \sigma)$ to all $S_i \in S$.

Signature Verification Upon receiving a message $(Verify, sid, m, \sigma, v')$ from some party P, give $(Verify, sid, m, \sigma, v')$ to $\mathcal{Z}$. Upon receiving $(Verified, sid, m, \sigma, \phi)$ from $\mathcal{Z}$, send $(Verified, sid, m, \sigma, f)$ to P, where f is determined as follows:

    1. If $v' = v$ and the entry $(m, \sigma, 1)$ is recorded, then set $f = 1$ (guarantees that if v' is the registered public key and σ is legitimately generated, then verification succeeds).

    2. Else, if $v' = v$ and no entry $(m, \sigma, 1)$ is recorded, set $f = 0$ (guarantees that if v' is the registered public key and m was not legitimately signed, then verification fails). Record the entry $(m, \sigma, 0)$.

    3. Else, if $v \neq v'$, set $f = \phi$.

Refreshment On input signaling that a refreshment phase starts in this round, record this and signal end of refreshment in the next round (this reflects that our protocol implementing the functionality takes one round to do the refreshment).

Note that all our functionalities receive initially a session id $sid = (S, sid')$ where S is the set of players who participate in realizing the functionality and sid' is a number identifying this particular instance of the functionality.

The functionality defines a player set S called the *servers* and a *verification key* v. Only the servers can ask $\mathcal{F}_{\mathtt{ThSig}}$ to sign messages, but any player with the correct key v can use $\mathcal{F}_{\mathtt{ThSig}}$ to verify a signature. For simplicity we also assume some external mechanism for the servers to agree on which message to sign and in which round. We model this by assuming throughout that all our environments behave such that if an honest player gets a message to sign as input, all honest players get the same message as input in the same round.

We note that the logic in Canetti's signature functionality is slightly more complicated than ours because it has to deal with the case where the signer is corrupted. In our case the single signer is replaced by the set of servers, and hence we can demand by bounding the number of corrupted servers that things will always work as if "the signer" is honest.

For a protocol π (in the $\mathcal{G}$-hybrid model) we can then say it is a *secure UC threshold signature scheme for the class* $\mathscr{Z}$ if π realizes $\mathcal{F}_{\mathtt{ThSig}}$ when quantifying over $\mathcal{Z} \in \mathscr{Z}$ in the definition of security.

3.1 Equivalence to a More Standard Notion

In [Can04] it was proved that for a (non-threshold) signature scheme, implementing the signature functionality in [Can04] is equivalent to the scheme being correct (i.e. signed messages are accepted by the verification algorithm), consistent (two verifications of the same message and signature give the same result) and unforgeable under chosen message attack.

THRESHOLD SIGNATURE SCHEME $\mathcal{F}_{\mathtt{ThSig}}$

Key Generation (well-formed) If all honest $S_i \in S$ receive the same message $(KeyGen, (S, sid'))$ in the same round, then after some rounds all honest parties $S_i \in S$ output one common message (sid, v).

Signature Generation (well-formed) If all honest $S_i \in S$ receive the same message $(Sign, sid, m)$ in the same round, then after some rounds all honest $S_i \in S$ output one common message $(Signature, sid, m, \sigma)$.

Signature Verification (well-formed) If an honest party P_i receives input $(Verify, sid, m, \sigma, v')$ in round r, then P_i outputs one corresponding message of the form $(Verified, sid, m, v', f)$ in round r.

No other messages (well-formed) No honest party outputs a message not described above.

Signature Verification (correct) If an honest party P_i receives input $(Verify, sid, m, \sigma, v)$ in round r and some honest party once output $(Signature, sid, m, \sigma)$, then P_i outputs $(Verified, sid, m, v, 1)$ in round r.

Signature Verification (consistent) If two honest parties P_i and P_j (not necessarily distinct) outputs $(Verified, sid, m, \sigma, v, f_i)$ respectively $(Verified, sid, m, \sigma, v, f_j)$, then $f_i = f_j$.

Signature Verification (unforgeable) If an honest party P_i outputs $(Verified, sid, m, v, 1)$ in round r, then in some round $r' \leq r$ an honest party received the input $(Sign, sid, m)$.

In this section we do a similar "sanity check" of our definition of a UC threshold signature scheme, by giving a property based definition of what it means for a protocol to be a secure threshold signature scheme and then proving that this notion is equivalent to the UC notion.

Let $\mathcal{Z}$ be a set of environments. We say that π has one of the properties in the figure above (relative to $\mathcal{Z}$) if for all $Z \in \mathcal{Z}$, the probability that the property fails when executing $\mathrm{HYB}^{\mathcal{G}}_{\pi, \mathcal{Z}}$ is negligible.

Theorem 1. *If π is well-formed, correct, consistent and unforgeable relative to $\mathcal{Z}$, then π is a secure UC threshold signature scheme for $\mathcal{Z}$.*

Briefly, this result is shown by constructing a UC simulator $\mathcal{T}$, which will generate on its own a set of keys for the signature scheme by executing internally an instance of π. Then, using the private key(s), it can trivially simulate $\mathcal{Z}$'s view of π by simply following the protocol to generate signatures. One then observes that the only way this could differ from actual executions is if $\mathcal{Z}$ can produce a valid signature that was not legally generated. Such a signature would be accepted in the hybrid process, but rejected in the ideal one. However, the unforgeability of π ensures that such events occur with negligible probability. The (tedious but straightforward) details can be found in [Alm05].

4   Passive Security

In this section we give a UC threshold signature scheme under the class $\mathcal{Z}$ of passive, adaptive environments which corrupts at most $t = n - 1$ players in each stage. In Section 5 we describe how to obtain active security. To simplify matters, we will assume a trusted dealer who distributes keys to the servers initially. The dealer is modeled as an ideal functionality $\mathcal{F}_{\mathrm{KeyGen}}$, in other words, we operate in the $\mathcal{F}_{\mathrm{KeyGen}}$-hybrid model.

FUNCTIONALITY $\mathcal{F}_{\mathrm{KeyGen}}$

Key Generation, initiate Having received the same message $(KeyGen, sid)$ from all honest parties in the same round, parse sid as (S, sid'), perform RSA key generation with security parameter k to obtain modulus N and exponents e, d. Next, for $i = 1, \ldots, n$, where n is the size of the set S, choose at random d_i in $[-nN^2 .. nN^2]$ and set $d_{public} = d - \sum_i d_i$. Then send $(KeyGen, sid, v, d_{public})$ to $\mathcal{Z}$, where v is the RSA public key (N, e).

Key Generation, finalize (To avoid having to specify how many rounds key generation will take, we let the environment decide when to return results) Upon receiving $(KeyGenFinish, sid)$ from $\mathcal{Z}$, send $(KeyGen, sid, v, d_{public}, d_i)$ to each $S_i \in S$.

The keys generated will be used in an RSA signature scheme, where we assume (as usual) that binary strings of length up to some polynomial in the security parameter k can be signed, where the signature on m is of form $H(m)^d \bmod N$, and where H is some preprocessing that typically involves a hash function. The details of this are left out of scope here.

We also assume secure point-to-point channels, i.e., we assume a functionality $\mathcal{F}_{\mathrm{SMT}}$ for secure message transmission. This functionality will accept inputs containing message and sender/receiver id, and will in the next round deliver the message to the intended receiver, revealing only the message length to the adversary. Whenever we speak about sending a message privately in the following, this refers to calling $\mathcal{F}_{\mathrm{SMT}}$.

PROTOCOL π

All parties in the protocol run the following code:

Key Generation, initiate On input $(KeyGen, sid)$, parse sid as (S, sid') and send $(KeyGen, sid)$ to $\mathcal{F}_{\mathtt{KeyGen}}$.

Key Generation, finalize Wait to receive $(KeyGen, sid, v, d_i, d_{public})$ from $\mathcal{F}_{\mathtt{KeyGen}}$, store this information, and output (sid, v). Here v is the RSA public key (N, e).

Signature Generation, initiate On input $(Sign, sid, m)$, send $(sid, m, H(m)^{d_i} \bmod N)$ to all $S_j \in S$.

Signature Generation, finalize Upon receiving $(sid, m, H(m)^{d_j} \bmod N)$ from all $S_j \in S$ (i.e. for $j = 1, \ldots, n$), compute the signature

$$\sigma = H(m)^{d_1} H(m)^{d_2} \cdots H(m)^{d_n} H(m)^{d_{public}} \bmod N,$$

and output $(Signature, sid, m, \sigma)$.

Signature Verification On input $(Verify, sid, m, \sigma, v')$, where v' is an RSA key (N', e'), define $f \in \{0,1\}$ by $f = 1$ iff $\sigma^{e'} \bmod N' = H(m)$, and output $(Verified, sid, m, \sigma, v', f)$.

Refreshment For each decorrupted P_i, his default state after round r is d_i, i.e., his actual share. Each $S_i \in S$ reshares d_i, i.e., chooses $d_{i,j}$ at random in $[-N^2..N^2]$, sets $d_{i,public} = d_i - \sum_j d_{i,j}$, sends $d_{i,public}$ to all $S_j \in S$ and $d_{i,j}$ privately to S_j. In the next round, each $S_i \in S$ computes $d_i^{new} = \sum_j d_{j,i}$. Finally, all $S_i \in S$ compute $d_{public}^{new} = d_{public} + \sum_i d_{i,public}$. Everyone signals end of refreshment. Only d_{public}^{new} and the private d_i^{new} are remembered in the next phase.

By Theorem 1 it is sufficient to show the following:

Theorem 2. *If the underlying signature scheme is unforgeable under chosen message attack, then the protocol π is well-formed, correct, consistent and unforgeable relative to the class $\mathscr{Z}$ of environments which corrupt, passively and adaptively, at most $n - 1$ players in each proactive stage.*

Proof. It is straight-forward to verify that the protocol is well-formed, correct and consistent (in fact these properties hold unconditionally). What remains is to prove that it is unforgeable. So, assume for the sake of contradiction that there exists $\mathcal{Z} \in \mathscr{Z}$ such that with some non-negligible probability $P(\mathcal{Z})$ it happens in $\mathrm{HYB}_{\pi, \mathcal{Z}}^{(\mathcal{F}_{\mathrm{SMT}}, \mathcal{F}_{\mathtt{KeyGen}})}$ that an honest party P_i outputs $(Verified, sid, m, \sigma, v, 1)$ in round r without m being signed in some round $r' \leq r$ (in the following we say that m *was signed* in round r' if $(Sign, sid, m)$ was input to all honest parties). We use this to construct a PPT reduction $\mathrm{Red}'(\mathcal{Z})$ which breaks the underlying signature scheme with some non-negligible probability P' related to $P(\mathcal{Z})$. It is given a random RSA verification key (N, e) and is given an oracle $\mathcal{O}(N, d) : m \mapsto H(m)^d \bmod N$. It then tries to compute a *forgery*, i.e. a value (m, σ) where $\sigma^e \bmod N = H(m)$ and where $\mathcal{O}(N, d)$ was not queried on m. The algorithm $\mathrm{Red}(\mathcal{Z})$ described on the next page is used as a sub-routine.

The strategy of $\mathrm{Red}(\mathcal{Z})$ is to run $\mathcal{Z}$ while simulating its view of the protocol. More precisely, $\mathrm{Red}(\mathcal{Z})$ runs $\mathrm{HYB}_{\pi, \mathcal{Z}}^{(\mathcal{F}_{\mathrm{SMT}}, \mathcal{F}_{\mathrm{KeyGen}})}$, but it simulates itself the actions of $(\mathcal{F}_{\mathrm{SMT}}, \mathcal{F}_{\mathrm{KeyGen}})$ and the (currently) honest players, using the verification key and oracle it is given, but of course without knowing the secret RSA key. The hope is that $\mathcal{Z}$ will behave (approximately) as in a real attack and will hence produce a forgery that can help us break the signature scheme.

The reduction $\mathrm{Red}(\mathcal{Z})$ uses the single inconsistent player (SIP) technique explained in the introduction. A new SIP is chosen at random after every refreshment phase. We use S_{j_u} to denote the SIP chosen after the opening refreshment of stage u. If the current SIP S_{j_u} is corrupted, $\mathrm{Red}(\mathcal{Z})$ rewinds to the beginning of stage u and tries again.

We now analyze $\mathrm{Red}(\mathcal{Z})$: Let an *attempt* for stage u be a run of $\mathrm{Red}(\mathcal{Z})$ from state State^{u-1} at the beginning of the opening refreshment of u, until S_{j_u} is corrupted or the closing refreshment of u begins. Let a *failed attempt* (*successful attempt*) be an attempt where S_{j_u} is (not) corrupted. Notice that $\mathrm{Red}(\mathcal{Z})$ is trying to create a sequence of successful attempts, closing with $\mathcal{Z}$ terminating or the unforgeability property being violated. Call such a sequence a *successful sequence*. Let d denote the signing key corresponding to the input verification key (N, e), and let $d_1^u, \ldots, d_n^u, d_{public}^u$ be the shares used by $\mathrm{Red}(\mathcal{Z})$ in successful attempt u, and similarly $d_{i,j}^u, d_{i,public}^u$ the values used in the refreshment in successful attempt u. These are called the *real shares* in the following. We first prove:

Claim 1: the view of $\mathcal{Z}$ in a successful sequence is statistically indistinguishable from its view in $\mathrm{HYB}_{\pi, \mathcal{Z}}^{(\mathcal{F}_{\mathrm{SMT}}, \mathcal{F}_{\mathrm{KeyGen}})}$.

Note that $\mathrm{Red}(\mathcal{Z})$, when it creates and updates the shares d_i, follows exactly the protocol, except that the secret is zero, instead of the correct d. We now want to argue that if we modify the shares generated by $\mathrm{Red}(\mathcal{Z})$ so they are consistent with d, $\mathcal{Z}$ will still see essentially the same view. To this end, define a new set of shares $d'^u_1, \ldots, d'^u_n, d'^u_{public}, d'^u_{i,j}, d'^u_{i,public}$ that are equal to the shares generated by $\mathrm{Red}(\mathcal{Z})$, except

$$d'^u_{j_u} = d^u_{j_u} + d, \quad d'_{j_u, j_{u+1}} = d_{j_u, j_{u+1}} + d .$$

We call these the *virtual* shares. Note that the new set of values is consistent with secret exponent d, but if we restrict to the subset seen by $\mathcal{Z}$ the virtual shares equal the real ones. Moreover, except with negligible probability, the virtual shares are legal, i.e., all shares are in the intervals specified in the protocol. This follows immediately from the fact that the size of the intervals is larger than

d by an exponential factor. Note also that when signatures are generated, the contribution from the SIP, σ_{j_u}, as generated by $\mathtt{Red}(\mathcal{Z})$ satisfies $\sigma_{j_u} = H(m)^{d'^u_{j_u}}$, since $\sigma = H(m)^d$ and $-d^u_{public} = \sum_{i \in S} d^u_i$. In other words, $\mathtt{Red}(\mathcal{Z})$ already generates signatures consistently with the virtual shares.

REDUCTION $\mathtt{Red}(\mathcal{Z})$

Run a copy of $\mathtt{HYB}^{(\mathcal{F}_{\mathrm{SMT}}, \mathcal{F}_{\mathrm{KeyGen}})}_{\pi, \mathcal{Z}}$ while simulating $(\mathcal{F}_{\mathrm{SMT}}, \mathcal{F}_{\mathrm{KeyGen}})$ and the honest parties as follows:

Key Generation, initiate On input $v = (N, e)$ and a set of players S, choose d_i at random in $[0..nN^2]$ for all $S_i \in S$ and set $d_{public} = -\sum_i d_i$. Then, choose a player S_{j_0} at random among the honest players in S and call S_{j_0} the *single inconsistent party* (SIP) for stage 0.

Key Generation, finalize When $\mathcal{Z}$ gives the command to generate keys, send $(KeyGen, sid, v, d_i, d_{public})$ to each S_i on behalf of $\mathcal{F}_{\mathrm{KeyGen}}$. Store the current state State^0 of $\mathtt{HYB}^{(\mathcal{F}_{\mathrm{SMT}}, \mathcal{F}_{\mathrm{KeyGen}})}_{\pi, \mathcal{Z}}$.

Refreshment On a signal that opening refreshment of u starts in this round, record state State^{u-1} of $\mathtt{HYB}^{(\mathcal{F}_{\mathrm{SMT}}, \mathcal{F}_{\mathrm{KeyGen}})}_{\pi, \mathcal{Z}}$ and set $\mathrm{State}^u := \mathrm{State}^{u-1}$. Then execute the refreshment on behalf of the honest players according to the protocol, using as input the current d_i. This results in a new set of d_i's for all the players, and a new d_{public}.[a] Update and record State^u. Finally $\mathtt{Red}(\mathcal{Z})$ picks a new SIP S_{j_u} among the $S_i \in S$ still honest after the refreshment phase.

Signature Generation, initiate When $\mathcal{Z}$ inputs $(Sign, sid, m)$ to all honest $S_i \in S$, call $\mathcal{O}(N, d)$ to obtain $\sigma = H(m)^d \bmod N$.

Signature Generation, finalize In the next round, for each $S_i \in S \setminus \{S_{j_p}\}$, set $\sigma_i = H(m)^{d_i} \bmod N$, and for the SIP S_{j_p}, compute

$$\sigma_{j_p} = \sigma \cdot H(m)^{-d_{public}} \cdot \prod_{S_i \in S \setminus \{S_{j_p}\}} \sigma_i^{-1}.$$

Then for all honest $S_i \in S$, send σ_i to all parties in S.

Corruption When $\mathcal{Z}$ corrupts a server S_i, $\mathtt{Red}(\mathcal{Z})$ sends the d_i it holds for S_i to $\mathcal{Z}$, or both d_i and its older share if corruption is made in opening refreshment of u. If $S_i = S_{j_u}$ (where S_{j_u} denotes the current SIP for stage u), then $\mathtt{Red}(\mathcal{Z})$ gives up this attempt to simulate stage u and restarts the simulation from the recorded state State^u at the beginning of the appropriate phase, using fresh randomness (notice that this involves choosing a new random SIP). To ensure that $\mathtt{Red}(\mathcal{Z})$ runs in PPT it will rerun each operational phase at most kn times and then give up the reduction completely.

Signature Verification $\mathtt{Red}(\mathcal{Z})$ does not need to do anything special here, since verification is just done as in the protocol using v'.

Termination If it ever happens that the unforgeability property is violated by some party P_i outputting $(Verified, sid, m, \sigma, v, 1)$, then $\mathtt{Red}(\mathcal{Z})$ terminates with output (m, σ). If $\mathcal{Z}$ terminates first, then $\mathtt{Red}(\mathcal{Z})$ terminates with an empty output.

---

[a] Notice that by inspecting the messages that $\mathcal{Z}$ sends privately on behalf of the corrupted parties in $\mathtt{HYB}^{(\mathcal{F}_{\mathrm{SMT}}, \mathcal{F}_{\mathrm{KeyGen}})}_{\pi, \mathcal{Z}}$, $\mathtt{Red}(\mathcal{Z})$ can also compute the d_i of all corrupted $S_i \in S$.

This means that the mapping from real to virtual shares creates (except for a negligibly small set of cases) a 1-1 correspondence between successful sequences generated by $\text{Red}(\mathcal{Z})$ and executions of $\text{HYB}_{\pi,\mathcal{Z}}^{(\mathcal{F}_{\text{SMT}},\mathcal{F}_{\text{KeyGen}})}$. Since $\mathcal{Z}$'s view is unchanged under this correspondence, Claim 1 follows.

Since no SIP is even defined in $\text{HYB}_{\pi,\mathcal{Z}}^{(\mathcal{F}_{\text{SMT}},\mathcal{F}_{\text{KeyGen}})}$ it follows from Claim 1 that all j_u are statistically independent of the view of $\mathcal{Z}$ in any attempt until S_{j_u} is corrupted. Since j_u is chosen uniformly at random and $\mathcal{Z}$ corrupts at most $n-1$ parties, it follows that in any given attempt, with probability at least statistically close to $1/n$ the environment $\mathcal{Z}$ does not corrupt S_{j_u}. From this it easily follows that after kn reruns we get a successful attempt except with negligible probability. Since the number of operational phases is polynomial it follows that $\text{Red}(\mathcal{Z})$ also gets a successful sequence, except with negligible probability. From Claim 1 it also follows that the unforgeability property fails with probability statistically close to $P(\mathcal{Z})$ in this successful sequence. Now, every time the unforgeability property fails in $\text{Red}(\mathcal{Z})$, by some party P_i outputting (*Verified, sid, m, σ, v*, 1), it by definition holds that $\sigma^e \bmod N = H(m)$ and that m was not signed in the the successful sequence. Therefore $\text{Red}(\mathcal{Z})$ never queried $\mathcal{O}(N,d)$ on m in the successful attempts used to produce the successful sequence.

It is tempting to believe that $\text{Red}(\mathcal{Z})$ could just output (m,σ) and break the signature scheme with probability statistically close to $P(\mathcal{Z})$. However, this may not work as $\text{Red}(\mathcal{Z})$ also makes queries to $\mathcal{O}(N,d)$ in the failed attempts. If m was queried in a failed attempt, $\text{Red}(\mathcal{Z})$ does not break the signature scheme by outputting (m,σ). Below we will say that a message on which the simulator queried $\mathcal{O}(N,d)$ during a failed attempt and for which $\mathcal{Z}$ did not request a signature generation in the successful sequence is a *dirty message*. When the environment outputs a forgery on a dirty message m in the final state, then we have a situation where $\mathcal{Z}$ produced a successful forgery, but where $\text{Red}(\mathcal{Z})$ cannot use this forgery as its own. Accordingly, successful forgeries by $\mathcal{Z}$ on dirty messages are called *useless forgeries*.

To see that useless forgeries are a real problem, consider the following environment $\mathcal{Z}$: it runs for k operational phases and in phase i picks a random message m_i from the set $\{0, 1, \ldots, k\}$ which was not signed already, and then inputs (*Sign, sid, m_i*) to all parties. After the signature is generated, $\mathcal{Z}$ corrupts all parties except one (at the end of any operational phase, it leaves all parties). After k phases it outputs a forgery (m_{k+1}, σ) on the single message $m_{k+1} \in \{0, 1, \ldots, k\}$ which was not signed yet. It is easy to see that $\text{Red}(\mathcal{Z})$ will have to rerun each operational phase an expected n times, and that the probability that m_{k+1} was not signed in any failed attempt thus is negligible. This shows that the reduction Red does not work for all $\mathcal{Z}$. So, we must come up with a better simulation strategy.

First of all we can assume that $\text{Red}(\mathcal{Z})$ never queries $\mathcal{O}(N,d)$ on the same message m twice by having it remember previous queries (here we use that RSA signatures are unique). For a run of $\text{Red}(\mathcal{Z})$ we then use $(m_1, \ldots, m_L)$ to denote the distinct messages on which $\text{Red}(\mathcal{Z})$ queried $\mathcal{O}(N,d)$, in the order of query. Furthermore, when $\text{Red}(\mathcal{Z})$ produces a useless forgery on some dirty message

m we define l_0 by $m_{l_0} = m$, and when $\mathrm{Red}(\mathcal{Z})$ produces a useful forgery or no forgery we let $l_0 = 0$. Clearly, given $\mathcal{Z}$ and the randomness $\mathfrak{r}$ used by $\mathrm{Red}(\mathcal{Z})$, the value l_0 is uniquely defined by some function $l_0 = l_0(\mathcal{Z}, \mathfrak{r})$.

Consider now the following reduction $\mathrm{Red}^{l_0}(\mathcal{Z})$ which has access to an oracle for the function l_0. When running with randomness $\mathfrak{r}$, $\mathrm{Red}^{l_0}(\mathcal{Z}; \mathfrak{r})$ runs $\mathrm{Red}(\mathcal{Z}; \mathfrak{r})$, but tries to keep m_{l_0} clean. First it queries $l_0 = l_0(\mathcal{Z}, \mathfrak{r})$ and proceeds as follows: When $l_0 = 0$ it just runs $\mathrm{Red}(\mathcal{Z}; \mathfrak{r})$ (so, $\mathrm{Red}^{l_0}(\mathcal{Z}; \mathfrak{r}) = \mathrm{Red}(\mathcal{Z}; \mathfrak{r})$ when $l_0 = 0$). When $l_0 > 0$ it runs $\mathrm{Red}(\mathcal{Z}; \mathfrak{r})$ with the following changes: initially it just counts on how many distinct messages it queried $\mathcal{O}(N, d)$, until it is about to query on the l_0'th message m_{l_0}. Then it remembers m_{l_0} and *does not query* $\mathcal{O}(N, d)$ on m_{l_0}. After m_{l_0} is defined $\mathrm{Red}^{l_0}(\mathcal{Z})$ still runs $\mathrm{Red}(\mathcal{Z}; \mathfrak{r})$, except that in addition to rerunning when the SIP S_{j_u} is corrupted it also reruns when it is about to query on m_{l_0}, *so that it never queries* $\mathcal{O}(N, d)$ *on* m_{l_0}. Notice that by definition of $l_0 > 0$ the message m_{l_0} would be dirty in $\mathrm{Red}(\mathcal{Z}; \mathfrak{r})$. So, if $\mathrm{Red}(\mathcal{Z}; \mathfrak{r})$ was run, the message m_{l_0} would by definition not be requested signed by $\mathcal{Z}$ in the successful sequence. So, all requests by $\mathcal{Z}$ to signed m_{l_0} would occur in failed attempts, because the SIP was corrupted. Therefore the modification in $\mathrm{Red}^{l_0}(\mathcal{Z}; \mathfrak{r})$ of aborting when m_{l_0} is requested signed only aborts attempts which would also have been aborted by $\mathrm{Red}(\mathcal{Z}; \mathfrak{r})$. In particular, the successful sequence of $\mathrm{Red}^{l_0}(\mathcal{Z}; \mathfrak{r})$ is identical to the successful sequence of $\mathrm{Red}(\mathcal{Z}; \mathfrak{r})$ (so, $\mathrm{Red}^{l_0}(\mathcal{Z}; \mathfrak{r}) = \mathrm{Red}(\mathcal{Z}; \mathfrak{r})$ when $l_0 > 0$). It follows that independent of l_0, $\mathrm{Red}^{l_0}(\mathcal{Z}; \mathfrak{r}) = \mathrm{Red}(\mathcal{Z}; \mathfrak{r})$. However, since $\mathrm{Red}^{l_0}(\mathcal{Z})$ by construction never queries $\mathcal{O}(N, d)$ on m_{l_0} it follows that $\mathrm{Red}^{l_0}(\mathcal{Z})$ produces no useless forgeries, so $\mathrm{Red}^{l_0}(\mathcal{Z})$ outputs a forgery (m, σ) with probability statistically close to $P(\mathcal{Z})$.

Consider finally the algorithm $\mathrm{Red}'(L, \mathcal{Z})$ which runs as follows: It first sample a uniformly random number $l_0' \in [L]$. Then it runs $\mathrm{Red}^{l_0}(\mathcal{Z}; \mathfrak{r})$ with uniformly random $\mathfrak{r}$, except that when $\mathrm{Red}^{l_0}(\mathcal{Z})$ queries the oracle l_0, $\mathrm{Red}'(L, \mathcal{Z})$ replies with l_0'. If $L > l_0$, then $l_0' = l_0(\mathcal{Z}, \mathfrak{r})$ with probability $1/L$. Since $\mathrm{Red}^{l_0}(\mathcal{Z})$ outputs a forgery with probability statistically close to $P(\mathcal{Z})$, it follows that when $L > l_0$, the algorithm $\mathrm{Red}'(L, \mathcal{Z})$ outputs a forgery with probability statistically close to $P(\mathcal{Z})/L$. Assume then that there exists a PPT environment $\mathcal{Z}$ which violates the unforgeability property in $\mathrm{HYB}_{\pi, \mathcal{Z}}^{(\mathcal{F}_{\mathrm{SMT}}, \mathcal{F}_{\mathrm{KeyGen}})}$ with non-negligible probability $P(\mathcal{Z})$. Then there also exists a polynomial bound $L(k)$ on the running time of $\mathcal{Z}$ and thus there exists a PPT algorithm $\mathrm{Red}' = \mathrm{Red}'(L(k), \mathcal{Z})$ which breaks the unforgeability under chosen message attack of the RSA signature scheme with probability P' statistically close to the non-negligible $P(\mathcal{Z})/L$, a contradiction. $\Diamond$

5   Active Security

In this section we sketch how to make the protocol robust. We follow the approach from [Rab98] and [CGJ+99] with some modifications to avoid share exposure.

5.1   The Protocol

Preliminaries. We need a statistically hiding integer commitment scheme *com*, where a commitment to integer a is denoted by $com(a)$ (we suppress here the random coins need to produce the commitment). We assume that the initial key setup generates parameters for such a scheme. We require that the scheme is *linear*. Informally this means that there exists a method to compute from two commitments $com(a)$ and $com(b)$ and an integer c a new commitment $com(a) + c \cdot com(b)$, and if one can open $com(a)$ to a and $com(b)$ to b, one can compute an opening of $com(a) + c \cdot com(b)$ to $z = a + cb$. This opening should reveal essentially no information about a and b except that $z = a + cb$. A commitment scheme with these properties exists, where binding is based on the factoring assumption [FD02].

Using this commitment scheme we can construct a statistically private VSS scheme as follows. Given a secret integer $s \in [0..B]$ in some known interval, pick a degree t polynomial f with $f(0) = sL$ by letting $a_0 = sL$, picking integer coefficients $a_1, \ldots, a_t$ as in [Rab98] and letting $f(x) = \sum_{j=0}^{t} a_j x^j$. Then for $j = 0, 1, \ldots, t$ compute a commitment $c_j = com(a_j)$ and broadcast $c_0, c_1, \ldots, c_t$. Then for $i = 1, \ldots, n$ compute $d_i = \sum_{j=0}^{t} i^j c_j$. Then compute an opening of d_i to $f(i)$ and send this opening to P_i. If P_i does not receive an opening of $\sum_{j=0}^{t} i^j c_j$ it complains and the dealer must broadcast an opening of $\sum_{j=0}^{t} i^j c_j$. If the dealer fails to do so, the VSS is rejected. It is straight-forward to verify that this is a secure integer VSS scheme that hides the shared value information theoretically.

A VSS to a secret s is given by the commitments d_i and we use $[s] = (com(f(1)), \ldots, com(f(n)))$ to denote a VSS to s. Given a VSS $[a] = (com(f(1)), \ldots, com(f(n)))$ and a VSS $[b] = (com(g(1)), \ldots, com(g(n)))$ and an integer c we can compute a VSS $[a] + c \cdot [b] = (com(f(1)) + c \cdot com(g(1)), \ldots, com(f(n)) + c \cdot com(g(n)))$. Clearly, from openings of $[a]$ and $[b]$ the parties can compute an opening of $[a] + c \cdot [b]$.

Generation of Challenges. We will be using several interactive proofs of the standard public coin 3-move form (Σ-protocols), where the prover must answer a challenge. For us, it will always be the case that all players have to give proofs simultaneously. After the opening messages of the proofs have been sent, the challenges are generated as follows: each party picks uniformly random k-bit values $a_i, b_i \in \{0, 1\}^k$, deals a VSS of a_i and then broadcasts b_i. The parties open all the VSS's (that were successfully generated) and compute the k-bit values $c_i = b_i \oplus \bigoplus_j a_j$. The string c_i is used as challenge in the proof given by P_i. The rationale for this method is that, although a corrupt P_i will not be able to predict the challenge ahead of time, a simulator can put itself in a position where it knows all a_j before b_i is chosen, and can therefore force c_i to be any desired value.

Key Setup. We have the following requirements on the key setup. First, let $p = 2p' + 1$ and $q = 2q' + 1$ be safe primes and let $N = pq$ and let SQ_N be the subgroup of squares in $\mathbb{Z}_N$ (which has order $p'q'$). We let $L = n!$ and require that $\gcd(e, L) = 1$.

The key generator now additionally broadcasts a random element g of order $p'q'$, for $i = 1, \ldots, n$ broadcasts the value $h_i = g^{d_i} \bmod N$, broadcasts public parameters for a commitment scheme as described above, and finally deals a VSS $\alpha_i = [d_i]$.

In the following, let EDL_N be the language for equality of discrete logarithms, where $(a, A, b, B) \in EDL_N$ iff $a, A, b, B \in SQ_N$ and there exists w such that $A = a^w \bmod N$ and $B = b^w \bmod N$.

Signature Generation, with Share Exposure. The only difference from the passive protocol is that after generating an alleged signature σ', the parties check whether $\sigma'^e \bmod N = H(m)$. If this is not the case, then each party P_i has to prove that $(\sigma_i^2 \bmod N, H(m)^2 \bmod N, h_i, g) \in EDL_N$. This is done using the same standard Σ-protocol that was used in [Rab98], but with the above method for generating the challenges. The protocol requires that the inputs are in SQ_N, but this is guaranteed by the key setup and the squarings done.

Let I be the set of i for which P_i failed this proof. The parties then compute the VSS $\alpha_I = \sum_{i \in I} \alpha_i$ and opens it to some value d_I. We have that $d_I = \sum_{i \in I} d_i$. Therefore, the correct signature σ satisfies $\sigma^2 = H(m)^{2(d_{public} + d_I)} \prod_{i \notin I} \sigma_i^2 \bmod N$. Finally, from $\sigma^2 \bmod N, H(m) = \sigma^e \bmod N$, we can easily compute σ, since 2 and e are relatively prime.

Refreshment. At the beginning of refreshment, decorrupted parties may not have reliable information determining their key shares. Therefore, each party P_j sends to each other party all the public information he holds on α_i, for all i. In other words, the commitments to the shares of d_i are sent. This means each player receives n suggestions for α_i, but since a majority will be correct, we may assume that all honest parties now agree on each α_i. Then for each P_i the key share d_i is privately reconstructed from the VSS α_i, i.e., players send the opening information for the commitments in α_i privately to P_i. Decorrupted players may not be able to send correct opening information, but a majority will be able to do so, and this is sufficient.

Next, the refreshment protocol proceeds as in the passive case, with the following changes:

1. In addition to sending $d_{i,j}$ to P_j, party P_i will also broadcast $h_{i,j} = g^{d_{i,j}} \bmod N$, deal a VSS $\alpha_{i,j} = [d_{i,j}]$ and give a zero-knowledge proof that $\alpha_{i,j}$ can be opened to a value $d_{i,j}$ for which $h_{i,j} = g^{d_{i,j}} \bmod N$. The details of this proof is given below.

2. If any of the proofs fails or $h_i \neq g^{d_{i,public}} \prod_{j=1}^{n} h_{i,j} \bmod N$, then P_i is detected as a cheater. Also, if $h_{i,j} \neq g^{d_{i,j}} \bmod N$, then P_j broadcasts a complaint. Then P_i must broadcast $d_{i,j}$ such that $h_{i,j} = g^{d_{i,j}} \bmod N$, and P_j adopts this values. If P_i fails to do so, then P_i is detected as a cheater.

3. For each party P_i which was detected as a cheater, the other parties simulate P_i, as follows. They define $d_{i,i} = d_i$ and let $d_{i,j} = 0$ for $j \neq i$ and let $d_{i,public} = 0$. Notice that $d_i = d_{i,public} + \sum_{j=1}^{n} d_{i,j}$ and that P_j knows $d_{i,j}$, as desired. Then they let $h_{i,i} = h_i$ and let $\alpha_{i,i} = \alpha_i$, and let $h_{i,j} = g^0 \bmod N$ and let $\alpha_{i,j}$ be a default secret sharing of 0. Notice that $h_{i,j} = g^{d_{i,j}} \bmod N$ and $\alpha_{i,j}$ is a secret sharing of $d_{i,j}$, for $j = 1, \ldots, n$, as desired.

4. Finally each P_i computes $d_i^{new} = \sum_{j=1}^{n} d_{j,i}$ and all parties compute $h_i^{new} = \prod_{j=1}^{n} h_{i,j}$ and $\alpha_i^{new} = \sum_{j=1}^{n} \alpha_{i,j}$.

The proof mentioned above proceeds by having each party run the following (in the role of prover) k times in parallel[1].

1. The prover knows some secret $s \in [0..B]$ and has broadcast $h = g^s \bmod N$ and dealt a VSS $\alpha = [s]$.
2. The prover broadcasts $H = g^r \bmod N$ for a uniformly random $r \in_R [0..(B + 2^k)]$ and deals a VSS $\beta = [r]$.
3. The prover is given a challenge $c \in \{0, 1\}$, generated as described earlier.
4. The parties open the VSS $c\alpha + \beta$ to some value z. If $h^c H \bmod N = g^z \bmod N$, then the parties accept the proof.

Using standard techniques, one can argue that this protocol is honest verifier zero-knowledge, and sound relative to the binding property of the commitments used.

5.2   Analysis

We give a sketch of the security analysis. We want to reprove Theorem 2 for the class of actively cheating adversaries. Except for unforgeability, the required properties are straight-forward to verify. In particular, the correctness follows directly from the binding property of the commitment scheme, the soundness of the applied proof systems and the observation that for a decorrupted party P_i, by virtue of the VSS α_i, there is always sufficient backup information at the beginning of refreshment in order to reconstruct the correct value d_i to P_i as his share. Formalizing this requires a rewinding argument to demonstrate that an adversary breaking correctness can break the binding property of the commitments. This rewinding does not cause any problems since first, the reduction is not part of the UC simulator and second, we may assume that we know the factorization of N (but not the trapdoor for the commitment scheme), and so we can simulate perfectly the actions of honest players in all cases, by just following the protocol.

The proof of unforgeability follows the proof from the passive case, with a few additions to the reduction to unforgeability of the signature scheme, as detailed below.

The key generation is simulated as in the passive case with the following addition. Pick a random square $h \bmod N$ (which will have order $p'q'$ except with negligible probability, by choice of N). Let $g = h^e \bmod N$. Then g is also a random element of order $p'q'$, as desired. Notice that $h = g^d \bmod N$. Now, for the consistent parties P_i, let $h_i = g^{d_i} \bmod N$ and let $\alpha_i = [d_i]$, and for the SIP P_{j_0}, let $h_{j_0} = h(\prod_{i \neq j_0} h_i)^{-1} \bmod N$ and let $\alpha_{j_0} = [d_{j_0}]$, such that $h_{j_0} = g^{d'_{j_0}} \bmod N$

---

[1]   We use parallel repetition of a standard protocol with a 1-bit challenge since this gives us soundness with no extra assumptions. If one is willing to make the strong RSA assumption, 1 repetition with a k-bit challenge is sufficient, this follows from results in [FD02].

where d'_{j_0} is the virtual share of the SIP. It can be seen that all additional values introduced in the simulation have the same distribution as in the protocol, except that α_i is a VSS of the incorrect share d_{j_0} instead of the virtual share d'_{j_0}; This is however unnoticeable as long as the SIP is honest, as the VSS is statistically hiding.

The refreshment protocol is simulated as in the passive case. Additionally, all parties broadcast the values $h^u_{i,j} = g^{d^u_{i,j}} \bmod N$ and deal VSS's $\alpha^u_{i,j} = [d^u_{i,j}]$. The value $h^u_{j_u-1,j_u} = g^{d'^u_{j_u-1,j_u}} \bmod N$ is computed using the virtual contribution. Since $d'^u_{j_u-1,j_u}$ is defined to be $d'^u_{j_u-1,j_u} = d^{u-1}_{j_u-1} - \sum_{i \neq j_u-1} d^u_{j_u-1,i}$, this can be computed as $h^u_{j_u-1,j_u} = h^{u-1}_{j_u}(\prod_{i \neq j_u} h^u_{j_u-1,i})^{-1} \bmod N$.

Notice that $\alpha^u_{j_u-1,j_u} = [d^u_{j_u-1,j_u}]$ is still computed using the incorrect contribution. This means that the simulator does not know a witness for the proof that $\alpha^u_{j_u-1,j_u}$ can be opened to a value x such that $h^u_{j_u-1,j_u} = g^x \bmod N$. Therefore this proof is simulated, as follows. Using the honest verifier zero-knowledge property, the first message in the k proofs are set up such that there exists exactly one string of challenges $c_{j_u-1} \in \{0,1\}^k$ which the simulator can answer. Then the simulator waits for the VSS's of the a_i values to be dealt, and using the shares of the honest parties it computes each a_i and broadcasts $b_{j_u-1} = c_{j_u-1} \oplus \bigoplus_i a_i$. Note that this simulation introduced *no new rewinding*. As a consequence of $\alpha^u_{j_u-1,j_u}$ being incorrect, the VSS $\alpha^u_{j_u}$ will be incorrect. Again this is not a problem as long as the SIP P_{j_u} is not corrupted.

The signature generation is simulated as in the passive protocol. Additionally, for the consistent parties a proof that $(\sigma_i, H(m), h_i, g) \in EDL_N$ is simulated by following the protocol (as d_i is known). Notice that the signature share of the SIP is computed as to make it $\sigma_{j_u} = H(m)^{-d'^u_{j_u}} \bmod N$. Therefore $(g, h^u_{j_u}, H(m)^2, \sigma^2_{j_u}) \in EDL_N$. So we can run the honest verifier simulator for the proof of membership in EDL_N, and in this way generate an opening message for which we can answer one challenge value. As above, we can make the challenge equal this value without rewinding, and hence complete the simulation. As long as the SIP is not corrupted this will give $\mathcal{Z}$ a view statistically close to that of the protocol. As for simulating the value α_I, notice that it is not a problem that the VSS $\alpha^u_{j_u}$ is not correct, as it will never enter the sum α_I when the SIP P_{j_u} is honest.

As in the passive case, the simulation is statistically close to the protocol until the SIP is corrupted, and as argued during the description, we introduced no more rewinding. Therefore the reduction goes through as in the passive case, using the same rewinding technique.

As for efficiency, note that although we introduced some changes compared to Rabin's original protocol, to make our proof go through, the performance is essentially the same: signature generation is constant round and requires broadcasting $O(n(k + \log n))$ bits.

5.3 Signature Generation, Without Share Exposure

Because of the model it is considered secure to open the VSS α_I to reveal the value d_I in the signing protocol, as the parties P_i for $i \in I$ are considered corrupted. In practice a party P_i might, however, end up in I just because a network plug was pulled or its network was congested because of a denial of service attack. In such a situation it might not be such a good idea to reveal d_i, as it constitutes a value which the 'adversary' does not know already. We can indeed do better.

Instead of opening the VSS α_I to the value d_I the parties notice that this VSS defines a polynomial f of degree at most t such that $f(0) = d_I L$ and the party P_i is holding an opening of a public commitment $com(f(i))$.

Each party P_i can therefore broadcast the value $h_i = H(m)^{f(i)} \bmod N$ and prove (using standard techniques similar to what we described above) that $com(f(i))$ can be opened to a value $f(i)$ such that $h_i = H(m)^{f(i)} \bmod N$.

This gives the parties at least $t + 1$ of the values $H(m)^{f(i)} \bmod N$. Therefore the parties can use interpolation as described in [Sho00] to compute $H(m)^{f(0)L} \bmod N = H(m)^{d_I L^2} \bmod N$. Then they compute $\sigma' = H(m)^{d_{public} L^2} H(m)^{\sigma_I L^2} \prod_{i \notin I} (\sigma_i)^{L^2} \bmod N = H(m)^{dL^2} \bmod N$. Using that $\gcd(e, L) = 1$ they then compute $H(m)^d \bmod N$ from $H(m)^{dL^2} \bmod N$.

It might seem puzzling that this is adaptively secure, given the similarity to the protocol from [Sho00] which is not known to be adaptively secure. The crucial point is that we applied the technique to compute $H(m)^{d_I L^2} \bmod N$ and not $H(m)^{dL^2} \bmod N$. Since the value d_I can be computed from the shares d_i of the corrupted parties, d_I is known to the simulator in the reduction (as opposed to d). Therefore it can 'simulate' the computation of $H(m)^{d_I L^2} \bmod N$ by simply running the protocol honestly.

References

[Alm05]   Jesús F. Almansa. *A Study for Cryptologic Protocols*. PhD thesis, BRICS, University of Aarhus, Department of Computer Science, IT-parken, Aabogade 34, DK-8200 Århus N, Denmark, 2005.

[Boy89]   C. Boyd. Digital multisignatures. In Oxford University Press, editor, *Cryptography and Coding*, pages 241–246, 1989.

[Can]   R. Canetti. Universally composable security: A new paradigm for cryptographic protocols. Cryptology ePrint Archive.

[Can01]   R. Canetti. Universally composable security: A new paradigm for cryptographic protocols. In *Proceedings of the 42nd IEEE Symposium on Foundations of Computer Science*, page 136, 2001. FOCS'01.

[Can04]   R. Canetti. Universally composable signature, certification, and authentication. Cryptology ePrint Archive, August 2004. Corrected version of the paper in Proceedings of the 17th IEEE Computer Security Foundations Workshop, pages 219–235, 2004.

[CGJ+99]   R. Canetti, R. Gennaro, S. Jarecki, H. Krawczyk, and T. Rabin. Adaptive security for threshold cryptosystems. In *LNCS*, volume 1666, pages 98–115, 1999. CRYPTO'99.

[Des87]   Yvo Desmedt. Society and group oriented cryptography: A new concept. In *LNCS*, volume 293, pages 120–127, 1987. CRYPTO'87.

[DN03]   Ivan Damgård and Jesper Buus Nielsen. Universally composable efficient
            multiparty computation from threshold homomorphic encryption. In
            LNCS, volume 2729, pages 247–264, 2003. CRYPTO'03.

[FD02]   E. Fujisaki and I. Damgård. A statistically-hiding integer commitment
            scheme based on groups with hidden order. In *Proceedings of Asiacrypt
            2002: 125-142*, pages 125–142, 2002. ASIACRYPT'02.

[FGMY97a]   Y. Frankel, P. Gemmell, P. Mackenzie, and Moti Yung. Optimal re-
            silience proactive public-key cryptosystems. In *Proceedings of the 38th
            IEEE Symposium on Foundations of Computer Science*, page 384, 1997.
            FOCS'97.

[FGMY97b]   Y. Frankel, P. Gemmell, P. Mackenzie, and Moti Yung. Proactive RSA.
            In *LNCS*, volume 1294, pages 440–454, 1997. CRYPTO'97.

[FGY96]   Yair Frankel, Peter Gemmell, and Moti Yung. Witness-based crypto-
            graphic program and robust function sharing. In *28th Annual ACM
            Symposium on Theory of Computing*, pages 499–508, 1996. STOC'96.

[FMY01]   Yair Frankel, Philip D. MacKenzie, and Moti Yung. Adaptive security
            for the additive-sharing based proactive RSA. In *LNCS*, volume 1992,
            pages 240–263, 2001. PKC'01.

[Fra89]   Yair Frankel. A practical protocol for large group oriented networks. In
            LNCS, volume 434, pages 56–61, 1989. EUROCRYPT'89.

[GJK96]   Rosario Gennaro, Stanislaw Jarecki, and Hugo Krawczyk. Robust and
            efficient sharing of RSA functions. In *LNCS*, volume 1109, pages 157–
            172, 1996. CRYPTO'96.

[JJKY95]   M. Jakobsson, S. Jarecki, H. Krawczyk, and Moti Yung. "proactive RSA
            for constant-size thresholds". Unpublished manuscript, 1995.

[JS05]   Stanislaw Jarecki and Nitesh Saxena. Further simplifications in proactive
            RSA signature schemes. In *LNCS*, volume 3378, pages 510–528, 2005.
            TCC'05.

[Nie04]   Jesper Buus Nielsen. *On Protocol Security in the Cryptographic Model*.
            PhD thesis, BRICS, University of Aarhus, Department of Computer Sci-
            ence, IT-parken, Aabogade 34, DK-8200 Århus N, Denmark, 2004.

[OY91]   R. Ostrovsky and M. Yung. How to withstand mobile virus attack. In
            *Proceedings of the 10th ACM Symposium on Principles of Distributed
            Computing*, pages 51–59, 1991. PODC'91.

[Rab98]   T. Rabin. A simplified approach to threshold and proactive RSA. In
            LNCS, volume 1462, pages 89–104, 1998. CRYPTO'98.

[SDFY94]   Alfredo De Santis, Yvo Desmedt, Yair Frankel, and Moti Yung. How to
            share a function securely. In *26th Annual ACM Symposium on Theory
            of Computing*, pages 522–533, 1994. STOC'94.

[Sho00]   Victor Shoup. Practical threshold signatures. In *LNCS*, volume 1807,
            pages 207–220, 2000. EUROCRYPT 2000.

Author Index

Lecture Notes in Computer Science

For information about Vols. 1–3899

please contact your bookseller or Springer

Vol. 3951: A. Leonardis, H. Bischof, A. Pinz (Eds.), Computer Vision – ECCV 2006, Part I. XXXV, 639 pages. 2006.

Vol. 3950: J.P. Müller, F. Zambonelli (Eds.), Agent-Oriented Software Engineering VI. XVI, 249 pages. 2006.

Vol. 3947: Y.-C. Chung, J.E. Moreira (Eds.), Advances in Grid and Pervasive Computing. XXI, 667 pages. 2006.

Vol. 3946: T.R. Roth-Berghofer, S. Schulz, D.B. Leake (Eds.), Modeling and Retrieval of Context. XI, 149 pages. 2006. (Sublibrary LNAI).

Vol. 3945: M. Hagiya, P. Wadler (Eds.), Functional and Logic Programming. X, 295 pages. 2006.

Vol. 3944: J. Quiñonero-Candela, I. Dagan, B. Magnini, F. d'Alché-Buc (Eds.), Machine Learning Challenges. XIII, 462 pages. 2006. (Sublibrary LNAI).

Vol. 3943: N. Guelfi, A. Savidis (Eds.), Rapid Integration of Software Engineering Techniques. X, 289 pages. 2006.

Vol. 3942: Z. Pan, R. Aylett, H. Diener, X. Jin, S. Göbel, L. Li (Eds.), Technologies for E-Learning and Digital Entertainment. XXV, 1396 pages. 2006.

Vol. 3941: S.W. Gilroy, M.D. Harrison (Eds.), Interactive Systems. XI, 267 pages. 2006.

Vol. 3940: C. Saunders, M. Grobelnik, S. Gunn, J. Shawe-Taylor (Eds.), Subspace, Latent Structure and Feature Selection. X, 209 pages. 2006.

Vol. 3939: C. Priami, L. Cardelli, S. Emmott (Eds.), Transactions on Computational Systems Biology IV. VII, 141 pages. 2006. (Sublibrary LNBI).

Vol. 3936: M. Lalmas, A. MacFarlane, S. Rüger, A. Tombros, T. Tsikrika, A. Yavlinsky (Eds.), Advances in Information Retrieval. XIX, 584 pages. 2006.

Vol. 3935: D. Won, S. Kim (Eds.), Information Security and Cryptology - ICISC 2005. XIV, 458 pages. 2006.

Vol. 3934: J.A. Clark, R.F. Paige, F.A. C. Polack, P.J. Brooke (Eds.), Security in Pervasive Computing. X, 243 pages. 2006.

Vol. 3933: F. Bonchi, J.-F. Boulicaut (Eds.), Knowledge Discovery in Inductive Databases. VIII, 251 pages. 2006.

Vol. 3931: B. Apolloni, M. Marinaro, G. Nicosia, R. Tagliaferri (Eds.), Neural Nets. XIII, 370 pages. 2006.

Vol. 3930: D.S. Yeung, Z.-Q. Liu, X.-Z. Wang, H. Yan (Eds.), Advances in Machine Learning and Cybernetics. XXI, 1110 pages. 2006. (Sublibrary LNAI).

Vol. 3929: W. MacCaull, M. Winter, I. Düntsch (Eds.), Relational Methods in Computer Science. VIII, 263 pages. 2006.

Vol. 3928: J. Domingo-Ferrer, J. Posegga, D. Schreckling (Eds.), Smart Card Research and Advanced Applications. XI, 359 pages. 2006.

Vol. 3927: J. Hespanha, A. Tiwari (Eds.), Hybrid Systems: Computation and Control. XII, 584 pages. 2006.

Vol. 3925: A. Valmari (Ed.), Model Checking Software. X, 307 pages. 2006.

Vol. 3924: P. Sestoft (Ed.), Programming Languages and Systems. XII, 343 pages. 2006.

Vol. 3923: A. Mycroft, A. Zeller (Eds.), Compiler Construction. XIII, 277 pages. 2006.

Vol. 3922: L. Baresi, R. Heckel (Eds.), Fundamental Approaches to Software Engineering. XIII, 427 pages. 2006.

Vol. 3921: L. Aceto, A. Ingólfsdóttir (Eds.), Foundations of Software Science and Computation Structures. XV, 447 pages. 2006.

Vol. 3920: H. Hermanns, J. Palsberg (Eds.), Tools and Algorithms for the Construction and Analysis of Systems. XIV, 506 pages. 2006.

Vol. 3918: W.K. Ng, M. Kitsuregawa, J. Li, K. Chang (Eds.), Advances in Knowledge Discovery and Data Mining. XXIV, 879 pages. 2006. (Sublibrary LNAI).

Vol. 3917: H. Chen, F.Y. Wang, C.C. Yang, D. Zeng, M. Chau, K. Chang (Eds.), Intelligence and Security Informatics. XII, 186 pages. 2006.

Vol. 3916: J. Li, Q. Yang, A.-H. Tan (Eds.), Data Mining for Biomedical Applications. VIII, 155 pages. 2006. (Sublibrary LNBI).

Vol. 3915: R. Nayak, M.J. Zaki (Eds.), Knowledge Discovery from XML Documents. VIII, 105 pages. 2006.

Vol. 3914: A. Garcia, R. Choren, C. Lucena, P. Giorgini, T. Holvoet, A. Romanovsky (Eds.), Software Engineering for Multi-Agent Systems IV. XIV, 255 pages. 2006.

Vol. 3911: R. Wyrzykowski, J. Dongarra, N. Meyer, J. Waśniewski (Eds.), Parallel Processing and Applied Mathematics. XXIII, 1126 pages. 2006.

Vol. 3910: S.A. Brueckner, G.D.M. Serugendo, D. Hales, F. Zambonelli (Eds.), Engineering Self-Organising Systems. XII, 245 pages. 2006. (Sublibrary LNAI).

Vol. 3909: A. Apostolico, C. Guerra, S. Istrail, P. Pevzner, M. Waterman (Eds.), Research in Computational Molecular Biology. XVII, 612 pages. 2006. (Sublibrary LNBI).

Vol. 3908: A. Bui, M. Bui, T. Böhme, H. Unger (Eds.), Innovative Internet Community Systems. VIII, 207 pages. 2006.

Vol. 3907: F. Rothlauf, J. Branke, S. Cagnoni, E. Costa, C. Cotta, R. Drechsler, E. Lutton, P. Machado, J.H. Moore, J. Romero, G.D. Smith, G. Squillero, H. Takagi (Eds.), Applications of Evolutionary Computing. XXIV, 813 pages. 2006.

Vol. 3906: J. Gottlieb, G.R. Raidl (Eds.), Evolutionary Computation in Combinatorial Optimization. XI, 293 pages. 2006.

Vol. 3905: P. Collet, M. Tomassini, M. Ebner, S. Gustafson, A. Ekárt (Eds.), Genetic Programming. XI, 361 pages. 2006.

Vol. 3904: M. Baldoni, U. Endriss, A. Omicini, P. Torroni (Eds.), Declarative Agent Languages and Technologies III. XII, 245 pages. 2006. (Sublibrary LNAI).

Vol. 3903: K. Chen, R. Deng, X. Lai, J. Zhou (Eds.), Information Security Practice and Experience. XIV, 392 pages. 2006.

Vol. 3902: R. Kronland-Martinet, T. Voinier, S. Ystad (Eds.), Computer Music Modeling and Retrieval. XI, 275 pages. 2006.

Vol. 3901: P.M. Hill (Ed.), Logic Based Program Synthesis and Transformation. X, 179 pages. 2006.

Vol. 3900: F. Toni, P. Torroni (Eds.), Computational Logic in Multi-Agent Systems. XVII, 427 pages. 2006. (Sublibrary LNAI).